W9-BXY-525

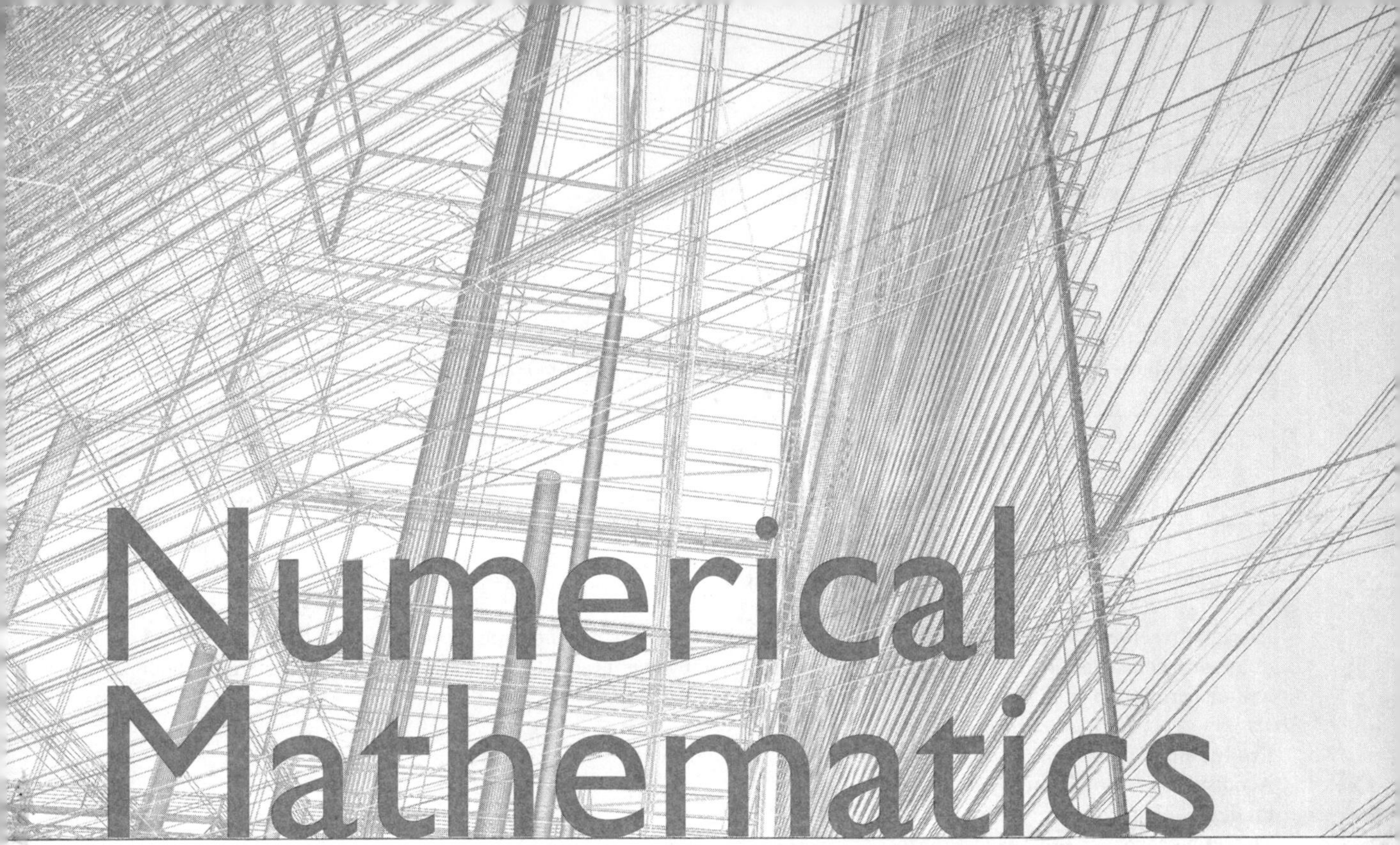

Numerical Mathematics

Matheus Grasselli
McMaster University
Dmitry Pelinovsky
McMaster University

JONES AND BARTLETT PUBLISHERS
Sudbury, Massachusetts
BOSTON TORONTO LONDON SINGAPORE

World Headquarters
Jones and Bartlett Publishers
40 Tall Pine Drive
Sudbury, MA 01776
978-443-5000
info@jbpub.com
www.jbpub.com

Jones and Bartlett Publishers Canada
6339 Ormindale Way
Mississauga, Ontario L5V 1J2
CANADA

Jones and Bartlett Publishers International
Barb House, Barb Mews
London W6 7PA
UK

Jones and Bartlett's books and products are available through most bookstores and online booksellers. To contact Jones and Bartlett Publishers directly, call 800-832-0034, fax 978-443-8000, or visit our website, www.jbpub.com.

Substantial discounts on bulk quantities of Jones and Bartlett's publications are available to corporations, professional associations, and other qualified organizations. For details and specific discount information, contact the special sales department at Jones and Bartlett via the above contact information or send an email to specialsales@jbpub.com.

Production Credits
Acquisitions Editor: Timothy Anderson
Production Director: Amy Rose
Marketing Manager: Andrea DeFronzo
Editorial Assistant: Laura Pagluica
Production Assistant: Sarah Bayle
Composition: Northeast Compositors, Inc.
Cover Design: Kristin E. Ohlin
Cover Image: © Ximagination/ShutterStock, Inc.
Cover Image: © Andreas Guskos/ShutterStock, Inc.
Printing and Binding: Malloy, Inc.
Cover Printing: Malloy, Inc.

Library of Congress Cataloging-in-Publication Data
Grasselli, Matheus.
Numerical mathematics / Matheus Grasselli and Dmitry Pelinovsky.
p. cm.
ISBN-13: 978-0-7637-3767-2
ISBN-10: 0-7637-3767-4
1. Numerical analysis. I. Pelinovsky, Dmitry. II. Title.
QA297.G7123 2008
518—dc22

2007009996

6048

Printed in the United States of America
11 10 09 08 07 10 9 8 7 6 5 4 3 2 1

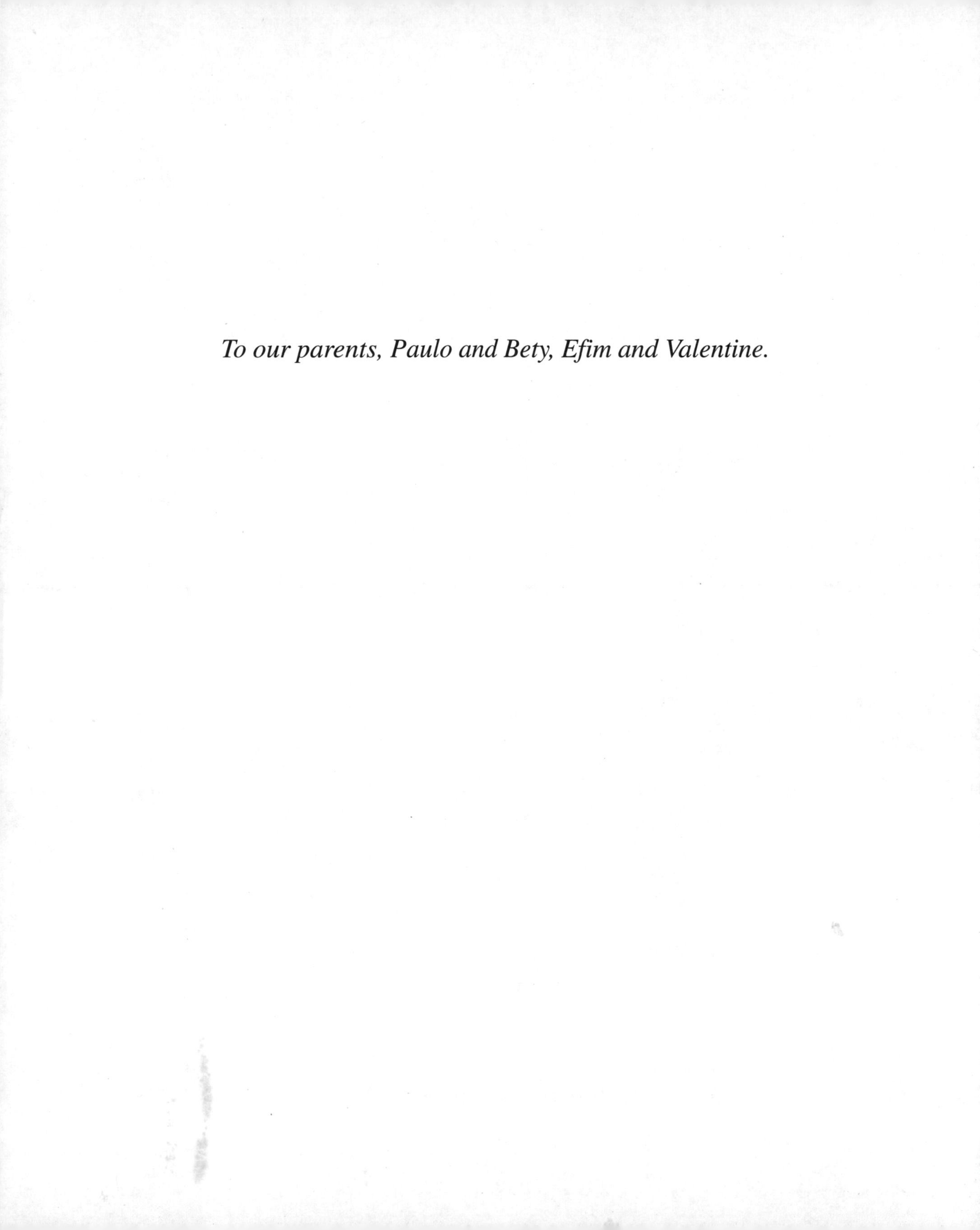

To our parents, Paulo and Bety, Efim and Valentine.

Preface

Laboratories are an essential part of most undergraduate scientific degrees. For example, students studying for a degree in physics will spend a significant amount of class time inside a laboratory, learning how to manipulate instruments, design experiments, collect data, and interpret results. This does not mean that all such students are going to become experimental physicists. Rather, such practice is based on the understanding that sciences derive their subject from Nature and that the best way to study her in a controlled way is through lab experiments. Therefore, even if not all students are going to become experimentalists, they should be exposed to the methods of the experimental work that serves as the motivation and ultimate validation of their fields. From a pedagogic point of view, lab classes reinforce and consolidate the concepts learned in other courses.

We believe that modern computers can play a similar role in mathematics. This book presents the innovative approach that numerical methods should be considered as a practical laboratory for undergraduate mathematics courses. We think this is innovative because it is not the state of affairs we currently encounter. On the one hand, first- and second-year students in mathematics, science, and engineering learn introductory mathematical concepts without making appropriate use of computer technology. On the other hand, upper-level courses on numerical methods put their emphases on specific topics such as computational algorithms, error analysis, convergence and stability, and coding and debugging procedures, with only passing references to the mathematical background, which is generally assumed to be known and understood beforehand.

Another way of explaning this is that numerical methods are traditionally taught at the undergraduate level as an isolated set of computational techniques that are not related to the main mathematical courses such as calculus and linear algebra. This is reflected in the standard textbooks on the subject. Those written by authors who are engineers or computational scientists tend to focus on specialized robust algorithms for hands-on experience in numerical computations and provide little explanation on the mathematical background linking these numerical methods with the theory of functions, linear algebra, and differential equations. Those written by mathematicians tend to focus on the theory of numerical analysis and are often too advanced for undergraduate students because they require a deep knowledge of analysis and provide fewer examples of practical computations.

We wrote this book to bridge the gap between numerical methods and undergraduate mathematics. It offers systematic and comprehensive implementations of numerical techniques in parallel with the discussions of the mathematical ideas involved. The central concept of the book is the *numerical laboratory*, where the emphasis is put on graphical visualization of mathematical properties, numerical verification of formal statements, and illustrations of the mathematical ideas.

With the overall goal of using numerical methods as pedagogical tools for learning mathematics, we had to make several choices regarding content and style. Any mathematical text must contain definitions, results, and applications. We have made effective use of the numerical laboratory already at the level of definitions by introducing computer examples of each new mathematical concept. With this in mind, we decided to include all the relevant definitions in the text, making it as self-contained as possible. Mathematical results are more delicate because establishing them might involve a long and hard sequence of logical steps—what mathematicians call a proof. Some of these logical steps are intuitive, directly related to the definitions, and can be discussed with the help of computer examples, whereas others are not. We decided to include the first type of argument in the text, while only outlining the second type of argument and making reference to other books. In any event, we use the numerical laboratory to motivate, verify, and visualize all the results presented in the text. As for applications, they generally fall into two categories: they can be related to other areas of mathematics, leading to new definitions and results, or they can be related to other disciplines, such as engineering or physical sciences. We decided to focus on the first type of application, making ample use of the numerical laboratory to show how results in one mathematics course can be applied to solve problems in another mathematics course. In this way, we can convey to the students a sense of unity for the mathematical material learned at the undergraduate level. Systematic applications in other disciplines would require developing the necessary concepts and background and would lead us too far from our purposes.

Regarding numerical methods, we concentrate on the description, implementation, and main usage of basic algorithms. This includes a discussion of their applicability, robustness, and efficiency. In this way, the book contains enough material to serve as the main textbook for a variety of undergraduate courses in numerical analysis. However, we refrained from covering specialized numerical methods or advanced versions of algorithms, which are best studied at the graduate level.

We therefore envision this book being used either as the main textbook for an introductory course in numerical methods or as a supplementary textbook for entry-level courses in calculus, linear algebra, and differential equations. Students using the book as a supplementary reference for basic mathematical courses are encouraged to develop their mathematical intuition with an effective component of technology, whereas students using it as the primary reference for a numerical course can aquire a broader and reinforced understanding of the subject. In the most optimistic scenario, students can keep and use this book throughout their entire undergraduate career, linking its contents to practically all undergraduate courses in mathematics.

Undergraduate Courses and the Numerical Laboratory

A typical sequence of mathematical courses required for undergraduate students in mathematics, science, and engineering programs includes the entry-level courses in calculus, linear algebra, and scientific computing, combined with more advanced courses in ordinary and partial differential equations, real and complex analysis, and numerical analysis. The material of our numerical laboratory is built around this type of standard undergraduate curriculum and can be roughly divided into three main parts.

These are preceded by Chapter 1, which is related to the material of an introductory course in scientific computing. It gives a self-contained overview of basic elements of the numerical laboratory, including matrix operations, programming, graphics and data files, error analysis, and floating-point arithmetic.

The first part consists of three chapters related to courses in elementary and advanced linear algebra. Chapter 2 starts with a discussion of the key theoretical concepts of finite-dimensional vector spaces, linear maps, vector and matrix norms, and systems of linear equations. It is followed by the numerical implementations of direct and iterative methods of solutions of the systems of linear equations. Chapter 3 discusses the theoretical properties of inner products, orthogonal projections, and orthonormal bases followed by the numerical methods of QR factorizations and least-square regressions. Chapter 4 covers properties of eigenvalues and eigenvectors in matrix eigenvalue problems and the numerical methods for singular value decompositions, power iterations for dominant eigenvalues, and simultaneous iterations for subspaces.

The second part consists of four chapters related to courses in calculus of one or several variables. Polynomial functions are described in Chapter 5 in the context of numerical problems of polynomial interpolations and approximations. The main subject of elementary calculus, derivatives, and integrals of scalar functions of a single variable, is covered in Chapter 6, in close ties with numerical problems of finite-difference approximations for derivatives and finite-sum representations for integrals. Functions of several variables are treated in Chapter 7, which covers partial derivatives, gradient vectors, vector fields and Jacobian matrices, line and surface integrals, and integral theorems of vector calculus, together with their numerical implementations. Chapter 8 presents the discussion of zeros and extrema of scalar functions of a single variable with root finding and minimization algorithms and techniques, and extends the same problems to functions of several variables.

The final part of the book consists of four chapters corresponding to advanced courses in ordinary and partial differential equations, Fourier analysis, and numerical analysis. Chapter 9 discusses properties of initial-value problems for ordinary differential equations (ODEs) followed by a discussion of numerical algorithms such as single-step Runge–Kutta methods, multistep Adams methods, and adaptive and implicit methods. Chapter 10 is devoted to the boundary-value problems both for ODEs and for partial differential equations (PDEs), including the three main classes of parabolic, hyperbolic, and elliptic PDEs. The central object of this chapter is discussion of the finite-difference method, which brings the main concepts of numerical calculus in action. Chapter 11 explains introductory concepts of spectral methods with a particular emphasis on the Fourier methods of trigonometric approximations and interpolations. Applications of these methods to various ODE and PDE problems establish close ties between this chapter and the previous chapters of the book. Chapter 12 explains introductory concepts of finite-element methods by studying spline and Hermite interpolations, variational methods, and their applications to ODEs.

Conventions

We use italics whenever we introduce a new concept, generally followed or preceeded by a formal definition and a specific example. We also use italics to refer to standard areas of mathematics. To avoid confusion, we combine the second usage of italics with capital letters, as in *Calculus* or *Functional Analysis*. *Very* occasionally we use italics simply to emphasize certain words. Mathematical results are motivated and described throughout the text and summarized in each chapter as *theorems*. In each chapter, we use the word *problem* to formulate a mathematical question of interest, which is solved in the book by analytical or computational methods of the numerical laboratory. Questions that students should solve are labeled as *exercises*. Some of these exercises are embedded in the main text, which means that they are directly linked with the discussion at hand. In addition, we offer a comprehensive collection of exercises at the end of each chapter. These exercises can be used for class tests and home assignments in the course of numerical analysis.

We divide each chapter into sections, as shown in the table of contents. Each section consists of several subsections, named in the margins of the page. These subsection headings simply indicate important parts of each section. At the end of each chapter, we present a summary of all important concepts and results. The summary is given in a form similar to the synopsis of an opera. If read before the performance, it gives an idea of the characters and the plot, but often sounds incomprehensible. Read afterward, it serves as a reminder of important moments, but might lack rhythm and dramatic flow.

The main body of the text consists of statements, discussions and formulas, inline MATLAB codes for numerical computations, and numerical and/or graphical outputs of the computations. The lists of topics, MATLAB functions, and scripts used throughout the book are given at the end of the book.

The standard version of the MATLAB package is the only software used in this text for all numerical computations.

MATLAB examples are inlined in the text and all exercises are expected to be implemented by the readers in the MATLAB Command window. The inline MATLAB code gives the necessary material for readers to generate the required output or graphical file. The inline code does not generally include additional elements such as labeling of axes, zooming and rotations of graphs, and modifications of embedded text and line widths, which can easily be achieved by using the MATLAB graphical user interface. The longer files containing scripts and functions developed throughout the text can be either typed by the students in the MATLAB Editor window or downloaded from the book website at `http://math.jbpub.com/catalog/0763737679`. This website also offers a password-protected instructor's area. Here, adopting professors have access to the exercise solutions and PDFs of the textbook's illustrations.

Acknowledgments

Students should always be the main motivation for writing a textbook, and we want to thank ours, whose enthusiasm toward the first incarnation of this text in the form of lecture notes propelled us to consider turning them into a book. We would have never ventured into the competitive and uncertain business of book writing without the zest and vision of Tim Anderson, our editor at Jones & Bartlett. Through him, we extend our gratitude to the whole team from J&B who worked with us for the past two years with extreme professionalism and patience.

Many professors participated as peer reviewers in the product development phase of this final text. We would like to thank and acknowledge the following: Bartek Protaz at McMaster University, Donald L. Hardcastle at Baylor University, Matt Davison at University of Western Ontario, Mary M. Hofle at Idaho State University, and Michelle L. Ghrist at the United States Air Force Acacdemy.

Close to home, we would like to thank our colleagues at McMaster University for creating the scholarly environment where this enterprise was possible. In particular, we thank Nicholas Kevlahan, Zlavek Kovarik, and Bartek Protaz for their careful reading of preliminary versions of the manuscript and valuable suggestions.

Moving even closer, we thank our families, partners, and children alike, for putting up with late night typing, grudging deadline complaints, and the general absent-mindedness of intense periods of authorship, all endured with love and understanding.

M. Grasselli
D. Pelinovsky

Contents

CHAPTER 1

Elements of the Laboratory

THIS CHAPTER IS A BRIEF INTRODUCTION to MATLAB that provides a student who has never before used the program with the knowledge necessary to follow the exposition in the rest of the book. It also serves to set up the notation we use in the book, in particular the way in which we intersperse MATLAB commands and outputs with the main body of the text. For this reason, we have intentionally kept it short, so that those students with previous experience with MATLAB can still read through and get acquainted with our notation without committing too much of their time.

Section 1.1 covers the basic way in which inputs and outputs appear in a MATLAB command window and is followed by a more detailed description of the type of variables used in MATLAB in Section 1.2 and some elementary operations that can be performed on them in Section 1.3. Part of the appeal of MATLAB to students in mathematical sciences (and professional scientists alike) is its ability to perform complicated mathematical tasks through its large collection of built-in functions, which are the subject of Section 1.4. A second source of MATLAB's appeal is the fact that it can also be used as a high-level language that students and scientists can use to write their own programs. The basics of programming in MATLAB are presented in Section 1.5, again assuming readers have no previous experience with other computer languages. Although it is perfectly possible to follow the simple programs presented in the rest of the book (and write new programs as required in some of the exercises) based only on the introduction given in this chapter, computer programming is an art and a science in its own right, and is certainly not the emphasis of our book. The remainder of the attraction of the scientific community to MATLAB rests on the program's flexible and powerful visualization tools, introduced in Section 1.6 and used repeatedly throughout the book.

The last two sections of this chapter are devoted to some fundamental concepts in *Numerical Analysis*. Section 1.7 discusses the way numbers are represented and manipulated in a computer, and Section 1.8 describes the different types of errors that are inherent to any numerical procedure.

1.1 Getting Started

When you start MATLAB, the first thing to do is to identify its command window on your screen. This is the window that includes a prompt like this:

```
>>
```

This is where you input simple statements and commands. When we deal with MATLAB examples, our convention in this book is to write down exactly what you should type in the command window followed by the actual MATLAB output. That way, you can see the results even if you are reading the book without a computer around. Of course, it is a lot more fun to adopt a hands-on approach and actually type the examples as you go along to see what happens. Needless to say, the fun can only increase if you modify the examples as much as you like to discover different results.

The simplest (and dullest!) way to use MATLAB is as a glorified desk calculator, such as follows:

```
>> (10+45*(-2))/(8-5)

ans =
      -26.6667
```

where, incidentally, we introduced the four basic arithmetic operations, as well as the use of brackets to override the order in which they are performed. In the absence of further instructions, MATLAB assigns any output to the variable `ans` (short for answer) and displays it in the command window. Also notice that the output in this example is displayed in four decimal places, which is the MATLAB format `short`.

Let us see how each of these features change in the next example. Suppose we want to calculate the numerical factor $1/(\sqrt{2\pi})^3$ with 15 significant digits and store it in the variable `num_factor`. In MATLAB, this is accomplished by the following:

```
>> format long
>> num_factor=1/(sqrt(2*pi)^3)

num_factor =
      0.06349363593424
```

To change MATLAB back to its short format, you should type the following:

```
>> format short
```

You can use the equal sign to assign values to variables without performing any calculations, as in:

```
>> x=13

x =
     13
```

where you can see MATLAB echoing what you have just done. You can prevent MATLAB from echo printing by ending a command line with a semicolon. For instance, simply typing

```
>> y=exp(12)*sqrt(log(31));
```

assigns the value $e^{12}\sqrt{\log 31}$ to the variable y without printing the result. To see the result, you need to type the name of the variable, for instance:

```
>> y

y =
    3.0160e+005
```

This example also shows how MATLAB displays scientific notation, that is, numbers of the form 3.0160×10^5.

To assign values to several variables in the same command line, you can use commas or semicolons, depending on whether you want the echo printing. For example, try

```
>> a=(-5)^(0.5);b=1+2*i;c=a+b,d=-1-3*i;

c =
   1.0000 + 4.2361i
```

which also shows how MATLAB handles complex numbers.

We end this introductory section with the command `who`, which returns a list of all current variables. For example, if you have typed all the examples in this section (and nothing else) since starting your current MATLAB session, you should obtain the following:

```
>> who

Your variables are:
a        b        c      d      num_factor  x y

>> clear
>> who
```

The command `clear` can be used to delete all current variables, so that the second use of the command `who` returns nothing.

1.2 Scalars, Vectors, and Matrices

If you have experience with computer programming, you are surely familiar with the fact that most languages require the programmer to declare in advance the types of variables to be used by a certain routine. Some even require the programmer to specify memory allocations. Not MATLAB! One of its most convenient features is that it creates variables as you need them, and in a format and type that are consistent with the calculation you are attempting to do. The result is that for most of the time the assignments of values to variables run smoothly, without requiring your attention. When MATLAB fails in this task, it displays an error message. For instance, when number 2 is erroneously assigned to be number 1, MATLAB complains:

```
>> 1 = 2
??? 1 = 2
      |
Error: Missing operator, comma, or semicolon.
```

Vectors The key to MATLAB's flexibility in handling variables is that it treats all of them as one kind of object: an *array*. An array is a collection of values labeled by indices. For example, an array with just one index is called a *vector*, as in

```
>> x=[1 -2 -1 3 0.5 1.7 3]

x =
    1.0000   -2.0000   -1.0000    3.0000    0.5000    1.7000    3.0000
```

Because the values in the preceding vector are displayed horizontally, it constitutes an example of a *row vector*. By contrast, you could have

```
>> y=[1;-2;-1]

y =
     1
    -2
    -1
```

which is an example of a *column vector*. When entering the values for row vectors, we can separate them by spaces (as in the previous example) or commas (try it!). For column vectors, we can use either a semicolon (as previously) or the carriage-return key (also known as the Enter key) to separate values.

As can be seen in the preceding examples, we use square brackets to enclose all the values of an array. In contrast, we use parentheses to access an individual value (also called a component of the vector) through its index.

For instance, for the previously entered vector you have

```
>> x(2)

x(2) =
    -2.0000

>> a=5*x(2)+x(5)

a =
    -9.5000
```

You should notice that indices in MATLAB always start at 1. If you want to know the number of components of a given vector, you can use the following MATLAB command:

```
>> length(x)

ans =
         7
```

In this book, when we are not mimicking MATLAB windows, we use standard mathematical notation and write row vectors as follows:

$$\mathbf{x} = [1, -2, -1, 3, 0.5, 1.7, 3],$$

and column vectors as

$$\mathbf{y} = \begin{bmatrix} 1 \\ -2 \\ -1 \end{bmatrix}.$$

We use subscripts to reference specific components, such as in $x_2 = -2$ and $y_3 = -1$.

Matrices

An array with two indices is called a *matrix*. The components of a matrix are input row by row, using either a semicolon or the Enter key to separate rows. For instance:

```
>> A=[1 -1; 3 4;0 2; -3 5]

A =
     1    -1
     3     4
     0     2
    -3     5
```

```
>> B=[1 -1 3 4
0 2 -3 5
2 -1 5 8]

B =
      1 -1  3 4
      0  2 -3 5
      2 -1  5 8
```

As before, you use parentheses to access individual components. The two possible indices are separated by a comma, and the convention is that the first index refers to the row whereas the second refers to the column where the component is located. For example, using the previous matrices, you have

```
>> a=A(3,2)

a =
   2

>> 4*B(1,2)+A(4,1)+10

ans =
      3
```

In standard mathematical notation, we denote a matrix as follows:

$$\mathbf{B} = \begin{bmatrix} 1 & -1 & 3 & 4 \\ 0 & 2 & -3 & 5 \\ 2 & -1 & 5 & 8 \end{bmatrix},$$

with individual components written as $b_{23} = -3$.

A matrix with m rows and n columns is said to be an $m \times n$ matrix, in tune with the convention that the first index always refers to rows and the second index refers to columns. To determine the number of rows and columns of a matrix in MATLAB, use the command `size`. For the previous examples, you have

```
>> size(A)

ans =
      4  2

>> s=size(B)

s =
    3  4
```

Because it consists of two values, the output for the command **size** is stored in a vector. In the preceding example, s_1 and s_2 store, respectively, the number of rows and columns for the matrix **B**.

We define the *transpose* of a matrix as the array obtained by interchanging the role of the indices of each component. In MATLAB, this is achieved as follows:

```
>> A=[1 -1; 3 4; 0 2; -3 5];
>> C=A'

C =
   1 3 0 -3
  -1 4 2  5
```

When the components of a matrix are complex numbers, the MATLAB operator introduced earlier returns the *complex conjugate* of the transpose matrix. For example:

```
>> A=[2+3i,1-i, 4; -1-2i,-3,5i]

A =
   2.0000 + 3.0000i   1.0000 - 1.0000i   4.0000
  -1.0000 - 2.0000i  -3.0000                  0 + 5.0000i

>> C=A'

C =
   2.0000 - 3.0000i  -1.0000 + 2.0000i
   1.0000 + 1.0000i  -3.0000
   4.0000                 0 - 5.0000i
```

Following this notation, we will always use $\mathbf{A}'$ for the complex conjugate of the transpose of **A**.

In its soul, MATLAB considers matrices to be its preferred objects—its own name stands for *Matrix Laboratory.* All other objects, such as vectors and scalars, are represented by matrices of a particular size. To see this, notice that a row vector such as the previous **x** is in fact interpreted by MATLAB as a 1×7 matrix, which can be confirmed by

```
>> x=[1 -2 -1 3 0.5 1.7 3];
>> size(x)

ans =
     1   7
```

```
>> x(1,5)

ans =
        0.5
```

Analogously, an $n \times 1$ matrix is identified with a column vector. Naturally, row vectors and column vectors are transposes of each other. For instance:

```
>> v=[1 -2 -1];
>> u=v'

u =
     1
    -2
    -1

>>size(u)

ans =
        3   1

>> u(2,1)

ans =
       -2
```

The colon operator

In the same vein, a single *scalar* is treated by MATLAB as a 1×1 matrix.

The single most useful and versatile sign in MATLAB is the *colon operator*. Used in an expression like `a:b:c`, it generates a sequence of numbers starting at a and ending on or before c, with step size b, while automatically storing the results in a row vector. If b is omitted, MATLAB takes it to be equal to 1. For example:

```
>> w=2:0.3:3

w =
    2.0000    2.3000    2.6000    2.9000

>> z=[30:34]

z =
    30    31    32    33    34
```

Notice that the use of square brackets in the preceding vectors is optional. Also notice how the final value can be left out if the step size is not chosen appropriately.

When used in place of an index, the colon operator acts as a wildcard, that is, it corresponds to *all* values of that index at once. In practice, this is used when we want to access an entire row or column of a matrix. For instance:

```
>> D=[1 -1 0; 3 2 4;0 -3 2];
>> D(:,2)

ans =
    -1
     2
    -3

>> row=D(3,:)

row =
       0 -3 2
```

Finally, you can use the colon operator to identify submatrices or subvectors. For instance, if you are interested in the first two rows and three columns of the matrix **B** given earlier, you could use the colon operator as follows:

```
>> block=B(1:2,1:3)

block =
      1     -1      3
      0      2     -3
```

Similarly, to select the components x_2, x_4, and x_6 of the following vector:

$$\mathbf{x} = [1, -2, -1, 3, 0.5, 1.7, 3],$$

you can write

```
>> selection=x(2:2:6)

selection =
   -2.0000     3.0000     1.7000
```

MATLAB also handles strings of characters in a vectorized way. That is, it stores each character as a component of a vector. The syntax to create a string is to enclose its contents in single quotation marks, as in:

```
>> subject='philosophy of mathematics'

subject =
          philosophy of mathematics

>> first_letter=subject(1)

first_letter =
              p

>> first_letter+x(2)

ans =
     110
```

You might be surprised by the numerical calculation done by MATLAB in this example. Because `x(2)=-2` is a number, an expression such as `first_letter+x(2)` makes sense only if MATLAB can assign a numerical value to the character p. Fortunately, the computer science community has a well-established way of assigning numerical values to characters, known as the American Standard Code for Information Interchange (ASCII). For instance, the ASCII code for the character p is 112, which justifies the preceding calculation.

General arrays

Scalars, vectors (both row and column), and matrices are the basic types of arrays considered in this book. Pictorially, we might call them, respectively, zero-, one-, and two-dimensional arrays, with an index playing the role of a spatial dimension. We refrain from doing so in this book because the word *dimension* is reserved for the more specific mathematical meaning of the dimension of a vector space, introduced in Section 2.1. The dimension analogy is convenient, however, when dealing with more general arrays. For instance, an array with three indices can be viewed as a three-dimensional cube. This is useful when you want to store multiple matrices under the same variable name. For instance, using the colon operator, the following command stores a 3×3 matrix as `cube(:,:,1)` and a different 3×3 matrix as `cube(:,:,2)`, therefore creating an array of size $3 \times 3 \times 2$.

```
>> cube(:,:,1)=[2 -3 0; 1 2 3; 0 0 1];
>> cube(:,:,2)=[1 0 0; 13 12 11; -1 -1 3];
>> cube

cube(:,:,1) =
              2    -3     0
              1     2     3
              0     0     1
```

```
cube(:,:,2) =
     1     0     0
    13    12    11
    -1    -1     3

>> size(cube)

ans =
     3     3     2
```

You might have noticed that MATLAB accepts a great variety of case-sensitive variable names. For instance, variable names can contain even certain special characters, such as the underscore. Certain words, however, are reserved for commands and functions, and MATLAB promptly warns you if you attempt to use them as the name of a variable.

```
>> end=5
??? end=5
     |
Error: Illegal use of reserved keyword "end".
```

1.3 Matrix Operations

We have demonstrated how the basic arithmetic operations on scalars are implemented in MATLAB by the operators +, -, *, and /. We have also shown that MATLAB actually interprets scalars as a 1×1 matrix. It should then come as no surprise that these operators are in fact used to implement matrix operations, for which the scalar and vector versions are special cases.

Addition and Subtraction

Matrix addition and subtraction are performed in a standard componentwise way. If you attempt to add or subtract matrices of different sizes, MATLAB displays an error message.

```
>> A=[2 -3 0; 1 2 3; 0 0 1];
>> B=[1 0 0; 13 12 11; -1 -1 3];
>> C=[5 -5; 2 1]; D=[1 15; -3 4];
>> A+B

ans =
     3    -3     0
    14    14    14
    -1    -1     4

>> C-D
```

```
ans =
     4   -20
     5    -3

>> A+C
??? Error using ==> plus
Matrix dimensions must agree.
```

Addition and subtraction of row and column vectors (that is, $1 \times n$ and $m \times 1$ matrices) as well as scalars (1×1 matrices), are special cases of these rules.

Multiplication

As for multiplication, given an $m \times n$ matrix $\mathbf{A}$ with entries (a_{ij}) and an $n \times p$ matrix $\mathbf{B}$ with entries (b_{ij}), the operator `*` returns the matrix product, that is, the $m \times p$ matrix $\mathbf{C} = \mathbf{A} * \mathbf{B}$ with entries given by

$$c_{ij} = \sum_{k=1}^{n} a_{ik} b_{kj}, \qquad i = 1, \ldots, m, \quad j = 1, \ldots, p. \tag{1.1}$$

For example,

```
>> A=[5 2 3; 1 -2 6];
>> B=[0 1 -1 3; 5 1 0 2; 4 -3 2 -1];
>> C=A*B

C =
    22    -2     1    16
    14   -19    11    -7

>> D=B*A
??? Error using ==> mtimes
Inner matrix dimensions must agree.
```

Observe how reversing the order of the matrices leads to an impossible operation because the product of two matrices $\mathbf{A}$ and $\mathbf{B}$ makes sense only if the number of *columns* of $\mathbf{A}$ equals the number of *rows* of $\mathbf{B}$. This is a rather dramatic way of showing that, for matrix multiplication, the order of factors alters the product! Less extreme examples are possible: for instance, if $\mathbf{A}$ is a 2×3 matrix and $\mathbf{B}$ is a 3×2 matrix, then both $\mathbf{A} * \mathbf{B}$ and $\mathbf{B} * \mathbf{A}$ are well defined, though the former has size 2×2, while the latter is a 3×3 matrix. More interestingly, even for square (that is $n \times n$) matrices, it is *not* generally true that $\mathbf{A} * \mathbf{B} = \mathbf{B} * \mathbf{A}$, as you can see in the next MATLAB example:

```
>>A=[1 0 -1;5 -3 6; 7 3 1];
>>B=[2 2 1; -1 4 1; 0 0 2];
>>A*B
```

```
ans =
     2     2    -1
    13    -2    14
    11    26    12

>>B*A

ans =
    19    -3    11
    26    -9    26
    14     6     2
```

Notice that if **A** is an $m \times n$ matrix and x is an $n \times 1$ column vector, then the matrix product $\mathbf{b} = \mathbf{A} * \mathbf{x}$ is an $m \times 1$ column vector. As we discuss in Section 2.2, this is a fruitful way to generate maps between column vectors. Analogously, if **y** is a $1 \times m$ row vector and **A** is an $m \times n$ matrix, then $\mathbf{c} = \mathbf{y} * \mathbf{A}$ is a $1 \times n$ row vector, although this kind of operation is used less often in practice:

```
>> A=[5 2 3; 1 -2 6];
>> x=[-1 1 0]';
>> b=A*x

b =
    -3
     3

>> y=[2 -17];
>> c=y*A

c =
  -7 38 -96
```

Inner and outer products

Moreover, it follows from equation (1.1) that multiplying the transpose of an $n \times 1$ column vector **a** with components $a_1, \ldots, a_n$ (that is, a $1 \times n$ row vector $\mathbf{a}'$) by an $n \times 1$ column vector **b** with components $b_1, \ldots, b_n$ results in a 1×1 matrix whose only component is

$$a_1 b_1 + a_2 b_2 + \cdots a_n b_n. \tag{1.2}$$

Because this is just a scalar, we call it the *scalar* product between the vectors **a** and **b**. We denote (1.2) by $\mathbf{a} \cdot \mathbf{b}$ and alternatively call it the *dot product* between **a** and **b**. In MATLAB, such a product can be obtained either using the multiplication operator `*` or using the command `dot`:

```
>> a=[7;2;-9]; b=[0;-1;3];
>> a'*b

ans =
        -29

>> dot(a,b)

ans =
   -29
```

The product of a row vector by a column vector leading to (1.2) is also called an *inner product*. This is because in the opposite order, according to equation (1.1), the product of a column vector by a row vector results in a full-blown $n \times n$ matrix, which is then called the *outer product* between two vectors:

```
>> a=[7;2;-9]; b=[0;-1;3];
>> a*b'

ans =
     0    -7    21
     0    -2     6
     0     9   -27
```

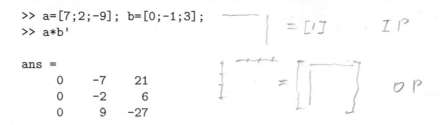

In Chapter 2, we identify vectors with three components with arrows in space. Under this identification, we say that two vectors are *perpendicular* or *orthogonal* when their inner product vanishes.

Cross product

Still on the subject of multiplication, let us define the *cross product* of two vectors with three components each as

$$\mathbf{a} \times \mathbf{b} = \begin{bmatrix} a_1 \\ a_2 \\ a_3 \end{bmatrix} \times \begin{bmatrix} b_1 \\ b_2 \\ b_3 \end{bmatrix} = \begin{bmatrix} a_2 b_3 - a_3 b_2 \\ a_3 b_1 - a_1 b_3 \\ a_1 b_2 - a_2 b_1 \end{bmatrix}. \tag{1.3}$$

You can immediately see from the preceding formula that the cross product is *anticommutative*, that is, $\mathbf{a} \times \mathbf{b} = -\mathbf{b} \times \mathbf{a}$. It is also nonassociative in the sense that $\mathbf{a} \times (\mathbf{b} \times \mathbf{c}) \neq (\mathbf{a} \times \mathbf{b}) \times \mathbf{c}$. The cross product is implemented with the MATLAB command `cross`, and you can verify the previous properties as follows:

```
>> a=[1;3;-4];b=[2;-1;0];c=[5;7;1];
>> [cross(a,b) cross(b,a)]
```

```
ans =
    -4     4
    -8     8
    -7     7
>> [cross(a,cross(b,c)),cross(cross(a,b),c)]

ans =
    49    41
   -15   -31
     1    12
```

Cross products are used extensively in Chapter 7. In particular, we use the fact that $\mathbf{a} \times \mathbf{b}$ is perpendicular to both $\mathbf{a}$ and $\mathbf{b}$, as the next example shows:

```
>> a=[1;3;-4];b=[2;-1;0];
>> [dot(a,cross(a,b)),dot(b,cross(a,b))]

ans =
     0     0
```

Other matrix operations

Matrix division is a much more delicate matter than matrix multiplication is. Recall that for real numbers a and b, what we really mean by $a\,/\,b$ is a number c such that $b \cdot c = a$. That is, division is defined as an inverse procedure. The same is true for matrices, only more so, because it is not clear *a priori* if the inverse procedure can be carried out at all. We therefore defer comments on the matrix division operator / until Section 1.4, where we introduce the inverse of a matrix and the MATLAB function that calculates it.

The last matrix operation in our list is exponentiation, implemented by the operator `^`. Its syntax is simply `A^p`, which stands for $\mathbf{A}^p$, where $\mathbf{A}$ is a square matrix and p is a scalar. If p is a positive integer, it is computed by repeated matrix multiplication, whereas if it is a negative integer, it requires the calculation of a matrix inverse:

```
>> A=[1 0; -1 2];
>> A^3

ans =
     1     0
    -7     8

>> A^(-2)
```

```
ans =
        1.0000         0
        0.7500    0.2500
```

Componentwise operations

Sometimes it is convenient to perform arithmetic operations on each component of a matrix, rather than following the rules of matrix multiplication, division, and exponentiation. To accommodate this need, MATLAB offers what its manual calls *array operations*, which we prefer to call *componentwise operations*. For instance, componentwise multiplication for matrices **A** and **B** consists of multiplying each component of **A** by the corresponding component of **B**. This is implemented by the MATLAB command `A.*B`, which is clearly only defined for matrices of the exact same size. A similar definition holds for componentwise division, through the command `A./B`. Componentwise exponentiation supports two different versions. If p is a scalar, then `A.^p` raises each component of the matrix **A** to the power p. Alternatively, for matrices of the same size, `A.^B` results in each component of **A** being raised by the corresponding component of **B**. Some examples should make it all clear.

```
>> A=[1 0; -1 2]; B=[-1 3; 4 1];
>> A.*B

ans =
       -1     0
       -4     2

>> A./B

ans =
       -1.0000         0
       -0.2500    2.0000

>> A.^2

ans =
        1     0
        1     4

>> B.^A

ans =
      -1.0000    1.0000
       0.2500    1.0000
```

Componentwise operations are similarly defined for vectors or arrays with more than two indices.

1.4 Built-in Functions

Many more interesting mathematical operations can be implemented in MATLAB using its large collection of functions. In Section 1.1, we used **sqrt**, **exp**, and **log**, which are simple MATLAB functions that implement, respectively, the square root, the exponential and the natural logarithm functions of *Calculus*. Each MATLAB function can take one or more arguments as an input and returns one or more variables as output. Both the input and output can be different types of arrays, depending on which function is being used. Additionally, optional arguments can be used to make the same function perform modified tasks. A quick way of obtaining information about a MATLAB function, while working in the command window, is to use the command **help** followed by the name of the function. For instance:

```
>> help sqrt
 SQRT   Square root.
    SQRT(X) is the square root of the elements of X. Complex
    results are produced if X is not positive.

    See also sqrtm.

    Overloaded functions or methods
    (ones with the same name in other directories)
       help sym/sqrt.m

    Reference page in Help browser
       doc sqrt
```

You learn from this that **sqrt** calculates the square root of the elements of an array A, that is, it is an example of a componentwise calculation. In fact, using sqrt(A) produces the same result as A.^(1/2). The help also suggests that you look at the command **sqrtm**. This implements the *matrix* square root, that is, X=sqrtm(A) produces the same result as X=A^(1/2), which in turn is a matrix satisfying X*X=A. Let us follow MATLAB's advice and see what the **help** command has to say about this function:

```
>> help sqrtm
SQRTM     Matrix square root.
    X = SQRTM(A) is the principal square root of the matrix A, i.e. X*X = A.

    X is the unique square root for which every eigenvalue has nonnegative
    real part.  If A has any real, negative eigenvalues then a complex
    result is produced.  If A is singular then A may not have a
    square root.  A warning is printed if exact singularity is detected.

    With two output arguments, [X, RESNORM] = SQRTM(A) does not print any
    warning, and returns the residual, norm(A-X^2,'fro')/norm(A,'fro').
```

```
With three output arguments, [X, ALPHA, CONDEST] = SQRTM(A) returns a
stability factor ALPHA and an estimate CONDEST of the matrix square root
condition number of X.  The residual NORM(A-X^2,'fro')/NORM(A,'fro') is
bounded approximately by N*ALPHA*EPS and the Frobenius norm relative
error in X is bounded approximately by N*ALPHA*CONDEST*EPS, where
N = MAX(SIZE(A)).

See also expm, logm, funm.

Reference page in Help browser
   doc sqrtm
```

This time, the help contains a great deal more information, quite possibly using terms that you haven't seen before and making several references to many more functions. Unlike a dictionary, which tries to use a small core of universally known words to explain the meaning of more complicated terms, the MATLAB `help` command makes no apologies in explaining its functions through references to advanced concepts and likely more sophisticated functions. It serves the purpose of accurately describing the syntax, uses, and variations of a MATLAB function. Implicitly, it assumes that the reader is familiar with a complete mathematical theory behind the numerical techniques being implemented.

A gentler way to gather information about a given function is to use MATLAB's interactive Help window, which can be started by clicking the question mark on the MATLAB screen or by typing `helpwin` in the command window. This leads to all the technical information displayed by the `help` command, but also provides links to MATLAB tutorials, which are rich sources of basic explanations and examples. Although not entirely self-contained, MATLAB tutorials are written in pedagogical style, with the typical hyperlink structure of electronic documents that encourages readers to spend hours navigating them.

Our overall strategy in this book is to use the computer as a learning tool for mathematical concepts. Our tactics for dealing with MATLAB functions reflect this. We freely use MATLAB functions as "black boxes" to illustrate conceptual points, even when their underlying algorithms are beyond your understanding at a given stage of the exposition. Quite often, we refer to later chapters in the book where the techniques used by a particular MATLAB function are further investigated. In these cases, you will learn how MATLAB implements certain functions long after having used them several times, and hopefully you will already be familiar with their different inputs, outputs, arguments, and the like. In some cases, we use functions that involve numerical techniques that are way beyond the scope of this book, but their effective use in examples and exercises serves the educational purpose we have in mind. The analogy of a successful pilot who does not necessarily understand the mechanics of a jet fighter is perhaps appropriate here, though in mathematics it should generally be employed with caution.

Matrix inverse

For an illustration of these tactics, consider the concept of a matrix inverse. We say that an $n \times n$ matrix $\mathbf{A}$ is *invertible* if there exists a matrix $\mathbf{B}$ such that

$$\mathbf{A} * \mathbf{B} = \mathbf{B} * \mathbf{A} = \mathbf{1}, \qquad (1.4)$$

where

$$\mathbf{1} = \begin{bmatrix} 1 & 0 & \cdots & 0 \\ 0 & 1 & \cdots & 0 \\ \vdots & & \ddots & \vdots \\ 0 & & \cdots & 1 \end{bmatrix}$$

is the n n identity matrix. If these equations are satisfied, we use the notation $\mathbf{B} = \mathbf{A}^{-1}$.

Suppose you want to find the inverse for the matrix

$$\mathbf{A} = \begin{bmatrix} 1 & 3 \\ -1 & 2 \end{bmatrix}.$$

Using the second equality in equation (1.4), you must find a matrix $\mathbf{B} = (b_{ij})$ such that

$$\begin{bmatrix} b_{11} & b_{12} \\ b_{21} & b_{22} \end{bmatrix} \begin{bmatrix} 1 & 3 \\ -1 & 2 \end{bmatrix} = \begin{bmatrix} 1 & 0 \\ 0 & 1 \end{bmatrix}. \qquad (1.5)$$

Using the rules of matrix multiplication, this is equivalent to the following two systems of equations:

$$\begin{aligned} b_{11} - b_{12} &= 1 \\ 3b_{11} + 2b_{12} &= 0 \end{aligned}$$

and

$$\begin{aligned} b_{21} - b_{22} &= 0 \\ 3b_{21} + 2b_{22} &= 1 \end{aligned}$$

You can immediately verify that the solutions to these systems are satisfied by $b_{11} = 0.4, b_{12} = -0.6, b_{21} = 0.2$, and $b_{22} = 0.2$. For $\mathbf{B}$ to be the inverse of $\mathbf{A}$, it must also satisfy the reversed order product $\mathbf{A} * \mathbf{B}$, which you can also verify using the preceding numbers (in Section 2.2, you learn why this second step is not necessary, since $\mathbf{B} * \mathbf{A} = \mathbf{1}$ is satisfied if and only if $\mathbf{A} * \mathbf{B} = \mathbf{1}$).

We can extrapolate this argument and conclude that finding the inverse of a general n n matrix is equivalent to solving n different systems, each of them consisting of n equations. It is easy to be convinced that this is a highly elaborate task. In MATLAB, it is implemented by the innocent-looking function `inv`. When the inverse does not exist, the matrix is said to be *singular* and MATLAB displays a warning:

```
>> A = [1 3; -1 2];
>> inv(A)

ans =
    0.4000   -0.6000
    0.2000    0.2000

>> B = [-1 3; 2  -6];
>> inv(B)
Warning: Matrix is singular to working precision.

ans =
       Inf    Inf
       Inf    Inf
```

If you read the MATLAB Help entry for the function `inv` (from the Help window), you find a rather discouraging passage, telling you that you should avoid explicitly calculating the matrix inverse at all costs. It indicates that whenever you encounter a matrix inverse in practice, it is often in connection with a simpler linear system, which should then be solved by different methods, such as those treated in Section 2.5. The actual algorithm used by MATLAB to calculate a matrix inverse invokes several routines from a lower-level FORTRAN library called LAPACK. All these are road signs for a place you don't want to go. You should content yourself with being a (moderate) user of the function `inv`, rather than dealing with its implementation details.

Matrix division Before we leave the subject, we give the definition for the matrix division operator announced in Section 1.3. The MATLAB command `A/C` gives, in principle, the same result as `A*inv(C)`, but is calculated in a different (and more efficient) way, in the spirit of solutions to linear systems mentioned earlier. Therefore, if you find yourself in the middle of a problem where it becomes necessary to calculate $\mathbf{A} * \mathbf{C}^{-1}$, you know which MATLAB command you should use. More precisely, you have that `A/C=(C'\A')'`, which makes use of the *backslash* matrix division operator `\`. We defer further discussion about this extremely powerful operator until Section 2.3, where it is introduced in connection with the solution of linear systems.

In the course of the text, we introduce many more MATLAB functions as we need them. Whenever you see a new function, you should read its corresponding Help entry and try to become familiar with its syntax. You are also encouraged to browse MATLAB's libraries and try out the many other functions that we do not mention in this text.

1.5 Programming with MATLAB

As you have seen in Section 1.4, the numerical implementation of mathematical tasks that are very simple to state, such as "calculate the inverse of this matrix," can actually involve a long series of nontrivial steps. These might be just *sequential steps* (executed in the order they appear), but could also involve *branch statements* (steps that are executed provided a certain condition is satisfied), *loops* (steps that are repeated in identical fashion or with slight variations for a certain number of times), or combinations of these.

For such complicated procedures, it is clear that you need more than the command window environment. If you already have experience with other computer languages, such as C or FORTRAN, you will recognize that what we are looking for is the MATLAB version of a *program*, or a *code*. In MATLAB, these are generally called *M-files* because they are saved with a **.m** extension. In this section, you learn the basics of constructing and using such files.

Judging by the questions we receive from our own students, the single most important thing to learn about M-files, if you are a beginner MATLAB user, is where to put them. We recommend that you create a new folder somewhere in your home directory, name it in an appropriate manner (say `C:\Documents\num_lab`), and then make sure that the `Current directory` field in the main MATLAB window shows this newly created folder. You can later create subfolders according to your own organizational tastes (say, one for each chapter of the book), but remember that MATLAB's command window only executes files located in the folder shown in `Current directory`. Having chosen the place to save M-files, you can use any text editor of your liking to create and modify them. To use MATLAB's own editor, go to `File > New > M-file` in the main menu, or type `edit` in the command window.

Functions

MATLAB distinguishes between two types of M-files: scripts and functions. A *function* is an M-file that takes one or more variables as input arguments, performs a certain task on them, and assigns values to one or more output variables. Typically, a function is designed to be used as a small tool in the course of a more elaborate project. In this way, a function is what other programming languages call subroutines.

Any function M–file must start with a line of the form

```
function [Y1,Y2,...,YM]=my_function_name(X1,...,XN),
```

where `Y1,...,` `YM` is a list of output variables, `my_function_name` indicates the name you want to give to the function, and `X1,...,XN` is a list of input arguments. This line is then usually followed by several lines beginning with the percent sign % and containing comments that can be read whenever a user types the command **help my_function_name**. This is then followed by the actual statements that constitute the function. These can be any of the MATLAB commands discussed so far, including the built-in functions, plus all the logical and loop statements that we introduce shortly. The function file should then be saved under the file name **my_function_name.m**.

For example, suppose you want to create a function that calculates the mean and standard deviation of your term marks in first year *Calculus*. Type the following as a new M-file and save it as **marks.m**.

```
function [m,s]=marks(x)
% [m,s]=marks(x)
% x is a vector, m and s are scalars
% the function computes the mean value m and standard deviation s
% of the data sample x

n=length(x);
m=sum(x)/n; % built-in function for the sum of components of a vector
s=sqrt(sum((x-m).^2)/n);  % componentwise operations
```

Observe the use of % to insert comments at any point in the code, which you should always use to document your files well. Not only does it make the code more readable for other users, it serves as a reminder of what you had intended if you read the code several months later (it is remarkable how time can obscure the clearest of ideas). In this book, we often use % as a device to introduce and explain new MATLAB on the fly.

You can now use your newly created function:

```
>> x=[80 70 85 65 70 75 90];
>> [m,s]=marks(x)

m =
    76.429

s =
    8.3299
```

Notice how the input argument is assigned to the variable **x** before calling the function **marks**. This is common practice because it makes it possible to use such variables as the input argument for other functions as well. Alternatively, you could have typed the vector directly inside the parentheses while calling the function.

In either case, a function sees only the variables passed to it as input arguments. That is, if inside a function you make reference to a variable that is being used in the workspace *without* passing that variable as an input argument first, MATLAB will produce an error message while executing the function. Similarly, after executing a function, only the variables listed as output become available for later use in the workspace. For instance, the command `n = length(x)` used in the function `marks` creates a variable that is available for use inside the function, but is not listed as an output. If you try to use it in the command window after running the function `marks`, the result will be another error message.

Scripts

By contrast, a *script* is an M-file with no input or output variables. It consists of a file with a list of commands that would otherwise be typed directly in the command window. Therefore, it shares all variables currently being used in the workspace, and any variable created during its execution also becomes available for later use in the command window. Scripts should be designed to contain the main sequence of steps for the execution of an entire project. These include the use of several different functions, as well as the display of results on the screen. It is also a good place to insert commands to systematically plot graphics or save data on files, as we describe in Section 1.6. Viewed in this way, scripts are the analogues of the main executable files in other programming languages.

As an example of a script, suppose that you want to use MATLAB to obtain your letter grade for first-year *Calculus*. What MATLAB has to do is (1) prompt you to type your term marks on the keyboard, (2) calculate their average, (3) convert the numerical average into a letter grade, and (4) display it on the screen. Step 1 can be handled by the command `input`; for Step 2, you can use the function `marks` created earlier; Step 4 can be achieved with the versatile command `disp`. For Step 3, assume that the final letter grade depends on which interval happens to contain the numerical average for the term marks. For instance, suppose that the letter grade is A if the average is above 80, B if it is between 70 and 79, C if it is between 60 and 69, D if it is between 50 and 59, and F if it is below 50. Therefore, the assignment of a letter grade is an example of a *branch statement*, that is, a statement that depends on certain conditions.

Branch statements

One way to implement this kind of statement in MATLAB is through the command `if`. The simplest syntax for this command is as follows:

```
if <condition>
    <statements>
end
```

where `<condition>` is a logical expression that is either true or false. If it is true, MATLAB executes all the commands listed in `<statements>`. If it is false, MATLAB skips these commands and proceeds to execute whatever comes after the command `end`. A slightly more complicated syntax uses the optional command `else`:

```
if <condition>
    <statements>
else
    <other_statements>
end
```

This time, if <condition> is true, MATLAB executes <statements> and moves to whatever comes after end, whereas if <condition> is false, MATLAB skips *statements*, executes <other_statements> instead, and moves to the next line after end. The full syntax for this kind of structure also uses the optional command elseif:

```
if <condition1>
    <statements1>
elseif <condition2>
    <statements2>

...

elseif <conditionN>
    <statementsN>
else
    <other_statements>
end
```

The dynamics for this is the following. If <condition1> is true, MATLAB executes <statements1> and moves to end. If <condition1> is false, MATLAB proceeds to verify <condition2>. If this turns out to be true, MATLAB executes <statements2> and moves to end. If it is false, MATLAB moves to the next elseif and repeats the procedure. If *all* the conditions turn out to be false, MATLAB executes <other_statements> and moves to end. If you are turned off by this kind of logical dynamics, return quickly to our example. Otherwise, read the MATLAB help on it for further joy.

Simple logical expressions are constructed with the following *relational* operators: == (equal), ~= (not equal), < (less than), > (greater than), <= (less than or equal to), and >= (greater than or equal to). For instance, the expression (51>10) is true, while the expression (13=20) is false.

More complicated logical expressions are obtained with the *logical* operators ~ (not), & (and), and | (or), in this order of precedence. For instance, in the letter grade example, if m denotes your average mark, you end up with a B grade if the logical expression (70<=m)&(m<80) is true, that is, if $70 \leq m < 80$. Here is the script that implements all of this, which you should save as letter_grade.m.

```
% letter_grade script
% calculates final letter grade based on marks entered by user

x=input('Enter your term marks as a vector: ');
[m,s]=marks(x);
disp('    Your numerical average is:')
disp(m)
if m>= 80
   disp('Your final grade is A - well done !')
elseif (70<=m)&(m<80)
   disp('Your final grade is B')
elseif (60<=m)&(m<70)
   disp('Your final grade is C')
elseif (50<=m)&(m<60)
   disp('Your final grade is D')
else
   disp('Sorry, but you failed the course!')
end
```

To run a script, simply type its file name in the command window.

```
>> letter_grade
Enter your term marks as a vector: [50 61 70 56 80 75]

Your numerical average is:
   65.3333

Your final grade is C
```

Loops

A different way to modify the flow of commands is through the use of *loops*, which can be implemented in MATLAB with the command `for`, whose syntax is as follows:

```
for <counts>
    <statements>
end
```

Such a loop executes `<statements>` as many times as indicated by `<counts>`. For example, the following function calculates the sum of squares of the first N integers:

```
function s=sum_squares(N)
% calculates the sum of squares for the first N integers
```

```
s=0;
for x=1:N
    s=s+x^2;
end
```

After saving this as `sum_squares.m`, you can try it in the command window:

```
>> sum_squares(1000)

ans =
      333833500
```

The variable used in `<counts>` can also be used as the index of an array appearing inside the loop. For instance, the next function creates a vector with squares of the first N integers and then calculates their sum:

```
function [s,x]=squares(N)
% squares of the first N integers and their sum

for k=1:N
    x(k)=k^2;
end
s=sum(x);
```

You then obtain the following:

```
>> [s,x]=squares(1000);
>> s

s =
    333833500
```

Of course, making use of componentwise operations, the same result could be achieved with the following much simpler command:

```
>> sum([1:1000].^2)

ans =
   333833500
```

This is an example of what MATLAB programmers called *vectorization*, that is, the use of MATLAB powers to operate on vectors and matrices in very compact commands. Quite often, this practice replaces entire loops with a single line, producing much shorter codes. Moreover, vectorized calculations run much faster because of the way MATLAB handles vectors internally.

In programming jargon, vectorized calculations run in *compiled* form, rather than in *interpreted* form, but you don't really need to worry about it for simple computations. The downside of using vectorization is that it sometimes results in codes that are much harder to understand at a later stage.

If you don't know exactly how many times to execute a loop, you can use the command while, with the following syntax:

```
while <condition>
     <statements>
end
```

This executes <statements> for as long as <condition> remains true. For instance, if you want to know how many terms of a geometric progression starting at a_0 with ratio q need to be added until you reach a number greater than a certain threshold M, you can use the following function:

```
function k=prog(a0,q,M)
% k is the number of steps necessary for the sum of
% terms of a geometric progression to become larger than a threshold M

x(1)=a0;
k=1;
while sum(x)<M
      x(k+1)=q*x(k);
      k=k+1;
end
```

You can use this function to estimate how many chessboard squares it would take for the proverbial Indian king to go beyond the entire world production of rice in a given year. In the legend, the king promises to pay a wise man one grain of rice for the first square, two for the second square, four for the third square, and so forth, but he soon reaches the limits of his wealth before covering all of the chessboard squares ($8 \times 8 = 64$). At 40 grains per teaspoon, 1,000 teaspoons per cup, 5 cups per kilo, and 400 million tons of rice produced worldwide in 2003 (ah, the Internet!), you obtain the following:

```
>> k=prog(1,2,40*1000*5*400*10^9)

k =
    57
```

Loops and branch statements can be nested into each other in several ways. A common practice is to use indentation while writing code containing nested structures, so that they can be visually identified by human readers. For example, the following function calculates the largest absolute row sum of a given matrix. As you will see in Section 2.4, this is one way of measuring the size of a matrix in a mathematically precise sense, called a *norm*.

```
function norm=my_norm(A)
% calculates the max absolute row sum for the matrix A

[m,n]=size(A);
largest=0;
for i=1:m          % loop along rows
    row_sum=0;
    for j=1:n      % loop along columns
        row_sum=row_sum+abs(A(i,j));
    end
    if row_sum>=largest
       largest=row_sum;
    end
end
norm=largest;
```

1.6 Graphics and Data Files

Visualization of data and results is at the core of any successful scientific enterprise. It also tremendously enhances the learning experience in mathematics. One of the explicit purposes of this book is that you become sufficiently familiar with MATLAB to use its graphing tools routinely in whatever subject you come across in your career as a student.

Plots

The simplest graphing command in MATLAB is `plot`, which creates a two-dimensional plot of one vector versus another. In other words, given two vectors $\mathbf{x} = (x_1, \ldots, x_n)$ and $\mathbf{y} = (y_1, \ldots, y_n)$, the command `plot` depicts the ordered pairs (x_i, y_i) on a cartesian plane. By default, it also uses a solid line to "connect the dots." Its most common use is to obtain the graph of functions of one variable. For example, suppose you want to graph the function $f(x) = \exp(-x)\sin(\pi x)$ for $x \in [0, 6]$. For this, you create a vector $\mathbf{x}$ of equally spaced values from the interval $[0, 6]$, say with step size 0.2, evaluate the function $f(x)$ at these points, and store the results in the components of another vector $\mathbf{y}$. You can then use the `plot` function for these two vectors:

```
>> x=0:0.2:6;
>> y=exp(-x).*sin(pi*x);
>> plot(x,y);
```

The graph shown in Figure 1.1 (*left*) should then appear in a separate window titled `Figure 1`.

You can discard the graph by simply closing the picture window or typing `close` in the MATLAB command window. Alternatively, you can save it in several different formats by using the menu option `File > Save as` in that same window. You can do a lot more in this window. For example, with the

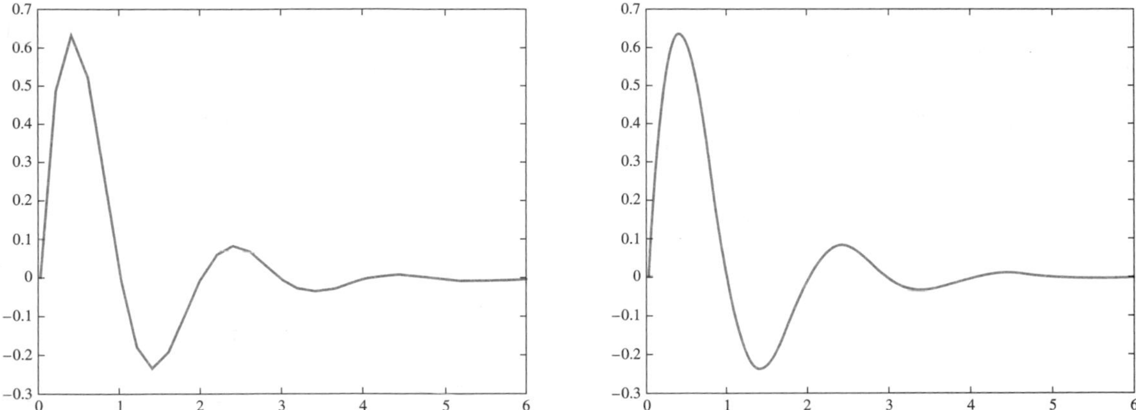

Figure 1.1 The graph of a one variable function with low resolution (*left*) and high resolution (*right*).

menu options listed under `Insert`, you can add names to the axis or a title to the entire figure, and the `Tools` menu options allow for several handy visualization tricks, such as zooming in and out or rotating the picture (more useful for the coming 3D plots). The important thing to remember here is that if you want to be able to add these modifications to your figure at a later time, you should save the figure as a MATLAB figure, using the extension `.fig` because MATLAB cannot edit figures in other formats (for instance, with an extension `.pdf`), even if it created them.

Say you are not happy with the graph and want to improve its resolution. You should then reduce the step size for the x–values, as in the following commands:

```
>> x=0:0.02:6;
>> y=exp(-x).*sin(pi*x);
>> plot(x,y);
```

If you have closed the picture window, the command `plot` reopens it with the new graph shown in Figure 1.1 (*right*). If you didn't close it, the command overwrites the new graph in the same window, regardless of whether the old graph was saved.

As learned in *Calculus*, a general way to obtain two-dimensional curves is to express them in *parametric form*. In the next example, you plot an ellipse centered at the point $(2, 0)$ with semiaxis of length 4 and $2\sqrt{3}$, which in parametric form is given by

$$\begin{aligned} x &= 4\cos(t) + 2 \\ y &= 2\sqrt{3}\sin(t) \end{aligned}$$

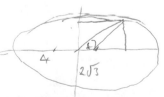

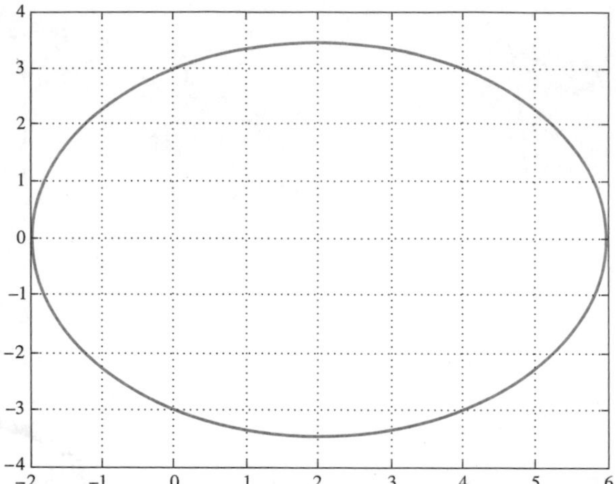

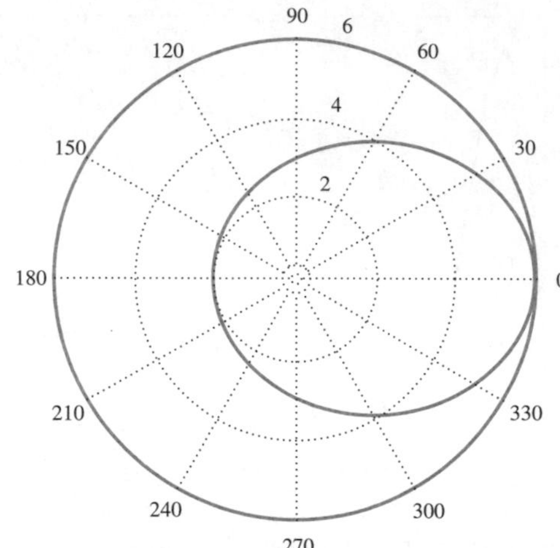

Figure 1.2 Ellipse in parametric form (*left*) and in polar coordinates (*right*).

for $0 \le t \le 2\pi$. The MATLAB commands for this are as follows:

```
>> t=[0:0.01:2*pi]; a=4; b=2*sqrt(3);
>> x=a*cos(t)+2;
>> y=b*sin(t);
>> grid on;
>> plot(x,y);
```

which result in the curve shown in Figure 1.2 (*left*).

A different way to express curves is through the use of *polar coordinates*. In MATLAB, this is implemented with the command `polar`. For example, the same ellipse as before can be obtained as follows:

```
>> t=[0:0.01:2*pi]; a=4; b=2*sqrt(3);
>> r=(b^2/a)./(1-sqrt(1-b^2/a^2)*cos(t));
>> polar(t,r);
```

The corresponding plot is shown in Figure 1.2 (*right*).

The `plot` function is extremely versatile, allowing you to use different colors and line styles, such as dashed or dotted lines. You can also depict data points in several different styles, such as stars, circles, diamonds, and the like, and even choose not to have any line connecting them. For example, the next commands produce a graph for the sine function using a red dashed line.

```
>> x=0:0.01:5; y=sin(x);
>> plot(x,y,'--r')
```

To see more than one graph in the same figure, you can use the command `hold on`. For example, the following commands produce a graph for the cosine

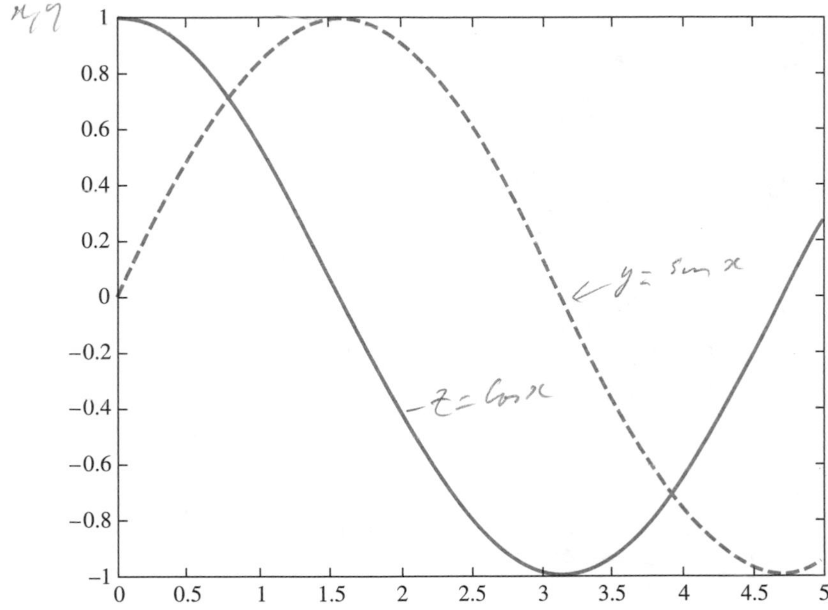

Figure 1.3 Sine and cosine functions.

function and depict it on the same figure as the graph just obtained for the sine function. The result is shown in Figure 1.3.

```
>> z=cos(x);
>> hold on;
>> plot(x,z)
```

To change MATLAB back to its default mode, that is, releasing old figures when new ones are created, type the command `hold off`.

Three-dimensional curves can be obtained by using the function `plot3`. For example, the following commands create the spiral depicted in Figure 1.4.

```
>> z=0:0.1:10;
>> x=exp(-0.2*z).*cos(pi*z);
>> y=exp(-0.2*z).*sin(pi*z);
>> plot3(x,y,z);
```

Surfaces

A standard way to generate surfaces of the form $z = f(x, y)$ in MATLAB is by using the command `surf(x,y,z)`, where **x**, **y** are vectors and **z** is a matrix. Before using it, you should complete a couple of preparatory steps. For example, suppose you want to plot the surface

$$z = xye^{-(x^2+y^2)}$$

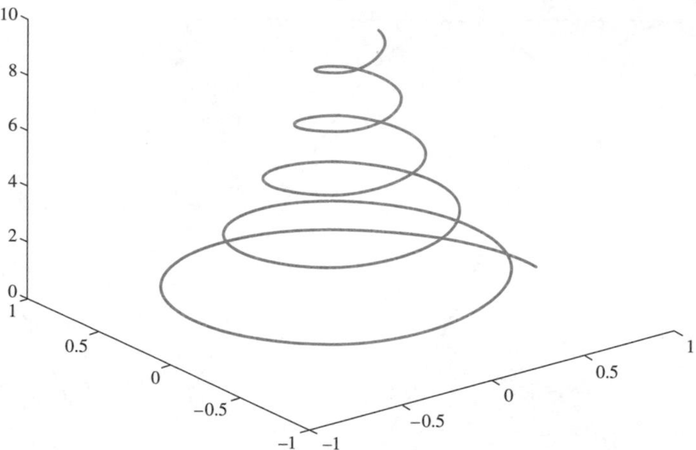

Figure 1.4 Example of a curve in three dimensions.

over the rectangular domain $-2 \leq x \leq 2$ and $-1 \leq y \leq 1$. Then, you should start by creating vectors for x and y, such as

```
>> x=-2:0.1:2;
>> y=-1:0.1:1;
```

These vectors are then used to compute the values for the matrix **z**. Notice in this example that, because **x** is a vector with 41 components and **y** is a vector with 21 components, the matrix **z** will contain the values for the function $z = f(x, y)$ over 21×41 points. Therefore, to make effective use of componentwise operations, you have to create a 21×41 matrix **X** whose rows are copies of the vector **x** and another 21×41 matrix **Y** whose columns are copies of the vector **y**. This is achieved by the command `meshgrid`:

```
>> [X,Y]=meshgrid(x,y);
```

The next two commands then compute the components for the matrix **z** and generate a surface with vertices at the points `(x(j),y(i),z(i,j))`. The result is shown in Figure 1.5

```
>> z=X.*Y.*exp(-X.^2-Y.^2);
>> surf(x,y,z);
```

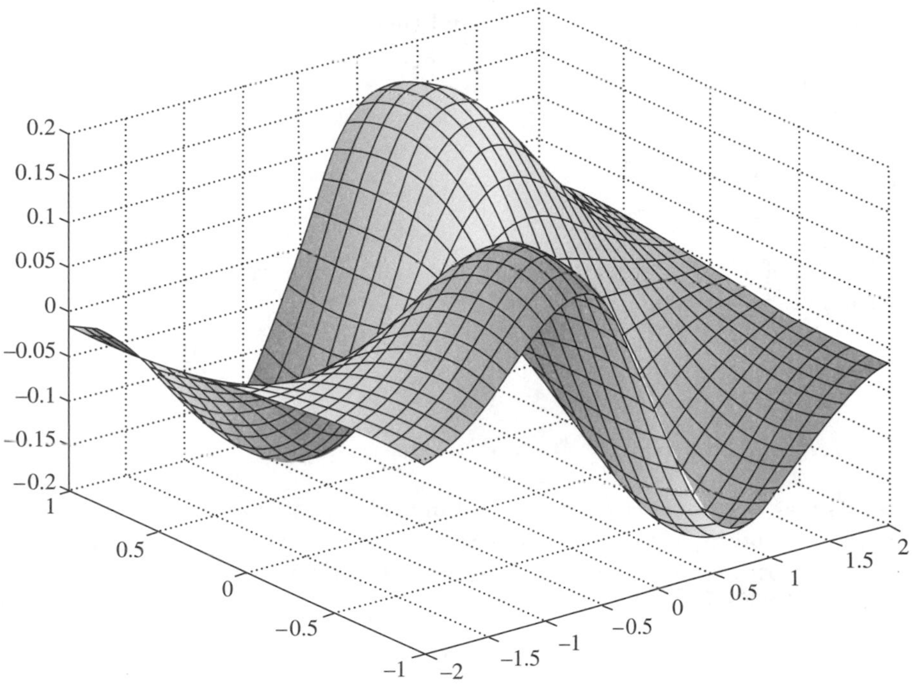

Figure 1.5 Example of a two-dimensional surface.

All figures in MATLAB can be supplemented with commands that manipulate their properties, such as axes, orientation, legends, displayed text, and so forth. Most of these effects can also be obtained interactively through MATLAB menus when a given figure is open. Our convention for the figures presented in this book is to include in the text only the commands that are necessary to generate them, leaving out all the commands used to change their displayed properties. In this way you can modify the appearance of any figure to suit your own visualization preferences.

Data files

We end this section by describing a simple and reliable way to store and retrieve MATLAB data. To save data currently available in the workspace, you can use the command `save`, which takes a file name and a list of variables as arguments. For instance, to save the variables `x,y,z`, used to generate the previous surface, into a file called `surface.mat`, use the command

```
>> save surface x y z
```

If you close MATLAB, or clear the content of its workspace, you can retrieve data previously stored in a file with the extension `.mat` by using the command `load`. For instance, close MATLAB, and then open it again and type

```
>> load surface
>> who

Your variables are:
   x  y  z
```

The variables `x y z` are returned to the system and can be used for further arithmetic operations.

1.7 Floating-Point Arithmetic

Let us now take a closer look at how numbers are manipulated by computers in general and by MATLAB in particular. For this, let us define the *floating-point number system* $\mathbb{F}$ with parameters (b, p, L, U) as the set of real numbers x that can be written in the form

$$x = \pm\left(a_0 + \frac{a_1}{b} + \frac{a_2}{b^2} + \cdots \frac{a_{p-1}}{b^{p-1}}\right) b^E, \tag{1.6}$$

for integers $0 \le a_j \le b-1$ and $L \le E \le U$. The positive integer parameter b is called the *base* for $\mathbb{F}$. For instance $b = 10$ for decimal numbers and $b = 2$ for binary numbers. The integers $L < 0 < U$ are called the *exponent range* of the system, while the positive integer p is called the *precision*. The system is said to be *normalized* when $a_0 = 0$ if and only if $x = 0$.

Our two primary examples of floating-point systems are the single-precision and double-precision IEEE systems (pronounced "I-triple-E", which stands for the Institute of Electrical and Electronics Engineers). For single-precision IEEE, we have

$$b = 2, \quad p = 24, \quad L = -126, \quad U = 127, \tag{1.7}$$

whereas for double-precision IEEE, the parameters are

$$b = 2, \quad p = 53, \quad L = -1022, \quad U = 1023. \tag{1.8}$$

Clearly, setting $L = -\infty$ and $U = p = \infty$ in (1.6) would allow for the representation of any conceivable real number. The fact that L, U, and p are finite is what distinguishes this type of *finite-precision* number system in which a small portion of real numbers can be represented. The following exercise tells you exactly how many different real numbers can be represented in a given normalized floating-point system.

Exercise 1.1 Show that the number of floating-point numbers for a normalized system with parameters (b, p, L, U) is

$$N = 2(b-1)b^{p-1}(U-L+1)+1.$$

For example, the single-precision IEEE system can represent about 4.26×10^9 (over four and a quarter billion) different numbers, while the double-precision IEEE supports about 1.84×10^{19} different numbers.

In MATLAB, all numerical values are stored and operated internally using double-precision IEEE, which is what you see when changing the display format to `long`. Keeping the display format `short`, however, only affects the way outputs are shown, but does not make MATLAB operate within the corresponding single-precision IEEE.

Special numbers

It is easy to see from Equation (1.6) that the smallest positive number that can be represented in a normalized floating-point system $\mathbb{F}$ is b^L, called the *underflow level* of the system. In double-precision IEEE, this is given by 2^{-1022}, or approximately 10^{-308}. In MATLAB, this minimal number is obtained by using the command `realmin`:

```
>> realmin

ans =
  2.2251e-308
```

Similarly, the largest number that can be represented in a normalized floating-point system $\mathbb{F}$ is called the *overflow level* of the system. From Equation (1.6), you see that this is given by $b^{U+1}(1-b^{-p})$, which for double-precision IEEE is $2^{1024}(1-2^{-53})$, or approximately 10^{308}. This can be retrieved with the MATLAB command `realmax`:

```
>> realmax

ans =
  1.7977e+308
```

Positive numbers that are smaller than `realmin` are automatically taken to be zero by MATLAB, while operations leading to positive numbers larger than `realmax` result in the output `Inf`, indicating an overflow. Similarly, operations resulting in negative numbers smaller than `-realmax` return the output `-Inf`. The only thing worse than obtaining `Inf` as the result of a calculation is the frightening outcome `NaN` (pronounced "not-a-number"), which corresponds to undefined numerical results such as `(Inf-Inf)` or `0/0`, and is MATLAB's way of letting you know that your numerical computations have gone astray.

Another important concept for floating-point number systems is what is called the *machine precision.* This is defined as the distance ε between 1 and the next larger number that can be represented in the system. Using Equation (1.6), you see that $\varepsilon = b^{1-p}$. In the single-precision IEEE system, this is 2^{-23}, or approximately 10^{-7}, whereas for double-precision IEEE the machine precision is 2^{-52}, or approximately 10^{-16}. In MATLAB, the machine precision can be retrieved by using the command `eps`:

```
>> eps

ans =
   2.2204e-16
```

At this point, you should make sure that you understand the difference between the underflow level and the machine precision. The fact that they are different originates from the fact that the numbers that can be represented in a given floating-point system are not equally spaced. Therefore, whereas the gap between zero and the next larger floating-point number is given by the underflow level, the (usually bigger) gap between 1 and the next larger floating-point number is given by the machine precision.

If a given number x cannot be exactly represented in a floating-point system $\mathbb{F}(b, p, L, U)$, then the process of choosing its floating-point approximation, denoted by $fl(x)$, is called *rounding.* Although many different rounding rules are available, the most commonly adopted (for instance, by the IEEE system) is called *rounding to nearest*, according to which $fl(x)$ is chosen as the floating-point number that is closest to x. In case of a tie, the rule dictates that $fl(x)$ should be the floating-point number whose last stored digit is even.

Round-off error

The distance between x and its floating-point representation $fl(x)$ is called the *round-off error.* It follows from the previous discussion that the underflow level is a measure of the round-off error for numbers close to zero, whereas the machine precision ε is a measure of the round-off error for numbers comparable to 1. More generally, we have that

$$|fl(x) - x| \leq \frac{\varepsilon}{2}|x|, \tag{1.9}$$

and so the machine precision can also be used to estimate the relative round-off error around any number x. In MATLAB, the estimate (1.9) for the absolute round-off error can also be obtained with the function `eps` using x as an argument:

```
>> eps(1000)

ans =
   1.1369e-13
```

Arithmetic operations on floating-point systems must be consistent with the way in which floating-point numbers are represented. For example, although the true sum for $x = 1.432 \times 10^4$ and $y = 3.598 \times 10^2$ is

$$x + y = 1.46798 \times 10^4,$$

in a system with $b = 10$ and $p = 4$, this sum is represented as 1.468×10^4. Therefore, such arithmetic operations lead to round-off errors even if the inputs are perfectly well-represented numbers. Moreover, they can lead to features that are not present when true arithmetic operations are considered.

To see this, let us use the MATLAB operators `+,-,*,/` to denote the floating-point arithmetic operations corresponding to true addition, subtraction, multiplication, and division. One of the unexpected features mentioned earlier is that floating-point addition is not necessarily associative. For instance, if **`delta`** denotes a floating-point number smaller than $\varepsilon/2$ but larger than $\varepsilon/4$, where ε denotes the machine precision, then we have that `(1+delta)+delta=1`, while `1+(delta+delta)>1`, as the next example shows:

```
>> delta=0.4*eps;
>> 1==(1+delta)+delta

ans =
     1

>> 1==1+(delta+delta)

ans =
     0
```

A similar unpleasant feature is provided by the following example:

```
>> delta=0.1*eps;
>> 0 == (1+delta)-(1-delta)

ans =
     1
```

You can see in this example that the true arithmetic result `2*delta`, which is a number much bigger than `realmin` and should therefore be easily distinguished from zero, got completely lost in the final result because `(1+delta)=1` and `(1-delta)=1` in the floating-point number system. That is, even though `delta` can be distinguished from zero, this distinction is lost when you operate with the relatively larger numbers `(1+delta)` and `(1-delta)`. This effect, known somewhat dramatically as *catastrophic cancellation*, should always be taken into consideration when dealing with the subtraction of two large numbers.

Because of these reservations, floating-point operations should be viewed with caution. In devising effective numerical methods, you should therefore try to keep the number of floating-point operations to a minimum, if only because they take computer time to be performed, let alone the rounding errors they produce. For this reason, the analysis of any numerical algorithm should include an estimate of the number of *fl*oating-point *op*erations it requires, what is called the *flop count* of the algorithm. In mathematics, as much as in any other enterprise, one should have as little flops as possible.

1.8 Error Analysis

To describe the different types of errors that occur in any numerical procedure, let us concentrate on the example of an input given by a single number $x \in \mathbb{R}$, upon which a theoretical procedure should be performed to produce another number $y \in \mathbb{R}$ as an output. To simplify the discussion even further, assume that the theoretical procedure corresponds to a function $f : \mathbb{R} \to \mathbb{R}$ so that ideally we have $y = f(x)$.

The input x is subject to the rounding error that appears when we represent it in the floating-point system, leading to an approximated input $\widehat{x}$. Also, the theoretical procedure f can rarely be implemented exactly, and so we must content ourselves with an approximated numerical procedure $\widehat{f}$, leading to an approximated output $\widehat{y}$. Therefore, the total error obtained by using the approximated procedure on the approximated input is

$$\widehat{f}(\widehat{x}) - f(x).$$

This total error can be further decomposed as

$$\widehat{f}(\widehat{x}) - f(x) = [\widehat{f}(\widehat{x}) - f(\widehat{x})] + [f(\widehat{x}) - f(x)]. \tag{1.10}$$

The first term in the preceding expression is generally called *computational error* because it measures the difference between the theoretical and the approximated procedures when applied to the same input. The second term is called *propagated data error* because it measures how the theoretical procedure alone is affected by the error in the input.

Truncation error

The computational error is usually caused by two independent components. One is the already familiar rounding error, caused by finite-precision arithmetic. A different one is the error caused by the fact that the approximated procedure $\widehat{f}$ is usually implemented as a truncation of the theoretical procedure f. For instance, we might decide to approximate a function f by a simpler polynomial of a given degree (according to Section 5.6). As another example, if f is the derivative of a given function, then we might want to approximate it by a *finite difference* (according to Section 6.1). This type of computational error is then called *truncation error*.

When the numerical method involves an iterative scheme, such as the methods for solving linear systems described in Section 2.6 or the methods for finding eigenvalues described in Section 4.6, the truncation errors may accumulate and play an essential role in computations. Typically, each iteration produces an approximated solution y_k, for the problem whose true solution is y, so that the truncation error is given by

$$e_k = |y_k - y|. \tag{1.11}$$

Rates of convergence

The iterative method is then said to be *convergent* if the truncation error e_k goes to zero as the number of iterations k gets large. Observe that the definition in Equation (1.11) ignores the round-off error to the extent that y_k is interpreted as the result of each iteration as if iterations were implemented using true arithmetic operations. Therefore, even a convergent method might not lead to any useful output if the round-off errors occurring at each step accumulate too fast. It then becomes paramount to know how fast an iterative method converges, which is referred to as the *rate of convergence*. We say that an iterative scheme converges at a rate r with constant $C > 0$ if the truncation errors satisfy

$$|e_{k+1}| \leq C|e_k|^r. \tag{1.12}$$

When $r = 1$, the convergence is said to be *linear*, whereas $r = 2$ and $r = 3$ are referred to as *quadratic* and *cubic* convergence rates, respectively. For instance, we show in Section 8.1 that root finding algorithms for zeros of nonlinear functions may converge both linearly (as in the bisection method) and quadratically (as in the Newton–Raphson method).

Moving forward with our definitions, the effect that the *relative input error*

$$\frac{|\widehat{x} - x|}{|x|} \tag{1.13}$$

has on the *relative output error*

$$\frac{|\widehat{y} - y|}{|y|} \tag{1.14}$$

is called the *sensitivity* of a mathematical procedure. One way to quantify this sensitivity is through the *condition number*, defined as the ratio of (1.14) by (1.13):

$$\text{cond} = \frac{|\widehat{y} - y|}{|\widehat{x} - x|}\frac{|x|}{|y|}.$$

In this way, a small condition number indicates a procedure for which the relative output error is comparable to the relative input error, which is then said to be *well-conditioned*. On the other hand, a procedure with a large condition number suggests that small relative input errors can be turned into quite large relative output errors, and the procedure is then termed *ill-conditioned.*

In practice, all that we are interested in is an estimate for the condition number. For example, assuming that $y = f(x)$ can be evaluated without any truncation error (that is, without replacing it with an approximated procedure $\widehat{f}$), we have that

$$\widehat{y} - y = f(\widehat{x}) - f(x) \approx f'(x)(\widehat{x} - x),$$

using the mean-value theorem. Therefore, the condition number for the problem of evaluating a differentiable function f at a given point x can be estimated as

$$\text{cond} = \frac{|\widehat{y} - y|}{|\widehat{x} - x|}\frac{|x|}{|y|} \approx \frac{|f'(x)(\widehat{x} - x)|}{|\widehat{x} - x|}\frac{|x|}{|f(x)|} = \left|\frac{xf'(x)}{f(x)}\right|.$$

Exercise 1.2 Show that the condition number for the problem of solving $y = f(x)$ for a given y and a monotone differentiable function f can be approximated by

$$\left|\frac{f(x)}{xf'(x)}\right|.$$

Looking back at Equation (1.10), you can see that the condition number quantifies the effect of propagated data errors because it refers to the exact procedure f, as opposed to the numerical approximation $\widehat{f}$. The concept of *stability*, on the other hand, refers to the effect that both computational and data errors have on the final output. That is, a numerical algorithm is *stable* if its output is relatively insensitive to the errors introduced during the computations, such as rounding and truncation.

1.9 Summary and Notes

We presented in this chapter the essential tools necessary to use MATLAB in the rest of the book, together with some elementary concepts in *Numerical Analysis*:

- **Section 1.1:** Simple arithmetic operations are performed in the MATLAB command window. The result of such operations can be assigned to variables and represented in different formats, such as `short` and `long`.

- **Section 1.2:** The basic objects handled by MATLAB are called *arrays*. An array with just one index is either a *column vector* or a *row vector*, depending on whether its components are arranged vertically or horizontally. An array with two indices is called a *matrix*, with the first index corresponding to a row and the second index to a column. A matrix with just one row and one column is called a *scalar*. The *colon* operator `:` is a versatile way to handle a sequence of indices. MATLAB can handle arrays with more than two indices, such as *cubes*.

- **Section 1.3:** Addition and subtraction of matrices are implemented with the operators `+` and `-` and are performed component by component. They are only defined for matrices of the same size. Matrix multiplication is implemented with the operator `*` and performed according to Equation (1.1), and requires the number of columns of the first matrix to be the same as the number of rows of the second matrix in the product. Vectors can be multiplied together as special cases of matrices. The product of a row vector by a column vector with the same number of components is a scalar, called the *inner product* between the vectors. A *cross product* is a special type of product valid only for vectors with three components. Matrix exponentiation is implemented with the `^` operator. Componentwise multiplication and exponentiation on matrices are performed by using the commands `.*` and `.^`.

- **Section 1.4:** MATLAB offers a large collection of built-in functions that implement more complicated numerical procedures. The syntax and use of each function can be learned by using the command `help` followed by the name of the function. The function `inv` finds the inverse of a *nonsingular* matrix and can be used to define a matrix division.

- **Section 1.5:** MATLAB can also be used as a high-level programming language. All programs in MATLAB are *M-files*, which can be either *functions* or *scripts*. Functions take several input arguments and return the result of the program in the form of output variables. Scripts are a list of commands with no input arguments or output variables. *Sequential statements* are parts of a program that are executed in sequence. *Branch statements* are parts of a program that are executed provided some conditions are satisfied by using the commands `if`, `elseif`, `else`, and `end`. *Loops* are parts of a program that are executed either a certain number of times, using the commands `for` and `end`, or while a certain condition remains true, using the commands `while` and `end`.

- **Section 1.6:** Simple two-dimensional graphics are generated by the MATLAB command `plot`, which takes several optional arguments specifying the properties of lines, points, colors, legends, axes, and so forth.

Curves in three-dimensions can be obtained by using the command `plot3`. Surfaces are obtained by using the command `surf`, which requires the specification of a two-dimensional grid obtained through `meshgrid`. Data can be saved and retrieved in MATLAB using the commands `save` and `load`.

- **Section 1.7:** Computers represent numbers in a *floating-point number system* characterized by its base, exponent range and precision. The two most popular systems are called single- and double-precision IEEE. The smallest and largest numbers that can be represented in MATLAB are given by the commands `realmax` and `realmin`. The *machine precision* is the distance between 1 and the next larger number that can be represented in the floating-point system. In MATLAB, this is given by `eps`. The error resulting from representing numbers in a finite-precision number system is called *round-off error*. Arithmetic operations are performed by a computer in a way that is consistent with the floating-point number systems and therefore do not satisfy the same properties as regular operations. The number of floating-point operations required by an algorithm is called its *flop count* and should be kept to a minimum.

- **Section 1.8:** The error incurred in numerical computations can be separated into *computational error*, caused by the difference between an approximated procedure and its exact counterpart, and *propagated data error*, measuring the effect of an approximated input. The computational error that arises by stopping a theoretically infinite sequence of iterations after a finite number of steps is called *truncation error*. The iterations are said to be *convergent* if the truncation error goes to zero as the number of iterations goes to infinity. The power law governing the decay of the error to zero determines the *rate of convergence* of the iterations. The *sensitivity* of a problem is the relative effect that perturbations on the input have on the output. Such effect is measured by the *condition number* of the problem and refers exclusively to the propagation of data errors. The concept of *stability* measures the effect that all errors introduced during a numerical procedure have on the final output.

A very complete source of information on MATLAB commands, built-in functions, and syntax is [11], which has the advantage of following closely the language used in the Help and tutorial pages of MATLAB. For more material on programming in MATLAB, we recommend [21], which uses an interplay between text and MATLAB expressions similar to our book. Our description of floating-point arithmetic and error analysis is based on [12].

1.10 Exercises

1. Let $z = x + iy$ be an arbitrary complex number.

   (a) Use MATLAB to evaluate the expression $e^x(\cos y + i \sin y)$ and compare it with the result obtained using the built-in function `exp`.

   (b) Write a MATLAB function whose input is z and whose output is the real numbers $r \geq 0$ and $-\pi \leq \theta \leq \pi$ in the polar representation $z = \rho e^{i\theta}$.

   (c) Evaluate *log(-1)* in MATLAB and interpret the result in terms of the complex logarithm of $z = -1$ using the polar representation given in the previous item.

2. Let

$$\mathbf{A} = \begin{bmatrix} -1 & 1 & -4 \\ 2 & 2 & 0 \\ 3 & 3 & 2 \end{bmatrix}, \qquad \mathbf{B} = \begin{bmatrix} 4 & 2 & 4 \\ 4 & -1 & 5 \\ 5 & 1 & -6 \end{bmatrix}.$$

   Perform the following operations in MATLAB and verify the result by hand:

   (a) Multiply the matrix $\mathbf{A}$ by the matrix $\mathbf{B}$.

   (b) Multiply each component of the matrix $\mathbf{A}$ by the corresponding component of the matrix $\mathbf{B}$.

   (c) Divide each component of the matrix $\mathbf{A}$ by the corresponding component of the matrix $\mathbf{B}$.

   (d) Perform the multiplications $\mathbf{AB}^{-1}$ and $\mathbf{A}^{-1}\mathbf{B}$.

   (e) Divide each component in the submatrix $\begin{bmatrix} 2 & 2 & 0 \\ 3 & 3 & 2 \end{bmatrix}$ by the corresponding component in the matrix $\mathbf{B}$.

   (f) Square each component in matrix $\mathbf{A}$.

3. Write a MATLAB function that calculates the angle between two vectors in $\mathbb{R}^3$ according to the formula

$$\cos\theta = \frac{\mathbf{x} \cdot \mathbf{y}}{\|\mathbf{x}\|_2 \|\mathbf{y}\|_2},$$

   where $\mathbf{x} \cdot \mathbf{y} = x_1 y_1 + x_2 y_2 + x_3 y_3$ and $\|\mathbf{x}\|_2 = \sqrt{x_1^2 + x_2^2 + x_3^2}$.

4. The *logistic map* is a simple way to generate chaotic behavior. It is defined by the iterations

$$x_{n+1} = r x_n (1 - x_n) \tag{1.15}$$

starting from an initial value $0 < x_0 < 1$. Write a MATLAB function with inputs r, x_0, and N that stores the sequence $x_1, x_2, \ldots, x_N$ given by Equation (1.15) as the components of a vector $\mathbf{x}$ and plots them with respect to n.

5. Use the function in the previous exercise to answer the following questions:
   (a) Analyze what happens to the sequence when $0 < r < 1$, using $N = 10$. Does the behavior depend on the initial value x_0? Does it depend on r?
   (b) Repeat the analysis for $1 < r < 2$, using $N = 10$. In particular, for fixed x_0, find an expression for the limit value in terms of r. Does the limit value depend on x_0?
   (c) Repeat the analysis for $2 < r < 3$, using $N = 100$. Observe what happens when r approaches 3.
   (d) Repeat the analysis for $3 < r < 3.4$.
   (e) What happens to the sequence when $r = 3.5$? Does the behavior depend on the initial value x_0?
   (f) What happens to the sequence when $r = 3.7$? Change the initial value x_0 by 0.001 and observe how the terms in the sequence change. Repeat this for several different initial values.

6. The *Lissajous curves* are given by the following parametric equations:

$$
\begin{aligned}
x &= A\cos(mt) \\
y &= B\sin(nt),
\end{aligned}
$$

   for $0 \le t \le 2\pi$. Write a MATLAB function that plots a Lissajous curve with parameters A, B, m, n and use it to answer the following questions.
   (a) For what parameter values is the Lissajous curve a circle?
   (b) For what parameter values is the Lissajous curve an ellipse?
   (c) Can the Lissajous curve be a parabola?
   (d) What happens to the Lissajous curves when the ratio m/n is irrational? Is the converse to this statement true?

7. The *hypotrochoids* are curves given by the following parametric equations:

$$
\begin{aligned}
x &= (R-r)\cos t + d\cos\left(\frac{R-r}{r}t\right) \\
y &= (R-r)\sin t - d\sin\left(\frac{R-r}{r}t\right),
\end{aligned}
$$

for $0 \le t \le K\pi$. Write a MATLAB function that plots a hypotrochoid with parameters R, r, d, K. Let $R = 1$, $d = 1.2$, and $K = 18$ and plot the hypotrochoids corresponding to $r = 0.9, 0.8, 0.7, 0.6, 0.5, 0.4, 0.3, 0.2,$ 0.1, and 0.05. Describe a mechanical way to trace a hypotrochoid with pen and paper and give a geometrical interpretation to the parameters R, r, d, K.

8. Use the command `polar` to graph the following curves given in polar coordinates:

   (a) $r = e^{\sin t} - 2\cos(4t)$, $0 \le t \le 2\pi$.

   (b) $r = 1 + 2\sin(t/2)$, $0 \le t \le 4\pi$.

   (c) $r = \sin^2(4\pi t) + \cos(4\pi t)$, $0 \le t \le 2\pi$.

9. Use the command `surf` to graph the following surfaces over intervals $-3 \le x \le 3$ and $-3 \le y \le 3$:

   (a) $z = |x| + |y|$.

   (b) $z = |xy|$.

   (c) $z = xy^2 - x^3$.

   (d) $z = xy^3 - yx^3$.

   (e) $z = x^2y^2e^{\ x^2-y^2}$.

   (f) $z = \sin(x^2 + y^2)$.

10. Consider a normalized floating-point number system with $b = 2$, $p = 3$, $L = -2$, and $U = 2$.

    (a) Find how many different numbers can be represented in this system.

    (b) Compute the underflow and overflow levels for the system, as well as its machine precision.

    (c) Plot all the numbers in this system in a horizontal line (in decimal format). Observe its granularity and the fact that the numbers are not equally spaced.

11. Calculate the *flop* count for performing the multiplication `A*B` of an $m \times n$ matrix $\mathbf{A}$ by an $n \times p$ matrix $\mathbf{B}$.

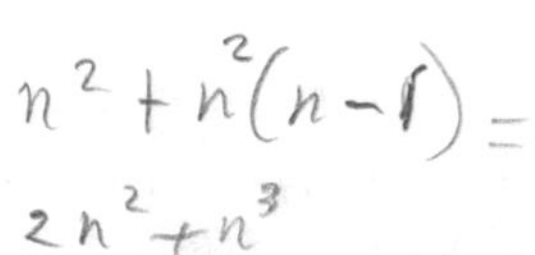

12. Using the following summation formulas:

$$\sum_{i=1}^{n} i = \frac{n(n+1)}{2}, \quad \sum_{i=1}^{n} i^2 = \frac{n(n+1)(2n+1)}{6},$$

find the *flop* count for the following sequence of commands:

```
a=0;
for p=1:n
     for q=p:n
          for r=q:n
               a=a+1;
          end
     end
end
```

What will be the value of a after the commands are executed?

13. One approximation for π consists of inscribing polygons in a circle of radius 1/2, starting with hexagons, and successively doubling the number of sides. The recurrence formula for the perimeters of such polygons is $p_{n+1} = 6 \cdot 2^n \cdot x_{n+1}$ where

$$x_{n+1} = \frac{\sqrt{x_n^2+1}-1}{x_n}, \qquad x_1 = 1/\sqrt{3}, \qquad n = 1, 2, \ldots \tag{1.16}$$

Observe that we can rewrite the preceding expression in the mathematically equivalent form

$$x_{n+1} = \frac{x_n}{\sqrt{x_n^2+1}+1}, \qquad x_1 = 1/\sqrt{3}, \qquad n = 1, 2, \ldots \tag{1.17}$$

Write a MATLAB function that calculates the first 26 approximations for π based on both Equations (1.16) and (1.17) and compare the results. Explain the differences in terms of rounding errors for floating-point operations.

14. Consider the function $f(x) = 1/\cos x$. Estimate the condition number for the problem of evaluating this function near the point $x = 1.5708$. Calculate the relative input and output errors when $\hat{x} = 1.57079$ and compare their ratio with your previous estimate for the condition number.

15. Estimate the condition number for the problem of solving the equation $x^3 - 3x^2 + 3x - 0.99 = 0$. Compare the solution to this equation with the solution to $x^3 - 3x^2 + 3x - 1 = 0$ and relate them with the condition number just calculated.

CHAPTER 2

Linear Systems

THIS CHAPTER CONSISTS OF the backbone for the entire book. Through the topic of linear systems, we explore the essential concepts of vector spaces and linear maps, which will then be used in every subsequent chapter. We give an overview of *Linear Algebra*, making full use of MATLAB to discuss its concepts, definitions, and main results.

The first four sections are of a more theoretical flavor. Section 2.1 contains the central definition of vector spaces, starting with the properties of familiar examples such as $\mathbb{R}^2$ and $\mathbb{R}^3$, and then presenting abstract vector spaces as a natural generalization. The most important technical concept from this section is linear independence, which is discussed in several examples, and leads to the definition of a basis, from where we can present the representation of vectors with respect to a given basis and the change of coordinates from one basis to another. Section 2.2 then proceeds with the equally important definition of linear maps, using MATLAB to explore some of their properties, such as null spaces and ranges. In Section 2.3, we show how linear systems can be written in matrix form, leading to a characterization of their solutions in terms of properties of the associated matrices. Finally, Section 2.4 introduces the concept of norms, which assign magnitudes to vectors and are therefore used to quantify the sensitivity of the solution of linear systems.

The remaining three sections have a more practical tone. Section 2.5 is centered around the Gaussian elimination technique for solving linear systems and the associated algorithms of backward and forward substitutions and LU factorization. We present step-by-step explanations, MATLAB codes and concrete examples of these methods. In Section 2.6, we turn to iterative methods for solving linear systems. We discuss the popular Jacobi and Gauss–Seidel methods, with particular emphasis on their convergence, analyzed through different MATLAB examples and counterexamples. We then present the more specialized technique of Cholesky factorization in Section 2.7. Besides being an important method for solving certain types of linear systems, this algorithm offers an opportunity to introduce the concepts of symmetric and positive definite matrices, which have repeated applications in later chapters.

Theoretical treatments of linear systems generally involve an alternative characterization of solutions formulated in terms of determinants, known as Cramer's rule. Because such characterization does not lead to efficient numerical methods, we present it in Section 2.8 only for completeness, in connection with the definition of determinants, which are important theoretical tools in their own right.

2.1 Vector Spaces

In Section 1.2, we used the term *vector* to represent a MATLAB array with only one index. In such context, vectors are usually referred to as "one-dimensional arrays," alluding to the fact that their components are displayed either as vertical or as horizontal lines (respectively, for column or row vectors). In this section, we provide the mathematical definition of a *vector space*—the arena for the development of linear algebra. Elements of vector spaces are naturally called vectors, and, at least at first encounter, a certain amount of care is necessary to distinguish between these two uses of the word. Vector spaces also carry an associated concept of "dimension," which differs significantly from the sense this word has when we refer to one-dimensional arrays. As usual in mathematics, after you gain confidence in dealing with these concepts, context alone will be enough for you to correctly identify the sense in which words are employed.

Vectors in $\mathbb{R}^2$

To help you develop intuition in the subject, we start with $\mathbb{R}^2$, the archetypical example of a vector space. It consists of the set of all elements of the form $\mathbf{x} = [x_1, x_2]$, where x_1 and x_2 are real numbers. The next step is to define operations on $\mathbb{R}^2$. If $\mathbf{x} = [x_1, x_2]$ and $\mathbf{y} = [y_1, y_2]$ are two such elements, then their sum is simply defined by

$$\mathbf{x} + \mathbf{y} = [x_1, x_2] + [y_1, y_2] := [x_1 + y_1, x_2 + y_2].$$

In other words, the sum of elements in $\mathbb{R}^2$ is defined componentwise and yields another element in $\mathbb{R}^2$. Using elementary properties of real numbers, it is now easy to show that the addition operation just defined is *commutative* ($\mathbf{x} + \mathbf{y} = \mathbf{y} + \mathbf{x}$), *associative* ($\mathbf{x} + (\mathbf{y} + \mathbf{z}) = (\mathbf{x} + \mathbf{y}) + \mathbf{z}$), admits an *additive identity* ($\mathbf{x} + [0, 0] = \mathbf{x}$), and that each element of $\mathbb{R}^2$ has an *additive inverse* ($\mathbf{x} + (-\mathbf{x}) = [0, 0]$).

Next, consider the multiplication of an element of $\mathbb{R}^2$ by a *scalar*. This is another term we used in Section 1.2, where it corresponded to a 1×1 MATLAB matrix, and it receives special treatment in this section. If $\lambda \in \mathbb{R}$ and $\mathbf{x} = [x_1, x_2] \in \mathbb{R}^2$, then their product is defined as

$$\lambda \mathbf{x} = \lambda [x_1, x_2] := [\lambda x_1, \lambda x_2].$$

You can see immediately that this multiplication operation is *associative* ($(\lambda\mu)\mathbf{x} = \lambda(\mu\mathbf{x})$), *distributive* ($(\lambda + \mu)\mathbf{x} = \lambda\mathbf{x} + \mu\mathbf{x}$ and $\lambda(\mathbf{x} + \mathbf{y}) = \lambda\mathbf{x} + \lambda\mathbf{y}$), and admits a *multiplicative identity* ($1\mathbf{x} = \mathbf{x}$).

These definitions conform with the geometric interpretation of an element $\mathbf{x} \in \mathbb{R}^2$ as a point in the plane with coordinates $[x_1, x_2]$. More specifically,

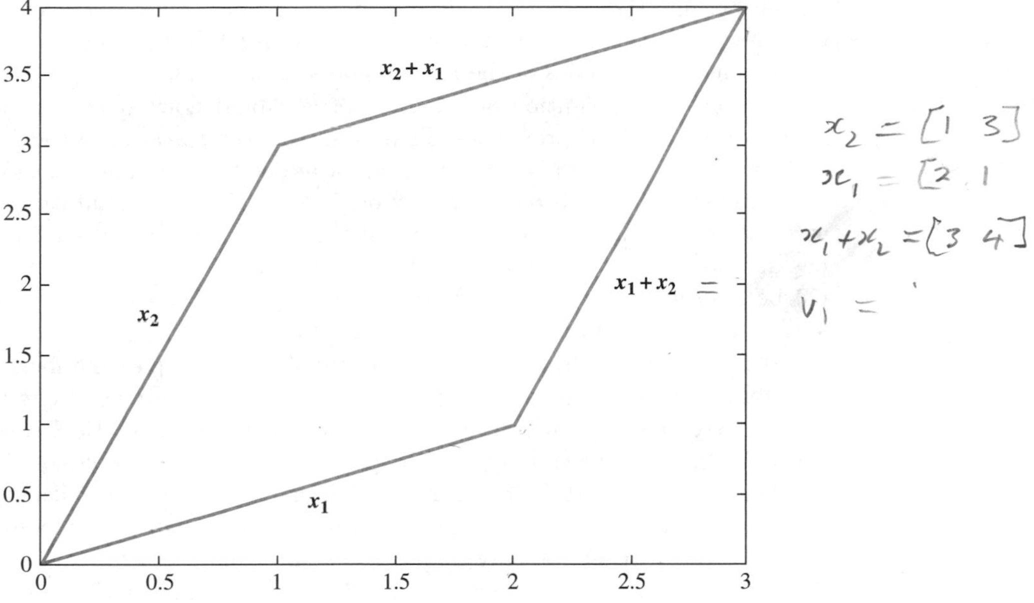

Figure 2.1 Parallelogram rule for vector addition.

they are well suited to the picture of $\mathbf{x}$ as an arrow based at the origin and "pointing" to the point $[x_1, x_2]$, which is the historical meaning of the term *vector* as used by physicists. For example, the vector $\lambda\mathbf{x}$ is an arrow pointing in the same direction as $\mathbf{x}$ if $\lambda > 0$ or in the opposite direction if $\lambda < 0$ but with length $|\lambda|$ times as big as the length of $\mathbf{x}$. The sum of two vectors in the plane obeys the parallelogram rule, which is illustrated in Figure 2.1, produced with the MATLAB commands

```
>> x1 = [2 1]; x2 = [1 3];
>> v1 = [0 x1(1) x1(1)+x2(1)];
>> w1 = [0 x1(2) x1(2)+x2(2)];
>> v2 = [0 x1(2) x1(2)+x1(1)];
>> w2 = [0 x2(2) x2(2)+x1(2)];
>> plot(v1,w1,'r',v2,w2,'b');
```

The generalization to the space $\mathbb{R}^n$ is straightforward: it consists of all n-tuples of the form $\mathbf{x} = [x_1, \ldots, x_n]$ with addition and multiplication by scalars defined by

$$\begin{aligned} \mathbf{x} + \mathbf{y} &= [x_1 + y_1, \ldots, x_n + y_n], \quad \forall \mathbf{x}, \mathbf{y} \in \mathbb{R}^n \\ \lambda\mathbf{x} &= [\lambda x_1, \ldots, \lambda x_n], \qquad \forall \lambda \in \mathbb{R}, \forall \mathbf{x} \in \mathbb{R}^n. \end{aligned}$$

Again, because these operations are defined componentwise, you can see immediately that they satisfy the properties listed before by simply appealing to elementary properties of the real numbers themselves.

Vector spaces

The abstract definition of a vector space should now come as a natural formalization of these properties. That is, a *real vector space* is a (nonempty) set V whose elements, called *vectors* , can be added together and multiplied by real numbers. The addition of two vectors should: (1) be commutative, (2) be associative, (3) admit an additive identity, and (4) be such that each element has an additive inverse. Moreover, the multiplication by real numbers should: (1) be associative, (2) be distributive with respect to sums of real numbers and vectors, and (3) admit a multiplicative inverse.

If we consider multiplication of elements of V by complex numbers, while maintaining all the properties listed earlier, then we obtain the definition of *complex vector spaces*, which are essential for many mathematical computations. An example of a complex vector space is the space of n-tuples of complex numbers, denoted by $\mathbb{C}^n$. In general, vector spaces are defined with a multiplication of elements in V by numbers in a field K, which are called scalars. In this book, however, we only consider real and complex numbers as scalars.

At this point, we can reconcile our different uses of the word *vector*. First, observe that pairs of the form (x_1, x_2) or triples of the form (x_1, x_2, x_3), depicted as arrows in two- and three-dimensional spaces and commonly referred to by physicists and engineers as vectors (recall the concepts of *force*, *momentum*, and *electric field*, for instance), are elements of the vector spaces $\mathbb{R}^2$ and $\mathbb{R}^3$, respectively, and are therefore vectors in the mathematical sense. The set of real numbers $\mathbb{R}$ is itself a vector space, implying that its elements can also be referred to as vectors (with just one component), even though no right-minded physicist would do so. More important, as we have just shown, the set $\mathbb{R}^n$ is a vector space for arbitrarily large integers n, although for $n > 3$ it is humanly impossible to visualize its elements as pointing arrows in space (some physicists have claimed that they can "see" objects in four-dimensional space, which just makes us wonder what they mean by "seeing"). Even more reassuring for the purposes of this book, observe that MATLAB row vectors with n components are just a concrete way to write down elements of $\mathbb{R}^n$. Specifically, the addition operator + when applied to two MATLAB row vectors with n components implements the componentwise addition of two elements of $\mathbb{R}^n$. Similarly, the multiplication operator * applied to a MATLAB scalar and a MATLAB row vector of n components just implements the multiplication of a vector in $\mathbb{R}^n$ by a scalar. The same is true for MATLAB column vectors, so that MATLAB vectors with their usual operations are specific examples of mathematical vector spaces. Their vector space properties can be verified with the following MATLAB commands involving three vectors in $\mathbb{R}^5$ and two scalars:

```
>> x=[1 -1 2 0 4]; y=[3 -2 5 8 11]; z=[7 1 -3 4 4.5];
>> lambda=2.5; mu=-3;
```

```
>> mu*(lambda*(x+y))-(mu*lambda)*(y+x)

ans =
     0     0     0     0     0

>> (lambda+mu)*((x+y)+z)-lambda*(x+(y+z))-mu*(x+(y+z))

ans =
     0     0     0     0     0
```

Before you think that row and column vectors are all that there is in the world of vector spaces, observe that the set of all $m \times n$ matrices, with componentwise addition and multiplication by scalars, is also a vector space, denoted by $\mathbb{M}^{m\times n}$. Therefore, in a mathematical sense, $m \times n$ matrices are also vectors. The same is also true for arrays with more than two indices, which are handled in a similar way by MATLAB, as we discussed in Section 1.2. To avoid confusion, we never willfully refer to matrices or arrays with more than two indices as vectors, unless we are forced to explicitly explore their vector space structure. Another example of vector space is the space $\mathcal{P}_n(\mathbb{R})$ of all polynomials of degree n with real coefficients, which is examined in Chapter 5.

Subspaces

Because vector spaces are sets, it is natural to explore the structure of their subsets. Consider, for example, the vector space $V = \mathbb{R}^3$. If we decide to look at the subset U of vectors having zero as their third component, we can verify that

$$[x_1, x_2, 0] + [y_1, y_2, 0] = [x_1 + y_1, x_2 + y_2, 0]$$

and

$$\lambda[x_1, x_2, 0] = [\lambda x_1, \lambda x_2, 0].$$

That is, the sum of two vectors in U is another vector in U, and so is the product of a vector in U by a real number. In mathematical terms, we say that U is *closed* with respect to addition and multiplication by scalars. Subsets U of vector spaces that have these closure properties are called *subspaces*, and it is a routine exercise to show that they are vector spaces in their own right, inheriting the same addition and multiplication by scalars of the original vector space V.

In the preceding example, if we decide to focus exclusively on vectors having a vanishing third component, we can concentrate on the vector space structure of the subspace U, which is naturally associated with a plane with elements $(x_1, x_2) \in \mathbb{R}^2$ and therefore simpler than the full space $\mathbb{R}^3$. This kind of reduction is crucial in applications. For instance, the solutions of an ordinary differential equation of order n form an n-dimensional subspace that is

much simpler than the original vector space of all functions with n continuous derivatives on which the ordinary differential equation is defined.

Often this simplification is not possible, in the sense that a subset of interest fails to be a subspace. For example, the sphere of radius 1 in $\mathbb{R}^3$, that is, the subset

$$S^1 = \{(x_1, x_2, x_3) \in \mathbb{R}^3 : x_1^2 + x_2^2 + x_3^2 = 1\}$$

is *not* a subspace of $\mathbb{R}^3$ because it is defined by a quadratic equation. To see this, try the following:

```
>> x=[1/sqrt(2) -1/sqrt(2) 0];
>> squares_x=x(1)^2+x(2)^2+x(3)^2

squares_x =
            1

>>  y=[1/sqrt(3) 1/sqrt(3) -1/sqrt(3)];
>> squares_y=y(1)^2+y(2)^2+y(3)^2

squares_y =
            1

>> z=x+y;
>> squares_z=z(1)^2+z(2)^2+z(3)^2

squares_z =
            2
```

Linear independence

Advancing in our study of general vector spaces, we say that a *linear combination* of vectors $\mathbf{v}_1, \mathbf{v}_2, \dots, \mathbf{v}_k \in V$ is a vector of the form

$$\lambda_1 \mathbf{v}_1 + \cdots + \lambda_k \mathbf{v}_k,$$

where $\lambda_1, \lambda_2, \dots, \lambda_k$ are arbitrary scalars. The set of all linear combinations of the vectors $\mathbf{v}_1, \mathbf{v}_2, \dots, \mathbf{v}_k$ is called their *linear span*, denoted by $\mathrm{Span}(\mathbf{v}_1, \dots, \mathbf{v}_k)$, and it is easy to see that this is a subspace of V.

The next definition is central to the entire subject of *Linear Algebra.* A set of vectors $\{\mathbf{v}_1, \dots, \mathbf{v}_k\}$ is said to be *linearly independent* if the identity

$$\lambda_1 \mathbf{v}_1 + \lambda_2 \mathbf{v}_2 + \cdots \lambda_k \mathbf{v}_k = \mathbf{0} \tag{2.1}$$

implies that there is no index i for which $\lambda_i \neq 0$. If a set of vectors is *not* linearly independent, then it is naturally said to be *linearly dependent.*

For a concrete understanding of linear independence, let us start with simple vector spaces. If $V = \mathbb{R}$, vectors are simply real numbers. Then, the only (nonempty) linearly independent sets are those containing exactly one vector. To see this, suppose that a set contains, for example, the two vectors $\mathbf{v}$ and $\mathbf{u}$ (with, say, $\mathbf{u} \neq 0$). Then, we can always write

$$\mathbf{v} + \lambda \mathbf{u} = 0$$

for a scalar λ. All we need to do is take $\lambda = -\mathbf{v}/\mathbf{u}$, because $\mathbf{v}$ and $\mathbf{u}$ are, after all, just real numbers. It is easy to see how this generalizes to a set with an arbitrary number of vectors. In other words, any set of two or more vectors in $\mathbb{R}$ is linearly dependent.

Moving up to $V = \mathbb{R}^2$, you can see that two vectors $\mathbf{v} = [v_1, v_2]$ and $\mathbf{u} = [u_1, u_2]$ are linearly dependent if and only if there exists a scalar λ such that $[v_1, v_2] = \lambda[u_1, u_2]$. But geometrically this means that $\mathbf{v}$ and $\mathbf{u}$ are co-linear. That is, two vectors in $\mathbb{R}^2$ are linearly independent if and only if they do *not* lie on the same line. But what about a set of three vectors?

Exercise 2.1 Let $\mathbf{v} = [v_1, v_2]$, $\mathbf{u} = [u_1, u_2]$, and $\mathbf{w} = [w_1, w_2]$ be three vectors in $\mathbb{R}^2$, assumed to be pairwise linearly independent. Show that it is *always* possible to find scalars λ and μ such that

$$\lambda \mathbf{v} + \mu \mathbf{u} + \mathbf{w} = \mathbf{0}.$$

In other words, any set of more than two vectors in $\mathbb{R}^2$ is linearly dependent. For example, for the vectors $\mathbf{v} = [-1, 2]$, $\mathbf{u} = [2, 5]$, and $\mathbf{w} = [1, 3]$ you have

```
>> v=[-1 2]; u=[2 6];  w=[2 1];
>> v-0.5*u+w

ans =
     0     0
```

Next, consider $V = \mathbb{R}^3$. Then, it is still true that any set of two vectors is linearly dependent if and only if the vectors lie on the same line. For sets of three vectors, however, there is no simple criterion to assert linear dependence or independence. That is not to say that there is no way to check whether a *particular* set of three vectors is linearly dependent or independent, and in fact later in Section 2.2, we describe a possible way of doing it. It is just that you can't determine their linear dependence just by looking at them. For instance, the example

```
>> v_1=[1 2 1]; v_2=[2 4 0]; v_3=[3 6 -1];
>> v_1-2*v_2+v_3

ans =
     0     0     0
```

shows that the vectors $\mathbf{v}_1 = [1, 2, 1]$, $\mathbf{v}_2 = [2, 4, 0]$, and $\mathbf{v}_3 = [3, 6, -1]$ are linearly dependent as $\mathbf{v}_1 = 2\mathbf{v}_2 - \mathbf{v}_3$. On the other hand, the next exercise shows that similar-looking vectors can be linearly independent.

Exercise 2.2 Show from the definition that the vectors $\mathbf{u}_1 = [1, 2, 1]$, $\mathbf{u}_2 = [2, 4, 0]$, and $\mathbf{u}_3 = [3, -6, -1]$ are linearly independent.

Basis

Looking back at Equation (2.1), you can see that an equivalent way of saying that a set of vectors $\{\mathbf{v}_1, \ldots, \mathbf{v}_k\}$ is linearly dependent is that at least one of them (say, $\mathbf{v}_k$) can be written as a linear combination of the others. In this case, $\text{Span}(\mathbf{v}_1, \mathbf{v}_2, ..., \mathbf{v}_k)$ is the same as $\text{Span}(\mathbf{v}_1, \mathbf{v}_2, ..., \mathbf{v}_{k-1})$. This suggests that the concept of linear independence is intimately related to the space spanned by a given set of vectors. Indeed, we say that a vector space V is *finite-dimensional* if there exists a finite set of vectors $\{\mathbf{v}_1, \ldots, \mathbf{v}_n\}$ such that $\text{Span}(\mathbf{v}_1, \ldots, \mathbf{v}_n) = V$. If, in addition to spanning V, the set of vectors $\{\mathbf{v}_1, \ldots, \mathbf{v}_n\}$ is linearly independent, then it is said to be a *basis* for a vector space V.

Several important results about linearly independent vectors can be proved at this point, all of which are extensively discussed in *Linear Algebra* courses. For instance, if a vector space V is spanned by a finite set of vectors, not necessarily linearly independent, then we can reduce the number of vectors in this set (by removing those that are linear combinations of the others) until we are left with a basis for V. In the opposite direction, if we know that V is finite-dimensional, then any linearly independent set of vectors in it can be extended to form a basis for V (by adding other linearly independent vectors to the set until they span the whole V). Moreover, such a set cannot be extended further because any set with more vectors than a basis will necessarily be linearly dependent. Finally, any two bases of a finite-dimensional vector space V have the same number of vectors, called the *dimension* of V and denoted by $\dim V$.

The culmination of these results is the following crucial theorem for finite-dimensional vector spaces:

Theorem 2.1 *If $\{\mathbf{v}_1, \ldots, \mathbf{v}_n\}$ is a basis for a vector space V, then each vector $\mathbf{x} \in V$ can be uniquely written as*

$$\mathbf{x} = x_1\mathbf{v}_1 + x_2\mathbf{v}_2 + \cdots + x_n\mathbf{v}_n. \tag{2.2}$$

The scalars $(x_1, \ldots, x_n)$ appearing in Equation (2.2) are called the *coordinates* of the vector $\mathbf{x}$ with respect to the basis $\{\mathbf{v}_1, \ldots, \mathbf{v}_n\}$. Although the decomposition (2.2) of a vector in terms of the elements of a given basis is unique, a finite-dimensional vector space possesses, in general, many different bases, leading to different ways to decompose the same vector.

Change of basis

Suppose, then, that $\{\mathbf{v}_1, \dots, \mathbf{v}_n\}$ and $\{\widetilde{\mathbf{v}}_1, \dots, \widetilde{\mathbf{v}}_n\}$ are two different bases for the same vector space V. We can then express each vector $\mathbf{v}_j$ as the unique linear combination

$$\mathbf{v}_j = c_{1j}\widetilde{\mathbf{v}}_1 + c_{2j}\widetilde{\mathbf{v}}_2 + \cdots c_{nj}\widetilde{\mathbf{v}}_n = \sum_{i=1}^{n} c_{ij}\widetilde{\mathbf{v}}_i. \tag{2.3}$$

The scalars $\{c_{ij}\}$ can then be collected together as the entries of an $n \times n$ matrix $\mathbf{C}$, called the *change of basis matrix* from $\{\widetilde{\mathbf{v}}_1, \dots, \widetilde{\mathbf{v}}_n\}$ to $\{\mathbf{v}_1, \dots, \mathbf{v}_n\}$. Therefore, if $(x_1, \dots, x_n)$ are the coordinates of a vector $\mathbf{x} \in V$ with respect to the basis $\{\mathbf{v}_1, \dots, \mathbf{v}_n\}$, we have

$$\begin{aligned}
\mathbf{v} &= x_1\mathbf{v}_1 + x_2\mathbf{v}_2 + \cdots + x_n\mathbf{v}_n \\
&= x_1\sum_{i=1}^{n} c_{i1}\widetilde{\mathbf{v}}_i + x_2\sum_{i=1}^{n} c_{i2}\widetilde{\mathbf{v}}_i + \cdots + x_n\sum_{i=1}^{n} c_{in}\widetilde{\mathbf{v}}_i \\
&= (x_1c_{11} + x_2c_{12} + \cdots + x_nc_{1n})\widetilde{\mathbf{v}}_1 + (x_1c_{21} + x_2c_{22} + \cdots + x_nc_{2n})\widetilde{\mathbf{v}}_2 + \\
&\quad \cdots + (x_1c_{n1} + x_2c_{n2} + \cdots + x_nc_{nn})\widetilde{\mathbf{v}}_n.
\end{aligned}$$

Thus, if $(\widetilde{x}_1, \dots, \widetilde{x}_n)$ are the coordinates of $\mathbf{v}$ with respect to the basis $\{\widetilde{\mathbf{v}}_1, \dots, \widetilde{\mathbf{v}}_n\}$, the preceding expression tells us that

$$\widetilde{x}_i = \sum_{j=1}^{n} c_{ij}x_j. \tag{2.4}$$

Comparing this expression with Equation (2.3), we say that the coordinates of vectors are transformed in a *contravariant* way under a change of basis.

We can reinterpret Equation (2.4) by saying that $(\widetilde{x}_1, \dots, \widetilde{x}_n)$ are the components of the column vector obtained by multiplying the matrix $\mathbf{C}$ by the column vector with components $(x_1, \dots, x_n)$. Because both bases consist of linearly independent vectors, we see that $(\widetilde{x}_1, \dots, \widetilde{x}_n) = (0, \dots, 0)$ if and only if $(x_1, \dots, x_n) = (0, \dots, 0)$. In other words, the change of basis matrix $\mathbf{C}$ is invertible. Conversely, given any basis $\{\widetilde{\mathbf{v}}_1, \dots, \widetilde{\mathbf{v}}_n\}$ and an invertible $n \times n$ matrix $\mathbf{C}$, we can obtain a new basis $\{\mathbf{v}_1, \dots, \mathbf{v}_n\}$ through Equation (2.3).

Let us see how these results fit together for example of the vector space $\mathbb{R}^3$. First of all, it is clear that the vectors

$$\mathbf{e}_1 = [1, 0, 0], \quad \mathbf{e}_2 = [0, 1, 0], \quad \mathbf{e}_3 = [0, 0, 1], \tag{2.5}$$

are linearly independent. Furthermore, they span $\mathbb{R}^3$, since we can write any vector $\mathbf{v} = [v_1, v_2, v_3] \in \mathbb{R}^3$ as

$$\mathbf{v} = v_1\mathbf{e}_1 + v_2\mathbf{e}_2 + v_3\mathbf{e}_3.$$

Therefore, $\{\mathbf{e}_1, \mathbf{e}_2, \mathbf{e}_3\}$ is a basis for $\mathbb{R}^3$, called the *canonical basis*. It follows from the previous discussion that any other basis for $\mathbb{R}^3$ must contain exactly three vectors, and any set of three linearly independent vectors in $\mathbb{R}^3$ is a basis for it. Also, any set of four or more vectors in $\mathbb{R}^3$ is necessarily linearly dependent. As a different example of a basis for $\mathbb{R}^3$, you can take the vectors

$$\mathbf{f}_1 = [1, 2, 1], \quad \mathbf{f}_2 = [1, 3, 2], \quad \mathbf{f}_3 = [0, 1, 2]. \tag{2.6}$$

Then, you can observe that

$$\begin{aligned} \mathbf{e}_1 &= 4\mathbf{f}_1 - 3\mathbf{f}_2 + \mathbf{f}_3 \\ \mathbf{e}_2 &= -2\mathbf{f}_1 + 2\mathbf{f}_2 - \mathbf{f}_3 \\ \mathbf{e}_3 &= \mathbf{f}_1 - \mathbf{f}_2 + \mathbf{f}_3, \end{aligned}$$

so that the change of basis matrix from $\{\mathbf{f}_1, \mathbf{f}_2, \mathbf{f}_3\}$ to $\{\mathbf{e}_1, \mathbf{e}_2, \mathbf{e}_3\}$ is

$$\mathbf{C} = \begin{bmatrix} 4 & -2 & 1 \\ -3 & 2 & -1 \\ 1 & -1 & 1 \end{bmatrix}.$$

Therefore, because in terms of the canonical basis $\{\mathbf{e}_1, \mathbf{e}_2, \mathbf{e}_3\}$ the vector $\mathbf{v} = [-1, 2.5, 7]$ can be decomposed as

$$\mathbf{v} = -\mathbf{e}_1 + 2.5\mathbf{e}_2 + 7\mathbf{e}_3,$$

its coordinates in terms of the basis $\{\mathbf{f}_1, \mathbf{f}_2, \mathbf{f}_3\}$ can be obtained by

```
>> [4 -2 1;-3 2 -1;1 -1 1]*[-1;2.5;7]

ans =
   -2.0000
    1.0000
    3.5000
```

That is, the vector $\mathbf{v}$ can be decomposed as

$$\mathbf{v} = -2\mathbf{f}_1 + \mathbf{f}_2 + 3.5\mathbf{f}_3,$$

which can be verified by

```
>> f_1=[1 2 1];f_2=[1 3 2]; f_3=[0 1 2];
>> v=-2*f_1+f_2+3.5*f_3

v =
      -1.0000     2.5000     7.0000
```

We can generalize this to the vector space $\mathbb{R}^n$, with a canonical basis of elementary vectors $\mathbf{e}_i$, that is, vectors with the ith component equal to 1 and all other components equal to zero. Because there are n such vectors, $\mathbb{R}^n$ is an

n-dimensional vector space. For a different example, the vector space $\mathbb{M}^{m\times n}$ also has a canonical basis of elementary matrices $\mathbf{E}_{ij}$, that is, matrices with the entry at the ith row and jth column equal to 1 and all other entries equal to zero. Because there are mn such matrices, we see that $\mathbb{M}^{m\times n}$ is a vector space of dimension mn. As another example, the *monomials* $\{1, x, x^2, \ldots, x^n\}$ form a canonical basis for the vector space $\mathcal{P}_n(\mathbb{R})$ of polynomials of degree n, from which we conclude that $\mathcal{P}_n(\mathbb{R})$ is an $(n+1)$-dimensional vector space.

If a vector space cannot be spanned by a finite set of vectors, then it is said to be *infinite dimensional.* An example of an infinite-dimensional vector space is the set of all polynomials with real (or complex) coefficients, denoted by $\mathcal{P}(\mathbb{R})$ (or $\mathcal{P}(\mathbb{C})$). Other popular examples are *function spaces*, that is, vector spaces of functions that satisfy certain defining properties. For instance, the space of all continuous real-valued functions on the interval $[0,1]$, denoted by $\mathcal{C}([0,1],\mathbb{R})$, is a vector space.

Exercise 2.3 Complete the definition of $\mathcal{P}(\mathbb{R})$ and $\mathcal{C}([0,1],\mathbb{R})$ by specifying their addition and multiplication by scalars, and then verify that the spaces satisfy the general properties of vector spaces. In each case, show that, given an arbitrary *finite* set of vectors, you can always find other vectors that cannot be spanned by them. Conclude that these spaces are infinite-dimensional.

The techniques for dealing with infinite-dimensional vector spaces are much harder than for finite-dimensional one, and are the subject of *Functional Analysis*. Although a full account of these techniques is outside the scope of this book, functional analytic ideas permeate most important applications, from the theory of polynomial interpolation of functions (Chapter 5) to the solutions of partial differential equations (Chapter 10). In these chapters, you will see how infinite-dimensional spaces can be systematically replaced by finite-dimensional ones through a *truncation* technique.

2.2 Linear Maps

The defining features of vector spaces are the properties of their addition and multiplication by scalars. We now turn to the study of maps preserving these properties. This is a general strategy in mathematics: whenever we encounter a type of set with some special structure, we are immediately led to the study of maps that preserve this structure. You will meet this idea again several times in your mathematical career, for example, when studying continuous maps between topological spaces, or differentiable maps between differentiable manifolds, or homomorphisms between groups.

A *linear map* from a vector space V to a vector space W is a map $\mathbf{A} : V \mapsto W$ satisfying

$$\mathbf{A}(\mathbf{u}+\mathbf{v}) = \mathbf{A}(\mathbf{u}) + \mathbf{A}(\mathbf{v}) \qquad \text{for all } \mathbf{u},\mathbf{v} \in V \tag{2.7}$$

and

$$\mathbf{A}(\lambda\mathbf{v}) = \lambda\mathbf{A}(\mathbf{v}) \qquad \text{for all scalars } \lambda \text{ and all } \mathbf{v} \in V. \tag{2.8}$$

Whenever there is no ambiguity, we denote the value of the map $\mathbf{A}$ at the vector $\mathbf{v}$ simply by $\mathbf{Av}$. If $\mathbf{A}$ and $\mathbf{B}$ are two linear maps acting on the same vector space V, then

$$\begin{aligned}(\mathbf{A}+\mathbf{B})\mathbf{v} &:= \mathbf{Av}+\mathbf{Bv}, \qquad \text{for all } \mathbf{v} \in V\\ (\lambda\mathbf{A})\mathbf{v} &:= \lambda\mathbf{Av}, \qquad \text{for all scalars } \lambda \text{ and all } \mathbf{v} \in V.\end{aligned}$$

You can immediately verify that the set of all linear maps on a vector space is itself a vector space. Similarly, given two linear maps $\mathbf{A} : V \mapsto W$ and $\mathbf{B} : W \mapsto Z$, their composition is the linear map given by

$$(\mathbf{AB})\mathbf{v} := \mathbf{A}(\mathbf{Bv}), \qquad \forall \mathbf{v} \in V.$$

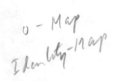

The two simplest linear maps are the *zero* and the *identity* maps, defined respectively by $\mathbf{0v} = \mathbf{0}$ and $\mathbf{1v} = \mathbf{v}$ for all $\mathbf{v} \in V$. Before dealing with the maps that will occupy our attention for the rest of the chapter, let us mention a few examples of linear maps on infinite-dimensional vector spaces, using the vector space $\mathcal{P}(\mathbb{R})$ of all polynomials with real coefficients as a prototype space. If $p \in \mathcal{P}(\mathbb{R})$, then the basic rules of *Calculus* dictate that the following are linear maps:

$$\begin{aligned}Dp(x) &:= \frac{dp(x)}{dx} \qquad \text{(differentiation)},\\ Tp(x) &:= \int p(x)dx \qquad \text{(integration)},\\ Mp(x) &:= xp(x) \qquad \text{(multiplication by } x\text{)}.\end{aligned}$$

From now on we concentrate on linear maps on finite-dimensional vector spaces. For example, consider the map from $\mathbb{R}^3$ to $\mathbb{R}^2$ given by

$$\mathbf{A}(x_1, x_2, x_3) = (2x_1 - x_2 + x_3, 3x_1 + x_2 - x_3). \tag{2.9}$$

That is, $\mathbf{A}$ maps the three-dimensional vector with components (x_1, x_2, x_3) into the two-dimensional vector $(2x_1 - x_2 + x_3, 3x_1 + x_2 - x_3)$. To see how this map acts on specific vectors, all you need to do is insert their components into Equation (2.9). For example,

$$\mathbf{A}(1, -1, 2) = (5, 0).$$

A more concise way of expressing Equation (2.9) makes use of matrices and column vectors. Observe that Equation (2.9) is equivalent to the matrix multiplication

$$\begin{bmatrix} 2 & -1 & 1 \\ 3 & 1 & -1 \end{bmatrix} \begin{bmatrix} x_1 \\ x_2 \\ x_3 \end{bmatrix} = \begin{bmatrix} 2x_1 - x_2 + x_3 \\ 3x_1 + x_2 - x_3 \end{bmatrix} \tag{2.10}$$

We can therefore identify the linear map $\mathbf{A}$ with the 2×3 matrix appearing in Equation (2.10) and rewrite its action on vectors in $\mathbb{R}^3$ as matrix multiplication of $\mathbf{A}$ by column vectors with three components. In MATLAB, the matrix multiplication in Equation (2.10) is implemented as follows:

```
>> A=[2 -1 1; 3 1 -1];
>> x=[1 -1 2]';
>> A*x

ans =
     5
     0
```

To show that the transformation $\mathbf{A}$ given by Equation (2.9) is linear, you need to perform a somewhat tedious calculation on general vectors and scalars, that is, start with vectors of the form (x_1, x_2, x_3) and (y_1, y_2, y_3) and a scalar λ, plug them into Equation (2.9), and verify that the result behaves according to the definition of a linear map. Alternatively, using the matrix notation in Equation (2.10), you can deduce the linearity of the transformation $\mathbf{A}$ from the linearity of matrix multiplication, that is, the fact that

$$\mathbf{A}(\mathbf{x} + \lambda\mathbf{y}) = \mathbf{A}\mathbf{x} + \lambda\mathbf{A}\mathbf{y},$$

for any 2×3 matrix $\mathbf{A}$ and any vectors $\mathbf{x}, \mathbf{y} \in \mathbb{R}^3$ and scalar λ. This is just a small instance of the usefulness of the matrix notation.

Matrix representation

Having used a convenient matrix notation for the linear map of the previous example, a natural question is to what extent other linear maps can also be represented by matrices. The answer is that *any* linear map between finite-dimensional vector spaces can be expressed in this way. This result is of such significance to our book that we are going to establish it in detail now.

The key idea is how to understand general linear maps in terms of bases. Suppose that $\mathbf{A} : V \mapsto W$ is a linear map on V and that $\{\mathbf{v}_1, \ldots, \mathbf{v}_n\}$ is a basis for V. Suppose further that we know how $\mathbf{A}$ acts on each of the basis vectors, that is, we know $\mathbf{A}\mathbf{v}_j$ for $j = 1, \ldots, n$. Recall from Equation (2.2) that an arbitrary vector $\mathbf{x} \in V$ can be uniquely written as

$$\mathbf{x} = x_1\mathbf{v}_1 + \cdots + x_n\mathbf{v}_n,$$

for scalars $\{x_1, \ldots, x_n\}$. Therefore, using the linearity of $\mathbf{A}$, we conclude that

$$\mathbf{A}\mathbf{x} = \mathbf{A}(x_1\mathbf{v}_1 + \cdots + x_n\mathbf{v}_n) = x_1\mathbf{A}\mathbf{v}_1 + \cdots + x_n\mathbf{A}\mathbf{v}_n.$$

That is, the values of $\mathbf{A}\mathbf{v}_1, \ldots, \mathbf{A}\mathbf{v}_n$ determine the values of $\mathbf{A}$ on arbitrary vectors in V.

Now let $\{\mathbf{w}_1, \ldots, \mathbf{w}_m\}$ be a basis for W (notice that m can be different from n). Then, each $\mathbf{A}\mathbf{v}_j$, being itself a vector in W, can be uniquely expressed as

$$\mathbf{A}\mathbf{v}_j = a_{1j}\mathbf{w}_1 + \cdots a_{mj}\mathbf{w}_m,$$

for scalars $\{a_{1j}, \ldots, a_{mj}\}$. Notice that the first index in this set of scalars runs from 1 to m, corresponding to the different elements in the basis for the m-dimensional vector space W. The second index in the preceding equation is fixed because we have decided to write down an expression for a fixed vector $\mathbf{A}\mathbf{v}_j$. The complete specification of the linear map $\mathbf{A}$, however, requires that we write down an expression for each $\mathbf{A}\mathbf{v}_j$, with j ranging from 1 to n.

Therefore, once we choose bases for the vector spaces V and W, a general linear map $\mathbf{A} : V \mapsto W$ is completely specified by the set of scalars a_{ij}, with $i = 1, \ldots, m$ and $j = 1, \ldots, n$. As you might have guessed by now, a convenient way to represent this set of scalars with two indices is through an $m \times n$ matrix. To conclude the identification between linear maps and matrices, observe that

$$\begin{aligned} \mathbf{A}\mathbf{x} &= x_1\mathbf{A}\mathbf{v}_1 + \cdots + x_n\mathbf{A}\mathbf{v}_n \\ &= x_1 \sum_{i=1}^{m} a_{i1}\mathbf{w}_i + \cdots + x_n \sum_{i=1}^{m} a_{in}\mathbf{w}_i \\ &= \sum_{i=1}^{m} \left(\sum_{j=1}^{n} a_{ij}x_j \right) \mathbf{w}_i. \end{aligned}$$

You should now recognize that the summands inside brackets in the preceding expression correspond exactly to the rule of matrix multiplication. That is, if we identify the linear map $\mathbf{A}$ with the $m \times n$ matrix with entries $\{a_{ij}\}$, and the vector $\mathbf{x} \in V$ with the column vector with components $(x_1, \ldots, x_n)$, then we can write

$$\mathbf{A}\mathbf{x} = \begin{bmatrix} a_{11} & \cdots & a_{1n} \\ \vdots & & \vdots \\ a_{m1} & \cdots & a_{mn} \end{bmatrix} \begin{bmatrix} x_1 \\ \vdots \\ x_n \end{bmatrix}.$$

Exercise 2.4 Find the matrix representation for the derivative map $D : \mathcal{P}_4(\mathbb{R}) \to \mathcal{P}_4(\mathbb{R})$ with respect to the canonical basis $\{1, x, x^2, x^3, x^4\}$ for $\mathcal{P}_4(\mathbb{R})$.

You should always remember that the entries $\{a_{ij}\}$ depend on the bases that are chosen for the vector spaces V and W. Different bases give rise to different matrices representing the same linear map.

As an important example of the effect of a change of basis, consider a linear map from $\mathbb{R}^n$ to $\mathbb{R}^n$ whose matrix representation with respect to the canonical basis $\{\mathbf{e}_1, \ldots, \mathbf{e}_n\}$ is $\mathbf{A}$ and let $\{\mathbf{v}_1, \ldots, \mathbf{v}_n\}$ be a new basis, which we can arrange as the columns of an invertible matrix $\mathbf{V} = [\mathbf{v}_1, \ldots, \mathbf{v}_n]$. Then, according to Theorem 2.1, any vector $\mathbf{x} = [x_1, \ldots, x_n] \in \mathbb{R}^n$ can be decomposed in the new basis as

$$\mathbf{x} = \lambda_1 \mathbf{v}_1 + \ldots + \lambda_n \mathbf{v}_n = \mathbf{V} \begin{bmatrix} \lambda_1 \\ \vdots \\ \lambda_n \end{bmatrix}.$$

Similarly, the vector $\mathbf{y} = \mathbf{A}\mathbf{x} = [y_1, \ldots, y_n] \in \mathbb{R}^n$ can also be decomposed as

$$\mathbf{y} = \mu_1 \mathbf{v}_1 + \ldots + \mu_n \mathbf{v}_n = \mathbf{V} \begin{bmatrix} \mu_1 \\ \vdots \\ \mu_n \end{bmatrix}.$$

Therefore,

$$\begin{bmatrix} \mu_1 \\ \vdots \\ \mu_n \end{bmatrix} = \mathbf{V}^{-1}\mathbf{A}\mathbf{V} \begin{bmatrix} \lambda_1 \\ \vdots \\ \lambda_n \end{bmatrix},$$

from which we conclude that $\mathbf{V}^{-1}\mathbf{A}\mathbf{V}$ is the matrix representation for the same linear map with respect to the basis $\{\mathbf{v}_1, \ldots, \mathbf{v}_n\}$.

For example, consider the basis for $\mathbb{R}^3$ introduced in (2.6) and the linear map whose matrix representation with respect to the canonical basis is

$$\mathbf{A} = \begin{bmatrix} 1 & 2 & 3 \\ 2 & 3 & 4 \\ 3 & 4 & 5 \end{bmatrix}.$$

Then, the matrix representation for this linear map with respect to the basis (2.6) is given by $\mathbf{V}^{-1}\mathbf{A}\mathbf{V}$, where

$$\mathbf{V} = \begin{bmatrix} 1 & 1 & 0 \\ 2 & 3 & 1 \\ 1 & 2 & 2 \end{bmatrix}.$$

```
>> v_1=[1 2 1]'; v_2=[1 3 2]'; v_3=[0 1 2]';
>> V = [v_1,v_2,v_3];
>> A = [1 2 3 ; 2 3 4; 3 4 5];
>> inv(V)*A*V

ans =
    24    39    24
   -16   -26   -16
    12    19    11
```

Null space and range

Let us now return to general concepts related to linear maps. The *null space* of a linear map $\mathbf{A} : V \mapsto W$ is the set

$$\text{null}(\mathbf{A}) = \{\mathbf{x} \in V : \mathbf{Ax} = \mathbf{0}\}. \tag{2.11}$$

The *range* of a linear map $\mathbf{A} : V \mapsto W$ is the set

$$\text{range}(\mathbf{A}) = \{\mathbf{b} \in W : \mathbf{Ax} = \mathbf{b} \text{ for some } \mathbf{x} \in V\}. \tag{2.12}$$

Exercise 2.5 Prove that null($\mathbf{A}$) and range($\mathbf{A}$) are subspaces of V and W, respectively.

For a better understanding of null($\mathbf{A}$) and range($\mathbf{A}$), let us explore the connection between linear maps and matrices a bit further and observe that we can write a general $m \times n$ matrix $\mathbf{A}$ with entries $\{a_{ij}\}$ as

$$\mathbf{A} = [\mathbf{a}_1 \cdots \mathbf{a}_n],$$

where each $\mathbf{a}_j$ is an m-dimensional column vector of the form

$$\mathbf{a}_j = \begin{bmatrix} a_{1j} \\ \vdots \\ a_{mj} \end{bmatrix}.$$

Therefore, if $\mathbf{x} \in \mathbb{R}^n$, the rules of matrix multiplication give

$$\mathbf{Ax} = [\mathbf{a}_1 \cdots \mathbf{a}_n] \begin{bmatrix} x_1 \\ \vdots \\ x_n \end{bmatrix} = x_1\mathbf{a}_1 + \cdots + x_n\mathbf{a}_n.$$

In other words, solving a matrix equation of the form $\mathbf{Ax} = \mathbf{b}$ is equivalent to finding scalars $(x_1, \ldots, x_n)$ for which the vector $\mathbf{b} \in \mathbb{R}^m$ can be written as a linear combination of the columns of $\mathbf{A}$.

For instance, finding the null space of a matrix is equivalent to finding all the solutions for the equation

$$\mathbf{Ax} = \mathbf{0}.$$

In MATLAB, the function `null` does this job for you. More precisely, because the null space is a subspace, MATLAB computes a suitable basis for it. The result is expressed in matrix form, where each column corresponds to a vector in the basis.

For the first example, consider the matrix

$$\mathbf{A} = \begin{bmatrix} 1 & 1 & 0 \\ 2 & 3 & 1 \\ 1 & 2 & 2 \end{bmatrix} \tag{2.13}$$

and try the following MATLAB commands:

```
>> A=[1 1 0; 2 3 1;1 2 2];
>> NA=null(A)

NA =
     Empty matrix: 3-by-0
```

In this example, the null space of the linear map $\mathbf{A}$ is the subspace $\{\mathbf{0}\}$, whose only possible basis is the empty set. In other words, the only way of writing the vector $\mathbf{0} \in \mathbb{R}^3$ as a linear combination of the columns of $\mathbf{A}$ is with zero scalars. That is, the columns of $\mathbf{A}$ form a linearly independent set of vectors.

Criterion for linear independence

Incidentally, this provides a general method for checking if a particular set of vectors is linearly independent: arrange them as the columns of a matrix and verify if its null space is trivial. The columns for the matrix $\mathbf{A}$ in this example are exactly the vectors $\{\mathbf{f}_1, \mathbf{f}_2, \mathbf{f}_3\}$ given in (2.6) as an alternative basis for $\mathbb{R}^3$.

Next, consider the matrix

$$\mathbf{B} = \begin{bmatrix} 1 & 2 & 3 \\ 2 & 4 & 6 \\ 1 & 0 & -1 \end{bmatrix} \tag{2.14}$$

for which MATLAB gives

```
>> B=[1 2 3;2 4  6;1 0 -1];
>> NB=null(B)

NB =
     0.4082
    -0.8165
     0.4082
```

This means that the scalars $(0.4082, -0.8165, 0.4082)$ can be used to make a linear combination of the columns of $\mathbf{B}$ equal to zero. Because the null space is a subspace, the same is true for any multiple of these scalars. For example, even without MATLAB, you can verify that

$$(1) \cdot \begin{bmatrix} 1 \\ 2 \\ 1 \end{bmatrix} + (-2) \cdot \begin{bmatrix} 2 \\ 4 \\ 0 \end{bmatrix} + (1) \cdot \begin{bmatrix} 3 \\ 6 \\ -1 \end{bmatrix} = \begin{bmatrix} 0 \\ 0 \\ 0 \end{bmatrix},$$

where the vector $[1, -2, 1]$ is proportional to the vector NB in the previous example.

For an example of a two-dimensional null space, consider the matrix

$$\mathbf{C} = \begin{bmatrix} 1 & 1 & 1 \\ -1 & -1 & -1 \\ 0 & 0 & 0 \end{bmatrix} \tag{2.15}$$

and the corresponding MATLAB commands

```
>> C=[1 1 1;-1 -1 -1; 0 0 0];
>> NC=null(C)

NC =
   -0.8165         0
    0.4082   -0.7071
    0.4082    0.7071
```

Because the null space is a subspace, any linear combination $\mathbf{x}$ of the columns of the preceding matrix $\mathbf{NC}$ also belongs to it, and therefore satisfies $\mathbf{Cx} = \mathbf{0}$. For example, it is clear that the first column of $\mathbf{NC}$ is proportional to $[-2, 1, 1]$, while its second column is proportional to $[0, -1, 1]$. Therefore, the scalars

$$(-2, -1, 3) = (-2, 1, 1) + 2 \cdot (0, -1, 1)$$

can be used to produce a vanishing linear combination of the columns of $\mathbf{C}$:

$$(-2) \cdot \begin{bmatrix} 1 \\ -1 \\ 0 \end{bmatrix} + (-1) \cdot \begin{bmatrix} 1 \\ -1 \\ 0 \end{bmatrix} + (3) \cdot \begin{bmatrix} 1 \\ -1 \\ 0 \end{bmatrix} = \begin{bmatrix} 0 \\ 0 \\ 0 \end{bmatrix}.$$

For the final example, let

$$\mathbf{D} = \begin{bmatrix} 0 & 0 & 0 \\ 0 & 0 & 0 \\ 0 & 0 & 0 \end{bmatrix}, \tag{2.16}$$

which can be easily created in MATLAB by using the command `zeros`:

```
>> D=zeros(3);
>> ND=null(D)

ND =

     1     0     0
     0     1     0
     0     0     1
```

This confirms the fact that the null space of a 3×3 matrix with vanishing entries is the entire space $\mathbb{R}^3$, because it is clear that $\mathbf{0x} = \mathbf{0}$ for any vector $\mathbf{x} \in \mathbb{R}^3$.

Having found the null space, what can be said about the range of a linear map? Recall that the range of $\mathbf{A}$ determines the subspace spanned by columns of $\mathbf{A}$. Suppose that we have found a basis $\{\mathbf{v}_1 \ldots, \mathbf{v}_k\}$ for the null space of a linear map $\mathbf{A} : V \mapsto W$ acting on a finite-dimensional space V. Recall from the discussion in Section 2.1 that this linearly independent set of vectors can then be extended to a basis $\{\mathbf{v}_1 \ldots, \mathbf{v}_k, \mathbf{u}_1, \ldots, \mathbf{u}_r\}$ for the whole vector space V, where $k + r = \dim V$. We can then write a general vector $\mathbf{v} \in V$ as the linear combination

$$\mathbf{v} = \lambda_1 \mathbf{v}_1 + \cdots + \lambda_k \mathbf{v}_k + \mu_1 \mathbf{u}_1 + \cdots + \mu_r \mathbf{u}_r.$$

Therefore, because the first k terms in the preceding sum all belong to the null space of $\mathbf{A}$, you obtain that a generic element $\mathbf{w} = \mathbf{Av}$ in the range of $\mathbf{A}$ can be written as

$$\mathbf{w} = \mathbf{Av} = \mu_1 \mathbf{Au}_1 + \cdots + \mu_r \mathbf{Au}_r,$$

which implies that the vectors $\{\mathbf{Au}_1, \ldots, \mathbf{Au}_r\}$ span the subspace range($\mathbf{A}$). To proceed with this argument, we need the following result.

Exercise 2.6 Prove that the set $\{\mathbf{Au}_1, \ldots, \mathbf{Au}_r\}$ is linearly independent.

Therefore, $\{\mathbf{Au}_1, \ldots, \mathbf{Au}_r\}$ constitutes a basis for the range of $\mathbf{A}$. What we have just sketched is the proof of the following theorem:

Theorem 2.2 (Dimension Formula) *If $\mathbf{A} : V \mapsto W$ is a linear map and V is finite-dimensional, then the range of $\mathbf{A}$ is also finite-dimensional and*

$$\dim \text{null}\ (\mathbf{A}) + \dim \text{range}\ (\mathbf{A}) = \dim V. \tag{2.17}$$

If the range of a linear map $\mathbf{A}$ is finite-dimensional, its dimension is called the *rank* of $\mathbf{A}$.

For illustrations, let us return to the matrices in previous examples. The MATLAB function that produces a suitable basis for the range of a matrix

A is **orth**, and the function that calculates its rank is **rank**. We discuss the algorithms behind these functions in Section 4.8; for now, let us just observe their work. For the matrix **A** in (2.13), you have:

```
>> A=[1 1 0;2 3 1;1 2 2];
>> r=rank(A), RA=orth(A)

r =
     3

RA =
   -0.2616    0.4891    0.8321
   -0.7659    0.4194   -0.4873
   -0.5873   -0.7648    0.2649
```

In this case, the range of $\mathbf{A} : \mathbb{R}^3 \mapsto \mathbb{R}^3$ is the entire space $\mathbb{R}^3$. The dimension formula is verified, since its null space has dimension zero.

Next, the matrix **B** in Equation (2.14) gives

```
>> B=[1 2 3;2 4 6;1 0 -1];
>> r=rank(B), RB=orth(B)

r =
     2

RB =
   -0.4463    0.0292
   -0.8925    0.0584
    0.0653    0.9979
```

Here the range of **B** is two-dimensional, while its null space is one-dimensional. Note that the columns of **RB** span the same subspace as the columns of **B**. This fact is sometimes hard to spot with the naked eye.

Now try the matrix **C** in Equation (2.15):

```
>> C=[1 1 1;-1 -1 -1;0 0 0];
>> r=rank(C), RC=orth(C)

r =
     1

RC =
   -0.7071
    0.7071
         0
```

You already know that the null space of **C** is two-dimensional, and this confirms that its range is one-dimensional. It is also evident that the single column of **RC** must be a multiple of the thrice-repeated column of **C**.

Finally, the matrix **D** in Equation (2.16) produces

```
>> D=zeros(3);
>> r=rank(D), RD=orth(C)

r =
    0

RC =
     Empty matrix: 3-by-0
```

For this degenerate case, the range of **D** is the zero-dimensional subspace $\{\mathbf{0}\}$, whose basis is the empty set, confirmed by MATLAB.

Surjective and injective maps

Bijective

We can introduce several secondary definitions now. For instance, a linear map $\mathbf{A} : V \mapsto W$ is *surjective* if its range is the entire space W. You can see that the matrix **A** in Equation (2.13) corresponds to a surjective linear map, whereas the matrices **B**, **C**, and **D** in Equations (2.14), (2.15), and (2.16) do not.

Similarly, a linear map $\mathbf{A} : V \mapsto W$ is *injective* if $\mathbf{u}_1 \neq \mathbf{u}_2$ implies that $\mathbf{A}\mathbf{u}_1 \neq \mathbf{A}\mathbf{u}_2$, that is, $\mathbf{A}(\mathbf{u}_1 - \mathbf{u}_2) \neq \mathbf{0}$. It is then clear that a linear map is injective if and only if its null space is empty. Therefore, the matrices **B**, **C**, and **D** in Equations (2.14), (2.15), and (2.16) correspond to linear maps that are not injective, whereas the matrix **A** in Equation (2.13) corresponds to an injective linear map. For confirmation, construct vectors $\mathbf{u}_1$ and $\mathbf{u}_2$ such that $\mathbf{u}_1 \neq \mathbf{u}_2$ but $\mathbf{B}\mathbf{u}_1 = \mathbf{B}\mathbf{u}_2$:

```
>> B=[1 2 3; 2 4 6; 1 0 -1];
>> u1=[1 0 1]'; u2=[2 -2 2]';
>> B*u1-B*u2

ans =
     0
     0
     0
```

For finite-dimensional vector spaces, the dimension formula from Theorem 2.2 quickly helps to settle many questions about surjective and injective linear maps. For instance, if the dimension of V is bigger than the dimension of W, then no linear map from V to W can be injective. Similarly, if the dimension of V is smaller than the dimension of W, then no linear map from V to W can be surjective. Finally, the same formula also implies the following remarkable property of finite-dimensional vector spaces:

Exercise 2.7 Show that a linear map between finite-dimensional vector spaces of the same dimension is injective if and only if it is surjective.

Invertible maps

A linear map $\mathbf{A} : V \mapsto W$ is *invertible* if there exists a unique map $\mathbf{A}^{-1} : W \mapsto V$, called the inverse of $\mathbf{A}$, such that

$$\mathbf{A}\mathbf{A}^{-1} = \mathbf{A}^{-1}\mathbf{A} = \mathbf{1}. \tag{2.18}$$

If a linear map does not admit an inverse, it is said to be *singular*.

It is easy to see that a linear map $\mathbf{A} : V \mapsto W$ is invertible if and only if, given any vector $\mathbf{b} \in W$, there exists a unique vector $\mathbf{x} \in V$ such that $\mathbf{A}\mathbf{x} = \mathbf{b}$.

Exercise 2.8 Prove that a linear map is invertible if and only if it is both surjective and injective.

For finite-dimensional spaces, this means that V and W must have the same dimension.

Two vector spaces are *isomorphic* if there is an invertible linear map from one to the other. Therefore, if two vector spaces are isomorphic and one of them is finite-dimensional, then so is the other, and their dimensions are the same. Using bases for both spaces, the brave reader can also prove the converse. That is, two finite-dimensional vector spaces are isomorphic if and only if they have the same dimension. Therefore, any n-dimensional vector space is isomorphic to $\mathbb{R}^n$.

To conclude this section, by comparing the definition of an invertible map given in Equation (2.18) with that of an invertible matrix given in Equation (1.4) we conclude that the matrix associated with the map $\mathbf{A}^{-1}$ is the inverse of the matrix associated with the map $\mathbf{A}$. Therefore, you can find the inverse of the linear map corresponding to the matrix $\mathbf{A}$ in Equation (2.13) as follows:

```
>>  A=[1 1 0;  2 3 1; 1 2 2];
>> Ainv=inv(A)

Ainv =
      4    -2     1
     -3     2    -1
      1    -1     1
```

As you have seen, none of the maps in the other examples are injective (or surjective). Therefore, they are not invertible either. This is what MATLAB has to say:

```
>> C=[1 1 1;  -1 -1 -1; 0 0 0];
>> inv(C)
```

```
Warning: Matrix is singular to working precision.

ans =
        Inf    Inf    Inf
        Inf    Inf    Inf
        Inf    Inf    Inf
```

The unpleasant outcome filled by `Inf` shows that MATLAB fails to find any inverse to the matrix **C**.

2.3 Systems of Linear Equations

We now turn to one of the central applications of *Linear Algebra*: the solution of systems of linear equations. Let us start with a simple example of a system of linear equations:

$$\begin{array}{rcrcrcl} x & + & y & & & = & 2 \\ 2x & + & 3y & + & z & = & 4 \\ x & + & 2y & + & 2z & = & 6 \end{array} \tag{2.19}$$

We call (2.19) a system because it consists of more than one equation, and say that it is linear because it contains only first-order powers of each of the unknowns x, y, and z. We soon neglect this use of the term *system*, by rewriting it as a single equation (albeit in matrix form), whereas the term *linear* will acquire a much more interesting meaning. A solution to such a system is a triple of numbers (x, y, z) that satisfy all the equations simultaneously. You can verify that, for this particular system, a solution is given by the triple $(6, -4, 4)$.

Sections 2.5 and 2.6 are devoted to numerical methods for solving linear systems. But before we turn to such methods, we must address several important mathematical questions. The obvious ones are: does a solution exist? If it does, is it unique? If it is not, what relates different solutions to the same system? All of these questions can be elegantly formulated and answered in the language of *Linear Algebra*.

Matrix representation

First, let us generalize the preceding example to a system of the form

$$\begin{array}{ccccccc} a_{11}x_1 & + & \cdots & + & a_{1k}x_k & = & b_1 \\ & & \vdots & & & & \\ a_{n1}x_1 & + & \cdots & + & a_{nk}x_k & = & b_k, \end{array} \tag{2.20}$$

where the scalars $\{a_{ij}\}$ and $\{b_j\}$ are interpreted as given coefficients and $\{x_1, \ldots, x_k\}$ are the unknowns that we need to determine. It is then clear that if we write

$$\mathbf{A} = \begin{bmatrix} a_{11} & \cdots & a_{1k} \\ & \vdots & \\ a_{n1} & \cdots & a_{nk} \end{bmatrix}, \quad \mathbf{x} = \begin{bmatrix} x_1 \\ \vdots \\ x_k \end{bmatrix}, \quad \mathbf{b} = \begin{bmatrix} b_1 \\ \vdots \\ b_n \end{bmatrix}, \tag{2.21}$$

then the system (2.20) can be expressed as

$$\mathbf{Ax} = \mathbf{b}. \tag{2.22}$$

Then, associate the set of all possible coefficients $b_1, \ldots, b_n$ appearing on the right-hand side of system (2.20) with the vector space $\mathbb{R}^n$ and the set of possible values for the unknowns $x_1, \ldots, x_k$ with the vector space $\mathbb{R}^k$. Following the discussion in Section 2.2, the matrix $\mathbf{A}$ appearing in (2.21) can be viewed as a linear map $\mathbf{A} : \mathbb{R}^k \to \mathbb{R}^n$. Notice that the same linear map $\mathbf{A}$ gives rise to infinitely many different linear systems, one for each choice of a vector $\mathbf{b}$. It is therefore the pair $(\mathbf{A}, \mathbf{b})$ that characterizes a linear system, and all meaningful discussions about the solution of a linear system should involve both.

Returning to the initial example, it can be written in the form $\mathbf{Ax} = \mathbf{b}$ with

$$\mathbf{A} = \begin{bmatrix} 1 & 1 & 0 \\ 2 & 3 & 1 \\ 1 & 2 & 2 \end{bmatrix}, \quad \mathbf{x} = \begin{bmatrix} x_1 \\ x_2 \\ x_3 \end{bmatrix}, \quad \mathbf{b} = \begin{bmatrix} 2 \\ 4 \\ 6 \end{bmatrix}.$$

One way of solving this system is by calculating the inverse of the matrix $\mathbf{A}$. It is clear that the vector $\mathbf{x} = \mathbf{A}^{-1}\mathbf{b}$ is a solution because

$$\mathbf{Ax} = \mathbf{A}(\mathbf{A}^{-1}\mathbf{b}) = \mathbf{1b} = \mathbf{b}.$$

Computing $\mathbf{A}^{-1}$ is a bad idea for two reasons. First, it does not generalize to nonsquare systems and second, it is numerically dangerous (i.e., ill-conditioned) and inefficient (i.e., slow). But for the moment, find the solution $\mathbf{x} = [6, -4, 4]$ for the previous linear system from the following MATLAB implementation:

```
>> A=[1 1 0; 2 3 1;1 2 2];
>> b=[2 4 6]';
>> x=inv(A)*b

x =
    6
   -4
    4
```

In matrix form, the problem of finding solutions for linear systems can be rephrased as follows:

Problem 2.1 (Linear Systems) Given a vector $\mathbf{b} \in \mathbb{R}^n$ and a matrix $\mathbf{A} \in M^{n\times k}$, find a vector $\mathbf{x} \in \mathbb{R}^k$ such that $\mathbf{Ax} = \mathbf{b}$.

For the case of square systems, that is, corresponding to $n \times n$ matrices, the solution of Problem 2.1 is summarized in the following theorem. The case of under- and overdetermined systems, where the number of equations differs from the number of unknowns, are treated in Section 3.4.

Theorem 2.3 *Consider a system of n linear equations with n unknowns $x_1, \ldots, x_n$, written in the form $\mathbf{Ax} = \mathbf{b}$, where the $n \times n$ matrix $\mathbf{A}$ and the vector $\mathbf{b} \in \mathbb{R}^n$ are given.*

1. *If $\mathbf{A}$ is invertible, then the system has a unique solution given by $\mathbf{x} = \mathbf{A}^{-1}\mathbf{b}$, regardless of the vector $\mathbf{b}$.*
2. *If $\mathbf{A}$ is singular and $\mathbf{b} \notin range(\mathbf{A})$, then the system has no solution.*
3. *If $\mathbf{A}$ is singular and $\mathbf{b} \in range(\mathbf{A})$, then the system has infinitely many solutions.*

The proofs for the first two items are mere rewritings of the concepts from Section 2.2. For the third item, it goes as follows: since $\mathbf{b} \in \mathrm{range}(A)$, there must exist a vector $\mathbf{x} \in \mathbb{R}^n$ such that $\mathbf{Ax} = \mathbf{b}$, which is therefore a solution to the system. Given that the matrix is singular, the null space of $\mathbf{A}$ must contain a nonzero vector $\mathbf{y} \in \mathbb{R}^n$ (otherwise the linear map would be injective, which in finite dimensions implies that it is invertible). But then any vector of the form $\mathbf{x} + \lambda\mathbf{y}$, for arbitrary scalars λ, is also a solution to the system, since

$$\mathbf{A}(\mathbf{x} + \lambda\mathbf{y}) = \mathbf{Ax} + \lambda\mathbf{Ay} = \mathbf{Ax} + \mathbf{0} = \mathbf{b}.$$

MATLAB has a powerful and versatile built-in function to solve linear systems. It is implemented through the operator \, called the *backslash* operator. The syntax to find the solution of a linear system of the form $\mathbf{Ax} = \mathbf{b}$ with this operator is `x=A\b`, which appeals to the intuitive meaning of "dividing" both sides of the system by the matrix $\mathbf{A}$. When a matrix is invertible, this operator is theoretically equivalent to $\mathbf{x} = \mathbf{A}^{-1}\mathbf{b}$. In practice, instead of inverting a matrix, it solves the system using a sophisticated implementation of the methods described in Section 2.5. You will see in Section 3.4 how it also implements the least-squares method for overdetermined linear systems.

To illustrate the use of the backslash operator \ in connection with the three possibilities predicted by Theorem 2.3, consider the MATLAB examples in Section 2.2. The system (2.19) corresponds to the matrix $\mathbf{A}$ in Equation (2.13), which we know is invertible. Therefore, you can use its inverse to calculate the solution to $\mathbf{Ax} = \mathbf{b}$ for whichever vector $\mathbf{b}$ you please, in particular for $\mathbf{b} = [2, 4, 6]$. With the backslash operator, this is implemented as

```
>> A=[1 1 0; 2 3 1;1 2 2];
>> b=[2 4 6]';
>> x=A\b

x =
     6
    -4
     4
```

which agrees with the previous calculation.

For the singular matrix $\mathbf{B}$ in Equation (2.14), the existence of a solution for $\mathbf{Bx} = \mathbf{b}$ depends on the choice of $\mathbf{b}$. As you have seen, the range of $\mathbf{B}$ is the two-dimensional subspace that spans any two of its columns, such as

$$\mathbf{v}_1 = \begin{bmatrix} 1 \\ 2 \\ 1 \end{bmatrix}, \quad \mathbf{v}_2 = \begin{bmatrix} 2 \\ 4 \\ 0 \end{bmatrix}.$$

Therefore, any $\mathbf{b}$ that is not a linear combination of these vectors produces a system with no solution. For instance:

```
>>B=[1 2 3;  2 4 6; 1 0 -1];
>>b=[1 0 0]';
>>x=B\b
Warning: Matrix is singular to working precision.

x =
   Inf
   Inf
   Inf
```

However, if $\mathbf{b}$ is a linear combination of the previous vectors $\mathbf{v}_1$ and $\mathbf{v}_2$, then the system has infinitely many solutions. For example, the vector $\mathbf{b} = [0, 0, 1]$ can be written as the linear combination $\mathbf{b} = \mathbf{v}_1 - 0.5\mathbf{v}_2$. Therefore, since $\mathbf{b} \in \text{range}(\mathbf{B})$, Theorem 2.3 guarantees that the system $\mathbf{Bx} = \mathbf{b}$ has infinitely many solutions. However, MATLAB cannot find any of them with the use of the backslash operator:

```
>>b=[0 0 1]';
>>x=B\b
Warning: Matrix is singular to working precision.

x =
   NaN
   NaN
   NaN
```

But you can easily verify that, for instance, the vector $\mathbf{x} = [1, -0.5, 0]$ is a solution to the system $\mathbf{Bx} = \mathbf{b}$:

```
>> x=[1 -0.5 0]';
>> B*x-b

ans =
     0
     0
     0
```

Moreover, recalling that the null space of $\mathbf{B}$ is the one-dimensional subspace of multiples of the vector $\mathbf{y} = [1, -2, 1]$, you can verify that any other vector of the form $\mathbf{x} + \lambda\mathbf{y}$ is also a solution:

```
>> x=[1 -0.5 0]'+17*[1 -2 1]';
>> B*x-b

ans =
     0
     0
     0
```

Thus, although the backslash operator \ fails to find a solution of the linear system with singular matrix $\mathbf{B}$, the solution may still exist if $\mathbf{b} \in \text{range}(\mathbf{B})$. The moral of this is that when MATLAB cannot find the solution of a square linear system, you better investigate the range of the matrix involved to see which of the outcomes of Theorem 2.3 is being observed.

2.4 Vector and Matrix Norms

When we discuss methods for solving linear systems, it is important to determine how much the final solution varies with the initial input and how the performance of the method depends on the particular linear system. Because both the initial data and the solution are given in terms of vectors, we now introduce an appropriate concept for measuring their size.

A *norm* on a vector space V is a function $\|\cdot\| : V \mapsto \mathbb{R}$ satisfying

1. $\|\mathbf{x}\| > 0$ if $\mathbf{x} \neq \mathbf{0}$.
2. $\|\lambda\mathbf{x}\| = |\lambda|\|\mathbf{x}\|, \qquad \forall\lambda \in \mathbb{R}, \quad \forall\mathbf{x} \in V.$
3. $\|\mathbf{x}+\mathbf{y}\| \leq \|\mathbf{x}\| + \|\mathbf{y}\|, \qquad \forall\mathbf{x},\mathbf{y} \in V.$

Notice that the first and second properties together imply that $\|\mathbf{x}\| = 0$ if and only if $\mathbf{x} = \mathbf{0}$. The third property is called the *triangle inequality.*

Vector norms

The most popular example of a norm in $\mathbb{R}^n$ is the *Euclidean* norm. For the vector $\mathbf{x} = [x_1, \cdots, x_n]$, it is defined as

$$\|\mathbf{x}\|_2 = \left(\sum_{i=1}^{n} |x_i|^2\right)^{1/2} \tag{2.23}$$

The subscript 2 indicates that this is a special case of what is called a *p-norm,* defined by

$$\|\mathbf{x}\|_p = \left(\sum_{i=1}^{n} |x_i|^p\right)^{1/p}, \tag{2.24}$$

where $1 \leq p < \infty$. You can see that the limiting case $p = 1$ takes the particularly simple form

$$\|\mathbf{x}\|_1 = \sum_{i=1}^{n} |x_i|. \tag{2.25}$$

The limit $p \to \infty$ is given by a different expression and deserves special attention, because the corresponding norm measures the maximal component of the vector, namely:

$$\|\mathbf{x}\|_\infty = \max_{1\leq i\leq n} |x_i|. \tag{2.26}$$

It is not completely straightforward to verify that each of these norms satisfies the properties of a norm, particularly the triangle inequality. After making an attempt, you might find it useful to check a *Real Analysis* book for the full argument.

In MATLAB, the p-norm of a vector is calculated by the function `norm`, which takes the value of $1 \leq p < \infty$ as an optional argument. If p is not specified, MATLAB computes the Euclidean norm by default. To obtain the norm $\|\cdot\|_\infty$ you must set the optional argument equal to `inf`.

```
>> x=[1 -1 0 2 4 -3];
>> norm(x,1)

ans =
        11

>> norm(x,2)

ans =
        5.5678

>> norm(x,inf)

ans =
        4
```

Observe that in the preceding example you have

$$\|\mathbf{x}\|_\infty \leq \|\mathbf{x}\|_2 \leq \|\mathbf{x}\|_1 \tag{2.27}$$

and

$$\|\mathbf{x}\|_1 \leq \sqrt{n}\|\mathbf{x}\|_2 \leq n\|\mathbf{x}\|_\infty, \tag{2.28}$$

where $n = 6$. It can be shown that these two inequalities hold for an arbitrary vector $\mathbf{x} \in \mathbb{R}^n$. The meaning behind inequalities (2.27) and (2.28) is that the three preceding norms give qualitatively similar estimates of the magnitude of vectors in $\mathbb{R}^n$. For instance, if any of them gets arbitrarily small or arbitrarily large when calculated along a particular sequence of vectors, then so do the other two. These inequalities are particular examples of the following deeper (and harder) result in *Analysis*:

Theorem 2.4 *Given any two norms* $\|\cdot\|_a$ *and* $\|\cdot\|_b$ *on a* finite-dimensional *vector space* V, *there exist constants* c *and* C *such that*

$$c\|\mathbf{x}\|_a \leq \|\mathbf{x}\|_b \leq C\|\mathbf{x}\|_a, \quad \forall \mathbf{x} \in V. \tag{2.29}$$

Two norms satisfying (2.29) are said to be *equivalent*. For *finite-dimensional* vector spaces, despite this equivalence, it is useful to consider several different norms because it is often the case that either a numerical calculation in a specific example or the statement and proofs of some general results are carried out more naturally in one norm than they are in another.

In what follows, we refer to norms on $\mathbb{R}^n$ simply as *vector norms*. By contrast, norms on the space $\mathbb{M}^{m\times n}$ of $m \times n$ matrices are called *matrix norms*, despite the fact that both are examples of norms on general vector spaces.

Matrix norms

Because an $m \times n$ matrix can be viewed as a linear map from $\mathbb{R}^n$ to $\mathbb{R}^m$, we can ask what its effect is on the size of the underlying vector. For example, suppose we use the norm $\|\cdot\|_p$ for $\mathbb{R}^n$ and the norm $\|\cdot\|_{p'}$ for $\mathbb{R}^m$. Then, for a given linear map $\mathbf{A} : \mathbb{R}^n \mapsto \mathbb{R}^m$, we want to know how $\|\mathbf{Ax}\|_{p'}$ (the size of the vector $\mathbf{Ax}$ measured with the p'-norm) compares with $\|\mathbf{x}\|_p$ (the size of the vector $\mathbf{x}$ measured with the p-norm). This kind of question is at the core of the sensitivity analysis for numerical methods. It can be handled by the concept of a matrix norm $\|\cdot\|_{pp'}$ *induced* by the underlying vector norms:

$$\|\mathbf{A}\|_{pp'} = \max_{\mathbf{x}\neq\mathbf{0}} \frac{\|\mathbf{Ax}\|_{p'}}{\|\mathbf{x}\|_p}. \tag{2.30}$$

The subscript indicates that this matrix norm depends on each of the vector norms used for $\mathbb{R}^n$ and $\mathbb{R}^m$. If $p = p'$, we denote it simply by $\|\mathbf{A}\|_p \equiv \|\mathbf{A}\|_{pp}$.

For example, when the 1-norm is used for both $\mathbb{R}^n$ and $\mathbb{R}^m$, the induced matrix norm $\|\cdot\|_1$ is given by the maximum absolute column sum of the matrix $\mathbf{A}$,

$$\|\mathbf{A}\|_1 = \max_{1\leq j\leq n} \sum_{i=1}^{m} |a_{ij}|. \tag{2.31}$$

Similarly, if we use the ∞-norm for both $\mathbb{R}^n$ and $\mathbb{R}^m$, the induced matrix norm $\|\cdot\|_\infty$ is the maximum absolute row sum of the matrix $\mathbf{A}$,

$$\|\mathbf{A}\|_\infty = \max_{1\leq i\leq m} \sum_{j=1}^{n} |a_{ij}|. \tag{2.32}$$

It is very easy to get confused with these two expressions. A convenient way to remember them is by observing that if $\mathbf{A}$ consists of a single $m \times 1$ column (i.e., $n = 1$), then the 1-matrix norm (2.31) reduces to the 1-vector norm (2.25), whereas the ∞-matrix norm (2.32) coincides with the ∞–vector norm (2.26). This is yet another example of consistency in the matrix notations, provided we always treat elements of $\mathbb{R}^m$ as column vectors.

The matrix norm induced by the 2-norm for vectors does not have an easy expression. It is given by the matrix's largest singular value, which is introduced in Section 4.8.

In MATLAB, the function used to calculate the matrix norm induced by the p–vector norm is also `norm`, but it can only take the arguments $p = 1, 2$, or ∞, with the default value being $p = 2$.

```
>> A=[-1 2 3; 0 1 -1; 2 -3 5];
>> norm(A,1)

ans =
     9

>>  norm(A,2)

ans =
    6.4426

>> norm(A,inf)

ans =
    10
```

Notice from this example that the inequalities in (2.27) and (2.28) do *not* hold for the induced matrix norms. However, since matrix spaces are finite-dimensional vector spaces, Theorem 2.4 still holds and all matrix norms are equivalent. In particular, inequalities of the form (2.29) hold for appropriate constants c and C, for example, for $a = 1$ and $b = 2$, you can choose $c = 0.5$ and $C = 1$ and observe that

```
>> norm(A,2)/norm(A,1)

ans =
    0.7158
```

Apart from the standard properties of norms, a matrix norm of the form (2.30) satisfies two important additional properties, which we mention for later use:

$$\|\mathbf{AB}\|_{pp'} \leq \|\mathbf{A}\|_{pp'}\|\mathbf{B}\|_{pp'}, \quad \forall \mathbf{A}, \mathbf{B} \in \mathbb{M}^{m\times n}, \tag{2.33}$$

$$\|\mathbf{Ax}\|_{p'} \leq \|\mathbf{A}\|_{pp'}\|\mathbf{x}\|_{p}, \quad \forall \mathbf{x} \in \mathbb{R}^{n}, \quad \forall \mathbf{A} \in \mathbb{M}^{m\times n}. \tag{2.34}$$

These properties are illustrated in the following example:

```
>> A=[-1 2 3; 0 1 -1; 2 -3 5];
>> B=[0 2 -1; 3 6 -2; 8 5 1];
>> x=[1 -1 0]';
>> norm(A*B) <= norm(A)*norm(B)

ans =
     1
```

```
>> norm(A*x,1) <= norm(A,1)*norm(x,1)

ans =
     1
```

Condition number

Let us now turn to a discussion of invertibility of square matrices. Because the only singular real number is zero, we can say that the absolute value of a real number is a good measure of how close to singular the number is: numbers with large absolute values are far from being singular. At first sight, you might be tempted to say the same for matrix norms, since they measure the "size" of a matrix. Quick reflection, however, should convince you that the size of a matrix has little to do with its invertibility. For example, a 2×2 matrix with all entries equal to a million will have a very large matrix norm (regardless of which one you choose) but is nevertheless singular.

An appropriate indicator that a square matrix is singular is given by its *condition number*, defined as

$$\mathrm{cond}_{pp'}(\mathbf{A}) = \begin{cases} \|\mathbf{A}\|_{pp'}\|\mathbf{A}^{-1}\|_{pp'} & , \quad \text{if } \mathbf{A} \text{ is invertible} \\ \infty & , \quad \text{otherwise.} \end{cases} \tag{2.35}$$

Large values of the condition number indicate that the matrix is close to singular, and any numerical procedure that depends on the inverse of such a matrix can pose severe difficulties. Therefore, it is recommended that you calculate the condition number of a matrix before attempting any method that depends on its invertibility.

At this point, it seems like we are in a loop because the definition of the condition number depends on the inverse of a matrix. The way out of the loop is to obtain estimates for the order of magnitude of the condition number of a matrix using steps that do not rely on its inverse. This is one instance where matrix norms induced by vector norms are better suited than general matrix norms. For if the norm is defined as in (2.30), then a simple rearrangement shows that

$$\mathrm{cond}_{pp'}(\mathbf{A}) = \frac{\left(\max\limits_{\mathbf{x}\neq\mathbf{0}} \frac{\|\mathbf{A}\mathbf{x}\|_{p'}}{\|\mathbf{x}\|_p}\right)}{\left(\min\limits_{\mathbf{x}\neq\mathbf{0}} \frac{\|\mathbf{A}\mathbf{x}\|_{p'}}{\|\mathbf{x}\|_p}\right)}, \tag{2.36}$$

that is, the condition number is the ratio of the maximum dilation to the maximum contraction that the matrix produces on nonzero vectors.

In MATLAB, the function `cond` calculates the condition number with respect to the 2-norm. As you will see in Section 4.8, in this case (2.36) reduces to the ratio between the largest and the smallest singular values of the matrix $\mathbf{A}$, which is also the implementation technique used by MATLAB.

```
>> A=[-4 11 3; 7 2 -1; 13 0 6];
>> cond_1=norm(A,1)*norm(inv(A),1)

cond_1 =
        8.9302

>> cond_inf=norm(A,inf)*norm(inv(A),inf)

cond_inf =
        6.6019

>> cond_2=norm(A,2)*norm(inv(A),2)

cond_2 =
        4.0443

>> cond(A)

ans =
        4.0443
```

A famous example producing large condition numbers is obtained by using *Hilbert matrices*, defined as square matrices with entries

$$\mathbf{H}_{ij}(n) = \frac{1}{(i+j-1)}, \qquad 1 \leq i, j \leq n,$$

which can be obtained with the MATLAB command `hilb`. For example, the 2×2 and 3×3 Hilbert matrices and their inverses are

```
>> H2=hilb(2)

H2 =
    1.0000    0.5000
    0.5000    0.3333

>> H3=hilb(3)

H3 =
    1.0000    0.5000    0.3333
    0.5000    0.3333    0.2500
    0.3333    0.2500    0.2000

>> inv(H2)

ans =
    4.0000   -6.0000
   -6.0000   12.0000
```

```
>> inv(H3)

ans =
    9.0000  -36.0000   30.0000
  -36.0000  192.0000 -180.0000
   30.0000 -180.0000  180.0000
```

As you can observe, the entries in the inverses are large integers, which become larger as the dimension of the Hilbert matrices increases. This leads very rapidly to extremely large condition numbers for matrices $\mathbf{H}(n)$ with $n \geq 10$.

Exercise 2.9 Plot the logarithm of the condition number for the first 20 Hilbert matrices $\mathbf{H}(n)$ and confirm that Hilbert matrices are ill-conditioned for $n > 10$.

Sensitivity

Let us now show how the concepts of matrix norms and condition numbers help to quantify the computational errors arising in the solution of linear systems. For this, suppose that we are given an $m \times n$ matrix $\mathbf{A}$ and an m-vector $\mathbf{b}$ and want to find an n-vector $\mathbf{x}$ satisfying

$$\mathbf{A}\mathbf{x} = \mathbf{b}.$$

Assume further that the matrix $\mathbf{A}$ is invertible, so that the system has a unique solution. In this problem, both the matrix $\mathbf{A}$ and the vector $\mathbf{b}$ should be viewed as input data, and therefore be subject to several sources of error. On a first attack of the problem, however, we consider the matrix $\mathbf{A}$ to be error free (for example, if it is given by a certain model, and we decide to ignore the possibility that the model itself is misspecified) and concentrate on the *measurement error* $\Delta\mathbf{b} = (\widehat{\mathbf{b}} - \mathbf{b})$ introduced by an imperfect input $\widehat{\mathbf{b}}$. This leads to a modified system

$$\mathbf{A}\widehat{\mathbf{x}} = \widehat{\mathbf{b}},$$

which induces an error in the solution given by $\Delta\mathbf{x} = (\widehat{\mathbf{x}} - \mathbf{x})$. Notice that both $\Delta\mathbf{b}$ and $\Delta\mathbf{x}$ are vectors, and the goal is to obtain the size of $\Delta\mathbf{x}$ in terms of the size of $\Delta\mathbf{b}$ and properties of the matrix $\mathbf{A}$. By taking norms on both sides of the original system $\mathbf{A}\mathbf{x} = \mathbf{b}$, we obtain

$$\|\mathbf{b}\| = \|\mathbf{A}\mathbf{x}\| \leq \|\mathbf{A}\|\|\mathbf{x}\|$$

so that

$$\|\mathbf{x}\| \geq \frac{\|\mathbf{b}\|}{\|\mathbf{A}\|}, \tag{2.37}$$

where the subscripts of vector and matrix norms are ommitted for ease of notations. Further, using the linearity of $\mathbf{A}$, we obtain

$$\mathbf{A}\Delta\mathbf{x} = \mathbf{A}(\widehat{\mathbf{x}} - \mathbf{x}) = \mathbf{A}\widehat{\mathbf{x}} - \mathbf{A}\mathbf{x} = \widehat{\mathbf{b}} - \mathbf{b} = \Delta\mathbf{b},$$

which implies that

$$\Delta \mathbf{x} = \mathbf{A}^{-1}(\Delta \mathbf{b}).$$

Taking norms on both sides of this equation then leads to

$$\|\Delta \mathbf{x}\| = \|\mathbf{A}^{-1}(\Delta \mathbf{b})\| \leq \|\mathbf{A}^{-1}\| \|\Delta \mathbf{b}\|. \tag{2.38}$$

From (2.37) and (2.38) we can conclude that

$$\frac{\|\Delta \mathbf{x}\|}{\|\mathbf{x}\|} \leq \text{cond}(\mathbf{A}) \frac{\|\Delta \mathbf{b}\|}{\|\mathbf{b}\|}. \tag{2.39}$$

That is, the condition number provides a bound for how much the relative error in the input $\mathbf{b}$ is magnified by the matrix $\mathbf{A}$ to produce a relative error in the output $\mathbf{x}$. Observe that this is based on idealized matrix calculations. For real computations, this is further increased by rounding errors (because of computer arithmetic) and truncation errors (for example, when iterative procedures are terminated after a certain number of steps), according to the discussion in Section 1.8.

To illustrate the sensitivity of linear systems, let $\mathbf{A}$ be the 5×5 Hilbert matrix, whose condition number is very large, as calculated in Exercise 2.9. Starting with the vector $\mathbf{b} = [1.767, 1.167, 0.912, 0.758, 0.652]$, you obtain the following solution to the linear system $\mathbf{Ax} = \mathbf{b}$:

```
>> b=[1.767 1.167 0.912 0.758 0.652]';
>> A-hilb(5);
>> x=A\b

x =
    1.2350
   -5.4600
   21.2100
  -28.8400
   17.0100
```

If you now slightly modify the input and consider the vector $\widehat{\mathbf{b}} = [1.77, 1.17, 0.91, 0.76, 0.65]$ instead, the solution you obtain is very different from the previous one:

```
>> bhat=[1.77 1.17 0.91 0.76 0.65]';
>> A=hilb(5);
>> xhat=A\bhat

xhat =
     -5.7500
    124.8000
   -539.7000
    817.6000
   -396.9000
```

Instead of perturbing the original linear system by modifying the vector $\mathbf{b}$, we can consider an error $\Delta\mathbf{A} = (\widehat{\mathbf{A}} - \mathbf{A})$ in the matrix itself. In this case, the condition number still provides a bound for the relative error in the final solution, according to the following exercise.

Exercise 2.10 Let $\widehat{\mathbf{A}}\widehat{\mathbf{x}} = \mathbf{b}$ denote a modified version of the linear system $\mathbf{A}\mathbf{x} = \mathbf{b}$, accounting for a measurement error $\Delta\mathbf{A} = \widehat{\mathbf{A}} - \mathbf{A}$ in the matrix of coefficients. Prove that

$$\frac{\|\Delta\mathbf{x}\|}{\|\widehat{\mathbf{x}}\|} \leq \mathrm{cond}(\mathbf{A})\frac{\|\Delta\mathbf{A}\|}{\|\mathbf{A}\|}. \tag{2.40}$$

2.5 Direct Methods

We now discuss numerical solutions to Problem 2.1 for the case of square linear systems, that is, with $k = n$. Nonsquare systems, that is, for general $n \times k$ matrices with $n \neq k$, are treated in Section 3.4 with the method of least squares.

There are two types of methods for solving linear systems: *direct* and *iterative*. If we disregard the effects of finite-precision arithmetics, direct methods, treated in this section, produce an exact solution in a finite number of steps. By contrast, iterative methods, considered in Section 2.6, produce approximate solutions that approach the exact solution as the number of steps becomes larger. It is perfectly legitimate of you to wonder why anyone would turn to iterative methods if direct methods are available. After all, what can be better than an exact solution? A partial answer to this is that direct methods can be computationally costly, in the sense that the number of *flops* necessary to obtain the final solution scales very rapidly with the size of the system at hand. If computer arithmetic were itself exact, this would not be a disadvantage of direct methods in relation to iterative ones because the latter require (by definition) infinitely many operations to reach the solution. However, after we take into account rounding errors introduced by finite-precision arithmetic, direct methods are not exact anymore, and it might happen that for a given system the desired accuracy can be achieved faster by iterative methods. For this reason, in what follows, we keep track of the nature and number of arithmetic operations involved in each method so that we can compare their merits and decide which one to use in a real problem.

Forward substitution

We start with a very special type of linear system defined by a *lower-triangular* matrix $\mathbf{L}$, which is an $n \times n$ matrix such that $l_{ij} = 0$ for all $i < j$. In other words, all entries above its diagonal are equal to zero. That is, the matrix equation $\mathbf{Lx} = \mathbf{b}$ has the form

$$\begin{bmatrix} l_{11} & 0 & \cdots & 0 \\ l_{21} & l_{22} & \cdots & 0 \\ \vdots & & & \vdots \\ l_{n1} & l_{n2} & \cdots & l_{nn} \end{bmatrix} \begin{bmatrix} x_1 \\ x_2 \\ \vdots \\ x_n \end{bmatrix} = \begin{bmatrix} b_1 \\ b_2 \\ \vdots \\ b_n \end{bmatrix}.$$

Therefore, if we start from the first equation of the system, corresponding to the first row of the matrix $\mathbf{L}$, we have

$$l_{11}x_1 = b_1 \qquad \Rightarrow \qquad x_1 = \frac{b_1}{l_{11}}.$$

We can then store this value for the variable x_1 and proceed directly to the second equation of the system, which reads

$$l_{21}x_1 + l_{22}x_2 = b_2 \quad \Rightarrow \quad x_2 = \frac{b_2 - l_{21}x_1}{l_{22}},$$

because x_1 is known for this step. Just to make sure that you understand the pattern, let us store the values of x_1 and x_2 and proceed with the third equation of the system:

$$l_{31}x_1 + l_{32}x_2 + l_{33}x_3 = b_3 \quad \to \quad x_3 = \frac{b_3 - (l_{31}x_1 + l_{32}x_2)}{l_{33}}.$$

It is now clear that at the ith step we would have already calculated the values for $x_1, \ldots, x_{i-1}$ and could use these values to obtain

$$x_i = \frac{1}{l_{ii}} \left(b_i - \sum_{j=1}^{i-1} l_{ij}x_j \right). \tag{2.41}$$

For obvious reasons, this algorithm for solving a lower-triangular system is called *forward substitution*. A possible MATLAB function implementing the forward substitution algorithm for solving lower-triangular systems can be coded as follows:

```
function x=forwsub(L,b)
% solves L*x=b by forward substitution
% L is an n-by-n lower triangular matrix
% b is an n-vector

[n,m]=size(L);
x=zeros(n,1);
for i=1:n
    x(i)=(b(i)-L(i,1:i-1)*x(1:i-1))/L(i,i);
end
```

Observe how we have used both the multiplication operator * and the colon operator : to rewrite the expression

$$\sum_{j=1}^{i-1} l_{ij}x_j,$$

which in matrix form simply becomes `L(i,1:i-1)*x(1:i-1)`. This leads to the question of how many *flops* are necessary for the entire algorithm, which is crucial in assessing its computational complexity. Given a row vector `a` and a column vector `b`, both with k components, we can see immediately that the product `a*b` requires k multiplications and $k-1$ additions. Using this count, we see for the algorithm of forward substitution that, for $i = 2, \ldots, n$, the number of *flops* required for the ith step of the for loop is

$$(i-1) + (i-2) + 1 + 1 = 2i - 1,$$

where the first two terms correspond to the vector multiplication for vectors of length $(i-1)$ and the last two terms correspond to a single subtraction and division. For the degenerate step $i = 1$, we have just one *flop*, corresponding to a single division, which also follows from the preceding formula for $i = 1$. Therefore, the total number of *flops* required by the MATLAB function `forwsub` is

$$\sum_{i=1}^{n}(2i-1) = 2\frac{n(n+1)}{2} - n = n^2,$$

where we have used the summation formulas

$$\sum_{1=1}^{n} i = \frac{n(n+1)}{2}, \qquad \sum_{i=1}^{n} i^2 = \frac{n(n+1)(2n+1)}{6}. \tag{2.42}$$

Let us apply the MATLAB function `forwsub` to the lower-triangular linear systems with $\mathbf{b} = [1, 2, 3]$ and the lower-triangular matrices

$$\mathbf{L}_1 = \begin{bmatrix} 2 & 0 & 0 \\ 1 & -1 & 0 \\ 3 & 4 & 5 \end{bmatrix}, \qquad \mathbf{L}_2 = \begin{bmatrix} 2 & 0 & 0 \\ 1 & 0 & 0 \\ 3 & 4 & 5 \end{bmatrix}.$$

The MATLAB function `forwsub` finds a unique solution of the linear system in the case of $\mathbf{L}_1$:

```
>> L1=[2 0 0;1 -1 0; 3 4 5];
>> b=[1 2 3]';
>> x = forwsub(L1,b)
```

```
x =
    0.5000
   -1.5000
    1.5000
```

However, things do not run so smoothly for the lower-triangular matrix $\mathbf{L}_2$:

```
>> L2=[2 0 0;1 0 0; 3 4 5];
>> x = forwsub(L2,b)
Warning: Divide by zero.
> In forwsub.m at line 9

x =
    0.5000
       Inf
      -Inf
```

The source of trouble is the vanishing diagonal entry appearing on the second row. As you can see from (2.41), this leads to a division by zero. The good news is that the same formula also shows that this is the only way in which forward substitution might fail. That is, provided all the diagonal entries are different from zero, forward substitution leads to the unique solution for the system. This is rephrased in the following theorem:

Theorem 2.5 *A lower-triangular matrix* $\mathbf{L}$ *is invertible if and only if all of its diagonal entries are different from zero.*

Backward substitution

The pictorial opposite of a lower-triangular matrix is an *upper-triangular* matrix, that is, a matrix $\mathbf{U}$ with the property that all the entries below its diagonal are equal to zero, that is, with $u_{ij} = 0$ for all $i > j$. Therefore, if $\mathbf{U}$ is upper triangular, the linear system $\mathbf{Ux} = \mathbf{b}$ has the form

$$\begin{bmatrix} u_{11} & u_{12} & \cdots & u_{1n} \\ 0 & u_{22} & \cdots & u_{2n} \\ \vdots & & & \vdots \\ 0 & 0 & \cdots & u_{nn} \end{bmatrix} \begin{bmatrix} x_1 \\ x_2 \\ \vdots \\ x_n \end{bmatrix} = \begin{bmatrix} b_1 \\ b_2 \\ \vdots \\ b_n \end{bmatrix}.$$

It is now hopeless to start from the first equation because it might contain terms in all variables. A more profitable approach is to start from the last equation, obtaining

$$u_{nn}x_n = b_n \qquad \Rightarrow \qquad x_n = \frac{b_n}{u_{nn}}.$$

We can then store this value for x_n and proceed backward to the equation before the last, which reads

$$u_{n-1,n-1}x_{n-1} + u_{n-1,n}x_n = b_n \quad \Rightarrow \quad x_{n-1} = \frac{b_{n-1} - u_{n-1,n}x_n}{u_{nn}},$$

because x_n is already known at this point. It is now easy to identify the general form for the step to calculate x_i. In it, we would have already calculated the values of $x_{i+1}, \ldots, x_n$ and could use these values to obtain

$$x_i = \frac{1}{u_{ii}}\left(b_i - \sum_{j=i+1}^{n} u_{ij}x_j\right). \tag{2.43}$$

This algorithm for solving an upper-triangular linear system is called *backward substitution.*

Exercise 2.11 Write a MATLAB function `backsub` that implements the backward substitution for an upper-triangular linear system and estimates its flop count as a function of the number of equations n of the upper-triangular linear system.

Exercise 2.12 Prove that an upper-triangular matrix $\mathbf{U}$ is invertible if and only if all of its diagonal entries are different from zero.

Gaussian elimination

Most linear systems encountered in practice are neither lower nor upper triangular, and we cannot apply the convenient forward and backward substitution methods directly. However, we can try to transform a given system into lower- or upper-triangular systems having the same solutions. To do that, we must know what kinds of transformations of linear systems have the property of preserving their solutions.

One such transformation is obtained by multiplying both sides of the matrix equation $\mathbf{Ax} = \mathbf{b}$ by a nonsingular matrix $\mathbf{M}$. To see this, observe that if $\mathbf{x}_0$ satisfies $\mathbf{MAx}_0 = \mathbf{Mb}$, then we have

$$\mathbf{Ax}_0 = \mathbf{M}^{-1}(\mathbf{MAx}_0) = \mathbf{M}^{-1}\mathbf{Mb} = \mathbf{b},$$

which shows that $\mathbf{x}_0$ is a solution to the original system $\mathbf{Ax} = \mathbf{b}$.

The method known as *Gaussian elimination* consists of successive transformations of a system through multiplication by nonsingular matrices $\mathbf{M}_k$ until the resulting system is upper triangular, which can then be solved by backward substitution. Start with the following example:

$$\mathbf{Ax} = \begin{bmatrix} 1 & 2 & 2 \\ 4 & 4 & 2 \\ 2 & 6 & 4 \end{bmatrix} \begin{bmatrix} x_1 \\ x_2 \\ x_3 \end{bmatrix} = \begin{bmatrix} 3 \\ 6 \\ 10 \end{bmatrix} = \mathbf{b}.$$

If you multiply both sides of this matrix equation by

$$\mathbf{M}_1 = \begin{bmatrix} 1 & 0 & 0 \\ -4 & 1 & 0 \\ -2 & 0 & 1 \end{bmatrix},$$

you obtain

$$\mathbf{M}_1\mathbf{A} = \begin{bmatrix} 1 & 0 & 0 \\ -4 & 1 & 0 \\ -2 & 0 & 1 \end{bmatrix} \begin{bmatrix} 1 & 2 & 2 \\ 4 & 4 & 2 \\ 2 & 6 & 4 \end{bmatrix} = \begin{bmatrix} 1 & 2 & 2 \\ 0 & -4 & -6 \\ 0 & 2 & 0 \end{bmatrix}$$

and

$$\mathbf{M}_1\mathbf{b} = \begin{bmatrix} 1 & 0 & 0 \\ -4 & 1 & 0 \\ -2 & 0 & 1 \end{bmatrix} \begin{bmatrix} 3 \\ 6 \\ 10 \end{bmatrix} = \begin{bmatrix} 3 \\ -6 \\ 4 \end{bmatrix}.$$

Observe what was achieved in this step: all the elements below the diagonal on the *first* column of the matrix $\mathbf{M}_1\mathbf{A}$ are zero. Next, multiply both sides of the new equation $\mathbf{M}_1\mathbf{A}\mathbf{x} = \mathbf{M}_1\mathbf{b}$ by

$$\mathbf{M}_2 = \begin{bmatrix} 1 & 0 & 0 \\ 0 & 1 & 0 \\ 0 & 0.5 & 1 \end{bmatrix},$$

obtaining

$$\mathbf{M}_2\mathbf{M}_1\mathbf{A} = \begin{bmatrix} 1 & 0 & 0 \\ 0 & 1 & 0 \\ 0 & 0.5 & 1 \end{bmatrix} \begin{bmatrix} 1 & 2 & 2 \\ 0 & -4 & -6 \\ 0 & 2 & 0 \end{bmatrix} = \begin{bmatrix} 1 & 2 & 2 \\ 0 & -4 & -6 \\ 0 & 0 & -3 \end{bmatrix}$$

and

$$\mathbf{M}_2\mathbf{M}_1\mathbf{b} = \begin{bmatrix} 1 & 0 & 0 \\ 0 & 1 & 0 \\ 0 & 0.5 & 1 \end{bmatrix} \begin{bmatrix} 3 \\ -6 \\ 4 \end{bmatrix} = \begin{bmatrix} 3 \\ -6 \\ 1 \end{bmatrix}.$$

Observe now that the modified matrix $\mathbf{M}_2\mathbf{M}_1\mathbf{A}$ is upper triangular, so that the system $\mathbf{M}_2\mathbf{M}_1\mathbf{A}\mathbf{x} = \mathbf{M}_2\mathbf{M}_1\mathbf{b}$ can be solved by backward substitution. Using the MATLAB function `backsub` from Exercise 2.11, you can find

```
>> U=[1 2 2; 0 -4 -6; 0 0 -3];
>> v=[3 -6 1]';
>> x=backsub(U,v)
```

```
x =
   -0.3333
    2.0000
   -0.3333
```

You can also use MATLAB to check that this is also the solution to the original system:

```
>> A=[1 2 2; 4 4 2; 2 6 4];
>> b=[3 6 10]';
>> x=inv(A)*b

x =
   -0.3333
    2.0000
   -0.3333
```

The general strategy for Gaussian elimination is to look at a matrix of the form

$$\mathbf{M}_k = \begin{bmatrix} 1 & \cdots & 0 & 0 & \cdots & 0 \\ \vdots & \ddots & \vdots & \vdots & & \vdots \\ 0 & \cdots & 1 & 0 & \cdots & 0 \\ 0 & \cdots & -m_{k+1} & 1 & \cdots & 0 \\ \vdots & & \vdots & \vdots & & \vdots \\ 0 & \cdots & -m_n & 0 & \cdots & 1 \end{bmatrix}. \tag{2.44}$$

The effect of this matrix on a general vector $\mathbf{v} \in \mathbb{R}^n$ is

$$\begin{bmatrix} 1 & \cdots & 0 & 0 & \cdots & 0 \\ \vdots & \ddots & \vdots & \vdots & & \vdots \\ 0 & \cdots & 1 & 0 & \cdots & 0 \\ 0 & \cdots & -m_{k+1} & 1 & \cdots & 0 \\ \vdots & & \vdots & \vdots & & \vdots \\ 0 & \cdots & -m_n & 0 & \cdots & 1 \end{bmatrix} \begin{bmatrix} v_1 \\ \vdots \\ v_k \\ v_{k+1} \\ \vdots \\ v_n \end{bmatrix} = \begin{bmatrix} v_1 \\ \vdots \\ v_k \\ v_{k+1} - m_{k+1}v_k \\ \vdots \\ v_n - m_n v_k \end{bmatrix}.$$

In particular, if the matrix $\mathbf{A}$ has already been made upper triangular up to its first $k-1$ columns and we choose

$$m_i = \frac{a_{ik}}{a_{kk}}, \quad i = k+1, \ldots, n, \tag{2.45}$$

then the matrix $\mathbf{M}_k$ leaves the first $k-1$ columns of $\mathbf{A}$ unchanged while transforming the kth column into

$$\begin{bmatrix} a_{1k} \\ \vdots \\ a_{kk} \\ 0 \\ \vdots \\ 0 \end{bmatrix}.$$

For this reason, the matrices $\mathbf{M}_k$ are called *elementary elimination matrices.* These elimination matrices are lower triangular with all of their diagonal entries being equal to 1. Hence, we conclude that each of them is invertible, and so is their product. Therefore, as long as none of the diagonal elements of $\mathbf{A}$ is zero, successive multiplication by exactly $n-1$ of such matrices leads to an upper triangular system of the form

$$\mathbf{M}_{n-1}\cdots\mathbf{M}_2\mathbf{M}_1\mathbf{A}\mathbf{x} = \mathbf{M}_{n-1}\cdots\mathbf{M}_2\mathbf{M}_1\mathbf{b}$$

having the same solution as the original system $\mathbf{A}\mathbf{x} = \mathbf{b}$.

A MATLAB function implementing Gaussian elimination is the following:

```
function [Aprime,bprime] = gausselim(A,b)
% reduces general system A*x=b to equivalent
% upper triangular system Aprime*x=Bprime
% A :  n x n input matrix
% b :  n x 1 vector
% Aprime : n x n output matrix
% bprime : n x 1 output vector
% WARNING: possibility of division by zero

[n,m]=size(A);
for p = 1:n-1                       % loop through columns
     for r=p+1:n                    % loop down rows
            M=A(r,p)/A(p,p);
            A(r,p)=0;
            A(r,p+1:n)=A(r,p+1:n)-A(p,p+1:n)*M;
            B(r)=B(r)-B(p)*M;
     end
end
Aprime=A;bprime=b;
```

Notice the warning about a possible division by zero in this code. It occurs whenever MATLAB encounters a vanishing entry on the diagonal of $\mathbf{A}$, due to the form of the multipliers in (2.45), where the term a_{kk} appearing in the denominator is commonly referred to as a *pivot.* The technique used to avoid division by a zero pivot, or the numerical instability created by a very small one, is called *pivoting* and is discussed later in this section.

The MATLAB function `gausselim` can be tested on the previous example:

```
>> A=[1 2 2; 4 4 2 ; 2 6 4];
>> b=[3 6 10]';
>> [U,v]=gausselim(A,b)

U =

     1     2     2
     0    -4    -6
     0     0    -3

v =

     3
    -6
     1
```

We see from the MATLAB code for `gausselim` that the *flop* count for Gaussian elimination for a system of n equations can be obtained from the expression

$$\sum_{p=1}^{n-1} \sum_{r=p+1}^{n} [1 + 2(n-p) + 2].$$

Because the final solution for the system still needs to be found by backward substitution, you must add to this the n^2 *flops* for this method.

Exercise 2.13 Show that the number of *flops* for the algorithm of Gaussian elimination is a cubic polynomial of the number of equations n in the linear system, and find the coefficient of the highest cubic power n^3.

LU factorization

Gaussian elimination can be summarized by the successive steps

$$\begin{aligned}
\mathbf{Ax} &= \mathbf{b} \\
\mathbf{M}_1\mathbf{Ax} &= \mathbf{M}_1\mathbf{b} \\
\mathbf{M}_2\mathbf{M}_1\mathbf{Ax} &= \mathbf{M}_2\mathbf{M}_1\mathbf{b} \\
&\vdots \\
\mathbf{M}_{n-1}\cdots\mathbf{M}_2\mathbf{M}_1\mathbf{Ax} &= \mathbf{M}_{n-1}\cdots\mathbf{M}_2\mathbf{M}_1\mathbf{b}
\end{aligned}$$

resulting in the equivalent system (that is, one with the same solution)

$$\mathbf{Ux} = \mathbf{v},$$

where $\mathbf{v} = \mathbf{M}_{n-1} \cdots \mathbf{M}_2\mathbf{M}_1\mathbf{b}$ is a column vector and

$$\mathbf{U} = \mathbf{M}_{n-1} \cdots \mathbf{M}_2\mathbf{M}_1\mathbf{A}$$

is an upper-triangular matrix. Recall that each of the matrices $\mathbf{M}_k$ is invertible, because they are lower triangular with non-vanishing diagonal entries. Therefore, if we denote its inverse by $\mathbf{L}_k = \mathbf{M}_k^{-1}$, we can use the preceding expression to rewrite $\mathbf{A}$ as

$$\mathbf{A} = (\mathbf{M}_{n-1} \cdots \mathbf{M}_2\mathbf{M}_1)^{-1}\mathbf{U} = \mathbf{M}_1^{-1} \cdots \mathbf{M}_{n-1}^{-1}\mathbf{U} = \mathbf{L}_1 \cdots \mathbf{L}_{n-1}\mathbf{U}. \quad (2.46)$$

The matrices $\mathbf{L}_k$ appearing in this expression are easy to obtain from the entries of the corresponding $\mathbf{M}_k$. To begin with, because of the following general result, they must be lower triangular:

Exercise 2.14 Prove that the inverse of a nonsingular lower-triangular matrix is also lower triangular.

More than just being lower triangular, it follows from the special form (2.44) of the matrices $\mathbf{M}_k$ that the inverse matrices $\mathbf{L}_k$ have exactly the same form but with multipliers of reversed signs, that is:

$$\mathbf{L}_k = \begin{bmatrix} 1 & \cdots & 0 & 0 & \cdots & 0 \\ \vdots & \ddots & \vdots & \vdots & & \vdots \\ 0 & \cdots & 1 & 0 & \cdots & 0 \\ 0 & \cdots & m_{k+1} & 1 & \cdots & 0 \\ \vdots & & \vdots & \vdots & & \vdots \\ 0 & \cdots & m_n & 0 & \cdots & 1 \end{bmatrix}. \quad (2.47)$$

For the matrices used in the previous example, you can verify (2.47) using MATLAB as follows:

```
>> M1= [1 0 0; -4 1 0; -2 0 1];
>> L1=inv(M1)

L1 =

     1     0     0
     4     1     0
     2     0     1

>> M2=[1 0 0; 0 1 0; 0 0.5 1];
>> L2=inv(M2)
```

```
L2 =
    1.0000         0         0
         0    1.0000         0
         0   -0.5000    1.0000
```

Let us now tackle the product $\mathbf{L}_1 \cdots \mathbf{L}_{n-1}$ appearing in (2.46), where each $\mathbf{L}_k$ is a lower-triangular matrix of the form (2.47). As the next exercise shows, such a product must itself be lower triangular:

Exercise 2.15 Show that the product of two lower-triangular matrices is also lower triangular.

Moreover, again because of the special form of the matrices $\mathbf{L}_k$, it is easy to show that, if $k < j$, then the product $\mathbf{L}_k\mathbf{L}_j$ consists of a lower-triangular matrix resembling the general form (2.47), but having the multipliers of $\mathbf{L}_k$ in its kth column and the multipliers of $\mathbf{L}_j$ in its jth column. For example, for matrices $\mathbf{L}_1$ and $\mathbf{L}_2$ in the previous example, you have

```
>> L=L1*L2

L =
    1.0000         0         0
    4.0000    1.0000         0
    2.0000   -0.5000    1.0000
```

Using the notation $\mathbf{L} = \mathbf{L}_1 \cdots \mathbf{L}_{n-1}$, we conclude that the original matrix $\mathbf{A}$ can be rewritten as

$$\mathbf{A} = \mathbf{LU}, \tag{2.48}$$

where $\mathbf{L}$ is lower triangular and $\mathbf{U}$ is upper triangular. This is appropriately called the *LU factorization* of the matrix $\mathbf{A}$. Moreover, the matrix $\mathbf{U}$ coincides with the outcome of Gaussian elimination for $\mathbf{A}$, whereas the matrix $\mathbf{L}$ can be easily obtained by adjoining together the columns of the individual matrices $\mathbf{L}_k$, which by their turn are obtained by changing the sign of the multipliers of the matrices $\mathbf{M}_k$.

The algorithm of LU factorization can be coded in the following MATLAB function:

```
function [L,U]=lufactor(A)
% LU factorization for n x n matrix A
% L is lower triangular
% U is upper triangular
% A= L*U
% WARNING: possible division by zero
```

```
n=length(A);
U=A;
L=eye(n);
for p=1:n
     for q=p+1:n
          L(q,p)=U(q,p)/U(p,p);
          U(q,p:n)=U(q,p:n)-U(p,p:n)*L(q,p);
     end
end
```

Again, observe the warning about a possible division by zero occurring whenever the original matrix $\mathbf{A}$ has a vanishing diagonal entry, which is dealt with by the *pivoting* technique described later in this section.

You can now try this function on the matrix $\mathbf{A}$ in the previous example:

```
>> A=[1 2 2; 4 4 2; 2 6 4];
>> [L,U]=lufactor(A)

L =
    1.0000         0         0
    4.0000    1.0000         0
    2.0000   -0.5000    1.0000

U =
     1     2     2
     0    -4    -6
     0     0    -3
```

which agrees with the previous calculations.

Once an LU factorization has been performed on a matrix $\mathbf{A}$, a system of the form $\mathbf{Ax} = \mathbf{b}$ can be easily solved by forward and backward substitution as follows. Since $\mathbf{A} = \mathbf{LU}$, the original system is equivalent to

$$\mathbf{LUx} = \mathbf{b}.$$

If we denote $\mathbf{Ux} = \mathbf{y}$, we obtain the intermediate system

$$\mathbf{Ly} = \mathbf{b},$$

which can be solved for $\mathbf{y}$ by forward substitution. After we find $\mathbf{y}$, we can then solve the upper-triangular system $\mathbf{Ux} = \mathbf{y}$ by backward substitution and produce the solution $\mathbf{x}$ for the original system.

Exercise 2.16 Using the MATLAB functions for forward and backward substitutions and for LU factorization, write a MATLAB function for solving a general linear system using LU factorization, followed by forward and then backward substitutions.

Gaussian elimination and LU factorization are obviously two ways of expressing the exact same idea. Not surprisingly, the *flop* count for LU factorization has the same behavior as the *flop* count for Gaussian elimination (see Exercise 2.13). The slight advantage of LU factorization is that, after the factors $\mathbf{L}$ and $\mathbf{U}$ for a given matrix $\mathbf{A}$ are found, they can be stored and used to solve many different systems, corresponding to different input vectors $\mathbf{b}$.

Pivoting

If the starting matrix $\mathbf{A}$ has a vanishing diagonal term, say, $a_{kk} = 0$, then both Gaussian elimination and LU factorization will lead to a division by zero as they reach the kth column of $\mathbf{A}$. This problem has an easy fix because the order in which equations appear on a system do not alter the solution of the system. Therefore, given a system in the form $\mathbf{Ax} = \mathbf{b}$, we can interchange any two rows of the matrix $\mathbf{A}$ without altering the solution to the system, provided we also interchange the corresponding rows for the vector $\mathbf{b}$. In the language of matrix transformations, the interchange of rows is achieved by multiplying $\mathbf{A}$ from the left by a permutation matrix $\mathbf{P}$. Because another interchange of the same rows recovers the original matrix, any permutation matrix is invertible and its inverse is $\mathbf{P}$ itself, which in turn confirms that $\mathbf{PAx} = \mathbf{Pb}$ and $\mathbf{Ax} = \mathbf{b}$ have the same solution.

Back to the problem. If we encounter a diagonal term $a_{kk} = 0$, all we need to do is to look for the first nonzero entry appearing below it, interchange the rows (for both the matrix $\mathbf{A}$ and the vector $\mathbf{b}$), and continue with the algorithm. In the extreme case where all entries below a_{kk} are zero, then there is nothing we need to do because this means that this kth column is already in diagonal form. We then just leave $a_{kk} = 0$ where it is and proceed to the next column, therefore producing a vanishing diagonal entry in the resulting upper-triangular matrix $\mathbf{U}$. In this way, we can complete the LU factorization for permutations of any matrix $\mathbf{A}$. When the matrix $\mathbf{A}$ is singular, this procedure still work, but results in a matrix $\mathbf{U}$ with some vanishing diagonal terms.

A far more delicate problem arises if small (as opposed to vanishing) pivots occur in the LU factorization. Suppose that ε is the precision accuracy level for some floating-point number system. Recall from Section 1.7 that ε is defined as the distance between 1 and the next larger floating-point number represented in the system. Now let $0 < \delta < 1/2$ and consider the matrix

$$\mathbf{A} = \begin{bmatrix} \delta & 1 \\ 1 & \varepsilon \end{bmatrix}.$$

Under exact arithmetic, the LU factorization algorithm leads to

$$\mathbf{U} = \mathbf{MA} = \begin{bmatrix} 1 & 0 \\ -1/\delta & 1 \end{bmatrix} \begin{bmatrix} \delta & 1 \\ 1 & \varepsilon \end{bmatrix} = \begin{bmatrix} \delta & 1 \\ 0 & \varepsilon - 1/\delta \end{bmatrix}$$

and

$$\mathbf{L} = \begin{bmatrix} 1 & 0 \\ 1/\delta & 1 \end{bmatrix}.$$

However, in floating-point arithmetic, the term $(\varepsilon - 1/\delta)$ is approximated by $-1/\delta$. With this approximation, the factorization leads to

$$\mathbf{LU} = \begin{bmatrix} 1 & 0 \\ 1/\delta & 1 \end{bmatrix} \begin{bmatrix} \delta & 1 \\ 0 & -1/\delta \end{bmatrix} = \begin{bmatrix} \delta & 1 \\ 1 & 0 \end{bmatrix} \neq \mathbf{A}.$$

That is, a small pivot produces a very large multiplier, leading to an LU factorization that corresponds to a matrix very different from the original matrix $\mathbf{A}$. This is yet another instance of "catastrophic cancellations" induced by very large numbers represented in floating-point arithmetic. For a concrete example, consider the following:

```
>> delta=1/4;
>> A=[delta 1; 1 eps(1)];
>> [L U]=lufactor(A);
>> (L*U==A)
ans =
     1     1
     1     0
```

The solution to the problem with small pivots is the same as before: on the kth step, apply permutations to bring the largest entry on or below the diagonal to the pivot position. Formally, each step involves multiplying the given matrix by $\mathbf{M}_k\mathbf{P}_k$, and so in the end we have

$$\mathbf{MA} = (\mathbf{M}_{n-1}\mathbf{P}_{n-1} \cdots \mathbf{M}_1\mathbf{P}_1)\mathbf{A} = \mathbf{U}, \tag{2.49}$$

where $\mathbf{U}$ is upper triangular. We can therefore write $\mathbf{A} = \mathbf{M}^{-1}\mathbf{U}$. The matrix $\mathbf{M}^{-1}$ is not, however, lower triangular. Instead, it is a permutation of a lower-triangular matrix. If we want to obtain a proper lower triangular matrix, we can reformulate the pivoting technique as in the next exercise.

Exercise 2.17 Prove that, if all the intermediate permutations are collected into a single matrix

$$\mathbf{P} = \mathbf{P}_{n-1} \cdots \mathbf{P}_1,$$

then (2.49) is equivalent to

$$\mathbf{PA} = \mathbf{LU},$$

where $\mathbf{L}$ is lower triangular.

Therefore, to solve a linear system $\mathbf{Ax} = \mathbf{b}$, we consider

$$\mathbf{PAx} = \mathbf{LUx} = \mathbf{Pb},$$

solve it first for the lower-triangular system $\mathbf{Ly} = \mathbf{Pb}$ by forward substitution, and then solve the upper-triangular system $\mathbf{Ux} = \mathbf{y}$ by backward substitution. The following MATLAB function provides the necessary matrices for this task.

```
function [P,L,U]=plufactor(A)
% PA=LU factorization for n x n A
% L is lower triangular
% U is upper triangular

n=length(A);
U=A;
L=eye(n);
P=L;
for p=1:n-1
       [y,k]=max(abs(U(p:n,p)));
       k=k+p-1;
       aux=U(p,p:n); U(p,p:n)=U(k,p:n); U(k,p:n)=aux;
       aux=L(p,1:p-1); L(p,1:p-1)=L(k,1:p-1); L(k,1:p-1)=aux;
       aux=P(p,:); P(p,:)=P(k,:); P(k,:)=aux;
       for q=p+1:n
              L(q,p)=U(q,p)/U(p,p);
              U(q,p:n)=U(q,p:n)-U(p,p:n)*L(q,p);
       end
end
```

You can now test this function on the previous example of small pivots:

```
>> delta=0.5*eps(1);
>> A=[delta 1; 1 eps(1)];
>> [P,L,U]=plufactor(A)

P =
     0     1
     1     0

L =
    1.0000         0
    0.0000    1.0000
```

```
U =
    1.0000    0.0000
         0    1.0000

>> (P*A==L*U)

ans =
     1     1
     1     1
```

The built-in MATLAB function `lu` implements LU factorization with the same type of pivoting technique just described. Variants of Gaussian elimination with pivoting are also the basis for the backslash operator \ when applied to square matrices.

2.6 Iterative Methods

The general idea of an iterative method for solving a linear system of the form $\mathbf{Ax} = \mathbf{b}$ is to formulate a scheme that calculates an approximate solution $\mathbf{x}^{(k+1)}$ at the $(k+1)$th step based on the approximation $\mathbf{x}^{(k)}$ obtained at the kth iteration. Starting with an initial guess $\mathbf{x}_0$, we then hope that the iterations produce better and better approximations. The method is said to be convergent if

$$\lim_{k\to\infty} \|\mathbf{x}^{(k)} - \mathbf{x}\| = 0, \tag{2.50}$$

where $\mathbf{x}$ is the solution of the system. Because of the equivalence of norms for finite-dimensional vector spaces, any convenient vector norm can be used in the limit (2.50). In possession of a convergent method, we obviously don't sit around and wait for an infinite number of steps until we reach the solution $\mathbf{x}$. Instead, we would like to stop the iterations once the error

$$\mathbf{e}^{(k)} = \mathbf{x} - \mathbf{x}^{(k)} \tag{2.51}$$

becomes sufficiently small (according to a given norm). But since $\mathbf{x}$ is not known, we can stop the iterations once the *residual*

$$\mathbf{r}^{(k)} = \mathbf{Ax}^{(k)} - \mathbf{b} \tag{2.52}$$

becomes sufficiently small.

The key to understanding why iterative methods are used at all, as opposed to the direct methods presented in Section 2.5, is that they generally produce an approximate solution within a desired tolerance level in just a few steps. For such small numbers of steps, iterative methods have lower storage and *flop* count requirements than direct methods do. In particular, they are much superior to direct methods if we have to deal with *sparse* matrices, that is, matrices with a large number of vanishing entries. For instance, as you will see in Section 10.5, iterative methods involving sparse matrices appear in the numerical solution of partial differential equations.

Jacobi method

The two most popular iterative algorithms for solutions of linear systems are the *Jacobi* and *Gauss–Seidel* methods. Given an approximate solution $\mathbf{x}^{(k)}$, the Jacobi method proceeds to calculate the first component of $\mathbf{x}^{(k+1)}$ from the first equation in the system, the second component of $\mathbf{x}^{(k+1)}$ from the second equation, and so on. For instance, for the system

$$\begin{bmatrix} 5 & -1 & 2 \\ 1 & -6 & 1 \\ 1 & 1 & 4 \end{bmatrix} \begin{bmatrix} x_1 \\ x_2 \\ x_3 \end{bmatrix} = \begin{bmatrix} 6 \\ -4 \\ 6 \end{bmatrix}, \tag{2.53}$$

the Jacobi iterations are

$$\begin{aligned} x_1^{(k+1)} &= \frac{6 + x_2^{(k)} - 2x_3^{(k)}}{5} \\ x_2^{(k+1)} &= \frac{4 + x_1^{(k)} + x_3^{(k)}}{6} \\ x_3^{(k+1)} &= \frac{6 - x_1^{(k)} - x_2^{(k)}}{4}. \end{aligned}$$

For a general system $\mathbf{Ax} = \mathbf{b}$, the Jacobi method is summarized in the expression

$$x_i^{(k+1)} = \frac{1}{a_{ii}} \left(b_i - \sum_{j \neq i} a_{ij} x_j^{(k)} \right), \tag{2.54}$$

which is easily implemented by the following MATLAB function:

```
function [x,iter] = jacobi(A,b,tol,maxit)
% uses Jacobi iteration to solve A*x=B
% A      square  n x n
% b      n x 1
% tol    tolerance
% maxit  maximal number of iterations
% iter   actual number of iterations required for convergence

[n,m]=size(A);
x = zeros(size(b));       %initial guess
i=1;
iter = maxit;
```

```
for i = 1: maxit
    for j=1:n
        y(j)=(b(j)-A(j,1:j-1)*x(1:j-1)-A(j,j+1:n)*x(j+1:n))/A(j,j);
    end
    if max(abs(A*y'-b)) < tol
        iter = i;
        break
    end
    x=y';
end
```

Notice that two termination criteria were implemented in the MATLAB function `jacobi`. First, it computes the norm of the residual (2.52) and compares it to the given tolerance. In addition, it compares the number of iterations to a maximal number `maxit`. Without this second condition, iterations could enter into an infinite loop in the situations when the Jacobi method does not converge and the tolerance might never be reached.

Applying the MATLAB function `jacobi` to the linear system (2.53), you can observe the rapid convergence of the method to the exact solution $\mathbf{x} = [1, 1, 1]$ such that the distance of less than 10^{-6} is reached in just 14 iterations.

```
>> A=[5 -1 2; 1 -6 1; 1 1 4];
>> b=[6 -4 6]';
>> [x,iter]=jacobi(A,b,10^-6,100)

x =
    1.0000
    1.0000
    1.0000

iter =
    14
```

As you can see from the commands in the function `jacobi`, the *flop* count for each iteration of the Jacobi method is

$$n\,(1+1+(n-1)+(n-2)) = 2n^2 - n.$$

For a reasonable number of iterations (say, 20), this is still much faster than the n^3 *flops* typically needed for direct methods. Unfortunately, the Jacobi method may fail to converge for some linear systems. To consider convergence of the Jacobi method, observe that the exact solution for the system satisfies

$$x_i = \frac{1}{a_{ii}}\left(b_i - \sum_{j\neq i} a_{ij}x_j\right). \tag{2.55}$$

Therefore, subtracting (2.55) from (2.54), we obtain

$$x_i^{(k+1)} - x_i = -\sum_{j\neq i} \frac{a_{ij}}{a_{ii}}(x_j^{(k)} - x_j).$$

That is, recalling the definition of the error $\mathbf{e}^{(k)}$ in (2.51), we have

$$e_i^{(k+1)} = -\sum_{j\neq i} \frac{a_{ij}}{a_{ii}} e_j^{(k)}.$$

Taking the absolute value on both sides leads to

$$|e_i^{(k+1)}| \leq \sum_{j\neq i} \left|\frac{a_{ij}}{a_{ii}}\right| |e_j^{(k)}| \leq \left(\sum_{j\neq i} \left|\frac{a_{ij}}{a_{ii}}\right|\right) \max_{1\leq j\leq n} |e_j^{(k)}|. \tag{2.56}$$

Defining

$$\mu = \max_i \sum_{j\neq i} \left|\frac{a_{ij}}{a_{ii}}\right|, \tag{2.57}$$

and taking the maximum over i on both sides of (2.56), we can see that

$$\max_{1\leq i\leq n} |e_i^{(k+1)}| \leq \mu \max_{1\leq j\leq n} |e_j^{(k)}|.$$

In other words, using the definition (2.26) for the ∞–norm, we have

$$\|\mathbf{e}^{(k+1)}\|_\infty \leq \mu \|\mathbf{e}^{(k)}\|_\infty.$$

Therefore, if $\mu < 1$, the error goes to zero according to the ∞–norm and the Jacobi method converges. Moreover, according to the definitions in Section 1.8, you can see that the convergence rate is at least linear.

To get a better understanding of the condition $\mu < 1$, observe that it is equivalent to

$$\sum_{j\neq i} |a_{ij}| < |a_{ii}|, \qquad \forall i = 1, \ldots, n. \tag{2.58}$$

In other words, $\mu < 1$ if and only if the diagonal entry of each row of $\mathbf{A}$ has an absolute value that is larger than the sum of the absolute values of the off-diagonal entries. A matrix satisfying this condition is said to be *strictly diagonally dominant.* Therefore, strict diagonal dominance is a sufficient condition for the convergence of the Jacobi method. For example, the matrix $\mathbf{A}$ in the previous example of the linear system (2.53) is strictly diagonally dominant, so that the Jacobi method is guaranteed to converge in this case.

However, condition (2.58) is far from being necessary for the convergence of the Jacobi method, as shown in the next exercise.

Exercise 2.18 Apply the Jacobi method to the linear system (2.53) with the second row of the matrix $\mathbf{A}$ replaced by $[4, -6, 4]$ and conclude that strict diagonal dominance is not necessary for the convergence of the Jacobi method.

Still on the topic of convergence for the Jacobi method, observe that it can depend on the order of the equations in a given linear system. For example, interchanging the first and last equations of the system in (2.53) (which clearly does not alter its solution) has the effect of producing a matrix that is no longer strictly diagonally dominant. Consequently, convergence of the Jacobi method is no longer guaranteed for this system. In fact, it diverges:

```
>> A=[1 1 4;1 -6 1; 5 -1 2];
>> b=[6 -4 6]';
>> [x,iter]=jacobi(A,b,10^-6,100)

x =
   1.0e+49 *

   -3.7612
    0.6869
   -7.9278

iter =
   100
```

Gauss–Seidel method

Looking back at the Jacobi method, you might be inclined to ask the following: once we find $x_1^{(k+1)}$ from the first equation, why not use this new component on the second equation, instead of the old $x_1^{(k)}$? After all, $\mathbf{x}^{(k+1)}$ is intended to be a better approximation of the solution than $\mathbf{x}^{(k)}$, and we expect that, in some sense, the same is true for individual components. The Gauss–Seidel method addresses exactly this criticism. In calculating each new component in the new approximation, it uses the newest available estimates of the other components. For example, for the system

$$\begin{bmatrix} 4 & 2 & 1 \\ -1 & 2 & 2 \\ 1 & 0 & 3 \end{bmatrix} \begin{bmatrix} x_1 \\ x_2 \\ x_3 \end{bmatrix} = \begin{bmatrix} 5 \\ 6 \\ 7 \end{bmatrix}, \tag{2.59}$$

the Gauss–Seidel iterations are

$$\begin{aligned} x_1^{(k+1)} &= \frac{5 - 2x_2^{(k)} - x_3^{(k)}}{4} \\ x_2^{(k+1)} &= \frac{6 + x_1^{(k+1)} - 2x_3^{(k)}}{2} \\ x_3^{(k+1)} &= \frac{7 - x_1^{(k+1)}}{3}. \end{aligned}$$

The general expression for these iterations, applied to a system $\mathbf{Ax} = \mathbf{b}$, is

$$x_i^{(k+1)} = \frac{1}{a_{ii}} \left(b_i - \sum_{j=1}^{i-1} a_{ij} x_j^{(k+1)} - \sum_{j=i+1}^{n} a_{ij} x_j^{(k)} \right). \tag{2.60}$$

Exercise 2.19 Write a MATLAB function that implements the Gauss–Seidel method. Obtain its *flop* count and compare it with the one for the Jacobi method.

We can adapt the steps of the previous argument for the Jacobi method and conclude that strict diagonal dominance is also a sufficient condition for convergence of the Gauss–Seidel iterations. Apart from this statement, a full comparison between the convergence properties of the two methods is more involved. For instance, it is possible to show that the Gauss–Seidel method converges for symmetric positive definite matrices (see Section 2.7 for the definitions), whereas the Jacobi method may diverge for such matrices.

Exercise 2.20 Use the Gauss–Seidel method to obtain a solution within a tolerance of 10^{-6} for the system

$$\begin{aligned} 6x - 3y + 2z &= 6 \\ -3x + 5y - 7z &= -4 \\ 2x - 7y + 14z &= 6 \end{aligned}$$

Repeat your calculations using the Jacobi method.

Although it is true that, for most linear systems, the Gauss–Seidel method is more likely to converge than the Jacobi method, there are cases when the opposite is true. For example, for the system

$$\begin{aligned} -x + 0.5y + 0.5z &= 1 \\ 0.7x - y + 1.1z &= -2 \\ 0.3x - y - z &= 1 \end{aligned}$$

the Jacobi method converges but the Gauss–Seidel method does not:

```
>> A=[-1 0.5 0.5; 0.7 -1 1.1; 0.3 -1 -1];
>> b=[1 -2 1]';
>> [x,iter]=jacobi(A,b,10^-6,100)

x =
   -1.7647
   -0.4370
   -1.0924

iter =
    76

>> [x,iter]=gaussseidel(A,b,10^-6,100)

x =
   1.0e+04 *

   -0.0002
   -1.8427
    1.8426

iter =
   100

>> norm (A*x-b)

ans =
   3.8696e+04
```

Convergence aside, a clear advantage of Gauss–Seidel is that it requires far less memory allocation because the new components can be written on top of the old ones as soon as they are calculated. On the other hand, the Gauss–Seidel method must be executed sequentially, whereas the Jacobi iterations can be implemented in parallel, with each component being calculated separately.

2.7 Cholesky Factorization

The LU factorization described in Section 2.5 is a general method that can be applied to any $n \times n$ matrix. Other types of factorizations, however, can be more efficient when dealing with matrices possessing some special properties. In this section, we review one such method, known as *Cholesky factorization.*

The special class of matrices for which this method is applicable is that of *symmetric positive definite* matrices.

We say that an $n \times n$ matrix $\mathbf{A}$ is *symmetric* if $\mathbf{A} = \mathbf{A}'$. For matrices with real entries, this simply means that $a_{ij} = a_{ji}$ for all i, j, while for matrices with complex numbers as entries, the symmetry condition means that $a_{ij} = \overline{a_{ji}}$ for all i, j. In either case, we immediately notice that if a matrix is symmetric, we don't need to specify all of its entries separately, leading to savings in memory allocation. This is just the tip of the iceberg concerning symmetric matrices, and in the next chapters we have the chance to investigate many of their fascinating features.

Positivity is a more delicate topic. In analogy with real numbers, we would like to say that a matrix is positive if it is, in some sense, "greater than zero." But because it is difficult to introduce an appropriate order relation for multidimensional objects such as matrices, the definition of positivity requires more work.

If we consider matrices not just as arrays of real (or complex) numbers, but as linear transformations between vector spaces, then it turns out that positivity can be defined according to their effect on vectors. More precisely, because $\mathbf{x}'$ and $\mathbf{Ax}$ are, respectively, a row and a column vector, we know that the product $\mathbf{x}' * (\mathbf{Ax})$, which can be denoted simply by $\mathbf{x}'\mathbf{Ax}$, is a real (or complex) number. As we discuss in Section 4.4, when $\mathbf{A}$ is symmetric, the scalar $\mathbf{x}'\mathbf{Ax}$ is always a real number. We then say that a symmetric matrix $\mathbf{A}$ is *positive definite* if

$$\mathbf{x}'\mathbf{Ax} > 0 \tag{2.61}$$

for all nonzero $\mathbf{x} \in \mathbb{R}^n$ (or $\mathbb{C}^n$).

We begin with Cholesky factorization for the simplest case of 2×2 matrices with real entries. A general 2×2 symmetric, positive definite real matrix $\mathbf{A}$ can be written as

$$\mathbf{A} = \begin{bmatrix} a_{11} & a_{21} \\ a_{21} & a_{22} \end{bmatrix}.$$

Applying an elementary elimination matrix to its first column gives

$$\mathbf{M}_1\mathbf{A} = \begin{bmatrix} 1 & 0 \\ -\frac{a_{21}}{a_{11}} & 1 \end{bmatrix} \begin{bmatrix} a_{11} & a_{21} \\ a_{21} & a_{22} \end{bmatrix} = \begin{bmatrix} a_{11} & a_{21} \\ 0 & a_{22} - \frac{a_{21}^2}{a_{11}} \end{bmatrix}.$$

Instead of being content with this upper-triangular matrix, we now try to obtain a diagonal matrix out of it. Namely, we apply an elementary elimination matrix to its transpose:

$$\mathbf{M}_1(\mathbf{M}_1\mathbf{A})' = \begin{bmatrix} 1 & 0 \\ -\frac{a_{21}}{a_{11}} & 1 \end{bmatrix} \begin{bmatrix} a_{11} & 0 \\ a_{21} & a_{22} - \frac{a_{21}^2}{a_{11}} \end{bmatrix} = \begin{bmatrix} a_{11} & 0 \\ 0 & a_{22} - \frac{a_{21}^2}{a_{11}} \end{bmatrix}.$$

That is

$$\mathbf{M}_1\mathbf{A}\mathbf{M}_1' = \begin{bmatrix} a_{11} & 0 \\ 0 & a_{22} - \frac{a_{21}^2}{a_{11}} \end{bmatrix},$$

which implies that

$$\mathbf{A} = \mathbf{L}_1\mathbf{D}\mathbf{L}_1', \tag{2.62}$$

where $\mathbf{L}_1 = \mathbf{M}_1^{-1}$ and $\mathbf{D}$ is the diagonal matrix

$$\mathbf{D} = \begin{bmatrix} a_{11} & 0 \\ 0 & a_{22} - \frac{a_{21}^2}{a_{11}} \end{bmatrix}.$$

Let us see how these calculations are implemented in a specific example:

```
>> A=[2 1; 1 3];
>> M1=[1 0; -1/2 1];
>> D=M1*A*M1'

D =
    2.0000         0
         0    2.5000
```

Observe how in the preceding example the matrix $\mathbf{D}$ has strictly positive diagonal entries. The next exercise shows that this is not a coincidence.

Exercise 2.21 Prove that, because $\mathbf{M}_1$ is invertible and $\mathbf{A}$ is positive definite, the matrix $\mathbf{M}_1\mathbf{A}\mathbf{M}_1'$ is positive definite. Use the basis vectors $\mathbf{e}_1$ and $\mathbf{e}_2$ to conclude that the positivity of $\mathbf{A}$ implies that

$$a_{11} > 0 \qquad \text{and} \qquad \left(a_{22} - \frac{a_{21}^2}{a_{11}}\right) > 0.$$

Armed with this result, we can then take the square root of the diagonal terms in $\mathbf{D}$ and define

$$\mathbf{L}_2 = \begin{bmatrix} \sqrt{a_{11}} & 0 \\ 0 & \sqrt{a_{22} - \frac{a_{21}^2}{a_{11}}} \end{bmatrix}.$$

Defining $\mathbf{L} = \mathbf{L}_1\mathbf{L}_2$, we see that (2.62) can be rewritten as

$$\mathbf{A} = (\mathbf{L}_1\mathbf{L}_2)(\mathbf{L}_2'\mathbf{L}_1') = \mathbf{L}\mathbf{L}'. \tag{2.63}$$

This is still a factorization of $\mathbf{A}$ into the product of a lower and an upper triangular matrix, but this time they are transposesof each other, which obviously reduces both storage and *flop* count requirements when compared to a simple LU factorization.

Exercise 2.22 Prove that $\mathbf{L}$ is the unique lower-triangular matrix with strictly positive diagonal entries for which (2.63) holds. That is, if $\mathbf{S}$ is a lower-triangular matrix with $s_{11}, s_{22} > 0$ and $\mathbf{SS}' = \mathbf{LL}'$, then you necessarily have $\mathbf{L} = \mathbf{S}$.

The steps involved in the Cholesky factorization of an $n \times n$ symmetric positive definite matrix $\mathbf{A}$ are careful generalizations of what we just did for the 2×2 case. We present them now for authorial peace of mind, but you are free to jump to Equation (2.65) if you find them too cumbersome.

Let us first identify some blocks within the matrix $\mathbf{A}$ and rewrite it as

$$\mathbf{A} = \begin{bmatrix} a_{11} & \mathbf{w}_1' \\ \mathbf{w}_1 & \mathbf{K}_1 \end{bmatrix},$$

where a_{11} is a scalar, $\mathbf{w}_1$ is an $(n-1) \times 1$ column vector, and $\mathbf{K}_1$ is an $(n-1) \times (n-1)$ symmetric matrix. For example, for the matrix

$$\mathbf{A} = \begin{bmatrix} 4 & -2 & 4 & 6 \\ -2 & 10 & -17 & 0 \\ 4 & -17 & 30 & 3 \\ 6 & 0 & 3 & 0 \end{bmatrix},$$

you have

$$a_{11} = 4, \quad \mathbf{w}_1 = \begin{bmatrix} -2 \\ 4 \\ 6 \end{bmatrix}, \quad \mathbf{K}_1 = \begin{bmatrix} 10 & -17 & 0 \\ -17 & 30 & 3 \\ 0 & 3 & 18 \end{bmatrix}.$$

Applying an elementary elimination matrix to the first column of $\mathbf{A}$, we obtain

$$\mathbf{M}_1\mathbf{A} = \begin{bmatrix} 1 & 0 \\ -\frac{\mathbf{w}_1}{a_{11}} & \mathbf{1}_{n-1} \end{bmatrix} \begin{bmatrix} a_{11} & \mathbf{w}_1' \\ \mathbf{w}_1 & \mathbf{K}_1 \end{bmatrix} = \begin{bmatrix} a_{11} & \mathbf{w}_1' \\ 0 & \mathbf{K}_1 - \frac{\mathbf{w}_1 * \mathbf{w}_1'}{a_{11}} \end{bmatrix}.$$

In the last matrix here, notice how the outer product $\mathbf{w}_1 * \mathbf{w}_1'$ of a column vector by its own transpose creates an $(n-1) \times (n-1)$ symmetric matrix. Now, like we did in the 2×2 case, let us apply the same elementary elimination matrix to the transpose of $\mathbf{M}_1\mathbf{A}$:

$$\begin{aligned} \mathbf{M}_1(\mathbf{M}_1\mathbf{A})' &= \begin{bmatrix} 1 & 0 \\ -\frac{\mathbf{w}_1}{a_{11}} & \mathbf{1}_{n-1} \end{bmatrix} \begin{bmatrix} a_{11} & 0 \\ \mathbf{w}_1' & \mathbf{K}_1 - \frac{\mathbf{w}_1 * \mathbf{w}_1'}{a_{11}} \end{bmatrix} \\ &= \begin{bmatrix} a_{11} & 0 \\ 0 & \mathbf{K}_1 - \frac{\mathbf{w}_1 * \mathbf{w}_1'}{a_{11}} \end{bmatrix}. \end{aligned}$$

Therefore,

$$\mathbf{M}_1\mathbf{A}\mathbf{M}_1' = \begin{bmatrix} a_{11} & 0 \\ 0 & \mathbf{A}^{(1)} \end{bmatrix},$$

where $\mathbf{A}^{(1)} = \mathbf{K}_1 - \frac{\mathbf{w}_1 * \mathbf{w}_1'}{a_{11}}$ is an $(n-1) \times (n-1)$ symmetric matrix, upon which the same procedure can be repeated. For our example, we obtain

```
>>A=[4  -2  4  6;-2 10 -17  0; 4 -17 30  3; 6 0 3 18];
>>M1=[1 0 0 0; 1/2 1 0 0; -1 0 1 0; -3/2 0 0 1];
>>M1*A*M1'

ans =
     4     0     0     0
     0     9   -15     3
     0   -15    26    -3
     0     3    -3     9
```

so that

$$\mathbf{A}^{(1)} = \begin{bmatrix} 9 & -15 & 3 \\ -15 & 26 & -3 \\ 3 & -3 & 9 \end{bmatrix}.$$

Continuing in this way, after $(n-1)$ steps, we obtain the diagonal matrix

$$\mathbf{M}_{n-1} \cdots \mathbf{M}_1\mathbf{A}\mathbf{M}_1' \cdots \mathbf{M}_{n-1}' = \begin{bmatrix} a_{11} & & \cdots & \\ & a_{11}^{(1)} & & \\ \vdots & & \ddots & \\ & \cdots & & a_{11}^{(n-1)} \end{bmatrix}. \quad (2.64)$$

At this point, we invoke the same crucial claim as before, namely, that the scalars

$$a_{11}, a_{11}^{(1)}, \ldots, a_{11}^{(n-1)}$$

are all strictly positive. For our example, the explicit steps are

```
>> M2=[1 0 0 0; 0 1 0 0; 0 5/3 1 0; 0 -1/3 0 1];
>> M2*M1*A*M1'*M2'
```

```
ans =
    4.0000         0         0         0
         0    9.0000         0         0
         0         0    1.0000    2.0000
         0    0.0000    2.0000    8.0000

>> M3=[1 0 0 0;0 1 0 0;0 0 1 0;0 0 -2 1];
>> M3*M2*M1*A*M1'*M2'*M3'

ans =
    4.0000         0         0         0
         0    9.0000         0         0
         0         0    1.0000    0.0000
         0   -0.0000   -0.0000    4.0000
```

We can then take the square root of the diagonal entries in (2.64) and define

$$\mathbf{L}_n = \begin{bmatrix} \sqrt{a_{11}} & & \cdots & \\ & \sqrt{a_{11}^{(1)}} & & \\ \vdots & & \ddots & \\ & \cdots & & \sqrt{a_{11}^{(n-1)}} \end{bmatrix}.$$

Using the same notation as before, that is, with $\mathbf{L}_i = \mathbf{M}_i^{-1}$ for $i = 1, \ldots, n-1$, and $\mathbf{L} = \mathbf{L}_1 \cdots \mathbf{L}_n$, we conclude from (2.64) that

$$\mathbf{A} = \mathbf{L}\mathbf{L}', \tag{2.65}$$

where $\mathbf{L}$ is the unique lower triangular $n \times n$ matrix with strictly positive diagonal entries for which this holds. In our example, we have

```
>>L1=inv(M1);L2=inv(M2);L3=inv(M3);L4=sqrt(M3*M2*M1*A*M1'*M2'*M3');
>>L=L1*L2*L3*L4

L =
    2.0000         0         0         0
   -1.0000    3.0000         0         0
    2.0000   -5.0000    1.0000         0
    3.0000    1.0000    2.0000    2.0000
```

After the existence of such decomposition is established, it is relatively easy to write a function to implement it.

Exercise 2.23 Write a MATLAB function for finding the lower-triangular matrix $\mathbf{L}$ in the factorization (2.65) of an $n \times n$ symmetric positive definite matrix $\mathbf{A}$.

Notice that MATLAB has a built-in function called `chol`, which you can use to see if your function is working properly. To conform with this notation, observe that the output of this function is the upper-triangular matrix $\mathbf{L}'$, instead of the lower-triangular matrix $\mathbf{L}$. As an application of this function, first use the matrix in the preceding example:

```
>>A=[4  -2  4  6;-2 10 -17  0; 4 -17 30  3; 6 0 3 18];
>>L=chol(A)'

L =
     2     0     0     0
    -1     3     0     0
     2    -5     1     0
     3     1     2     2
```

which agrees with the previous calculations.

As a further application of this function, here is an example where it fails to produce the desired result:

```
>> B=[-4  -2  4  6;-2 10 -17  0; 4 -17 30  3; 6 0 3 18];
>> chol(B)
??? Error using ==> chol Matrix must be positive definite.
```

This happened because the matrix $\mathbf{B}$, albeit symmetric, is not positive definite (indeed, $\mathbf{e}_1'\mathbf{B}\mathbf{e}_1 = -4 < 0$ for the first basis vector in $\mathbb{R}^4$). Because this is the only way in which Cholesky decomposition fails, it can be used to check whether a given matrix is positive definite.

2.8 Determinants

The determinant of an $n \times n$ matrix $\mathbf{A}$ is a scalar $\det(\mathbf{A})$ that can be used to characterize the invertibility of $\mathbf{A}$. For $n = 1$, the determinant of $\mathbf{A} = [a_{11}]$ is simply $\det(\mathbf{A}) = a_{11}$, and we immediately see that a 1×1 matrix $\mathbf{A}$ is invertible if and only if $\det(\mathbf{A}) \neq 0$.

For $n = 2$, the determinant is defined as

$$\det\left(\begin{bmatrix} a_{11} & a_{12} \\ a_{21} & a_{22} \end{bmatrix}\right) = a_{11}a_{22} - a_{12}a_{21}. \tag{2.66}$$

We can see from this definition that if we exchange the columns of a 2×2 matrix, then its determinant simply changes sign, that is,

$$\det\left(\begin{bmatrix} a_{12} & a_{11} \\ a_{22} & a_{21} \end{bmatrix}\right) = -\det\left(\begin{bmatrix} a_{11} & a_{12} \\ a_{21} & a_{22} \end{bmatrix}\right).$$

Moreover, it is also easy to see that

$$\det\left(\begin{bmatrix} \alpha a_{11} + b_1 & a_{12} \\ \alpha a_{12} + b_2 & a_{22} \end{bmatrix}\right) = \alpha \det\left(\begin{bmatrix} a_{11} & a_{12} \\ a_{21} & a_{22} \end{bmatrix}\right) + \det\left(\begin{bmatrix} b_1 & a_{12} \\ b_2 & a_{22} \end{bmatrix}\right).$$

In other words, the determinant is an *alternating multilinear* function of the columns of **A**. It therefore follows that if one column of **A** is a multiple of the other, then the determinant of **A** must be zero. Recalling that a matrix is invertible if and only if its columns are linearly independent, we see that a 2×2 matrix **A** is invertible if and only if $\det(\mathbf{A}) \neq 0$.

To obtain the determinant of a general $n \times n$ matrix **A**, let us define a *permutation* as a bijective map σ from a finite set $\{1, 2, \ldots, n\}$ to itself. For instance:

$$\sigma(1, 2, 3, 4) = (2, 3, 1, 4)$$

is a permutation of the set $\{1, 2, 3, 4\}$. The result of the permutation σ on a particular element i is denoted by $\sigma(i)$. For instance, for the preceding permutation we have $\sigma(1) = 2, \sigma(2) = 3, \sigma(3) = 1$, and $\sigma(4) = 4$. Every permutation can be implemented as a finite sequence of exchange between two elements in the set, called *transpositions*. For example, the previous permutation can be implemented as an exchange between 1 and 3 followed by an exchange between 2 and 3. The number of transpositions necessary to implement a permutation determines if the permutation is *even* or *odd*. For example, the previous permutation is even because it can be implemented with two transpositions, whereas the permutation

$$\sigma(1, 2, 3, 4) = (2, 1, 3, 4)$$

is odd because it consists of just one transposition.

Determinants of $n \times n$ matrices

The *determinant* of an $n \times n$ matrix **A** with entries a_{ij} is then defined as

$$\det(\mathbf{A}) = \sum_{\sigma} \operatorname{sign}(\sigma) \prod_{i=1}^{n} a_{i\sigma(i)}, \tag{2.67}$$

where σ ranges over all the permutations of the set $\{1, 2, \ldots, n\}$ and $\operatorname{sign}(\sigma) = 1$ if the permutation is even and $\operatorname{sign}(\sigma) = -1$ if the permutation is odd. For example, the only permutations of the set $\{1, 2\}$ are $(1, 2)$, which is even, and $(2, 1)$, which is odd. Applying the definition (2.67) for $n = 2$ then recovers the formula (2.66).

For $n = 3$, the even permutations of the set $\{1, 2, 3\}$ are $(1, 2, 3)$, $(3, 1, 2)$, and $(2, 3, 1)$, whereas the odd permutations are $(1, 3, 2)$, $(3, 2, 1)$, and $(2, 1, 3)$. You can then verify that (2.67) gives

$$\begin{aligned} \det(\mathbf{A}) &= a_{11}a_{22}a_{33} + a_{13}a_{21}a_{32} + a_{12}a_{23}a_{31} \\ &\quad - a_{13}a_{22}a_{31} - a_{11}a_{23}a_{32} - a_{12}a_{21}a_{33}. \end{aligned} \tag{2.68}$$

A numerical algorithm implementing formula (2.67) for an $n \times n$ matrix would have a *flop* count of order at least $n!$ terms (because this is the number of permutations of a set of n elements), which is clearly impractical. Instead of calculating the determinant directly from (2.67), we can use some of its properties and the factorization methods introduced earlier in this chapter.

First of all, observe that (2.67) implies that the determinant of an $n \times n$ matrix $\mathbf{A}$ is also an alternating multilinear function of the columns of $\mathbf{A}$. It then follows that the determinant vanishes whenever one of the columns of $\mathbf{A}$ is a linear combination of the others. Therefore, an $n \times n$ matrix $\mathbf{A}$ is invertible if and only if $\det(\mathbf{A}) \neq 0$.

The definition (2.67) also implies that the determinant of an upper triangular matrix is simply the product of its diagonal entries, as can be readily verified in the special cases (2.66) and (2.68). Moreover, the determinant of a product is given by

$$\det(\mathbf{AB}) = \det(\mathbf{A})\det(\mathbf{B}). \tag{2.69}$$

Finally, the determinant of a matrix $\mathbf{A}$ and of its transpose are equal, which implies that the determinant is also an alternating multilinear function of the rows of $\mathbf{A}$.

Using these properties, you can calculate the determinant of a general matrix $\mathbf{A}$ by performing its LU factorization and then multiplying the diagonal entries of the matrix $\mathbf{U}$. This is illustrated in the following example, where we also introduce the MATLAB built-in function `det`:

```
>> A=[3 2 -1 4 0; -1 2 7 5 1; 0 3 9 5 -2;8 -3 -2 1 2; 5 -4 3 0 -1];
>> [L,U]=lufactor(A);
>> D=prod(diag(U))

D =
  -1.7690e+03

>> det(A)

ans =
       -1769
```

Because the multiplication of diagonal elements can be done in $(n-1)$ operations, you can see that the *flop* count for this procedure for finding the determinant of an $n \times n$ matrix is of the same order as that of LU factorization, namely, n^3, which is a significant economy when compared to the original $n!$ terms in (2.67).

Cramer's rule

Let us now return to the problem of solving the linear system $\mathbf{Ax} = \mathbf{b}$. Denoting the columns of $\mathbf{A}$ by $\mathbf{a}_1, \ldots, \mathbf{a}_n$, we can use the properties of determinants to obtain that, for any $j \neq 1$,

$$\begin{aligned}[\det([\alpha \mathbf{a}_1 + \beta \mathbf{a}_j \quad \mathbf{a}_2 \quad \cdots \quad \mathbf{a}_n]) &= \alpha \det(\mathbf{A}) + \beta \det([\mathbf{a}_j \quad \mathbf{a}_2 \quad \cdots \quad \mathbf{a}_n]) \\ &= \alpha \det(\mathbf{A}).\end{aligned}$$

The same argument holds if instead of $\beta \mathbf{a}_j$ we add any linear combinations of the columns $\mathbf{a}_2, \ldots, \mathbf{a}_n$ to the first column of $\mathbf{A}$. In particular, we can choose the components of the solutions vector $\mathbf{x} = [x_1, \ldots, x_n]$ as the scalars multiplying each corresponding column of $\mathbf{A}$, resulting in

$$\det\left(\begin{bmatrix} a_{11}x_1 + a_{12}x_2 + \cdots a_{1n}x_n & a_{12} & \cdots & a_{1n} \\ a_{21}x_1 + a_{22}x_2 + \cdots a_{2n}x_n & a_{22} & \cdots & a_{2n} \\ \vdots & & \vdots & \\ a_{n1}x_1 + a_{n2}x_2 + \cdots a_{nn}x_n & a_{n2} & \cdots & a_{nn} \end{bmatrix}\right) = x_1 \det(\mathbf{A}).$$

But because $\mathbf{Ax} = \mathbf{b}$, the matrix on the left side of the preceding equation is simply the matrix $\mathbf{A}$ with its first column replaced by the vector $\mathbf{b}$, which we denote by $\mathbf{A}_1$. Repeating the same argument for the other columns of $\mathbf{A}$ leads to *Cramer's rule* for linear systems:

$$x_i = \frac{\det(\mathbf{A}_i)}{\det(\mathbf{A})}, \quad i = 1, \ldots, n \tag{2.70}$$

where $\mathbf{A}_i$ is the matrix obtained by replacing the ith column of $\mathbf{A}$ by the vector $\mathbf{b}$.

It follows from (2.70) that the linear system $\mathbf{Ax} = \mathbf{b}$ has a unique solution if and only if the determinant of $\mathbf{A}$ is different from zero, which is equivalent to the matrix $\mathbf{A}$ being invertible and is therefore in agreement with the first item of Theorem 2.3. As an example in this case, consider the following:

```
>> A=[3 2 -1 4 0; -1 2 7 5 1; 0 3 9 5 -2;8 -3 -2 1 2; 5 -4 3 0 -1];
>> det(A)

ans =

       -1769

>> b=[1 -1 2 -5 3]';
>> cramer=[det([b A(:,2:5)])
det([A(:,1) b A(:,3:5)])
det([A(:,1:2) b A(:,4:5)])
```

```
det([A(:,1:3) b A(:,5)])
det([A(:,1:4) b])]'

cramer =
       2226         3981         1425        -3746         4788

>> x=cramer/det(A)

x =
   -1.2583   -2.2504   -0.8055    2.1176   -2.7066

>> (A\b)'

ans =
   -1.2583   -2.2504   -0.8055    2.1176   -2.7066
```

As you can see, the solution obtained from Cramer's rule coincides with that produced by MATLAB using the backslash operator.

When $\det(\mathbf{A}) = 0$ and $\det(\mathbf{A}_i) \neq 0$ for at least one i, then (2.70) implies that the system $\mathbf{Ax} = \mathbf{b}$ does not have a solution, which corresponds to the case of a singular matrix $\mathbf{A}$ with $\mathbf{b} \notin \text{range}(\mathbf{A})$, as in the second item of Theorem 2.3. You can verify this for the matrix $\mathbf{B}$ given in Equation (2.14) and the vector $\mathbf{b} = [1, 0, 0]$, which is not in the range of $\mathbf{B}$:

```
>>  B=[1 2 3;2 4 6;1 0 -1];
>> det(B)

ans =
     0

>>  cramer=[det([b B(:,2:3)])
det([B(:,1) b B(:,3)])
det([B(:,1:2) b ])]'

cramer =
    -4     8    -4

>> x=cramer/det(B)
Warning: Divide by zero.

x =
   Inf   Inf   Inf
```

which agrees with the outcome obtained from the backslash operator `B\b`, as you have seen when you used this matrix as an example for item 2 of Theorem 2.3.

When $\det(\mathbf{A}) = 0$ and $\det(\mathbf{A}_i) = 0$ for all i, then (2.70) is ambiguous regarding the existence of solutions for the system $\mathbf{Ax} = \mathbf{b}$. When $\det(\mathbf{A}) = 0$, $\mathbf{A}$ is singular, so that according to Theorem 2.3, the system $\mathbf{Ax} = \mathbf{b}$ either has no solutions at all or has infinitely many solutions, depending or whether $\mathbf{b}$ is in the range $\mathbf{A}$. The problem with Cramer's rule is that both cases might result in vanishing $\det(\mathbf{A}_i)$ for all i. Continuing with the matrix $\mathbf{B}$ of Example (2.14), but this time using the $\mathbf{b} = [0, 0, 1]$, which lies in the range of $\mathbf{B}$, you obtain:

```
>> B=[1 2 3;  2 4 6; 1 0 -1];
>> b=[0 0 1]';
>> cramer=[det([b B(:,2:3)])
det([B(:,1) b B(:,3)])
det([B(:,1:2) b ])]'

cramer =
     0     0     0

>> x=cramer/det(B)
Warning: Divide by zero.

x =
   NaN   NaN   NaN
```

As in the example used for item 3 of Theorem 2.3, you can observe that Cramer's rule fails to find any of the solutions to this system, despite the fact that any vector of the form $\mathbf{x} + \lambda\mathbf{y}$, with $\mathbf{x} = [1, -0.5, 0]$ and $\mathbf{y} = [1, -2, 1]$ is a solution. Once more, the lesson to be drawn is that the outcome `NaN` merely indicates an undetermined situation requiring further analysis.

For an example of a system with no solutions leading to $\det(\mathbf{A}_i) = 0$ for all i, consider the following:

```
>> A=[1 1 1;2 2 2;3 3 3];
>> b=[2 4 -7]'
>> cramer=[det([b A(:,2:3)])
det([A(:,1) b A(:,3)])
det([A(:,1:2) b ])]'
```

Injection, surjection, Bijection

alternate signs

$$\begin{vmatrix} a_{11} & a_{12} & a_{13} \\ a_{21} & a_{22} & a_{23} \\ a_{31} & a_{32} & a_{33} \end{vmatrix} = a_{11}(a_{22}a_{33} - a_{23}a_{32}) + a_{12}(a_{21}a_{33} - a_{31}a_{23}) + a_{13}(a_{21}a_{32} - a_{31}a_{22})$$

$a_{12}a_{23}a_{31}$

$$= a_{11}a_{22}a_{33} - a_{11}a_{23}a_{32} - a_{12}a_{21}a_{33} + a_{12}a_{31}a_{23}$$

$$+ a_{13}a_{21}a_{32} - a_{13}a_{31}a_{22} -$$

$a_{12}a_{23}a_{31}$

ex. $a_{13}a_{31}a_{22} = - a_{13}a_{22}a_{31}$

$- a_{12}a_{31}a_{23} = + a_{12}a_{23}a_{31}$

```
cramer =
     0     0     0

>> x=cramer/det(A)
Warning: Divide by zero.

x =
   NaN   NaN   NaN
```

In this example, it is easy to see that the range of $\mathbf{A}$ is the one-dimensional subspace spanned by the vector $\mathbf{v} = [1, 2, 3]$, which clearly does not contain the vector $\mathbf{b} = [2, 4, -7]$, implying that the system $\mathbf{Ax} = \mathbf{b}$ does not have a solution. Nevertheless, you see that $\det(\mathbf{A}) = \det(\mathbf{A}_1) = \det(\mathbf{A}_2) = \det(\mathbf{A}_3) = 0$. This confirms that Cramer's rule alone is not enough to decide whether a system has infinitely many solutions or no solutions at all when $\det(\mathbf{A}) = 0$ and $\det(\mathbf{A}_i) = 0$ for all i.

2.9 Summary and Notes

We reviewed in this chapter the main concepts and results of *Linear Algebra* in the context of numerical methods for solving linear systems:

- **Section 2.1:** *Vector spaces* are abstract generalizations of $\mathbb{R}^2$ and $\mathbb{R}^3$. They are defined through the properties of addition of vectors (commutativity, associativity, additive identity, and additive inverse) and multiplication by scalar (associativity, distributivity, and multiplicative identity). *Subspaces* are subsets of vector spaces that are themselves vector spaces. A set of vectors is said to be *linearly dependent* if one of them can be written as a linear combination of the others. A linearly independent set of vectors spanning the entire space is called a *basis*. A vector space is *finite-dimensional* if it has a basis with a finite number of vectors.

- **Section 2.2:** *Linear maps* are maps from one vector space to another that preserve the structure of addition of vectors and multiplication by scalars. Linear maps between finite-dimensional vector spaces are identified with *matrices*. The *null space* of a linear map is the set of vectors that are mapped to $\mathbf{0}$. The *range* of a linear map is the set of vectors that arise as the result of the linear map applied to other vectors. The dimension of the range is called the *rank* of the linear map, which equals the number of linearly independent columns of the corresponding matrix. Any n-dimensional vector space is *isomorphic* to $\mathbb{R}^n$.

- **Section 2.3:** A system of n linear equations with k unknowns can be written in matrix form as $\mathbf{Ax} = \mathbf{b}$. Problem 2.1 consists of finding the

k-dimensional vector $\mathbf{x}$ given the $n \times k$ matrix $\mathbf{A}$ and the n-dimensional vector $\mathbf{b}$. For $n \times n$ matrices, the solution to this problem is characterized in the three items of Theorem 2.3.

- **Section 2.4:** The size of vectors is measured by *norms*. For finite-dimensional vector spaces, any two norms are equivalent, in the sense defined in Theorem 2.4. Norms on the underlying finite-dimensional vector spaces induce a *matrix norm* for linear maps between them. With these matrix norms we can define the *condition number* of a matrix, which is used to measure the sensitivity of solutions of $\mathbf{Ax} = \mathbf{b}$ with respect to perturbation in $\mathbf{A}$ or $\mathbf{b}$.

- **Section 2.5:** *Direct methods* for solving $n \times n$ linear systems are those that produce a solution in a finite number of steps. These include *forward substitution* for lower-triangular systems and *backward substitution* for upper-triangular ones, both with a *flop* count of order n^2. General systems are transformed into upper-triangular ones by *Gaussian elimination*, and then solved by backward substitution, with a total *flop* count of order n^3. A rearrangement of Gaussian elimination produces the *LU factorization* of a matrix. The technique of *pivoting* is used to circumvent the problem of dividing by small numbers during this factorization. With this technique, you can accomplish the LU factorization of permutations of any matrix $\mathbf{A}$ in the form $\mathbf{PA} = \mathbf{LU}$.

- **Section 2.6:** *Iterative methods* for solving $n \times n$ linear systems are those that produce a solution through an infinite sequence of approximations. The method is said to be *convergent* if the error for the approximation tends to zero, and the iterations are stopped in practice when this error becomes smaller than a given tolerance. The *Jacobi* and *Gauss–Seidel* methods are the most popular iterative methods for linear systems. A sufficient (but not necessary) condition for convergence of either method is *strict diagonal dominance* for the matrix $\mathbf{A}$.

- **Section 2.7:** A symmetric, positive definite matrix can be decomposed as the product of a lower-triangular matrix $\mathbf{L}$ and its transpose $\mathbf{L}'$. This is known as the *Cholesky* decomposition.

- **Section 2.8:** *Determinants* are alternating multilinear functions of the columns of an $n \times n$ matrix $\mathbf{A}$, and $\mathbf{A}$ is invertible if and only if its determinant $\det(\mathbf{A})$ is not zero. Determinants can be used to provide an alternative characterization for the solutions of a linear system $\mathbf{Ax} = \mathbf{b}$ in what is known as *Cramer's rule.*

The concepts and results on vector spaces, linear maps, and linear systems are standard and can be found in any undergraduate book on *Linear Algebra.* The classic introductory book in the subject is [10], which is still very readable.

A more modern treatment written for a mixed audience of mathematics and physics students is presented in [13]. The material on norms, including the proof of Theorem 2.4, can be found in [14]. The numerical methods for this chapter are all covered in [12]. Determinants and the Cramer rule are treated in [13].

2.10 Exercises

1. In each of the following items, determine if the given vectors are linearly dependent or independent in the corresponding vector space:

   (a) $V = \mathbb{R}^4$, $\mathbf{v}_1 = [2, -1, 3, 0]$, $\mathbf{v}_2 = [1, 0, -2, 4]$, $\mathbf{v}_3 = [-3, 1, 1, 2]$.

   (b) $V = \mathbb{M}^{3\times 3}$, $\mathbf{A}_1 = \begin{bmatrix} 1 & 0 & 3 \\ -2 & 4 & 1 \\ 5 & 1 & 6 \end{bmatrix}$, $\mathbf{A}_2 = \begin{bmatrix} -1 & 2 & -1 \\ 0 & 3 & 7 \\ 1 & 2 & 3 \end{bmatrix}$,

   $$\mathbf{A}_3 = \begin{bmatrix} 10 & -14 & 16 \\ -6 & -9 & -46 \\ 8 & -11 & -3 \end{bmatrix}.$$

   (c) $V = \mathcal{P}_3(\mathbb{R})$, $p_1(x) = 1 + x^2$, $p_2(x) = 3x^2 + x^3$, $p_3(x) = -2 - 4x + 9x^2 + 3x^3$.

2. Consider the first five Legendre polynomials:

   $$P_k(x) = \frac{1}{k!2^k} \frac{d^k}{dx^k}(x^2 - 1)^k, \qquad k = 0, 1, 2, 3, 4.$$

   (a) Show that these polynomials form a basis for the vector space $\mathcal{P}_4(\mathbb{R})$.

   (b) Obtain the coordinates of the polynomial $p(x) = 2 - x^2 + 3x^4$ with respect to this basis.

   (c) Find the matrix representation for the derivative map $D : \mathcal{P}_4(\mathbb{R}) \to \mathcal{P}_4(\mathbb{R})$ with respect to this basis.

3. Consider the linear map $\mathbf{T}_\mathbf{A} : \mathbb{M}^{2\times 2} \to \mathbb{M}^{2\times 2}$ defined by $\mathbf{T}_\mathbf{A}(\mathbf{B}) = \mathbf{AB}$, where

   $$\mathbf{A} = \begin{bmatrix} -1 & 2 \\ 0 & 3 \end{bmatrix} \tag{2.71}$$

   is a fixed 2×2 matrix. Obtain the 4×4 matrix representation of this linear map with respect to the canonical basis for $\mathbb{M}^{2\times 2}$. Use this matrix to verify that this linear map is invertible and express its inverse in terms of the matrix $\mathbf{A}$.

4. Consider the linear map $\mathbf{T_A} : \mathbb{M}^{2\times 2} \to \mathbb{M}^{2\times 2}$ defined by $\mathbf{T_A}(\mathbf{B}) = \mathbf{AB} - \mathbf{BA}$, where $\mathbf{A}$ is the fixed matrix defined in (2.71). Obtain the null space and the range of $\mathbf{T_A}$ and verify that it is not invertible.

5. For each of the following items, find the null space and the range of the linear map and verify the dimension formula (2.17):

   (a) $\mathbf{A}(x_1, x_2, x_3) = (2x_1 - x_2 + x_3, 3x_1 + x_2 - x_3)$

   (b) $\mathbf{A}(x_1, x_2) = (2x_1 + 4x_2, x_1 - x_2, 3x_2)$

   (c) $\mathbf{A}(x_1, x_2, x_3, x_4) = (x_1 - x_2, -x_1 + x_2, x_3 + x_4, -x_3 - x_4)$

6. Let $\mathbf{A} = \begin{bmatrix} 3 & 2 & -1 \\ 0 & 2 & 5 \\ 1 & -1 & 4 \end{bmatrix}$ and $\mathbf{B} = \begin{bmatrix} 2 & -1 \\ 3 & -8 \end{bmatrix}$ and $\mathbf{C} = \begin{bmatrix} \mathbf{A} & \mathbf{0} \\ \mathbf{0} & \mathbf{B} \end{bmatrix}$.

   Find $\mathbf{C}^{-1}$ and express it in terms of $\mathbf{A}^{-1}$ and $\mathbf{B}^{-1}$.

7. Let $\mathbf{A} = \begin{bmatrix} 1 & 1 & 2 \\ 1 & -2 & 2 \\ 1 & 2 & -1 \end{bmatrix}$ and $\mathbf{b} = \begin{bmatrix} 1 \\ 4 \\ 2 \end{bmatrix}$. Solve the system $\mathbf{Ax} = \mathbf{b}$

   by finding the inverse of $\mathbf{A}$ using the function `inv`. Compare it with the solution obtained using the backslash operator \.

8. Let $\mathbf{B} = \begin{bmatrix} -1 & 2 & -3 \\ 0 & 4 & 1 \\ 3 & -6 & 9 \end{bmatrix}$ and $\mathbf{b} = \begin{bmatrix} -12 \\ -1 \\ 36 \end{bmatrix}$.

   (a) Verify that $\mathbf{x} = [1, -1, 3]$ is a solution to this system.

   (b) Verify that $\mathbf{v} = [-14, -1, 4]$ is in the null space of $\mathbf{B}$.

   (c) Verify that $\mathbf{x} + \lambda\mathbf{v}$, for an arbitrary real number λ, is also a solution to this system.

   (d) Try to solve the system $\mathbf{Bx} = \mathbf{b}$ using both `x=inv(B)*b` and `x=B\b`.

   (e) Find the scalar λ for which $\mathbf{x} + \lambda\mathbf{v}$ equals the solution obtained with the backslash operator \.

9. Let $\mathbf{C} = \begin{bmatrix} 1 & -1 & 2 \\ -1 & 1 & -2 \\ 2 & -2 & 4 \end{bmatrix}$ and $\mathbf{b} = \begin{bmatrix} 1 \\ 0 \\ 0 \end{bmatrix}$.

   (a) Show that $\mathbf{b} \notin \text{range}(\mathbf{C})$.

   (b) Try to solve the system $\mathbf{Cx} = \mathbf{b}$ using both `x=inv(C)*b` and `x=C\b`.

10. Let $\mathbf{A} = \begin{bmatrix} \cos\theta & \sin\theta \\ -\sin\theta & \cos\theta \end{bmatrix}$. Find expressions for $\text{cond}_1(\mathbf{A})$, $\text{cond}_\infty(\mathbf{A})$ and $\det(\mathbf{A})$ as functions of θ and verify them for specific values of θ using the corresponding MATLAB functions. Give a geometric interpretation for $\mathbf{A}$ and its inverse.

11. Calculate by hand the LU factorization of the matrices and verify your answer using MATLAB:

(a)

$$\mathbf{A} = \begin{bmatrix} -1 & 1 & -4 \\ 2 & 2 & 0 \\ 3 & 3 & 2 \end{bmatrix}$$

(b)

$$\mathbf{B} = \begin{bmatrix} 4 & 2 & 4 \\ 4 & 0 & 5 \\ 5 & 1 & -6 \end{bmatrix}$$

12. Let $\mathbf{A}$ and $\mathbf{B}$ be invertible $n \times n$ matrices and let $\mathbf{b}$ be an $n \times 1$ vector. Write a MATLAB function with inputs $(\mathbf{A}, \mathbf{B}, \mathbf{b})$ to solve the equation

$$\mathbf{x} = \mathbf{B}^{-1}(2\mathbf{A}^{-1} + \mathbf{1})\mathbf{b},$$

making use of the functions `plufactor`, `forwsub`, and `backsub`, but *without* calculating any matrix inverses. Verify that your function gives the correct answer for a concrete example.

13. The determinant of a triangular matrix is the product of its diagonal entries. Also, the determinant of a product of matrices is the product of the determinants of each matrix. Use these facts in the following items:

    (a) Write a MATLAB function that calculates the determinant of a general $n \times n$ matrix $\mathbf{A}$ using its *LU* factorization with pivoting. Find the leading order term in the *flop* count of your function.

    (b) Use your function to calculate the determinants of an $n \times n$ Hilbert matrix $\mathbf{H}(n)$ and its inverse, for $n = 2, \ldots, 20$.

    (c) To avoid underflow and overflow, modify your function so that it computes the logarithm of the determinant of a matrix.

    (d) Use the modified function to calculate the determinants of an $n \times n$ Hilbert matrix and its inverse, for $n = 2, \ldots, 20$.

14. (a) Calculate by hand the first two iterations of the Jacobi method for solving the system

$$\begin{aligned} 4x + y &= 4 \\ 2x + 5y &= 5, \end{aligned}$$

starting from $\mathbf{x}_0 = [0, 0]$.

(b) Show that the iterations converge and estimate the linear rate of convergence.

(c) Use the function `jacobi` for this system with a 10^{-6} tolerance.

15. (a) Prove that the Jacobi method converges when used to solve the linear system

$$\begin{aligned} 3x - y + z &= 3 \\ 2x + 4y - z &= 5 \\ -y + 2z &= 1. \end{aligned}$$

(b) Knowing that the exact solution to this system is $\mathbf{x} = [1, 1, 1]$, estimate how many iterations of the Jacobi method are sufficient to make the error less than 10^{-6} in all components.

(c) Calculate by hand the first two iterations for the Gauss–Seidel method for solving the same system.

(d) Use the functions `jacobi` and `gaussseidel` for this system with a 10^{-6} tolerance.

16. Calculate by hand the Cholesky factorization of the following matrices and verify the results using MATLAB:

(a)

$$\mathbf{A} = \begin{bmatrix} 4 & -2 & 6 \\ -2 & 17 & -11 \\ 6 & -11 & 22 \end{bmatrix}$$

(b)

$$\mathbf{B} = \begin{bmatrix} 9 & -3 & 1 \\ -3 & 5 & 2 \\ 1 & 2 & 9 \end{bmatrix}$$

17. Calculate the Cholesky factorization for the matrix $\mathbf{A}'\mathbf{A}$ where

$$\mathbf{A} = \begin{bmatrix} 1 & -2 & 3 \\ -1 & 2 & 1 \\ 0 & 1 & -1 \end{bmatrix}.$$

Use the factorization just obtained to solve the system $\mathbf{A}'\mathbf{A}\mathbf{x} = \mathbf{b}$, where $\mathbf{b} = [2, -1, 1]$.

CHAPTER 3

Orthogonality

IN THIS CHAPTER, we explore the notion of orthogonality in general vector spaces. This is built around the concept of an inner product, which is introduced in Section 3.1 as a generalization of the familiar dot product for vectors in $\mathbb{R}^3$. With it, other geometric ideas such as the Euclidean distance and the Pythagoras theorem can be developed in full generality, even for infinite-dimensional vector spaces. We can then use the intuition built from the geometry of three-dimensional space to understand much more complicated spaces. Section 3.2 is devoted to orthogonal projections and the related minimization of Euclidean distance. In this section, we also describe the Gram–Schmidt orthogonalization procedure, which has great practical and theoretical importance. As always in this book, all these abstract notions are illustrated by MATLAB examples.

Section 3.3 introduces the QR factorization method, whereby a general matrix is written as the product of an orthogonal matrix and an upper-triangular matrix. This allows us to explore the consequences of orthogonality even when the original problem is not presented in orthogonal form. Three different implementations of this factorization are discussed through detailed MATLAB codes.

In Section 3.4, the full machinery of orthogonality is employed in the popular least-squares method, where a linear system with more equations than unknowns is solved according to a minimization criterion. The most important application of this method is to find the best fit of a set of data points to a specified model, in what is known as linear regression, which we discuss in detail through a series of MATLAB examples.

3.1 Inner Product Spaces

In Chapter 2, we repeatedly used our basic intuition of vectors in $\mathbb{R}^2$ and $\mathbb{R}^3$ as motivation for abstract concepts in vector spaces. For instance, the definition of a general vector space given in Section 2.1 is based on the properties of algebraic operations of vectors (addition and multiplication by scalar) while the concept of a norm given in Section 2.4 is based on the properties of the geometric length of vectors (positivity, homogeneity, and the triangle inequality). In this section, we generalize yet another fundamental concept pertaining to the geometry of two- and three-dimensional vectors: orthogonality, that is, the property of two vectors forming a right angle between them.

In Section 1.3 we showed how to use the matrix multiplication operation between a row vector and a column vector to obtain a scalar. Using MATLAB notation, this *dot product* of two vectors in $\mathbb{R}^n$ is given by

$$\mathbf{x} \cdot \mathbf{y} := \mathbf{x}' * \mathbf{y} = x_1 y_1 + \cdots + x_n y_n. \tag{3.1}$$

The Euclidean norm introduced in Section 2.4 is related to the dot product (3.1) by

$$\|\mathbf{x}\|_2 = \sqrt{x_1^2 + \cdots + x_n^2} = \sqrt{\mathbf{x} \cdot \mathbf{x}}. \tag{3.2}$$

Using this Euclidean norm to represent the lengths of vectors in $\mathbb{R}^3$, you can apply elementary properties of Euclidean geometry to solve the next exercise.

Exercise 3.1 Let $0 \leq \theta \leq \pi$ be the smallest angle between two non-zero vectors $\mathbf{x}, \mathbf{y} \in \mathbb{R}^3$. Prove that

$$\cos\theta = \frac{\mathbf{x} \cdot \mathbf{y}}{\|\mathbf{x}\|_2 \|\mathbf{y}\|_2}. \tag{3.3}$$

Because $\cos(\pi/2) = 0$, you can see from Equation (3.3) that two vectors in $\mathbb{R}^3$ are perpendicular to each other if and only if their dot product is zero. The same notion applies to higher-dimensional vectors: we say that two vectors $\mathbf{x}$ and $\mathbf{y}$ in $\mathbb{R}^n$ are *orthogonal* if and only if $\mathbf{x} \cdot \mathbf{y} = 0$. The next example illustrates two orthogonal vectors in $\mathbb{R}^5$:

```
>> x = [1,-2,1,3,4];
>> y = [2,1,-1,-1,1];
>> x*y'

ans =
     0
```

Inner products

For a general vector space, where vectors cannot be easily identified with arrows in space, it is hard to give a meaning for the angle between vectors. To define orthogonality in this case, we first must obtain the appropriate generalized version for the dot product on general vector spaces. Let us define an *inner product* on a real vector space V as a map

$$\langle \cdot, \cdot \rangle : V \times V \mapsto \mathbb{R} \tag{3.4}$$

satisfying:

1. $\langle \mathbf{x}, \mathbf{y} + \lambda\mathbf{z} \rangle = \langle \mathbf{x}, \mathbf{y} \rangle + \lambda \langle \mathbf{x}, \mathbf{z} \rangle, \qquad \forall \mathbf{x}, \mathbf{y}, \mathbf{z} \in V, \quad \forall \lambda \in \mathbb{R}.$
2. $\langle \mathbf{x}, \mathbf{y} \rangle = \langle \mathbf{y}, \mathbf{x} \rangle, \qquad \forall \mathbf{x}, \mathbf{y} \in V.$
3. $\langle \mathbf{x}, \mathbf{x} \rangle > 0, \quad \forall \mathbf{x} \neq \mathbf{0} \in V.$

You can immediatly verify that the dot product $\mathbf{x} \cdot \mathbf{y}$ in $\mathbb{R}^n$ satisfies the preceding properties. But several other examples of inner products can be constructed for different vector spaces. For instance, starting with the dot product in $\mathbb{R}^n$, we can generate infinitely many different inner products by setting

$$\langle \mathbf{x}, \mathbf{y} \rangle_{\mathbf{A}} := \mathbf{A}\mathbf{x} \cdot \mathbf{A}\mathbf{y}, \tag{3.5}$$

where $\mathbf{A}$ is any nonsingular $n \times n$ matrix. Recalling that any n-dimensional vector space is isomorphic to $\mathbb{R}^n$, this gives a recipe for producing inner products on finite-dimensional vector spaces.

Analogously, an inner product on a complex vector space V is defined as a map $\langle \cdot, \cdot \rangle : V \times V \mapsto \mathbb{C}$ satisfying the preceding first and third properties but with the second property replaced by $\langle \mathbf{x}, \mathbf{y} \rangle = \overline{\langle \mathbf{y}, \mathbf{x} \rangle}$. It is easy to verify that these properties are satisfied for the complex dot product of two column vectors in $\mathbb{C}^n$:

$$\mathbf{x} \cdot \mathbf{y} := \mathbf{x}' * \mathbf{y} = \bar{x}_1 y_1 + \cdots + \bar{x}_n y_n. \tag{3.6}$$

For an example of an inner product in infinite dimensions, consider the next exercise.

Exercise 3.2 Let $V = \mathcal{C}([0,1], \mathbb{R})$ be the vector space of continuous real-valued functions on the interval $[0,1]$. Show that

$$\langle f, g \rangle = \int_0^1 f(x) g(x) dx$$

defines an inner product on V.

Given any inner product $\langle \cdot, \cdot \rangle$ on a vector space V (real or complex), we can obtain an analogue for the Euclidean norm by setting

$$\|\mathbf{x}\| = \sqrt{\langle \mathbf{x}, \mathbf{x} \rangle}, \tag{3.7}$$

which generalizes the 2-norm (3.2). It follows from property 3 of inner products that $\|\mathbf{x}\| > 0$ if $\mathbf{x} \neq \mathbf{0}$. Moreover, we obtain from properties 1 and 2 that $\|\lambda \mathbf{x}\| = |\lambda| \|\mathbf{x}\|$. That is, the first two properties of a vector norm introduced in Section 2.4 are satisfied by (3.7). To prove that the third property in

Section 2.4 (the triangle inequality) is also satisfied, we need to establish an important property of inner products, known as the *Cauchy–Schwarz inequality*. Although it is hard to overemphasize its importance (for instance, it is used to prove the *uncertainty principle* in *Quantum Mechanics*!), the Cauchy–Schwarz inequality has a remarkably simple proof. For real vector spaces, it goes as follows: given two arbitrary vectors $\mathbf{x}, \mathbf{y} \in V$ and a scalar $\lambda \in \mathbb{R}$, we start by expanding $\|\mathbf{y} - \lambda\mathbf{x}\|^2$ as

$$\|\mathbf{y} - \lambda\mathbf{x}\|^2 = \langle \mathbf{y} - \lambda\mathbf{x}, \mathbf{y} - \lambda\mathbf{x} \rangle = \|\mathbf{y}\|^2 - 2\lambda\langle \mathbf{x}, \mathbf{y} \rangle + \lambda^2\|\mathbf{x}\|^2, \tag{3.8}$$

obtaining a quadratic expression in the variable λ. It then follows from the positivity of an inner product that this expression is always greater than or equal to zero. Therefore, its discriminant must be negative, implying that

$$4\langle \mathbf{x}, \mathbf{y} \rangle^2 - 4\|\mathbf{x}\|^2\|\mathbf{y}\|^2 \leq 0.$$

Taking square roots, we arrive at the Cauchy–Schwarz inequality:

$$|\langle \mathbf{x}, \mathbf{y} \rangle| \leq \|\mathbf{x}\|\|\mathbf{y}\|. \tag{3.9}$$

The following MATLAB commands illustrate it:

```
>> x=[-1 2 0 3 1 -2];
>> y=[ 3 2 5 4 -1.5 1];
>> norm(x,2)*norm(y,2)-abs(x*y')

ans =
   23.4811
```

Exercise 3.3 Use the Cauchy–Schwarz inequality to prove that the norm defined in (3.7) satisfies

$$\|\mathbf{x} + \mathbf{y}\| \leq \|\mathbf{x}\| + \|\mathbf{y}\|, \qquad \text{for any } \mathbf{x}, \mathbf{y} \in V.$$

Orthogonality

Having defined an inner product and the norm induced by it, we say that two vectors $\mathbf{x}$ and $\mathbf{y}$ in a general vector space V are *orthogonal* according to the inner product $\langle \cdot, \cdot \rangle$ if and only if

$$\langle \mathbf{x}, \mathbf{y} \rangle = 0. \tag{3.10}$$

The reason why this is an appropriate definition for orthogonality is that we can generalize several important results of elementary Euclidean geometry in terms of it. For instance, if $\mathbf{x}$ and $\mathbf{y}$ are orthogonal vectors in V, then setting $\lambda = -1$ in the quadratic expression (3.8) gives

$$\|\mathbf{x} + \mathbf{y}\|^2 = \|\mathbf{x}\|^2 + \|\mathbf{y}\|^2, \tag{3.11}$$

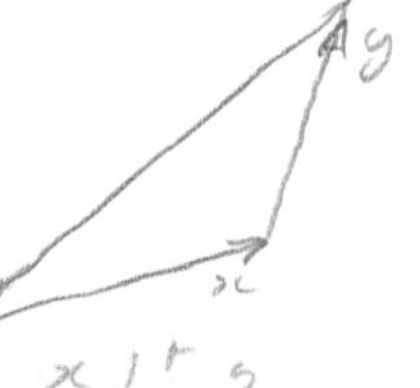

which is nothing but the *Pythagorean Theorem* for a triangle formed by the orthogonal vectors $\mathbf{x}$ and $\mathbf{y}$. The following MATLAB code illustrates the Pythagorean Theorem (3.11):

```
>> x = [1,-2,1,3,4];
>> y = [2,1,-1,-1,1];
>> norm(x+y,2)^2-norm(x,2)^2-norm(y,2)^2

ans =
  1.7764e-015
```

where the small number of the order of 10^{-15} is at the level of machine precision for MATLAB computations.

To generalize yet another identity of elementary Euclidean geometry, we can set $\lambda = 1$ and $\lambda = -1$ in (3.8) and conclude that

$$\|\mathbf{y} + \mathbf{x}\|^2 + \|\mathbf{y} - \mathbf{x}\|^2 = 2(\|\mathbf{y}\|^2 + \|\mathbf{x}\|^2), \tag{3.12}$$

which is known as the *parallelogram law.* Because it is a direct consequence of the definition (3.7), the parallelogram law can be used as a criterion to determine whether a given norm is induced by an inner product. Consider the following MATLAB commands:

```
>> x=[-1 2 0 3 1 -2];
>> y=[ 3 2 5 4 -1.5 1];
>> norm(x+y,1)^2+norm(x-y,1)^2-2*(norm(x,1)^2+norm(y,1)^2)

ans =
   -86

>> norm(x+y,2)^2+norm(x-y,2)^2-2*(norm(x,2)^2+norm(y,2)^2)

ans =
     0

>> norm(x+y,inf)^2+norm(x-y,inf)^2-2*(norm(x,inf)^2+norm(y,inf)^2)

ans =
     6
```

These examples show that (at least for $\mathbb{R}^6$) neither $\|\cdot\|_1$ nor $\|\cdot\|_\infty$ satisfy the parallelogram law (3.12), implying that there is no inner product in $\mathbb{R}^6$ for which (3.7) holds for these norms. In fact, for $p \neq 2$, it can be shown that there is no inner product in $\mathbb{R}^n$ such that the norms $\|\cdot\|_p$ introduced in Section 2.4 satisfy (3.7).

Conversely, if a norm does satisfy the parallelogram law, then the *polarization identity* introduced in the next exercise can be used to uniquely define an inner product between two vectors such that (3.7) holds.

Exercise 3.4 Suppose that a given norm $\|\cdot\|$ on a real vector space V satisfies the parallelogram law (3.12). Prove that

$$\langle \mathbf{x}, \mathbf{y} \rangle := \frac{1}{2}\left(\|\mathbf{x}+\mathbf{y}\|^2 - \|\mathbf{x}\|^2 - \|\mathbf{y}\|^2\right) \tag{3.13}$$

defines an inner product on V.

Orthogonal sets

We say that a set of vectors $\{\mathbf{x}_1, \dots, \mathbf{x}_k\}$ in an inner product space V is an *orthogonal set* if $\langle \mathbf{x}_i, \mathbf{x}_j \rangle = 0$, for all $i \neq j$. Orthogonal sets of vectors define convenient bases in the vector space V, with extensive applications. For instance, such bases are used for trigonometric approximations in Chapter 11 or for the finite elements in Chapter 12. Among many remarkable properties of orthogonal sets of vectors, we can easily verify that any set of nonzero orthogonal vectors is automatically linearly independent. To see this, let $\{\mathbf{x}_1, \dots, \mathbf{x}_k\}$ be an orthogonal set and suppose that we write down the linear combination

$$\lambda_1 \mathbf{x}_1 + \cdots \lambda_k \mathbf{x}_k = \mathbf{0}.$$

Then, taking the inner product of both sides of this equality with any of the vectors $\mathbf{x}_i$ gives

$$0 = \langle \lambda_1 \mathbf{x}_1 + \cdots \lambda_k \mathbf{x}_k, \mathbf{x}_i \rangle = \lambda_i \|\mathbf{x}_i\|^2,$$

which implies that $\lambda_i = 0$. The last expression can be simplified if each vector in the set $\{\mathbf{x}_1, \dots, \mathbf{x}_k\}$ has norm equal to one. Orthogonal sets with this property, that is, such that $\langle \mathbf{x}_i, \mathbf{x}_i \rangle = \|\mathbf{x}_i\|^2 = 1$, for all i, are called *orthonormal sets*.

Exercise 3.5 Suppose that $\{\mathbf{x}_1, \dots, \mathbf{x}_k\}$ is an orthonormal set of vectors. Use the Pythagorean theorem to show that

$$\|\lambda_1 \mathbf{x}_1 + \cdots \lambda_k \mathbf{x}_k\|^2 = \lambda_1^2 + \cdots \lambda_k^2.$$

Adjoint maps

We now want to explore the relationship between linear maps and inner products. For motivation, consider a linear map $\mathbf{A}$ from the vector space $\mathbb{R}^n$ to the vector space $\mathbb{R}^m$ equipped with their respective dot products. In MATLAB matrix notation, the dot product $\mathbf{x} \cdot \mathbf{A}\mathbf{y}$ in $\mathbb{R}^m$ can be written as

$$\mathbf{x} \cdot \mathbf{A}\mathbf{y} = \mathbf{x}' * (\mathbf{A} * \mathbf{y}) = (\mathbf{A}' * \mathbf{x})' * \mathbf{y} = \mathbf{A}'\mathbf{x} \cdot \mathbf{y}, \tag{3.14}$$

where we have used the properties of the transpose of a matrix product. Therefore, to *move* a linear map from one factor of a dot product to another, all we need to do is take its transpose.

Following this idea, for general finite-dimensional inner product spaces V and W, we define the *adjoint* of a linear map $\mathbf{A} : V \to W$ as the *unique* map $\mathbf{A}^* : W \to V$ satisfying

$$\langle \mathbf{A}^*\mathbf{w}, \mathbf{v}\rangle = \langle \mathbf{w}, \mathbf{A}\mathbf{v}\rangle, \qquad \forall \mathbf{w} \in W, \forall \mathbf{v} \in V. \tag{3.15}$$

We say that a map $\mathbf{A} : V \to V$ is *self-adjoint* if $\mathbf{A}^* = \mathbf{A}$.

If $V = \mathbb{R}^n$, then the adjoint of a linear map from $V = \mathbb{R}^n$ to $W = \mathbb{R}^m$ with respect to the dot product (3.1) is obtained by taking the transpose of its corresponding matrix with respect to the canonical bases for $\mathbb{R}^n$ and $\mathbb{R}^m$, as in the identity (3.14). In this case, a self-adjoint map from $V = \mathbb{R}^n$ to itself corresponds to a *symmetric* matrix with the property $\mathbf{A}' = \mathbf{A}$. Similarly, when $V = \mathbb{C}^n$, the same identity shows that the adjoint of a linear map $\mathbf{A}$ with respect to the complex dot product (3.6) is obtained by taking the conjugate-transpose of its corresponding matrix. In this case, a self-adjoint map from $V = \mathbb{C}^n$ satisfies $\mathbf{A}' = \mathbf{A}$ and is called a *Hermitian* matrix.

In general, the null space of a linear map $\mathbf{A}$ is related to the range of its adjoint $\mathbf{A}^*$ through the equation

$$\text{null}(\mathbf{A}) = \text{range}(\mathbf{A}^*)^{\perp}. \tag{3.16}$$

To establish (3.16), suppose that $\mathbf{v} \in V$ and let $\mathbf{y} = \mathbf{A}^*\mathbf{x}$ be an arbitrary vector in the range of $\mathbf{A}^*$. Then

$$\langle \mathbf{v}, \mathbf{y}\rangle = \langle \mathbf{v}, \mathbf{A}^*\mathbf{x}\rangle = \langle \mathbf{A}\mathbf{v}, \mathbf{x}\rangle,$$

which implies that $\mathbf{v}$ is in the null space of $\mathbf{A}$ if and only if it is orthogonal to $\mathbf{y}$. For a concrete example of (3.16), consider the following:

```
>> A=[1 2 3; 2 4 6; 1 0 -1]

A =

     1     2     3
     2     4     6
     1     0    -1

>> v=null(A), B=orth(A')

v =

    0.4082
```

```
    -0.8165
     0.4082


B =
    -0.2583    0.8756
    -0.5323    0.2237
    -0.8062   -0.4282

>> [v'*B(:,1),v'*B(:,2)]

ans =
    1.0e-15 *

     0.1258    0.0195
```

Observe in this example that the null space of $\mathbf{A}$ is spanned by the single vector $[1, -2, 1]$, which is orthogonal, within machine precision, to the two vectors that span the range of $\mathbf{A}'$.

Isometries

For the next concept, recall how the matrix norm was defined in Section 2.4 in terms of the effect that a linear map had on the size of the underlying vectors. Similarly, we can ask what effect linear maps have on the inner product of the underlying vectors. That is, if we know $\langle \mathbf{x}, \mathbf{y} \rangle$, what can we say about $\langle \mathbf{Ax}, \mathbf{Ay} \rangle$? For example:

```
>> e1 = [1 0 0]'; e2 = [0 1 0]'; e3 = [0 0 1]';
>> A = [0 -1 2; 2 3 4; -3 2 5];
>> (A*e1)'*(A*e2)

ans =
     0

>> (A*e1)'*(A*e3)

ans =
    -7

>> (A*e2)'*(A*e3)

ans =
    20
```

In this example, starting with three orthogonal vectors $\{\mathbf{e}_1, \mathbf{e}_2, \mathbf{e}_3\}$, the linear map $\mathbf{A}$ produces three vectors $\{\mathbf{Ae}_1, \mathbf{Ae}_2, \mathbf{Ae}_3\}$ with very different properties with regard to their inner products : the vectors $\{\mathbf{Ae}_1, \mathbf{Ae}_2\}$ are still orthogonal, while the angle enclosed by $\{\mathbf{Ae}_1, \mathbf{Ae}_3\}$ is greater than $\pi/2$ and the one enclosed by $\{\mathbf{Ae}_2, \mathbf{Ae}_3\}$ is smaller than $\pi/2$. That is, in general, linear maps modify inner products in many diverse ways.

On the other hand, suppose that a linear map $\mathbf{U} : V \to V$ does not alter the inner products between arbitrary vectors in V, that is, suppose it satisfies

$$\langle \mathbf{Ux}, \mathbf{Uy} \rangle = \langle \mathbf{x}, \mathbf{y} \rangle, \quad \forall \mathbf{x}, \mathbf{y} \in V. \tag{3.17}$$

Such a map is called an *isometry* because it also preserves the Euclidean norm on V, in the sense that $\|\mathbf{Ux}\| = \|\mathbf{x}\|$ for all $\mathbf{x} \in V$. Moreover, using the definition of an adjoint map, it is easy to see that an isometry satisfies $\mathbf{U}^*\mathbf{U} = \mathbf{1}$, where $\mathbf{1}$ is the identity map on V. Typical examples of isometries are rotations in $\mathbb{R}^3$. For instance, the matrices implementing counterclockwise rotations around the z-, y-, and x-axis are

$$\mathbf{U}_1 = \begin{bmatrix} \cos\theta_1 & -\sin\theta_1 & 0 \\ \sin\theta_1 & \cos\theta_1 & 0 \\ 0 & 0 & 1 \end{bmatrix},$$

$$\mathbf{U}_2 = \begin{bmatrix} \cos\theta_2 & 0 & -\sin\theta_2 \\ 0 & 1 & 0 \\ \sin\theta_2 & 0 & \cos\theta_2 \end{bmatrix},$$

$$\mathbf{U}_3 = \begin{bmatrix} 1 & 0 & 0 \\ 0 & \cos\theta_3 & -\sin\theta_3 \\ 0 & \sin\theta_3 & \cos\theta_3 \end{bmatrix}.$$

The following MATLAB commands implement counterclockwise rotations by an angle $\pi/4$ around the z-axis, $-\pi/3$ around the y-axis, and $\pi/6$ around the x-axis. You can then observe that the dot product of two vectors $\mathbf{x}$ and $\mathbf{y}$ in $\mathbb{R}^3$ is invariant under these rotations:

```
>> U1 = [cos(pi/4) -sin(pi/4) 0 ; sin(pi/4) cos(pi/4) 0; 0 0 1];
>> U2 = [cos(pi/3) 0 sin(pi/3); 0 1 0; -sin(pi/3) 0 cos(pi/3)];
>> U3 = [1 0 0 ; 0 cos(pi/6) -sin(pi/6); 0 sin(pi/6) cos(pi/6)];
>> U = U3*U2*U1

U =

    0.3536   -0.3536    0.8660
    0.9186    0.3062   -0.2500
   -0.1768    0.8839    0.4330
```

```
>> x = [1;2;-1]; y = [0;1;-3];
>> (U*x)'*(U*y)-x'*y

ans =
     0
```

In particular, notice that the product of the isometries $\mathbf{U}_1$, $\mathbf{U}_2$, and $\mathbf{U}_3$ is itself an isometry.

Orthogonal and unitary matrices

In Section 2.3, we showed how to arrange a given set of vectors as the columns of a matrix in order to verify their linear independence. The same type of construction turns out to be useful for checking the orthogonality of a set of vectors. Let $\{\mathbf{u}_1, \ldots, \mathbf{u}_k\}$ be a set of vectors in $\mathbb{R}^n$ and consider the $n \times k$ matrix

$$\mathbf{U} = [\mathbf{u}_1, \cdots, \mathbf{u}_k].$$

Then we can verify that

$$\mathbf{U}'\mathbf{U} = \begin{bmatrix} \mathbf{u}_1' \\ \vdots \\ \mathbf{u}_k' \end{bmatrix} [\mathbf{u}_1 \cdots \mathbf{u}_k] = \begin{bmatrix} \mathbf{u}_1 \cdot \mathbf{u}_1 & \mathbf{u}_1 \cdot \mathbf{u}_2 & \cdots & \mathbf{u}_1 \cdot \mathbf{u}_k \\ \mathbf{u}_2 \cdot \mathbf{u}_1 & \mathbf{u}_2 \cdot \mathbf{u}_2 & \cdots & \mathbf{u}_2 \cdot \mathbf{u}_k \\ \vdots & & \ddots & \vdots \\ \mathbf{u}_k \cdot \mathbf{u}_1 & \mathbf{u}_k \cdot \mathbf{u}_2 & \cdots & \mathbf{u}_k \cdot \mathbf{u}_k \end{bmatrix}.$$

Therefore, the vectors $\{\mathbf{u}_1, \ldots, \mathbf{u}_k\}$ form an orthogonal set if and only if $\mathbf{U}'\mathbf{U}$ is a diagonal matrix, and an orthonormal set if and only if $\mathbf{U}'\mathbf{U}$ is the k–dimensional identity matrix $\mathbf{1}_k$. For example, you can immediately verify that any subset of the canonical basis $\{\mathbf{e}_1, \mathbf{e}_2, \ldots, \mathbf{e}_n\}$ is an orthonormal set in $\mathbb{R}^n$. For a different example of an orthonormal set, consider the following:

```
>> u1 = [1/2 1/2 1/2 1/2]';
>> u2 = [1/2 1/2 -1/2 -1/2]';
>> u3 = [-1/2 1/2 -1/2 1/2]';
>> U = [u1 u2 u3];
>> U'*U

ans =
     1     0     0
     0     1     0
     0     0     1
```

In general, an $n \times k$ matrix $\mathbf{U}$ of rank k satisfying $\mathbf{U}'\mathbf{U} = \mathbf{1}_k$ is called *left orthogonal* if it has real entries, or *left unitary* if it has complex entries. When $n = k$, such a matrix is simply called *orthogonal* if its entries are real, or *unitary* if its entries are complex. In either case, we have $\mathbf{U}'\mathbf{U} = \mathbf{U}\mathbf{U}' = \mathbf{1}_n$, which implies that every orthogonal or unitary matrix is invertible and satisfies the property $\mathbf{U}^{-1} = \mathbf{U}'$. Because this is equivalent to

$$\mathbf{U}\mathbf{x} \cdot \mathbf{U}\mathbf{x} = \mathbf{x} \cdot \mathbf{x}, \tag{3.18}$$

we see that orthogonal matrices correspond to isometries in $\mathbb{R}^n$, preserving the inner product (3.1), whereas unitary matrices correspond to isometries in $\mathbb{C}^n$, preserving the inner product (3.6).

3.2 Orthogonal Projections

Given a nonzero vector $\mathbf{x}$ and an arbitrary vector $\mathbf{y}$ in $\mathbb{R}^n$, consider the problem of finding the multiple $\lambda\mathbf{x}$ of the vector $\mathbf{x}$ that is closest to $\mathbf{y}$ according to the Euclidean norm, that is, finding λ such that the distance $\|\mathbf{y} - \lambda\mathbf{x}\|$ is as small as possible. According to (3.8), $\|\mathbf{y} - \lambda\mathbf{x}\|^2$ is a quadratic function of the variable λ, which achieves its only minimum at

$$\lambda = \frac{\langle \mathbf{x}, \mathbf{y} \rangle}{\langle \mathbf{x}, \mathbf{x} \rangle}. \tag{3.19}$$

The following MATLAB example plots this quadratic function of λ in Figure 3.1 with its only minimum at $\lambda = 0.5$.

```
>> x = [-1 2 0 3 1 -2]';
>> y = [ 3 2 5 4 -1.5 1]';
>> lambda = [-2:0.1:2];
>> for i = 1:length(lambda)
```

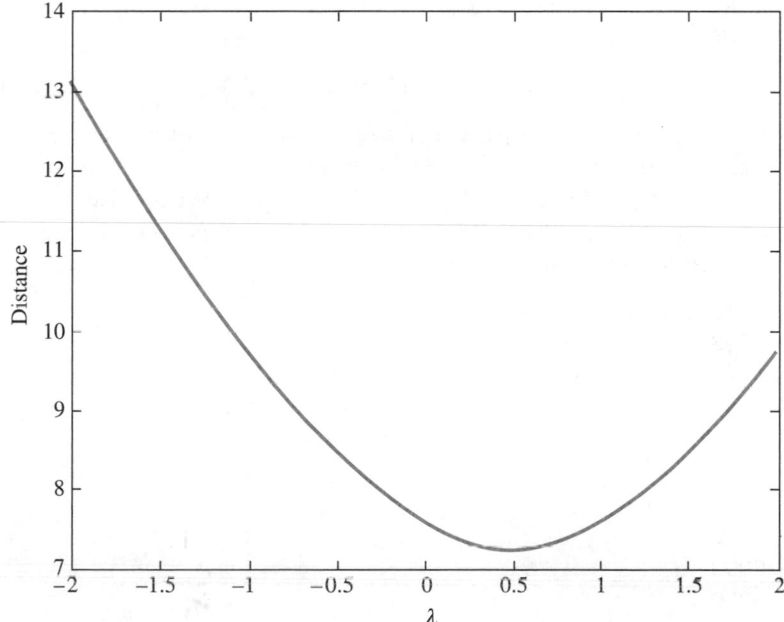

Figure 3.1 Distance $\|\mathbf{y} - \lambda\mathbf{x}\|$ as a function of λ.

```
>>       dist(i) = norm(y-lambda(i)*x);
>> end
>> plot(lambda,dist);
>> lambda=(x'*y)/(x'*x)

lambda =
                  0.5000
```

Moreover, observe that λ given by (3.19) is the only scalar for which the difference $\mathbf{y} - \lambda\mathbf{x}$ is orthogonal to $\mathbf{x}$ because it is the only solution to the equation

$$\langle \mathbf{x}, \mathbf{y} - \lambda\mathbf{x} \rangle = 0.$$

The vector

$$\mathbf{u} = \frac{\langle \mathbf{x}, \mathbf{y} \rangle}{\langle \mathbf{x}, \mathbf{x} \rangle}\mathbf{x} \tag{3.20}$$

is then called the *orthogonal projection* of $\mathbf{y}$ in the direction of $\mathbf{x}$ and, as we have just shown, is the multiple of $\mathbf{x}$ that is closest to $\mathbf{y}$ according to the Euclidean norm.

Orthogonal decomposition

Observe that if $\mathbf{y}$ is already orthogonal to $\mathbf{x}$, then (3.19) gives $\lambda = 0$, and so the orthogonal projection of $\mathbf{y}$ in the direction of $\mathbf{x}$ is the null vector. On the other hand, if $\mathbf{y}$ is already in the direction of $\mathbf{x}$, that is, if $\mathbf{y} = \mu\mathbf{x}$ for some scalar μ, then (3.19) gives $\lambda = \mu$ and the orthogonal projection of $\mathbf{y}$ in the direction of $\mathbf{x}$ is the vector $\mathbf{y}$ itself. This indicates that any vector $\mathbf{y}$ can be uniquely written as

$$\mathbf{y} = \mathbf{u} + \mathbf{v}, \tag{3.21}$$

where $\mathbf{u}$ is the orthogonal projection of $\mathbf{y}$ in the direction of $\mathbf{x}$ and $\mathbf{v} = \mathbf{y} - \mathbf{u}$ is its orthogonal complement, with the property $\langle \mathbf{x}, \mathbf{v} \rangle = 0$. The decomposition (3.21) is called the *orthogonal decomposition* of $\mathbf{y}$ given $\mathbf{x}$ with respect to the inner product $\langle \cdot, \cdot \rangle$. The following MATLAB commands implement this decomposition for an example in $\mathbb{R}^6$:

```
>> x = [-1 2 0 3 1 -2]';
>> y = [ 3 2 5 4 -1.5 1]';
>> u = ((x'*y)/(x'*x))*x

u =
   -0.5000
    1.0000
         0
    1.5000
    0.5000
   -1.0000
```

```
>> v=y-u

v =
     3.5000
     1.0000
     5.0000
     2.5000
    -2.0000
     2.0000

>> x'*v

ans =
         0
```

A graphical illustration of the decomposition (3.21) in $\mathbb{R}^2$ is useful for understanding the geometric meaning of the orthogonal projection $\mathbf{u}$ and the orthogonal complement $\mathbf{v}$. This is shown in Figure 3.2, generated by the following MATLAB commands:

```
>> x = [5 1]'; y = [1 3]';
```

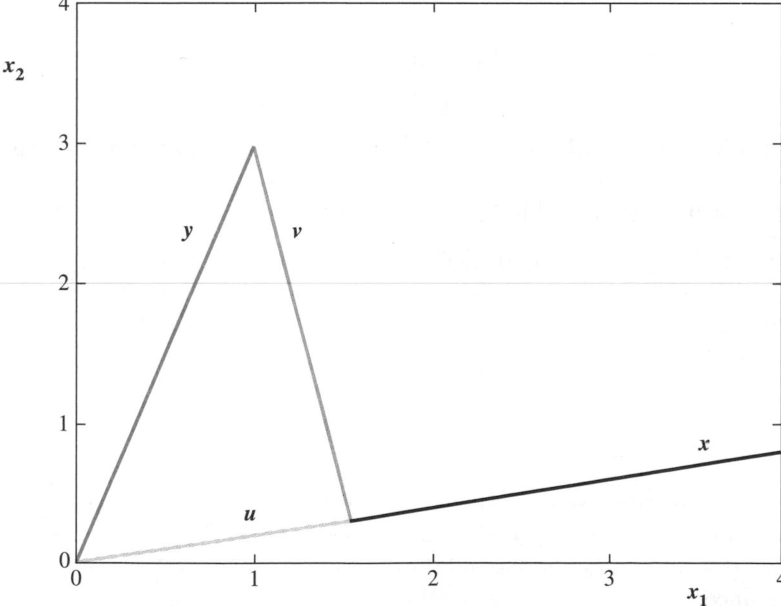

Figure 3.2 Orthogonal projection of vector $\mathbf{y}$ in the direction of the vector $\mathbf{x}$.

```
>> u = ((x'*y)/(x'*x))*x; v = y-u;
>> v1 = [0 x(1)]; w1 = [0 x(2)];
>> v2 = [0 y(1)]; w2 = [0 y(2)];
>> v3 = [0 u(1) u(1) + v(1)]; w3 = [0 u(2) u(2) + v(2)];
>> plot(v1,w1,'r',v2,w2,'b',v3,w3,'g');
```

Projections

In general, a linear map $\mathbf{P} : V \to V$ with the property that $\mathbf{P}^2 = \mathbf{P}$ is called a *projection* because it maps a general vector $\mathbf{x} \in V$ onto the subspace $U = \text{range}(\mathbf{P})$ and leaves unchanged any vector that is already in this subspace. To confirm this, suppose that $\mathbf{x} \in U$. Then, by definition of the range, there must exist a vector $\mathbf{y} \in V$ such that $\mathbf{x} = \mathbf{Py}$. But then,

$$\mathbf{Px} = \mathbf{P}(\mathbf{Py}) = \mathbf{P}^2\mathbf{y} = \mathbf{Py} = \mathbf{x}.$$

If V is an inner product space, a projection $\mathbf{P}$ satisfying the additional property $\mathbf{P} = \mathbf{P}^*$ is said to be an *orthogonal projection.* To see the reason for this name, consider the difference between a general vector $\mathbf{y} \in V$ and its projection $\mathbf{Py}$ onto range($\mathbf{P}$), that is, consider the *residual* vector $\mathbf{r} = \mathbf{y} - \mathbf{Py}$. We can then verify that this vector is orthogonal to range($\mathbf{P}$), in the sense that, if $\mathbf{x} \in \text{range}(\mathbf{P})$, then

$$\langle \mathbf{x}, \mathbf{r} \rangle = \langle \mathbf{x}, \mathbf{y} - \mathbf{Py} \rangle = \langle \mathbf{x}, \mathbf{y} \rangle - \langle \mathbf{x}, \mathbf{Py} \rangle = \langle \mathbf{x}, \mathbf{y} \rangle - \langle \mathbf{Px}, \mathbf{y} \rangle = 0.$$

As an easy example of an orthogonal projection, consider the matrix

$$\mathbf{P} = \begin{bmatrix} 1 & 0 & 0 \\ 0 & 1 & 0 \\ 0 & 0 & 0 \end{bmatrix}, \tag{3.22}$$

which projects a general vector $\mathbf{y} \in \mathbb{R}^3$ onto the subspace spanned by the two first canonical basis vectors $\{\mathbf{e}_1, \mathbf{e}_2\}$. The orthogonality property mentioned earlier is easily shown in this example:

```
>> P = [1 0 0; 0 1 0; 0 0 0];
>> y = [2 -1 3]';
>> x = [4 5 0]';
>> x'*(y-P*y)

ans =
     0
```

More interesting, we can ask which is the vector in the subspace range($\mathbf{P}$) that is closest to a vector $\mathbf{y} \in \mathbb{R}^3$ with respect to the Euclidean norm. In the previous example, this corresponds to finding the scalars λ, μ that minimize the distance

$$\|\mathbf{y} - \lambda\mathbf{e}_1 - \mu\mathbf{e}_2\| = \sqrt{(2 - \lambda)^2 + (1 + \mu)^2 + 9)}.$$

The following commands compute this function of two variables and plot the result in Figure 3.3, using the techniques described in Section 1.6.

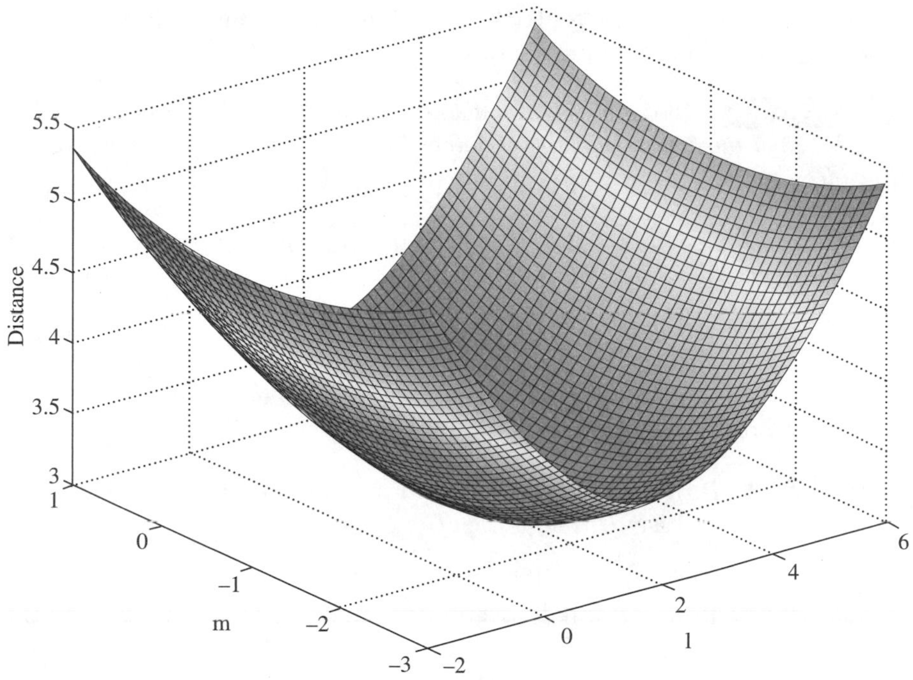

Figure 3.3 Distance $\|\mathbf{y} - \lambda\mathbf{e}_1 - \mu\mathbf{e}_2\|$ as a function of λ and μ.

```
>> lambda=-2:0.1:6;
>> mu=-3:0.1:1;
>> [Lambda,Mu]=meshgrid(lambda,mu);
>> dist=sqrt((2-Lambda).^2+(1+Mu).^2+9);
>> surf(lambda,mu,dist);
```

You can see from the figure that this distance is minimized for the scalar $\lambda = 2$ and $\mu = -1$, which correspond exactly to the vector $\mathbf{Py} = (2, -1, 0)$, that is, the orthogonal projection of $\mathbf{y}$ onto the subspace $\mathrm{Span}(\mathbf{e}_1, \mathbf{e}_2)$. This property can be demonstrated in general as follows. Let $\mathbf{y} \in V$ and suppose that $\mathbf{x} \in \mathrm{range}(\mathbf{P})$ is any vector in the range of $\mathbf{P}$. Then,

$$\begin{aligned}
\|\mathbf{x} - \mathbf{y}\|^2 - \|\mathbf{Py} - \mathbf{y}\|^2 &= \langle \mathbf{x} - \mathbf{y}, \mathbf{x} - \mathbf{y} \rangle - \langle \mathbf{Py} - \mathbf{y}, \mathbf{Py} - \mathbf{y} \rangle \\
&= \langle \mathbf{x}, \mathbf{x} \rangle - 2\langle \mathbf{x}, \mathbf{y} \rangle + \langle \mathbf{y}, \mathbf{y} \rangle - \langle \mathbf{Py}, \mathbf{Py} \rangle + 2\langle \mathbf{y}, \mathbf{Py} \rangle - \langle \mathbf{y}, \mathbf{y} \rangle \\
&= \langle \mathbf{x}, \mathbf{x} \rangle - 2\langle \mathbf{Px}, \mathbf{y} \rangle + \langle \mathbf{Py}, \mathbf{Py} \rangle \\
&= \langle \mathbf{x} - \mathbf{Py}, \mathbf{x} - \mathbf{Py} \rangle \\
&= \|\mathbf{x} - \mathbf{Py}\|^2,
\end{aligned}$$

which shows that $\|\mathbf{x} - \mathbf{y}\| \geq \|\mathbf{P}\mathbf{y} - \mathbf{y}\|$ with equality if and only if $\mathbf{x} = \mathbf{P}\mathbf{y}$. The following theorem summarizes this result.

Theorem 3.1 *Given an orthogonal projection* $\mathbf{P}$ *and an arbitrary vector* $\mathbf{y} \in V$*, then the vector in range*($\mathbf{P}$) *that is closest to* $\mathbf{y}$ *with respect to the Euclidean norm* (3.7) *is given by* $\mathbf{P}\mathbf{y}$*.*

We can also generalize the orthogonal decomposition given in (3.21). For that, we define the orthogonal complement of a subspace $U \in V$ as the set

$$U^{\perp} := \{\mathbf{y} \in V : \langle \mathbf{y}, \mathbf{x} \rangle = 0 \text{ for all } \mathbf{x} \in U\}. \tag{3.23}$$

Now try the following simple exercises concerning orthogonal complements and orthogonal projections.

Exercise 3.6 Prove that $U^{\perp}$ is itself a subspace. Moreover, show that if two subspaces U_1, U_2 satisfy $U_1 \subset U_2$, then $U_2^{\perp} \subset U_1^{\perp}$.

Exercise 3.7 Prove that if $\mathbf{P}$ is an orthogonal projection, then so is the linear map $\mathbf{P}^{\perp} = \mathbf{1} - \mathbf{P}$, where $\mathbf{1}$ is the identity operator. Moreover, show that range$(\mathbf{P})^{\perp}$ = range$(\mathbf{P}^{\perp})$. Observe that $\mathbf{P}^{\perp}\mathbf{y} = \mathbf{y} - \mathbf{P}\mathbf{y}$ coincides with the residual vector $\mathbf{r}$ associated with an orthogonal projection $\mathbf{P}$.

For example, if $\mathbf{P}$ is the orthogonal projection defined in (3.22), then

$$\mathbf{P}^{\perp} = \begin{bmatrix} 0 & 0 & 0 \\ 0 & 0 & 0 \\ 0 & 0 & 1 \end{bmatrix}, \tag{3.24}$$

whose range is the subspace of $\mathbb{R}^3$ spanned by the basis vector $\mathbf{e}_3$. It is then easy to see that any vector $\mathbf{y} = [y_1, y_2, y_3] \in \mathbb{R}^3$ can be uniquely decomposed as

$$\mathbf{y} = \begin{bmatrix} y_1 \\ y_2 \\ 0 \end{bmatrix} + \begin{bmatrix} 0 \\ 0 \\ y_3 \end{bmatrix},$$

with $[y_1, y_2, 0] \in$ range$(\mathbf{P})$ and $[0, 0, y_3] \in$ range$(\mathbf{P})^{\perp}$.

Another example is the orthogonal projection $\mathbf{P}_{\mathbf{x}}$ in the direction of a vector $\mathbf{x}$ in an inner product space V, which is given by the linear map

$$\mathbf{P}_{\mathbf{x}}\mathbf{y} = \frac{\langle \mathbf{x}, \mathbf{y} \rangle}{\langle \mathbf{x}, \mathbf{x} \rangle}\mathbf{x}, \tag{3.25}$$

as we established at the beginning of this section. In this case,

$$\mathbf{P}_{\mathbf{x}}^{\perp}\,\mathbf{y} = \mathbf{y} - \frac{\langle \mathbf{x}, \mathbf{y} \rangle}{\langle \mathbf{x}, \mathbf{x} \rangle}\mathbf{x} \tag{3.26}$$

defines an orthogonal projection on the subspace $\text{Span}(\mathbf{x})^{\perp}$, which is appropriately called the hyperplane orthogonal to $\mathbf{x}$.

Without specifying a basis for the inner product space V, it is difficult to get a concrete grasp of the linear map $\mathbf{P}_{\mathbf{x}}$. If $V = \mathbb{R}^n$ (or equivalently, when we specify a basis for a finite-dimensional inner product space V), then the orthogonal projection $\mathbf{P}_{\mathbf{x}}$ associated with the column-vector $\mathbf{x}$ acquires the following interesting matrix form:

$$\mathbf{P}_{\mathbf{x}}\,\mathbf{y} = \frac{\mathbf{x}' * \mathbf{y} * \mathbf{x}}{\mathbf{x}' * \mathbf{x}} = \frac{\mathbf{x} * \mathbf{x}' * \mathbf{y}}{\mathbf{x}' * \mathbf{x}} = \frac{\mathbf{x} * \mathbf{x}'}{\mathbf{x}' * \mathbf{x}}\mathbf{y},$$

where you can observe that the inner product $\mathbf{x}' * \mathbf{x} = \|\mathbf{x}\|^2$ is a scalar, while the outer product $\mathbf{x} * \mathbf{x}'$ is an $n \times n$ matrix. It is now an easy task to verify that the linear map $\mathbf{P}_x = (\mathbf{x} * \mathbf{x}')/\|\mathbf{x}\|^2$ is self-adjoint in $\mathbb{R}^n$.

As an example, you can calculate the matrix for the orthogonal projection in the direction of the vector $\mathbf{x} = [-1, 0, 2]$, as well as the projection onto its orthogonal hyperplane:

```
>> x = [-1 0 2]';
>> P = (x*x')/(x'*x)

P =
    0.2000         0   -0.4000
         0         0         0
   -0.4000         0    0.8000

>> P_perp = eye(3)-P

P_perp =
    0.8000         0    0.4000
         0    1.0000         0
    0.4000         0    0.2000
```

In general, given an orthogonal projection $\mathbf{P}$ and an arbitrary vector $\mathbf{y} \in V$, we can define the vectors

$$\mathbf{x} = \mathbf{P}\mathbf{y} \in \text{range}(\mathbf{P}), \quad \mathbf{z} = \mathbf{P}^{\perp}\mathbf{y} \in \text{range}(\mathbf{P})^{\perp},$$

with the property that

$$\mathbf{x} + \mathbf{z} = \mathbf{P}\mathbf{y} + \mathbf{P}^{\perp}\mathbf{y} = \mathbf{y}. \tag{3.27}$$

As you can see, this is the anticipated generalization of the decomposition (3.21). Using Theorem 3.1, you can prove the following uniqueness result for the decomposition (3.27).

Theorem 3.2 *If* $\mathbf{P}$ *is an orthogonal projection and*

$$\mathbf{y} = \mathbf{u} + \mathbf{v} \tag{3.28}$$

for some $\mathbf{u} \in \text{range}(\mathbf{P})$ *and* $\mathbf{v} \in \text{range}(\mathbf{P})^{\perp}$, *then*

$$\mathbf{u} = \mathbf{P}\mathbf{y}, \qquad \mathbf{v} = \mathbf{P}^{\perp}\mathbf{y}. \tag{3.29}$$

In finite dimensions, the same type of decomposition holds for an arbitrary subspace U and its orthogonal complement $U^{\perp}$. Assume for a moment that we are given a set of orthonormal vectors $\{\mathbf{q}_1, \ldots, \mathbf{q}_k\}$ such that $\text{Span}(\mathbf{q}_1, \ldots, \mathbf{q}_k) = U$. Then, the linear map $\mathbf{P}_U$ defined by

$$\mathbf{P}_U\mathbf{y} = \langle \mathbf{y}, \mathbf{q}_1 \rangle \mathbf{q}_1 + \cdots + \langle \mathbf{y}, \mathbf{q}_k \rangle \mathbf{q}_k \tag{3.30}$$

is an orthogonal projection with the property that $\text{range}(\mathbf{P}_U) = U$. Moreover, we have that

$$\mathbf{P}_U^{\perp}\mathbf{y} = \mathbf{y} - \langle \mathbf{y}, \mathbf{q}_1 \rangle \mathbf{q}_1 - \cdots - \langle \mathbf{y}, \mathbf{q}_k \rangle \mathbf{q}_k, \tag{3.31}$$

with $\text{range}(\mathbf{P}_U^{\perp}) = U^{\perp}$.

Orthogonalization

In other words, all that we need in order to obtain the orthogonal decomposition of V into U and $U^{\perp}$ is an orthonormal set of vectors spanning the subspace U. But because U is itself a vector space, it must have a basis of linearly independent vectors $\{\mathbf{u}_1, \ldots, \mathbf{u}_k\}$. We are then led to the following problem.

Problem 3.1 (Orthogonalization) Given a set of linearly independent vectors $\{\mathbf{u}_1, \ldots, \mathbf{u}_k\}$ in an inner product space V, find a set of *orthonormal* vectors $\{\mathbf{q}_1, \ldots, \mathbf{q}_k\}$ such that $\text{Span}(\mathbf{q}_1, \ldots, \mathbf{q}_k) = \text{Span}(\mathbf{u}_1, \ldots, \mathbf{u}_k)$.

We now describe the *Gram–Schmidt orthogonalization procedure* for solving Problem 3.1. For the first vector, we take

$$\mathbf{q}_1 = \frac{\mathbf{u}_1}{\|\mathbf{u}_1\|}, \tag{3.32}$$

so that $\text{Span}(\mathbf{u}_1) = \text{Span}(\mathbf{q}_1)$ and $\|\mathbf{q}_1\| = 1$. Next, define the auxiliary vector

$$\mathbf{v}_2 = \mathbf{u}_2 - \langle \mathbf{u}_2, \mathbf{q}_1 \rangle \mathbf{q}_1$$

and take the second vector of the target set to be

$$\mathbf{q}_2 = \frac{\mathbf{v}_2}{\|\mathbf{v}_2\|}. \tag{3.33}$$

We can then verify that $\|\mathbf{q}_2\| = 1$ and

$$\langle \mathbf{q}_1, \mathbf{q}_2 \rangle = \frac{1}{\|\mathbf{v}_2\|} \langle \mathbf{q}_1, \mathbf{u}_2 - \langle \mathbf{u}_2, \mathbf{q}_1 \rangle \mathbf{q}_1 \rangle = 0,$$

so that $\{\mathbf{q}_1, \mathbf{q}_2\}$ is an orthonormal set with $\text{Span}(\mathbf{q}_1, \mathbf{q}_2) = \text{Span}(\mathbf{u}_1, \mathbf{u}_2)$. Next, define

$$\begin{aligned} \mathbf{v}_3 &= \mathbf{u}_3 - \langle \mathbf{u}_3, \mathbf{q}_1 \rangle \mathbf{q}_1 - \langle \mathbf{u}_3, \mathbf{q}_2 \rangle \mathbf{q}_2 \\ \mathbf{q}_3 &= \frac{\mathbf{v}_3}{\|\mathbf{v}_3\|} \end{aligned} \tag{3.34}$$

and verify that $\|\mathbf{q}_3\| = 1$ and

$$\begin{aligned} \langle \mathbf{q}_1, \mathbf{q}_3 \rangle &= \frac{1}{\|\mathbf{v}_3\|} \langle \mathbf{q}_1, \mathbf{u}_3 - \langle \mathbf{u}_3, \mathbf{q}_1 \rangle \mathbf{q}_1 - \langle \mathbf{u}_3, \mathbf{q}_2 \rangle \mathbf{q}_2 \rangle = 0 \\ \langle \mathbf{q}_2, \mathbf{q}_3 \rangle &= \frac{1}{\|\mathbf{v}_3\|} \langle \mathbf{q}_2, \mathbf{u}_3 - \langle \mathbf{u}_3, \mathbf{q}_1 \rangle \mathbf{q}_1 - \langle \mathbf{u}_3, \mathbf{q}_2 \rangle \mathbf{q}_2 \rangle = 0, \end{aligned}$$

so that $\{\mathbf{q}_1, \mathbf{q}_2, \mathbf{q}_3\}$ is an orthonormal set with $\text{Span}(\mathbf{q}_1, \mathbf{q}_2, \mathbf{q}_3) = \text{Span}(\mathbf{u}_1, \mathbf{u}_2, \mathbf{u}_3)$. We can then proceed inductively by setting

$$\begin{aligned} \mathbf{v}_i &= \mathbf{u}_i - \langle \mathbf{u}_i, \mathbf{q}_1 \rangle \mathbf{q}_1 - \cdots - \langle \mathbf{u}_i, \mathbf{q}_{i-1} \rangle \mathbf{q}_{i-1} \\ \mathbf{q}_i &= \frac{\mathbf{v}_i}{\|\mathbf{v}_i\|} \end{aligned} \tag{3.35}$$

until we obtain the desired set $\{\mathbf{q}_1, \ldots, \mathbf{q}_k\}$. You can see how the procedure works in practice in the following MATLAB example:

```
>> v1 = [-1 -1 1 1]'; v2 = [-5 -5 7 7]'; v3 = [4 -2 0 6]';
>> q1 = v1/norm(v1);
>> q2 = (v2-(v2'*q1)*q1)/norm(v2-(v2'*q1)*q1);
>> q3 = (v3-(v3'*q1)*q1-(v3'*q2)*q2)/norm(v3-(v3'*q1)*q1-(v3'*q2)*q2);
>> Q = [q1 q2 q3]
```

```
Q =
   -0.5000    0.5000    0.5000
   -0.5000    0.5000   -0.5000
    0.5000    0.5000   -0.5000
    0.5000    0.5000    0.5000

>> Q'*Q

ans =
     1     0     0
     0     1     0
     0     0     1
```

Finally, observe that if $\{\mathbf{u}_1, \dots, \mathbf{u}_n\}$ is a basis for an n-dimensional inner product space V, then applying the Gram–Schmidt orthogonalization algorithm to these vectors produces an orthonormal set $\{\mathbf{q}_1, \dots, \mathbf{q}_n\}$ that spans the entire space V. An orthonormal set with this property is called an *orthonormal basis*, and the Gram–Schmidt orthogonalization procedure tells us that such a basis always exists.

3.3 QR Factorization

You have already seen in Section 2.5 how writing a matrix as a product of special types of matrices can be useful in the numerical solution of square linear systems. For instance, LU factorization reduces the solution of a general linear system to the solution of much simpler systems with upper- and lower-triangular matrices. In this section, we present another type of factorization you can use to better exploit the properties of orthogonal vectors.

Recall from Section 3.2 that, given linearly independent vectors $\{\mathbf{u}_1, \dots, \mathbf{u}_k\}$ in $\mathbb{R}^n$, we can use the Gram–Schmidt algorithm to produce a set of orthonormal vectors $\{\mathbf{q}_1, \dots, \mathbf{q}_k\}$ spanning the same subspace. In particular, it follows that we can rewrite the original vectors $\{\mathbf{u}_1, \dots, \mathbf{u}_k\}$ as linear combinations of the orthonormal vectors $\{\mathbf{q}_1, \dots, \mathbf{q}_k\}$. In matrix notation, this means that if the vectors $\{\mathbf{u}_1, \dots, \mathbf{u}_k\}$ are arranged as the columns of an $n \times k$ matrix $\mathbf{A}$ and the vectors $\{\mathbf{q}_1, \dots, \mathbf{q}_k\}$ are arranged as the columns of another $n \times k$ matrix $\mathbf{Q}$, then there must exist a $k \times k$ matrix $\mathbf{R}$ such that

$$\mathbf{A} = \mathbf{QR}. \tag{3.36}$$

To find the entries of the matrix $\mathbf{R}$, all we need to do is rewrite the Equations (3.35) used in the Gram–Schmidt procedure. For example, looking at steps of the Gram–Schmidt orthogonalization algorithm (3.32)–(3.34) for three linearly independent vectors $\{\mathbf{u}_1, \mathbf{u}_2, \mathbf{u}_3\}$, we find that

$$\begin{aligned}
\mathbf{u}_1 &= \|\mathbf{u}_1\|\mathbf{q}_1 \\
\mathbf{u}_2 &= \langle \mathbf{q}_1, \mathbf{u}_2\rangle \mathbf{q}_1 + \|\mathbf{u}_2 - \langle \mathbf{u}_2, \mathbf{q}_1\rangle \mathbf{q}_1\|\mathbf{q}_2 \\
\mathbf{u}_3 &= \langle \mathbf{q}_1, \mathbf{u}_3\rangle \mathbf{q}_1 + \langle \mathbf{q}_2, \mathbf{u}_3\rangle \mathbf{q}_2 + \|\mathbf{u}_3 - \langle \mathbf{u}_3, \mathbf{q}_1\rangle \mathbf{q}_1 - \langle \mathbf{u}_3, \mathbf{q}_2\rangle \mathbf{q}_2\|\mathbf{q}_3.
\end{aligned}$$

For the vectors $\mathbf{u}_1 = [-1, -1, 1, 1]$, $\mathbf{u}_2 = [-5, -5, 7, 7]$, and $\mathbf{u}_3 = [4, -2, 0, 6]$ in $\mathbb{R}^4$, this gives

$$\begin{bmatrix} -1 & -5 & 4 \\ -1 & -5 & -2 \\ 1 & 7 & 0 \\ 1 & 7 & 6 \end{bmatrix} = \begin{bmatrix} -1/2 & 1/2 & 1/2 \\ -1/2 & 1/2 & -1/2 \\ 1/2 & 1/2 & -1/2 \\ 1/2 & 1/2 & 1/2 \end{bmatrix} \begin{bmatrix} 2 & 12 & 2 \\ 0 & 2 & 4 \\ 0 & 0 & 6 \end{bmatrix}. \tag{3.37}$$

In general, the matrix $\mathbf{R}$ obtained in this way is upper triangular with entries

$$\begin{aligned}
r_{ij} &= \langle \mathbf{q}_i, \mathbf{u}_j\rangle, \qquad \text{if } i < j \\
r_{jj} &= \left\| \mathbf{u}_j - \sum_{i=1}^{j-1} r_{ij}\mathbf{q}_i \right\| \\
r_{ij} &= 0, \qquad \text{if } i > j.
\end{aligned} \tag{3.38}$$

Reduced QR factorization

The factorization (3.36) of a full-rank $n \times k$ matrix $\mathbf{A}$ into the product of a left-orthogonal $n \times k$ matrix $\mathbf{Q}$ and a $k \times k$ upper-triangular matrix $\mathbf{R}$ with strictly positive diagonal entries is called a *reduced QR factorization.* Its existence is provided by the Gram–Schmidt orthogonalization procedure. For uniqueness, suppose that $\mathbf{S}$ and $\mathbf{Q}$ are two left-orthogonal $n \times k$ matrices and $\mathbf{T}$ and $\mathbf{R}$ are two $k \times k$ upper-triangular matrices with strictly positive diagonal entries. If $\mathbf{QR} = \mathbf{ST}$, then the matrix $\mathbf{Q}'\mathbf{S}$ must be both upper and lower triangular, implying that $\mathbf{Q}'\mathbf{S} = \mathbf{D}$ for some diagonal matrix $\mathbf{D}$. Moreover, in this case, $\mathbf{D}^2 = \mathbf{1}_k$, so that the diagonal entries of $\mathbf{D}$ must be equal to ± 1. Finally, using the fact that both $\mathbf{R}$ and $\mathbf{T}$ have strictly positive diagonal entries, it follows that $\mathbf{D} = \mathbf{1}_k$, which then implies that $\mathbf{Q} = \mathbf{S}$ and $\mathbf{R} = \mathbf{T}$. These results are summarized in the next theorem.

Theorem 3.3 *Any $n \times k$ matrix $\mathbf{A}$ of rank $k \leq n$ admits a unique reduced QR factorization $\mathbf{A} = \mathbf{QR}$, where $\mathbf{Q}$ is an $n \times k$ left-orthogonal matrix and $\mathbf{R}$ is a $k \times k$ upper-triangular matrix with strictly positive diagonal entries.*

The reduced QR factorization can be implemented by the following MATLAB code:

```
function [Q,R]=QRfactor(A)
% QR factorization for n x k matrix A
% Q is n x k left-orthogonal matrix
% R is k x k upper triangular matrix with positive diagonal entries
% A = Q*R

[n,k] = size(A);
for j = 1:k
    Q(:,j) = A(:,j);
    for i = 1:j-1
        R(i,j) = Q(:,i)'*A(:,j);
        Q(:,j) = Q(:,j)-R(i,j)*Q(:,i);
    end
    R(j,j) = norm(Q(:,j));
    if  R(j,j) == 0
        error('matrix is not of full rank')
    end
    Q(:,j) = Q(:,j)/R(j,j);
end
```

You can now test this function on the matrix of Example (3.37):

```
>> A = [-1 -5 4; -1 -5 -2; 1 7 0; 1 7 6];
>> [Q,R] = QRfactor(A)

Q =
   -0.5000    0.5000    0.5000
   -0.5000    0.5000   -0.5000
    0.5000    0.5000   -0.5000
    0.5000    0.5000    0.5000

R =
     2    12     2
     0     2     4
     0     0     6
```

The reduced QR factorization is implemented following the steps of the Gram–Schmidt algorithm presented in Section 3.2. In particular, observe that it generates the matrix $\mathbf{R}$ column by column because each step of the Gram–Schmidt procedure expresses the vector $\mathbf{v}_i$ as a linear combination of the vectors $\{\mathbf{q}_1, \ldots, \mathbf{q}_i\}$ obtained up to that step. Such implementation, however, presents numerical shortcomings, mainly related to the fact that the orthogonality between the vectors $\mathbf{q}_i$ can be lost by the successive floating-point operations executed on them in the inner loop of the preceding function. For an illustration, consider the following pathological example:

```
>> A = [1 1 1; 10^-8 0 0; 0 10^-8 0; 0 0 10^-8];
>> [Q,R] = QRfactor(A);
>> Q'*Q

ans =
    1.0000   -0.0000   -0.0000
   -0.0000    1.0000    0.5000
   -0.0000    0.5000    1.0000
```

Observe that the matrix $\mathbf{Q}$ is not left-orthogonal since $\mathbf{Q}'\mathbf{Q} \neq \mathbf{1}_3$. To overcome this difficulty, we can consider an alternative formulation of the Gram–Schmidt algorithm that, although mathematically equivalent to the classical one, achieves numerical superiority by generating the matrix $\mathbf{R}$ row by row, therefore avoiding the execution of deleterious floating-point operations once the vectors $\mathbf{q}_i$ are computed and stored. The following MATLAB code implements the reduced QR factorization based on the modified Gram–Schmidt procedure:

```
function [Q,R]=modifiedQR(A)
% modified QR factorization for n x k matrix A
% Q is n x k left-orthogonal matrix
% R is k x k upper-triangular matrix with positive diagonal entries
% A = Q*R

[n,k] = size(A);
for i = 1:k
    R(i,i) = norm(A(:,i));
    if  R(i,i) == 0
        error('matrix is not of full rank')
    end
    Q(:,i) = A(:,i)/R(i,i);
    for j = i+1:k
        R(i,j) = Q(:,i)'*A(:,j);
        A(:,j) = A(:,j)-R(i,j)*Q(:,i);
    end
end
```

Testing the MATLAB function `modifiedQR` on the previous pathological example gives a better outcome with the left-orthogonal matrix $\mathbf{Q}$:

```
>> [Q,R] = modifiedQR(A);
>> Q'*Q
```

```
ans =
    1.0000   -0.0000   -0.0000
   -0.0000    1.0000    0.0000
   -0.0000    0.0000    1.0000
```

Exercise 3.8 Compute the *flop* count for the modified QR factorization algorithm for an $n \times k$ matrix and show that the leading order term is $2nk^2$.

Left inverse

Recall that an $n \times n$ matrix is invertible if and only if its columns are linearly independent. For an $n \times k$ matrix with $k \leq n$, having linearly independent columns is equivalent to having a full rank k. In this case, we have the following analogous result:

Theorem 3.4 *Any $n \times k$ matrix $\mathbf{A}$ of rank $k \leq n$ admits a left-inverse $\widetilde{\mathbf{A}}$, that is, a $k \times n$ matrix such that $\widetilde{\mathbf{A}}\mathbf{A} = \mathbf{1}_k$.*

To prove this theorem, simply consider the reduced QR factorization of $\mathbf{A}$ and put

$$\widetilde{\mathbf{A}} := \mathbf{R}^{-1}\mathbf{Q}',$$

which must exist because $\mathbf{R}$ is invertible (being upper triangular with strictly positive diagonal entries). We then verify that

$$\widetilde{\mathbf{A}}\mathbf{A} = \mathbf{R}^{-1}\mathbf{Q}'\mathbf{Q}\mathbf{R} = \mathbf{R}^{-1}\mathbf{R} = \mathbf{1}_k,$$

implying that $\widetilde{\mathbf{A}}$ is a left inverse for $\mathbf{A}$. Moreover, for the reversed-order product $\mathbf{A}\widetilde{\mathbf{A}}$, we have

$$\mathbf{A}\widetilde{\mathbf{A}} = \mathbf{Q}\mathbf{R}\mathbf{R}^{-1}\mathbf{Q}' = \mathbf{Q}\mathbf{Q}',$$

which is not necessarily the identity matrix $\mathbf{1}_n$ since $\mathbf{Q}$ is only left-orthogonal. However, we observe that $\mathbf{Q}\mathbf{Q}'$ is a self-adjoint matrix satisfying $(\mathbf{Q}\mathbf{Q}')^2 = \mathbf{Q}\mathbf{Q}'$, thus being an orthogonal projection. Finally, since $\mathbf{Q}\mathbf{Q}'\mathbf{v} = \mathbf{A}\widetilde{\mathbf{A}}\mathbf{v} \in \text{range}(\mathbf{A})$ for any $\mathbf{v} \in \mathbb{R}^n$, we see that

$$\mathbf{P} = \mathbf{Q}\mathbf{Q}' \tag{3.39}$$

is the orthogonal projection onto range($\mathbf{A}$). The following MATLAB commands confirm the preceding properties:

```
>> A = [-1 -5 4; -1 -5 -2; 1 7 0; 1 7 6];
>> [Q,R] = QRfactor(A);
>> Atilde = inv(R)*Q'
```

```
Atilde =
   -0.8333   -2.6667   -2.1667   -0.3333
    0.0833    0.4167    0.4167    0.0833
    0.0833   -0.0833   -0.0833    0.0833

>> Atilde*A

ans =
    1.0000    0.0000   -0.0000
   -0.0000    1.0000    0.0000
         0    0.0000    1.0000

>> P = Q*Q'

P =
    0.7500    0.2500   -0.2500    0.2500
    0.2500    0.7500    0.2500   -0.2500
   -0.2500    0.2500    0.7500    0.2500
    0.2500   -0.2500    0.2500    0.7500
```

Full QR factorization

For another type of QR factorization, we follow the steps of the LU factorization from Section 2.5, except that the elementary transformations $\mathbf{M}_k$ are now performed by orthogonal matrices. That is, our general strategy is to find successive orthogonal matrices $\mathbf{H}_k$ each with the property of inserting zeros under the diagonal of the kth columns of a given matrix $\mathbf{A}$.

To find such transformations, let $\mathbf{x} = [x_1, \dots, x_n] \in \mathbb{R}^n$ be a given vector and consider another vector $\mathbf{y} \in \mathbb{R}^n$, with the same norm as $\mathbf{x}$, having all but its first component equal to zero. It is easy to see that there are only two possibilities for this vector, namely,

$$\mathbf{y} = \pm\|\mathbf{x}\|\mathbf{e}_1,$$

where $\mathbf{e}_1 = [1, 0, \dots, 0] \in \mathbb{R}^n$. To choose between these two possibilities, we consider numerical issues and require $\mathbf{y}$ to be as far from $\mathbf{x}$ as possible, which then leads to the unique choice:

$$\mathbf{y} = -\text{sign}(x_1)\|\mathbf{x}\|\mathbf{e}_1. \tag{3.40}$$

Having determined the vector $\mathbf{y}$, we now want to find an orthogonal matrix $\mathbf{H}_1$ satisfying

$$\mathbf{H}_1\mathbf{x} = \mathbf{y}. \tag{3.41}$$

The problem (3.41) is a special case of a general problem: given two column-vectors $\mathbf{x}$ and $\mathbf{y}$ in $\mathbb{R}^n$ with $\|\mathbf{x}\| = \|\mathbf{y}\|$, find an orthogonal matrix $\mathbf{H}$ such that $\mathbf{Hx} = \mathbf{y}$. One solution to this problem is given by the matrix

$$\mathbf{H} = \mathbf{1}_n - 2\frac{(\mathbf{x}-\mathbf{y}) * (\mathbf{x}-\mathbf{y})'}{\|\mathbf{x}-\mathbf{y}\|^2}, \tag{3.42}$$

where the outer product of the column-vector $(\mathbf{x}-\mathbf{y})$ by its own transpose produces an $n \times n$ matrix. An explicit calculation shows that

$$\begin{aligned}
\mathbf{H}' * \mathbf{H} &= \left(\mathbf{1}_n - 2\frac{(\mathbf{x}-\mathbf{y}) * (\mathbf{x}-\mathbf{y})'}{\|\mathbf{x}-\mathbf{y}\|^2}\right) * \left(\mathbf{1}_n - 2\frac{(\mathbf{x}-\mathbf{y}) * (\mathbf{x}-\mathbf{y})'}{\|\mathbf{x}-\mathbf{y}\|^2}\right) \\
&= \mathbf{1}_n - 4\frac{(\mathbf{x}-\mathbf{y}) * (\mathbf{x}-\mathbf{y})'}{\|\mathbf{x}-\mathbf{y}\|^2} + 4\frac{(\mathbf{x}-\mathbf{y}) * (\mathbf{x}-\mathbf{y})' * (\mathbf{x}-\mathbf{y}) * (\mathbf{x}-\mathbf{y})'}{\|\mathbf{x}-\mathbf{y}\|^4} \\
&= \mathbf{1}_n
\end{aligned}$$

and

$$\begin{aligned}
\mathbf{Hx} &= \mathbf{x} - 2\frac{(\mathbf{x}-\mathbf{y}) * (\mathbf{x}-\mathbf{y})' * \mathbf{x}}{\|\mathbf{x}-\mathbf{y}\|^2} = \mathbf{x} - \frac{(\mathbf{x}-\mathbf{y}) * (2\mathbf{x}' * \mathbf{x} - 2\mathbf{y}' * \mathbf{x})}{\|\mathbf{x}-\mathbf{y}\|^2} \\
&= \mathbf{x} - (\mathbf{x}-\mathbf{y})\frac{\mathbf{x}' * \mathbf{x} - 2\mathbf{y}' * \mathbf{x} + \mathbf{y}'\mathbf{y}}{\|\mathbf{x}-\mathbf{y}\|^2} = \mathbf{x} - (\mathbf{x}-\mathbf{y}) = \mathbf{y}.
\end{aligned}$$

Returning to the original problem, if we insert the vector $\mathbf{y}$ from (3.40) into (3.42), we obtain

$$\mathbf{H}_1 = \mathbf{1}_n - 2\frac{\mathbf{w}_1 * \mathbf{w}_1'}{\|\mathbf{w}_1\|^2},$$

where

$$\mathbf{w}_1 = \mathbf{x} - \mathbf{y} = \mathbf{x} + \text{sign}(x_1)\|\mathbf{x}\|\mathbf{e}_1. \tag{3.43}$$

The next example illustrates the steps taken up to this point.

```
>> x = [2 1 2]'; e1 = [1 0 0]';
>> w1 = x+sign(x(1))*norm(x)*e1

w1 =
     5
     1
     2

>>H1 = eye(3)-2*(w1*w1')/(w1'*w1)

H1 =
   -0.6667   -0.3333   -0.6667
   -0.3333    0.9333   -0.1333
   -0.6667   -0.1333    0.7333
```

```
>> y = H1*x

y =
  -3.0000
   0.0000
   0.0000
```

Householder transformations

Therefore, if $\mathbf{x}$ is the first column of a matrix $\mathbf{A}$, we can use the orthogonal matrix $\mathbf{H}_1$ to produce zeros below the diagonal entries of this column. To proceed further, let us define the *Householder transformation* associated with a nonzero column vector $\mathbf{w} \in \mathbb{R}^n$ as

$$\mathbf{H}_\mathbf{w} = \mathbf{1} - 2\frac{\mathbf{w} * \mathbf{w}'}{\|\mathbf{w}\|^2}. \tag{3.44}$$

You can easily verify that $\mathbf{H}_\mathbf{w}$ is both self-adjoint *and* orthogonal. Although $\mathbf{H}_\mathbf{w}$ is not itself an orthogonal projection, the next exercise shows how it is related to the orthogonal projection $\mathbf{P}_\mathbf{w}$.

Exercise 3.9 Show that

$$\frac{\mathbf{H}_\mathbf{w}\mathbf{x} - \mathbf{x}}{2} = -\mathbf{P}_\mathbf{w}\mathbf{x},$$

where $\mathbf{P}_\mathbf{w}$ is the orthogonal projection of $\mathbf{x} \in \mathbb{R}^n$ in the direction of $\mathbf{w}$.

Suppose then that we have already obtained zero entries below the diagonal for the first $(k-1)$ columns of a given matrix $\mathbf{A}$. If $\mathbf{x}$ is its kth column, we need to find an orthogonal matrix that annihilates the last $(n-k)$ components of $\mathbf{x}$. If we denote

$$\widetilde{\mathbf{x}} = \begin{bmatrix} 0 \\ \vdots \\ 0 \\ x_k \\ \vdots \\ x_n \end{bmatrix} \tag{3.45}$$

we can verify that this is achieved by the Householder transformation $\mathbf{H}_k$ associated with the vector

$$\mathbf{w}_k = \widetilde{\mathbf{x}} + \text{sign}(x_k)\|\widetilde{\mathbf{x}}\|\mathbf{e}_k. \tag{3.46}$$

Because the first $(k-1)$ components of the vector $\mathbf{w}_k$ are zero, it is easy to see that $\mathbf{H}_k$ leaves the first $(k-1)$ components of any vector unchanged. As a result of this fact, applying $\mathbf{H}_k$ to the (already upper-triangular) first $(k-1)$

columns of the matrix $\mathbf{A}$ does not change them at all. Continuing the previous example, let us construct the transformation $\mathbf{H}_2$:

```
>> x = [2 1 2]'; e2 = [0 1 0]'; xtilde = [0 1 2]';
>> w2 = xtilde+sign(x(2))*norm(xtilde)*e2

w2 =
         0
    3.2361
    2.0000

>> H2 = eye(3)-2*(w2*w2')/(w2'*w2)

H2 =
    1.0000         0         0
         0   -0.4472   -0.8944
         0   -0.8944    0.4472

>> y = H2*x

y =
    2.0000
   -2.2361
    0.0000
```

Applying successive Householder transformations to the columns of a full-rank $n \times k$ matrix $\mathbf{A}$ we obtain

$$\mathbf{H}_{k-1} \cdots \mathbf{H}_1 \mathbf{A} = \begin{bmatrix} \widetilde{\mathbf{R}} \\ \mathbf{O} \end{bmatrix},$$

where $\widetilde{\mathbf{R}}$ is a $k \times k$ upper-triangular matrix and $\mathbf{O}$ denotes an $(n-k) \times k$ matrix with zero entries. But since the product of orthogonal matrices is another orthogonal matrix, we can introduce the $n \times n$ orthogonal matrix $\mathbf{Q}' = \mathbf{H}_{k-1} \cdots \mathbf{H}_1$ and obtain

$$\mathbf{A} = \mathbf{Q} \begin{bmatrix} \widetilde{\mathbf{R}} \\ \mathbf{O} \end{bmatrix}, \tag{3.47}$$

which is called the *full QR factorization* (or the *Householder QR factorization*) of the matrix **A**. This factorization can be implemented in MATLAB as follows:

```
function [Q,R]=householderQR(A)
% Householder method for QR factorization
% A is n x k matrix , k <= n full rank
% Q is n x n orthogonal
% R is n x k upper triangular
% A = Q * R

[n,k] = size(A); R = A; Q = eye(n);
for p = 1:k
    x = R(p:n,p);
    if x(1) ~= 0
        x(1) = x(1)+sign(x(1))*norm(x);
        x = x./norm(x);
    else
        x(1) = x(1)+norm(x);
        x = x./norm(x);
    end
    R(p:n,p:k) = R(p:n,p:k)-2*x*(x'*R(p:n,p:k));
    Q(p:n,1:n) = Q(p:n,1:n)-2*x*(x'*Q(p:n,1:n));
end
Q = Q';
```

The following MATLAB commands compute the full QR factorization for the previous example:

```
>> A = [-1 -5 4; -1 -5 -2; 1 7 0; 1 7 6]
>> [Q,R] = householderQR(A)

Q =
   -0.5000   -0.5000    0.5000    0.5000
   -0.5000   -0.5000   -0.5000   -0.5000
    0.5000   -0.5000   -0.5000    0.5000
    0.5000   -0.5000    0.5000   -0.5000

R =
    2.0000   12.0000    2.0000
   -0.0000   -2.0000   -4.0000
    0.0000   -0.0000    6.0000
    0.0000   -0.0000    0.0000
```

Observe the differences between the reduced and full QR factorizations applied to the same 4×3 matrix. For the reduced factorization, **Q** is a 4×3

left-orthogonal matrix and $\mathbf{R}$ is a 3×3 upper-triangular matrix with strictly positive diagonal entries. From the full factorization, you obtain a 4×4 orthogonal matrix $\mathbf{Q}$ and a 4×3 matrix $\mathbf{R}$ with zeros below the diagonal and possibly negative diagonal entries.

The full QR factorization is implemented by using the MATLAB function `qr`. In its simplest form, the function `qr` takes the matrix $\mathbf{A}$ as the only input and returns the unitary matrix $\mathbf{Q}$ and the upper-triangular matrix $\mathbf{R}$ such that $\mathbf{A} = \mathbf{QR}$. With three output arguments, the same function also returns a permutation matrix $\mathbf{E}$ so that $\mathbf{AE} = \mathbf{QR}$. The columns of the permutation matrix $\mathbf{E}$ are chosen so that the absolute values of the diagonal entries of the matrix $\mathbf{R}$ are decreasing. You can check that the MATLAB code

```
>> [Q,R] = qr(A);
```

with the matrix $\mathbf{A}$ of the previous example returns the identical output for the matrices $\mathbf{Q}$ and $\mathbf{R}$.

3.4 The Least-Squares Method

We have shown in Sections 2.5 and 2.6 how to numerically solve linear systems with square matrices. We now turn to overdetermined linear systems, where the number of equations exceeds the number of variables.

Let $\mathbf{A}$ be an $n \times k$ matrix of full-rank $k < n$ and $\mathbf{b}$ be a vector in $\mathbb{R}^n$. Consider the overdetermined linear system $\mathbf{Ax} = \mathbf{b}$ for the vector $\mathbf{x}$ in $\mathbb{R}^k$. Recalling the discussion in Section 2.3, we see that this system has a solution if and only if $\mathbf{b}$ can be written as a linear combination of the columns of $\mathbf{A}$, or in other words, if and only if $\mathbf{b} \in \text{range}(\mathbf{A})$.

In applications, however, it is very rare that $\mathbf{b}$ lies in the range of $\mathbf{A}$, and the linear system $\mathbf{Ax} = \mathbf{b}$ typically has no solutions at all. In this case, we can still try to find a vector $\widetilde{\mathbf{b}}$ in range($\mathbf{A}$) that best approximates $\mathbf{b}$. We refer to the solution of $\mathbf{Ax} = \widetilde{\mathbf{b}}$ as the best approximation to the solution of the overdetermined linear system and denote it by $\mathbf{Ax} \simeq \mathbf{b}$. If the Euclidean norm is used to measure the distance between the two vectors $\mathbf{b}$ and $\widetilde{\mathbf{b}}$, then the problem can be formulated as follows:

Problem 3.2 (Least-Squares) Given a full-rank $n \times k$ matrix $\mathbf{A}$ and a vector $\mathbf{b} \in \mathbb{R}^n$, find the vector $\mathbf{x} \in \mathbb{R}^k$ that minimizes the Euclidean norm $\|\mathbf{Ax} - \mathbf{b}\|$.

Normal equations

In other words, find the vector $\widetilde{\mathbf{b}} = \mathbf{Ax} \in \text{range}(\mathbf{A})$ such that the norm of the residual vector $\mathbf{r} = \mathbf{Ax} - \mathbf{b}$ is as small as possible. It follows from Theorem 3.1 that this vector must be given by $\widetilde{\mathbf{b}} = \mathbf{Pb}$, where $\mathbf{P}$ is the *unique* orthogonal projection onto range($\mathbf{A}$). Therefore, the solution $\mathbf{x}$ to Problem 3.2 must satisfy

$$\mathbf{A}\mathbf{x} = \mathbf{P}\mathbf{b}. \tag{3.48}$$

Multiplying both sides of this equation by $\mathbf{A}'$, we obtain

$$\mathbf{A}'\mathbf{A}\mathbf{x} = \mathbf{A}'\mathbf{P}\mathbf{b} = \mathbf{A}'\mathbf{P}'\mathbf{b} = (\mathbf{P}\mathbf{A})'\mathbf{b} = \mathbf{A}'\mathbf{b} \tag{3.49}$$

Conversely, if $\mathbf{x} \in \mathbb{R}^k$ satisfies (3.49), then we obtain that, for any other vector $\mathbf{v} \in \mathbb{R}^k$,

$$\langle \mathbf{A}\mathbf{x} - \mathbf{b}, \mathbf{A}\mathbf{v} \rangle = \langle \mathbf{A}'\mathbf{A}\mathbf{x} - \mathbf{A}'\mathbf{b}, \mathbf{v} \rangle = \langle \mathbf{0}, \mathbf{v} \rangle = 0,$$

which means that $(\mathbf{A}\mathbf{x} - \mathbf{b}) \in \text{range}(\mathbf{A})^{\perp}$, implying that $\mathbf{x}$ minimizes the norm $\|\mathbf{A}\mathbf{x} - \mathbf{b}\|$ and is therefore a solution to Problem 3.2. In other words, $\mathbf{x} \in \mathbb{R}^k$ is a solution of Problem 3.2 if and only if it satisfies the *normal equations*

$$\mathbf{A}'\mathbf{A}\mathbf{x} = \mathbf{A}'\mathbf{b}. \tag{3.50}$$

In view of this result, we can reduce the least-squares problem $\mathbf{A}\mathbf{x} \simeq \mathbf{b}$ to the square linear system $\mathbf{A}'\mathbf{A}\mathbf{x} = \mathbf{A}'\mathbf{b}$, where $\mathbf{A}'\mathbf{A}$ is a $k \times k$ matrix.

Exercise 3.10 Use inner product properties to show that the $k \times k$ matrix $\mathbf{A}'\mathbf{A}$ is nonsingular if and only if the $n \times k$ matrix $\mathbf{A}$ has full rank $k \leq n$.

Therefore, provided that the $n \times k$ matrix $\mathbf{A}$ is of full-rank $k \leq n$, the least-squares problem $\mathbf{A}\mathbf{x} \simeq \mathbf{b}$ has a unique solution $\mathbf{x}$ obtained by solving the normal equations (3.50). Observe that if $k = n = \text{rank}(\mathbf{A})$, then $\mathbf{A}$ is invertible, in which case the unique solution $\mathbf{x} = \mathbf{A}^{-1}\mathbf{b}$ shows that $\widetilde{\mathbf{b}} = \mathbf{b}$. On the other hand, when $\text{rank}(\mathbf{A}) < k \leq n$, the matrix $\mathbf{A}$ is said to be *rank deficient*, in which case the $k \times k$ matrix $\mathbf{A}'\mathbf{A}$ is singular. In this case, the least-squares problem $\mathbf{A}\mathbf{x} \simeq \mathbf{b}$ does not have a unique solution, and is best handled by using the singular value decomposition introduced in Section 4.8.

Let us see how the least-squares method works in a simple example. Take

$$\mathbf{A} = \begin{bmatrix} 1 & 2 \\ 3 & 4 \\ 5 & 6 \end{bmatrix}, \qquad \mathbf{b} = \begin{bmatrix} 1 \\ -1 \\ 0 \end{bmatrix}.$$

Then

$$\mathbf{A}'\mathbf{A} = \begin{bmatrix} 35 & 44 \\ 44 & 56 \end{bmatrix}, \qquad \mathbf{A}'\mathbf{b} = \begin{bmatrix} -2 \\ -2 \end{bmatrix},$$

so that the normal equations correspond to the linear system

$$\begin{bmatrix} 35 & 44 \\ 44 & 56 \end{bmatrix} \begin{bmatrix} x_1 \\ x_2 \end{bmatrix} = \begin{bmatrix} -2 \\ -2 \end{bmatrix},$$

whose solution is easily found to be $\mathbf{x} = [-1, 0.75]$. You can then compute the residual vector

$$\mathbf{r} = \mathbf{A}\mathbf{x} - \mathbf{b} = \begin{bmatrix} -0.5 \\ 1 \\ -0.5 \end{bmatrix}$$

and verify that it is indeed orthogonal to each column of $\mathbf{A}$ (and therefore to range($\mathbf{A}$), which is spanned by the columns of $\mathbf{A}$).

Recall that we used the MATLAB backslash operator \ in Section 2.3 to solve a linear system of the form $\mathbf{A}\mathbf{x} = \mathbf{b}$ in the case where $\mathbf{A}$ was an $n \times n$ matrix. As it turns out, when $\mathbf{A}$ is an $n \times k$ matrix, the same command implements the solution of the least-squares problem $\mathbf{A}\mathbf{x} \simeq \mathbf{b}$.

```
>> A = [1 2; 3 4; 5 6]; b = [1 -1 0]';
>> x = A\b

x =
   -1.0000
    0.7500
```

QR and least-squares

Now observe that, although we used the orthogonal projection $\mathbf{P}$ in the derivation of the normal equations, we do not really need to find $\mathbf{P}$ explicitly in order to write them down. But recall from Equation (3.39) that such a projection is given by

$$\mathbf{P} = \mathbf{Q}\mathbf{Q}',$$

where $\mathbf{Q}$ is the left-orthogonal matrix obtained in the reduced QR factorization of $\mathbf{A}$. Recall also that the matrix $\widetilde{\mathbf{A}} = \mathbf{R}^{-1}\mathbf{Q}'$ is the left inverse of $\mathbf{A}$, so that we can multiply both sides of (3.48) by $\widetilde{\mathbf{A}}$ to obtain

$$\mathbf{x} = \widetilde{\mathbf{A}}\mathbf{P}\mathbf{b} = \mathbf{R}^{-1}\mathbf{Q}'\mathbf{Q}\mathbf{Q}'\mathbf{b} = \mathbf{R}^{-1}\mathbf{Q}'\mathbf{b}.$$

That is, if you know the QR factorization of $\mathbf{A}$, then you don't need to solve the normal equations to find the solution to the least-squares problem $\mathbf{A}\mathbf{x} \simeq \mathbf{b}$

lar system $\mathbf{Rx} = \mathbf{Q}'\mathbf{b}$

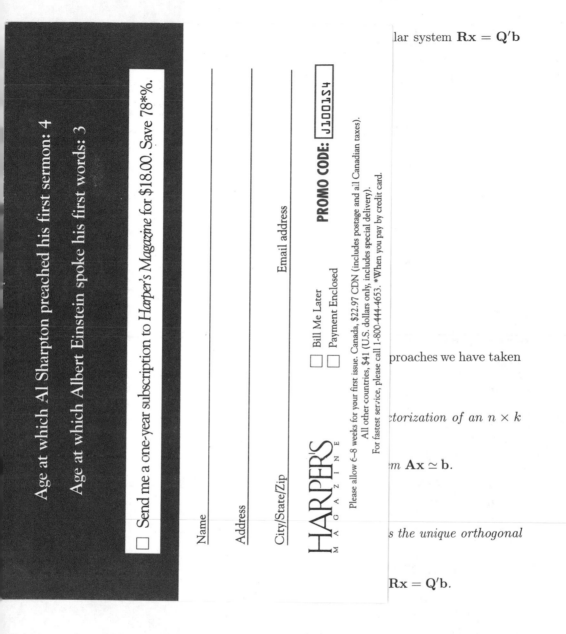

proaches we have taken

ctorization of an $n \times k$

m $\mathbf{Ax} \simeq \mathbf{b}$.

s the unique orthogonal

$\mathbf{Rx} = \mathbf{Q}'\mathbf{b}$.

If instead of the reduced QR factorization we use the full QR factorization (3.47), then we can obtain some more detailed information in the solution of the least-squares problem. Namely, let us write

$$\mathbf{Q}'\mathbf{b} = \begin{bmatrix} \mathbf{c}_1 \\ \mathbf{c}_2 \end{bmatrix},$$

for vectors $\mathbf{c}_1 \in \mathbb{R}^k$ and $\mathbf{c}_2 \in \mathbb{R}^{n-k}$ and observe that

$$\begin{aligned} \|\mathbf{r}\|^2 &= \|\mathbf{A}\mathbf{x} - \mathbf{b}\|^2 = \left\| \mathbf{Q} \begin{bmatrix} \mathbf{R} \\ \mathbf{O} \end{bmatrix} \mathbf{x} - \mathbf{b} \right\|^2 \\ &= \left\| \mathbf{Q}'\mathbf{Q} \begin{bmatrix} \mathbf{R} \\ \mathbf{O} \end{bmatrix} \mathbf{x} - \mathbf{Q}'\mathbf{b} \right\|^2 = \left\| \begin{bmatrix} \mathbf{R} \\ \mathbf{O} \end{bmatrix} \mathbf{x} - \begin{bmatrix} \mathbf{c}_1 \\ \mathbf{c}_2 \end{bmatrix} \right\|^2 \\ &= \|\mathbf{R}\mathbf{x} - \mathbf{c}_1\|^2 + \|\mathbf{c}_2\|^2. \end{aligned}$$

Therefore, the minimal residual norm is obtained for $\mathbf{R}\mathbf{x} = \mathbf{c}_1$ and satisfies $\|\mathbf{r}\| = \|\mathbf{c}_2\|$. These formulas are illustrated with the MATLAB computations of the previous example:

```
>> [Q,R] = householderQR(A)

Q =
   -0.1690    0.8971    0.4082
   -0.5071    0.2760   -0.8165
   -0.8452   -0.3450    0.4082

R =
   -5.9161   -7.4374
         0    0.8281
         0    0.0000

>> c = Q'*b

c =
    0.3381
    0.6211
    1.2247

>> x = R(1:2,1:2)\c(1:2)

x =
   -1.0000
    0.7500

>> norm(A*x-b)

ans =
    1.2247
```

Data fitting

The most common use of the least-squares method is the *linear regression* technique for data fitting, which can be expressed as the following problem.

Problem 3.3 (Linear Regression) Given a set of data points $(t_1, y_1), \ldots,$ (t_n, y_n) and a set of functions $f_1(t), \ldots, f_k(t)$, $k \leq n$, find the parameters $x_1, \cdots, x_k$ for which the functional relation

$$f(t) = x_1 f_1(t) + \cdots x_k f_k(t) \tag{3.51}$$

describes the data with minimal square error

$$E = \sum_{i=1}^{n} (y_i - f(t_i))^2. \tag{3.52}$$

Therefore, this technique *regresses* the potentially complicated and unknown relation between the given data points to the simpler functional relation provided by $f(t)$. Moreover, it is called a *linear* regression because the function $f(t)$ is assumed to be a linear combination of the predetermined functions $f_1(t), \ldots, f_k(t)$. Observe, however, that this does not mean that the data points are linearly related since each of the functions $f_i(t)$ can be highly nonlinear in the variable t. For instance, we can have

$$f_1(t) = t^2, \quad f_2(t) = \exp(t/20), \quad f_3(t) = \sin(2t).$$

All that is required is that the overall functional relation $f(t)$ be linear with respect to the *parameters*, that is, for the preceding functions we would have

$$f(t) = x_1 t^2 + x_2 \exp(t/20) + x_3 \sin(2t). \tag{3.53}$$

If we now insert (3.51) into the error function (3.52), we obtain (taking proper care of the different indices)

$$E = \sum_{i=1}^{n} \left[y_i - \sum_{j=1}^{k} x_j f_j(t_i) \right]^2 = \|\mathbf{y} - \mathbf{A}\mathbf{x}\|^2,$$

where $\mathbf{y} = [y_1, \ldots, y_n] \in \mathbb{R}^n$ and $\mathbf{A}$ is the $n \times k$ matrix with entries

$$a_{ij} = f_j(t_i).$$

Therefore, we see that the parameters $x_1, \ldots, x_k$ solving Problem 3.3 consist of the vector $\mathbf{x} \in \mathbb{R}^k$ solving Problem 3.2 for these specific $\mathbf{b} = \mathbf{y}$ and $\mathbf{A}$.

For illustration, suppose that you are given the following set of data points in the form (t, y):

$$\{(16, 12), (18, 9), (20, 8), (22, 7), (24, 8), (26, 4), (28, 2)\}. \tag{3.54}$$

Let us try to relate these points with three functions: (1) a straight line, (2) a parabola, and (3) a linear combination of elementary functions given in (3.53). For the straight line, consider a linear function

$$f(t) = x_1 + x_2 t.$$

Then, putting $f_1(t) = 1$ and $f_2(t) = t$ you obtain the following overdetermined system:

$$\begin{bmatrix} 1 & 16 \\ 1 & 18 \\ 1 & 20 \\ 1 & 22 \\ 1 & 24 \\ 1 & 26 \\ 1 & 28 \end{bmatrix} \begin{bmatrix} x_1 \\ x_2 \end{bmatrix} = \begin{bmatrix} 12 \\ 9 \\ 8 \\ 7 \\ 8 \\ 4 \\ 2 \end{bmatrix},$$

whose least-squares solution can be obtained by

```
>> A = [1 16;1 18; 1 20; 1 22; 1 24; 1 26; 1 28];
>> y = [12 9 8 7 8 4 2]';
>> x = A\y

x =
   23.7857
   -0.7500

>> E1 = norm(A*x-y)

E1 =
    2.5355
```

That is, the best straight line describing the data points (3.54) is

$$f(t) = 23.7857 - 0.75t.$$

For the parabola, consider a quadratic function

$$f(t) = x_1 + x_2 t + x_3 t^2.$$

Then, putting $f_1(t) = 1$, $f_2(t) = t$, and $f_3(t) = t^2$ leads to the overdetermined system

$$\begin{bmatrix} 1 & 16 & 256 \\ 1 & 18 & 324 \\ 1 & 20 & 400 \\ 1 & 22 & 484 \\ 1 & 24 & 576 \\ 1 & 26 & 676 \\ 1 & 28 & 784 \end{bmatrix} \begin{bmatrix} x_1 \\ x_2 \\ x_3 \end{bmatrix} = \begin{bmatrix} 12 \\ 9 \\ 8 \\ 7 \\ 8 \\ 4 \\ 2 \end{bmatrix}.$$

whose least-squares solution is given by

```
>> A = [1 16 16^2;1 18 18^2; 1 20 20^2; 1 22 22^2;
        1 24 24^2; 1 26 26^2; 1 28 28^2];
>> y = [12 9 8 7 8 4 2]'; x = A\y

x =
   15.4286
    0.0357
   -0.0179

>> E2 = norm(A*x-y)

E2 =
    2.4495
```

That is, the best parabola describing the data points (3.54) is

$$f(t) = 15.4286 + 0.0357t - 0.0179t^2.$$

For the linear combination of the elementary functions (3.53), you need to calculate in MATLAB the matrix **A** associated with these functions and the data points (3.54), as well as the corresponding least-squares solution to the linear regression problem as follows:

```
>> t = [16 18 20 22 24 26 28]';
>> y = [12 9 8 7 8 4 2]';
>> A = [t.^2 exp(t/20) sin(2*t)]

A =
  256.0000    2.2255    0.5514
  324.0000    2.4596   -0.9918
  400.0000    2.7183    0.7451
  484.0000    3.0042    0.0177
```

```
  576.0000    3.3201   -0.7683
  676.0000    3.6693    0.9866
  784.0000    4.0552   -0.5216

>> x = A\y

x =
   -0.0558
   11.4275
   -0.1782

>> E3 = norm(A*x-y)

E3 =
    2.7488
```

Therefore, the best function of the form (3.53) describing the data points (3.54) is

$$f(t) = -0.0558t^2 + 11.4275\exp(t/20) - 0.1782\sin(2t).$$

Comparing the final errors E1, E2, and E3 obtained in the three different regressions, you can see that they all have comparable (and rather bad) performances. Plotting all regressions and the original data points by using the MATLAB commands:

```
>> t = [16 18 20 22 24 26 28]';
>> y = [12 9 8 7 8 4 2]';
>> x = [16:0.01:28];
>> f1 = 23.7857-0.75*x;
>> f2 = 15.4286+0.0357*x-0.0179*x.^2;
>> f3 = -0.0558*x.^2+11.4275*exp(x/20)-0.1782*sin(2*x);
>> plot(t,y,'d',x,f1,x,f2,'--r',x,f3,'-.k')
>> legend('data','line','parabola','non-polynomial')
```

results in Figure 3.4, which shows how far the three regressions are from the given set of data points.

The choice of the elementary functions $f_j(t)$ depends on the underlying model that is assumed to have generated the data points in the first place. After the model is selected, the least-squares method provides a well-defined way of finding the best linear combination of the elementary functions. The appropriateness of the model, however, including the assumption that the final function $f(t)$ depends linearly on the parameters $x_1, \ldots, x_k$ is outside the scope of *Linear Algebra*, it is the subject of *Mathematical Statistics* instead.

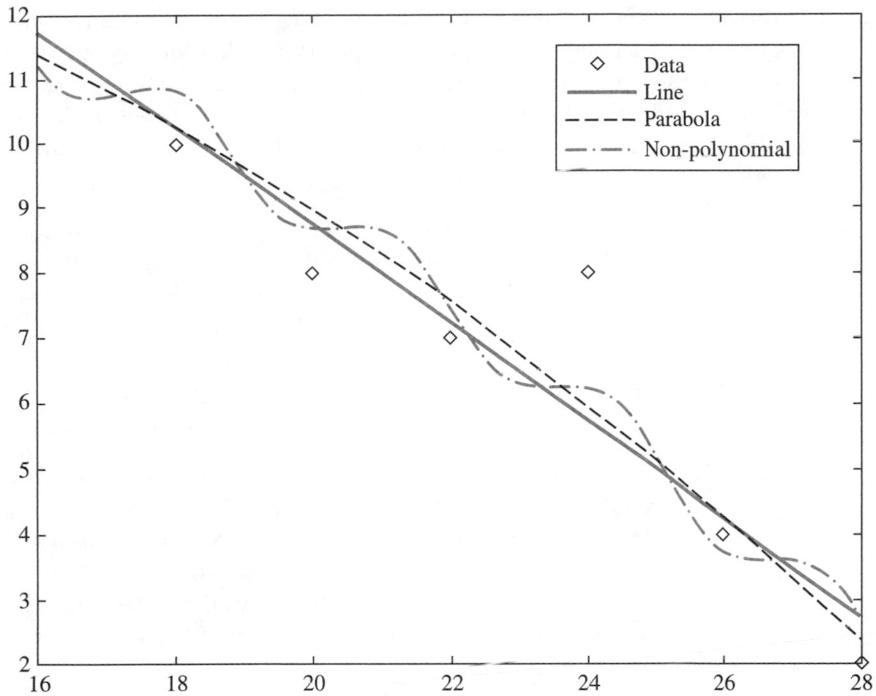

Figure 3.4 Three different linear regressions for the same data points.

3.5 Summary and Notes

This chapter discusses concepts related to the notion of orthogonality for general vector spaces and their main consequences and applications:

- **Section 3.1:** *Inner products* are generalizations of the dot product for abstract vector spaces. From an inner product, we can define a *Euclidean norm* satisfying the *Cauchy–Schwarz inequality.* Two vectors are *orthogonal* if their inner product is zero. The *adjoint* of a linear map $\mathbf{A}$ is a linear map $\mathbf{A}^*$ satisfying $\langle \mathbf{A}^*\mathbf{w}, \mathbf{v}\rangle = \langle \mathbf{w}, \mathbf{A}\mathbf{v}\rangle$. Maps satisfying $\mathbf{A}^* = \mathbf{A}$ are said to be *self-adjoint.* Maps that preserve an inner product are called *isometries.* They correspond to *orthogonal matrices* in $\mathbb{R}^n$ and *unitary matrices* in $\mathbb{C}^n$.

- **Section 3.2:** A *projection* is a linear map $\mathbf{P}$ satisfying $\mathbf{P}^2 = \mathbf{P}$. If in addition $\mathbf{P}^* = \mathbf{P}$, then the projection is said to be *orthogonal.* An orthogonal projection $\mathbf{P}$ gives rise to a unique decomposition of V into the direct sum of the range of $\mathbf{P}$ and its orthogonal complement. When V is finite-dimensional, the same decomposition holds for any subspace U and its orthogonal complement $U^\perp$. Given a linearly independent set of vectors, Problem 3.1 consists of finding an orthonormal set of vectors spanning the same subspace. This is solved with the Gram–Schmidt procedure.

- **Section 3.3:** An $n \times k$ real matrix $\mathbf{A}$ with rank $k \leq n$ can be uniquely written as $\mathbf{A} = \mathbf{QR}$, where $\mathbf{Q}$ is an $n \times k$ left-orthogonal matrix and $\mathbf{R}$ is a $k \times k$ upper-triangular matrix with strictly positive diagonal entries. This is called the *reduced QR factorization* of $\mathbf{A}$ to distinguish it from the *full QR factorization*, where $\mathbf{Q}$ is an $n \times n$ orthogonal matrix and $\mathbf{R}$ is an $n \times k$ matrix with $r_{ij} = 0$ whenever $i > j$. This last type of factorization is obtained through the use of *Householder transformations.*
- **Section 3.4:** Given an overdetermined linear system $\mathbf{Ax} = \mathbf{b}$ for an $n \times k$ matrix $\mathbf{A}$ and a vector $\mathbf{b} \in \mathbb{R}^n$, Problem 3.2 consists of finding a vector $\mathbf{x} \in \mathbb{R}^k$ such that the residual norm $\|\mathbf{Ax} - \mathbf{b}\|$ is as small as possible. Theorem 3.5 characterizes the solution to this problem in terms of the *normal equations* $\mathbf{A}'\mathbf{Ax} = \mathbf{A}'\mathbf{b}$, as well as using the QR factorization of the matrix $\mathbf{A}$. The main application of the least-squares method consists of Problem 3.3, where we try to find the best fit of a set of data points to a specified model.

Our suggested reference for inner product spaces and orthogonal projections is [13], in particular for the motivation we gave for the Gram–Schmidt procedure as a necessary step to establish the orthogonal decomposition. The QR factorization and the least-squares problem are treated in [12].

3.6 Exercises

1. Consider the inner product $\langle \mathbf{x}, \mathbf{y} \rangle := \mathbf{Ax} \cdot \mathbf{Ay}$ on $\mathbb{R}^3$, where

$$\mathbf{A} = \begin{bmatrix} 1 & 2 & 2 \\ 4 & 4 & 2 \\ 2 & 6 & 4 \end{bmatrix}.$$

Replace the dot product in expression (3.3) with this inner product (as well as the Euclidean norm with the corresponding norm induced by the new inner product) and use the new formula to calculate the angle between the canonical basis vectors $\{\mathbf{e}_1, \mathbf{e}_2, \mathbf{e}_3\}$.

2. Use the parallelogram law (3.12) to show that there is no inner product $\langle \cdot, \cdot \rangle$ in $\mathbb{R}^5$ such that the p-norm (2.24) for $p = 3$ can be written as $\|x\|_3 = \sqrt{\langle x, x \rangle}$.

3. Show that the functions $\left\{\frac{1}{\sqrt{2\pi}}, \frac{\sin x}{\sqrt{\pi}}, \frac{\cos x}{\sqrt{\pi}}\right\}$ are orthonormal vectors in the space $\mathcal{C}([-\pi,\pi],\mathbb{R})$ of continuous, real-valued functions, on the interval $[-\pi,\pi]$ with respect to the inner product

$$\langle f, g\rangle = \int_{-\pi}^{\pi} f(x)g(x)dx.$$

Extend the preceding set by adding two more orthonormal functions with respect to the same inner product.

4. Let $\mathbf{A} : V \mapsto W$ be a linear map between finite-dimensional vector spaces V and W. Use (3.16) and the dimension formula (2.17) to show that

$$\dim \text{null}(\mathbf{A}^*) = \dim \text{null}(\mathbf{A}) + \dim W - \dim V$$

and

$$\dim \text{range}(\mathbf{A}^*) = \dim \text{range}(\mathbf{A}).$$

Verify these formulas for the linear map $\mathbf{A} : \mathbb{R}^4 \to \mathbb{R}^5$, where

$$\mathbf{A} = \begin{bmatrix} -2 & 3 & 4 & 1 \\ 6 & 3 & -2 & -3 \\ 4 & -6 & -8 & -2 \\ 0 & 3 & 7 & 0 \end{bmatrix}.$$

5. Consider the linear map $\mathbf{A} : \mathbb{R}^3 \to \mathbb{R}^3$ given by

$$\mathbf{A}(x_1, x_2, x_3) = (x_1 - x_2, -x_1 + x_2, x_3).$$

(a) Find the adjoint map $\mathbf{A}^*$.

(b) Obtain the matrix representations of $\mathbf{A}$ and $\mathbf{A}^*$ with respect to the basis

$$\mathbf{f}_1 = [1,2,1], \quad \mathbf{f}_2 = [1,3,2], \quad \mathbf{f}_3 = [0,1,2].$$

(c) Verify that the matrix corresponding to $\mathbf{A}^*$ is *not* the transpose of the matrix corresponding to $\mathbf{A}$ with respect to this preceding basis and explain why this is not a contradiction.

6. Consider the inner product

$$\langle p(x), q(x)\rangle = \int_{-1}^{1} p(x)q(x)dx, \tag{3.55}$$

in the five-dimensional vector space $\mathcal{P}_4([-1,1])$ of polynomials of degree 4 on the interval $[-1,1]$. Define the linear map $\mathbf{A} : \mathcal{P}_4([-1,1]) \to \mathcal{P}_4([-1,1])$ by

$$\mathbf{A}(c_1x^4 + c_2x^3 + c_3x^2 + c_4x + c_5) = c_3x^2. \tag{3.56}$$

(a) Using definition (3.15) directly, verify that $\mathbf{A}$ is not self-adjoint.

(b) Verify that the matrix representation of $\mathbf{A}$ with respect to the canonical basis $\{1, x, x^2, x^3, x^4\}$ is symmetric and explain why this is not a contradiction.

7. Apply the Gram–Schmidt orthogonalization procedure to the canonical basis $\{1, x, x^2, x^3, x^4\}$ to find an orthonormal basis for the space $\mathcal{P}_4([-1,1])$ with respect to the inner product (3.55). Use this basis to find the adjoint of the linear map $\mathbf{A}$ defined in (3.56).

8. For each of the following matrices, calculate the reduced QR factorization, the modified QR factorization, and the full QR factorization by hand and verify your answer using the functions `QRfactor`, `modifiedQR`, and `householderQR`:

(a)

$$\mathbf{A} = \begin{bmatrix} 1 & 2 & 1 \\ 1 & 1 & 1 \\ 0 & 3 & 1 \\ 1 & 0 & -1 \end{bmatrix},$$

(b)

$$\mathbf{B} = \begin{bmatrix} 2 & 4 & 2 \\ 6 & 0 & 8 \\ 3 & 2 & -8 \\ 4 & -2 & 1 \end{bmatrix}.$$

9. Find the left inverses and the orthogonal projections onto the range of the matrices in the previous exercise.

10. Try to apply the Gram–Schmidt orthogonalization procedure to the columns of the matrix:

$$\mathbf{C} = \begin{bmatrix} 2 & 4 & 2 \\ 6 & 12 & 8 \\ 3 & 6 & -8 \\ 1 & 2 & -1 \end{bmatrix}.$$

Find the QR factorization for the matrix $\mathbf{C}$ using the function `QRfactor`, verify that the matrix $\mathbf{Q}$ obtained is *not* left orthogonal, and explain why this is not a contradiction. Then, find the full QR factorization for $\mathbf{C}$, either by using the function `householderQR` or the MATLAB built-in function `qr`, and verify that the matrix $\mathbf{Q}$ obtained is orthogonal.

11. A matrix is said to be in *upper Hessenberg form* if it has zeros below each *subdiagonal* entry. Given the matrix

$$\mathbf{A} = \begin{bmatrix} 4 & 2 & 4 \\ 4 & 0 & 5 \\ 5 & 1 & -6 \end{bmatrix},$$

find a Householder matrix $\mathbf{H_w}$ such that $\mathbf{H'_w A H_w}$ is in upper Hessenberg form.

12. Consider the points $(1, 3)$, $(2, 3)$, $(3, 4)$, and $(4, 7)$, which lie near a parabola $f(t) = x_1 + x_2 t + x_3 t^2$.

(a) Set up the overdetermined linear system $\mathbf{Ax} = \mathbf{b}$ for the best fit for these points to the parabola by least-squares.

(b) Verify that the matrix $\mathbf{A}$ has full rank and find its reduced QR factorization $\mathbf{A} = \mathbf{QR}$.

(c) Verify that the matrix $\mathbf{A'A}$ is invertible and solve the normal equations for this problem.

(d) Verify that the residual vector $\mathbf{r} = \mathbf{b} - \mathbf{Ax}$ is orthogonal to the range of $\mathbf{A}$.

(e) Verify that $\mathbf{x} = \mathbf{R}^{-1}\mathbf{Q'b}$ coincides with the solution to the normal equations.

(f) Plot the four points together with the parabola corresponding to the solution to the least-squares problem.

CHAPTER **4**

Eigenvalues and Eigenvectors

THIS CHAPTER COMPLETES the part of our book dedicated to *Linear Algebra*. We study here the eigenvalues and eigenvectors of matrices, both theoretically and numerically. The basic definitions are provided in Section 4.1, together with Theorem 4.1, which establishes the existence of eigenvalues for any matrix with complex entries. This theorem is based on a reformulation of the definition of eigenvalues in terms of the characteristic polynomial of a matrix, which makes use of the determinant function. In this formulation, the existence of eigenvalues is a simple consequence of the Fundamental Theorem of Algebra. Besides not being numerically stable, the approach to eigenvalues using the characteristic polynomial provides little information about the eigenvectors of a matrix. For this reason, numerical algorithms for finding eigenvalues and eigenvectors are developed along entirely different routes.

Based on this observation, we present the essential theoretical properties of eigenvalues in Section 4.2 independently of determinants and characteristic polynomials. In particular, we outline the proof for the Schur decomposition presented in Theorem 4.2, which is repeatedly used in this chapter, both in the proof of other theoretical properties and in justifying the steps of numerical algorithms. Other properties of eigenvalues, such as their algebraic multiplicity and the location of eigenvalues in the complex plane, are discussed with the aid of MATLAB examples.

Section 4.3 is dedicated to the properties of eigenvectors, in particular linear independence and the related concept of geometric multiplicity. We then proceed to show how linearly independent eigenvectors can be used to put a matrix in diagonal form, and how this result needs to be modified in the presence of linearly dependent eigenvectors.

The strongest results available for eigenvalues and eigenvectors concern the special class of normal matrices, studied in Section 4.4. These are precisely the type of matrices whose eigenvectors form an orthonormal basis, as established in Theorem 4.6. Particular examples of normal matrices are self-adjoint and unitary matrices, whose eigenvalues have very special and useful properties.

The next three sections are devoted to numerical methods for finding eigenvectors and eigenvalues. Section 4.5 characterizes the overall sensitivity of the eigenvalue problem in terms of the condition number of the matrix of eigenvectors, together with some more specific results concerning the sensitivity of individual eigenvalues. Section 4.6 consists of a detailed discussion of general algorithms for finding individual eigenvalues and eigenvectors. These are necessarily iterative because it is theoretically impossible to find the eigenvalues of an $n \times n$ matrix for $n \geq 4$ in a finite number of steps. Section 4.7 then deals with algorithms that produce several eigenvalues and eigenvectors simultaneously, culminating with the elegant QR iterations algorithm. The chapter ends with Section 4.8 with the singular value decomposition of a matrix, which has deep theoretical significance but also important practical applications, in particular to problems involving rank-deficient matrices.

4.1 Matrix Eigenvalue Problems

Let V be a finite-dimensional vector space. We say that a scalar λ is an *eigenvalue* of the linear map $\mathbf{A} : V \mapsto V$ if there exists a vector $\mathbf{v} \neq \mathbf{0} \in V$ such that

$$\mathbf{A}\mathbf{v} = \lambda\mathbf{v}. \tag{4.1}$$

Observe that this equation is equivalent to

$$(\mathbf{A} - \lambda\mathbf{1})\mathbf{v} = \mathbf{0}, \tag{4.2}$$

so that λ is an eigenvalue of $\mathbf{A}$ if and only if the map $(\mathbf{A} - \lambda\mathbf{1})$ has a nontrivial null space, which is equivalent to the map $(\mathbf{A} - \lambda\mathbf{1})$ being singular (noninvertible).

If λ is an eigenvalue of $\mathbf{A}$, then any vector $\mathbf{x} \in V$ satisfying $\mathbf{A}\mathbf{x} = \lambda\mathbf{x}$ is said to be an *eigenvector* of $\mathbf{A}$ associated with the eigenvalue λ. In other words, the set of eigenvectors of $\mathbf{A}$ associated with the eigenvalue λ equals the subspace $\text{null}(\mathbf{A} - \lambda\mathbf{1})$.

Easy examples of eigenvalues are provided by the diagonal entries of a matrix $\mathbf{A} : \mathbb{R}^n \to \mathbb{R}^n$, with the unit vectors in the canonical basis of $\mathbb{R}^n$ as the corresponding eigenvectors. For instance, the eigenvalues of

$$\mathbf{A} = \begin{bmatrix} 2 & 0 \\ 0 & -3 \end{bmatrix},$$

are $\lambda_1 = 2$, with eigenvector $\mathbf{e}_1 = [1, 0]$, and $\lambda_2 = -3$, with eigenvector $\mathbf{e}_2 = [0, 1]$.

Up to this point in the book, all the results derived for real vector spaces have simple extensions to the complex case, obtained by adjusting the notations and slightly modifying the arguments. For eigenvalues and eigenvectors, however, this is no longer true, and the distinction between real and complex vector spaces becomes essential. As an example, consider the real vector space $V = \mathbb{R}^2$ and define the linear map

$$\mathbf{A} = \begin{bmatrix} 0 & -1 \\ 1 & 0 \end{bmatrix}, \tag{4.3}$$

which is just the orthogonal matrix of a counterclockwise rotation by $\pi/2$ around the origin in $\mathbb{R}^2$ (recall the matrices of rotations introduced in Section 3.1). Since the result of such rotation on any nonzero vector in $\mathbf{v} \in \mathbb{R}^2$ is *never* a multiple of $\mathbf{v}$ itself, you can see that this map has no eigenvalues in $\mathbb{R}^2$. Algebraically, finding an eigenvalue for $\mathbf{A}$ corresponds to solving the equation

$$\begin{bmatrix} 0 & -1 \\ 1 & 0 \end{bmatrix} \begin{bmatrix} x_1 \\ x_2 \end{bmatrix} = \lambda \begin{bmatrix} x_1 \\ x_2 \end{bmatrix}, \tag{4.4}$$

for some vector $[x_1, x_2] \neq [0, 0]$. But this corresponds to

$$\begin{aligned} -x_2 &= \lambda x_1 \\ x_1 &= \lambda x_2 \end{aligned} \tag{4.5}$$

which is therefore equivalent to

$$\lambda^2 = -1. \tag{4.6}$$

Since there is no real number λ satisfying Equation (4.6) you can confirm that $\mathbf{A}$ has no eigenvalues when acting on the real vector space $\mathbb{R}^2$. However, let the same $\mathbf{A}$ be a linear map on the complex vector space $V = \mathbb{C}^2$, and you can conclude that it has the eigenvalues $\lambda = \pm i$. Moreover, you can easily recognize the eigenvectors associated with these eigenvalues, simply noticing that

$$\begin{bmatrix} 0 & -1 \\ 1 & 0 \end{bmatrix} \begin{bmatrix} 1 \\ \mp i \end{bmatrix} = \pm i \begin{bmatrix} 1 \\ \mp i \end{bmatrix},$$

which implies that $\mathbf{v}_{\pm} = [1, \mp i] \in \mathbb{C}^2$ are eigenvectors for eigenvalues $\lambda = \pm i$. In view of this example, for the rest of this chapter we assume that V is a finite-dimensional complex vector space, and therefore isomorphic to $\mathbb{C}^n$.

Some elementary properties of eigenvalues can be obtained directly from the definition (4.1) or its equivalent formulation (4.2), as in the next exercise.

Exercise 4.1 Use (4.2) and (3.16) to show that λ is an eigenvalue of a linear map $\mathbf{A}$ if and only if its complex conjugate $\bar{\lambda}$ is an eigenvalue of the adjoint map $\mathbf{A}^*$.

For more elaborate properties of eigenvalues, it is convenient to rephrase (4.2) in terms of the determinant function. For this, recall from Section 2.8 that a matrix is singular if and only if its determinant equals zero. Therefore, you can immediately see that λ is an eigenvalue of $\mathbf{A} : \mathbb{C}^n \to \mathbb{C}^n$ if and only if it is a solution to the *characteristic equation*

$$\det(\mathbf{A} - \lambda \mathbf{1}) = 0. \tag{4.7}$$

Characteristic polynomial

The function $D(\lambda) = \det(\mathbf{A} - \lambda\mathbf{1})$ is a polynomial of degree n in the variable λ, called the *characteristic polynomial* of $\mathbf{A}$. Invoking the *Fundamental Theorem of Algebra* (see Section 5.1), such a polynomial can be factorized as

$$D(\lambda) = (\lambda_1 - \lambda)^{m_1}(\lambda_2 - \lambda)^{m_2} \cdots (\lambda_p - \lambda)^{m_p}, \tag{4.8}$$

where $\lambda_1, \ldots, \lambda_p$ are all the roots of $D(\lambda)$, and hence the eigenvalues of $\mathbf{A}$ and $m_1 + m_2 + \cdots m_p = n$ are their corresponding *algebraic multiplicities.* By combining these observations together, we establish the following theorem:

Theorem 4.1 *Every $n \times n$ complex matrix has exactly n eigenvalues, repeated according to their algebraic multiplicities.*

This theorem is not true if $V = \mathbb{R}^n$ since linear maps on real vector spaces may not have any eigenvalues, according to the counterexample in (4.3). However, a modification of the argument using the characteristic polynomial can be used to prove the result in the following exercise.

Exercise 4.2 Show that, when n is an odd number, every linear map on $\mathbb{R}^n$ has at least one real eigenvalue.

Based on the result of Theorem 4.1, we formulate the central problem for this chapter:

Problem 4.1 (Eigenvalues and eigenvectors) Find all the eigenvalues and eigenvectors of an $n \times n$ complex matrix $\mathbf{A}$.

In MATLAB, the solution for this problem is obtained with the function `eig`, as in the following example:

```
>> A = [2 4 2; 6 0 8; 3 2 -8];
>> [V,lambda] = eig(A)  % V is the matrix of eigenvectors
                        % lambda is the diagonal matrix of eigenvalues

V =
   -0.6300   -0.5547    0.1132
   -0.7451    0.8321   -0.6838
   -0.2191    0.0000    0.7208

lambda =
    7.4261         0         0
         0   -4.0000         0
         0         0   -9.4261
```

The approach to Problem 4.1 based on the characteristic polynomial $D(\lambda)$, although popular and well established, has several disadvantages. Pedagogically, it relies on the determinant function, which provides no insight on eigenvectors corresponding to eigenvalues. In particular, two different matrices with different eigenvectors may have the same determinant function $D(\lambda)$ and the same eigenvalues. For instance, the following two matrices

$$\mathbf{A} = \begin{bmatrix} 3 & 0 \\ 0 & 3 \end{bmatrix}, \qquad \mathbf{B} = \begin{bmatrix} 3 & 2 \\ 0 & 3 \end{bmatrix} \tag{4.9}$$

have the same polynomial $D(\lambda) = (3 - \lambda)^2$ with a double eigenvalue $\lambda = 3$, but have very different eigenvectors. Because $\text{null}(\mathbf{A} - 3\mathbf{1})$ is two-dimensional, any vector in $\mathbb{R}^2$ is an eigenvector of $\mathbf{A}$, as can be easily verified. However, you can also directly verify that the only eigenvectors of $\mathbf{B}$ are multiples of $\mathbf{e}_1$, corresponding to the fact that $\text{null}(\mathbf{B} - 3\mathbf{1})$ is a one-dimensional subspace.

In practical terms, obtaining the characteristic polynomial $D(\lambda)$ for a given $n \times n$ matrix $\mathbf{A}$ is an expensive and numerically unstable task. Therefore, even assuming that you can compute the roots of such polynomials (for example, using the methods described in Chapter 8), there is little guarantee that these roots correspond to the actual eigenvalues of the original matrix $\mathbf{A}$ because of a large computational error.

Accordingly, our approach in this book is to avoid the determinant function $D(\lambda)$ and the characteristic equation (4.7), both computationally and as tools for deriving properties of eigenvalues and eigenvectors. Instead, we introduce iterative numerical methods for Problem 4.1 without computing the roots of the characteristic equation (4.7).

4.2 Properties of Eigenvalues

We now introduce some theoretical properties of eigenvalues that are important both for numerical computations and for applications of *Linear Algebra*. For brevity of notation, we always use $V = \mathbb{C}^n$ as the complex vector space and an $n \times n$ matrix $\mathbf{A}$ with complex entries as the linear map from V to itself. It is only a matter of mathematical culture to rephrase all statements and arguments to the general case of an n-dimensional complex vector space and a linear map acting on it. Indeed, when a particular basis is chosen for V, a linear map $\mathbf{A}$ becomes an $n \times n$ matrix in $\mathbb{C}^n$.

Existence of eigenvalues

Let $\mathbf{x} \in \mathbb{C}^n$ be a nonzero vector and consider the $(n + 1)$ vectors

$$\mathbf{x}, \mathbf{A}\mathbf{x}, \mathbf{A}^2\mathbf{x}, \ldots, \mathbf{A}^n\mathbf{x}$$

obtained by successive applications of $\mathbf{A}$ to $\mathbf{x}$. Since these $(n + 1)$ vectors cannot be linearly independent in $\mathbb{C}^n$ (because the dimension of $\mathbb{C}^n$ is n), and since $\mathbf{x}$ is a nonzero vector, there must exist complex numbers $\mu_0, \ldots, \mu_n$, with at least one $\mu_j \neq 0$, for $0 < j \leq n$, such that

$$\mathbf{0} = \mu_0 \mathbf{x} + \mu_1 \mathbf{A}\mathbf{x} + \cdots + \mu_n \mathbf{A}^n \mathbf{x}. \tag{4.10}$$

If the righthand side of (4.10) is associated with the polynomial

$$P(z) = \mu_0 + \mu_1 z + \cdots \mu_n z^n,$$

then the degree of $P(z)$ is at least $j > 0$. According to the *Fundamental Theorem of Algebra* (see Section 5.1), there exist complex roots of the polynomial $P(z)$, denoted by $\lambda_1, \ldots, \lambda_k$, such that

$$\mu_0 + \mu_1 z + \cdots + \mu_n z^n = c(z - \lambda_1)^{m_1} \cdots (z - \lambda_k)^{m_k}, \tag{4.11}$$

for a nonzero constant c and nonzero integers $m_1 + \cdots + m_k \geq j$. Substituting this back into (4.10) then gives

$$\begin{aligned} \mathbf{0} &= \mu_0 \mathbf{x} + \mu_1 \mathbf{A}\mathbf{x} + \cdots + \mu_n \mathbf{A}^n \mathbf{x} \\ &= (\mu_0 \mathbf{1} + \mu_1 \mathbf{A} + \cdots + \mu_n \mathbf{A}^n)\mathbf{x} \\ &= c(\mathbf{A} - \lambda_1 \mathbf{1})^{m_1} \cdots (\mathbf{A} - \lambda_k \mathbf{1})^{m_k} \mathbf{x}. \end{aligned}$$

But this implies that $(\mathbf{A} - \lambda_i \mathbf{1})$ is not injective for at least one value of λ_i, $i = 1, \ldots, k$. Therefore, such λ_i must be an eigenvalue of $\mathbf{A}$. This argument, which uses the *Fundamental Theorem of Algebra* but avoids using the determinant of $(\mathbf{A} - \lambda \mathbf{1})$, is the proof that at least one eigenvalue (and its associated eigenvector) of the square matrix $\mathbf{A}$ exists.

Suppose now that $\mathbf{A}$ is a 2×2 complex matrix and λ is one of its eigenvalues (which must exist, according to the argument just presented). We then know that the subspace range$(\mathbf{A} - \lambda \mathbf{1})$ must have dimension zero or one because otherwise $(\mathbf{A} - \lambda \mathbf{1})$ would be invertible, contradicting the fact that λ is an eigenvalue. If the dimension of range$(\mathbf{A} - \lambda \mathbf{1})$ is zero, then the dimension formula (2.17) tells us that null$(\mathbf{A} - \lambda \mathbf{1})$ is two-dimensional so that any vector in $\mathbb{C}^2$ is an eigenvector of $\mathbf{A}$ associated with the eigenvalue λ, implying that

$$\mathbf{A} = \begin{bmatrix} \lambda & 0 \\ 0 & \lambda \end{bmatrix}. \tag{4.12}$$

On the other hand, if the dimension of null$(\mathbf{A} - \lambda \mathbf{1})$ is one, then it must coincide with the subspace spanned by a nonzero vector $\mathbf{u}$, which we might take to satisfy $\|\mathbf{u}\| = 1$. In this case, we can find another nonzero vector $\mathbf{w}$ such that $\{\mathbf{u}, \mathbf{w}\}$ form an orthonormal basis for $\mathbb{C}^2$ (for example, by applying the Gram–Schmidt algorithm to $\mathbf{u}$ and any other linearly independent vector $\mathbf{v}$, which must exist because $\mathbb{C}^2$ is two-dimensional). Therefore, if $\mathbf{U}$ is the unitary matrix having $\mathbf{u}$ and $\mathbf{w}$ as its columns, we obtain that

$$\mathbf{U}'\mathbf{A}\mathbf{U} = \begin{bmatrix} \mathbf{u}' * \mathbf{A}\mathbf{u} & \mathbf{u}' * \mathbf{A}\mathbf{w} \\ \mathbf{w}' * \mathbf{A}\mathbf{u} & \mathbf{w}' * \mathbf{A}\mathbf{w} \end{bmatrix}.$$

But since $\mathbf{Au} = \lambda\mathbf{u}$, it is easy to see that $\mathbf{w}' * \mathbf{Au} = 0$, implying that

$$\mathbf{U}'\mathbf{AU} = \begin{bmatrix} \mathbf{u}' * \mathbf{Au} & \mathbf{u}' * \mathbf{Aw} \\ 0 & \mathbf{w}' * \mathbf{Aw} \end{bmatrix}. \tag{4.13}$$

Whichever the case, we see from either (4.12) or (4.13) that, given any 2×2 complex matrix $\mathbf{A}$, we can always find a unitary matrix $\mathbf{U}$ such that $\mathbf{U}'\mathbf{AU}$ is upper triangular.

A similar type of proof, based on the existence of at least one eigenvalue and using an induction argument, leads to the following general result:

Schur decomposition

Theorem 4.2 *If $\mathbf{A}$ is an $n \times n$ complex matrix, then there exists a unitary $n \times n$ matrix $\mathbf{U}$ such that $\mathbf{U}'\mathbf{AU}$ is upper triangular.*

Another way of expressing the result of this theorem is that any $n \times n$ complex matrix $\mathbf{A}$ can be expressed as

$$\mathbf{A} = \mathbf{UBU}', \tag{4.14}$$

where $\mathbf{U}$ is an $n \times n$ unitary matrix and $\mathbf{B}$ is an $n \times n$ upper-triangular matrix. In this form, Theorem 4.2 is known as the *complex Schur decomposition* of a matrix $\mathbf{A}$. In MATLAB, such decomposition can be obtained from the built-in function `schur` with the optional argument `'complex'` (the argument `'real'`, which is unfortunately the default MATLAB value, produces the 'real' Schur decomposition of $\mathbf{A}$, which is different from the factorization formula (4.14)). As an example, consider the following:

```
>> A = [1 2 2 -2; 0 2 0 -3; 3 -6 -4 6; -1 2 2 -3];
>> [U,B] = schur(A,'complex')

U =
   -0.2828   -0.1234    0.9512    0.0000
   -0.1414    0.7098    0.0501   -0.6882
    0.8485    0.3703    0.3004    0.2294
   -0.4243    0.5864   -0.0501    0.6882

B =
   -7.0000   -2.6186    0.4036    8.6302
         0   -1.0000    0.2920   -1.1894
         0         0    2.0000    0.0000
         0         0         0    2.0000
```

The Schur decomposition is significant to Problem 4.1 because the eigenvalues of an $n \times n$ upper-triangular matrix coincide with the diagonal entries of the matrix, as the next exercise shows.

Exercise 4.3 Use Exercise 2.12 to show that a complex number λ is an eigenvalue of an $n \times n$ upper-triangular matrix if and only if it coincides with at least one of its diagonal entries.

For the matrix from the previous example, you have:

```
>> eig(B)'

ans =
   -7.0000   -1.0000    2.0000    2.0000
```

Similarity transformations

We say that two $n \times n$ matrices $\mathbf{A}$ and $\mathbf{B}$ are related by a *similarity transformation* if and only if there exists an invertible $n \times n$ matrix $\mathbf{T}$ such that

$$\mathbf{B} = \mathbf{T}^{-1}\mathbf{A}\mathbf{T}. \tag{4.15}$$

It is then easy to see that

$$\mathbf{B}\mathbf{y} = \lambda\mathbf{y} \Leftrightarrow \mathbf{T}^{-1}\mathbf{A}\mathbf{T}\mathbf{y} = \lambda\mathbf{y} \Leftrightarrow \mathbf{A}(\mathbf{T}\mathbf{y}) = \lambda(\mathbf{T}\mathbf{y}),$$

which implies that λ is an eigenvalue of $\mathbf{B}$ with nonzero eigenvector $\mathbf{y}$ if and only if it is an eigenvalue of $\mathbf{A}$ with nonzero eigenvector $\mathbf{T}\mathbf{y}$.

Therefore, since a unitary matrix $\mathbf{U}$ is invertible with $\mathbf{U}^{-1} = \mathbf{U}'$, the eigenvalues of a general $n \times n$ complex matrix $\mathbf{A}$ coincide with the diagonal entries of the upper-triangular matrix $\mathbf{B} = \mathbf{U}'\mathbf{A}\mathbf{U}$ obtained in Theorem 4.2. For instance, the eigenvalues for the matrix in the previous example are

```
>> A = [1 2 2 -2; 0 2 0 -3; 3 -6 -4 6; -1 2 2 -3];
>> eig(A)'

ans =
   -7.0000    2.0000   -1.0000    2.0000
```

which coincide with the diagonal entries of the upper-triangular matrix $\mathbf{B}$ obtained from the Schur decomposition (4.14).

The Schur decomposition, however, is not unique. Consider the following operations:

```
>> A = [1 2 2 -2; 0 2 0 -3; 3 -6 -4 6; -1 2 2 -3];
>> [V,lambda] = eig(A);
>> v1 = V(:,1); v2 = V(:,3); % eigenvectors associated with -7,-1
```

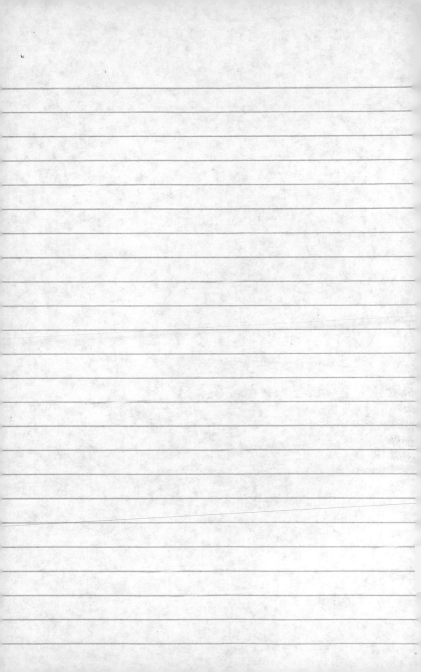

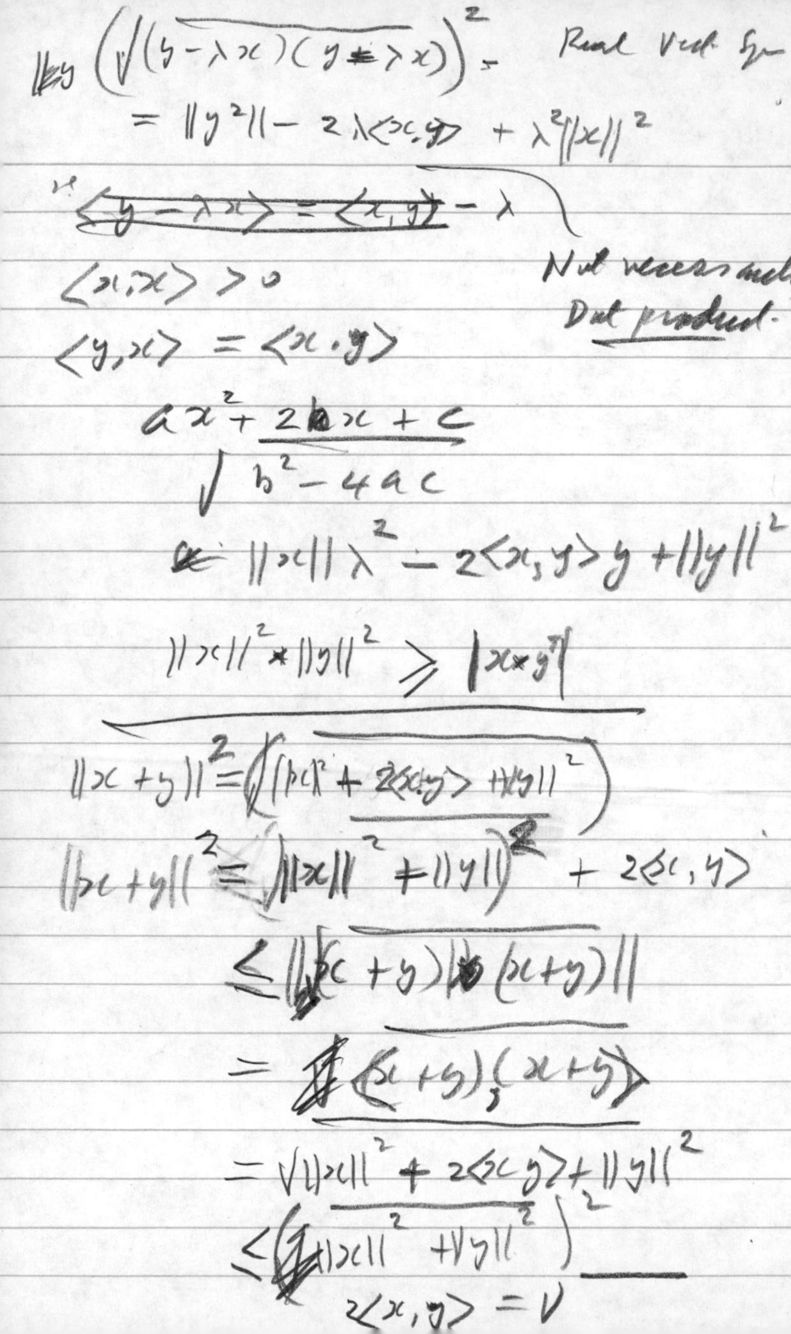

$\left(\sqrt{(y-\lambda x)(y-\lambda x)}\right)^2 =$ Real Vect Sp

$= \|y^2\| - 2\lambda\langle x, y\rangle + \lambda^2\|x\|^2$

~~$\langle y - \lambda x\rangle = \langle x, y\rangle$~~ $- \lambda$

$\langle x, x\rangle > 0$

Not necessarily Dot product.

$\langle y, x\rangle = \langle x \cdot y\rangle$

$ax^2 + 2bx + c$

$\sqrt{b^2 - 4ac}$

$\|x\|\lambda^2 - 2\langle x, y\rangle y + \|y\|^2$

$\|x\|^2 * \|y\|^2 \geq |x * y|$

$\|x+y\|^2 = \left(\sqrt{\|x\|^2 + 2\langle x, y\rangle + \|y\|^2}\right)$

$\|x+y\|^2 = \left(\sqrt{\|x\|^2} + \|y\|\right)^2 + 2\langle x, y\rangle$

$\leq \|(x+y)(x+y)\|$

$= \langle (x+y), (x+y)\rangle$

$= \sqrt{\|x\|^2 + 2\langle x, y\rangle + \|y\|^2}$

$\leq \left(\sqrt{\|x\|^2 + \|y\|^2}\right)^2$

$2\langle x, y\rangle = \sqrt{\ }$

```
>> v3 = [1 0 0 0]'; v4 = [0 1 0 0]'; % arbitrary linearly independent vectors
>> q1 = v1/norm(v1); % Gram-Schmidt orthogonalization procedure
>> q2 = (v2-(v2'*q1)*q1)/norm(v2-(v2'*q1)*q1);
>> q3 = (v3-(v3'*q1)*q1-(v3'*q2)*q2); q3 = q3/norm(q3);
>> q4 = (v4-(v4'*q1)*q1-(v4'*q2)*q2-(v4'*q3)*q3); q4 = q4/norm(q4);
>> U = [q1,q2,q3,q4]

U =
   [illegible]    0.1234    0.9512         0
   -0.1414    0.7095    [illegible]    0.6882
   -0.8485    0.3702    0.2004   -0.2294
   -0.4243    0.5864   -0.0501   -0.6882

>> B=U'*A*U

B =
   -1.0000    2.6186   -0.4036   -8.6302
   -0.0000   -1.0000    0.2920    1.1894
   -0.0000   -0.0000    2.0000    0.0000
   [illegible]   -0.0000    [illegible]    2.0000

>> U*B*U'

ans =
    1.0000    2.0000    2.0000   -2.0000
   -0.0000    2.0000   -0.0000   -3.0000
    3.0000   -6.0000   -4.0000    6.0000
   -1.0000    2.0000    2.0000   -3.0000
```

As you can see, the unitary matrix **U** and the upper triangular matrix **B** obtained in this example, although different from the ones obtained in the MATLAB example following Theorem 4.2, also satisfy **A** = **UBU**′. But despite this nonuniqueness of the Schur decomposition, you can see in the example that the diagonal entries of the two different upper triangular matrices are the same. Now, if we ignore repeated appearances, the fact that these diagonal entries are the same just follows from the fact that they are, after all, eigenvalues of the original matrix **A**, and therefore should not depend on which particular unitary matrix **U** is used to produce the similarity transformation leading to the upper-triangular matrix **B**. However, we observe that even the number of times a given value appears on the diagonal of **B** seems to be the same for different similarity transformations, which suggests a deeper property of eigenvalues, summarized in the next theorem.

Algebraic multiplicity

THEOREM [illegible] *If* **A** = **UBU**′ *for an* $n \times n$ *unitary matrix* **U** *and an upper-triangular matrix* **B**, *then the number of times an eigenvalue* λ *appears on the diagonal of* **B** *equals the algebraic multiplicity of the eigenvalue.*

An immediate consequence of this theorem is that the sum of the algebraic multiplicity of all eigenvalues of an $n \times n$ complex matrix is always equal to n (because the corresponding upper-triangular matrix always has n diagonal entries), which agrees with the result of Theorem 4.1.

The proof of Theorem 4.3, which we do not present here, relies on properties of the subspaces null$(\mathbf{A} - \lambda\mathbf{1})^n$, where λ is an eigenvalue of $\mathbf{A}$. In particular, it can be shown that the dimension of null$(\mathbf{A} - \lambda\mathbf{1})^n$ coincides with the algebraic multiplicity of λ, as can be verified in the next example:

```
>> A = [1 2 2 -2; 0 2 0 -3; 3 -6 -4 6; -1 2 2 -3];
>> lambda = eig(A)'

lambda =
   -7.0000    2.0000   -1.0000    2.0000

>> null((A-lambda(1)*eye(4))^4)

ans =
    0.2828
    0.1414
   -0.8485
    0.4243

>> null((A-lambda(2)*eye(4))^4)

ans =
    0.9428   -0.0028
    0.2336   -0.7078
    0.2378    0.7064
   -0.0000    0.0000

>> null((A-lambda(3)*eye(4))^4)

ans =
   -0.0000
   -0.7071
   -0.0000
   -0.7071
```

You can see from this example that the eigenvalue $\lambda = 2$ has algebraic multiplicity equal to two because the subspace null$(\mathbf{A} - 2\mathbf{1})^4$ is two-dimensional, whereas the other eigenvalues $\lambda = -7$ and $\lambda = -1$ each have algebraic multiplicity equal to one. This confirms the result in Theorem 4.3 since such algebraic multiplicities are exactly the number of times that these eigenvalues appear on the diagonal of the upper-triangular matrix $\mathbf{B}$ in the Schur decomposition $\mathbf{A} = \mathbf{UBU}'$.

Block-triangular matrices

As a generalization of an upper-triangular matrix, we say that an $n \times n$ matrix $\mathbf{C}$ is *block triangular* if it can be written as

$$\begin{bmatrix} \mathbf{C}_{11} & \cdots & \cdots & \mathbf{C}_{1p} \\ \mathbf{0} & \mathbf{C}_{22} & \cdots & \mathbf{C}_{2p} \\ \mathbf{0} & \vdots & \ddots & \vdots \\ \mathbf{0} & \mathbf{0} & \cdots & \mathbf{C}_{pp} \end{bmatrix}, \tag{4.16}$$

where each diagonal block $\mathbf{C}_{ii}$ is an $n_i \times n_i$ matrix and $n_1 + \cdots n_p = n$. For example, the matrix

$$\mathbf{C} = \begin{bmatrix} 4 & 3 & -2 & 5 & 7 \\ 0 & 5 & 3 & 6 & 9 \\ 0 & -5 & 2 & 1 & -7 \\ 0 & 0 & 0 & 4 & 3 \\ 0 & 0 & 0 & -2 & -1 \end{bmatrix}$$

is block triangular because it can be written in the form (4.16) with

$$\mathbf{C}_{11} = [4], \quad \mathbf{C}_{22} = \begin{bmatrix} 5 & 3 \\ -5 & 2 \end{bmatrix}, \quad \mathbf{C}_{33} = \begin{bmatrix} 4 & 3 \\ -2 & -1 \end{bmatrix}.$$

You can now apply the same argument used in connection with upper-triangular matrices and prove generalizations of Exercises 2.12 and 4.3.

Exercise 4.4 Show that a block-triangular matrix of the form (4.16) is invertible if and only if all of its diagonal blocks $\mathbf{C}_{ii}$ are invertible.

Exercise 4.5 Show that λ is an eigenvalue of a block-triangular matrix of the form (4.16) if and only if it is an eigenvalue of at least one of its diagonal block $\mathbf{C}_{ii}$.

For example, you can verify that the eigenvalues of matrices $\mathbf{C}_1$, $\mathbf{C}_2$, and $\mathbf{C}_3$ coincide with the eigenvalues of the matrix $\mathbf{C}$ in the previous example:

```
>> C = [4 3 -2 5 7; 0 5 3 6 9; 0 -5 2 1 -7;0 0 0 4 3; 0 0 0 -2 -1];
>> C22 = [5 3; -5 2];
>> C33 = [4 3; -2 -1];
>> eig(C)
```

```
ans =
   4.0000
   3.5000 + 3.5707i
   3.5000 - 3.5707i
   2.0000
   1.0000

>> eig(C22)

ans =
   3.5000 + 3.5707i
   3.5000 - 3.5707i

>> eig(C33)

ans =
    2.0000
    1.0000
```

Spectral radius

The subset of $\mathbb{C}$ consisting of the eigenvalues of an $n \times n$ complex matrix $\mathbf{A}$ is called the *spectrum* of $\mathbf{A}$ and is denoted by $\sigma(\mathbf{A})$. The *spectral radius* of $\mathbf{A}$ is then defined as

$$\rho(\mathbf{A}) = \max\{|\lambda| : \lambda \in \sigma(\mathbf{A})\}. \tag{4.17}$$

Therefore, as points on the complex plane, all the eigenvalues of $\mathbf{A}$ must be located inside a disk of radius $\rho(\mathbf{A})$. Now, if $\mathbf{x}$ is a nonzero eigenvector associated with the eigenvalue λ and $\|\cdot\|$ denotes any vector norm on $\mathbb{C}^n$ from Section 2.4, then, for any integer $k \geq 1$, we have

$$|\lambda|^k = \frac{\|\lambda^k \mathbf{x}\|}{\|\mathbf{x}\|} = \frac{\|\mathbf{A}^k \mathbf{x}\|}{\|\mathbf{x}\|} \leq \max \frac{\|\mathbf{A}^k \mathbf{x}\|}{\|\mathbf{x}\|} = \|\mathbf{A}^k\|,$$

which implies that

$$\rho(\mathbf{A}) \leq \|\mathbf{A}^k\|^{1/k} \tag{4.18}$$

for any induced matrix norm $\|\cdot\|$ and any integer $k \geq 1$. In fact, the following expression for the spectral radius, known as *Gelfand's formula*, is valid for any induced matrix norm:

$$\rho(\mathbf{A}) = \lim_{k \to \infty} \|\mathbf{A}^k\|^{1/k}. \tag{4.19}$$

In the following MATLAB script, we calculate finite approximations for the Gelfand's formula (4.19) for three different matrix norms, using the matrix

$$\mathbf{A} = \begin{bmatrix} 4 & 2 & 4 \\ 2 & 0 & 5 \\ 5 & 1 & -6 \end{bmatrix}. \tag{4.20}$$

```
% radius_script
A = [4 2 4; 2 0 5; 5 1 -6];
lambda = eig(A)'
for k = 1:150
    a(k) = norm((A^k)^(1/k),1);
    b(k) = norm((A^k)^(1/k),2);
    c(k) = norm((A^k)^(1/k),inf);
end
k = 1:150; plot(k,a,k,b,'--r',k,c,'-.k')
legend('1-norm','2-norm','\infty-norm')
```

When the script is run, it computes eigenvalues of $\mathbf{A}$, from which you can conclude that $\rho(\mathbf{A}) = 7.7850$, and plots the numerical approximations of $\rho(\mathbf{A})$, as shown in Figure 4.1.

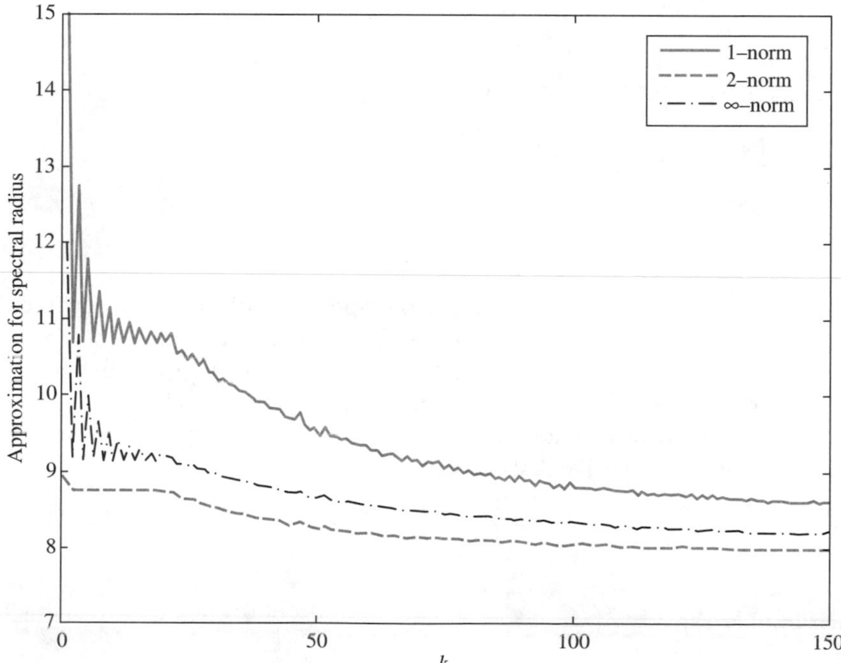

Figure 4.1 Gelfand's formula for the spectral radius.

```
>> radius_script

lambda =
    6.9336   -7.7850   -1.1486
```

As you can see in Figure 4.1, depending on the norm you use, the upper bound for $\rho(\mathbf{A})$ provided by the theoretical formula (4.18) might require a large k to become sharp. However, the entries of $\mathbf{A}^k$ with a large k can quickly approach the largest number available in the floating-point system so that the numerical implementation of Gelfand's formula (4.19) becomes problematic.

If $\mathbf{A}$ is a self-adjoint matrix, this unpleasant situation does not arise because in this case the Euclidean matrix norm satisfies

$$\|\mathbf{A}^{2k}\|_2 = \|\mathbf{A}\|_2^{2k},$$

for any positive integer k, so that Gelfand's formula reduces to

$$\rho(\mathbf{A}) = \|\mathbf{A}\|_2. \tag{4.21}$$

The validity of this formula is illustrated with the following MATLAB code for a symmetric matrix $\mathbf{A}$:

```
>> A = [4 2 4; 2 0 5; 4 5 -6];
>> lambda = eig(A)'

lambda =
   -9.3410   -0.0579    7.3988

>> norm(A)

ans =
    9.3410
```

Gershgorin's disks

For a more detailed localization of the spectrum of an $n \times n$ matrix $\mathbf{A}$, suppose that λ is one of its eigenvalues with corresponding eigenvector $\mathbf{v}$ and assume that $\|\mathbf{v}\|_\infty = 1$ (which can always be obtained by dividing any nonzero eigenvector associated with λ by its ∞-norm). Then, the vector $\mathbf{v}$ must have at least one of its components, say v_k, satisfying $|v_k| = 1$, by definition of the ∞-norm. From the identity $\mathbf{Av} = \lambda\mathbf{v}$ we obtain

$$\sum_{j=1}^{n} a_{kj}x_j = \lambda x_k,$$

which implies that

$$\sum_{j\neq k} a_{kj}x_j = (\lambda - a_{kk})x_k.$$

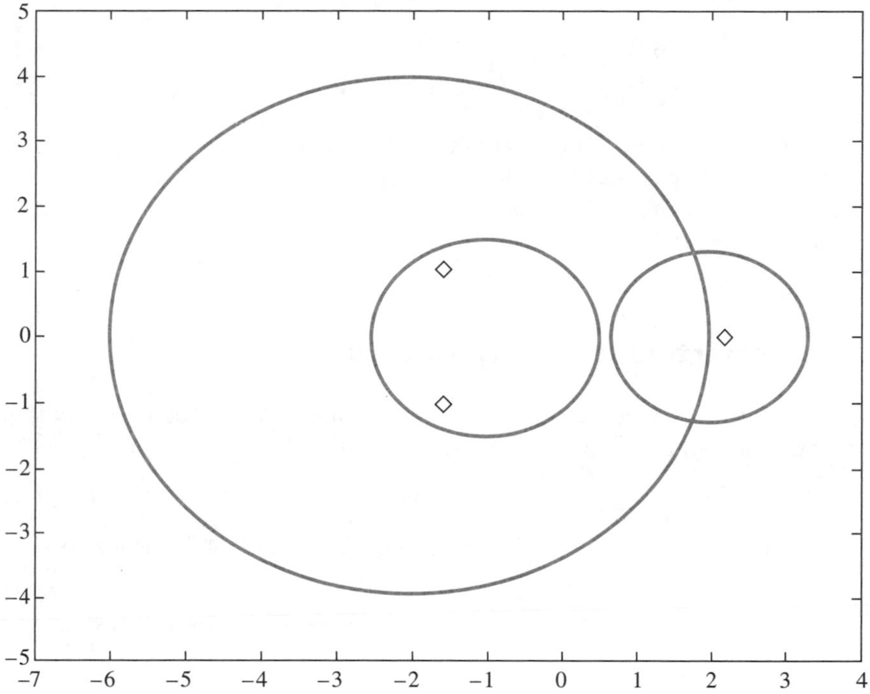

Figure 4.2 Gershgorin's disks and eigenvalues.

Taking the absolute value of both sides of this equation then gives

$$|\lambda - a_{kk}| \leq \sum_{j \neq k} |a_{kj}||x_j| \leq \sum_{j \neq k} |a_{kj}|. \tag{4.22}$$

In other words, the eigenvalue λ is contained in a disk centered at a_{kk} with radius $\sum_{j \neq k} |a_{kj}|$. Repeating this argument for the other eigenvalues, we conclude that the spectrum of an $n \times n$ matrix $\mathbf{A}$ is contained in the union of n disks centered around the diagonal entry of each row with radius equal to the sum of the absolute values of the off-diagonal entries in the same row. This result is called *Gershgorin's theorem* and is illustrated in Figure 4.2, corresponding to the matrix in the following example:

```
>> A=[2 -0.3 1; 4 -2 0; 0.5 1 -1];
>> lambda=eig(A)

lambda =
```

```
   2.1722
  -1.5861 + 1.0368i
  -1.5861 - 1.0368i

>> plot(real(lambda),imag(lambda),'dr')
>> t=[0:0.01:2*pi];hold on;
>> plot(2+1.3*sin(t),1.3*cos(t))
>> plot(-2+4*sin(t),4*cos(t))
>> plot(-1+1.5*sin(t),1.5*cos(t));axis([-7 4 -5 5]);
```

4.3 Properties of Eigenvectors

Linear independence

Suppose that $\lambda_1 \neq \lambda_2$ are two distinct eigenvalues of $\mathbf{A}$ and let $\mathbf{v}_1, \mathbf{v}_2$ be their corresponding nonzero eigenvectors. If we assume that

$$\mathbf{v}_1 = \mu \mathbf{v}_2, \tag{4.23}$$

for some scalar $\mu \neq 0$, then applying $\mathbf{A}$ to both sides of (4.23) leads to

$$\lambda_1 \mathbf{v}_1 = \mu \lambda_2 \mathbf{v}_2. \tag{4.24}$$

Multiplying equation (4.23) by λ_1 and subtracting the result from Equation (4.24) leads to the equality

$$(\lambda_1 - \lambda_2)\mu \mathbf{v}_2 = \mathbf{0}.$$

Since $\lambda_1 \neq \lambda_2$, $\mu \neq 0$, and $\mathbf{v}_2 \neq \mathbf{0}$, we have a contradiction, which implies that (4.23) cannot hold. Recalling the definition of linear dependence given in Section 2.1, we conclude that the eigenvectors $\mathbf{v}_1, \mathbf{v}_2$ must be linearly independent.

Next, suppose that $\lambda_1, \lambda_2, \lambda_3$ are three distinct eigenvalues with their corresponding nonzero eigenvectors $\mathbf{v}_1, \mathbf{v}_2, \mathbf{v}_3$. Then, assuming that these vectors are linearly dependent implies that you can write

$$\mathbf{v}_3 = \mu_1 \mathbf{v}_1 + \mu_2 \mathbf{v}_2, \tag{4.25}$$

with $\mu_1 \neq 0$, $\mu_2 \neq 0$. When the matrix $\mathbf{A}$ is applied to Equation (4.25), we obtain

$$\lambda_3 \mathbf{v}_3 = \mu_1 \lambda_1 \mathbf{v}_1 + \mu_2 \lambda_2 \mathbf{v}_2. \tag{4.26}$$

Multiplying (4.25) by λ_3 and subtracting the result from (4.26) then leads to

$$0 = \mu_1(\lambda_1 - \lambda_3)\mathbf{v}_1 + \mu_2(\lambda_2 - \lambda_3)\mathbf{v}_2,$$

which is a contradiction because we already know that $\mathbf{v}_1, \mathbf{v}_2$ are linearly independent. Therefore, the three vectors $\mathbf{v}_1, \mathbf{v}_2, \mathbf{v}_3$ must be linearly independent, too. Continuing in this way, we can prove the next theorem in exactly $(k-1)$ steps.

Theorem 4.4 *If $\lambda_1, \ldots, \lambda_k$ are distinct eigenvalues of an $n \times n$ matrix $\mathbf{A}$, then their corresponding nonzero eigenvectors $\mathbf{v}_1, \ldots, \mathbf{v}_k$ are linearly independent.*

An easy consequence of this theorem is the property that an $n \times n$ matrix has at most n distinct eigenvalues, which agrees with Theorem 4.1. In the next example, the matrix $\mathbf{A}$ has three distinct eigenvalues and linearly independent eigenvectors in $\mathbb{R}^3$, meaning that the corresponding matrix of eigenvectors has full rank equal to three:

```
>> A = [2 4 2; 6 0 8; 3 2 -8];
>> [V,lambda] = eig(A);
>> diag(lambda)'

ans =
    7.4261   -4.0000   -9.4261

>> rank(V)

ans =
     3
```

Geometric multiplicity

Having established that eigenvectors associated with different eigenvalues must be linearly independent, we are led to consider the linear independence of eigenvectors associated with the same eigenvalue. Given an eigenvalue λ, the number of linearly independent eigenvectors associated with it is given by the dimension of the subspace $\text{null}(\mathbf{A} - \lambda\mathbf{1})$ and is called the *geometric multiplicity* of λ. Recalling the discussion following Theorem 4.3 and using the fact that $\text{null}(\mathbf{A} - \lambda\mathbf{1}) \subset \text{null}(\mathbf{A} - \lambda\mathbf{1})^n$, we conclude that the geometric multiplicity of an eigenvalue is always smaller than or equal to its algebraic multiplicity.

For an illustration of these concepts, consider the following example:

```
>> A=[2 6 -15; 1 1 -5; 1 2 -6]

A =
     2     6   -15
     1     1    -5
     1     2    -6

>> eig(A)'
```

```
ans =
  -1.0000 - 0.0000i  -1.0000 + 0.0000i  -1.0000

>> null((A+eye(3))^3)

ans =
     1     0     0
     0     1     0
     0     0     1

>> null(A+eye(3))

ans =
    0.1366   -0.9737
   -0.9289   -0.0618
   -0.3442   -0.2195
```

We observe that $\lambda = -1$ is the only eigenvalue of $\mathbf{A}$, with algebraic multiplicity equal to 3 and geometric multiplicity equal to 2, meaning that $\mathbf{A}$ has only two linearly independent eigenvectors.

Diagonalization

We can use linearly independent eigenvectors in the diagonalization of a matrix $\mathbf{A}$. For this, suppose that $\mathbf{v}_1, \ldots, \mathbf{v}_k$ are linearly independent eigenvectors of $\mathbf{A}$ associated with the (not necessarily distinct) eigenvalues $\lambda_1, \ldots,$ λ_k. Let $\mathbf{V}$ be the full-rank $n \times k$ matrix having these eigenvectors as its columns, and let $\widetilde{\mathbf{V}}$ be its left inverse (which exists according to Theorem 3.4). Then,

$$\begin{aligned}
\widetilde{\mathbf{V}}\mathbf{A}\mathbf{V} &= \widetilde{\mathbf{V}}\mathbf{A}\,[\mathbf{v}_1, \mathbf{v}_2, \cdots, \mathbf{v}_k] = \widetilde{\mathbf{V}}[\lambda_1\mathbf{v}_1, \lambda_2\mathbf{v}_2, \cdots, \lambda_k\mathbf{v}_k] \\
&= \widetilde{\mathbf{V}}\mathbf{V}\begin{bmatrix} \lambda_1 & & & 0 \\ & \lambda_2 & & \\ & & \ddots & \\ 0 & & & \lambda_k \end{bmatrix} = \begin{bmatrix} \lambda_1 & & & 0 \\ & \lambda_2 & & \\ & & \ddots & \\ 0 & & & \lambda_k \end{bmatrix}.
\end{aligned}$$

We have therefore established the following result:

Theorem 4.5 *If $\mathbf{v}_1, \ldots, \mathbf{v}_k$ are linearly independent eigenvectors of an $n \times n$ matrix $\mathbf{A}$ associated with the eigenvalues $\lambda_1, \ldots, \lambda_k$, then there exists* *an* $n \times k$ *matrix* $\mathbf{V}$ *with left inverse* $\widetilde{\mathbf{V}}$ *such that*

$$\widetilde{\mathbf{V}}\mathbf{A}\mathbf{V} = \begin{bmatrix} \lambda_1 & & & 0 \\ & \lambda_2 & & \\ & & \ddots & \\ 0 & & & \lambda_k \end{bmatrix}. \tag{4.27}$$

In the following example, the 4×4 matrix $\mathbf{A}$ has eigenvalues $\lambda_1 = 4$, $\lambda_2 = \lambda_4 = 2$, and $\lambda_3 = 3$. Using the linearly independent eigenvectors $\mathbf{v}_2$, $\mathbf{v}_3$, $\mathbf{v}_4$, you can compute the matrices $\mathbf{V}$, $\widetilde{\mathbf{V}}$ and illustrate the diagonalization formula (4.27):

```
>> A = [4 -2 4/3 4/3; 0 2 1/3 -1/3; 0 0 2 0; 0 0 -1 3];
>> [V,lambda] = eig(A);
>> diag(lambda)'

ans =
     4     2     3     2

>> V = V(1:4,2:4);
>> [Q,R] = QRfactor(V);
>> Vtilde = inv(R)*Q';
>> Vtilde*A*V

ans =
    2.0000    0.0000    0.0000
   -0.0000    3.0000    0.0000
    0.0000   -0.0000    2.0000
```

In particular, if $\mathbf{v}_1, \ldots, \mathbf{v}_n$ form a basis in $\mathbb{R}^n$, then $\widetilde{\mathbf{V}} = \mathbf{V}^{-1}$ and the matrix $\mathbf{A}$ is said to be *diagonalizable*. In this case, the similarity transformation $\mathbf{V}^{-1}\mathbf{A}\mathbf{V}$ reduces $\mathbf{A}$ to a diagonal matrix of eigenvalues:

$$\mathbf{V}^{-1}\mathbf{A}\mathbf{V} = \begin{bmatrix} \lambda_1 & & & 0 \\ & \lambda_2 & & \\ & & \ddots & \\ 0 & & & \lambda_n \end{bmatrix}. \tag{4.28}$$

The following example shows that the diagonal matrix of eigenvalues in the diagonalization formula (4.28) is precisely the second output of the MATLAB function `eig`:

```
>> A = [2 4 5; 3 6 7; -1 4 -2];
>> [V,lambda] = eig(A);
>> inv(V)*A*V

ans =
   -3.8137    0.0000    0.0000
    0.0000   -0.1578   -0.0000
    0.0000    0.0000    9.9715

>> lambda

lambda =
   -3.8137         0         0
         0   -0.1578         0
         0         0    9.9715
```

Invariant subspaces

The result in Theorem 4.5 does not hold when the eigenvectors $\mathbf{v}_1, \ldots, \mathbf{v}_k$ are linearly dependent. To cover this case, let us say that a subspace S of a vector space V is an *invariant subspace* for the linear map $\mathbf{A}$ if

$$\mathbf{u} \in S \Rightarrow \mathbf{A}\mathbf{u} \in S. \tag{4.29}$$

Exercise 4.6 Let $\mathbf{v}_1, \ldots, \mathbf{v}_k$ be (not necessarily linearly independent) eigenvectors of a linear map $\mathbf{A}$. Show that the subspace $S = \text{Span}(\mathbf{v}_1, \ldots, \mathbf{v}_k)$ is a p-dimensional invariant subspace for $\mathbf{A}$, with $p \leq k$.

Suppose that $S \subset \mathbb{C}^n$ is a p-dimensional invariant subspace of an $n \times n$ matrix $\mathbf{A}$, and denote by $\mathbf{V}_1$ the $n \times p$ matrix whose columns are an orthonormal basis for S. Then, applying $\mathbf{A}$ to a column of $\mathbf{V}_1$ must result in a vector in S, which we can write as a linear combination of the columns of $\mathbf{V}_1$. This means that there must exist a $p \times p$ matrix $\mathbf{B}_{11}$ such that

$$\mathbf{A}\mathbf{V}_1 = \mathbf{V}_1\mathbf{B}_{11}. \tag{4.30}$$

On the other hand, denote by $\mathbf{V}_2$ the $n \times (n-p)$ matrix whose columns are an orthonormal basis for $S^\perp$, the orthogonal complement of S, which must be an $(n-p)$-dimensional subspace of $\mathbb{C}^n$, but not necessarily an invariant subspace for $\mathbf{A}$. Writing

$$\mathbf{V} = \begin{bmatrix} \mathbf{V}_1 & \mathbf{V}_2 \end{bmatrix},$$

where $\mathbf{V}$ is an $n \times n$ unitary matrix, we obtain the following similarity transformation:

$$\begin{aligned} \mathbf{V}'\mathbf{A}\mathbf{V} &= \begin{bmatrix} \mathbf{V}_1' \\ \mathbf{V}_2' \end{bmatrix} \mathbf{A} \begin{bmatrix} \mathbf{V}_1 & \mathbf{V}_2 \end{bmatrix} = \begin{bmatrix} \mathbf{V}_1'\mathbf{A}\mathbf{V}_1 & \mathbf{V}_1'\mathbf{A}\mathbf{V}_2 \\ \mathbf{V}_2'\mathbf{A}\mathbf{V}_1 & \mathbf{V}_2'\mathbf{A}\mathbf{V}_2 \end{bmatrix} \\ &= \begin{bmatrix} \mathbf{V}_1'\mathbf{V}_1\mathbf{B}_{11} & \mathbf{V}_1'\mathbf{A}\mathbf{V}_2 \\ \mathbf{V}_2'\mathbf{V}_1\mathbf{B}_{11} & \mathbf{V}_2'\mathbf{A}\mathbf{V}_2 \end{bmatrix} = \begin{bmatrix} \mathbf{B}_{11} & \mathbf{V}_1'\mathbf{A}\mathbf{V}_2 \\ \mathbf{0} & \mathbf{V}_2'\mathbf{A}\mathbf{V}_2 \end{bmatrix}. \end{aligned} \tag{4.31}$$

When $S = \text{Span}(\mathbf{v}_1, \ldots, \mathbf{v}_k)$ is the invariant subspace associated with the (not necessarily linearly independent) eigenvectors $\mathbf{v}_1, \ldots, \mathbf{v}_k$, with $p \leq k$, the block-triangular matrix obtained in Equation (4.31) can be viewed as a weaker version of (4.28) and constitutes an essential ingredient for the numerical methods for finding eigenvalues presented in Section 4.6.

4.4 Normal Matrices

Self-adjoint matrices play an essential role in most practical applications of linear algebra. For instance, one of the postulates of *Quantum Mechanics* is that the observable quantities in finite-dimensional quantum mechanical systems, such as the spin of a particle in a given direction, correspond to self-adjoint matrices. As another example, most matrices arising in discrete approximations of partial differential equations are self-adjoint. Arguably, the theoretical reason behind their importance lies in the properties of their eigenvalues and eigenvectors.

Similarly, because of the isometry property (3.18), unitary matrices play an important role in situations where an inner product is preserved, which generally involve some type of *symmetry*. For example, another postulate of *Quantum Mechanics* says that the time evolution of a finite-dimensional quantum system is given by unitary matrices, corresponding to a symmetry with respect to translations in time. Accordingly, properties of their eigenvalues and eigenvectors have both theoretical and practical importance.

Both self-adjoint and unitary matrices are special cases of *normal* matrices, defined as matrices that satisfy the identity $\mathbf{A}\mathbf{A}' = \mathbf{A}'\mathbf{A}$. The importance of normal matrices relies on the fact that they are exactly the class of matrices for which the diagonalization expressed in (4.28) holds for a unitary matrix $\mathbf{V}$. The next theorem states this fact, together with properties of eigenvectors of normal matrices.

Theorem 4.6 *Let $\mathbf{A}$ be an $n \times n$ complex matrix. Then:*

1. *There exists a unitary matrix $\mathbf{U}$ such that $\mathbf{U}'\mathbf{A}\mathbf{U}$ is diagonal if and only if $\mathbf{A}$ is a normal matrix.*

2. *The eigenvectors of* $\mathbf{A}$ *form an orthonormal basis for* $\mathbb{C}^n$ *if and only if* $\mathbf{A}$ *is normal.*

3. *The eigenvectors of a normal matrix* $\mathbf{A}$ *associated with distinct eigenvalues are orthogonal.*

To see why the first part of this theorem is true, suppose that (4.28) holds for a unitary matrix $\mathbf{U}$, that is, suppose that

$$\mathbf{U}'\mathbf{A}\mathbf{U} = \begin{bmatrix} \lambda_1 & & & 0 \\ & \lambda_2 & & \\ & & \ddots & \\ 0 & & & \lambda_n \end{bmatrix}. \tag{4.32}$$

Taking the conjugate transpose of both sides of (4.32) then gives

$$\mathbf{U}'\mathbf{A}'\mathbf{U} = \begin{bmatrix} \bar{\lambda}_1 & & & 0 \\ & \bar{\lambda}_2 & & \\ & & \ddots & \\ 0 & & & \bar{\lambda}_n \end{bmatrix}. \tag{4.33}$$

We then have

$$\mathbf{U}'\mathbf{A}\mathbf{A}'\mathbf{U} = \begin{bmatrix} \bar{\lambda}_1\lambda_1 & & & 0 \\ & \bar{\lambda}_2\lambda_2 & & \\ & & \ddots & \\ 0 & & & \bar{\lambda}_n\lambda_n \end{bmatrix} = \mathbf{U}'\mathbf{A}'\mathbf{A}\mathbf{U},$$

which implies that $\mathbf{A}\mathbf{A}' = \mathbf{A}'\mathbf{A}$, so that $\mathbf{A}$ is normal.

Conversely, suppose that $\mathbf{A}$ is normal and let $\mathbf{A} = \mathbf{U}\mathbf{B}\mathbf{U}'$ be its Schur decomposition (4.14). Then, the identity $\mathbf{A}\mathbf{A}' = \mathbf{A}'\mathbf{A}$ implies that $\mathbf{B}\mathbf{B}' = \mathbf{B}'\mathbf{B}$ as well. But because $\mathbf{B}$ is upper triangular, equating the first diagonal entry of both sides of this equation gives

$$|b_{11}|^2 + |b_{12}|^2 + \cdots + |b_{1n}|^2 = |b_{11}|^2,$$

which can only hold provided $b_{12} = \cdots = b_{1n} = 0$. Similarly, equating the second diagonal entry of both sides of $\mathbf{B}\mathbf{B}' = \mathbf{B}'\mathbf{B}$ leads to

$$|b_{22}|^2 + |b_{23}|^2 + \cdots + |b_{2n}|^2 = |b_{22}|^2,$$

which can only hold provided $b_{23} = \cdots = b_{2n} = 0$. Continuing in this way, we conclude that $\mathbf{B}$ must be diagonal, which completes the proof of the first part of Theorem 4.6.

For the second part of Theorem 4.6, observe that the entries of the diagonal matrix $\mathbf{B}$ must be eigenvalues of $\mathbf{A}$. It then follows from the identity $\mathbf{AU} = \mathbf{UB}$ that the columns of $\mathbf{U}$, which form an orthonormal basis for $\mathbb{C}^n$, must be eigenvectors of $\mathbf{A}$.

As for the third part of Theorem 4.6, it requires the following exercise, which can be seen as a stronger version of Exercise 4.1 for the case of normal matrices:

Exercise 4.7 Show that if $\mathbf{A}$ is an $n \times n$ normal matrix, then $\|\mathbf{Ax}\| = \|\mathbf{A}'\mathbf{x}\|$ for all $\mathbf{x} \in \mathbb{C}^n$. Use this fact to show that for a normal matrix $\mathbf{A}$ we have $\mathbf{Ax} = \lambda\mathbf{x}$ if and only if $\mathbf{A}'\mathbf{x} = \bar{\lambda}\mathbf{x}$.

Suppose then that $\mathbf{v}_1$ and $\mathbf{v}_2$ are nonzero eigenvectors associated with two distinct eigenvalues λ_1 and λ_2. Using the result of Exercise 4.7, we obtain

$$0 = \langle \mathbf{v}_1, \mathbf{Av}_2 \rangle - \langle \mathbf{A}'\mathbf{v}_1, \mathbf{v}_2 \rangle = \lambda_2 \langle \mathbf{v}_1, \mathbf{v}_2 \rangle - \lambda_1 \langle \mathbf{v}_1, \mathbf{v}_2 \rangle = (\lambda_2 - \lambda_1) \langle \mathbf{v}_1, \mathbf{v}_2 \rangle,$$

which implies that $\langle \mathbf{v}_1, \mathbf{v}_2 \rangle = 0$.

For example, consider the following normal matrix and its Schur decomposition:

```
>> A=[i 2+i; -2-i 3-5i];
>> [U,B]=schur(A,'complex')

U =

     -0.4280 + 0.8560i       -0.1003 - 0.2715i
     -0.2592 - 0.1296i       -0.8979 + 0.3312i

B =

     -0.3028 + 1.6056i        0.0000 - 0.0000i
      0                       3.3028 - 5.6056i
```

The diagonal entries of $\mathbf{B}$ are the eigenvalues of $\mathbf{A}$, whose corresponding eigenvectors are the two columns of the unitary matrix $\mathbf{U}$, which therefore form a basis for $\mathbb{C}^2$.

Eigenvalues of self-adjoint matrices

Despite all the information provided by Theorem 4.6 regarding their eigenvectors, not much can be said about the eigenvalues of general normal matrices. For the special cases of self-adjoint and unitary matrices, however, much more can be said. Suppose that λ is an eigenvalue of a self-adjoint matrix $\mathbf{A}$, with a corresponding nonzero eigenvector $\mathbf{v} \in \mathbb{C}^n$. Then, we have

$$\lambda\|\mathbf{v}\|^2 = \langle \mathbf{v}, \mathbf{A}\mathbf{v} \rangle = \langle \mathbf{A}\mathbf{v}, \mathbf{v} \rangle = \bar{\lambda}\|\mathbf{v}\|^2,$$

which implies that λ must be a real number, since $\lambda = \bar{\lambda}$. Notice that this is true even if the self-adjoint matrix $\mathbf{A}$ happens to have complex entries, as can be seen in the next example:

```
>> A = [2 4+i 2; 4-i 0 8; 2 8 -8];
>> eig(A)'
ans =

  -12.9759   -1.2709    8.2468
```

If, in addition to being self-adjoint, the matrix $\mathbf{A}$ happens to be *positive*, in the sense that $\langle \mathbf{x}, \mathbf{A}\mathbf{x} \rangle \geq 0$ for all $\mathbf{x} \in \mathbb{C}^n$ (see Section 2.7), then its eigenvalues are *positive real numbers.* This is because we can take the vector $\mathbf{x}$ to be a nonzero eigenvector associated with the eigenvalue λ and conclude that

$$0 \leq \langle \mathbf{x}, \mathbf{A}\mathbf{x} \rangle = \lambda\|\mathbf{x}\|^2,$$

which implies that $\lambda \geq 0$. In practice, the most important examples occur for matrices of the form $\mathbf{A} = \mathbf{C}'\mathbf{C}$ where $\mathbf{C}$ is any $m \times n$ matrix, which are clearly self-adjoint and positive, since

$$\langle \mathbf{x}, \mathbf{A}\mathbf{x} \rangle = \langle \mathbf{x}, \mathbf{C}'\mathbf{C}\mathbf{x} \rangle = \langle \mathbf{C}\mathbf{x}, \mathbf{C}\mathbf{x} \rangle \geq 0.$$

The following example constructs a positive symmetric 3×3 matrix $\mathbf{A}$ by using a 4×3 matrix $\mathbf{C}$:

```
>> C = [2 1 -1; 3 0 4; 2 -3 1; 0 4 3];
>> eig(C'*C)'

ans =
     5.8958   26.9255   37.1787
```

which confirms that eigenvalues of $\mathbf{A}$ are positive numbers.

Eigenvalues of unitary matrices

Similarly, suppose that λ is an eigenvalue of a unitary matrix $\mathbf{U}$, with a corresponding nonzero eigenvector $\mathbf{v} \in \mathbb{C}^n$. Then,

$$\langle \mathbf{v}, \mathbf{v} \rangle = \langle \mathbf{U}\mathbf{v}, \mathbf{U}\mathbf{v} \rangle = \langle \lambda\mathbf{v}, \lambda\mathbf{v} \rangle = |\lambda|^2\langle \mathbf{v}, \mathbf{v} \rangle,$$

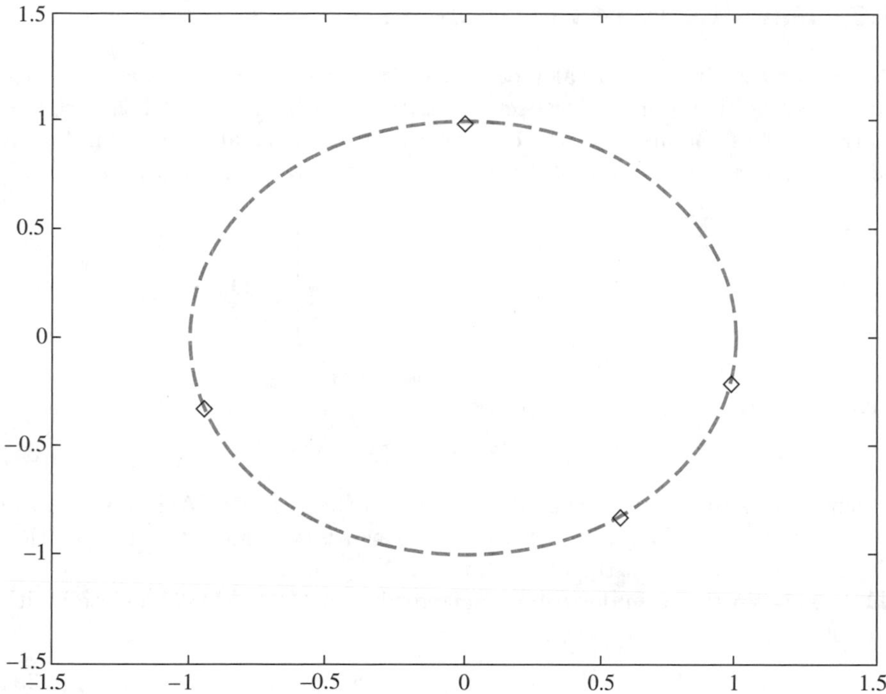

Figure 4.3 Eigenvalues of a unitary matrix

which implies that $|\lambda| = 1$. In other words, the eigenvalues of unitary matrices are located on the circle of radius one in the complex plane, as shown in Figure 4.3, obtained from the following example:

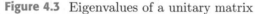

```
>> A=[-1 3 4 -2; 0 5 i -3;1-3*i  0 -2 1;3 2 -2*i 0];
>> [U,B]=schur(A,'complex'); % unitary matrix U obtained from Schur
>> lambda=eig(U)

lambda =
  -0.0151 + 0.9999i
  -0.9438 - 0.3305i
   0.5765 - 0.8171i
   0.9796 - 0.2011i

>> plot(real(lambda),imag(lambda),'dr'); hold on;
>> t=[0:0.01:2*pi];
>> plot(sin(t),cos(t),'--')
```

4.5 Sensitivity of Eigenvalues

Before introducing numerical methods for finding eigenvalues, let us study how sensitive they are with respect to small perturbations $\Delta\mathbf{A}$ of the original matrix $\mathbf{A}$. Assume first that the eigenvectors $\mathbf{v}_1, \ldots, \mathbf{v}_n$ are linearly independent so that according to (4.28) the matrix $\mathbf{A}$ can be diagonalized as

$$\mathbf{V}^{-1}\mathbf{A}\mathbf{V} = \begin{bmatrix} \lambda_1 & & & 0 \\ & \lambda_2 & & \\ & & \ddots & \\ 0 & & & \lambda_n \end{bmatrix} = \mathbf{D}. \tag{4.34}$$

Denoting $\Delta\mathbf{D} = \mathbf{V}^{-1}(\Delta\mathbf{A})\mathbf{V}$, we see that

$$\mathbf{V}^{-1}(\mathbf{A} + \Delta\mathbf{A})\mathbf{V} = \mathbf{D} + \Delta\mathbf{D}, \tag{4.35}$$

which implies that the eigenvalues of the perturbed matrix $(\mathbf{A} + \Delta\mathbf{A})$ are the same as the eigenvalues of $\mathbf{D} + \Delta\mathbf{D}$. In other words, we can obtain information about perturbations on the eigenvalues of $\mathbf{A}$ by analyzing the eigenvalues of $(\mathbf{D} + \Delta\mathbf{D})$. In this vein, let μ be an eigenvalue of $(\mathbf{D} + \Delta\mathbf{D})$ associated with a nonzero eigenvector $\mathbf{w}$, so that

$$(\mathbf{D} + \Delta\mathbf{D})\mathbf{w} = \mu\mathbf{w}. \tag{4.36}$$

If μ is also an eigenvalue of $\mathbf{A}$, then it is unperturbed and there is nothing more to say about it. Otherwise, if μ is not an eigenvalue of $\mathbf{A}$, then it is not an eigenvalue of $\mathbf{D}$ either, which means that the matrix $(\mathbf{D} - \mu\mathbf{1})$ is invertible. In this case, we can rewrite (4.36) as

$$\mathbf{w} = (\mu\mathbf{1} - \mathbf{D})^{-1}\Delta\mathbf{D}\mathbf{w}, \tag{4.37}$$

which after taking norms on both sides and dividing by $\|\mathbf{w}\|_2\|(\mu\mathbf{1} - \mathbf{D})^{-1}\|_2$ leads to

$$\|(\mu\mathbf{1} - \mathbf{D})^{-1}\|_2^{-1} \leq \|\Delta\mathbf{D}\|_2. \tag{4.38}$$

Moreover, because $(\mu\mathbf{1} - \mathbf{D})$ is a diagonal matrix, it is easy to see that

$$\|(\mu\mathbf{1} - \mathbf{D})^{-1}\|_2 = \frac{1}{|\mu - \lambda_j|},$$

where λ_j is the eigenvalue of $\mathbf{A}$ closest to μ. Therefore, inserting this into (4.38) gives

$$\begin{aligned} |\mu - \lambda_j| &\leq \|\Delta\mathbf{D}\|_2 = \|\mathbf{V}^{-1}(\Delta\mathbf{A})\mathbf{V}\|_2 \\ &\leq \|\mathbf{V}^{-1}\|_2\|\mathbf{V}\|_2\|\Delta\mathbf{A}\|_2 \\ &= \mathrm{cond}_2(\mathbf{V})\|\Delta\mathbf{A}\|_2. \end{aligned} \tag{4.39}$$

Comparing this with (2.39), we see that while the sensitivity of solutions to the linear system $\mathbf{A}\mathbf{x} = \mathbf{b}$ is governed by the condition number of $\mathbf{A}$, the sensitivity of the eigenvalue problem depends on the condition number for the matrix $\mathbf{V}$ of eigenvectors of $\mathbf{A}$. In other words, the eigenvalue problem for $\mathbf{A}$ is ill-conditioned whenever $\mathbf{V}$ is close to singular, which means that the eigenvectors of $\mathbf{A}$ are nearly linearly dependent. Conversely, if the eigenvectors of $\mathbf{A}$ are far from being linearly dependent, then $\mathbf{V}$ is far from being singular and the eigenvalue problem for $\mathbf{A}$ is well-conditioned.

As an example of an ill-conditioned eigenvalue problem, consider the matrix

$$\mathbf{A} = \begin{bmatrix} 5 & 9 & 9 \\ -72 & -79 & -72 \\ 90 & 90 & 80 \end{bmatrix}, \tag{4.40}$$

for which you have:

```
>> A=[5 9 9; -72 -79 -72; 90 90 80];
>> [V,lambda]=eig(A);
>> cond(V)

ans =
  374.5806
```

Because the condition number of $\mathbf{V}$ is large, you can expect that small perturbations on $\mathbf{A}$ will result in very different eigenvalues, which is confirmed in the following examples:

```
>> B=[5 9.1 9; -72 -79 -72; 90 90 80];
>> C=[5 9 9; -72 -78.9 -72; 90 90 80];
>> F=[5 9 9; -72 -79 -72; 89.9 90 80];
>> mu=[eig(A) eig(B) eig(C) eig(F)]

mu =
    5.0000     4.2783 + 3.7110i   4.0030 + 3.2764i   4.0247
    2.0000     4.2783 - 3.7110i   4.0030 - 3.2764i   3.2294
   -1.0000    -2.5565            -1.9059            -1.2541
```

By contrast, the eigenvalue problem is always well-conditioned when $\mathbf{A}$ is a normal matrix, because Theorem 4.6 guarantees that the matrix of eigenvectors $\mathbf{V}$ can be chosen to be unitary, in which case $\text{cond}_2(\mathbf{V}) = 1$. For instance, consider the following self-adjoint matrix:

```
>> A=[2 -1 4; -1 1 -3; 4 -3 5];
>> [V,lambda]=eig(A);
>> cond(V)

ans =
     1.0000
```

You can then verify that small perturbations on the matrix **A** correspond to small perturbations on its eigenvalues:

```
>> B=[2.1 -1 4; -1 1 -3; 4 -3 5];
>> C=[2 -1 4; -1 0.9 -3; 4 -3 5];
>> F=[2 -1 4; -1 1 -3; 3.9 -3 5];
>> mu=[eig(A) eig(B) eig(C) eig(F)]

mu =
   -1.3560   -1.3241   -1.3863    8.9040
    0.4123    0.4547    0.3554    0.4167
    8.9438    8.9694    8.9309   -1.3207
```

In order to derive (4.39), we had to assume that the eigenvalues of **A** were linearly independent. Moreover, (4.39) provides an overall bound for the sensitivity of all eigenvalues. For the sensitivity of individual eigenvalues, including the case of linearly dependent eigenvectors, we need to resort to a more detailed analysis that makes use of the eigenvectors of the adjoint matrix $\mathbf{A}'$. Let λ be an eigenvalue of **A** associated with the eigenvector **x**. According to Exercise 4.1, we know that $\bar{\lambda}$ is an eigenvalue of $\mathbf{A}'$. Therefore, there must exist a nonzero vector **y** satisfying

$$\mathbf{A}'\mathbf{y} = \bar{\lambda}\mathbf{y}.$$

Left-eigenvectors

Taking the conjugate transpose of both sides of this equation gives

$$\mathbf{y}' * \mathbf{A} = \lambda\mathbf{y}', \tag{4.41}$$

which justifies calling **y** a *left-eigenvector* of **A**.

Back to the sensitivity analysis, consider the modified eigenvalue problem

$$(\mathbf{A} + \Delta\mathbf{A})(\mathbf{x} + \Delta\mathbf{x}) = (\lambda + \Delta\lambda)(\mathbf{x} + \Delta\mathbf{x}). \tag{4.42}$$

Expanding both sides of this equation we obtain

$$\mathbf{Ax} + \mathbf{A}\Delta\mathbf{x} + (\Delta\mathbf{A})\mathbf{x} + \Delta\mathbf{A}\Delta\mathbf{x} = \lambda\mathbf{x} + \lambda\Delta\mathbf{x} + (\Delta\lambda)\mathbf{x} + \Delta\lambda\Delta\mathbf{x}.$$

Dropping the second-order terms $\Delta\mathbf{A}\Delta\mathbf{x}$ and $\Delta\lambda\Delta\mathbf{x}$ (because they are products of small perturbations and therefore are much smaller than the other terms in the equation) and using the fact that $\mathbf{Ax} = \lambda\mathbf{x}$, gives the following approximation

$$\mathbf{A}\Delta\mathbf{x} + (\Delta\mathbf{A})\mathbf{x} \approx \lambda\Delta\mathbf{x} + (\Delta\lambda)\mathbf{x}.$$

Multiplying both sides from the left by the row vector $\mathbf{y}'$ and using the fact that $(\mathbf{y}' * \mathbf{A}) * \Delta\mathbf{x} = (\lambda\mathbf{y}') * \Delta\mathbf{x}$ leads to

$$(\mathbf{y}' * \Delta\mathbf{A}) * \mathbf{x} \approx (\Delta\lambda)\mathbf{y}' * \mathbf{x}.$$

Therefore, provided that $\mathbf{y}' * \mathbf{x} \neq 0$, we have

$$\Delta\lambda = \frac{(\mathbf{y}' * \Delta\mathbf{A}) * \mathbf{x}}{\mathbf{y}' * \mathbf{x}}.$$

Taking norms of both sides gives the following approximate bound:

$$|\Delta\lambda| \lesssim \frac{\|\mathbf{y}\|_2\|\mathbf{x}\|_2}{|\mathbf{y}' * \mathbf{x}|}\|\Delta\mathbf{A}\|_2 = \frac{1}{\cos\theta}\|\Delta\mathbf{A}\|_2, \tag{4.43}$$

where θ is the angle between $\mathbf{x}$ and $\mathbf{y}$. In other words, the sensitivity of an individual eigenvalue λ depends on the orthogonality of the corresponding eigenvector $\mathbf{x}$ and left-eigenvector $\mathbf{y}$.

For instance, you can calculate the factors $(\cos\theta)^{-1}$ for the individual eigenvalues of the matrix (4.40) as follows:

```
>> A=[5 9 9; -72 -79 -72; 90 90 80];
>> [V,lambda]=eig(A),[Y,mu]=eig(A');
>> c_eig=[norm(Y(:,1))*norm(V(:,1))/abs(Y(:,1)'*V(:,1)),
norm(Y(:,2))*norm(V(:,2))/abs(Y(:,2)'*V(:,2)),
norm(Y(:,3))*norm(V(:,3))/abs(Y(:,3)'*V(:,3))]

c_eig =
  118.4662
  172.8872
   55.6709
```

These values coincide with the results obtained from the MATLAB built-in function `condeig`:

```
>> condeig(A)

ans =
  118.4662
  172.8872
   55.6709
```

You then conclude that the most sensitive eigenvalue for this matrix is $\lambda_2 = 2$, and the least sensitive is $\lambda_3 = -1$, as can be verified using the same perturbations as before:

```
>> B=[5 9.1 9; -72 -79 -72; 90 90 80];
>> C=[5 9 9; -72 -78.9 -72; 90 90 80];
>> F=[5 9 9; -72 -79 -72; 89.9 90 80];
>> Delta=[abs(eig(B)-eig(A)) abs(eig(C)-eig(A)) abs(eig(F)-eig(A))]
```

```
Delta =
    3.7805    3.4248    0.9753
    4.3545    3.8402    1.2294
    1.5565    0.9059    0.2541
```

When the eigenvector $\mathbf{x}$ and left-eigenvector $\mathbf{y}$ associated with an eigenvalue λ happen to be orthogonal, then the analysis leading to (4.43) fails to apply. Such a situation never arises when λ is a simple eigenvalue (i.e., with algebraic multiplicity equal to 1), but occur when λ is a *defective* eigenvalue, that is, an eigenvalue with geometric multiplicity strictly smaller than its algebraic multiplicity. Such eigenvalues are generally extremely sensitive to perturbations, as the next example shows.

Consider a 16×16 matrix $\mathbf{A}$ with all the diagonal entries equal to 2, all the entries immediately above the diagonal equal to 1, the lower-left entry equal to δ, and all other entries equal to 0. Using the properties of the determinant function, it is easy to see that the characteristic equation for this matrix is

$$\det(\mathbf{A} - \lambda\mathbf{1}) = (\lambda - 2)^{16} - \delta = 0.$$

When $\delta = 0$, $\lambda = 2$ is an eigenvalue of $\mathbf{A}$ with algebraic multiplicity equal to 16 and geometric multiplicity equal to 1 (because there is only one possible eigenvector for this eigenvalue). In other words, $\lambda = 2$ is a highly defective eigenvalue. For an extremely small perturbation corresponding to $\delta = 10^{-16}$, which is of the order of machine precision, you can see that the 16 different eigenvalues of $\mathbf{A}$ now satisfy the equation

$$(\lambda - 2)^{16} = 10^{-16},$$

that is, they are complex numbers on a circle centered at 2 with radius $r = 0.1$. This can be verified in MATLAB as follows:

```
>>A=[2 1 0 0 0 0 0 0 0 0 0 0 0 0 0 0; 0 2 1 0 0 0 0 0 0 0 0 0 0 0 0 0
0 0 2 1 0 0 0 0 0 0 0 0 0 0 0 0; 0 0 0 2 1 0 0 0 0 0 0 0 0 0 0 0
0 0 0 0 2 1 0 0 0 0 0 0 0 0 0 0; 0 0 0 0 0 2 1 0 0 0 0 0 0 0 0 0
0 0 0 0 0 0 2 1 0 0 0 0 0 0 0 0; 0 0 0 0 0 0 0 2 1 0 0 0 0 0 0 0
0 0 0 0 0 0 0 0 2 1 0 0 0 0 0 0; 0 0 0 0 0 0 0 0 0 2 1 0 0 0 0 0
0 0 0 0 0 0 0 0 0 0 2 1 0 0 0 0; 0 0 0 0 0 0 0 0 0 0 0 2 1 0 0 0
0 0 0 0 0 0 0 0 0 0 0 0 2 1 0 0; 0 0 0 0 0 0 0 0 0 0 0 0 0 2 1 0
0 0 0 0 0 0 0 0 0 0 0 0 0 0 2 1; 10^-16 0 0 0 0 0 0 0 0 0 0 0 0 0 0 2];
>> lambda=eig(A);
>> plot(real(lambda),imag(lambda),'dr');hold on
>> t=[0:0.01:2*pi];
>> plot(2+0.1*sin(t),0.1*cos(t),'--');axis([1.85 2.15 -.15 .15])
```

which produces the eigenvalues shown in Figure 4.4.

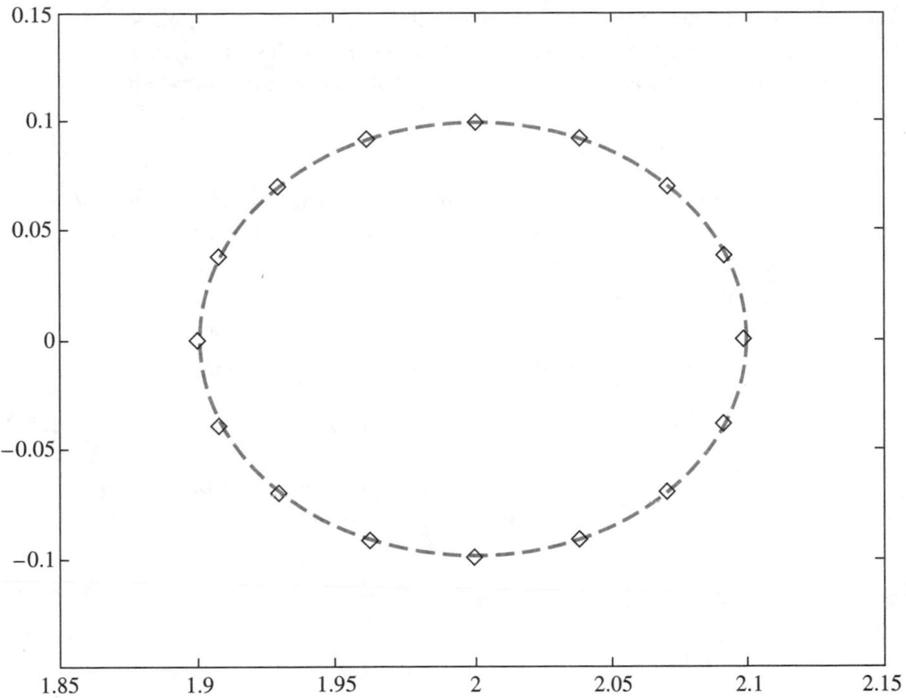

Figure 4.4 Perturbations of a highly defective eigenvalue $\lambda = 2$.

The sensitivity of eigenvectors is an even more delicate matter because it depends on the sensitivity of the corresponding eigenvalues as well as on the distance between the eigenvalues. For our purposes, it is enough to observe that when a matrix has well-conditioned eigenvalues that are not too close to each other, then the problem of computing eigenvectors is well-conditioned.

4.6 Power Iterations

Let us now consider numerical methods for finding an eigenvector $\mathbf{x}$ and its associated eigenvalue λ of an $n \times n$ complex matrix $\mathbf{A}$. In general, any method for computing eigenvalues *necessarily* involves an infinite number of steps. This is because finding eigenvalues of an $n \times n$ matrix is equivalent to finding the roots of its characteristic polynomial of degree n, and for $n > 4$ such roots cannot be found, in general, in a finite number of steps. In other words, we must consider iterative methods producing an approximated eigenvector $\mathbf{x}^{(k)}$ associated with an approximated eigenvalue $\lambda^{(k)}$ that converge to the desired eigenvector $\mathbf{x}$ and eigenvalue λ as the number of iterations becomes larger.

We then need to use some criterion to decide when the iterations should stop. Similar to the iterative methods for linear systems described in Section 2.6, we specify a desired tolerance level δ and stop the iterations when

$$\|\mathbf{A}\mathbf{x}^{(k)} - \lambda^{(k)}\mathbf{x}^{(k)}\| < \delta, \tag{4.44}$$

according to some vector norm. Although any norm produces equivalent results (see Section 2.4), the ∞-norm $\|\cdot\|_\infty$ renders itself more convenient for establishing the convergence rates of the iterative methods and is the preferred norm in the subsequent presentation.

Largest eigenvalue

Let us first assume that the $n \times n$ matrix $\mathbf{A}$ has a dominant eigenvalue λ_1. That is, suppose that the eigenvalues of $\mathbf{A}$ can be ordered as

$$|\lambda_1| > |\lambda_2| \geq \cdots \geq |\lambda_n|. \tag{4.45}$$

Assume further that their associated normalized eigenvectors $\mathbf{v}_1, \mathbf{v}_2, \ldots, \mathbf{v}_n$ form a basis in $\mathbb{C}^n$, so that we can write any initial vector $\mathbf{x}^{(0)}$ as

$$\mathbf{x}^{(0)} = \mu_1\mathbf{v}_1 + \cdots \mu_n\mathbf{v}_n \tag{4.46}$$

for some scalars $\mu_1, \ldots, \mu_n$. Finally, assume that the initial vector $\mathbf{x}_0$ is such that $\mu_1 \neq 0$ in (4.46). Then, for $k \geq 1$, define the iterative scheme

$$\begin{cases} \mathbf{y}^{(k)} = \mathbf{A}\mathbf{x}^{(k-1)} \\ \mathbf{x}^{(k)} = \frac{\mathbf{y}^{(k)}}{\|\mathbf{y}^{(k)}\|_\infty} \end{cases}. \tag{4.47}$$

At first sight, it seems strange to define these iterations in two separate steps. After all, we could have just inserted $\mathbf{y}^{(k)}$ directly into the expression for $\mathbf{x}^{(k)}$ and reduce it to just one step. In fact, doing exactly that leads to

$$\mathbf{x}^{(k)} = \frac{\mathbf{A}\mathbf{x}^{(k-1)}}{\|\mathbf{A}\mathbf{x}^{(k-1)}\|_\infty} = \frac{\mathbf{A}^2\mathbf{x}^{(k-2)}}{\|\mathbf{A}^2\mathbf{x}^{(k-2)}\|_\infty} = \cdots = \frac{\mathbf{A}^k\mathbf{x}^{(0)}}{\|\mathbf{A}^k\mathbf{x}^{(0)}\|_\infty}. \tag{4.48}$$

The problem with this expression is that on each term we have the same power of $\mathbf{A}$, which prevents us from exploring the fact that λ_1 has a larger absolute value than all the other eigenvalues. However, inserting the same expression in the definition of $\mathbf{y}^{(k)}$ leads to

$$\mathbf{y}^{(k)} = \mathbf{A}\mathbf{x}^{(k-1)} = \mathbf{A}\frac{\mathbf{A}^{k-1}\mathbf{x}^{(0)}}{\|\mathbf{A}^{k-1}\mathbf{x}^{(0)}\|_\infty}, \tag{4.49}$$

so that

$$\|\mathbf{y}^{(k)}\|_\infty = \frac{\|\mathbf{A}^k\mathbf{x}^{(0)}\|_\infty}{\|\mathbf{A}^{k-1}\mathbf{x}^{(0)}\|_\infty}. \tag{4.50}$$

Therefore, using the expansion (4.46), we find that

$$\|\mathbf{y}^{(k)}\|_\infty = |\lambda_1| \frac{\left\| \mu_1 \mathbf{v}_1 + \sum_{i=2}^{n} \mu_i \left(\frac{\lambda_i}{\lambda_1}\right)^k \mathbf{v}_i \right\|_\infty}{\left\| \mu_1 \mathbf{v}_1 + \sum_{i=2}^{n} \mu_i \left(\frac{\lambda_i}{\lambda_1}\right)^{k-1} \mathbf{v}_i \right\|_\infty}. \tag{4.51}$$

Since $|\lambda_i|/|\lambda_1| < 1$ for all $i > 1$ and $\mu_1 \neq 0$, we see that

$$\lim_{k\to\infty} \|\mathbf{y}^{(k)}\|_\infty = |\lambda_1|. \tag{4.52}$$

Moreover, because the preceding limit is equivalent to the convergence of the summands in (4.51), its convergence rate can be obtained from the error estimate

$$\|\mathbf{e}^{(k)}\|_\infty = \left\| \sum_{i=2}^{n} \mu_i \left(\frac{\lambda_i}{\lambda_1}\right)^k \mathbf{v}_i \right\|_\infty \leq \left|\frac{\lambda_2}{\lambda_1}\right|^k \sum_{i=2}^{n} |\mu_i| \|\mathbf{v}_i\|_\infty.$$

In the notation of Section 1.8, we see that $\|\mathbf{e}^{(k+1)}\|_\infty \leq C\|\mathbf{e}^{(k)}\|_\infty$ so that the convergence rate for power iterations is *linear* and has a constant

$$C = \left|\frac{\lambda_2}{\lambda_1}\right|. \tag{4.53}$$

That is, the smaller the second eigenvalue is compared to the first, the faster the method will converge. Conversely, the convergence can be quite slow if the first and second eigenvalues have comparable sizes.

In the same vein, inserting the expansion (4.46) into (4.48) leads to

$$\mathbf{x}^{(k)} = \frac{\lambda_1^k \left(\mu_1 \mathbf{v}_1 + \sum_{i=2}^{n} \mu_i \left(\frac{\lambda_i}{\lambda_1}\right)^k \mathbf{v}_i \right)}{|\lambda_1|^k \left\| \mu_1 \mathbf{v}_1 + \sum_{i=2}^{n} \mu_i \left(\frac{\lambda_i}{\lambda_1}\right)^k \mathbf{v}_i \right\|_\infty}. \tag{4.54}$$

That is, defining $\widetilde{\mathbf{v}}_1 = \frac{\mu_1}{|\mu_1|}\mathbf{v}_1$, we see that (1) if $\lambda_1 > 0$, the iterations converge to $\widetilde{\mathbf{v}}_1$, (2) if $\lambda_1 < 0$, they oscillate between the limits $\pm\widetilde{\mathbf{v}}_1$, and (3) if λ_1 is a general complex number, the iterations jump between different vectors of the form $\alpha\widetilde{\mathbf{v}}_1$ with $|\alpha| = 1$. Since the eigenvector is defined only up to multiplication by a nonzero scalar, the vector $\mathbf{x}^{(k)}$ for large enough k is an approximation for the eigenvector associated with λ_1 in all three cases.

Up to this point, we have produced only a sequence of approximations $\mathbf{x}^{(k)}$ for the eigenvector $\mathbf{v}_1$ and another sequence of approximations $\|\mathbf{y}^{(k)}\|_\infty$ for the absolute value of the eigenvalue λ_1. However, any eigenvalue λ associated with a nonzero eigenvector $\mathbf{v}$ must satisfy the following *Rayleigh quotient*:

$$\lambda = \frac{\langle \mathbf{v}, \mathbf{A}\mathbf{v} \rangle}{\langle \mathbf{v}, \mathbf{v} \rangle}. \tag{4.55}$$

Therefore, we can use the approximation $\mathbf{x}^{(k)}$ for the eigenvector $\mathbf{v}_1$ and define the corresponding approximation $\lambda^{(k)}$ for eigenvalue λ_1 as

$$\lambda^{(k)} = \frac{\langle \mathbf{x}^{(k)}, \mathbf{A}\mathbf{x}^{(k)} \rangle}{\langle \mathbf{x}^{(k)}, \mathbf{x}^{(k)} \rangle}. \tag{4.56}$$

An alternative explanation for the approximation (4.56) is that it gives the orthogonal projection of $\mathbf{A}\mathbf{x}^{(k)}$ in the direction of $\mathbf{x}^{(k)}$ (according to Section 3.2). In other words, $\lambda^{(k)}\mathbf{x}^{(k)}$ is the multiple of $\mathbf{x}^{(k)}$, which is closest to $\mathbf{A}\mathbf{x}^{(k)}$ in the Euclidean norm.

Power iterations can be implemented in MATLAB through the following code:

```
function Y = powerit(A,x0,max,tol)
% power iteration to find the largest eigenpair of A
% x0 initial column vector, length 1
% tol is the tolerance bound
% use infinity norm throughout computations

x = x0; error = 1; i = 0;
while (i <= max) & (error > tol)
    w = A*x;
    x = w./norm(w,inf);
    lambda = x'*A*x/(x'*x);
    error = norm(A*x-lambda*x,inf);
    i = i + 1;
    Y(i,:) = [x;lambda;error].'; % real transpose
end
```

Notice that we have introduced a maximum number of iterations in the MATLAB function `powerit` in addition to the stopping criterion (4.44). This modification is required because the convergence of the power iterations relies on the number of assumptions such as the existence of the dominant eigenvalue in (4.45) and the nonzero projection $\mu_1 \neq 0$ in (4.46). If the assumptions are not satisfied, the method does not converge and the only way to break the resulting infinite loop of computations is to limit the maximal number of iterations.

The output of the MATLAB function `powerit` is a matrix whose rows contain the approximated eigenvector, the approximated eigenvalue, and the error for each iteration. To illustrate the different behaviors of power iterations, consider the matrix

$$\mathbf{A} = \begin{bmatrix} \lambda_1 & 2 & -1 \\ 0 & \lambda_2 & 3 \\ 0 & 0 & \lambda_3 \end{bmatrix}, \tag{4.57}$$

so that you have exact control of the eigenvalues $\lambda_1, \lambda_2, \lambda_3$. Moreover, you also know that the eigenvector associated with λ_1 is $\mathbf{v}_1 = [1, 0, 0]$, making the convergence properties of $\mathbf{x}^{(k)}$ easier to identify.

For the first example, take

$$\lambda_1 = 4, \quad \lambda_2 = 1, \quad \lambda_3 = 0.8,$$

with an initial guess $\mathbf{x}^{(0)} = [1,\ 1,\ 1]'$, so that all assumptions for convergence are satisfied:

```
>> A = [4 2 -1; 0 1 3; 0 0 0.8];
>> Y = powerit(A,[1 1 1]',100,10^(-4))

Y =
    1.0000    0.8000    0.1600    3.8932    1.8345
    1.0000    0.2353    0.0235    4.2801    0.7012
    1.0000    0.0688    0.0042    4.1194    0.2019
    1.0000    0.0197    0.0008    4.0375    0.0574
    1.0000    0.0055    0.0002    4.0107    0.0160
    1.0000    0.0015    0.0000    4.0029    0.0044
    1.0000    0.0004    0.0000    4.0008    0.0012
    1.0000    0.0001    0.0000    4.0002    0.0003
    1.0000    0.0000    0.0000    4.0001    0.0001
```

Observe how the iterations converge to the correct eigenvalue $\lambda_1 = 4$ and eigenvector $\mathbf{v}_1 = [1, 0, 0]$. Also notice the linear convergence of the error (last column) to zero.

You can change the example to

$$\lambda_1 = -4, \quad \lambda_2 = 2, \quad \lambda_3 = 0.8$$

and make the same initial guess, and still satisfy the earlier assumptions, but this time you obtain:

```
>> A = [-4 2 -1; 0 2 3; 0 0 0.8];
>> Y = powerit(A,[1 1 1]',100,10^(-4))
```

```
Y =
   -0.6000    1.0000    0.1600   -0.0314    4.2212
    1.0000    0.5849    0.0302   -1.5803    2.1847
   -1.0000    0.4406    0.0084   -3.7458    2.5571
    1.0000    0.1861    0.0014   -3.4402    1.0163
   -1.0000    0.1037    0.0003   -4.1409    0.6376
    1.0000    0.0495    0.0001   -3.8866    0.2916
   -1.0000    0.0254    0.0000   -4.0469    0.1538
    1.0000    0.0126    0.0000   -3.9739    0.0750
   -1.0000    0.0063    0.0000   -4.0124    0.0380
    1.0000    0.0032    0.0000   -3.9936    0.0189
   -1.0000    0.0016    0.0000   -4.0031    0.0095
    1.0000    0.0008    0.0000   -3.9984    0.0047
   -1.0000    0.0004    0.0000   -4.0008    0.0024
    1.0000    0.0002    0.0000   -3.9996    0.0012
   -1.0000    0.0001    0.0000   -4.0002    0.0006
    1.0000    0.0000    0.0000   -3.9999    0.0003
   -1.0000    0.0000    0.0000   -4.0000    0.0001
    1.0000    0.0000    0.0000   -4.0000    0.0001
```

Observe that the convergence to the correct eigenvalue and eigenvector happens now at a slower rate because the ratio of the second to the first eigenvalue is now larger. Observe also that, because $\lambda_1 < 0$, the approximated eigenvector oscillates between $\pm\mathbf{v}_1$ in each iteration.

Exercise 4.8 Consider the same example with $\lambda_2 = 3.9$ and find the number of iterations required to achieve the accuracy of 10^{-4}. Change to $\lambda_2 = 4$, and observe and explain why the iterations fail to converge.

Finally, take

$$\lambda_1 = 10 - 3i, \quad \lambda_2 = 1, \quad \lambda_3 = 0.8$$

and modify the initial guess to the vector $\mathbf{x}^{(0)} = [0, 1, 1]$, for which $\mu_1 = 0$. Although in principle the method should fail, in practice round-off errors create the nonzero value for the projection to $\mathbf{v}_1$ and the method eventually converges to the solution:

```
>> A = [10-3i 2 -1; 0 1 3; 0 0 0.8];
>> Y = powerit(A,[0 1 1]',100,10^(-4))
```

```
Y =
   0.2500              1.0000       0.2000       2.4553 - 0.1701i    3.7534
   0.9851 - 0.1718i    0.3666       0.0367       9.5639 - 2.5362i    3.1687
   0.9065 - 0.4223i    0.0431       0.0026      10.0591 - 2.9592i    0.4028
   0.7504 - 0.6610i    0.0049       0.0002      10.0069 - 2.9936i    0.0455
   0.5295 - 0.8483i    0.0005       0.0000      10.0005 - 2.9991i    0.0049
   0.2635 - 0.9647i    0.0001       0.0000      10.0000 - 2.9999i    0.0005
  -0.0248 - 0.9997i    0.0000       0.0000      10.0000 - 3.0000i    0.0001
```

Smallest eigenvalue

With a small modification, we can use power iterations to find the eigenvalue of a matrix $\mathbf{A}$ with the smallest absolute value. It follows directly from (4.2) that if a matrix $\mathbf{A}$ is not invertible, then $\lambda = 0$ is its eigenvalue with the smallest absolute value. Otherwise, suppose that $\mathbf{A}$ is invertible and observe that

$$\begin{aligned} \mathbf{Av} = \lambda\mathbf{v} \quad &\Leftrightarrow \quad \mathbf{A}^{-1}\mathbf{Av} = \mathbf{A}^{-1}(\lambda\mathbf{v}) \\ &\Leftrightarrow \quad \mathbf{v} = \lambda\mathbf{A}^{-1}\mathbf{v} \\ &\Leftrightarrow \quad \mathbf{A}^{-1}\mathbf{v} = \lambda^{-1}\mathbf{v}. \end{aligned}$$

Therefore, λ is an eigenvalue of $\mathbf{A}$ associated with an eigenvector $\mathbf{v}$ if and only if λ^{-1} is an eigenvalue of $\mathbf{A}^{-1}$ associated with the eigenvector $\mathbf{v}$. It then follows that applying the power iterations algorithm to the matrix $\mathbf{A}^{-1}$ will produce the eigenvalue λ_n^{-1}, where λ_n denotes the eigenvalue of $\mathbf{A}$ with the smallest absolute value. Accordingly, now assume that

$$|\lambda_1| \geq |\lambda_2| \geq \cdots > |\lambda_n| \tag{4.58}$$

and that the projection of the initial guess $\mathbf{x}^{(0)}$ onto the eigenvector $\mathbf{v}_n$ for the smallest eigenvalue is nonzero, that is, $\mu_n \neq 0$ in (4.46).

The iteration scheme for finding the smallest eigenvalue is then

$$\begin{cases} \mathbf{Ay}^{(k)} = \mathbf{x}^{(k-1)} \\ \mathbf{x}^{(k)} = \frac{\mathbf{y}^{(k)}}{\|\mathbf{y}^{(k)}\|_\infty} \end{cases}, \tag{4.59}$$

which is equivalent to

$$\begin{cases} \mathbf{y}^{(k)} = \mathbf{A}^{-1}\mathbf{x}^{(k-1)} \\ \mathbf{x}^{(k)} = \frac{\mathbf{y}^{(k)}}{\|\mathbf{y}^{(k)}\|_\infty} \end{cases}. \tag{4.60}$$

Observe that the first step in this scheme corresponds to solving a linear system for the variable $\mathbf{y}^{(k)}$, which should be done without explicitly computing the inverse of $\mathbf{A}$ for better computational efficiency. In particular, since the matrix remains the same throughout the iterations, an efficient way of implementing them is to do an LU factorization of $\mathbf{A}$ before the iterations start, and then use backward and forward substitutions for the solution of the linear systems as it is described in Section 2.5.

The rest of the analysis done for the power iterations (4.47) carries over to these *inverse power iterations.* In particular, the convergence rate to the smallest eigenvalue is still linear, but is now determined by the constant

$$C = \left| \frac{\lambda_n}{\lambda_{n-1}} \right|. \tag{4.61}$$

An implementation of this method is given in the following MATLAB function:

```
function Y = inversepowerit(A,x0,maxit,tol)
% inverse power iteration to find the
% minimal eigenpair of A
% x0 initial column vector
% tol is the tolerance bound

[L,U] = lufactor(A); x = x0; error = 1; i = 0;
 while (i <= maxit) & (error > tol)
    w = backsub(U,forwsub(L,x));
    x = w./norm(w,inf);
    lambda = (x'*A*x)/(x'*x);
    error = norm(A*x-lambda*x,inf);
    i = i + 1;
    Y(i,:) = [x;lambda;error].';
end
```

For illustration, we use this function to calculate the smallest eigenvalue for the matrix (4.57) with $\lambda_1 = -4$, $\lambda_2 = 2$, and $\lambda_3 = 0.8$:

```
>> A = [-4 2 -1; 0 2 3; 0 0 0.8];
>> Y = inversepowerit(A,[1,1,1]',100,10^(-4))

Y =
   -0.9091   -1.0000    0.9091   -0.2741    0.9765
   -0.5258   -1.0000    0.5155    0.5726    0.1190
   -0.5202   -1.0000    0.4394    0.6988    0.0445
   -0.5055   -1.0000    0.4149    0.7647    0.0146
   -0.5026   -1.0000    0.4058    0.7854    0.0059
   -0.5009   -1.0000    0.4023    0.7943    0.0023
   -0.5004   -1.0000    0.4009    0.7977    0.0009
   -0.5002   -1.0000    0.4004    0.7991    0.0004
   -0.5001   -1.0000    0.4001    0.7996    0.0001
   -0.5000   -1.0000    0.4001    0.7999    0.0001
```

Intermediate eigenvalues

If we are interested in approximating intermediate eigenvalues, we first must specify a number $b \in \mathbb{C}$, and then consider the problem of finding the eigenvalue of $\mathbf{A}$ that is closest to b in absolute value. Since

$$\mathbf{A}\mathbf{v} = \lambda\mathbf{v} \Leftrightarrow (\mathbf{A} - b\mathbf{1})\mathbf{v} = (\lambda - b)\mathbf{v},$$

it follows that λ is an eigenvalue of $\mathbf{A}$ with eigenvector $\mathbf{v}$ if and only if $(\lambda - b)$ is an eigenvalue of $(\mathbf{A} - b\mathbf{1})$ with the same eigenvector. Therefore if $(\lambda - b)$ is the smallest eigenvalue for the matrix $(\mathbf{A} - b\mathbf{1})$, then λ is the eigenvalue of $\mathbf{A}$ that is closest to the complex number b, and we can use inverse power iterations to find it. The convergence rate in this case is still linear, but is now determined by the modified constant

$$C = \left|\frac{\lambda - b}{\mu - b}\right|, \tag{4.62}$$

where μ is the next eigenvalue nearest to b. As an example, take $b = 2.2$ and the matrix (4.57) with $\lambda_1 = -4$, $\lambda_2 = 2$, and $\lambda_3 = 0.8$:

```
>> A = [-4 2 -1; 0 2 3; 0 0 0.8];
>> Y = inversepowerit(A-2.2*eye(3),[1,1,1]',100,10^(-4))

Y =
   -0.3255   -1.0000   -0.0455    0.0787    0.0601
    0.3312    1.0000    0.0059   -0.1820    0.0072
   -0.3330   -1.0000   -0.0008   -0.1974    0.0010
    0.3333    1.0000    0.0001   -0.1996    0.0001
   -0.3333   -1.0000   -0.0000   -0.1999    0.0000
```

The resulting eigenvalue $\lambda - b = -0.2$ corresponds to the intermediate eigenvalue $\lambda_2 = 2$ of the matrix $\mathbf{A}$. Notice how the convergence in this example was accelerated because of the fact that the constant $C = 0.2/1.4 \approx 0.143$ in (4.62) is very small.

Rayleigh quotient iterations

The last observation is the base for the improved iterative method known as *Rayleigh quotient iterations*. The idea for this method is that, since the Rayleigh quotient (4.56) gives an approximation for the eigenvalue λ, we can use it as a choice of b in order to obtain a small constant C in (4.62), therefore enhancing the convergence of power iterations to that eigenvalue. We are then led to the following iterative scheme, for $k \geq 1$ starting from an initial guess $\mathbf{x}^{(0)}$:

$$\begin{cases} \lambda^{(k-1)} = \frac{\langle \mathbf{x}^{(k-1)}, \mathbf{A}\mathbf{x}^{(k-1)}\rangle}{\langle \mathbf{x}^{(k-1)}, \mathbf{x}^{(k-1)}\rangle} \\ (\mathbf{A} - \lambda^{(k-1)}\mathbf{1})\mathbf{y}^{(k)} = \mathbf{x}^{(k-1)} \\ \mathbf{x}^{(k)} = \frac{\mathbf{y}^{(k)}}{\|\mathbf{y}^{(k)}\|_2} \end{cases}. \tag{4.63}$$

Note that the second step in (4.63) requires solving a linear system with a different matrix $(\mathbf{A}-\lambda^{(k-1)}\mathbf{1})$ at each iteration so that we cannot use the same factorization argument used in the implementation of inverse power iterations.

The convergence rate for the iterative method (4.63) is at least quadratic and in most cases even cubic. This type of extraordinarily fast convergence more than compensates for the necessity of solving a linear system with a different matrix at each iteration. The following MATLAB function implements the iterative scheme (4.63):

```
function Y = rayleigh(A,x0,maxit,tol)
% Rayleigh quotient iteration

[m,n] = size(A); x = x0; i = 1;
lambda = x'*A*x;
error=norm(A*x-lambda*x);
while (i < maxit) & (error > tol)
    w = (A-lambda*eye(m))\x;
    x = w/norm(w);
    lambda = x'*A*x;
    error = norm(A*x-lambda*x);
    Y(i,:) = [x;lambda;error].';
    i=i+1;
end
```

The iterations may converge to different eigenvectors depending on the initial guess $\mathbf{x}^{(0)}$, and so in principle we can try to find all eigenvalues of $\mathbf{A}$ by modifying the initial guess $\mathbf{x}^{(0)}$. For example, take the matrix (4.57) with $\lambda_1 = -4$, $\lambda_2 = 3.9$, and $\lambda_3 = 0.8$. The initial guess $\mathbf{x}^{(0)} = [-1, -1, 1]$ leads to:

```
>> format long
>> A = [-4 2 -1; 0 3.9 3; 0 0 0.8];
>> Y = rayleigh(A,[-1,-1,1]',100,10^-15);
>> Y(:,4:5)

ans =
   0.81170298516373   0.02343426920442
   0.80001924747192   0.00005753802531
   0.80000000013598   0.00000000022846
   0.80000000000000   0.00000000000000
```

where we choose to print the approximations to the eigenvalue $\lambda_3 = 0.8$ as well as the error to highlight the extraordinary rate of convergence, resulting in an error smaller than 10^{-15} in just four iterations! A different initial guess $\mathbf{x}^{(0)} = [1, 2, 3]$ now gives

```
>> Y = rayleigh(A,[1,1,1]',100,10^-15);
>> Y(:,4:5)

ans =
   3.21021581232051   3.45836387934243
   3.44315511994641   0.35411059036995
   3.98778294099921   0.09034671964631
   3.90246535983193   0.00239984660074
   3.90000197443248   0.00000192238823
   3.90000000000126   0.00000000000123
   3.90000000000000   0.00000000000000
   3.90000000000000   0.00000000000000
```

where you can again observe the fast and accurate way you arrive at the eigenvalue $\lambda_2 = 3.9$. Finally, the initial guess $\mathbf{x}^{(0)} = [-1, 1, 0]$ produces

```
>> Y = rayleigh(A,[-1,1,0]',100,10^-15);
>> Y(:,4:5)

ans =
  -3.12854607306611   1.66927969235320
  -4.05522840964635   0.25123742978047
  -4.00043467397882   0.00171853336554
  -4.00000002393457   0.00000009454157
  -4.00000000000000   0.00000000000000
```

showing convergence to the eigenvalue $\lambda_1 = -4$ in five iterations. If you find the requirement of changing the initial guess unpleasant, recall from Exercise 4.8 that it takes a large number of iterations to converge to $\lambda_1 = -4$ using the power method because $\lambda_2 = 3.9$ is close in absolute value to $\lambda_1 = -4$ and the linear convergence of the power method is slow. Therefore, a little hindrance in dealing with the choice of $\mathbf{x}^{(0)}$ is more than outweighed by the effectiveness of this method. Moreover, the Rayleigh iterations converge even if some eigenvalues have the same absolute value, as you can verify in the next exercise.

Exercise 4.9 *Compute all three eigenvalues of the matrix* $\mathbf{A}$ *given by* (4.57) *with* $\lambda_1 = -5$, $\lambda_2 = 5$, *and* $\lambda_3 = 2$.

Deflation

Suppose that we have already calculated an eigenvalue λ_1 and eigenvector $\mathbf{v}_1$ of an $n \times n$ complex matrix $\mathbf{A}$. For instance, λ_1 could be the dominant eigenvalue of $\mathbf{A}$, obtained using power iterations. We now describe a *deflation* method producing an $(n-1) \times (n-1)$ matrix whose eigenvalues are the remaining eigenvalues of $\mathbf{A}$.

Let the eigenvector $\mathbf{v}_1 \in \mathbb{C}^n$ be normalized with respect to the Euclidean norm and let $\mathbf{H}_1$ be the Householder transformation such that

$$\mathbf{H}_1\mathbf{v}_1 = \mathbf{e}_1,$$

according to the algorithm described in Section 3.3. Because $\mathbf{H}_1$ is a unitary matrix, we obtain that

$$\mathbf{H}_1\mathbf{A}\mathbf{H}_1'\mathbf{e}_1 = \mathbf{H}_1\mathbf{A}\mathbf{v}_1 = \lambda_1\mathbf{H}_1\mathbf{v}_1 = \lambda_1\mathbf{e}_1,$$

so that the matrix $\mathbf{H}_1\mathbf{A}\mathbf{H}_1'$ must be of the form

$$\mathbf{H}_1\mathbf{A}\mathbf{H}_1' = \begin{bmatrix} \lambda_1 & \mathbf{b}' \\ \mathbf{0} & \mathbf{B} \end{bmatrix} \tag{4.64}$$

for some column vector $\mathbf{b} \in \mathbb{C}^{n-1}$ and some $(n-1) \times (n-1)$ matrix $\mathbf{B}$. Because $\mathbf{H}_1\mathbf{A}\mathbf{H}_1'$ is a similarity transformation, the eigenvalues of $\mathbf{A}$ must be the same as the eigenvalues of the block-triangular matrix given by (4.64) (see (4.16)). Therefore, the eigenvalues of $\mathbf{B}$ must coincide with the remaining eigenvalues of $\mathbf{A}$ (see Exercise 4.5). For example, consider the following 4×4 case:

```
>> A = [1 2 2 -2; 0 2 0 -3; 3 -6 -4 6; -1 2 2 -3];
>> eig(A)'

ans =
   -7.0000    2.0000   -1.0000    2.0000

>> Y = powerit(A,[1,1,1,1]',100,10^(-10));
>> [n,m] = size(Y);
>> lambda1 = Y(n,m-1)

lambda1 =
   -7.0000

>> v1 = Y(n,1:m-2)'/norm(Y(n,1:m-2));
>> e1 = [1 0 0 0]';
>> H = eye(4)-2*((v1-e1)*(v1-e1)')/((v1-e1)'*(v1-e1));
>> C = H*A*H'

C =
   -7.0000   -7.8561    1.5991    4.1502
   -0.0000    1.0107   -0.6275   -2.0303
    0.0000   -0.3948    1.7496   -0.8102
   -0.0000   -0.8577   -0.5441    0.2397
```

```
>> B = C(2:4,2:4);
>> eig(B)'

ans =
   -1.0000      2.0000      2.0000
```

You can then proceed to calculate an eigenvalue λ_2 and its associated eigenvector $\mathbf{u}_2$ for the smaller matrix $\mathbf{B}$, say, by using power iterations again. As we have just established, λ_2 is automatically an eigenvalue of the original matrix $\mathbf{A}$. Moreover, provided $\lambda_2 \neq \lambda_1$, we can verify that the vector

$$\mathbf{v}_2 = \mathbf{H}' \begin{bmatrix} \alpha \\ \mathbf{u}_2 \end{bmatrix}, \qquad \alpha = \frac{\mathbf{b}' * \mathbf{u}_2}{\lambda_2 - \lambda_1}$$

is an eigenvector of $\mathbf{A}$ associated with the eigenvalue λ_2, since it satisfies

$$\begin{aligned} \mathbf{A}\mathbf{v}_2 &= \mathbf{H}' \begin{bmatrix} \lambda_1 & \mathbf{b}' \\ \mathbf{0} & \mathbf{B} \end{bmatrix} \begin{bmatrix} \alpha \\ \mathbf{u}_2 \end{bmatrix} = \mathbf{H}' \begin{bmatrix} \alpha\lambda_1 + \mathbf{b}' * \mathbf{u}_2 \\ \mathbf{B}\mathbf{u}_2 \end{bmatrix} \\ &= \mathbf{H}' \begin{bmatrix} \lambda_2\alpha \\ \lambda_2\mathbf{u}_2 \end{bmatrix} = \lambda_2\mathbf{H}' \begin{bmatrix} \alpha \\ \mathbf{u}_2 \end{bmatrix} = \lambda_2\mathbf{v}_2. \end{aligned}$$

Continuing the previous example, we obtain:

```
>> W = powerit(B,[1,1,1]',100,10^(-10));
>> [n,m] = size(W);
>> lambda2 = W(n,m-1)

lambda2 =
     2.0000

>> u2 = W(n,1:3)';
>> b = C(1,2:4);
>> alpha = (b*u2)/(lambda2-lambda1);
>> v2 = H'*[alpha; u2]

v2 =
     0.8123
    -0.3840
     0.7902
    -0.0000

>> norm(A*v2-lambda2*v2)

ans =
   7.2614e-011
```

The last output confirms that $\mathbf{v}_2$ is an eigenvector of $\mathbf{A}$ associated with the eigenvalue λ_2, up to computational error.

Having calculated the eigenvalue λ_2, we can apply the deflation method to the matrix $\mathbf{B}$ to obtain an $(n-2) \times (n-2)$ matrix whose eigenvalues are the remaining eigenvalues of $\mathbf{B}$. Continuing in this way, we see that we could in principle find all the eigenvalues of $\mathbf{A}$ (and their associated eigenvectors, provided the eigenvalues are simple), one by one, through deflation to smaller and smaller matrices. In the interest of efficiency, it might be better to consider numerical methods that compute all the eigenvalues and eigenvectors of $\mathbf{A}$ at the same time, and these are the topics of the next section.

4.7 Simultaneous Iterations

We now move to numerical methods that compute several different eigenvalues and their associated eigenvectors at once. Recall that the power iterations method relies on the fact that, provided a single dominant eigenvalue exists for the $n \times n$ matrix $\mathbf{A}$, successive applications of $\mathbf{A}$ to an arbitrary initial vector $\mathbf{x}^{(0)}$ converge to the subspace spanned by its dominant eigenvector. Simultaneous iterations extend this idea in the case of an initial $n \times p$ matrix $\mathbf{X}^{(0)}$, exploring the convergence to the subspace $S \in \mathbb{C}^n$ generated by the first p eigenvectors of $\mathbf{A}$.

Similar to the hypotheses made for power iterations, assume that the eigenvalues of an $n \times n$ matrix $\mathbf{A}$ can be ordered as

$$|\lambda_1| \geq |\lambda_2| \geq \cdots \geq |\lambda_p| > |\lambda_{p+1}| \geq |\lambda_{p+2}| \geq |\lambda_n|, \tag{4.65}$$

that is, assume that $\mathbf{A}$ has p dominant eigenvalues. Notice how in the preceding ordering we only assumed strict dominance from λ_p to λ_{p+1}, whereas equality of absolute value is allowed to occur among the first p eigenvalues. We also assume that the associated eigenvectors $\mathbf{v}_1, \ldots, \mathbf{v}_n$ are linearly independent in $\mathbb{C}^n$ and let $\mathbf{X}_0$ be an initial $n \times p$ matrix of rank p. Analogous to the condition that the initial guess $\mathbf{x}^{(0)}$ for power iterations should have a nonzero component in the direction of the dominant eigenvector $\mathbf{v}_1$, let us assume also that no vector in the invariant subspace $S = \text{Span}(\mathbf{v}_1, \ldots, \mathbf{v}_p)$ is orthogonal to all columns of $\mathbf{X}_0$.

Orthogonal iterations

Under these conditions, define the $n \times p$ matrix

$$\mathbf{X}_k = \mathbf{A}^k \mathbf{X}_0. \tag{4.66}$$

Then, a modification of the analysis presented for power iterations in Section 4.6 shows that, as k becomes large, each column of $\mathbf{X}_k$ converges to a vector in the invariant subspace S. In the same way as we normalized each of the vectors in the power iterations, we obtain orthonormal vectors by applying a QR factorization to each previous matrix $\mathbf{X}_k$. This leads to the following *orthogonal iterations*:

$$\begin{cases} \widehat{\mathbf{Q}}_{k-1}\widehat{\mathbf{R}}_{k-1} = \mathbf{X}_{k-1} \\ \mathbf{X}_k = \mathbf{A}\widehat{\mathbf{Q}}_{k-1} \end{cases}. \tag{4.67}$$

Since by definition $\widehat{\mathbf{Q}}_k$ is an $n \times p$ left-orthogonal matrix whose columns span the same subspace as the columns of $\mathbf{X}_k$, we obtain that the sequence $\widehat{\mathbf{Q}}_k$ must converge to an $n \times p$ left-orthogonal matrix $\widehat{\mathbf{Q}}$ whose columns span the invariant subspace S.

Therefore, there must exist a $p \times p$ matrix $\mathbf{B}$ such that

$$\mathbf{A}\widehat{\mathbf{Q}} = \widehat{\mathbf{Q}}\mathbf{B}. \tag{4.68}$$

But this shows that

$$\mathbf{B}\mathbf{y} = \lambda\mathbf{y} \Leftrightarrow \widehat{\mathbf{Q}}\mathbf{B}\mathbf{y} = \lambda\widehat{\mathbf{Q}}\mathbf{y} \Leftrightarrow \mathbf{A}(\widehat{\mathbf{Q}}\mathbf{y}) = \lambda(\widehat{\mathbf{Q}}\mathbf{y}),$$

which implies that the eigenvalues of $\mathbf{B}$ must be the first p eigenvalues of $\mathbf{A}$.

If this was all that we could conclude, then the iterations (4.67) would not have taken us very far because we would still have to find a way to compute all the eigenvalues of $\mathbf{B}$. The key point is that, if there is any $1 \le j \le p$ for which $|\lambda_j| > |\lambda_{j+1}|$, then the preceding matrix $\widehat{\mathbf{Q}}$ can be partitioned as

$$\widehat{\mathbf{Q}} = \begin{bmatrix} \mathbf{Q}^{(j)} & \mathbf{Q}^{(p-j)} \end{bmatrix},$$

where $\mathbf{Q}^{(j)}$ is the $n \times j$ left-orthogonal matrix that would arise if the iterations were restricted to the first j columns of $\mathbf{X}_0$. In other words, the columns of $\mathbf{Q}^{(j)}$ form an orthonormal basis for the invariant subspace of $\mathbf{A}$ spanned by its first j eigenvectors. Therefore, from the same argument used in Section 4.3 in connection to invariant subspaces, it follows that $\mathbf{B}$ must be block triangular. In particular, if the strict inequality $|\lambda_j| > |\lambda_{j+1}|$ holds for all $j = 1, \dots, p$, then repeating this argument for each pair of consecutive eigenvalues shows that $\mathbf{B}$ must be upper triangular, in which case the first p eigenvalues of $\mathbf{A}$ can be readily obtained from the diagonal entries of $\mathbf{B}$.

Here is a MATLAB implementation for the method of orthogonal iterations:

```
function [Q,B] = orthogit(A,X0,maxit)
% orthogonal iterations for the first p eigenvalues of A
% X0 is nxp inital matrix
% Q  is nxp left-orthogonal matrix
% B is pxp block-triangular matrix
```

```
X = X0;
for i = 1:maxit
    [Q,R] = QRfactor(X);
    X = A*Q;
end
B = Q'*A*Q;
```

For the first example, consider a 4×4 matrix that has two dominant eigenvalues with the same absolute value:

```
>> A = [-1 2 3 1; 2 -1 1 3; 3 1 -1 2; 1 3 2 -1];
>> eig(A)'

ans =
   -5.0000   -3.0000   -1.0000    5.0000

>> X0 = [1 0 0; 0 1 0; 0 0 1; 1 -1 2];
>> [Q,B] = orthogit(A,X0,25)

Q =
    0.7071    0.0000   -0.5000
         0    0.7071   -0.5000
         0    0.7071    0.5000
    0.7071   -0.0000    0.5000


B =
         0    5.0000    0.0000
    5.0000   -0.0000    0.0000
    0.0000    0.0000   -3.0000
```

Notice how the matrix $\mathbf{B}$ in this example turns out to be block triangular, with the eigenvalue $\lambda_3 = -3$ as one of its diagonal entries, while clearly $\lambda_1 = 5$ and $\lambda_2 = -5$ are the two eigenvalues of the remaining block. Also notice how the third column of $\mathbf{Q}$ is indeed the eigenvector of $\mathbf{A}$ associated with λ_3, while the other two columns of $\mathbf{Q}$ are not eigenvectors of $\mathbf{A}$ themselves, but span the same subspace as the eigenvector associated with λ_1 and λ_2.

Consider now an example where all the three dominant eigenvalues of a 4×4 matrix have distinct absolute values:

```
>> A = [2 3 5 6; 3 -1 -2 -3; 5 -2 0 2; 6 -3 2 22];
>> eig(A)'

ans =
   -6.9114    0.7956    4.7923   24.3235
```

```
>> [Q,B] = orthogit(A,X0,50)

Q =
    0.2740    0.5969   -0.7391
   -0.0909   -0.5631   -0.3370
    0.1417   -0.5531   -0.5222
    0.9469   -0.1440    0.2597

B =
   24.3235   -0.0000    0.0000
   -0.0000   -6.9114   -0.0000
   -0.0000   -0.0000    4.7923
```

Notice now that **B** is upper triangular as it should be (because the three dominant eigenvalues have distinct absolute values) and has the eigenvalues of **A** as its diagonal entries. Moreover, each column of **Q** is an eigenvector of **A** associated with the eigenvalue appearing in the corresponding column of **B**, because **B** turns out to be diagonal in this example.

To summarize, the orthogonal iterations (4.67) starting from an initial $n \times p$ matrix $\mathbf{X}_0$ lead us to the expression $\mathbf{A}\widehat{\mathbf{Q}} = \widehat{\mathbf{Q}}\mathbf{B}$ where $\mathbf{B}$ is a $p \times p$ block-triangular matrix. Furthermore, each diagonal block in **B** is a $k \times k$ matrix whenever **A** has k eigenvalues with the same absolute value among its first p dominant eigenvalues. In particular, if all the first p eigenvalues have different absolute values, then **B** is an upper-triangular matrix.

As the analysis of orthogonal iterations suggests, we don't need to restrict ourselves to the first p eigenvectors of an $n \times n$ matrix **A**. Provided the eigenvectors of **A** form a basis for $\mathbb{C}^n$, simply putting $n = p$ in the orthogonal iterations (4.67) will produce an $n \times n$ block-triangular matrix **B** with the same eigenvalues as **A**.

The *QR iterations* method gives an alternative and elegant reformulation of this scheme: **QR iterations**

$$\begin{cases} \mathbf{Q}_k\mathbf{R}_k = \mathbf{A}_{k-1} \\ \mathbf{A}_k = \mathbf{R}_k\mathbf{Q}_k \end{cases} \tag{4.69}$$

starting with an $n \times n$ matrix $\mathbf{A}_0 = \mathbf{A}$. We claim that this iterative scheme is equivalent to (4.67) under the correspondence:

$$\begin{aligned} \mathbf{X}_0 &= \mathbf{1} \\ \widehat{\mathbf{Q}}_k &= \mathbf{Q}_1\mathbf{Q}_2\cdots\mathbf{Q}_k \\ \widehat{\mathbf{R}}_k &= \mathbf{R}_k. \end{aligned} \tag{4.70}$$

To show this equivalence, observe that for the step $k = 1$, we have

$$\begin{aligned}\widehat{\mathbf{Q}}_1\widehat{\mathbf{R}}_1 &= \mathbf{X}_1 = \mathbf{A}\widehat{\mathbf{Q}}_0 = \mathbf{A} \\ \mathbf{Q}_1\mathbf{R}_1 &= \mathbf{A}_0 = \mathbf{A},\end{aligned}$$

so that

$$\begin{aligned}\widehat{\mathbf{Q}}_1 &= \mathbf{Q}_1 \\ \widehat{\mathbf{R}}_1 &= \mathbf{R}_1.\end{aligned}$$

Assuming now that (4.70) is true for $(n-1)$, we have, for the nth step:

$$\begin{aligned}\widehat{\mathbf{Q}}_n\widehat{\mathbf{R}}_n &= \mathbf{X}_n = \mathbf{A}\widehat{\mathbf{Q}}_{n-1} = \mathbf{A}\mathbf{Q}_1\mathbf{Q}_2\cdots\mathbf{Q}_{n-1} \\ &= \mathbf{Q}_1(\mathbf{R}_1\mathbf{Q}_1)\mathbf{Q}_2\cdots\mathbf{Q}_{n-1} = \mathbf{Q}_1\mathbf{Q}_2(\mathbf{R}_2\mathbf{Q}_2)\cdots\mathbf{Q}_{n-1} \\ &= \mathbf{Q}_1\mathbf{Q}_2\cdots(\mathbf{R}_{n-1}\mathbf{Q}_{n-1}) = \mathbf{Q}_1\mathbf{Q}_2\cdots\mathbf{Q}_n\mathbf{R}_n,\end{aligned}$$

which, by the uniqueness of the QR factorization, implies that the correspondence (4.70) is true for all k. A similar induction then shows that

$$\mathbf{A}_k = (\mathbf{Q}_1\mathbf{Q}_2\cdots\mathbf{Q}_k)'\mathbf{A}(\mathbf{Q}_1\mathbf{Q}_2\cdots\mathbf{Q}_k) = \widehat{\mathbf{Q}}_k'\mathbf{A}\widehat{\mathbf{Q}}_k,$$

so that the matrix $\mathbf{A}_k$ produced by QR iterations converges to a block-triangular matrix $\mathbf{B}$ having the same eigenvalues as the original matrix $\mathbf{A}$.

The iterations (4.69) can be easily implemented as follows:

```
function [U,B] = QRit(A,maxit)
% pure QR algorithm for eigenpairs
% U is an orthogonal matrix
% B is a block triangular matrix

[n,m] = size(A); U = eye(n);
for i = 1 : maxit
    [Q,R] = householderQR(A);
    A = R*Q;
    U = U*Q;
end
B = A;
```

For example, consider first a matrix with complex conjugate eigenvalues:

```
>> A = [1 0 3; 1 -1 -3; -1 2 -1];
>> eig(A)
```

```
ans =
   1.0000
  -1.0000 + 3.0000i
  -1.0000 - 3.0000i

>> [U,B] = QRit(A,50)

U =
   -0.6114   -0.2116    0.7625
    0.7407   -0.4919    0.4575
    0.2783    0.8445    0.4575

B =
   -1.2517   -3.5388   -2.1810
    2.5611   -0.7483   -0.1842
    0.0000   -0.0000    1.0000
```

Because two of the eigenvalues have the same absolute value, we see that **B** is block triangular as opposed to upper triangular. This means that the remaining eigenvalues still need to be determined as the eigenvalues of a 2×2 block.

For a relatively large matrix with distinct eigenvalues, consider the following:

```
>> A = [-1 2 3 -2 5; 2 4 -2 3 1; 3 -2 1 5 7; -2 3 5 2 -9; 5 1 7 -9 3];
>> [U,B] = QRit(A,100)

U =
   -0.3562   -0.0265    0.2401   -0.1615   -0.8881
    0.0611    0.2056    0.4150   -0.8520    0.2365
   -0.2919    0.5391    0.6155    0.4600    0.1838
    0.4953   -0.5792    0.6189    0.1840   -0.0476
   -0.7341   -0.5752    0.0910   -0.0513    0.3455

B =
   14.1996   -0.0000   -0.0000    0.0000    0.0000
   -0.0000  -12.7498    0.0000    0.0000   -0.0000
    0.0000    0.0000    6.8847    0.0000   -0.0000
    0.0000    0.0000    0.0000    4.8712   -0.0000
   -0.0000    0.0000    0.0000   -0.0000   -4.2057
```

Observe now that **B** is actually diagonal, because of the fact that **A** is self-adjoint. Also notice the large number of iterations used previously to produce the desired result. If you repeat these examples with fewer iterations, you will observe that the smaller the ratio $|\lambda_{i+1}|/|\lambda_i|$, for a given pair of successive eigenvalues, the faster the iterations produce zeros in their corresponding columns for the matrix $\mathbf{B}$.

QR iterations with a shift

This last observation suggests that a way to improve the convergence for QR iterations is to use a shift of the form $(\mathbf{A} - b\mathbf{1})$, where $b \in \mathbb{C}$ is close to an eigenvalue. For simplicity, assume first that the absolute values for the eigenvalues of $\mathbf{A}$ are all distinct. An efficient choice of shift $(\mathbf{A} - b\mathbf{1})$ to speed up the convergence for QR iterations is when $b \in \mathbb{C}$ is an approximated eigenvalue of $\mathbf{A}$. Use the same notations as earlier, and write

$$\mathbf{A}_k = \widehat{\mathbf{Q}}_k' \mathbf{A} \widehat{\mathbf{Q}}_k,$$

where $\widehat{\mathbf{Q}}_k$ converges to the orthogonal matrix whose columns span the same space as the eigenvectors of $\mathbf{A}$. Therefore, denoting the last column of $\widehat{\mathbf{Q}}_k$ by $\widehat{\mathbf{q}}_n^{(k)}$, we have that

$$a_{nn}^{(k)} = \frac{\langle \widehat{\mathbf{q}}_n^{(k)}, \mathbf{A}\widehat{\mathbf{q}}_n^{(k)} \rangle}{\langle \widehat{\mathbf{q}}_n^{(k)}, \widehat{\mathbf{q}}_n^{(k)} \rangle} \tag{4.71}$$

is a Rayleigh quotient approximation for λ_n, where $a_{nn}^{(k)}$ denotes the last diagonal entry of the matrix $\mathbf{A}_k$. This leads to the following iterative scheme for calculating the last eigenvalue of $\mathbf{A}$ based on *QR iterations with shifts*:

$$\begin{cases} b_k = a_{nn}^{(k-1)} \\ \mathbf{Q}_k \mathbf{R}_k = (\mathbf{A}_{k-1} - b_k \mathbf{1}) \\ \mathbf{A}_k = \mathbf{R}_k \mathbf{Q}_k + b_k \mathbf{1} \end{cases} . \tag{4.72}$$

We then run these iterations until this eigenvalue has been determined within a certain tolerance. For instance, since $\mathbf{A}_k$ converges to an upper-triangular matrix, all the off-diagonal entries of its last row must converge to zero. Therefore, we can use the norm of the $(n-1)$-dimensional vector consisting of these off-diagonal entries as the stopping condition for the iterations. Suppose we decide to stop the iterations at the step k_n, meaning that the last eigenvalue of $\mathbf{A}$ is declared to be the bottom-right entry of the matrix $\mathbf{A}_{k_n}$. We can then concentrate on finding the remaining eigenvalues of $\mathbf{A}$ as the eigenvalues for the $(n-1) \times (n-1)$ matrix obtained by removing the last row and last column of $\mathbf{A}_{k_n}$, which we do by running the iterations (4.72) again with this reduced matrix as the initial input. We then repeat the procedure and continue in this way until all the eigenvalues of $\mathbf{A}$ are deemed to be found.

The following MATLAB function implements all these steps. For comparison with the pure QR algorithm, we record the number of iterations necessary to achieve the desired precision for each eigenvalue.

```
function [V,B,iterations] = QRshift(A,tol)
% QR algorithm with shifts
% V orthogonal matrix
% B vector of eigenvalues

[n,m] = size(A); V = eye(n); iterations = zeros(1,n);
for i=n:-1:2
    while norm(A(i,1:i-1)) > tol
        shift = A(i,i);
        [Q,R] = householderQR(A-shift*eye(i));
        A = R*Q+shift*eye(i);
        V(1:n,1:i) = V(1:n,1:i)*Q;
        iterations(i) = iterations(i)+1;
    end
    B(i) = A(i,i);
    A = A(1:i-1,1:i-1);
end
B(1) = A(1,1);
```

You can test this function on the matrix of the last example:

```
>> A = [-1 2 3 -2 5; 2 4 -2 3 1; 3 -2 1 5 7; -2 3 5 2 -9; 5 1 7 -9 3];
>> [V,B,iterations] = QRshift(A,10^-5)

V =
   -0.0265    0.8881   -0.3562   -0.2401    0.1615
    0.2056   -0.2365    0.0611   -0.4150    0.8520
    0.5391   -0.1838   -0.2919   -0.6155   -0.4600
   -0.5792    0.0476    0.4953   -0.6189   -0.1840
   -0.5752   -0.3455   -0.7341   -0.0910    0.0513

B =
  -12.7498   -4.2057   14.1996    6.8847    4.8712

iterations =
     0     3     3     1     4
```

Observe how few iterations are necessary to achieve a desired tolerance level, especially when compared with the pure QR algorithm. Notice also that the eigenvalues are not exactly ordered according to their absolute values. This is because the order in which they are found depends on how close to the shift they happened to be. Because of this fact, the method also works for the case of different eigenvalues with the same absolute value, as the next example shows,

```
>> A = [-1 2 3 1; 2 -1 1 3; 3 1 -1 2; 1 3 2 -1];
>> [V,B,iterations] = QRshift(A,10^-5)

V =
    0.5000   -0.5000   -0.5000    0.5000
    0.5000    0.5000   -0.5000   -0.5000
    0.5000    0.5000    0.5000    0.5000
    0.5000   -0.5000    0.5000   -0.5000

B =
    5.0000   -5.0000   -3.0000   -1.0000

iterations =
     0     2     3     1
```

The method also works equally well when $\mathbf{A}$ has complex entries and complex eigenvalues, as shown in the next example.

```
>> A = [2+i 2 -1; -1 i+4 3; 2 3i-1 5];
>> eig(A)

ans =
   2.4041 - 2.8284i
   2.5498 + 2.1146i
   6.0461 + 2.7138i

>> [V,B,iterations] = QRshift(A,10^-5)

V =
   0.2805 - 0.3148i   0.7924 + 0.3375i  -0.2835 - 0.0094i
  -0.5218 - 0.2529i  -0.0410 + 0.4944i   0.2185 + 0.6082i
   0.0484 + 0.6954i   0.1072 - 0.0253i  -0.4722 + 0.5281i

B =
   2.4041 - 2.8284i   2.5498 + 2.1146i   6.0461 + 2.7138i

iterations =
     0     2     5
```

It fails, however, when applied to a matrix $\mathbf{A}$ with real entries but complex-conjugate eigenvalues because all the operations in the algorithm will then involve only real-valued arithmetic and will never produce the desired eigenvalues. For these cases, extensions of the method, such as the *Francis shift*, are available but we do not introduce them here.

4.8 Singular Value Decomposition

The singular value decomposition is a general and powerful way to rewrite an arbitrary $n \times k$ complex matrix $\mathbf{A}$ of any rank $r \leq k \leq n$ in terms of unitary and diagonal matrices. The *singular values* $\sigma_1 \geq \sigma_2 \geq \ldots \geq \sigma_k$ of an $n \times k$ complex matrix $\mathbf{A}$ are the nonnegative square roots of the eigenvalues of the $k \times k$ matrix $\mathbf{A}'\mathbf{A}$. Notice that singular values are always well defined because the eigenvalues $\lambda_1, \cdots, \lambda_k$ for the self-adjoint positive matrix $\mathbf{A}'\mathbf{A}$ are non-negative real numbers. Moreover, the normalized eigenvectors $\mathbf{v}_1, \ldots, \mathbf{v}_k$ of $\mathbf{A}'\mathbf{A}$ are called the *right singular vectors* of $\mathbf{A}$ and form an orthonormal basis for $\mathbb{C}^k$. Similarly, the normalized eigenvectors $\mathbf{u}_1, \ldots, \mathbf{u}_n$ for the $n \times n$ self-adjoint matrix $\mathbf{A}\mathbf{A}'$ are called the *left singular vectors* of $\mathbf{A}$ and form an orthonormal basis for $\mathbb{C}^n$.

For example, consider the 4×3 matrix

$$\mathbf{A} = \begin{bmatrix} 1 & 2 & 3 \\ -1 & -3 & -2 \\ 1 & 4 & 3 \\ 2 & 1 & 4 \end{bmatrix} \tag{4.73}$$

with

$$\mathbf{A}'\mathbf{A} = \begin{bmatrix} 7 & 5 & 16 \\ 5 & 30 & 16 \\ 16 & 16 & 38 \end{bmatrix}, \qquad \mathbf{A}\mathbf{A}' = \begin{bmatrix} 14 & -1 & 18 & 16 \\ -1 & 14 & 5 & -7 \\ 18 & 5 & 26 & 18 \\ 16 & -7 & 18 & 21. \end{bmatrix},$$

whose eigenvalues and eigenvectors can be obtained with the following MATLAB commands:

```
>> A = [1 2 3;  -1 3 -2; 1 4 3; 2 1 4];
>> [V,lambda]=eig(A'*A)

V =
   -0.9083    0.2789    0.3119
   -0.0690   -0.8350    0.5459
    0.4127    0.4743    0.7776

lambda =
    0.1102         0         0
         0   19.2412         0
         0         0   55.6486

>> [U,mu]=eig(A*A')
```

```
U =
    0.6444   -0.5777    0.0073    0.5009
    0.3682    0.3734   -0.8509   -0.0308
   -0.6444   -0.1623   -0.3735    0.6473
    0.1841    0.7074    0.3693    0.5738

mu =
    0.0000         0         0         0
         0    0.1102         0         0
         0         0   19.2412         0
         0         0         0   55.6486
```

As expected, the eigenvalues of the positive definite matrices $\mathbf{A}'\mathbf{A}$ and $\mathbf{A}\mathbf{A}'$ are non-negative real numbers. Moreover, in accordance with Theorem 4.6, the eigenvectors of $\mathbf{A}'\mathbf{A}$ form an orthonormal basis for $\mathbb{C}^3$, while the eigenvectors of $\mathbf{A}\mathbf{A}'$ form an orthonormal basis for $\mathbb{C}^4$. In this example, the singular values of $\mathbf{A}$ are the numbers $\sigma_1 = \sqrt{55.6486} = 7.4598$, $\sigma_2 = \sqrt{19.2412} = 4.3865$, and $\sigma_3 = \sqrt{0.1102} = 0.3320$, and the right and left singular vectors of $\mathbf{A}$ are the columns of the matrices $\mathbf{V}$ and $\mathbf{U}$. Apart from sign changes, these are exactly the outcomes of the MATLAB function `svd` applied to the matrix $\mathbf{A}$:

```
>> A = [1 2 3;  -1 3 -2; 1 4 3; 2 1 4];
>> [U,S,V] = svd(A)

U =
   -0.5009   -0.0073   -0.5777   -0.6444
    0.0308    0.8509    0.3734   -0.3682
   -0.6473    0.3735   -0.1623    0.6444
   -0.5738   -0.3693    0.7074   -0.1841

S =
    7.4598         0         0
         0    4.3865         0
         0         0    0.3320
         0         0         0

V =
   -0.3119   -0.2789    0.9083
   -0.5459    0.8350    0.0690
   -0.7776   -0.4743   -0.4127
```

With these notations, let $\mathbf{u}_1, \ldots, \mathbf{u}_n$ be the columns of an $n \times n$ unitary matrix $\mathbf{U}$ and let $\mathbf{v}_1, \ldots, \mathbf{v}_k$ be the columns of a $k \times k$ unitary matrix $\mathbf{V}$. The *singular value decomposition* of a matrix $\mathbf{A}$ consists of writing it as the product

$$\mathbf{A} = \mathbf{U}\boldsymbol{\Sigma}\mathbf{V}', \tag{4.74}$$

where $\boldsymbol{\Sigma}$ is the $n \times k$ diagonal matrix with entries

$$\sigma_{ij} = \begin{cases} \sigma_i, & i = j \\ 0, & i \neq j. \end{cases} \tag{4.75}$$

This can be verified for the matrix of the previous example:

```
>> A = [1 2 3;  -1 3 -2; 1 4 3; 2 1 4];
>> [U,S,V] = svd(A);
>> U*S*V'

ans =
    1.0000    2.0000    3.0000
   -1.0000    3.0000   -2.0000
    1.0000    4.0000    3.0000
    2.0000    1.0000    4.0000
```

Although the existence of at least one singular value decomposition for any matrix $\mathbf{A}$ (real or complex) is an absolute pinnacle of *Linear Algebra*, its proof would take us too far adrift from our purposes. Numerically, the computation of the eigenvalues and eigenvector of $\mathbf{A}'\mathbf{A}$ and $\mathbf{A}\mathbf{A}'$ is not accurate or stable with respect to the entries of the original matrix $\mathbf{A}$, so that a practical algorithm for the singular value decomposition works directly with $\mathbf{A}$. Instead of attempting to describe these advanced algorithms in an introductory textbook, we content ourselves with invoking the MATLAB built-in function `svd`.

Economy-size singular value decomposition

A convenient way to rewrite (4.74) is

$$\mathbf{A} = \mathbf{U}\boldsymbol{\Sigma}\mathbf{V}' = [\mathbf{U}_1 \quad \mathbf{U}_2] \begin{bmatrix} \boldsymbol{\Sigma}_1 \\ \mathbf{0} \end{bmatrix} \mathbf{V}' = \mathbf{U}_1\boldsymbol{\Sigma}_1\mathbf{V}', \tag{4.76}$$

for an $n \times k$ matrix $\mathbf{U}_1$ and a $k \times k$ matrix Σ_1. This is called the *economy-size singular value decomposition* of $\mathbf{A}$ and can be obtained using the optional argument `econ` in the MATLAB function `svd`. For instance, for the matrix of the previous example, we have the following economy-size singular value decomposition:

```
>> A = [1 2 3;  -1 3 -2; 1 4 3; 2 1 4];
>> [U1,S1,V]=svd(A,'econ')

U1 =
   -0.5009   -0.0073   -0.5777
    0.0308    0.8509    0.3734
   -0.6473    0.3735   -0.1623
   -0.5738   -0.3693    0.7074

S1 =
    7.4598         0         0
         0    4.3865         0
         0         0    0.3320

V =
   -0.3119   -0.2789    0.9083
   -0.5459    0.8350    0.0690
   -0.7776   -0.4743   -0.4127
```

Rank-deficient least squares

As mentioned in Section 3.4, the singular value decomposition can be used to solve the least-squares problem $\mathbf{Ax} \simeq \mathbf{b}$. Because the diagonal entries for the $k \times k$ matrix $\mathbf{\Sigma}_1$ in (4.76) are the square roots of the eigenvalues of $\mathbf{A}'\mathbf{A}$, it follows from Exercise 3.10 that $\mathbf{\Sigma}_1$ is invertible if and only if $\text{rank}(\mathbf{A}) = k \leq n$. For such a full-rank matrix, you can verify that the vector

$$\mathbf{x} = \mathbf{V}\mathbf{\Sigma}_1^{-1}\mathbf{U}_1'\mathbf{b} \tag{4.77}$$

satisfies the normal equations $\mathbf{A}'\mathbf{Ax} = \mathbf{A}'\mathbf{b}$, being therefore the unique solution to the least-squares problem $\mathbf{Ax} \simeq \mathbf{b}$.

The following MATLAB commands confirm that (4.77) coincides with the solution to the least squares obtained in MATLAB using the backslash operator:

```
>> A = [1 2 3;  -1 3 -2; 1 4 3; 2 1 4];
>> b=[-1 2 -3 1]';
>> [U1,S1,V]=svd(A,'econ');
>> x=V*inv(S1)*U1'*b

x =
    6.7966
    0.4237
   -3.3559

>> A\b
```

```
ans =
    6.7966
    0.4237
   -3.3559
```

In general, we can use the properties of the unitary matrices **U** and **V** to show that the vector

$$\mathbf{x} = \sum_{\sigma_j \neq 0} \frac{\mathbf{u}_i' * \mathbf{b}}{\sigma_j} \mathbf{v_j} \qquad (4.78)$$

satisfies the normal equations, even when **A** is rank deficient. For example, the 4×3 matrix

$$\begin{bmatrix} -1 & 0 & 1 \\ 2 & 4 & 6 \\ 1 & 3 & 5 \\ -2 & 1 & 4 \end{bmatrix}$$

is clearly rank deficient because the second column is the average of the other two. As a result, the matrix $\mathbf{A}'\mathbf{A}$ is singular, which implies that the diagonal matrix $\mathbf{\Sigma}_1$ in (4.76) has at least one vanishing diagonal entry, as can be verified in MATLAB:

```
>> A=[-1 0 1; 2 4 6; 1 3 5; -2 1 4];
>>   [U1,S1,V]=svd(A,'econ');
>> S1

S1 =
   10.1802         0         0
         0    3.2192         0
         0         0    0.0000
```

Letting $\mathbf{b} = [2, 0, 3, -1]$, you can calculate the vector (4.78) and verify that it satisfies the normal equations $\mathbf{A}'\mathbf{A} = \mathbf{A}'\mathbf{b}$ as follows:

```
>> b=[2 0 3 -1]';
>> x=(U1(:,1)'*b/S1(1,1))*V(:,1)+(U1(:,2)'*b/S1(2,2))*V(:,2)

x =
    0.1266
    0.1099
    0.0931

>> A'*A*x-A'*b
```

```
ans =
   1.0e-14 *

   -0.2665
   -0.6217
   -0.8882
```

According to Theorem 2.3, however, the solution to the normal equations found in (4.78) cannot be unique in the rank-deficient case because the matrix $\mathbf{A}'\mathbf{A}$ is singular. In terms of the least-squares problem, this means that there are infinitely many vectors that minimize the residual norm $\|\mathbf{Ax} - \mathbf{b}\|_2$. In the previous example, this can be verified by adding any nonzero vector $\boldsymbol{\eta}$ in the null space of $\mathbf{A}'\mathbf{A}$ to the solution found in (4.77):

```
>> eta=null(A'*A)

eta =
   -0.4082
    0.8165
   -0.4082

>> lambda=[-10:0.1:10];
>> for i=1:length(lambda)
residual_norm(i)=norm(A*(x+lambda(i)*eta)-b);
solution_norm(i)=norm(x+lambda(i)*eta);
end
>> plot(lambda,residual_norm,'b',lambda,solution_norm,'r')
```

As you can see in Figure 4.5, for a rank-deficient matrix $\mathbf{A}$, any of the infinitely many solutions to the least-squares problem $\mathbf{Ax} \simeq \mathbf{b}$ yields the same minimal value for the residual norm $\|\mathbf{Ax} - \mathbf{b}\|_2$. However, the solutions have different Euclidean norms, the smallest one corresponding to the vector $\mathbf{x}$ in (4.78).

Rank determination

The discussion on the least-squares problem suggests that the singular value decomposition can be useful in determining the rank of a given matrix $\mathbf{A}$, which is the number of its linearly independent columns. Because multiplication by a unitary matrix preserves linear independence, we see that the rank of $\mathbf{A}$ coincides with the rank of $\boldsymbol{\Sigma}$ in the decomposition formula (4.74), which in turn is simply given by the number of nonzero singular values of $\mathbf{A}$. The following MATLAB code illustrates these computations with a simple example:

```
>> A = [1 3 2 5; -1 -3 3 8; 0 0 -2 4; 2 6 1 9; -3 -9 1 2];
>> rank(A)
```

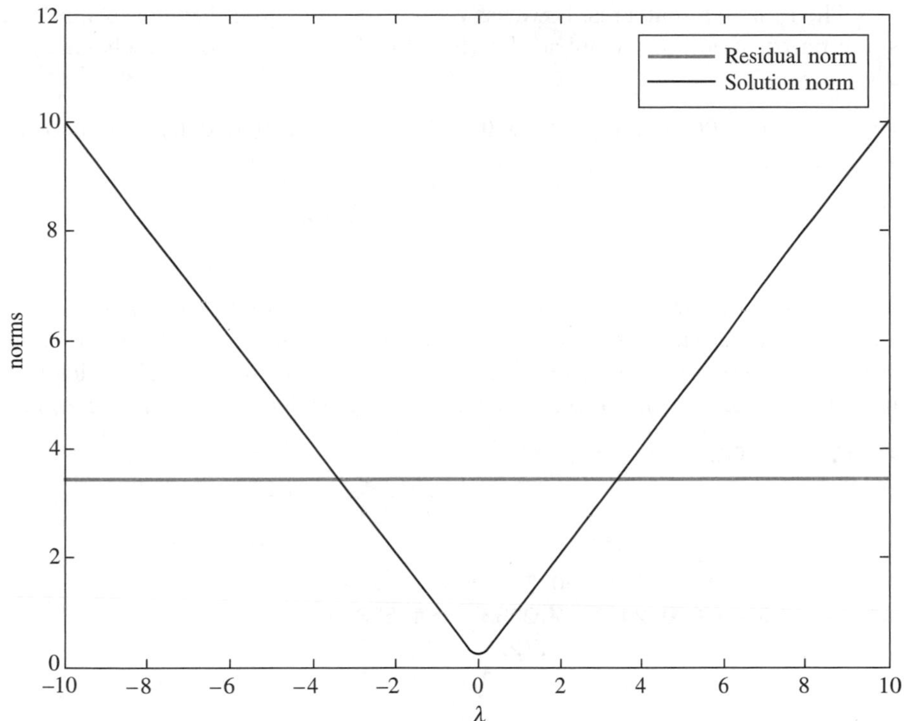

Figure 4.5 The residual norm $\|\mathbf{A}\mathbf{x} - \mathbf{b}\|_2$ and the solution norm $\|\mathbf{x}\|_2$ for different solutions to the least-squares problem $\mathbf{A}\mathbf{x} \simeq \mathbf{b}$ for a rank-deficient matrix $\mathbf{A}$.

```
ans =
     3

>> [U,S,V] = svd(A);
>> S

S =
   14.4795         0         0         0
         0   11.7923         0         0
         0         0    3.2069         0
         0         0         0    0.0000
         0         0         0         0
```

The rank of a matrix is, however, a very unstable quantity. For example, a small perturbation of the entries for the preceding matrix completely changes its rank:

```
>> B = A + 0.001*[1 0 0 0; 0 0 0 0; 0 0 0 0; 0 0 0 0; 0 0 0 0];
>> rank(B)

ans =
     4
```

This situation characterizes what is known as *near-rank deficiency*, that is, the full-rank matrix **B** is very close to a rank-deficient matrix **A**. As a result, numerical methods that require **B** to be a full-rank matrix might lead to unstable outcomes when applied to a near-rank deficient matrix. For example:

```
>> [Q,R] = QRfactor(B);
>> R

R =
    3.8732   11.6189   -0.5158    2.3249
         0    0.0029   -2.2083   -4.5539
         0         0    3.7225    7.5602
         0         0         0   10.3296
```

which shows that the matrix **R** in the QR factorization of a near-rank deficient matrix **B** is very close to singular, hence en route to all sorts of ill-conditioned problems.

Therefore, when dealing with a matrix **A** representing some real data subject to measurement error, it is better to determine its *numerical rank* as the number of singular values of **A** that are greater than a given tolerance. For instance, for the previous example you would have

```
>> [U,S,V] = svd(B);
>> s = diag(S)'

s =
   14.4796   11.7924    3.2070    0.0008
```

in which case a tolerance of, say, 10^{-3} gives a numerical rank equal to 3, as it should be since the perturbation in the matrix is smaller than the tolerance level is. This is exactly the procedure used by the MATLAB built-in function `rank`, which takes as an optional argument the desired tolerance level:

```
>> rank(B,10^-3)

ans =
     3
```

Along similar lines, an alternative way to express the singular value decomposition (4.74) is

$$\mathbf{A} = \sigma_1 \mathbf{u}_1 * \mathbf{v}_1' + \sigma_2 \mathbf{u}_2 * \mathbf{v}_2' + \cdots + \sigma_k \mathbf{u}_k * \mathbf{v}_k', \tag{4.79}$$

where each outer product $\mathbf{u}_i * \mathbf{v}_i'$ is a rank-1 $n \times k$ matrix. Truncating this sum up to the first p nonzero singular values produces an approximation of $\mathbf{A}$ with rank p. Moreover, it can be shown that this is the best rank-p approximation of $\mathbf{A}$ with respect to the Euclidean norm on the (nk)-dimensional space of $n \times k$ matrices. Therefore, this defines a systematic procedure for approximating a given matrix $\mathbf{B}$ by a lower-rank matrix of a desired rank p. Returning to the previous example, you can find the best rank-3 approximation for the near-rank deficient 5×4 matrix $\mathbf{B}$:

```
>> S(4,4) = 0;
>> U*S*V'

ans =
    1.0004    3.0002    2.0000    5.0000
   -0.9998   -3.0001    3.0000    8.0000
   -0.0002    0.0001   -2.0000    4.0000
    2.0003    5.9999    1.0000    9.0000
   -3.0001   -9.0000    1.0000    2.0000
```

Euclidean matrix norm and condition number

The singular value decomposition is also used for computing the Euclidean matrix norm and condition numbers introduced in Section 2.4. Recall that the matrix 2-norm induced by the Euclidean vector norm $\|\mathbf{x}\|_2$ is defined as

$$\|\mathbf{A}\|_2 = \max_{\mathbf{x} \neq \mathbf{0}} \frac{\|\mathbf{A}\mathbf{x}\|_2}{\|\mathbf{x}\|_2}. \tag{4.80}$$

Using the definition of the Euclidean norm $\|\mathbf{x}\|_2 = \langle \mathbf{x}, \mathbf{x} \rangle^{1/2}$, we obtain

$$\|\mathbf{A}\|_2 = \max_{\mathbf{x} \neq \mathbf{0}} \frac{\langle \mathbf{A}\mathbf{x}, \mathbf{A}\mathbf{x} \rangle^{1/2}}{\langle \mathbf{x}, \mathbf{x} \rangle^{1/2}} = \frac{\langle \mathbf{x}, \mathbf{A}'\mathbf{A}\mathbf{x} \rangle^{1/2}}{\langle \mathbf{x}, \mathbf{x} \rangle^{1/2}}. \tag{4.81}$$

Using the orthonormal basis of eigenvectors $\mathbf{v}_1, \ldots, \mathbf{v}_k$ of $(\mathbf{A}'\mathbf{A})$ associated with the eigenvalues $\lambda_1, \ldots, \lambda_k$, we can expand any vector $\mathbf{x} \in \mathbb{C}^k$ as

$$\mathbf{x} = \mu_1 \mathbf{v}_1 + \cdots \mu_k \mathbf{v}_k.$$

Inserting this into (4.81) then gives

$$\|\mathbf{A}\|_2 = \max_{\mathbf{x} \neq \mathbf{0}} \frac{\left(\sum_{i-1}^{k} \lambda_i \mu_i^2\right)^{1/2}}{\left(\sum_{i=1}^{k} \mu_i^2\right)^{1/2}}. \tag{4.82}$$

It is then clear that, since the eigenvalues of $\mathbf{A}'\mathbf{A}$ are arranged in decreasing order, the maximum occurs at $\mathbf{x} = \mathbf{v}_1$ and has value $\lambda_1^{1/2}$. Therefore,

$$\|\mathbf{A}\|_2 = \lambda_1^{1/2} = \sigma_1. \tag{4.83}$$

Similarly, using expression (2.36) for the condition number, exactly the same argument leads us to conclude that

$$\text{cond}_2(\mathbf{A}) = \frac{\sigma_1}{\sigma_k}. \tag{4.84}$$

Equations (4.83) and (4.84) can be rephrased as follows:

Theorem 4.7 *The matrix norm $\|\mathbf{A}\|_2$ induced by the Euclidean vector norm is given by the largest singular value of $\mathbf{A}$, whereas the Euclidean condition number $\text{cond}_2(\mathbf{A})$ is given by the ratio between the largest and smallest singular values of $\mathbf{A}$.*

As mentioned in Section 2.4, this is exactly the technique used by MATLAB to compute the Euclidean condition number. Here is an example of both concepts:

```
>> A = [-4 11 3; 7 2 -1; 13 0 6];
>> [U,S,V] = svd(A);
>> s = diag(S)'

s =
   15.9474   11.6246    3.9432

>> s(1)/s(3)

ans =
    4.0443

>> norm(A)

ans =
   15.9474

>> cond(A)

ans =
    4.0443
```

The final application for the singular value decomposition is a construction of orthonormal bases for the null space and the range of a given matrix $\mathbf{A}$. Observe from

Orthonormal bases

$$\mathbf{AV} = \mathbf{U}\Sigma,$$

that the columns of $\mathbf{V}$ corresponding to singular values equal to zero form an orthonormal basis for the subspace null($\mathbf{A}$). Similarly, the columns of $\mathbf{U}$ corresponding to nonzero singular values form an orthonormal basis for range($\mathbf{A}$), while those corresponding to singular values equal to zero form an orthonormal basis for range$(\mathbf{A})^{\perp}$.

True to Vedic form, we end this *Linear Algebra* part of our book by returning to the MATLAB built-in functions `null` and `orth`, the first MATLAB functions we used in Chapter 2. Both functions use the singular value decomposition and the columns of $\mathbf{U}$ and $\mathbf{V}$ to find orthonormal bases for null($\mathbf{A}$) and range($\mathbf{A}$):

```
>> B = [1 2 3;2 4 6;1 0 -1];
>> [U,S,V] = svd(B)

U =
   -0.4463    0.0292   -0.8944
   -0.8925    0.0584    0.4472
    0.0653    0.9979   -0.0000

S =
    8.3841         0         0
         0    1.3066         0
         0         0    0.0000

V =
   -0.2583    0.8756    0.4082
   -0.5323    0.2237   -0.8165
   -0.8062   -0.4282    0.4082

>> null(B)

ans =
    0.4082
   -0.8165
    0.4082
```

```
>> orth(B)

ans =
   -0.4463    0.0292
   -0.8925    0.0584
    0.0653    0.9979
```

4.9 Summary and Notes

This chapter introduces the concepts of eigenvalues and eigenvectors of a matrix, their main properties, and numerical algorithms for computing them:

- **Section 4.1:** An *eigenvalue* of an $n \times n$ matrix $\mathbf{A}$ is a scalar satisfying $\mathbf{A}\mathbf{v} = \lambda\mathbf{v}$ for some vector $\mathbf{v} \neq \mathbf{0}$. Eigenvalues are the roots of the *characteristic polynomial* of a matrix, from which it follows that any matrix acting on $\mathbb{C}^n$ has n (possible multiple) eigenvalues, according to Theorem 4.1. Problem 4.1 consists of finding all the eigenvalues and eigenvectors of such matrices.

- **Section 4.2:** Any $n \times n$ complex matrix $\mathbf{A}$ can be written as $\mathbf{A} = \mathbf{U}\mathbf{B}\mathbf{U}'$, where $\mathbf{U}$ is unitary and $\mathbf{B}$ is upper triangular. This is known as the (complex) *Schur decomposition.* Because this is a particular case of a *similarity transformation*, $\mathbf{A}$ and $\mathbf{B}$ have the same eigenvalues. Moreover, Theorem 4.3 says that the algebraic multiplicity of an eigenvalue equals the number of times it appears on the diagonal of $\mathbf{B}$. The *spectral radius* of a matrix is the absolute value of its largest eigenvalue. A more detailed localization of eigenvalues in the complex plane is provided by the *Gershgorin disks.*

- **Section 4.3:** Eigenvectors associated with distinct eigenvalues are linearly independent, according to Theorem 4.4. The *geometric multiplicity* of an eigenvalue is the number of linearly independent eigenvectors associated with it and is always less than or equal to its algebraic multiplicity. An $n \times n$ matrix with n linearly independent eigenvectors can be diagonalized through a similarity transformation using the matrix of eigenvectors.

- **Section 4.4:** An $n \times n$ complex matrix is said to be normal if $\mathbf{A}\mathbf{A}' = \mathbf{A}'\mathbf{A}$. Particular cases of normal matrices are self-adjoint and unitary matrices. Eigenvectors of a normal matrix associated with distinct eigenvalues are orthogonal. Moreover, the eigenvectors of a matrix form an orthonormal basis for $\mathbb{C}^n$ if and only if the matrix is normal. The eigenvalues of a self-adjoint matrix are real numbers, and they are positive real numbers if the matrix is positive. The eigenvalues of a unitary matrix have absolute value equal to one.

- **Section 4.5:** The overall sensitivity of the eigenvalue problem is determined by the condition number of the matrix of eigenvectors. Eigenvectors that are close to being linearly dependent lead to an ill-conditioned eigenvalue problem. The sensitivity of an individual simple eigenvalue depends on the angle between its associated eigenvector and the corresponding *left-eigenvector*. Multiple eigenvalues are highly sensitive with respect to perturbations.

- **Section 4.6:** All numerical methods for Problem 4.1 are necessarily iterative. The largest eigenvalue of a matrix and its associated eigenvector can be found through *power iterations*, whose convergence rate is linear and depends on the ratio between the second largest and the largest eigenvalues. The smallest eigenvalue of a matrix and its associated eigenvector can be found through *inverse power iterations*, which also converge linearly and depend on the ratio between the smallest and the second smallest eigenvalues. Intermediate eigenvalues can be found by introducing a *shift* in the inverse power iterations. These can be significantly improved using the *Rayleigh quotients* as the values for the shifts, resulting in a quadratic (and sometimes cubic) convergence rate. After an eigenvalue of a matrix $\mathbf{A}$ is found, the *deflation* technique can be used to find a smaller matrix $\mathbf{B}$ whose eigenvalues coincide with the remaining eigenvalues of $\mathbf{A}$.

- **Section 4.7:** Several eigenvalues and eigenvectors of a matrix $\mathbf{A}$ can be found simultaneously by using *orthogonal iterations* or the equivalent *QR iterations*. These can be improved through the introduction of *shifts*.

- **Section 4.8:** Any complex $n \times k$ matrix $\mathbf{A}$ can be written as $\mathbf{A} = \mathbf{U}\Sigma\mathbf{V}'$, where $\mathbf{U}$ and $\mathbf{V}$ are $n \times n$ and $k \times k$ unitary matrices and Σ is an $n \times k$ diagonal matrix. In this decomposition, the diagonal entries of Σ are the *singular values* of $\mathbf{A}$, defined as the square roots of the eigenvalues of $\mathbf{A}'\mathbf{A}$. This *singular value decomposition* can be used to provide a general solution for the least-squares problem when the matrix $\mathbf{A}$ is *rank deficient*. It can also be used to determine the rank of $\mathbf{A}$, its Euclidean matrix norm and condition number, and bases for its null space and range.

The material for Section 4.1 is standard and can be found, for example, in [13]. Our treatment of the properties of eigenvalues and eigenvectors without the use of determinants is inspired by [2], where the complete proofs of Theorems 4.2 and 4.3 can be found. Sensitivity of eigenvalues and the iterative methods developed in Sections 4.6 and 4.7, are treated in [12]. These and many more advanced methods for computing eigenvalues and eigenvectors are described in the classic [30]. For an account of the singular value decomposition from a theoretical point of view, we recommend [2]. The different uses of the singular value decomposition in numerical applications are described in [12].

4.10 Exercises

1. Find the eigenvalues of the following matrices and compare them with the eigenvalues of their conjugate transposes:

   (a) $\mathbf{A}_1 = \begin{bmatrix} 1 & 2 & 1 & -2 \\ 2 & 1 & 1 & -1 \\ 1 & 1 & -3 & 0 \\ -2 & -1 & 0 & 4 \end{bmatrix}$

   (b) $\mathbf{A}_2 = \begin{bmatrix} 6 & 1 & -3 & 3 \\ 0 & 3 & 2 & -1 \\ 1 & 9 & -4 & 1 \\ 0 & -3 & 2 & 7 \end{bmatrix}$

   (c) $\mathbf{A}_3 = \begin{bmatrix} i & 2 & -1 & 0 \\ 2 & 9 & 4 & 2i \\ 3 & -2 & 1 & -1 \\ 5 & -3i & 0 & 1 \end{bmatrix}$

   (d) $\mathbf{A}_4 = \begin{bmatrix} 0 & 0 & 0 & 2 \\ 1 & 0 & 0 & -5 \\ 0 & 1 & 0 & 3 \\ 0 & 0 & 1 & 1 \end{bmatrix}$

   (e) $\mathbf{A}_5 = \begin{bmatrix} 1 & 0 & 1 & 0 \\ 1 & 1 & 1 & 0 \\ 0 & 0 & 1 & 0 \\ 1 & 1 & 1 & 1 \end{bmatrix}$

2. Obtain the characteristic polynomial $p(x)$ associated with each of the preceding matrices and find its roots with corresponding multiplicity. Evaluate each characteristic polynomial on its corresponding matrix, that is, calculate $p(\mathbf{A})$.

3. Use the MATLAB built-in function `schur` with the optional argument `'complex'` to obtain the Schur decomposition $\mathbf{A} = \mathbf{UBU}'$ for each of the matrices in the first exercise. In each case, compare the diagonal entries of the matrix $\mathbf{A}$ with the eigenvalues for the matrix $\mathbf{B}$.

4. Find the spectral radius for each of the matrices in the first exercise and verify Gelfand's formula (4.19) for the norms $\|\cdot\|_1$, $\|\cdot\|_2$ and $\|\cdot\|_\infty$ with $k = 10, 100, 1000$.

5. Repeat the following items for each of the matrices in the first exercise.

   (a) Compare the eigenvectors of $\mathbf{A}$ associated with an eigenvalue λ with the eigenvectors of $\mathbf{A}'$ associated with the corresponding eigenvalue $\bar{\lambda}$.

   (b) Verify that eigenvectors associated with distinct eigenvalues are linearly independent.

   (c) Calculate the geometric multiplicity of each eigenvalue.

   (d) Compute $\mathbf{V}^{-1}\mathbf{AV}$, where $\mathbf{V}$ is the matrix of eigenvectors of $\mathbf{A}$, and identify which matrices are diagonalizable.

6. Consider the matrix

$$\mathbf{A} = \begin{bmatrix} -i & -i & 0 \\ -i & i & 0 \\ 0 & 0 & 1 \end{bmatrix}.$$

   (a) Show that $\mathbf{A}$ is an example of a normal matrix that is neither self-adjoint nor unitary.

   (b) Use the MATLAB built-in function `schur` with optional argument `'complex'` to find its Schur decomposition $\mathbf{A} = \mathbf{UBU}'$.

   (c) Verify that the columns of $\mathbf{U}$ are eigenvectors of $\mathbf{A}$ and form an orthonormal basis for $\mathbb{C}^3$.

   (d) Compare the eigenvalues and eigenvectors of $\mathbf{A}$ with those of $\mathbf{A}'$.

7. Implement the power iteration method starting with $\mathbf{x}_0 = [1/2, 1/2, 1/2, 1/2]$ to find the maximal eigenvalue of

$$\mathbf{A} = \begin{bmatrix} 2 & 3 & 5 & 6 \\ 3 & -1 & -2 & -3 \\ 5 & -2 & 0 & 2 \\ 6 & -3 & 2 & 22 \end{bmatrix},$$

accurate to 10^{-6}, as well as the corresponding normalized eigenvector. Using the eigenvalues of $\mathbf{A}$ calculated by MATLAB, determine the rate of convergence for the method in this example.

8. Consider the matrix

$$\mathbf{A} = \begin{bmatrix} 10 & 15 & 0 \\ 15 & 20 & 5 \\ 0 & 5 & 45 \end{bmatrix}$$

and the vector $\mathbf{v} = [4, -3, , 0]$.

(a) Calculate the Rayleigh quotient resulting from $\mathbf{A}$ and $\mathbf{v}$.

(b) Knowing that the negative eigenvalue of $\mathbf{A}$ is close to -1, set up one step of the inverse iterations with a shift that could be used to provide a better estimate of that eigenvalue.

(c) Use the function `rayleigh` to find the negative eigenvalue of $\mathbf{A}$.

9. Use the function `rayleigh` to find the eigenvalue of the 10×10 Hilbert matrix $\mathbf{H}(10)$ that is nearest to $b = 10^{-5}$, accurate to machine precision. Identify the rate of convergence from the successive approximations for the eigenvalue.

10. Consider the matrix

$$\mathbf{A} = \begin{bmatrix} 6 & 2 & 1 \\ 2 & 3 & 1 \\ 3 & 1 & -1 \end{bmatrix}.$$

(a) Compute the matrices $\widehat{\mathbf{Q}}_1$ and $\widehat{\mathbf{Q}}_2$ for the simultaneous orthogonal iterations:

$$\begin{aligned} \widehat{\mathbf{Q}}_1\widehat{\mathbf{R}}_1 &= \mathbf{X}_0 \\ \mathbf{X}_1 &= \mathbf{A}\widehat{\mathbf{Q}}_1 \\ \widehat{\mathbf{Q}}_2\widehat{\mathbf{R}}_2 &= \mathbf{X}_1 \end{aligned}$$

(b) Estimate the rate of convergence for the method applied to this example, knowing that the eigenvalues of $\mathbf{A}$ are $\{7.63, 1.88, 0.49\}$.

(c) Use the function `orthogit` to find the eigenvectors of $\mathbf{A}$ associated with its largest two eigenvalues.

11. Consider the matrix

$$\mathbf{A} = \begin{bmatrix} 2 & 4 & 2 \\ 6 & 0 & 8 \\ 3 & 2 & -8 \end{bmatrix}.$$

(a) Compute first the matrices $\mathbf{A}_1$ and $\mathbf{A}_2$ for the QR iterations:

$$\begin{aligned} \mathbf{Q}_1\mathbf{R}_1 &= \mathbf{A}_0 \\ \mathbf{A}_1 &= \mathbf{R}_1\mathbf{Q}_1 \\ \mathbf{Q}_2\mathbf{R}_2 &= \mathbf{A}_1 \\ \mathbf{A}_2 &= \mathbf{R}_2\mathbf{Q}_2, \end{aligned}$$

starting with $\mathbf{A}_0 = \mathbf{A}$.

(b) Knowing that the eigenvalues of $\mathbf{A}$ are $\{-9.4, 7.4, -4\}$, what can you say about the rate of convergence of the QR algorithm applied to $\mathbf{A}$?

(c) Use the function `QRit` to find the eigenvectors of $\mathbf{A}$.

12. (a) Use `QRshift` to determine the eigenvalues of the 10×10 Hilbert matrix $\mathbf{H}(10)$, accurate to machine precision.

    (b) Apply Householder transformations to $\mathbf{H}(10)$ to obtain an upper Hessenberg matrix $\mathbf{B}$ (that is, with zeros below the first subdiagonal) that has the same eigenvalues as $\mathbf{H}(10)$.

    (c) Use `QRshift` to find the eigenvalues of $\mathbf{B}$, accurate to machine precision, and compare the number of iterations with the number of iterations necessary for calculating the eigenvalues of $\mathbf{H}(10)$.

13. Write a MATLAB function to find the singular value decomposition of a matrix $\mathbf{A}$ using only functions designed to compute eigenvalues and eigenvectors of matrices. That is, you can use functions such as `powerit` and `QRit`, but should not use the built-in function `svd`.

CHAPTER 5

Polynomial Functions

POLYNOMIALS ARE THE SIMPLEST FUNCTIONS of elementary *Calculus*. They can be used for numerical representations of other functions (such as trigonometric, hyperbolic, or logarithmic) or for the representation of an arbitrary set of data points. Examples of such representations are Taylor polynomials (partial sums of a Taylor series), polynomial interpolants (polynomials fitting the data points), and orthogonal polynomials (satisfying orthogonality relations with respect to an inner product).

The advantages of representations of functions by polynomials stem from the fact that polynomials are well suited for differential and integral calculus. For instance, derivatives and integrals of polynomial functions are themselves polynomial functions that can be easily computed.

This chapter covers properties of polynomial functions; polynomial interpolation using Vandermonde, Lagrange, and Newton polynomials; least-squares fit; and orthogonal polynomial approximations, as well as the errors of the polynomial interpolation and approximation. It combines concepts of *Calculus*, *Linear Algebra*, and *Numerical Analysis*.

5.1 Properties of Polynomials

A general polynomial of degree n is a function of the form

$$p_n(x) = c_1 x^n + c_2 x^{n-1} + \ldots + c_n x + c_{n+1}, \tag{5.1}$$

where $c_1, c_2, ..., c_{n+1}$ are scalar coefficients. Our choice of enumeration of coefficients from the highest-order term x^n to the lowest-order term x^0 reflects the MATLAB notation: the polynomial $p_n(x)$ in Equation (5.1) is identified with the row vector $\mathbf{c} = [c_1, c_2, ..., c_{n+1}]$ in the finite-dimensional vector space $V = \mathbb{R}^{n+1}$. As you have seen in Section 2.1, the space $\mathcal{P}_n(\mathbb{R})$ of polynomials of degree n is an $(n+1)$-dimensional vector space, which is therefore isomorphic to $\mathbb{R}^{n+1}$.

This representation can also be used to operate with polynomials of different degrees. For instance, if you want to construct a linear combination of two polynomials

$$p_2(x) = 2x^2 + 3x + 1, \qquad p_4(x) = x^4 - 3x^3 - x^2 - 2x - 4,$$

you build two vectors of coefficients in the largest space $\mathcal{P}_4(\mathbb{R})$ and form a linear combination of them. For example,

$$3p_2(x) + 2p_4(x) = 2x^4 - 6x^3 + 4x^2 + 5x - 5,$$

which can be immediately implemented in MATLAB as

```
>> p1 = [0,0,2,3,1];
>> p2 = [1,-3,-1,-2,-4];
>> pComb = 3*p1 + 2*p2

pComb =
     2    -6     4     5    -5
```

Roots and factorization

Zeros of polynomial functions satisfy the following property:

Theorem 5.1 (Fundamental Theorem of Algebra) *A polynomial of degree n has exactly n (possibly multiple) complex roots.*

An explicit restatement of this theorem is that any polynomial $p_n(x)$ in the form (5.1) can be factorized as

$$p_n(x) = c_1(x - x_1)^{m_1}(x - x_2)^{m_2}...(x - x_k)^{m_k}, \tag{5.2}$$

where $x_1, x_2, ..., x_k$ are distinct roots with *multiplicities* $m_1, m_2, ..., m_k$ such that $m_1 + m_2 + ... + m_k = n$. We have already used the factorization (5.2) and Theorem 5.1 in the analysis of eigenvalues of an $n \times n$ matrix in Section 4.1. We will now have another opportunity to explore the connection between eigenvalues and roots of polynomials.

The numerical search for roots of a general function is performed by general *root finding algorithms*, which we describe in Section 8.1. In MATLAB, roots of polynomial functions can be found by using the command `roots`, which takes the vector of coefficients of a polynomial and returns the vector of its roots. You might wonder if the function `roots` can be used to compute eigenvalues of square matrices from their determinant equation. As explained in Section 4.1, finding the characteristic polynomial associated with a given matrix is not a stable procedure. However, starting with a given polynomial $p_n(x)$ in the form (5.1), it is easy to see that the $n \times n$ matrix

$$\mathbf{A} = \frac{1}{c_1}\begin{bmatrix} 0 & 0 & \cdots & 0 & -c_{n+1} \\ c_1 & 0 & \cdots & 0 & -c_n \\ 0 & c_1 & \cdots & 0 & -c_{n-1} \\ \vdots & \vdots & \ddots & \vdots & \vdots \\ 0 & 0 & \cdots & c_1 & -c_2 \end{bmatrix} \tag{5.3}$$

has $p_n(x)$ as its characteristic polynomial. You can then use the numerical algorithms in Sections 4.6 and 4.7 to find the eigenvalues of this *companion matrix* $\mathbf{A}$, which then coincide with the roots of $p_n(x)$. This is exactly the procedure used by the MATLAB function `roots`. Stability and efficiency aside, this procedure has the advantage of producing complex as well as real-valued roots, a generality that is not always shared by other root finding algorithms.

We illustrate the use of `roots` on the polynomials

$$\begin{aligned} p_4^{(1)}(x) &= x^4 + 2x^3 - 7x^2 - 8x + 12 \\ p_4^{(2)}(x) &= 2x^4 + 3x^2 + 5x^2 + 9x + 5. \end{aligned}$$

All four roots of the first polynomial are real and *simple* (that is, have multiplicity equal to one), while the second polynomial has two complex-conjugate roots and a double real root:

```
>> p1 = [1,2,-7,-8,12];
>> r1 = roots(p1)'

r1 =
   -3.0000    -2.0000     2.0000     1.0000

>> p2 = [2,3,5,9,5];
>> r2 = roots(p2)'

r2 =
   0.2500 - 1.5612i   0.2500 + 1.5612i  -1.0000                -1.0000
```

For comparison, you can calculate the eigenvalues of the corresponding companion matrices:

```
>>  A1=[0 0 0 -p1(5)/p1(1)
1 0 0 -p1(4)/p1(1)
0 1 0 -p1(3)/p1(1)
0 0 1 -p1(2)/p1(1)];

>> eig(A1)'

ans =
    2.0000     1.0000    -2.0000    -3.0000

>> A2=[0 0 0 -p2(5)/p2(1)
1 0 0 -p2(4)/p2(1)
0 1 0 -p2(3)/p2(1)
0 0 1 -p2(2)/p2(1)];

>> eig(A2)
```

```
ans =
   0.2500 - 1.5612i
   0.2500 + 1.5612i
  -1.0000 - 0.0000i
  -1.0000 + 0.0000i
```

It is sometimes necessary to find the coefficients $c_1, c_2, ..., c_{n+1}$ of the polynomial $p_n(x)$ in the representation (5.1) from its roots $x_1, x_2, ..., x_n$. This converse operation with respect to the function `roots` is performed by the function `poly`, which takes the vector of roots, expands the factors in (5.2), and returns the vector of coefficients of polynomials. The coefficient c_1 of the highest power is normalized by one.

```
>> q1 = poly(r1)

q1 =
    1.0000    2.0000   -7.0000   -8.0000   12.0000

>> q2 = poly(r2)

q2 =
    1.0000    1.5000    2.5000    4.5000    2.5000
```

Exercise 5.1 The roots of the Chebyshev polynomial $T_n(x)$ are located at the points

$$x_k = \cos\frac{\pi(2k-1)}{2n}, \qquad k = 1, 2, ..., n.$$

Find coefficients of the first 10 Chebyshev polynomials $T_n(x)$.

Derivatives and integrals

As we explained in Section 2.2, derivatives and integrals of polynomials are linear operations. Therefore, the representation (5.1) can be differentiated and integrated term by term to yield the following general formulas:

$$p_n'(x) = nc_1x^{n-1} + (n-1)c_2x^{n-2} + \ldots + c_n, \tag{5.4}$$

$$\int p_n(x)dx = \frac{c_1}{n+1}x^{n+1} + \frac{c_2}{n}x^n + \ldots + \frac{c_n}{2}x^2 + c_{n+1}x + c_{n+2}, \tag{5.5}$$

where the coefficient c_{n+2} is arbitrary. You see that for a polynomial $p_n(x)$ represented by a row vector of coefficients in $\mathbb{R}^{n+1}$, the derivative $p_n'(x)$ is represented by a row vector in $\mathbb{R}^n$, while the integral $\int p_n(x)dx$ is represented by a row vector in $\mathbb{R}^{n+2}$. Derivatives and integrals of polynomials are computed by the MATLAB functions `polyder` and `polyint`. With the second input argument in `polyint`, you can assign any specific number to c_{n+2}, while $c_{n+2} = 0$ is used by default.

```
>> p = [1,2,-7,-8,12];
>> Pder = polyder(p)

Pder =
     4     6   -14    -8

>> Pint1 = polyint(p)

Pint1 =
    0.2000    0.5000   -2.3333   -4.0000   12.0000         0

>>Pint2 = polyint(p,10)

Pint2 =
    0.2000    0.5000   -2.3333   -4.0000   12.0000   10.0000
```

Products

Nonlinear operations on polynomial functions are more difficult because they involve changes in the order of summation of power terms. For example, the product

$$p_{n+m}(x) = p_n(x)p_m(x)$$

of the polynomials $p_n(x)$ and $p_m(x)$ is obtained through the following convolution sum:

$$\left(\sum_{i=1}^{n+1} a_i x^{n+1-i}\right)\left(\sum_{j=1}^{m+1} b_j x^{m+1-j}\right) = \sum_{l=1}^{n+m+1} c_l x^{n+m+1-l}. \tag{5.6}$$

In MATLAB, this is implemented by the function `conv`, which operates on the vectors of coefficients. For instance, for the polynomials

$$\begin{aligned} p_4^{(1)}(x) &= x^4 + 2x^3 - 7x^2 - 8x + 12 \\ p_4^{(2)}(x) &= 2x^4 + 3x^3 + 5x^2 + 9x + 5 \end{aligned}$$

you have

```
>> a = [ 1,2,-7,-8,12];
>> b = [2,3,5,9,5];
>> c = conv(a,b)

c =
     2     7    -3   -18   -12   -57   -47    68    60
```

corresponding to the product

$$p_8(x) = 2x^8 + 7x^7 - 3x^6 - 18x^5 - 12x^4 - 57x^3 - 47x^2 + 68x + 60.$$

Rational functions

Ratios of two polynomials are called *rational functions.* A rational function of the form $p_n(x)/p_m(x)$ has finitely many *zeros* at the roots of the numerator $p_n(x)$ and finitely many *poles* at the roots of the denominator $p_m(x)$. For instance, the rational function

$$f(x) = \frac{p_2(x)}{p_3(x)} = \frac{x(x-3)}{(x-1)(x+1)^2}$$

has two poles at $x = 1$ and $x = -1$ and two zeros at $x = 0$ and $x = 3$. The graph of $f(x)$ shows that the *simple* pole at $x = -1$ leads to the infinite jump of $f(x)$ while the *double* pole at $x = 1$ leads to the infinite singularity but no jump of $f(x)$.

```
>> x=linspace(-2,4,1001)+0.0001; % to avoid division by zero
>> y=(x.*(x-3))./((x+1).*(x-1).^2);
>> plot(x,y);
>> axis([-2,4,-10,10])
```

Partial fractions

You can gain a better understanding of rational functions through their *partial fraction decomposition.* For example, the preceding rational function can be rewritten as

$$f(x) = \frac{x(x-3)}{(x-1)(x+1)^2} = \frac{1}{x+1} - \frac{1}{(x-1)^2}.$$

In this form, the contributions of the poles $x = -1$ and $x = 1$ to the graph in Figure 5.1 can be clearly identified. Moreover, the integral of $f(x)$ is easily calculated as sum of the integrals of the partial fractions.

For a general rational function $p_n(x)/p_m(x)$ with $n \leq m$, the partial fraction decomposition takes the form:

$$\frac{p_n(x)}{p_m(x)} = \sum_{i=1}^{k} \left(\frac{c_{i,1}}{(x-x_i)} + \frac{c_{i,2}}{(x-x_i)^2} + \ldots + \frac{c_{i,m_i}}{(x-x_i)^{m_i}} \right), \tag{5.7}$$

where k is the number of distinct roots x_i of the denominator $p_m(x)$, each with multiplicity m_j. In MATLAB, this is implemented by the function `residue`, which takes the coefficients of $p_n(x)$ and $p_m(x)$ as input and returns the coefficients $c_{i,j}$ and the vector of roots x_i. For example, the decomposition

$$\frac{x^2 - x + 3}{x^2 + x - 6} = -\frac{3}{x+3} + \frac{1}{x-2}$$

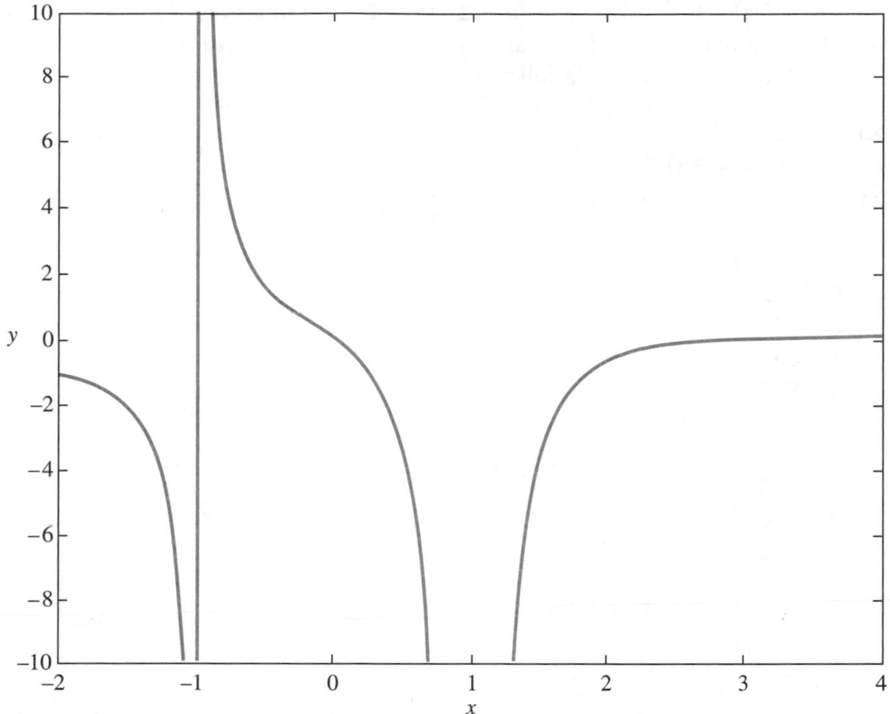

Figure 5.1 A rational function with two roots and two poles.

corresponds to a rational function with two simple poles and is implemented by the following MATLAB commands:

```
>>a=[1 -1 3];
>>b=[1 1 -6];
>>[c,x]=residue(a,b)

c =
    -3
     1

x =
    -3
     2
```

For a different example, consider the rational function

$$g(x) = \frac{3x - 2}{x^4 - 5x^3 + 6x^2 + 4x - 8}.$$

Since $x^4 - 5x^3 + 6x^2 + 4x - 8 = (x-2)^3(x+1)$, the function $g(x)$ has a *triple* pole at $x = 2$ and a simple pole at $x = -1$. Its partial fraction decomposition is computed by MATLAB as follows:

```
>>a=[3 -2];
>>b=[1 -5 6 4 -8];
>>[c,x]=residue(a,b)

c =
   -0.1852
    0.5556
    1.3333
    0.1852


x =
    2.0000
    2.0000
    2.0000
   -1.0000
```

which corresponds to the decomposition

$$\frac{3x+2}{x^4 - 5x^3 + 6x^2 + 4x - 8} = -\frac{5}{27(x-2)} + \frac{5}{9(x-2)^2} + \frac{4}{3(x-2)^3} + \frac{5}{27(x+1)}.$$

Long division

If $n > m$, the rational function $p_n(x)/p_m(x)$ can be represented in *long division* form:

$$\frac{p_n(x)}{p_m(x)} = p_{n-m}(x) + \frac{p_{m-1}(x)}{p_m(x)}, \tag{5.8}$$

where $p_{n-m}(x)$ is called the *quotient* and $p_{m-1}(x)$ is called the *remainder*. In component form this becomes

$$\frac{\sum_{i=1}^{n+1} a_i x^{n+1-i}}{\sum_{j=1}^{m+1} b_j x^{m+1-j}} = \sum_{l=1}^{n-m+1} c_l x^{n-m+1-l} + \frac{\sum_{i=1}^{m} d_i x^{m-i}}{\sum_{j=1}^{m+1} b_j x^{m+1-j}}. \tag{5.9}$$

In MATLAB, long division is implemented by the function `deconv`, which takes the coefficients of $p_n(x)$ and $p_m(x)$ as input and returns the coefficients of $p_{n-m}(x)$ and p_{m-1} as output. For instance, for the polynomials

$$\begin{aligned} p_4(x) &= x^4 + 2x^3 - 7x^2 - 8x + 12 \\ p_2(x) &= 2x^2 + 3x + 5 \end{aligned}$$

you obtain

```
>> a = [ 1,2,-7,-8,12];
>> b = [2,3,5];
>> [c,d] = deconv(a,b)

c =
    0.5000    0.2500   -5.1250

d =
         0         0         0    6.1250   37.6250
```

which corresponds to

$$\frac{x^4 + 2x^3 - 7x^2 - 8x + 12}{2x^2 + 3x + 5} = 0.5x^2 + 0.25x - 5.125 + \frac{6.125x + 37.625}{2x^2 + 3x + 5}.$$

You can use this example to check that long division (5.9) is the inverse operation to multiplication (5.6):

```
>> aa = conv(c,b)+d

aa =
     1     2    -7    -8    12
```

If $n \geq m$, the MATLAB function **residue** executes a long division of ...rm (5.8) to obtain the quotient polynomial $p_{n-m}(x)$ first, and then ... calculate the partial fraction decomposition of the remainder term ... The outputs in this case are the coefficients $c_{i,j}$ and the roots ...l fraction decomposition, plus a third vector with the ... naturally, if $n < m$, this third output returns an

...ion (5.7) is also valid when the denom- ... the function **residue** treats them

```
x =
          0 + 2.0000i
          0 - 2.0000i

r =
      []
```

corresponding to the decomposition

$$\frac{x+3}{x^2+4} = \frac{0.5 - 0.75i}{x - 2i} + \frac{0.5 + 0.75}{x + 2i}.$$

However, you learn in *Calculus* that irreducible quadratic terms in the denominator (leading to complex conjugate roots) result in the *irreducible* partial fraction decomposition:

$$\frac{p_n(x)}{p_m(x)} = \sum_{i=1}^{k_c} \left(\frac{a_{i,1}x + b_{i,1}}{(x^2 + \alpha_i x + \beta_i)} + \frac{a_{i,2}x + b_{i,2}}{(x^2 + \alpha_i x + \beta_i)^2} + \dots + \frac{a_{i,m_i}x + b_{i,m_i}}{(x^2 + \alpha_i x + \beta_i)^{m_i}} \right)$$
$$+ \sum_{i=1}^{k_r} \left(\frac{c_{i,1}}{(x - x_i)} + \frac{c_{i,2}}{(x - x_i)^2} + \dots + \frac{c_{i,m_i}}{(x - x_i)^{m_i}} \right),$$

where $2k_c$ roots are complex conjugated and k_r roots are real. As an example of such decomposition, you have

$$\frac{2x^2 + 5x + 10}{(x+2)(x^2+4)} = \frac{1}{x+2} + \frac{x+3}{x^2+4}.$$

As you can see from the previous MATLAB example, this type of irreducible decomposition cannot be obtained by a simple application of the function `residue`. In the next exercise, you are asked to remedy this situation.

Exercise 5.2 Use the MATLAB function `residue` and write a new function that performs partial fraction decomposition with irreducible quadratic factors from the partial fraction decomposition (5.7).

5.2 Vandermonde Interpolation

As you learn in *Calculus*, one can use a Taylor series expansion to app differentiable functions by polynomials up to a certain error.

Theorem 5.2 (Taylor Series) *Let $f(x)$ be $(n+1)$-times di the interval $[a, b]$. Then, for any $x_0, x \in (a, b)$*

$$f(x) = f(x_0) + f'(x_0)(x - x_0) + \frac{1}{2!} f''(x_0)(x - x_0$$
$$\dots + \frac{1}{n!} f^{(n)}(x_0)(x - x_0)^n +$$

where

$$R_n(x) = \frac{1}{(n+1)!} f^{(n+1)}(\xi)(x - x_0)^{n+1} \tag{5.11}$$

and $x_0 \le \xi \le x$.

In general terms, the higher the degree of differentiability for the function, the better you can control the approximation error $R_n(x)$. In particular, if the function is *smooth*, meaning that it can be differentiated infinitely often, then the error in its Taylor series can be made arbitrarily small. The class of infinitely differentiable function is, however, too small for many practical applications. It is therefore reassuring that an approximation by polynomials with an arbitrarily small error is available for a much wider class of functions because of the following deep result from *Real Analysis*.

Theorem 5.3 (Weierstrass Approximation) *If $f(x)$ is a real-valued continuous function on the interval $[a, b]$ and if $\epsilon > 0$ is given, then there exists a polynomial $p_n(x)$ such that*

$$\max_{a \le x \le b} |f(x) - p_n(x)| \le \epsilon. \tag{5.12}$$

In many practical problems, you are given a discrete set of data points with an unknown functional relation between them. For instance, you can try to use the table of the population of Canada in the years 1991, 1996, 2001, 2006 (the census years) and approximate the size of the Canadian population in all other intermediate years, for example, the year 2004. By Theorem 5.3, you know that if the functional relation between two variables (say, x and y) is at least continuous, then it can be approximated arbitrarily well by some polynomial. This motivates the following problem:

Problem 5.1 (Polynomial Interpolation) Given a set of $(n+1)$ data points (x_1, y_1), (x_2, y_2), ..., (x_{n+1}, y_{n+1}), find a polynomial $p_n(x)$ of degree n satisfying

$$p_n(x_j) = y_j, \qquad j = 1, 2, ..., n+1. \tag{5.13}$$

Linear interpolation

When $n = 1$, Problem 5.1 is called a *linear interpolation*. In other words, you try to connect two data points (x_1, y_1) and (x_2, y_2) by a straight line:

$$p_1(x) = y_1 + \left(\frac{y_2 - y_1}{x_2 - x_1}\right)(x - x_1) = \frac{x - x_2}{x_1 - x_2} y_1 + \frac{x - x_1}{x_2 - x_1} y_2. \qquad (5.14)$$

The first representation uses the slope of the *secant line* between two points, while the second representation is based on the explicit factorization of grid points (x_1, x_2) with the weights (y_1, y_2). Both formulas in (5.14) are equivalent and represent the unique solution of the linear interpolation problem. The first is known as the *Newton forward difference interpolation* (see Section 5.4), whereas the second is known as the *Lagrange interpolation* (see Section 5.3).

Polynomials that represent an unknown functional dependence on the discrete set of data points are called *interpolants.* Polynomial interpolation is used in numerical approximations of solutions of differential and integral equations. Working on a vector space of polynomials of degree n, rather than on a space of continuous functions, differential and integral equations are reduced to finite–dimensional linear and nonlinear algebraic systems, where the numerical matrix laboratory is particularly useful. In many other practical problems, interpolating polynomials helps researchers visualize continuous functions from the discrete set of data points, collocate these functions at selected intermediate points, as well as understand a unknown functional dependence.

General interpolation

The general analytical solution of the polynomial interpolation problem can be found in a closed form involving Vandermonde determinants. Although this solution is not computationally efficient (similar to other applications of determinants), it is still a simple theoretical tool that enables us to establish conditions under which Problem 5.1 has a unique solution (see Theorem 5.4).

Since each polynomial of degree n is determined by $(n+1)$ coefficients c_1, c_2, ..., c_{n+1} in the representation (5.1), we can rewrite problem (5.13) as the linear system:

$$x_j^n c_1 + x_j^{n-1} c_2 + ... + x_j c_n + c_{n+1} = y_j, \qquad j = 1, 2, ..., n+1, \qquad (5.15)$$

or in the matrix form:

$$\begin{bmatrix} x_1^n & x_1^{n-1} & \cdots & x_1 & 1 \\ x_2^n & x_2^{n-1} & \cdots & x_2 & 1 \\ \vdots & \vdots & \cdots & \vdots & \vdots \\ x_{n+1}^n & x_{n+1}^{n-1} & \cdots & x_{n+1} & 1 \end{bmatrix} \begin{bmatrix} c_1 \\ c_2 \\ \vdots \\ c_{n+1} \end{bmatrix} = \begin{bmatrix} y_1 \\ y_2 \\ \vdots \\ y_{n+1} \end{bmatrix}. \qquad (5.16)$$

According to Theorem 2.3, the linear system (5.16) has a unique solution if and only if the coefficient matrix is invertible, which is in turn equivalent to its determinant being different from zero. The determinant of the coefficient matrix is the *Vandermonde determinant* with explicit representation:

$$\begin{vmatrix} x_1^n & x_1^{n-1} & \cdots & x_1 & 1 \\ x_2^n & x_2^{n-1} & \cdots & x_2 & 1 \\ \vdots & \vdots & \cdots & \vdots & \vdots \\ x_{n+1}^n & x_{n+1}^{n-1} & \cdots & x_{n+1} & 1 \end{vmatrix} = \prod_{i=1}^{n+1} \prod_{j=i+1}^{n+1} (x_i - x_j). \tag{5.17}$$

If $x_i \neq x_j$ for $i \neq j$ (i.e., all grid points x_1, x_2, ..., x_{n+1} are distinct), there exists a unique solution for the coefficient vector $[c_1, c_2, ..., c_{n+1}] \in \mathbb{R}^{n+1}$. What we have just proved can be summarized in the following theorem:

Theorem 5.4 *There exists a unique solution $p_n(x)$ of Problem 5.1 if and only if the grid points $x_1, x_2, ..., x_{n+1}$ are all distinct.*

Vandermonde matrices in the form of (5.16) are built with the MATLAB function `vander`, defined on a vector of grid points $[x_1, x_2, ..., x_{n+1}]$. By using the MATLAB linear algebra solver \ (introduced in Section 2.3), you can immediately find a numerical solution of the polynomial interpolation problem for small values of n. The following example uses a data set on an equally spaced grid, generated by a small random perturbation of the polynomial $p_3(x) = 2x^3 + 3x^2 + x + 2$.

```
>> x = [0 0.2 0.4 0.6 0.8 1.0];
>> y = [2.0079, 2.3454, 3.0197, 4.1274, 5.7655, 8.0315];
>> V = vander(x)

V =

         0         0         0         0         0    1.0000
    0.0003    0.0016    0.0080    0.0400    0.2000    1.0000
    0.0102    0.0256    0.0640    0.1600    0.4000    1.0000
    0.0778    0.1296    0.2160    0.3600    0.6000    1.0000
    0.3277    0.4096    0.5120    0.6400    0.8000    1.0000
    1.0000    1.0000    1.0000    1.0000    1.0000    1.0000

>> c = V \ y'

c =

    0.0026
    0.0052
    2.0036
    3.0060
    1.0061
    2.0079
```

Alternatively, the exact same procedure for finding the coefficients of this type of Vandermonde interpolating polynomial is executed by the MATLAB function **polyfit**, which operates on two input vectors for grid points $[x_1, x_2, ..., x_{n+1}]$ and data points $[y_1, y_2, ..., y_{n+1}]$ and an input integer that specifies the degree of the interpolating polynomial $p_n(x)$. To visualize an interpolating polynomial $p_n(x)$ from the given vector of coefficients $[c_1, c_2, ..., c_{n+1}]$, you can use MATLAB two-dimensional graphics described in Section 1.6 and the MATLAB function **polyval**. The function **polyval** is called with two input vectors for the coefficients of the polynomial $p_n(x)$ and the grid points where the polynomial is computed. In the following example, the interpolating polynomial $p_n(x)$ is drawn in Figure 5.2 with the given set of data points. You need to use a dense grid of x values to show a continuous function $y = p_n(x)$ in between the given set of interpolated points:

```
>> x = [0 0.2 0.4 0.6 0.8 1.0];
>> y = [2.0079, 2.3454, 3.0197, 4.1274, 5.7655, 8.0315];
>> c = polyfit(x,y,length(x)-1)

c =
    0.0026    0.0052    2.0036    3.0060    1.0061    2.0079

>> xInt = 0 : 0.01 : 1;
>> yInt = polyval(c,xInt);
>> plot(xInt,yInt,'g',x,y,'b.')
```

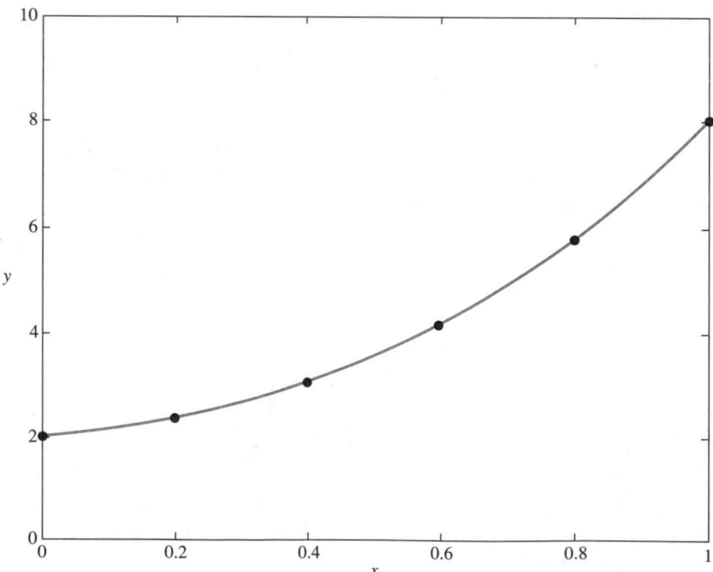

Figure 5.2 Example of Vandermonde polynomial interpolation.

If there are repeated values of $x_1, x_2, \ldots, x_{n+1}$, Problem 5.1 becomes singular since no polynomial function is multivalued. Depending on the source of data points, repeated values of x_j may represent multivalued functions, functions with vertical jump singularities, or functions defined in other forms, such as parametrically or implicitly.

Ill-posedness of Vandermonde interpolation

Albeit mathematically perfect, the solution to the polynomial interpolation problem using Vandermonde matrices is not computationally effective because Vandermonde matrices are ill-conditioned for large values of n. You can see this by calculating their 2-norm condition number. As explained in Section 2.4, the condition number of the coefficient matrix determines the accuracy in computations of a numerical solution of a linear system. Large condition numbers indicate a nearly singular matrix, the inverse of which is determined with large round-off error. In the following example, you can construct a simple loop that calculates the condition number of Vandermonde matrices of increasing dimension:

```
>> for n = 10 : 4 : 30
       x = 0.1*(1 : n);
       A = vander(x);
       c = cond(A);
       fprintf('n = %2.0f c = %3.1e \n',n,c);
   end

 n = 10 c = 5.6e+007
 n = 14 c = 2.6e+011
 n = 18 c = 6.4e+015
 n = 22 c = 3.4e+020
 n = 26 c = 2.4e+024
 n = 30 c = 2.8e+028
```

To understand why the Vandermonde matrices are ill-conditioned for large values of n, plot the monomials x^k, $0 \le k \le n$ on the interval $[0, 1]$:

```
>> x = linspace(0,1,101);
>> A = vander(x);
>> plot(x',A(:,91:101))
```

The monomials x^k, $k = 0, 1, ..., n$ form a basis for the vector space $\mathcal{P}_n$ of all polynomials of degree n. Therefore, according to Theorem 2.1, any polynomial $p_n(x)$ can be written as a linear combination of the monomials. Figure 5.3 shows that the basis functions tend to look the same for larger values of k. This phenomenon is similar to vectors in a finite-dimensional vector space that, despite being linearly independent, point in nearly the same direction. As a result, it is harder to identify projections of a particular polynomial $p_n(x)$ into the nearly collinear basis of monomials x^k for large k.

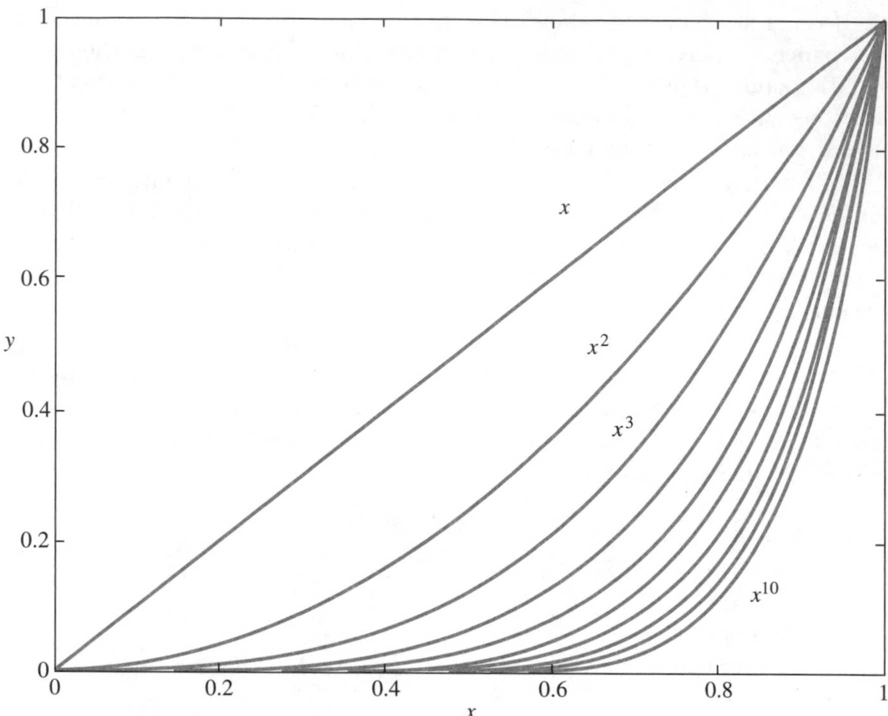

Figure 5.3 Basis functions for Vandermonde interpolation.

Because the Vandermonde matrices are ill-conditioned for large values of n, the linear algebra solvers may produce inaccurate solutions for the corresponding linear systems. To ensure that the coefficients of the interpolating polynomial $p_n(x)$ are found with low round-off error, in Sections 5.3 and 5.4 we reformulate the analytical solution of Problem 5.1 in terms of two computational algorithms that avoid such ill-conditioned linear systems.

5.3 Lagrange Interpolation

An elementary alternative to Vandermonde interpolation, avoiding the numerical solution of an ill-conditioned system, is known as *Lagrange interpolation.* This algorithm consists of two main steps. First, for a given set of distinct grid points $x_1, \ldots, x_{n+1}$, we prepare a set of *Lagrange polynomials* $L_{1,n}(x), \ldots, L_{n+1,n}(x)$ given by the formula

$$L_{j,n}(x) = \frac{(x - x_1)(x - x_2)...(x - x_{j-1})(x - x_{j+1})...(x - x_{n+1})}{(x_j - x_1)(x_j - x_2)...(x_j - x_{j-1})(x_j - x_{j+1})...(x_j - x_{n+1})}, \quad (5.18)$$

for $j = 1, \ldots, n+1$. By construction, it immediately follows that, for any j,

- Each polynomial $L_{j,n}(x)$ has degree n.
- $L_{j,n}(x_i) = 0$ for $i \neq j$.
- $L_{j,n}(x_j) = 1$.

In the second step, given any set of data values $y_1, \ldots, y_{n+1}$, the polynomial interpolant $p_n(x)$ is obtained as the scalar product of the vector $[y_1, ..., y_{n+1}]$ and the vector of Lagrange polynomials $[L_{1,n}(x), ..., L_{n+1,n}(x)]$ evaluated at any x value:

$$p_n(x) = y_1 L_{1,n}(x) + y_2 L_{2,n}(x) + ... + y_{n+1} L_{n+1,n}(x). \tag{5.19}$$

From the properties of the Lagrange polynomials, it is clear that the polynomial $p_n(x)$ solves Problem 5.1. The Lagrange interpolation form (5.19) looks different from the Vandermonde interpolation form (5.16). However, by Theorem 5.4, it is the same polynomial $p_n(x)$, as can be checked if all factors of the Lagrange polynomials are expanded in the representation (5.1).

In practice, the x values often consist of a fixed set of grid points, while polynomial interpolation needs to be performed for several different sets of y values on this grid. Therefore, even though the calculation and storage of the set of Lagrange polynomials associated with an array of x values (the first step) looks laborious, Lagrange interpolation is computationally competitive and the interpolant for any given set of y values is obtained in a single vector operation (the second step).

The MATLAB function **Lagrange** performs both steps of Lagrange interpolation.

```
function yInt = Lagrange(x,y,xInt)
% performs Lagrange interpolation
% x,y - row vectors of the same size for data points
% xInt - row vector of any size for interpolated values
% yInt - row vector of the same size as xInt

n = length(x) - 1;
ni = length(xInt);
L = ones(ni,n+1); % prepare a table of Lagrange polynomials
for j = 1 : (n+1)
    for i = 1 : (n+1)
        if (i ~= j)
            L(:,j) = L(:,j).*(xInt' - x(i))/(x(j)-x(i));
        end
    end
end
yInt = y*L'; % compute the values of interpolations
```

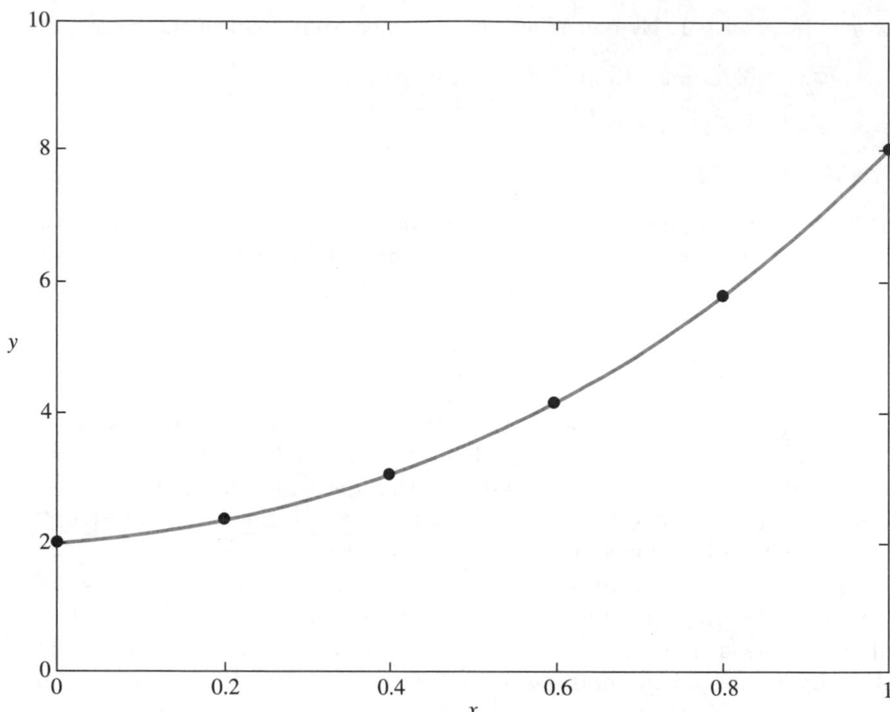

Figure 5.4 Example of Lagrange polynomial interpolation.

You can try the MATLAB function `Lagrange` for the same data points as in the example of Section 5.2. As a result of the uniqueness of the interpolating polynomial, the graphical output in Figure 5.4 shows no differences from the output of the Vandermonde interpolation in Figure 5.2.

```
>> x = [0 0.2 0.4 0.6 0.8 1.0];
>> y = [2.0079, 2.3454, 3.0197, 4.1274, 5.7655, 8.0315];
>> xInt = 0 : 0.01 : 1;
>> yInt = Lagrange(x,y,xInt);
>> plot(xInt,yInt,'g',x,y,'b*');
```

Exercise 5.3 Prove that

$$L_{1,n}(x) + L_{2,n}(x) + \ldots + L_{n+1,n}(x) = 1, \qquad \forall x \in \mathbb{R}.$$

Illustrate this property numerically on the grid of integers $x_k = k$, $k = 1, \ldots, 10$.

The advantage of Lagrange polynomial interpolation is that you can avoid the ill-conditioned Vandermonde matrices and solutions of large algebraic systems. As discussed, this interpolation algorithm is convenient if the grid of x values is fixed, while the set of y values changes because of, for instance, a dynamic evolution of a time-dependent solution. However, Lagrange interpolation is not convenient if a fast change in the grid of x values is required (for instance, if a new grid point is added or if the step size of a grid is changed). In those cases, the entire set of Lagrange polynomials needs to be recomputed and the results of the previous interpolation become unusable. A better solution to this problem is known as the Newton interpolation.

5.4 Newton Interpolation

Newton interpolation is an alternative algorithm for the polynomial interpolation that replaces the ill-conditioned algorithm of Vandermonde interpolation (in Section 5.2) and complements the well-conditioned algorithm for Lagrange interpolation (in Section 5.3). The Newton interpolating polynomial $p_n(x)$ that passes through the given set of data points (x_1, y_1), (x_2, y_2), ..., (x_{n+1}, y_{n+1}) is prepared by a recursive procedure based on the following special form of the interpolating polynomial $p_n(x)$:

$$p_n(x) = a_0 + a_1(x - x_1) + a_2(x - x_1)(x - x_2) + \dots \\ + a_n(x - x_1)(x - x_2)\cdots(x - x_n), \tag{5.20}$$

where $a_0, a_1, \dots, a_n$ are unknown coefficients. The algorithm consists of finding the coefficients a_0, a_1, ..., a_n by inserting the data points into (5.20) one by one and using the conditions of the polynomial interpolation.

Divided differences

We start by inserting x_1 into (5.20) and using the condition that $y_1 = p_n(x_1)$ to obtain

$$y_1 = p_n(x_1) = a_0.$$

Next, inserting x_2 into (5.20) and using the condition that $y_2 = p_n(x_2)$ and the value just obtained for a_0, we obtain

$$y_2 = p_n(x_2) = y_1 + a_1(x_2 - x_1) \Rightarrow a_1 = \frac{y_2 - y_1}{x_2 - x_1}.$$

For the next step, we insert x_3 into (5.20), and use the values just obtained for a_0 and a_1, together with the fact that $y_3 = p_n(x_3)$ to get

$$y_3 = p_n(x_3) = y_1 + \frac{y_2 - y_1}{x_2 - x_1}(x_3 - x_1) + a_2(x_3 - x_1)(x_3 - x_2),$$

which implies that

$$a_2 = \frac{1}{x_3 - x_1}\left(\frac{y_3 - y_2}{x_3 - x_2} - \frac{y_2 - y_1}{x_2 - x_1}\right).$$

These formulas suggest a reformulation of the algorithm in terms of what are called *divided differences.* To do that, notice that if $p[x; x_1, x_2, ..., x_k]$ is the Newton interpolating polynomial through the points (x_1, y_1), (x_2, y_2), ..., (x_k, y_k) and $p[x; x_2, x_3, ..., x_{k+1}]$ is the Newton interpolating polynomial through the points (x_2, y_2), (x_3, y_3), ..., (x_{k+1}, y_{k+1}), then the Newton interpolating polynomial through the points (x_1, y_1), (x_2, y_2), ..., (x_{k+1}, y_{k+1}) is

$$p[x; x_1, x_2, ..., x_{k+1}] = \frac{p[x; x_2, x_3, ..., x_{k+1}](x - x_1) - p[x; x_1, x_2, ..., x_k](x - x_{k+1})}{x_{k+1} - x_1}.$$

Comparing the coefficient in front of the largest term x^k, we obtain that the coefficient a_k of the Newton interpolating polynomial can be computed from the following recursive procedure.

Theorem 5.5 *Let $D_k[x_1, x_2, ..., x_{k+1}]$ be the kth divided difference, defined by the recursive formula:*

$$D_k[x_1, x_2, ..., x_{k+1}] = \frac{D_{k-1}[x_2, x_3, ..., x_{k+1}] - D_{k-1}[x_1, x_2, ..., x_k]}{x_{k+1} - x_1}, \quad (5.21)$$

starting with the zero-order difference $D_0[x_1] = y_1$. Then, $a_k = D_k[x_1, x_2, ..., x_{k+1}]$ is the coefficient of the interpolating polynomial (5.20).

The first divided difference in the recursive formula (5.21) corresponds to the slope approximation of the first derivatives,

$$D_1[x_k, x_{k+1}] = \frac{y_{k+1} - y_k}{x_{k+1} - x_k}.$$

The second, third, and higher-order divided differences in (5.21) are related to the difference approximations of the second, third, and higher-order derivatives, as we show in Section 6.2 for uniform grids. In the general case, the hierarchy of divided differences can be organized in the following table.

Grid Points	Data Points	1st Differences	2nd Differences	3rd Differences
x_1	y_1			
x_2	y_2	$D_1[x_1, x_2]$		
x_3	y_3	$D_1[x_2, x_3]$	$D_2[x_1, x_2, x_3]$	
x_4	y_4	$D_1[x_3, x_4]$	$D_2[x_2, x_3, x_4]$	$D_3[x_1, x_2, x_3, x_4]$

With the use of divided differences, the Newton interpolating polynomial $y = p_n(x)$ can be rewritten in the explicit form:

Nested multiplications

$$p_n(x) = \sum_{k=0}^{n} D_k[x_1, x_2, \ldots, x_{k+1}] \prod_{j=1}^{k} (x - x_j). \tag{5.22}$$

The Newton polynomial in the form (5.22) can be coded effectively by using the algorithm of nested multiplications. Nested multiplications for polynomials are based on the equivalent representation of the general polynomial $p_n(x)$ in the form:

$$\begin{aligned} p_n(x) &= c_1 x^n + c_2 x^{n-1} + \ldots + c_n x + c_{n+1} \\ &= (\ldots(((c_1 x + c_2)x + c_3)x + c_4)\ldots + c_n)x + c_{n+1}. \end{aligned} \tag{5.23}$$

A simple example illustrates nested multiplications for the polynomial

$$\begin{aligned} p_4(x) &= x^4 - 2x^3 + x^2 - 3x + 5 \\ &= 5 + x(-3 + x(1 + x(-2 + x))), \end{aligned}$$

computed on the uniform grid on $[1, 2]$ with the step size 0.2:

```
>> p = [1,-2,1,-3,4];
>> x = 1 : 0.2 : 2;
>> y = p(1)*ones(size(x));
>> for k = 2 : length(p)
>>     y = y.*x + p(k);
>> end
>> y

y =
    1.0000    0.4576    0.1136    0.1216    0.6736    2.0000
```

Exercise 5.4 Use nested multiplications for evaluation of polynomials of even degree

$$p_{2n}(x) = a_1 x^{2n} + a_2 x^{2n-2} + \ldots + a_n x^2 + a_{n+1}.$$

Use the MATLAB function `cputime` to check how much faster the nested multiplication algorithm works compared to the standard summation formula for a polynomial with $n = 1000$ and a vector x of 100 data points.

The MATLAB function `Newton` computes the table of divided differences and the Newton interpolating polynomials by using the recursive algorithm (5.21) and nested multiplications.

```
function [yInt,a] = Newton(x,y,xInt)
% computes the Newton interpolating polynomial
% x,y - row vectors of the same size for the data points
% xInt - row vector for grid points
% yInt - row vector for interpolated values at xInt
% a - row vector for coefficients of Newton polynomial

n = length(x)-1;
D = zeros(n+1,n+1); % the matrix for Newton divided differences
D(:,1) = y'; % zero-order divided differences
for k = 1 : n
    D(k+1:n+1,k+1) = (D(k+1:n+1,k)-D(k:n,k))./(x(k+1:n+1)-x(1:n-k+1))';
end
a = diag(D)';
nInt = length(xInt); % computation of interpolated values
yInt = a(n+1)*ones(1,nInt); % initialization
for k = 1 : n
    yInt = a(n+1-k)+yInt.*(xInt-x(n+1-k)); % nested multiplications
end
```

You can apply the MATLAB function `Newton` to the data points obtained from the function $f(x) = e^{-x^2}$ on the interval $[0, 3]$. Figure 5.5 shows that the interpolating polynomial $y = p_n(x)$ (solid curve) through the given set of data points (dots) is undistinguished from the exact function $y = f(x)$ (dotted curve).

```
>> x = 0 : 0.3 : 3;
>> y = exp(-x.^2);
>> xInt = 0 : 0.001 : 3;
>> [yInt,aCoef] = Newton(x,y,xInt);
>> a = aCoef

a =
    1.0000   -0.2869   -0.7233    0.5779   -0.0626   -0.1295
    0.0859   -0.0224   -0.0013    0.0035   -0.0015

>> yExact = exp(-xInt.^2);
>> plot(xInt,yInt,'g',x,y,'b.',xInt,yExact,':r');
```

The values for coefficients of the Newton polynomial $p_n(x)$ can be obtained from the MATLAB linear algebra solver \. The result is identical to the recursive algorithm of divided differences, but involves more operations and has no advantages of direct recursive algorithms.

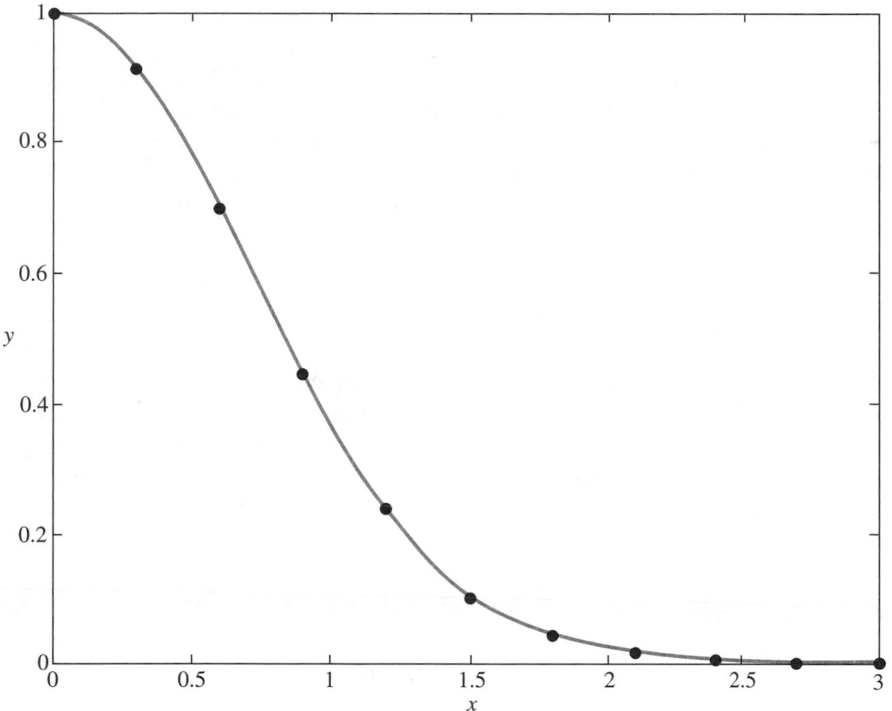

Figure 5.5 Example of Newton polynomial interpolation.

```
>> N = ones(n+1,1); % the coefficient matrix for linear system Na = y
>> for j = 1 : n
>>     N = [ N, N(:,j).*(x'-x(j))];
>> end
>> a = (N\(y'))'

a =
    1.0000   -0.2869   -0.7233    0.5779   -0.0626   -0.1295
    0.0859   -0.0224   -0.0013    0.0035   -0.0015
```

Because the interpolating polynomial $y = p_n(x)$ is unique when the x values of data points are all different (see Theorem 5.4), it is only the computational algorithm that distinguishes Newton interpolation from Vandermonde and Lagrange interpolations.

Exercise 5.5 Write a function that converts coefficients $a_0, a_1, \ldots, a_{n+1}$ of the Newton polynomial (5.20) to coefficients $c_1, c_2, \ldots, c_{n+1}$ of the standard polynomial form (5.1). Demonstrate that the Vandermonde and Newton interpolants are the same on a numerical example.

The Newton polynomial interpolation is convenient if a few data points are added or deleted from the given set of data points. In such problems, the table of divided differences in the recursive formula (5.21) can be expanded or truncated and the previously computed divided differences can be reused. In addition, the Newton interpolating polynomials are useful for constructing finite differences for first-order and higher-order numerical derivatives by means of derivatives of interpolating polynomials (see Section 6.2).

5.5 Errors of Polynomial Interpolation

In many cases, the set of data points in Problem 5.1 represent a sample of experimentally detected or numerically approximated values satisfying a theoretical dependence $y = f(x)$. The difference between the interpolation polynomial $p_n(x)$ and the function $f(x)$ is referred to as *the truncation error* $E_n(x)$ of the polynomial interpolation. The next theorem gives the truncation error in explicit form:

Theorem 5.6 *Let $p_n(x)$ be an interpolating polynomial for a given set of data points $y_k = f(x_k)$, for distinct grid points x_k, $k = 1, \ldots, n+1$. Let $a = \min_{1 \le j \le n+1}(x_j)$ and $b = \max_{1 \le j \le n+1}(x_j)$ and suppose that $f(x)$ is $(n+1)$ times continuously differentiable on $[a, b]$. Then, for each $x \in [a, b]$ there exists a $\xi \in [a, b]$, such that*

$$
\begin{aligned}
E_n(x) &= f(x) - p_n(x) \\
&= \frac{1}{(n+1)!} f^{(n+1)}(\xi)(x - x_1)(x - x_2)\ldots(x - x_{n+1}).
\end{aligned}
\tag{5.24}
$$

For the proof of this theorem, please see [7]. It follows from the error formula (5.24) that the error $E_n(x)$ vanishes at the grid points $x = x_k$, $k = 1, \ldots, n+1$ and the error $E_n(x)$ is identically zero if $f(x)$ is a polynomial of degree n or less.

We illustrate the error of polynomial interpolation for the function $f(x) = \sin x$ on the interval $[0, \pi]$ interpolated with a polynomial $p_6(x)$ on the equally spaced grid with step size $h = \pi/6$. The error $E_n(x)$ has local maxima at the midpoints of the grid of x values, and it becomes larger near the ends of the interpolation interval (see Figure 5.6). This property of polynomial interpolation on an equally spaced grid follows from the behavior of the product $\prod_{k=1}^{n+1}(x - x_k)$ appearing in the error formula (5.24).

```
>> x = linspace(0,pi,7);
>> y = sin(x);
```

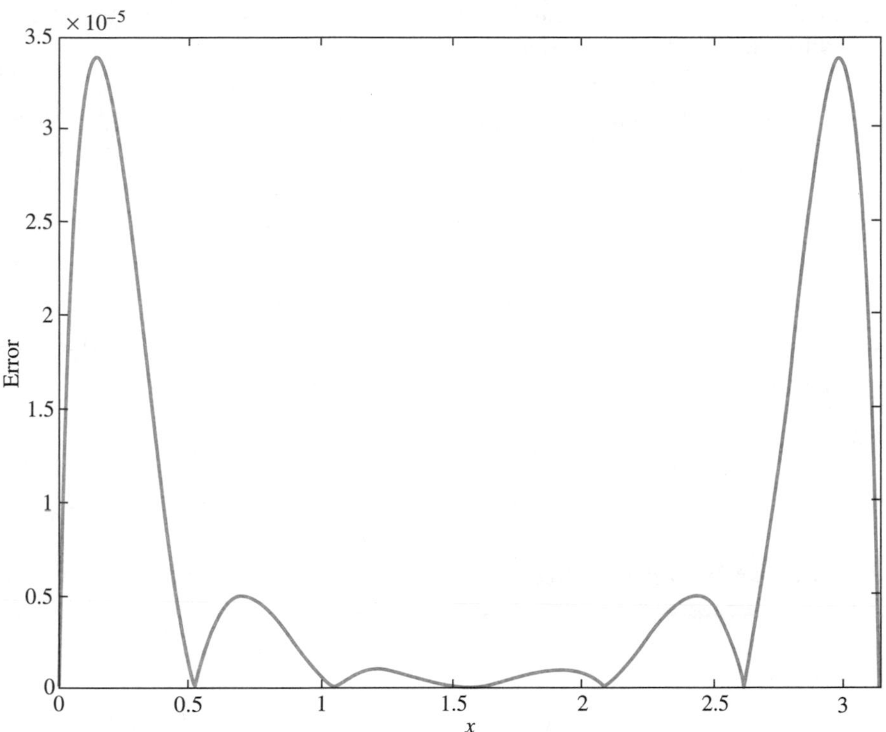

Figure 5.6 Error of the polynomial interpolation.

```
>> c = polyfit(x,y,6);
>> xInt = linspace(0,pi,1000);
>> yInt = polyval(c,xInt);
>> yExact = sin(xInt);
>> Error = abs(yExact-yInt);
>> ErrorAbsolute = max(Error)

ErrorAbsolute =
      3.3873e-005

>> plot(xInt,Error)
```

Equally spaced interpolation

For equally spaced grid points, the upper bound of the error $E_n(x)$ can be found from a maximization of product $\prod_{k=1}^{n+1}(x - x_k)$ appearing in (5.24).

Let $x_1 = a$, $x_{n+1} = b$, and the data points be equally spaced on the interval $[a, b]$ with the constant *step size* h such that

$$x_k = a + (k-1)h, \qquad k = 1, 2, \ldots, n+1, \qquad h = \frac{b-a}{n}. \tag{5.25}$$

Let

$$M_n = \max_{a \le x \le b} |f^{(n)}(x)|, \qquad C_n = \max_{0 \le z \le n} \prod_{k=0}^{n} |z-k|. \tag{5.26}$$

By using the scaling transformation $x = a + hz$, it follows from the error formula (5.24) that the upper bound for the truncation error $E_n(x)$ is

$$E_{\max} = \max_{a \le x \le b} |E_n(x)| \le \frac{1}{(n+1)!} C_n M_{n+1} h^{n+1}. \tag{5.27}$$

Exercise 5.6 By computing the maximum in the product (5.26), show that

$$C_1 = \frac{1}{4}, \qquad C_2 = \frac{2}{3\sqrt{3}}, \qquad C_3 = 1.$$

Illustrate the validity of C_1, C_2, and C_3 with an example of the polynomial interpolation for $f(x) = x^2$, $f(x) = x^3$, and $f(x) = x^4$, respectively.

The error bound (5.27) suggests a method to reduce the truncation error by decreasing the step size h. You have to be careful, however, because the degree of interpolating polynomial n could be related to the step size h when the interpolation interval $[a, b]$ is fixed. There exist two possibilities in the change of the interpolating polynomials: (1) fixed n and (2) fixed $[a, b]$.

Convergence rates

The case of fixed values of n and reduced values of h implies that we interpolate the same functional dependence on a shorter interpolation interval (assuming that the new data points can be found without additional limitations). According to the upper bound (5.27), the maximum error $E_{\max}$ must converge to zero as the function

$$E_{\max} = ch^p \qquad \text{as} \qquad h \to 0 \tag{5.28}$$

for some $c > 0$ and $p = n + 1$. You can check this behavior for the function $f(x) = \sin x$ on the interpolation interval $[0, 6h]$ with $n = 6$. In this case, you should have $p = 7$ in the power function (5.28). To verify this, calculate the errors for different values of the step size $h = \pi/m$, for $10 \le m \le 30$ and plot them, as shown in Figure 5.7. To obtain the functional relation implied by this graph, take logarithms on both sides of (5.28) and use the linear regression from Section 5.6 to find the power p and the constant c. The corresponding MATLAB computations are collected in the script `error_interp`:

```
% error_interp
m1 = 10;
m2 = 30;
for m = m1 : m2
    x = linspace(0,6*pi/m,7);
    y = sin(x);
    h(m-m1+1) = x(2)-x(1);
    c = polyfit(x,y,6);
    xInt = linspace(0,6*pi/m,100);
    yInt = polyval(c,xInt);
    yExact = sin(xInt);
    Error(m-m1+1) = max(abs(yExact-yInt));
end
a = polyfit(log(h),log(Error),1);
power = a(1)
ErrorApr = exp(a(2))*exp(power*log(h));
plot(h,Error,'b.',h,ErrorApr,':g')
```

The output of the MATLAB script is close to the theoretical value $p = 7$ for the variable `power`:

```
>> error_interp

power =
    6.6990
```

Runge phenomenon

The case of a fixed interpolation interval $[a, b]$ and reduced values of step size h implies that the degree of interpolating polynomial n grows. If there exists a uniform bound M_∞ for all higher derivatives of $f(x)$, the truncation error decreases sufficiently fast as $n \to \infty$ because

$$\lim_{n\to\infty} h^{n+1} = \lim_{n\to\infty} \left(\frac{b-a}{n}\right)^{n+1} = \lim_{n\to\infty} \exp\left(-(n+1)\log\frac{n}{b-a}\right) = 0.$$

However, the round-off error increases with larger values of n, as more arithmetic operations are performed. As a result, there exists an optimal number n (or equivalently, an optimal step size h) that gives the minimum of the total sum of the truncation and round-off errors.

The story becomes worse if the upper bound M_n for the nth derivative of $f(x)$ grows with larger values of n because the truncation error may decay slowly or even diverge as $n \to \infty$. The following example illustrates both cases. The balance between the truncation and round-off errors occurs in the case

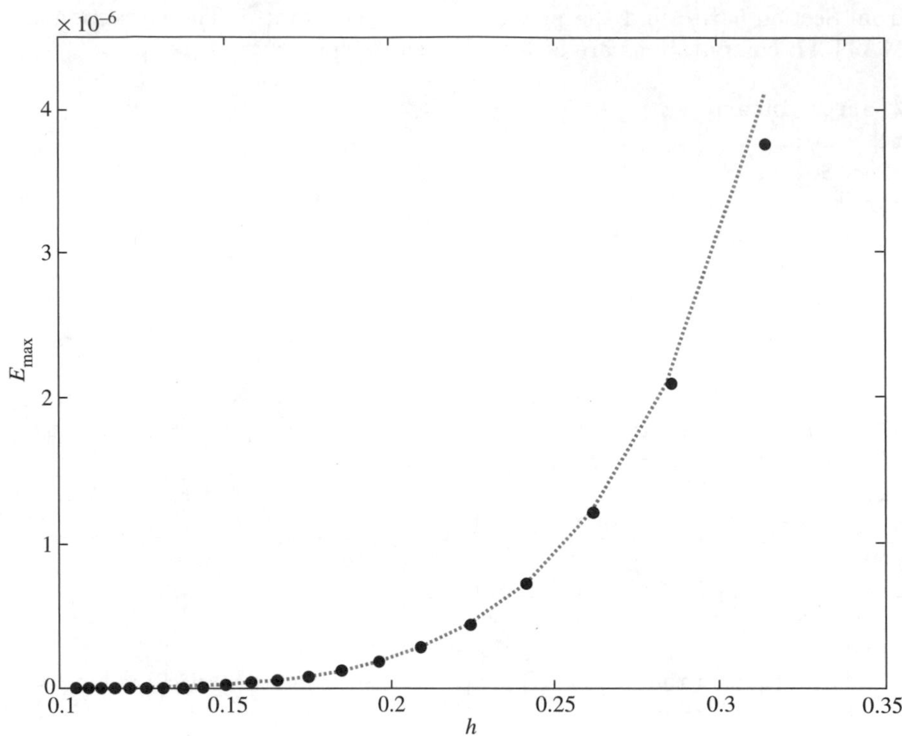

Figure 5.7 Maximum error versus step size h at the fixed value of $n = 6$.

of the function $f(x) = \sin x$ on the interval $[0, \pi]$ since all derivatives of $f(x)$ are bounded by $M_\infty = 1$. However, the truncation error diverges in the case of the Runge example $f(x) = 1/(1+25x^2)$ on the interval $[-1, 1]$ because the bounds M_n for the higher derivatives of $f(x)$ diverge as $n \to \infty$. The output of the MATLAB script `Runge_phenomenon` is shown in Figure 5.8.

```
% Runge_phenomenon
for n = 2 : 2 : 40
    x = linspace(0,pi,n+1);
    y = sin(x);
    c = polyfit(x,y,n);
    xInt = linspace(0,pi,1000);
    yInt = polyval(c,xInt);
    yExact = sin(xInt);
    Error(n) = max(abs(yExact-yInt));
end
figure(1); semilogy(Error,'.');
```

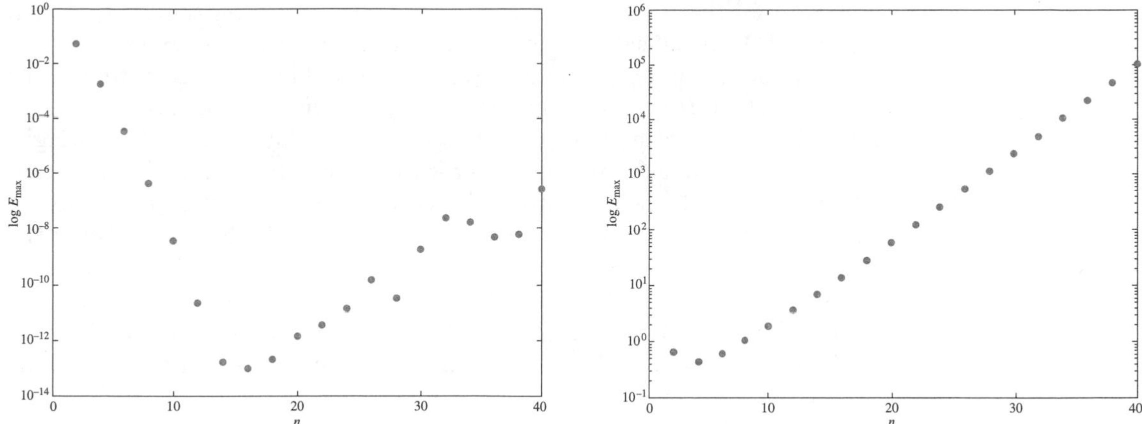

Figure 5.8 Maximum error versus the degree n at the fixed value of the interpolation interval. Left: $f(x) = \sin x$ on $[0, \pi]$. Right: $f(x) = 1/(1 + 25x^2)$ on $[-1, 1]$.

```
for n = 2 : 2 : 40
    x = linspace(-1,1,n+1);
    y = 1./(1+25*x.^2);
    c = polyfit(x,y,n);
    xInt = linspace(-1,1,1000);
    yInt = polyval(c,xInt);
    yExact = 1./(1+25*xInt.^2);
    Error(n) = max(abs(yExact-yInt));
end
figure(2); semilogy(Error,'.');
```

The polynomial interpolation fails for the Runge example $f(x) = 1/(1 + 25x^2)$ if the interpolation interval is fixed. This disaster for polynomial interpolation is called *polynomial wiggle* or the *Runge phenomenon.* As the degree of the interpolating polynomial $p_n(x)$ grows, the polynomial may display large swings between its extremal values. The truncation error measured at the extremal points grows with larger values of n.

Exercise 5.7 Plot the maximum error of the polynomial interpolation versus the step size h for the Runge example $f(x) = 1/(1 + 25x^2)$ on the interpolation interval $[-3h, 3h]$ at the fixed value of $n = 6$. Confirm that the maximum error reduces like $E_{\max} = ch^7$ and no polynomial wiggle occurs if the degree of the interpolating polynomial $n = 6$ is fixed.

The next computations display three interpolating polynomials $p_n(x)$ with

$n = 4, 8, 12$ for the Runge example $f(x) = 1/(1 + 25x^2)$ on the interpolation interval $[-1, 1]$. Artificial extremal values of interpolating polynomials $p_n(x)$ caused by the Runge phenomenon destroy the graphical representation of the original function $f(x)$. Meantime, the interpolating polynomials $p_n(x)$ on a smaller interpolation interval $[-0.1, 0.1]$ do not display a polynomial wiggle. According to a criterion in [28], the growth of the truncation error occurs if the length of the interpolation interval exceeds the distance from the midpoint of the interpolation interval to the first singularity of the function $f(z)$ in the complex plane $z = x + iy$. For the function $f(z) = 1/(1 + 25z^2)$ on $[-1, 1]$, the distance from the origin to the symmetric simple poles at $z = \pm i/5$ is $R = 0.2$. Therefore, the interval $[-0.1, 0.1]$ fits in the disk of radius $R = 0.2$, whereas the interval $[-1, 1]$ is beyond the disk. The output of the MATLAB script `polynomial_wiggle` is shown on Figure 5.9.

```
% polynomial_wiggle
figure(1); hold on;
for n = 4 : 4 : 12
    x = linspace(-1,1,n+1);
    y = 1./(1+25*x.^2);
    c = polyfit(x,y,n);
    xInt = linspace(-1,1,1000);
    yInt = polyval(c,xInt);
    plot(xInt,yInt,x,y,'*');
end
yExact = 1./(1+25*xInt.^2);
plot(xInt,yExact,':r'); hold off;
figure(2); hold on;
for n = 4 : 4 : 20
    x = linspace(-0.1,0.1,n+1);
    y = 1./(1+25*x.^2);
    c = polyfit(x,y,n);
    xInt = linspace(-0.1,0.1,1000);
    yInt = polyval(c,xInt);
    plot(xInt,yInt,x,y,'*');
end
yExact = 1./(1+25*xInt.^2);
plot(xInt,yExact,':r'); hold off;
```

Because of the polynomial wiggle and the growth of the round-off error, polynomial interpolation is unsuccessful in many cases. It does not provide accurate approximations and adequate visualizations of given functional dependencies. More advanced algorithms for numerical interpolation (and other problems in numerical analysis) are based on trigonometric functions and other orthogonal functions (as in the spectral methods of Chapter 11) and

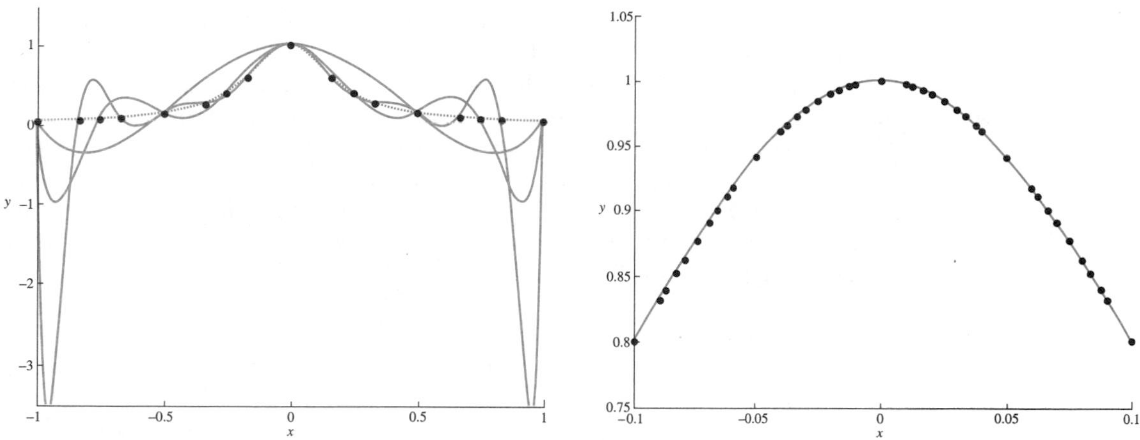

Figure 5.9 Left: Polynomial wiggle of the Runge function on $[-1, 1]$. Right: smooth polynomial interpolation of the same function on $[-0.1, 0.1]$.

piecewise polynomials and splines (as in the finite-element methods of Chapter 12).

Exercise 5.8 Compute polynomial interpolation of the Runge example $f(x) = 1/(1 + 25x^2)$ on $[-1, 1]$ on the grid of Chebyshev nodes:

$$x_k = \cos\frac{\pi k}{n}, \qquad k = 0, 1, 2, \ldots, n.$$

Show that no polynomial wiggle occurs for large values of n.

5.6 Polynomial Approximation

When the data points are corrupted by round-off errors of experimental measurements or numerical computations, solutions of the interpolation problem generate stiff interpolants with sharp corners and large truncation errors. In many cases, the corrupted data points represent a deterministic dependence between two variables x and y, which could sometimes be predicted from a theoretical study. Even if the deterministic dependence is fully predicted by the corresponding theory, it makes sense to ask whether and with what error the *real data* correspond to the *theoretical curve*. The problem becomes even more interesting if the theory has some limitations and some coefficients of the theoretical dependence remain unknown.

The preceding motivations change the problem of interest from *interpolation* to *approximation*. Instead of requiring that the function $y = f(x)$ *pass through* each point of the given data set, we shall look for a function $y = f(x)$ that is *sufficiently close* to the set of data points. There are multiple ways to define the *measure of closeness* between a smooth function $y = f(x)$ and the set of data points (x_1, y_1), (x_2, y_2), ..., (x_{n+1}, y_{n+1}), which is referred to as the *approximation error*. If x is an independent variable and y is a dependent variable, it may make sense to define the approximation error as the total square error between values of y_k and $f(x_k)$ for $k = 1, 2, \ldots, n+1$:

$$E = \sum_{k=1}^{n+1} (y_k - f(x_k))^2 \geq 0. \tag{5.29}$$

We shall in particular look for a polynomial function $y = f(x)$, which is referred to as *polynomial approximation*. Polynomial approximation becomes effective when the data values fit a simple theoretical curve that corresponds to polynomial or power functions. In some other cases, nonlinear functions predicted by the theory can be effectively transformed into polynomials after a change of dependent and independent variables.

Problem 5.2 (Polynomial Approximation) Given a set of data points (x_1, y_1),(x_2, y_2), ..., (x_{n+1}, y_{n+1}), find the polynomial function $y = p_m(x)$, where $m < n$, such that the total error (5.29) is minimal.

In this formulation, the approximation problem is equivalent to the minimization problem for quadratic functions. Similar minimizations are used in the general least-square methods (see Section 3.4) as well as in the finite element method for solutions of differential equations (see Section 12.3). Note that Problem 5.2 is equivalent to Problem 5.1 when $m = n$. When $m > n$, Problem 5.2 is ill-posed because more parameters of the polynomial functions exist than the given data points. Such an underdetermined system generally has infinitely many solutions.

Minimization of square error

We know that each polynomial $p_m(x)$ is completely defined by the set of $(m+1)$ coefficients:

$$p_m(x) = c_1 x^m + c_2 x^{m-1} + \ldots + c_m x + c_{m+1}. \tag{5.30}$$

Therefore, the problem can be reformulated as the algebraic system of linear equations for $[c_1, c_2, \ldots, c_{m+1}] \in \mathbb{R}^{m+1}$. To do so, we substitute the polynomial $p_m(x)$ into (5.29) so that the approximation error becomes a quadratic nonnegative function $E = E(c_1, c_2, \ldots, c_{m+1})$. Minimization of the quadratic function leads to the search for critical points, where the first derivatives of

$E(c_1, c_2, \ldots, c_{m+1})$ vanish. Computing the first derivatives of $E(c_1, c_2, \ldots, c_{m+1})$, we obtain a system of linear equations:

$$\frac{\partial E}{\partial c_j} = 2\sum_{k=1}^{n+1}\left(c_1 x_k^m + c_2 x_k^{m-1} + \ldots + c_m x_k + c_{m+1} - y_k\right) x_k^{m-j+1} = 0, \tag{5.31}$$

where $j = 1, 2, \ldots, m+1$. If the coefficient matrix of the linear system (5.31) is invertible, the linear system has a unique solution. In this case, the unique critical point is the point of global minimum of $E(c_1, c_2, \ldots, c_{m+1})$ since E is a non-negative function. If the coefficient matrix of the linear system (5.31) is singular, no solutions or infinitely many solutions may exist. In the former case, no critical points (and minima) occur for finite values of $c_1, c_2, \ldots, c_{m+1}$. In the latter case, the set of critical points form a line, or a higher dimensional set, but all critical points correspond to the same value of E, that is, they represent a one- or multidimensional set of global minima of $E(c_1, c_2, \ldots, c_{m+1})$.

The linear system (5.31) can be rewritten in matrix form as follows:

$$\begin{bmatrix} \sum_{k=1}^{n+1} x_k^{2m} & \sum_{k=1}^{n+1} x_k^{2m-1} & \ldots & \sum_{k=1}^{n+1} x_k^{m} \\ \sum_{k=1}^{n+1} x_k^{2m-1} & \sum_{k=1}^{n+1} x_k^{2m-2} & \ldots & \sum_{k=1}^{n+1} x_k^{m-1} \\ \vdots & \vdots & \ldots & \vdots \\ \sum_{k=1}^{n+1} x_k^{m} & \sum_{k=1}^{n+1} x_k^{m-1} & \ldots & \sum_{k=1}^{n+1} x_k^{0} \end{bmatrix} \begin{bmatrix} c_1 \\ c_2 \\ \vdots \\ c_{m+1} \end{bmatrix} = \begin{bmatrix} \sum_{k=1}^{n+1} y_k x_k^m \\ \sum_{k=1}^{n+1} y_k x_k^{m-1} \\ \vdots \\ \sum_{k=1}^{n+1} y_k \end{bmatrix}. \tag{5.32}$$

This seemingly complicated expression can be simplified using a $(n+1) \times (m+1)$ *rectangular Vandermonde matrix*:

$$\mathbf{A} = \begin{bmatrix} x_1^m & x_1^{m-1} & \ldots & 1 \\ x_2^m & x_2^{m-1} & \ldots & 1 \\ \vdots & \vdots & \ldots & \vdots \\ x_{n+1}^m & x_{n+1}^{m-1} & \ldots & 1 \end{bmatrix} : \mathbb{R}^{m+1} \mapsto \mathbb{R}^{n+1}.$$

You can verify that the system (5.32) is equivalent to

$$A'Ac = A'y, \tag{5.33}$$

where $\mathbf{y} = [y_1, y_2, \ldots, y_{n+1}] \in \mathbb{R}^{n+1}$. You can recognize this as the normal equations (3.50) of Section 3.4 corresponding to the *overdetermined* linear system

$$c_1 x_k^m + c_2 x_k^{m-1} + \ldots + c_m x_k + c_{m+1} = y_k, \qquad k = 1, 2, \ldots, n+1, \tag{5.34}$$

where $n > m$.

It follows from Section 3.4 that the matrix $\mathbf{A}'\mathbf{A}$ is invertible if and only if the matrix $\mathbf{A}$ has full rank $(m+1)$, that is, it has $(m+1)$ linearly independent rows (because $m < n$). Therefore, we have found the condition for uniqueness of the global minimum of $E(c_1, c_2, \ldots, c_{m+1})$.

Theorem 5.7 *There exists a unique solution* $\mathbf{c} = (\mathbf{A}'\mathbf{A})^{-1}\mathbf{A}'\mathbf{y}$ *of the linear system (5.33) if and only if there exist at least* $m+1$ *distinct values of* x_k *for* $k = 1, 2, \ldots, n+1$.

We illustrate the solution of Problem 5.2 with an example, where a set of 200 data points is generated from a cubic polynomial $y = f(x) = x^3 - 6x^2 + 8x - 10$ corrupted by a small random perturbation $f_{\text{rand}}(x)$. Random values of $f_{\text{rand}}(x)$ are chosen from a uniform distribution on the interval $[0, 1]$, which is generated by the MATLAB function `rand`. The example is coded in the MATLAB script `polynomial_approximation`.

```
% polynomial_approximation
x = linspace(-0.5,4.5,201);
y = x.^3-6*x.^2+8*x-10+2*(0.5-rand(size(x))); % given function
n = length(x)-1; m = 3;
A = vander(x); A = A(:,n-m+1:n+1); % Vandermonde matrix
c = ((A'*A)\(A'*y'))' % solution of the linear system
xApr = linspace(-0.5,4.5,201);
yApr = polyval(c,xApr);
yExact=xApr.^3 - 6*xApr.^2 + 8*xApr - 10;
plot(x,y,'.g',xApr,yApr,'b',xApr,yExact,':m');
```

When the script is executed, you find that the least-square approximation with a cubic polynomial function $p_3(x)$ recovers the parameters of the original function $y = f(x)$ within a small error. The polynomial approximation to the given set of data points is shown on Figure 5.10.

```
>> polynomial_approximation

c =

    0.9966   -5.9691    7.9779  -10.0682
```

The MATLAB function `polyfit` computes the coefficients of the polynomial $p_m(x)$, from the inputs $\mathbf{x} = [x_1, \ldots, x_{n+1}]$, $\mathbf{y} = [y_1, \ldots, y_{n+1}]$, and $m < n$. Equivalently, you can use the backslash operator \ to solve the

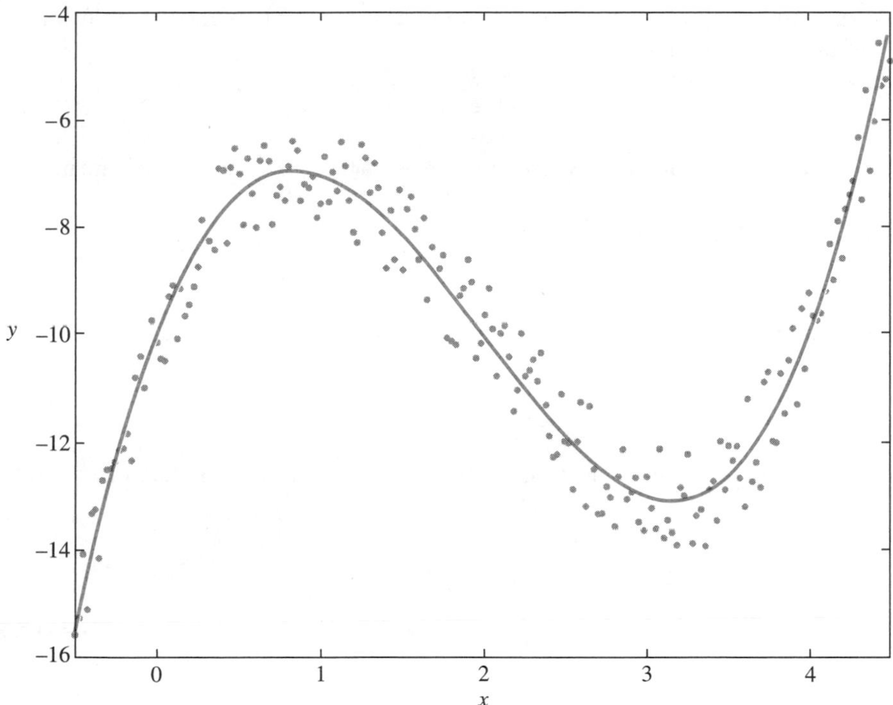

Figure 5.10 Example of cubic polynomial approximation.

overdetermined linear system $\mathbf{Ac} = \mathbf{y}$ and obtain the same solution, as shown next using the data from the previous example:

```
>> c = polyfit(x,y,m) % Variant with the polyfit function

c =
    0.9966   -5.9691    7.9779  -10.0682

>> c = (V\y')'  % Variant with the linear algebra solver

c =
    0.9966   -5.9691    7.9779  -10.0682
```

When $m = 1$, Problem 5.2 is called *linear regression.* It is particularly important because many physical laws (e.g., Hookes' law, Newton's law) can be approximated by linear functions for small perturbations. In addition, many other approximations (power functions, logarithmic and exponential functions) can be reduced to a linear regression. Although the linear regression

Linear regression

is explained in detail in Section 3.4, we can recover the same results from an algebraic point of view. Letting

$$p_1(x) = c_1 x + c_2, \tag{5.35}$$

the system (5.33) for linear regression takes the particularly simple form

$$\begin{pmatrix} \sum_{k=1}^{n+1} x_k^2 & \sum_{k=1}^{n+1} x_k \\ \sum_{k=1}^{n+1} x_k & \sum_{k=1}^{n+1} x_k^0 \end{pmatrix} \begin{pmatrix} c_1 \\ c_2 \end{pmatrix} = \begin{pmatrix} \sum_{k=1}^{n+1} y_k x_k \\ \sum_{k=1}^{n+1} y_k \end{pmatrix}. \tag{5.36}$$

This linear system has an exact solution, which can be expressed as

$$c_1 = \frac{\sigma_{xy}}{\sigma_x^2}, \quad c_2 = \bar{y} - \frac{\sigma_{xy}}{\sigma_x^2}\bar{x}, \tag{5.37}$$

where, using statistical notation, $\bar{x}$ and $\bar{y}$ are the sample means of x and y values,

$$\bar{x} = \frac{1}{n+1}\sum_{k=1}^{n+1} x_k, \quad \bar{y} = \frac{1}{n+1}\sum_{k=1}^{n+1} y_k,$$

σ_x^2 and σ_y^2 are the sample variances

$$\sigma_x^2 = \frac{1}{n+1}\sum_{k=1}^{n+1}(x_k - \bar{x})^2, \quad \sigma_y^2 = \frac{1}{n+1}\sum_{k-1}^{n+1}(y_k - \bar{y})^2$$

and σ_{xy} is the sample covariance

$$\sigma_{xy} = \frac{1}{n+1}\sum_{k=1}^{n+1}(x_k - \bar{x})(y_k - \bar{y}).$$

In this notation, the total square error defined by (5.29) takes the form:

$$E = (n+1)(1-\rho^2)\sigma_y^2, \qquad \rho = \frac{\sigma_{xy}}{\sigma_x \sigma_y}, \tag{5.38}$$

where $-1 \le \rho \le 1$ is the correlation coefficient. When $\rho^2 = 1$ ($\sigma_{xy}^2 = \sigma_x^2 \sigma_y^2$), the total square error E vanishes in (5.38), since all data points are perfectly correlated and belong to the straight line

$$y = \bar{y} + \frac{\sigma_{xy}}{\sigma_x^2}(x - \bar{x}).$$

When $\rho^2 = 0$ ($\sigma_{xy} = 0$), the total square error E is maximal in (5.38) because the data points are highly uncorrelated and are scattered around the constant value $y = \bar{y}$.

Linear regression can be used for problems that are not *a priori* given in linear form. For instance, the power function

Linearization

$$y = cx^p \tag{5.39}$$

with unknown parameters c and p is not a linear function for $p \neq 1$. Recall that this function was used in Section 5.5 for analysis of the truncation error of polynomial interpolation, and that linear regression was constructed in the MATLAB script `error_interp` (see Figure 5.7). To linearize the power (5.39), we take logarithms of both sides of (5.39), which results in

$$\log y = p \log x + \log c. \tag{5.40}$$

Therefore, the transformed data points $Y_i = \log y_i$ and $X_i = \log x_i$ can be fit to a linear function

$$Y = p_1(X) = c_1 X + c_2,$$

where $c_1 = p$ and $c_2 = \log c$. The data points in the next example are generated by the random perturbation of the function $y = 5x^{1/4}$. The MATLAB script `nonlinear_regression` computes the values for c and p from the linear regression and plots the given set of data points and the power function $f(x) = cx^p$ in log–log and linear scales.

```
% nonlinear_regression
x = linspace(0.05,2.5,101);
y = 5*x.^(0.25) + 2*(0.5-rand(size(x))); % given function
xBar = mean(log(x)); yBar = mean(log(y)); % sample mean
xSigma = mean((log(x)-xBar).^2);
ySigma = mean((log(y)-yBar).^2);
xySigma = mean((log(x)-xBar).*(log(y)-yBar)); % sample covariance
c1 = xySigma/xSigma; c2 = yBar - xBar*c1;  % Linear regression
c = exp(c2), p = c1, yApr = c*x.^(p);  % power function
r = xySigma/sqrt(xSigma*ySigma) % correlation
E = length(y)*ySigma*(1-r^2) % approximation error
figure(1); loglog(x,y,'g.',x,yApr,'b'); % linear regression in (log x, log y)
figure(2); plot(x,y,'g.',x,yApr,'b');   % nonlinear regression in (x,y)
```

When the script is executed, you find parameters c and p close to the unperturbed function $y = 5x^{1/4}$ and a high correlation between samples of x and y values (ρ is close to 1). Log–log and linear plots of the data points and the nonlinear regression $y = cx^p$ are shown on Figure 5.11.

```
>> nonlinear_regression

c =

    4.9314
```

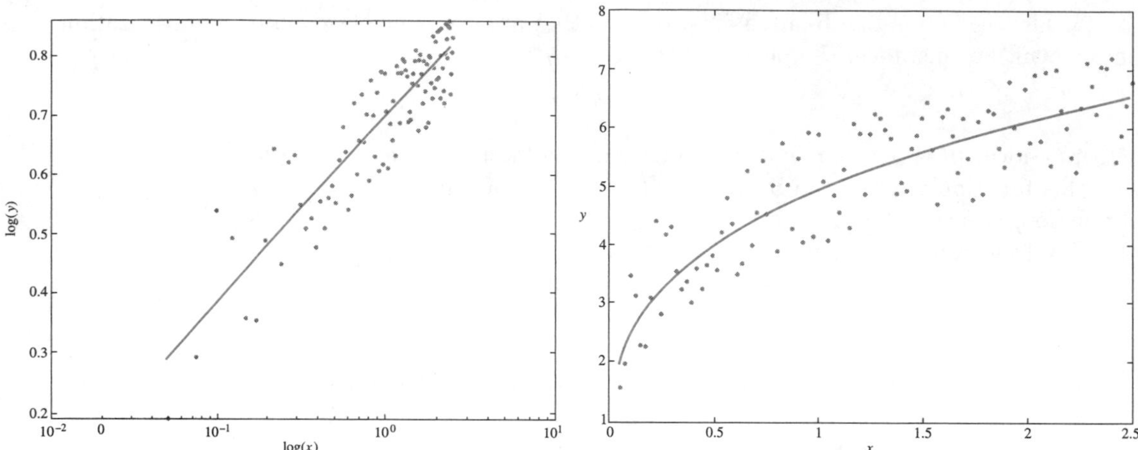

Figure 5.11 Linear regression in log-log scale (*left*) and nonlinear regression in linear scale (*right*).

```
p =
     0.2678

r =
     0.8936

E =
     1.3253
```

Similar techniques of polynomial approximation can be used for some other nonlinear functions such as exponential and logarithmic functions.

Exercise 5.9 Generate a set of data points from a random perturbation of the function $y = ce^{-px^2}$ for $c = 2$ and $p = 1$ on the interval $[-1, 3]$. Use the method of linearization, and compute the polynomial approximation of the data points. Show that the correlation coefficient ρ increases when the amplitude of the random noise reduces to zero and that parameters c and p of the Gaussian function approach their unperturbed values $c = 2$ and $p = 1$ in this limit.

Approximation error

The solution to Problem 5.2 generally has a nonzero approximation error (5.29). The approximation error becomes smaller when the degree m of the polynomial $p_m(x)$ increases but generally vanishes only when $m = n$ and the polynomial $p_n(x)$ gives a unique solution to Problem 5.1 (provided that all values of x_k are distinct). When the approximation error is zero, other sources of errors may arise such as the truncation and round-off errors. When it is nonzero, it makes no sense to study other errors because the approximation error is the dominant source for discrepancy between the set of data points and the polynomial $y = p_m(x)$.

The next example illustrates how the approximation error reduces with larger values of m. The function $f(x) = e^{-x}\cos x$ is used to generate a set of 51 equally spaced data points on $[0, 5]$ and the data set is approximated with the polynomials $p_m(x)$, where $1 \leq m \leq 11$. The example is coded in the MATLAB script approximation_error.

```
% approximation_error
x = 0 : 0.1 : 5; n = length(x)-1;
y = exp(-x).*cos(x); figure(1);
hold on;
for m = 1 : 11
    c = polyfit(x,y,m);
    yy = polyval(c,x);
    E = sum((y-yy).^2) ;
    fprintf('m = %d, E = %6.8f\n',m,E);
    if ((m == 2) | (m == 4) )
        xApr = 0 : 0.001 : 5;
        yApr = polyval(c,xApr);
        plot(x,y,'g.',xApr,yApr,'b');
    end
end
```

When the script is executed, it shows reduction of the approximation error E in (5.29) for larger degrees of the polynomial $p_m(x)$. Figure 5.12 shows that the polynomial $p_4(x)$ is sufficiently close to the given data set with $m = 4 < n = 50$.

```
>> approximation_error

 m = 1, E = 1.92849730
 m = 2, E = 0.38518965
 m = 3, E = 0.01422757
 m = 4, E = 0.00254079
 m = 5, E = 0.00097279
 m = 6, E = 0.00007298
 m = 7, E = 0.00000117
 m = 8, E = 0.00000001
 m = 9, E = 0.00000000
 m = 10, E = 0.00000000

Warning: Polynomial is badly conditioned. Remove repeated data
points or try centering and scaling as described in HELP POLYFIT.

 m = 11, E = 0.00000000
```

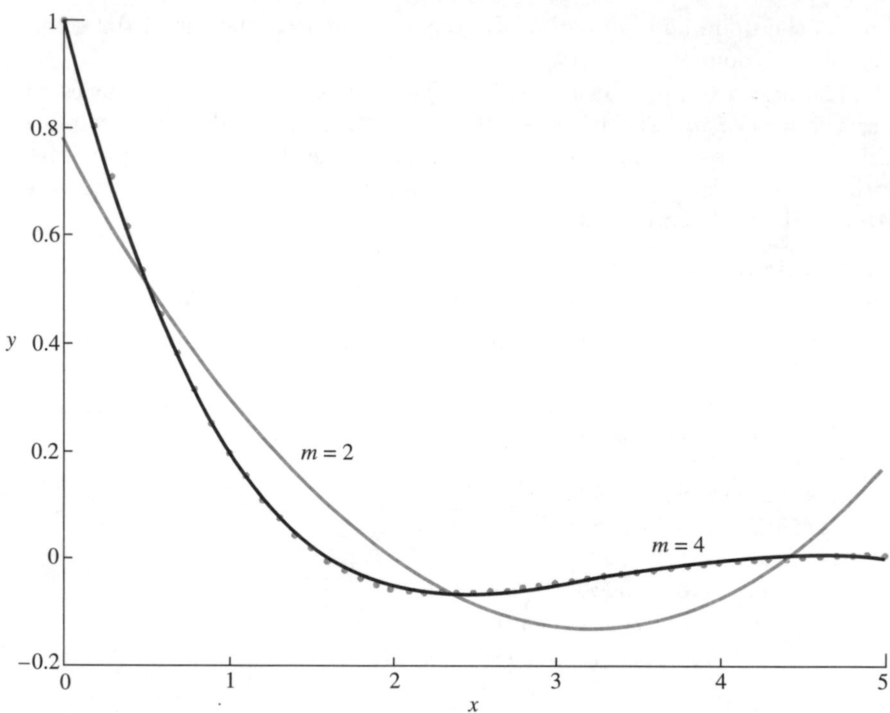

Figure 5.12 Polynomial approximations $p_m(x)$ with $m = 2$ and $m = 4$ for function $f(x) = e^{-x} \cos x$.

You can see from the outcome that MATLAB produces a warning at $m = 11$. This warning indicates that the polynomial approximation in the Vandermonde form is an ill-conditioned problem similar to the polynomial interpolation in the Vandermonde form in Section 5.2. This difficulty can be overcome using the set of orthogonal polynomials introduced in Section 5.7.

5.7 Approximation with Orthogonal Polynomials

As you have seen in Section 5.6, approximating a set of data points using a polynomial $p_m(x)$ in the form (5.30) leads to the system (5.32), which is ill-conditioned for large values of m. The derivation of (5.32) rests upon the use of the canonical basis $\{1, x, x^2, \ldots, x^m\}$ for the $(m+1)$-dimensional vector space $\mathcal{P}_m$ of all real-valued polynomials of degree m (see Section 2.1). Therefore, we would like to revise the structure of power terms in $p_m(x)$, or in other words use a different basis for it so that the matrix $\mathbf{A}'\mathbf{A}$ in the linear system (5.33) becomes as simple as possible. For example, if $\mathbf{A}'\mathbf{A}$ is a diagonal matrix with nonzero diagonal entries, its inverse consists of the inverse of each diagonal entry, and a general solution to the system (5.33) yielding the vector $\mathbf{c}$ can be obtained without any significant numerical error.

Discrete orthogonal polynomials

For this, observe first that, given a set of $(n+1)$ data points in Problem 5.2, we can identify each polynomial $p(x) \in \mathcal{P}_m$ with the $(n+1)$-dimensional vector $[p(x_1), p(x_2), \ldots, p(x_{n+1})] \in \mathbb{R}^{n+1}$ and consider the inner product

$$\langle p(x), q(x)\rangle := \sum_{l=1}^{n+1} p(x_l)q(x_l), \qquad \forall p(x), q(x) \in \mathcal{P}_m. \tag{5.41}$$

Having defined the inner product (5.41), we know from Section 3.2 that the Gram–Schmidt orthogonalization procedure applied to the basis $\{1, x, x^2, \ldots, x^m\}$ produces another basis $\{\pi_0(x), \pi_1(x), \ldots, \pi_m(x)\}$, with the following properties:

- Each $\pi_k(x)$ is a polynomial of degree k, for $k = 0, 1, \ldots, m$.
- $\{\pi_0(x), \pi_1(x), \ldots, \pi_m(x)\}$ is an orthogonal set, that is, $\langle \pi_k(x), \pi_n(x)\rangle = 0$, whenever $k \neq n$.
- Each $\pi_k(x)$ is normalized, that is, $\|\pi_k\| = \sqrt{\langle \pi_k, \pi_k\rangle} = 1$, for $k = 0, 1, \ldots, m$.

Using the orthonormal basis $\{\pi_k(x)\}_{k=0}^m$, we can uniquely represent any polynomial $p_m(x)$ in the form

$$p_m(x) = c_0\pi_0(x) + c_1\pi_1(x) + \ldots + c_m\pi_m(x), \tag{5.42}$$

where the coefficients $c_0, c_1, \ldots, c_m$ are given by

$$c_k = \langle p_m, \pi_k\rangle.$$

Since the polynomial $p_m(x)$ is unknown in Problem 5.2, these coefficients must be found by minimizing the total square error $E = E(c_0, c_1, \ldots, c_m)$ computed from (5.29) and (5.42). Using the orthogonality of the polynomials $\{\pi_0(x), \pi_1(x), \ldots, \pi_m(x)\}$, the total square error can be written as

$$\begin{aligned} E &= \sum_{l=1}^{n+1}\left(y_l - \sum_{k=0}^{m} c_k\pi_k(x_l)\right)^2 \\ &= \langle \mathbf{y}, \mathbf{y}\rangle - 2\sum_{k=0}^{m} c_k\langle \mathbf{y}, \pi_k\rangle + \sum_{k_1=0}^{m}\sum_{k_1=0}^{m} c_{k_1}c_{k_2}\langle \pi_{k_1}, \pi_{k_2}\rangle \\ &= \langle \mathbf{y}, \mathbf{y}\rangle - 2\sum_{k=0}^{m} c_k\langle \mathbf{y}, \pi_k\rangle + \sum_{k=0}^{m} c_k^2 \end{aligned} \tag{5.43}$$

Taking derivatives of this expression then leads to

$$\frac{\partial E}{\partial c_j} = 2\left(c_j - \langle \mathbf{y}, \pi_j\rangle\right), \quad j = 0, 1, 2, \ldots, m. \tag{5.44}$$

It then follows immediately that

$$c_j = \sum_{l=1}^{n+1} y_l \pi_j(x_l) = \langle \mathbf{y}, \pi_j \rangle \tag{5.45}$$

give the unique critical point for the function $E(c_0, c_1, \ldots, c_m)$. Moreover, inserting (5.45) into (5.43), we obtain the following expression for the minimal square error:

$$E_{min} = \sum_{l=1}^{n+1} y_l^2 - \sum_{k=0}^{m} c_k^2. \tag{5.46}$$

Instead of the Gram–Schmidt orthogonalization procedure, we can define the polynomials $\{\pi_0(x), \pi_1(x), \ldots, \pi_m(x)\}$ as

$$\pi_k(x) = \frac{\widetilde{\pi}_k(x)}{\|\widetilde{\pi}_k\|}, \qquad k = 0, 1, \ldots, m, \tag{5.47}$$

where $\widetilde{\pi}_{-1}(x) = 0$, $\widetilde{\pi}_0(x) = 1$, and the remaining polynomials $\widetilde{\pi}_k(x)$ are found through the three-term recurrence relation

$$\widetilde{\pi}_{k+1}(x) = (x - \alpha_k)\widetilde{\pi}_k(x) - \beta_k \widetilde{\pi}_{k-1}(x), \qquad k = 0, 1, \ldots, m-1, \tag{5.48}$$

where

$$\alpha_k = \frac{(\widetilde{\pi}_k, x\widetilde{\pi}_k)}{(\widetilde{\pi}_k, \widetilde{\pi}_k)} \tag{5.49}$$

and

$$\beta_k = \frac{(\widetilde{\pi}_{k-1}, x\widetilde{\pi}_k)}{(\widetilde{\pi}_{k-1}, \widetilde{\pi}_{k-1})} = \frac{(x\widetilde{\pi}_{k-1}, \widetilde{\pi}_k)}{(\widetilde{\pi}_{k-1}, \widetilde{\pi}_{k-1})} = \frac{(\widetilde{\pi}_k, \widetilde{\pi}_k)}{(\widetilde{\pi}_{k-1}, \widetilde{\pi}_{k-1})}. \tag{5.50}$$

Exercise 5.10 Prove that $\{\pi_0(x), \pi_1(x), \ldots, \pi_m(x)\}$ defined in (5.47)–(5.50) form an orthonormal basis with respect to the inner product (5.41). Compare it with the basis obtained from the Gram–Schmidt procedure applied to $\{1, x, x^2, \ldots, x^m\}$.

The procedure for obtaining the set of discrete orthogonal polynomials $\{\widetilde{\pi}_k(x)\}_{k=0}^m$ from the given set of data points $\{x_l\}_{l=1}^{n+1}$ and finding the coefficients $\{c_k\}_{k=0}^m$ of the polynomial $p_m(x)$ from the given set of data points $\{y_l\}_{l=1}^{n+1}$ is illustrated using the function $f(x) = e^{-x}\cos x$ in the MATLAB script `orthogonal_polynomials`. As in Section 5.6, you can construct a set of 51 equally spaced data points on $[0, 5]$ and compute the polynomial function $p_m(x)$ for $1 \leq m \leq 41$. The family of discrete orthogonal polynomials $\{\pi_k(x)\}_{k=0}^m$ can be extended up to $m = n$ and can be computed prior to computations of the polynomial functions $p_m(x)$ with $m \leq n$.

```
% orthogonal_polynomials
x = 0 : 0.1 : 5; n = length(x)-1; y = exp(-x).*cos(x);
p(1,:) = ones(size(x)); % discrete orthogonal polynomials
for k = 1 : n
    alpha = (p(k,:)*(x.*p(k,:))')/(p(k,:)*p(k,:)');
    if k == 1
        p(k+1,:) = (x-alpha).*p(k,:);
    else
        beta = (p(k,:)*p(k,:)')/(p(k-1,:)*p(k-1,:)');
        p(k+1,:) = (x-alpha).*p(k,:)- beta*p(k-1,:);
        % three-term recurrence relation
    end
end
for k = 1 : n+1
    p(k,:) = p(k,:)/sqrt(p(k,:)*p(k,:)');  % normalization
end
for m = 1 : 41  % computations of approximations p_m(x)
    yy = zeros(size(x));
    for k = 1 : m+1
        c(k) = y*p(k,:)';
        yy = yy + c(k)*p(k,:);
    end
    E(m) = y*y' - c*c'; % approximation error
end
semilogy(E,'.');
```

The MATLAB script `orthogonal_polynomials` computes the approximation error (5.46) and plots it versus m in the semi-log scale in Figure 5.13. You can see that the approximation error converges quickly to negligibly small values and no obstacles in computations of the polynomial approximation $p_m(x)$ arise for larger values of m. You may also observe that the approximation error starts to grow for $m > 20$ because of the effects of the round-off errors.

Exercise 5.11 Consider the least-square approximation of the polynomial function $f(x) = x^4 - x^2 + 3x - 5$ by using the discrete orthogonal polynomials $\{\pi_k(x)\}_{k=0}^{m}$ on an equally spaced grid $\{x_l\}_{l=1}^{n+1}$ on $[-1, 1]$ with $n = 10$ and $m = 2, 4, 6$. Plot the original function $f(x)$ and the three approximations $p_m(x)$ on the same graph. Show that the error E is identically zero for $m = 4$ and the approximations are identical for $m = 4$ and $m = 6$.

Continuous orthogonal polynomials

The problem of approximating a given set of data points by a polynomial can be extended to a more abstract problem of approximation of an arbitrary function $y = f(x)$ on a given interval $[a, b]$. You have already seen one instance

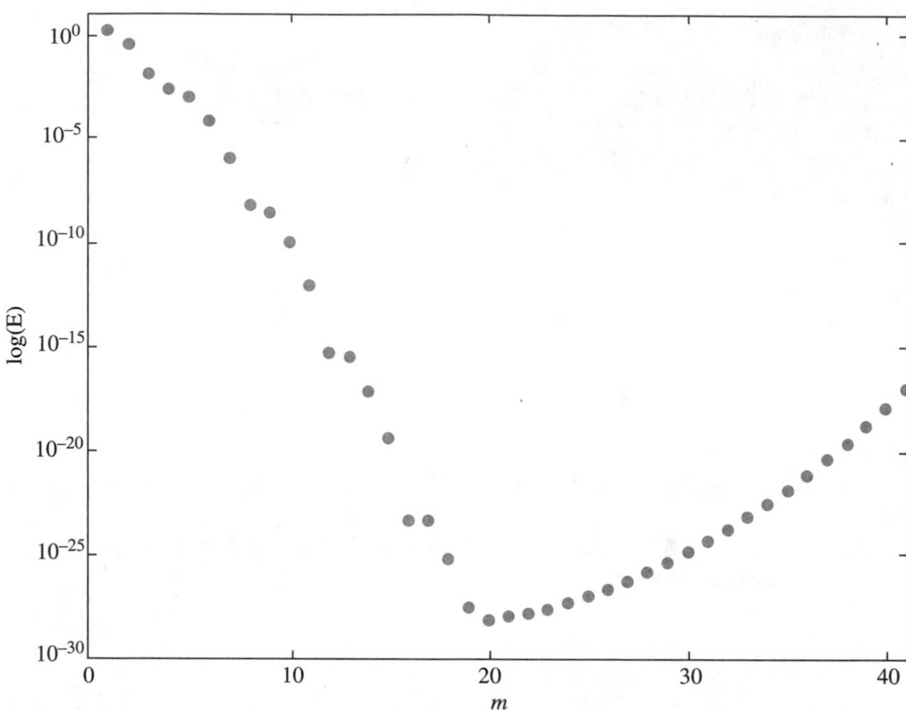

Figure 5.13 Error of the polynomial approximation.

of this idea in Theorem 5.3, according to which any continuous function $f(x)$ can be approximated by the polynomial function $p_m(x)$, where the degree m depends on the tolerance ε used in expression (5.12) to measure the error between $f(x)$ and $p_m(x)$. In general, such a problem requires the techniques of *Functional Analysis* because the vector spaces of functions are typically infinite–dimensional, such as the space $\mathcal{C}[a, b]$ of continuous functions on $[a, b]$. In what follows, however, we treat it as much as possible as an extension of Problem 5.2 and stay on the realm of finite–dimensional vector spaces.

Problem 5.3 (Function Approximation) Given a continuous function $f(x)$ on the interval $[a, b]$, find the polynomial function $y = p_m(x)$ of degree $m \in \mathbb{N}$ that minimizes the total square error

$$E = \int_a^b (f(x) - p_m(x))^2 \, dx. \tag{5.51}$$

This problem is solved by introducing a set of orthogonal polynomials $\{\pi_k(x)\}_{k=0}^m$ with respect to the inner product:

$$\langle p(x), q(x)\rangle := \int_a^b p(x)q(x)dx, \qquad \forall p(x), q(x) \in \mathcal{P}_m. \tag{5.52}$$

One intuitive way to interpret (5.52) is to think of it as a limit of (5.41) when the number n of grid points $a \leq x_1 < x_2 < ... < x_n < x_{n+1} \leq b$ goes to infinity.

The formal construction of the set of orthogonal polynomials $\{\pi_k(x)\}_{k=0}^m$ with respect to the inner product (5.52) follows from the same orthogonalization procedure as before. After we find these polynomials, we can represent any polynomial $p_m(x)$ as (5.42). Following the same argument leading to (5.45), we conclude that the coefficients $c_0, c_1, \ldots, c_m$ are now given by

$$c_k = \langle \pi_k, f\rangle, \qquad k = 0, 1, \ldots, m. \tag{5.53}$$

There are, however, three important differences. The first is that the continuous inner product (5.52) must be used for the computations of $\langle \pi_k, f\rangle$ and, therefore, continuous integrals must be computed either exactly or with the technique of numerical integration (see Section 6.5). The second difference is that the degree of polynomial m can be increased without any bound. In the limit $m \to \infty$, the orthogonal polynomials $\{\pi_k(x)\}_{k=0}^\infty$ provide an orthonormal basis for the infinite-dimensional vector space of *square integrable functions* on $[a, b]$, denoted by $L^2([a, b])$ (see [24] for further details). The third difference is that the polynomials $\{\pi_k(x)\}_{k=0}^\infty$ can be computed for any $x \in [a, b]$ in terms of the orthogonal Legendre polynomials introduced next.

Legendre polynomials

Legendre polynomials on the interval $[-1, 1]$, denoted by $P_k(x)$ for non-negative integers k, are defined as orthogonal polynomials with respect to the inner product (5.52) for $a = -1$, $b = 1$ and satisfy the normalization condition $\langle P_k, P_k\rangle = \frac{2}{2k+1}$. Observe that this normalization condition is different from the standard normalization $\langle \pi_k(x), \pi_k(x)\rangle = 1$ used for most orthogonal polynomials. Despite this difference, Legendre polynomials can still be obtained from the canonical basis $\{1, x, x^2, \ldots\}$ through a simple modification of the Gram–Schmidt orthogonalization procedure, taking into account the required normalization. This is equivalent to using the recurrence relation

$$P_{k+1}(x) = \frac{2k+1}{k+1} x P_k(x) - \frac{k}{k+1} P_{k-1}(x), \qquad k = 0, 1, \ldots, \tag{5.54}$$

starting with $P_{-1}(x) = 0$ and $P_0(x) = 1$. The first few Legendre polynomials are as follows:

$$P_1 = x$$
$$P_2 = \frac{1}{2}(3x^2 - 1)$$
$$P_3 = \frac{1}{2}(5x^3 - 3x)$$
$$P_4 = \frac{1}{8}\left(35x^4 - 30x^2 + 3\right).$$

Alternatively, Legendre polynomials can be calculated using the derivative formula

$$P_k(x) = \frac{1}{k!2^k}\frac{d^k}{dx^k}(x^2 - 1)^k. \tag{5.55}$$

It follows from linear independence that any polynomial of degree m can be represented by a unique linear combination of the Legendre polynomials $\{P_k(x)\}_{k=0}^m$. For instance,

$$\begin{aligned} f(x) &= x^4 - x^3 + 2x^2 + 3x - 5 \\ &= \frac{1}{105}\left(24P_4(x) - 42P_3(x) + 200P_2(x) + 252P_1(x) - 434P_0(x)\right). \end{aligned} \tag{5.56}$$

The function LegPoly computes this decomposition based on a direct comparison of power terms of different degrees:

```
function c = LegPoly(f)
% computes coefficients of the decomposition of the given polynomial
% in terms of the orthogonal set of Legendre polynomials
% f - vector for coefficients of the given polynomial f(x)
% c - vector for coefficients of the orthogonal decomposition

m = length(f)-1;
for k = m : -1 : 1
    p = poly([ones(1,k),-ones(1,k)]); % computations of P_k(x)
    for kk = 1 : k
        p = polyder(p);
    end
    p = p/(factorial(k)*2^k);
    for kk = 1 : m-k
        p = [0,p];  % extension of the polynomial to the degree m
    end
    c(m-k+1) = f(m-k+1)/p(m-k+1);
    f = f - c(m-k+1)*p;  % recursive procedure
end
c = [c,f(m+1)]
```

For example, we can use `LegPoly` to obtain the representation (5.56).

```
>> format rat, c = LegPoly([1,-1,2,3,-5])

c =

     8/35        -2/5        40/21        12/5        -62/15
```

Starting with Legendre polynomials, we can map an arbitrary interval $[a, b]$ into the interval $[-1, 1]$ by the transformation

$$x \longrightarrow \frac{2x - (a+b)}{b-a} \tag{5.57}$$

and define

$$\pi_k(x) = \sqrt{\frac{2k+1}{b-a}} P_k\left(\frac{2x-a-b}{b-a}\right), \qquad k \in \mathbb{N}. \tag{5.58}$$

We can then immediately show that $\{\pi_k(x)\}_{k \in \mathbb{N}}$ is a set of orthogonal polynomials with respect to (5.52).

These polynomials can then be used to solve Problem 5.3 for an arbitrary continuous function $f(x)$ on $[a, b]$. Unfortunately, the integrals $\langle P_k, f \rangle$ cannot be computed exactly for a general function $f(x)$. Therefore, the theoretical tool for solving Problem 5.3 does not have many practical applications, unless a discretization of the continuous interval $[a, b]$ is made. For instance, such discretizations can be used to approximate the integrals numerically (see Section 6.5) or to reduce Problem 5.3 to Problem 5.2, in which case, the set of data points $\{y_l\}_{l=1}^{n+1}$ is obtained from the values of the function $y = f(x)$ on the set of data points $\{x_l\}_{l=1}^{n+1}$.

The method of least square approximation relies on the technique of the minimization of the total square error. Similar techniques are employed widely in numerical methods, for instance, in the Rayleigh–Ritz variational method that contributes to the finite-element method. Details of these applications are explained in Section 12.3. Other expressions for the approximation errors can also be employed, but the minimization technique of the total error would become more technically involved. Therefore, the standard treatment of analysis uses exclusively (5.51) as the simplest expression for the approximation error.

5.8 Summary and Notes

This chapter covers properties of the polynomial functions and their applications. These include various analytic and numerical techniques for polynomial interpolation and approximation of a given set of discrete data points or given continuous functions on finite intervals.

- **Section 5.1:** Among the properties of polynomial functions, we cover the fundamental theorem of algebra on roots and factorizations of polynomials (Theorem 5.1) and the algorithms of computations of derivatives and integrals, products, long divisions, and partial fraction decompositions of the polynomial functions.

- **Section 5.2:** To consider the numerical problem of the polynomial interpolation (Problem 5.2), we review Taylor series for continuously differentiable and smooth functions (Theorem 5.2), Weierstrass Theorem for polynomial approximation of continuous functions (Theorem 5.3), and the main theorem on existence and uniqueness of interpolating polynomials (Theorem 5.4).

- **Sections 5.3–5.4:** We show that the Vandermonde representation of the interpolating polynomials from a direct solution of the linear system is ill-posed numerically in the sense that the numerical solution for many interpolation points is inaccurate because of a nearly singular coefficient matrix. The same polynomials can be expressed in two alternative forms that are well-posed numerically, namely, in Lagrange and Newton interpolations.

- **Section 5.5:** When the polynomial interpolation is well-posed numerically, for example, in the case of Lagrange and Newton interpolations, the round-off errors are extremely small, on the level of floating-arithmetic accuracy. In this case, the error of polynomial interpolation is merely the truncation error described in Theorem 5.6. We discuss convergence rates of the truncation error on the example of equally spaced interpolation points. We show that the truncation error converges to zero if the interpolation interval is small enough, but it may diverge to infinity for larger interpolation intervals. The latter phenomenon (Runge phenomenon) is associated with large-amplitude swings of the interpolating polynomials between the values at the interpolation points.

- **Section 5.6:** Polynomial approximation is studied in Problem 5.2 (compare this section with Section 3.4). The least-square approximation is based on the minimization of quadratic functions and is summarized in Theorem 5.7. Linear and nonlinear regressions are studied as particular cases of the polynomial approximation. Convergence of the approximation error for larger degrees of the approximating polynomials is explained.

- **Section 5.7:** We develop robust solutions of Problem 5.2 by using discrete orthogonal polynomials, which are built for the specific set of grid points. The same solutions are extended to the function approximation problem (Problem 5.3) using orthogonal Legendre polynomials.

Elementary operations with polynomial functions such as long divisions, partial fractions, multiplications, derivatives, and integrals are studied in *Calculus* (the suggested text is [26]). Polynomial functions as examples of finite-dimensional vector spaces, roots and factorizations of polynomials, linear systems for polynomial interpolation and approximations are considered in *Linear Algebra* (the suggested text is [20]). Algorithms and errors of the polynomial interpolation and approximation are studied in *Numerical Analysis* (the suggested text is [7]). Advanced interpolation methods that involve orthogonal polynomials and Chebyshev grid points are covered, for example, in [24].

5.9 Exercises

1. Consider a set of Legendre polynomials:

$$P_n(x) = \frac{1}{2^n n!} \frac{d^n}{dx^n} \left(x^2 - 1\right)^n, \qquad n = 0, 1, 2, ...$$

   and find roots of the first 10 Legendre polynomials. Plot black dots for the locations of the roots in the complex plane for $n = 5$ and $n = 10$.

2. The extremal values of the Chebyshev polynomial $T_n(x)$ are located at the points

$$x_k = \cos\frac{\pi k}{n}, \qquad k = 0, 1, 2, \ldots, n.$$

   Find coefficients of the first 10 Chebyshev polynomials $T_n(x)$.

3. Compute the partial fraction decomposition of the rational functions

$$\frac{x^3 + 2x + 2}{x^2 + 5x - 6}, \qquad \frac{x^4 - x + 3}{x^4 + 2x^2 + 1},$$

   by using the MATLAB functions.

4. Find roots and write the following polynomial in the factorized form:

$$p(x) = 2x^6 + x^5 + 4x^4 + 3x^3 + 2x^2 + 1.$$

   Write the same polynomial in the nested multiplication form. Evaluate $p(x)$, $p'(x)$, and $p''(x)$ at $x = 2$ by using the nested multiplication algorithm.

5. Use polynomial interpolation and prove numerically the summation formulas:

$$\sum_{k=1}^{n} k = \frac{n(n+1)}{2},$$
$$\sum_{k=1}^{n} k^2 = \frac{n(n+1)(2n+1)}{6},$$
$$\sum_{k=1}^{n} k^3 = \frac{n(n+1)(2n+1)(3n+1)}{24}.$$

6. Use the nested multiplication algorithm and compute the values $p(z) + p(-z)$ and $p(z)/p(-z)$ for a polynomial $p(z) = a_1 + a_2 z + ... + a_n z^n$ defined by the coefficient vector $\mathbf{a} = [a_1, a_2, \ldots, a_n]$.

7. Find the quartic polynomial that interpolates the function $f(x) = \sqrt[4]{x}$ on $[0, 2]$ with the equally spaced grid. Repeat the exercise with the function $f(x) = x^4$.

8. Find the polynomial interpolation on $[-1, 2]$ of the four data points $(-1, 2)$, $(0, 5)$, $(1, 1)$, and $(2, -1)$ by using (a) the MATLAB function `polyfit`, (b) the Vandermonde matrices, (c) the Lagrange polynomials, and (d) the Newton divided differences.

9. Find the Lagrange and Newton polynomials for the data points $(0, 1)$, $(2, 5)$, $(3, -2)$, $(6, 0)$, and $(10, -4)$ and plot the polynomial interpolants and the given data points on the same graph. Expand these polynomials to the standard form and show that they are identical. Are the data points related to a polynomial function of degree (a) three, (b) four, or (c) five?

10. Write the MATLAB script that computes the derivative of the polynomial interpolant for the given set of data points at any point of the interpolation interval. Compute the total error between the derivative of the polynomial interpolant and the analytical derivative for the function $f(x) = \tan x$ on a dense grid on the interpolation interval $[0, \pi/4]$. Plot the total error versus the number n of the given data points.

11. Consider the polynomial interpolation of $f(x) = \sqrt{x}$ on $[1, 2]$ with an equally spaced grid. Compute and plot the total error versus the degree of the interpolating polynomial in log–log scale. Repeat the exercise on $[0, 1]$.

12. Consider the special function $\Gamma(x)$, called the Gamma function. Interpolate the known values $\Gamma(k+1) = k!$ for $k \in \mathbb{N}$ with a polynomial $p_n(x)$ on $[1, n+1]$ for $n = 1, 2, \ldots, 5$.

13. Consider the linear interpolation of $f(x) = \log(x)$ on the interval $[1-h, 1+h]$ with two grid nodes at $x_0 = 1 - h/\sqrt{3}$ and $x_1 = 1 + h/\sqrt{3}$. Plot the total error versus h and find the convergence rate of the computational error as $h \to 0$.

14. Consider the polynomial interpolation of the function

$$f(x) = \frac{1}{2 + \cos x}$$

on the equally spaced grid points. Determine if the Runge phenomenon occurs for the interpolation on the intervals $[-\pi, \pi]$ and $[0, 2\pi]$.

15. Consider the logistic model for population growth

$$f(t) = \frac{a}{1 + be^{-ct}},$$

where (a, b, c) are positive parameters and $f(t)$ means the population in the area. Use the following data points $(0, 75)$, $(1, 110)$, $(2, 140)$, $(3, 180)$, and $(4, 225)$ and compute the parameters (b, c) from the linear regression after the linearization. Compute the parameter a from a point estimate at $t = 0$. Predict the value of $f(t)$ at $t = 5$. Plot the data points and the least-square approximation on the same graph.

16. Consider the set of 100 data points from the polynomial function $f(x) = x^6 - 2x^4 + 4x^2 - 5x + 1$. Construct the polynomial approximation with $p_n(x)$ for $n = 1, 2, \ldots, 6$ and show that the approximation error reduces with n and becomes identically zero for $n = 6$.

17. Write the MATLAB script for modeling a random walk of K particles from $\mathbf{x}_0 = 0$ at $n = 0$ to $\mathbf{x}_N$ at $n = N$, where $\mathbf{x}_k \in \mathbb{R}^K$. Each kth particle moves to the right $x_{k,n+1} = x_{k,n} + 1$ and to the left $x_{k,n+1} = x_{k,n} - 1$ with the equal probability $\frac{1}{2}$. Plot the mean value μ_n and variance σ_n^2 versus n for $n \in \mathbb{N}$, where

$$\mu_n = \frac{1}{K}\sum_{k=0}^{K} x_{k,n}, \qquad \sigma_n^2 = \frac{1}{K}\sum_{k=0}^{K}(x_{k,n} - \mu_n)^2.$$

Compute the linear regression for μ_n and σ_n^2 versus n and confirm applicability of the theoretical laws $\mu_n = 0$ and $\sigma_n^2 = Dn$, which are valid for large $n \in \mathbb{N}$. Approximate the coefficient of diffusion D.

18. Compute the least-square polynomial approximation of the function $f(x) = \sin(\pi x)$ on $[0, 1]$ with the discrete orthogonal polynomials $\{\pi_k(x)\}_{k=0}^{m}$ for the equally spaced grid with $n = 100$ and $m \leq k$. Plot the approximations for $m = 4$ and $m = 8$ together with the original function $f(x)$, and compute the absolute values of the total square error.

19. Compute the function approximation of $f(x) = \sin(\pi x)$ on $[-1, 1]$ with the Legendre polynomials $\{P_k(x)\}_{k=0}^{m}$ by analytic computations of integrals for $m = 4$. Plot the approximation together with the original function and compute the approximation error.

20. Represent the polynomial $f(x) = x^{10} - 2x^8 + 3x^5 + 2x^4 + 10x - 1$ as a linear combination of the Legendre polynomials $\{P_k(x)\}_{k=0}^{m}$ with $m = 2$, $m = 6$, and $m = 10$. Plot the original function on $[-2, 2]$ and the three approximations on the same graph. Compute the total square error of the polynomial approximation through the equally spaced discrete grid on the interval $[-2, 2]$.

21. Consider the function $f(x) = \sqrt{1 - x^2}$ on $[-a, a]$ with the discrete orthogonal polynomials $\{\pi_k(x)\}_{k=0}^{m}$ for the equally spaced grid with $n = 100$ and $m \leq k$. Plot the approximation error versus m for $a = \frac{1}{2}$ and $a = 1$.

CHAPTER **6**

Differential and Integral Calculus

IN *CALCULUS,* you learn how to find derivatives by memorizing elementary derivatives and using the Newton–Leibniz differentiation rules, whereas integrals are found by memorizing antiderivatives and tricks of substitutions. What you are not told is that all the formulas and tricks are enough to pass the course but fail to work in many practical cases. For instance, the special functions of mathematical physics are not given in analytical form suitable to explicit differentiation, and many integrals, such as the statistical error function given by the integral of the Gaussian function e^{-x^2}, cannot be computed in a closed form. Therefore, we would like to consider numerical strategy for computing derivatives and integrals that will remain robust in many practical cases.

This chapter covers the finite differences for numerical derivatives and finite sums for numerical integrals, as well as algorithms and errors in numerical differentiation and integration.

6.1 Derivatives and Finite Differences

Recall from *Calculus* that a continuous function $f(x)$ is said to be *differentiable* at the point x_0 if any of the following three equivalent limits exists:

$$\begin{aligned} f'(x_0) &= \lim_{h\to 0} \frac{f(x_0+h)-f(x_0)}{h} \\ &= \lim_{h\to 0} \frac{f(x_0)-f(x_0-h)}{h} \\ &= \lim_{h\to 0} \frac{f(x_0+h)-f(x_0-h)}{2h}. \end{aligned} \tag{6.1}$$

Using discrete values in the numerical representation of continuous functions leads to a numerical approximation to the limits (6.1) for some nonzero value of h. We call such approximation for the derivative $f'(x_0)$ the *numerical derivative* of $f(x)$ at x_0 and denote it by $Df[x_0]$.

In most numerical problems, it is assumed that a continuous function $f(x)$ is defined on the given set of data points. You then have to approximate its derivatives at a specific point x_0 using other data points located to the left, to the right, or on both sides of x_0.

Problem 6.1 (Numerical Derivative) Assume that the points (x_k, y_k), $k \in \mathbb{Z}$, represent the continuous function $y = f(x)$ on a discrete grid. Given m points to the right, left, or both sides of a fixed grid point x_0, find the numerical derivative $Df[x_0]$.

Forward, backward, central differences

Because each of the limits (6.1) uses exactly two points, the simplest solution of Problem 6.1 can be constructed for $m = 1$. We replace continuous limits with discrete quotients and define the *forward*, *backward*, and *central difference* approximations for numerical derivatives, respectively:

$$\text{(forward)} \quad Df[x_0] = \frac{y_1 - y_0}{x_1 - x_0}, \tag{6.2}$$

$$\text{(backward)} \quad Df[x_0] = \frac{y_0 - y_{-1}}{x_0 - x_{-1}}, \tag{6.3}$$

$$\text{(central)} \quad Df[x_0] = \frac{y_1 - y_{-1}}{x_1 - x_{-1}}. \tag{6.4}$$

The forward difference approximation is the slope of the linear interpolation between the selected point (x_0, y_0) and the right adjacent point (x_1, y_1) (see Section 5.2). The backward difference approximation is the slope of the linear interpolation between the selected point (x_0, y_0) and the left adjacent point (x_{-1}, y_{-1}). The central difference approximation is the slope of the linear interpolation between the left and right adjacent points (x_{-1}, y_{-1}) and (x_1, y_1) to the selected point (x_0, y_0).

The three difference approximations are illustrated in the next example, where the derivative of the function $f(x) = e^{10x}$ is approximated at $x = 0.5$. For the given value $h = 0.01$ of the distance between the adjacent grid points, the forward and backward differences are highly inaccurate: the forward difference overestimates the correct slope, whereas the backward underestimates it. The central difference gives a better result, which indicates that the errors of the forward and backward differences cancel out in the central difference approximation:

```
>> x0 = 0.5; x_1 = 0.49; x1 = 0.51;
>> y0 = exp(10*x0); y_1 = exp(10*x_1); y1 = exp(10*x1);
>> yDexact = 10*exp(10*x0)

yDexact =
  1.4841e+003

>> yDforward = (y1-y0)/(x1-x0)

yDforward =
  1.5609e+003
```

```
>> yDbackward = (y0-y_1)/(x0-x_1)

yDbackward =
  1.4123e+003

>> yDcentral = (y1-y_1)/(x1-x_1)

yDcentral =
  1.4866e+003
```

Numerical derivatives can be applied to many problems, most importantly for numerical solutions of differential equations. We mention a few related applications of numerical derivatives and give references to where these applications are discussed in detail:

- Forward differences are useful in *explicit solvers of the initial-value problems for ordinary differential equations (ODE)* (Section 9.2). Given the values of $f(x_0)$ and $f'(x_0)$, the forward difference approximates the value of $f(x_1)$ at the *next* data point.
- Backward differences are useful in *implicit ODE solvers* of the initial-value problems (Section 9.5). Given the value of $f(x_0)$, the backward difference for $f'(x_1)$ relates it to the unknown value of $f(x_1)$. Other applications of backward differences occur in the *multistep ODE solvers* for initial-value problems (Section 9.4) and the *secant method* for approximations of zeros of nonlinear functions (Section 8.1).
- Central differences are useful in *finite-difference ODE solvers* of the boundary-value problems (Section 10.1). When the values of $f'(x_0)$ and $f''(x_0)$ are approximated by the central differences, the unknown values of $f(x_{-1})$, $f(x_0)$, and $f(x_1)$ become related to each other by a system of algebraic equations.

Errors of finite differences

Because the linear interpolation between two data points has a nonzero truncation error, the slope approximations of forward, backward, and central differences are also subject to a truncation error. A useful tool to understand the truncation error for numerical derivatives is the Taylor series described by Theorem 5.2. Applying the Taylor series approximation (5.10) with $n = 1$ to the forward and backward difference approximations (6.2) and (6.3), we obtain the exact representation for truncation errors:

$$f'(x_0) - \frac{f(x_1) - f(x_0)}{h} = -\frac{1}{2}f''(\xi)h, \qquad \xi \in [x_0, x_1], \tag{6.5}$$

where $h = x_1 - x_0$, and

$$f'(x_0) - \frac{f(x_0) - f(x_{-1})}{h} = \frac{1}{2}f''(\xi)h, \qquad \xi \in [x_{-1}, x_0], \tag{6.6}$$

where $h = x_0 - x_{-1}$. On the other hand, applying the Taylor series approximation (5.10) with $n = 2$ to the central difference approximation (6.4), we obtain a modified representation for the truncation error:

$$f'(x_0) - \frac{f(x_1) - f(x_{-1})}{2h} = -\frac{1}{6}f^{(3)}(\xi)h^2, \qquad \xi \in [x_{-1}, x_1], \tag{6.7}$$

where $h = x_1 - x_0 = x_0 - x_{-1}$. The cancellation of the first power of h in the truncation error (6.7) explains the better accuracy of the central difference approximation compared to the forward and backward difference approximations. Note that this cancellation occurs only in the case of an equally spaced grid, when $x_1 - x_0 = x_0 - x_{-1}$.

Exercise 6.1 Prove from the Taylor series approximation (5.10) with $n = 2$ that the error of the central difference approximation (6.4) is linear in $h = (x_1 - x_{-1})/2$ if it is computed on a non-uniform grid with $x_1 - x_0 \neq x_0 - x_{-1}$. Illustrate the linear convergence of this numerical derivative with a numerical example.

In many problems, the truncation error of the numerical derivative is a power function of the step size h. As a result, we can classify different approximations for the same derivative and develop modifications of numerical derivatives with better accuracy. We say that the numerical approximation is of order n if the truncation error E is $O(h^n)$, that is, if

$$\lim_{h \to 0} \frac{E(h)}{h^n} = \alpha_n, \tag{6.8}$$

for some constant $\alpha_n \neq 0$. This is equivalent to saying that

$$E(h) \approx \alpha_n h^n,$$

for h close to zero. From (6.5), (6.6), and (6.7), we see that the forward and backward difference approximations for derivatives are of order $n = 1$, whereas the central difference approximation is of order $n = 2$.

The next example shows the total numerical error in the three difference approximations of the first derivative for the function $f(x) = e^{10x}$ at the point $x = 0.5$ versus the step size h. The output of the MATLAB script `numerical_derivatives` is shown in Figure 6.1.

```
% numerical_derivatives
h = logspace(-1,-15,15*5); x0 = 0.5*ones(size(h));
xn1 = x0-h; x1=x0+h;
y0 = exp(10*x0); yn1 = exp(10*xn1); y1 = exp(10*x1);
yDe=10*exp(10*x0);                          % exact derivative
yDf = (y1-y0)./h; eDf = abs(yDf-yDe);       % forward difference
```

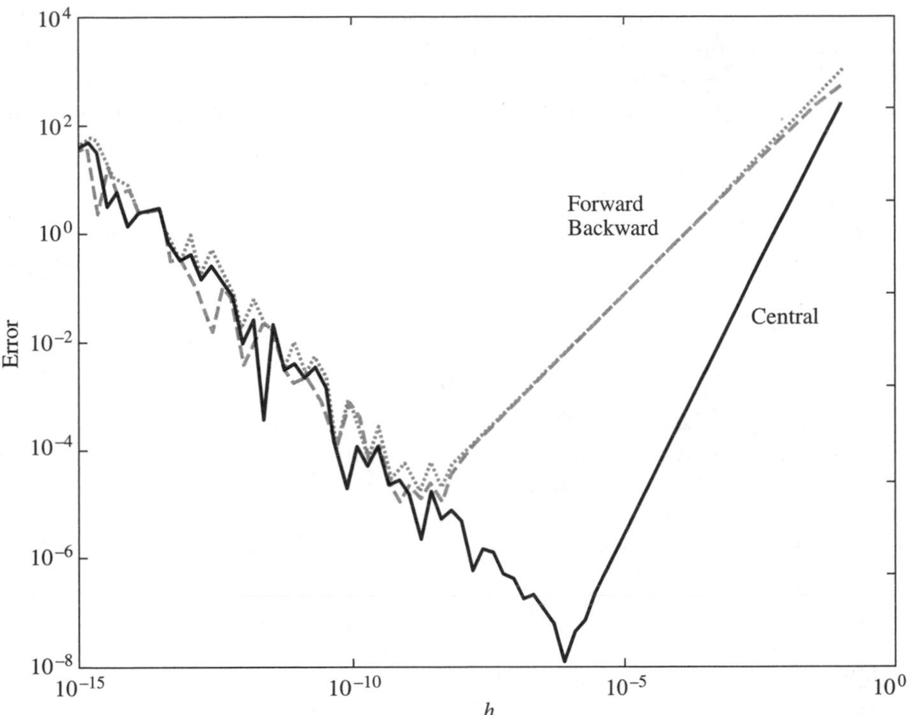

Figure 6.1 Total error of forward, backward, and central difference approximations for the first derivative versus the step size.

```
yDb = (y0-yn1)./h; eDb = abs(yDb-yDe); % backward difference
yDc=(y1-yn1)./(2*h); eDc=abs(yDc-yDe); % central difference
loglog(h,eDf,'g:',h,eDb,'b--',h,eDc,'r');
```

Exercise 6.2 Consider the function $f(x) = x^{3/2}$ and compute the derivative $f'(0) = 0$ using forward differences for different values of h. Using the least squares technique from Section 5.6, find the power law $E = ch^p$ for the dependence of the truncation error E on the step size h and show that the convergence occurs for $p \sim 1/2 < 1$.

Round-off error of differentiation

When the step size h decreases, the numerical error drops faster for the central difference approximation. Figure 6.1 shows that the slope of $\log E$ versus $\log h$ for central differences is twice as much as that for forward and backward differences. When the step size h becomes too small, the numerical error increases with further decrease of h. The latter phenomenon is explained by an increasing round-off error, which occurs when the ratio of two small numbers is computed in floating-point arithmetic, leading to a catastrophic cancellation (see Section 1.7). The minimum error at the optimal step size h represents the minimum value in the total sum of truncation and round-off errors,

$$E_{\text{total}}(h) = \alpha_n h^n + \frac{2\varepsilon}{h},$$

where n is the order of the truncation error and ε is the precision of the floating-point system. The minimum value for $E_{\text{total}}(h)$ occurs for an optimal value

$$h_{\text{opt}} = \left(\frac{2\epsilon}{n\alpha_n}\right)^{1/(n+1)}.$$

You can see in Figure 6.1 that the minimum error occurs for a larger value h_{opt} in the central difference approximation and the minimum value of the error $E_{\text{total}}(h_{\text{opt}})$ is smaller. However, for very small values of h, the difference between the three numerical approximations disappears and the numerical error E for all three approximations grows as the inverse power of h.

Numerical derivatives based on finite differences represent an ill-conditioned numerical problem because computations with smaller-than-optimal step sizes decrease the accuracy of numerical approximations. Higher-order difference approximations and alternative methods of numerical differentiations need to be developed for better accuracy of numerical derivatives. For instance, the solution of Problem 6.1 for $m > 1$ gives a numerical derivative of higher accuracy (Section 6.3). Additionally, trigonometric interpolation allows us to avoid large round-off errors of numerical differentiation (Section 11.2).

6.2 Higher-Order Numerical Derivatives

Higher-order numerical derivatives can be defined in terms of higher-order derivatives of the interpolation polynomials described in Section 5.2. We shall construct a *hierarchy* of higher-order finite difference formulas at a selected point x_0, assuming that the original function $f(x)$ is infinitely smooth.

Derivative of interpolant

Let $p_n(x)$ be the interpolation polynomial, say, in the Newton form of Section 5.4, through n points to the right, left, or both sides of the point x_0 and define the nth-order numerical derivative $D^n f[x_0]$ as the nth-order derivative of the interpolating polynomial $p_n(x)$:

$$D^n f[x_0] = p_n^{(n)}(x_0). \tag{6.9}$$

Because $p_n(x)$ is a polynomial of degree n, it follows that $p_n^{(n)}(x)$ is constant in x, and the Newton representation (5.22) for $p_n(x)$ implies that

$$D^n f[x_0] = n! D_n[x_0, x_1, \ldots, x_n], \tag{6.10}$$

where $D_n[x_0, x_1, \dots, x_n]$ is the nth-order divided difference defined in Section 5.4.

Using the representation (6.10), note that the Newton interpolating polynomial $p_n(x)$ through the data points (x_0, y_0), (x_1, y_1),..., (x_n, y_n) takes the form of a discrete Taylor polynomial:

$$p_n(x) = y_0 + Df[x_0](x - x_0) + \frac{1}{2!}D^2 f[x_0](x - x_0)(x - x_1) + \dots$$
$$+\frac{1}{n!}D^n f[x_0](x - x_0)(x - x_1)\dots(x - x_{n-1}). \quad (6.11)$$

Higher-order forward differences

Let the set of grid points $x_1, x_2, \dots, x_n$ be defined to the right of the given point x_0 and the data points be equally spaced with a constant step size h:

$$x_j = x_0 + jh, \qquad j = 1, 2, \dots, n. \quad (6.12)$$

In this case, the higher-order forward numerical derivatives are recursively defined from the recursive formula (5.21) of Theorem 5.5:

$$D^n f[x_0] = n!\frac{D_{n-1}[x_1, x_2, \dots, x_n] - D_{n-1}[x_0, x_1, \dots, x_{n-1}]}{nh} = DD^{n-1} f[x_0],$$

so that

$$D^n f[x_0] = DD^{n-1} f[x_0] = D^2 D^{n-2} f[x_0] = \dots = D^{n-1} Df[x_0]. \quad (6.13)$$

The recursive algorithm is expressed with the repeated matrix multiplication of a large matrix of first-order forward differences. The next example computes the first nine forward differences.

```
>> n = 100;
>> A = diag(ones(n-1,1),1)-diag(ones(n,1));
>> B = A;
>> for k = 1 : 9
       D(k,:) = B(1,1:10);
       B = A*B;
   end
>> Der = D

Der =
    -1     1     0     0     0     0     0     0     0     0
     1    -2     1     0     0     0     0     0     0     0
    -1     3    -3     1     0     0     0     0     0     0
     1    -4     6    -4     1     0     0     0     0     0
    -1     5   -10    10    -5     1     0     0     0     0
     1    -6    15   -20    15    -6     1     0     0     0
    -1     7   -21    35   -35    21     7     1     0     0
     1    -8    28   -56    70   -56    28    -8     1     0
    -1     9   -36    84  -126   126   -84    36    -9     1
```

Observe how the coefficients of the higher-order forward numerical derivatives resemble those in the expansion of the binomial $(p-q)^n$ in powers of p and q:

$$\begin{aligned} Df[x_0] &= \frac{y_1 - y_0}{h} \\ D^2 f[x_0] &= \frac{y_2 - 2y_1 + y_0}{h^2} \\ D^3 f[x_0] &= \frac{y_3 - 3y_2 + 3y_1 - y_0}{h^3} \\ D^4 f[x_0] &= \frac{y_4 - 4y_3 + 6y_2 - 4y_1 + y_0}{h^4}. \end{aligned}$$

We call the numerators of the preceding expressions the nth-order *forward differences* for the function f. For example, the first-order forward difference for f is simply $(y_1 - y_0)$, in accordance with (6.2), the second-order forward difference for f is $(y_2 - 2y_1 + y_0)$, and so on.

Exercise 6.3 Let the set of grid points $x_{-n}, \ldots, x_{-2}, x_{-1}$ be defined to the left of the given point x_0 and the data points be equally spaced with a constant step size h:

$$x_j = x_0 + jh, \qquad j = -1, -2, \ldots, -n. \tag{6.14}$$

Compute recursively the higher-order backward numerical derivatives and write explicitly the first four *backward differences*.

Forward and backward differences for a given set of data points are computed with the MATLAB function `diff`, which takes the vector of data points and the order of the numerical derivative and outputs the vector of differences. The vector of differences can be used both for forward and backward differences. The next example shows computations of the first order and second order numerical derivatives for the function $y = \sin x$ on the interval $[0, 2\pi]$. Because the forward differences require data points to the right of a specific data point, the first forward difference cannot be computed at the last data point, whereas the second forward difference cannot be computed at the last two data points. Similarly, the first backward difference cannot be computed at the first data point, whereas the second backward difference cannot be computed at the first two data points. As a result, the output of the MATLAB function `diff` returns the vector of smaller length compared to the vector of data points. The output of MATLAB script `forward_backward_differences` is shown in Figure 6.2. Comparing the graphs of forward (dark solid curves) and backward (bright solid curves) numerical derivatives with the exact derivatives (dotted curves), you can see that the forward differences shift the derivatives to the left of the exact derivatives, whereas the backward differences shift the derivatives to the right of the exact derivatives.

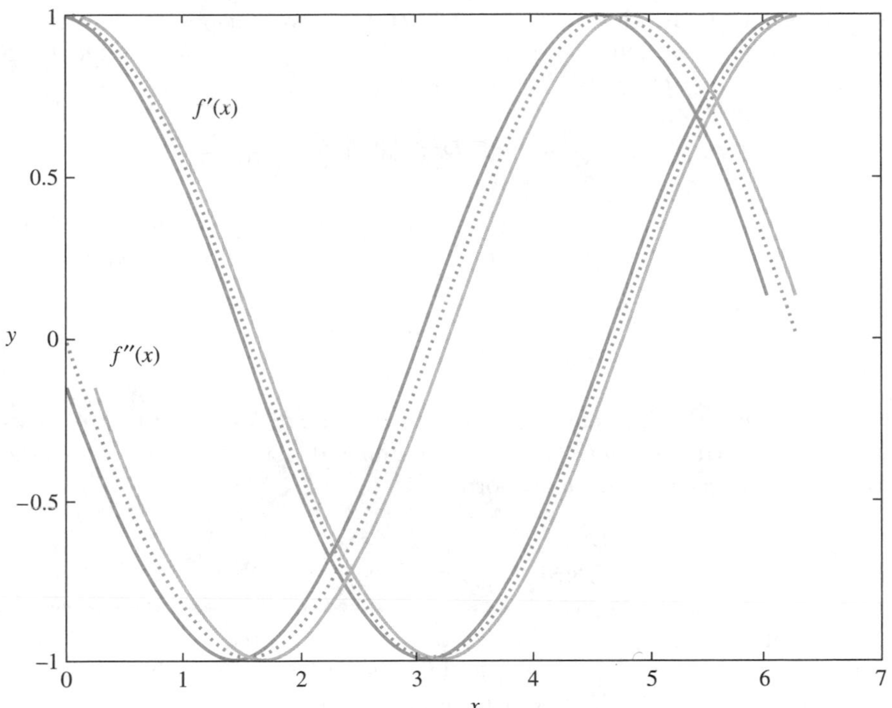

Figure 6.2 First-order and second-order forward and backward numerical derivatives for $f(x) = \sin x$.

```
% forward_backward_differences
n = 50; x = linspace(0,2*pi,n); h = x(2)-x(1); y = sin(x);
y1 = diff(y)/h; y2 = diff(y,2)/h^2; % numerical derivatives
y1ex = cos(x); y2ex = -sin(x);      % exact derivatives
plot(x(1:n-1),y1,'b',x(2:n),y1,'g',x,y1ex,':r');
plot(x(1:n-2),y2,'b',x(3:n),y2,'g',x,y2ex,':r');
```

Higher-order central differences

To construct the hierarchy of central differences, we consider an equally spaced grid extended symmetrically from the given point $x = x_0$ to the left and to the right:

$$x_j = x_0 + jh, \qquad j = \pm 1, \pm 2, \ldots, \pm n. \tag{6.15}$$

By analogy with the case of forward differences, we start with the central difference approximation of the first derivative (6.4) and apply the recursion formula $D^2 f[x_0] = DDf[x_0]$. We then obtain

$$Df[x_0] = \frac{y_1 - y_{-1}}{2h}, \qquad D^2 f[x_0] = \frac{y_2 - 2y_0 + y_{-2}}{(2h)^2}.$$

It is clear that the second-order numerical derivative $D^2 f[x_0]$ can be rewritten for the grid with the half step size $h \to h/2$ when $D^2 f[x_0]$ is defined in terms of adjacent values y_1 and y_{-1} to the central value y_0:

$$D^2 f[x_0] = \frac{y_1 - 2y_0 + y_{-1}}{h^2}.$$

As a result, central difference approximations for odd-numbered derivatives can be computed differently from even-numbered derivatives, by using the corresponding recursion formulas for $n \geq 1$:

$$D^{2n+1} f[x_0] = DD^{2n} f[x_0], \tag{6.16}$$
$$D^{2n+2} f[x_0] = D^2 D^{2n} f[x_0]. \tag{6.17}$$

Central differences can be computed with the repeated matrix multiplication of a large matrix of first-order and second-order central differences. The next example computes the first eight central differences.

```
>> n = 100; m = 5;
>> A = diag(ones(n-1,1),1)-diag(ones(n-1,1),-1);
>> A2=diag(ones(n-1,1),1)+diag(ones(n-1,1),-1)-2*diag(ones(n,1));
>> B = A; C = A2; k = 1;
>> while (k < (2*m-2))
       D(k,:) = B(m,1:2*m-1);
       D(k+1,:) = C(m,1:2*m-1);
       B = B*A2;
       C = C*A2;
       k = k+2;
   end
>> Der = D

Der =
     0     0     0    -1     0     1     0     0     0
     0     0     0     1    -2     1     0     0     0
     0     0    -1     2     0    -2     1     0     0
     0     0     1    -4     6    -4     1     0     0
     0    -1     4    -5     0     5    -4     1     0
     0     1    -6    15   -20    15    -6     1     0
    -1     6   -14    14     0   -14    14    -6     1
     1    -8    28   -56    70   -56    28    -8     1
```

The coefficients of the first four central numerical derivatives are written explicitly as follows:

$$\begin{aligned} Df[x_0] &= \frac{y_1 - y_{-1}}{2h} \\ D^2f[x_0] &= \frac{y_1 - 2y_0 + y_{-1}}{h^2} \\ D^3f[x_0] &= \frac{y_2 - 2y_1 + 2y_{-1} - y_{-2}}{2h^3} \\ D^4f[x_0] &= \frac{y_2 - 4y_1 + 6y_0 - 4y_{-1} + y_{-2}}{h^4}. \end{aligned}$$

We call the numerators of the preceding expressions the nth-order *central differences* for the function f. For instance, the first central difference for f is $(y_1 - y_{-1})$, which coincides with the expression in (6.4); the second central difference for f is given by $(y_1 - 2y_0 + y_{-1})$, and so on.

We illustrate the computations of first-order and second-order central numerical derivatives in the next example for the function $y = \sin x$ on the interval $[0, 2\pi]$. The output of the MATLAB script `central_differences` is shown in Figure 6.3. Comparison of numerical derivatives (solid curves)

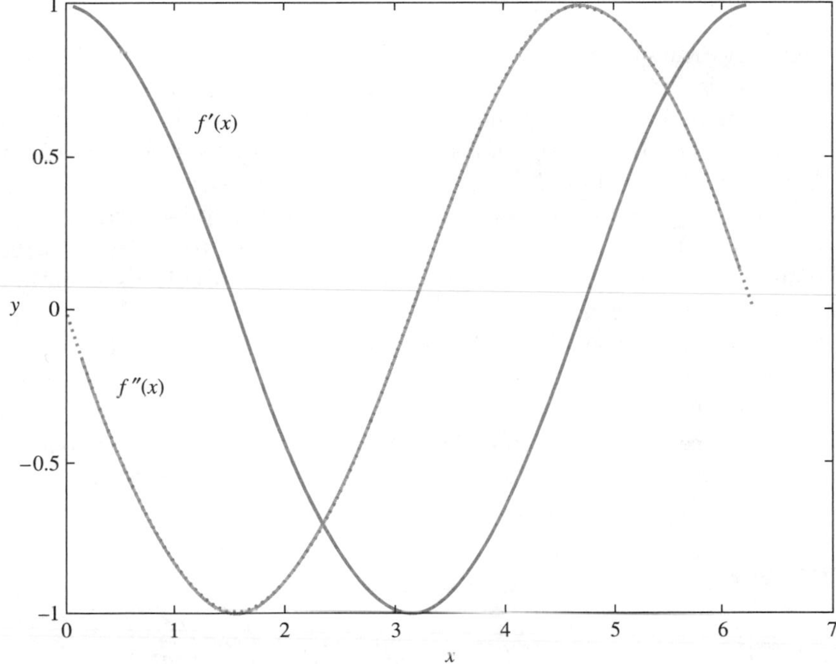

Figure 6.3 First-order and second-order central numerical derivatives for $f(x) = \sin x$.

with the exact derivatives (dotted curves) shows that the central differences accurately reproduce the behavior of the exact derivatives.

```
% central_differences
n = 50; x = linspace(0,2*pi,n); h = x(2)-x(1); y = sin(x);
y1 = (diff(y(2:n))+diff(y(1:n-1)))/(2*h); % numerical derivatives
y2 = (diff(y1(2:n-2))+diff(y1(1:n-3)))/(2*h);
y1ex = cos(x); y2ex = -sin(x); % exact derivatives
plot(x(2:n-1),y1,'b',x,y1ex,':r',x(3:n-2),y2,'g',x,y2ex,':r');
```

Error of higher-order derivatives

As you have seen in Section 6.1, the numerical approximation for derivatives based on first-order forward and backward differences is of order $n = 1$ since the truncation error is $O(h)$, whereas the approximation based on first-order central differences is of order $n = 2$ since the truncation error is $O(h^2)$ (see (6.8) for the definition of the $O(h^n)$ notation). Using the preceding construction, the nth-order forward numerical derivatives are defined as the nth-order derivative of the Newton polynomial $p_n(x)$, which is just a constant term. Because of Theorem 5.6, the truncation error of the constant polynomial is $O(h)$. Therefore, it should not come as a surprise that the approximations for derivatives based on higher-order forward differences all remain of order $n = 1$.

Similarly, using the preceding construction, the nth-order central numerical derivative is defined as the nth-order derivative of $p_{n+1}(x)$ for odd n and of $p_n(x)$ for even n. Therefore, when n is odd, since the nth derivative of $p_{n+1}(x)$ is a linear polynomial, whose truncation error is $O(h^2)$ (see Section 5.5), we see that the approximation for numerical derivatives based on central differences is of order $n = 2$. When n is even, because the nth derivative of $p_n(x)$ is a just constant, we would expect the truncation error to be $O(h)$, but the remarkable cancellation of odd powers of h (see Section 6.1) leads again to an approximation of order $n = 2$.

Finally, the error formula (5.24) for polynomial interpolation implies that the truncation error for higher-order derivatives is identically zero if $f(x)$ is a polynomial of degree n for forward differences and of degree $(n + 1)$ for central differences. With these observations, we have just proved the following important result on the errors of higher-order numerical derivatives.

Theorem 6.1

1. *Let $f(x)$ be an $(n+1)$-times continuously differentiable function on an interval $[a, b]$ that contains the grid points (6.12). Let forward differences for nth-order numerical derivatives be defined recursively by*

(6.13). There is a constant $C_n > 0$*, such that*

$$|f^{(n)}(x_0) - D^n f[x_0]| \le C_n M_{n+1} h, \tag{6.18}$$

where $M_{n+1} = \max_{x\in[a,b]} |f^{(n+1)}(x)|$.

2. *Let* $f(x)$ *be an* $(n+2)$*–times continuously differentiable function on an interval* $[a, b]$ *that contains the grid points (6.15). Let central differences for nth-order numerical derivatives be defined recursively by (6.16)–(6.17). There is a constant* $C_n > 0$*, such that*

$$|f^{(n)}(x_0) - D^n f[x_0]| \le C_n M_{n+2} h^2, \tag{6.19}$$

where $M_{n+2} = \max_{x\in[a,b]} |f^{(n+2)}(x)|$.

The exact expressions for the truncation errors of numerical derivatives can be obtained from the Taylor series (in Theorem 5.2). For example, the truncation errors of the forward and central difference approximations of the first derivative are given by (6.5) and (6.7), respectively. The truncation errors of the forward and central difference approximations of the second derivative are given by [25]:

$$f''(x_0) - \frac{f(x_2) - 2f(x_1) + f(x_0)}{h^2} = -f^{(3)}(\xi)h, \quad \xi \in [x_0, x_2] \tag{6.20}$$

and

$$f''(x_0) - \frac{f(x_1) - 2f(x_0) + f(x_{-1})}{h^2} = -\frac{1}{12} f^{(4)}(\xi)h^2, \quad \xi \in [x_{-1}, x_1]. \tag{6.21}$$

Exercise 6.4

1. Verify numerically that the truncation error of the nth-order forward numerical derivatives for the power function $f(x) = x^{n+1}$ converges to zero linearly in h and that $C_n = \frac{n}{2}$ in the upper bound (6.18).

2. Verify numerically that the truncation error of the nth-order central numerical derivatives for the power function $f(x) = x^{n+2}$ converges to zero quadratically in h, according to the upper bound (6.19).

6.3 Multipoint First-Order Numerical Derivatives

The nth-order numerical derivative of a function $f(x)$ is computed from values of $f(x)$ at data points to the right, left, or both sides of a central point $x = x_0$ (see Section 6.2). In a different development, we can use more data points to improve the accuracy of the first-order numerical derivative $Df(x_0)$. We shall refer to these numerical derivatives of higher accuracy as *multipoint first-order numerical derivatives.*

Derivative of interpolants

Let $p_m(x)$ be the Newton interpolation polynomial through m points to the right, left, or both sides of the point $x = x_0$ and define the m-point first-order numerical derivative $D_m f[x_0]$ of the function $f(x)$ as the derivative of the interpolating polynomial

$$D_m f[x_0] = p'_m(x_0), \tag{6.22}$$

where $D_1 f[x_0] \equiv Df[x_0]$. The m-point formula (6.22) is a useful approximation that shows an important role of polynomial interpolation in the context of numerical differentiation. The expression (6.22) remains valid for nonuniform grids with variable step size, and it can be coded algorithmically for any practical computations. It is still an implicit formula that expresses $D_m f[x_0]$ in terms of the derivative of the polynomial interpolant $p_m(x)$. Therefore, we seek to obtain explicit formulas for m-point first-order numerical derivatives at least for uniform grids. In these cases, such formulas can be implemented directly to a numerical solution, for example, in the finite-difference method for boundary-value ODE problems (see Section 10.1). To find the explicit m-point first-order numerical derivatives, we refer to the following result from *Numerical Analysis.*

Theorem 6.2

1. *Let $f(x)$ be an $(m+1)$-times continuously differentiable function on an interval $[a, b]$ that contains the grid points (6.12). Let forward differences for the m-point first-order numerical derivatives be defined by the derivative of the interpolating polynomial (6.22). There is a constant $C_m > 0$, such that*

$$|f'(x_0) - D_m f[x_0]| \le C_m M_{m+1} h^m, \tag{6.23}$$

where $M_{m+1} = \max_{x \in [a,b]} |f^{(m+1)}(x)|$.

2. *Let $f(x)$ be an $(2m+1)$-times continuously differentiable function on an interval $[a, b]$ that contains the grid points (6.15). Let central differences for the m-point first-order numerical derivatives be defined by the derivative of the interpolating polynomial (6.22). There is a constant*

$C_m > 0$, such that

$$|f'(x_0) - D_m f[x_0]| \le C_m M_{2m+1} h^{2m}, \tag{6.24}$$

where $M_{2m+1} = \max_{x \in [a,b]} |f^{(2m+1)}(x)|$.

Exercise 6.5 Use the error formula (5.24) in Theorem 5.6 and prove that the upper bound (6.23) can be replaced by the exact equality with $C_m = \frac{1}{m+1}$ and $M_{m+1} = f^{(m+1)}(\xi)$, where $\xi \in [x_0, x_m]$.

Multipoint forward derivatives

We shall use Theorem 6.6.2 to develop a general explicit solution for Problem 6.1 on the uniform grids (6.12) and (6.15). Let the m-point first-order forward numerical derivative be represented in the form of a weighted sum:

$$D_m f[x_0] = \frac{1}{h} \sum_{k=0}^{m} w_k y_k. \tag{6.25}$$

Using Theorem 6.6.2, we postulate that the m-point forward numerical derivative must be exact for the power functions $y = x^j$, $j = 0, 1, \dots, m$. This condition at $x = x_0$ leads to the linear system

$$\frac{1}{h} \sum_{k=0}^{m} w_k x_k^j = j x_0^{j-1}, \qquad j = 0, 1, \dots, m. \tag{6.26}$$

For equally spaced grid points (6.12), the linear system (6.26) has a unique solution. This solution is not computationally stable for large m because the Vandermonde matrix for the linear system (6.26) is ill-conditioned (recall the discussion from Section 5.2). Nevertheless, a solution can still be found for $m \le 10$, as the next example shows:

```
>> format rat;
>> for m = 1 : 5
       x = linspace(1,2,m+1); h = 1/m;  % x_0 = 1
       A = fliplr(vander(x))';
       b = (0:nm)'; w = h*(A\b)'
   end

w =

    -1            1

w =

   -3/2           2            -1/2
```

```
w =
   -11/6        3        -3/2        1/3
w =
   -25/12       4        -3          4/3        -1/4
w =
   -137/60      5        -5          10/3       -5/4        1/5
```

In particular, the 2-point, 3-point, and 4-point first-order forward numerical derivatives are given by

$$D_2 f[x_0] = \frac{-y_2 + 4y_1 - 3y_0}{2h}$$
$$D_3 f[x_0] = \frac{2y_3 - 9y_2 + 18y_1 - 11y_0}{6h}$$
$$D_4 f[x_0] = \frac{-3y_4 + 16y_3 - 36y_2 + 48y_1 - 25y_0}{12h},$$

in addition to the 1-point forward derivative (6.2).

Multipoint central derivatives

Similarly, let the m-point first-order central numerical derivative be represented in the form of the weighted sum:

$$D_m f[x_0] = \frac{1}{h} \sum_{k=-m}^{m} w_k y_k. \qquad (6.27)$$

Using Theorem 6.6.2, we again require that the weighted sum (6.27) becomes exact for the power functions $y = x^j$, $j = 0, 1, \ldots, 2m$, which results at $x = x_0$ in the linear system:

$$\frac{1}{h} \sum_{k=-m}^{m} w_k x_k^j = j x_0^{j-1}, \qquad j = 0, 1, \ldots, 2m. \qquad (6.28)$$

Again, a unique solution exists for the equally spaced grid points (6.15), as the next example shows:

```
>> format rat;
>> for m = 1 : 3
       x = linspace(0,2,2*m+1); h = 1/m; % x_0 = 1
       A = fliplr(vander(x))';
       b = (0:2*m)'; w = h*(A\b)'
   end

w =
    -1/2        0        1/2
w =
     1/12      -2/3       0         2/3       -1/12
w =
    -1/60       3/20     -3/4       0          3/4       -3/20       1/60
```

In particular, the 2-point and 3-point first-order central numerical derivatives are given by

$$D_2 f[x_0] = \frac{-y_2 + 8y_1 - 8y_{-1} + y_{-2}}{12h}$$
$$D_3 f[x_0] = \frac{y_3 - 9y_2 + 45y_1 - 45y_{-1} + 9y_{-2} - y_{-3}}{60h},$$

in addition to the 1-point central numerical derivative (6.4).

The exact expressions for the truncation errors of numerical derivatives can be obtained from the Taylor series (in Theorem 5.2). For example, the truncation errors of the forward and central difference approximations $D_2 f[x_0]$ of the first derivative are given by [25]:

$$f'(x_0) - \frac{-f(x_2) + 4f(x_1) - 3f(x_0)}{2h} = \frac{1}{3} f^{(3)}(\xi) h^2, \tag{6.29}$$

where $\xi \in [x_0, x_2]$, and

$$f'(x_0) - \frac{-f(x_2) + 8f(x_1) - 8f(x_{-1}) + f(x_{-2})}{12h} = \frac{1}{30} f^{(5)}(\xi) h^4, \tag{6.30}$$

where $\xi \in [x_{-2}, x_2]$.

Open and closed differentiation

So far, we have been considering *open* numerical derivatives, that is, when as many data points as needed were available to the right, left, or both sides of the given point $x = x_0$. Suppose now that you are solving a numerical differentiation problem on a bounded grid, and you need to find derivatives at each point of the grid. You will soon encounter the problem of how to define numerical derivatives close to the end points of the bounded interval. For example, when you use the 1-point central-difference approximation for the first derivative, you will need one point to the left of the left end point, whereas if you use the 2-point approximation, you will need two points. This modification of Problem 6.1 is known as the *closed* numerical differentiation.

Two possible algorithms incorporate finite-difference approximations in the solution of the closed numerical differentiation problem. One consists of specifying boundary conditions for the end points on the bounded interval. If you model a closed ring, you can use *periodic* boundary conditions, which imply that all points located to the left of the left end point have the same values as the grid points located to the left of the right end point. The numerical derivatives are defined for any $m \geq 1$, and the derivative operation is equivalent to the multiplication of a matrix to the vector of data values. Another set of boundary conditions would be zero values outside of the grid, which correspond to solutions on compact intervals. For instance, Dirichlet boundary conditions at the end points represent the solution on a compact closed interval.

Another technique would be to combine forward, backward, and central difference approximations of the same accuracy under the closed differentiation matrix. For instance, you could use the 1-point central differences for all interior points of the grid, 2-point forward difference for the left end point, and 2-point backward difference for the right end point, which all have the same accuracy $O(h^2)$. Closed numerical derivatives are discussed in more detail in Section 10.1.

6.4 Richardson Extrapolation

After we have constructed higher-order numerical derivatives and multipoint first-order numerical derivatives, we can ask if the accuracy of numerical derivatives can be improved. As a matter of fact, an effective recursive algorithm known as the *Richardson extrapolation* can be developed for this purpose. The algorithm is based on the characterization of the order of the truncation error as a power of the step size h. Richardson extrapolation is a fundamental trick, when a better multipoint approximation can be obtained by cancelling the truncation error of the previous multipoint approximation up to the leading order.

Richardson extrapolation for forward differences

The zeroth extrapolation, denoted by R_1, is defined as the 1-point first-order numerical derivative $Df[x_0]$. Considering the grid point (x_1, y_1) with $x_1 = x_0 + h$, this gives

$$R_1(h) = \frac{y_1 - y_0}{h}. \tag{6.31}$$

If we now consider another grid point (x_2, y_2) with double the step size, that is, $x_2 = x_0 + 2h$, we obtain

$$R_1(2h) = \frac{y_2 - y_0}{2h}.$$

The error formula (6.5) suggests that the truncation error for each of the preceding numerical derivatives is $O(h)$, although it is also clear that $R_1(2h)$ is less accurate than $R_1(h)$. The idea now is to go beyond the information provided in (6.5) by using the Taylor series (5.10) with $n = 2$. Namely, using the fact that

$$\begin{aligned} y_1 &= y_0 + f'(x_0)h + \frac{1}{2}f''(x_0)h^2 + O(h^3) \\ y_2 &= y_0 + f'(x_0)(2h) + \frac{1}{2}f''(x_0)(2h)^2 + O(h^3), \end{aligned}$$

we find that $R_1(h)$ and $R_1(2h)$ are related to the exact derivative $f'(x_0)$ by

$$\begin{aligned} R_1(h) &= f'(x_0) + \frac{1}{2}f''(x_0)h + O(h^2) \\ R_1(2h) &= f'(x_0) + f''(x_0)h + O(h^2). \end{aligned}$$

Therefore, we can cancel the O(h) term by setting

$$R_2(h) := 2R_1(h) - R_1(2h) = \frac{-y_2 + 4y_1 - 3y_0}{2h}, \tag{6.32}$$

which leads to

$$R_2(h) = f'(x_0) + O(h^2).$$

This confirms the result obtained in Theorem 6.2 with $m = 2$ because you can see that $R_2(h)$ defined in (6.32) coincides with $D_2 f[x_0]$ in Section 6.3, whose error is $O(h^2)$. That is, the first Richardson extrapolation $R_2(h)$ coincides with the 2-point forward numerical derivative. Using a Taylor series with $n = 3$, we can obtain even more precise information regarding the error in R_2. Begin by observing that, using the definition (6.32) for a step size $2h$, we obtain

$$R_2(2h) = \frac{-y_4 + 4y_2 - 3y_0}{4h}.$$

Now, using the fact that

$$\begin{aligned}
y_1 &= y_0 + f'(x_0)h + \frac{1}{2}f''(x_0)h^2 + \frac{1}{6}f^{(3)}(x_0)h^3 + O(h^4)\\
y_2 &= y_0 + f'(x_0)(2h) + \frac{1}{2}f''(x_0)(2h)^2 + \frac{1}{6}f^{(3)}(x_0)(2h)^3 + O(h^4)\\
y_4 &= y_0 + f'(x_0)(4h) + \frac{1}{2}f''(x_0)(4h)^2 + \frac{1}{6}f^{(3)}(x_0)(4h)^3 + O(h^4),
\end{aligned}$$

we find that

$$\begin{aligned}
R_2(h) &= f'(x_0) - \frac{1}{3}f^{(3)}(x_0)h^2 + O(h^3)\\
R_2(2h) &= f'(x_0) - \frac{4}{3}f^{(3)}(x_0)h^2 + O(h^3).
\end{aligned}$$

Therefore, we can cancel the $O(h^2)$ by setting

$$R_3(h) := \frac{4R_2(h) - R_2(2h)}{3} = \frac{y_4 - 12y_2 + 32y_1 - 21y_0}{12h}, \tag{6.33}$$

which leads to

$$R_3(h) = f'(x_0) + O(h^3).$$

Observe that the numerical derivative $R_3(h)$ defined in (6.33) is not the same as $D_3 f[x_0]$ from Section 6.3 since it involves the point (x_4, y_4), which does not occur in the formula for $D_3 f[x_0]$. The derivative $R_3(h)$ does not coincide with $D_4 f[x_0]$ either, since the point (x_3, y_3) is excluded from the definition of $R_3(h)$. Therefore, we have found a new numerical derivative of order 3, which we call the second Richardson extrapolation.

To generalize the definition of Richardson extrapolations, consider grid points (x_k, y_k) with

$$x_k = x_0 + 2^k h, \qquad k \in \mathbb{N}.$$

The $(k-1)$th Richardson extrapolation $R_k(h)$ gives a numerical approximation of order k for the exact derivative $f'(x_0)$, that is, it satisfies

$$f'(x_0) = R_k(h) + \alpha_k h^k. \tag{6.34}$$

As before, by introducing a higher-order term in the Taylor series for $f(x)$, we obtain that α_k remains the same for h and $2h$ so that we can cancel the truncation error of order k using the recursive formula:

$$R_{k+1}(h) = \frac{2^k R_k(h) - R_k(2h)}{2^k - 1} = R_k(h) + \frac{R_k(h) - R_k(2h)}{2^k - 1}, \tag{6.35}$$

starting with R_1 given by (6.31). The second formula is preferable to the first because it reduces the rounding error related to subtraction of a small number from a very large number. The recursive Richardson extrapolations can be organized in the following table.

Step Size	Data Point	Forward Difference	First Extrapolation	Second Extrapolation	Third Extrapolation
h	(x_1, y_1)	$R_1(h)$	$R_2(h)$	$R_3(h)$	$R_4(h)$
$2h$	(x_2, y_2)	$R_1(2h)$	$R_2(2h)$	$R_3(2h)$	
$4h$	(x_4, y_4)	$R_1(4h)$	$R_2(4h)$		
$8h$	(x_8, y_8)	$R_1(8h)$			

The forward Richardson extrapolations are coded in the following MATLAB function:

```
function Dnum = ForwRichExtrap(x0,y0,x,y)
% evaluates forward Richardson extrapolations
% x0,y0 - selected data point
% x,y - row-vectors of the same size for data points
% Dnum - row-vector of forward Richardson extrapolations
```

```
maxRich = length(x);
D = ones(maxRich,maxRich);
D(:,1) = (y' - y0)./(x' - x0); % forward numerical derivatives
for k = 2 : maxRich    % Richardson extrapolations
    for kk = 1 : (maxRich-k+1)
        D(kk,k) = D(kk,k-1)+(D(kk,k-1) - D(kk+1,k-1))/(2^(k-1)-1);
    end
end
Dnum = D(1,:);
```

The next example shows an application of the MATLAB function `ForwRichExtrap` to the logarithmic function $f(x) = 2\log x$ at the point $x_0 = 2$. The exact derivative is $f'(x_0) = 1$ and we use four extrapolations with the step size $h = 0.1$. The output suggests that the algorithm quickly converges to the exact value, subject to a small round-off error.

```
>> x0 = 2;
>> y0 = 2*log(x0);
>> h = 0.1;
>> x = x0+h*2.^(0:3);
>> y = 2*log(x);
>> format long;
>> Dnum = ForwRichExtrap(x0,y0,x,y)

Dnum =
   0.97580328338864   0.99850476873403   0.99980775427313   0.99995418240356
```

Exercise 6.6 Write a MATLAB function for backward Richardson extrapolations. Use the function to plot the computational error versus the number of the Richardson extrapolations for the function $f(x) = x^3 \sin(1/x)$ at $x = 0$, where $f'(0) = 0$.

Richardson extrapolation for central differences

The recursive Richardson extrapolations for central numerical derivatives can be obtained in a similar manner. The only difference is that the truncation error of the central difference approximation (6.7) is expanded in even powers of h, as it follows from Theorem 5.2 for $x_1 - x_0 = x_0 - x_{-1} = h$.

More precisely, the zeroth extrapolation for central differences is defined as the 1-point first-order numerical derivative $Df[x_0]$ based on central derivatives, that is,

$$R_1(h) = \frac{y_1 - y_{-1}}{2h}. \tag{6.36}$$

Using double the step size, we obtain

$$R_1(2h) = \frac{y_2 - y_{-2}}{4h}.$$

We can then proceed as before, using a Taylor series with $n = 4$, to show that

$$R_1(h) = f'(x_0) + \frac{1}{6} f^{(3)}(x_0)h^2 + O(h^4)$$
$$R_1(2h) = f'(x_0) + \frac{2}{3} f^{(3)}(x_0)h^2 + O(h^4).$$

As a result, we can cancel the term $O(h^2)$ by defining

$$R_2(h) := \frac{1}{3}\left(4R_1(h) - R_1(2h)\right) = \frac{-y_2 + 8y_1 - 8y_{-1} + y_{-2}}{12h}, \tag{6.37}$$

which leads to

$$R_2(h) = f'(x_0) + O(h^4).$$

Once more, this confirms the result of Theorem 6.6.2 because you can recognize $R_2(h)$ defined in (6.37) as the 2-point central numerical derivative $D_2 f[x_0]$ (see Section 6.3).

To cancel out the previous $O(h^4)$ term, observe first that using a step size of $2h$ in (6.37) gives

$$R_2(2h) = \frac{-y_4 + 8y_2 - 8y_{-2} + y_{-4}}{24h}.$$

We can use a Taylor series with $n = 6$ to show that

$$R_2(h) = f'(x_0) - \frac{1}{30} f^{(5)}(x_0)h^4 + O(h^6)$$
$$R_2(2h) = f'(x_0) - \frac{16}{30} f^{(3)}(x_0)h^4 + O(h^6).$$

Therefore, we can cancel the $O(h^6)$ term by defining

$$\begin{aligned} R_3(h) &= \frac{16R_2(h) - R_2(2h)}{15} \\ &= \frac{y_4 - 40y_2 + 256y_1 - 256y_{-1} + 40y_{-2} - y_{-4}}{360h}, \end{aligned} \tag{6.38}$$

which leads to

$$R_3(h) = f'(x_0) + O(h^6).$$

The numerical derivative $R_3(h)$ is not the same as $D_3 f[x_0]$ from Section 6.3 since it involves the points (x_4, y_4) and (x_{-4}, y_{-4}), which do not occur in the formula for $D_3 f[x_0]$. As before, the derivative $R_3(h)$ does not coincide with $D_4 f[x_0]$ either since the points (x_3, y_3) and (x_{-3}, y_{-3}) are also excluded from the definition of $R_3(h)$. We have therefore found a new numerical derivative of order $n = 6$.

For higher-order Richardson extrapolations for central differences, consider a grid of data points (x_k, y_k) with

$$x_k = x_0 + 2^k h, \qquad k \in \mathbb{Z}.$$

The $(k-1)$th Richardson extrapolation $R_k(h)$ gives a numerical approximation of order $2k$ for the exact derivative $f'(x_0)$, that is, it satisfies

$$f'(x_0) = R_k(h) + \alpha_k h^{2k}, \qquad k \in \mathbb{N}. \tag{6.39}$$

As we have seen previously, by introducing higher-order terms in the Taylor expansion for $f(x)$, we obtain that α_k remains the same for h and $2h$ so that we cancel the truncation error of the order $2k$ using the recursive formula:

$$R_{k+1}(h) = \frac{4^k R_k(h) - R_k(2h)}{4^k - 1} = R_k(h) + \frac{R_k(h) - R_k(2h)}{4^k - 1}. \tag{6.40}$$

The second formula is again preferable to the first because it reduces the rounding error related to subtraction of a small number from a very large number. The recursive Richardson extrapolations can be organized exactly in the same table as the one used for forward numerical derivatives. There are two differences, however: (1) the first-order numerical derivative $R_1(h)$ is computed from the central difference approximation (6.36) and (2) the recursive formula (6.40) is used for Richardson extrapolations instead of (6.35).

The central Richardson extrapolations are coded in the following MATLAB function:

```
function Dnum = CentRichExtrap(x0,y0,x,y,xNeg,yNeg)
% evaluates central Richardson extrapolations
% x0,y0 - selected data point
% x,y - row-vectors of the same size for right data points
% xNeg,yNeg - row-vectors of the same size for left data points
% Dnum - row-vector of central Richardson extrapolations

maxRich = length(x);
D = ones(maxRich,maxRich);
D(:,1) = (y' - yNeg')./(x' - xNeg'); % central numerical derivatives
```

```
for k = 2 : maxRich    % Richardson extrapolations
    for kk = 1 : (maxRich-k+1)
        D(kk,k) = D(kk,k-1)+(D(kk,k-1) - D(kk+1,k-1))/(4^(k-1)-1);
    end
end
Dnum = D(1,:);
```

The next example gives an alternative solution to the previous example with the central Richardson extrapolations. The output suggests that the algorithm based on central numerical derivatives recovers the exact value $f'(2) = 1$ much faster than a similar algorithm based on forward numerical derivatives.

```
>> x0 = 2;
>> y0 = 2*log(x0);
>> h = 0.1;
>> x = x0+h*2.^(0:3);
>> y = 2*log(x);
>> xNeg = x0-h*2.^(0:3);
>> yNeg = 2*log(xNeg);
>> Dnum = CentRichExtrap(x0,y0,x,y,xNeg,yNeg)

Dnum =
   1.00083458556983   0.99999495498952   1.00000014890054   0.99999997861220
```

Multipoint difference approximations for higher-order numerical derivatives can be obtained by two alternative routes. The same algorithm of Richardson extrapolations can be repeated starting with the lowest-order forward and central difference approximations for the second-order and higher-order derivatives of a function $y = f(x)$ at the point $x = x_0$. For instance, the second-order numerical derivatives

$$R_1(h) = D^2 f[x_0] = \frac{y_2 - 2y_1 + y_0}{h^2} \tag{6.41}$$

and

$$R_1(h) = D^2 f[x_0] = \frac{y_1 - 2y_0 + y_{-1}}{h^2} \tag{6.42}$$

have truncation errors that are $\mathrm{O}(h)$ and $\mathrm{O}(h^2)$, respectively. The Richardson extrapolations are computed from the cancellation of the corresponding terms in the truncation error as follows:

$$R_2(h) = \frac{-y_4 + 10y_2 - 16y_1 + 7y_0}{4h^2} \tag{6.43}$$

and

$$R_2(h) = \frac{-y_2 + 16y_1 - 30y_0 + 16y_{-1} - y_{-2}}{12h^2}. \tag{6.44}$$

The new numerical derivatives have truncation errors that are $O(h^2)$ and $O(h^4)$, respectively.

On the other hand, the multipoint difference approximations can be obtained with the repeated matrix multiplications of the differentiation matrix that represents the first-order numerical derivatives.

Exercise 6.7 Apply a repeated matrix multiplication to the 2-point first-order numerical derivatives $D_2^2 f[x_0] = D_2 D_2 f[x_0]$ and obtain new forward and central difference approximations for the second-order derivatives. Compare the resulting answers with the formulas (6.43) and (6.44) and explain any differences.

6.5 Integrals and Finite Sums

Let $f(x)$ be a continuous function and consider its definite integral over a bounded interval $[a, b]$. The most elementary definition of the integral is given in *Calculus* as the limit of Riemann sums:

$$\int_a^b f(x)dx = \lim_{n\to\infty} \sum_{i=1}^{n} f(\xi_i)(x_{i+1} - x_i), \tag{6.45}$$

where

$$a \le x_1 < x_2 < x_3 < \ldots < x_n < x_{n+1} \le b \tag{6.46}$$

is a partition of the interval $[a, b]$ and $\xi_1, \xi_2, \ldots, \xi_n$ are points such that $x_i \le \xi_i \le x_{i+1}$, $i = 1, 2, \ldots, n$. The partition of the interval $[a, b]$ into n subintervals can be done arbitrarily since the limit $n \to \infty$ is independent of it, as long as the distinct interior points $x_1, x_2, \ldots, x_{n+1}$ fill the interval $[a, b]$ densely in the sense that

$$\lim_{n\to\infty} \max_{1\le k\le n} |x_{k+1} - x_k| = 0.$$

An equivalent definition of the integral of $f(x)$ on $[a, b]$ is given graphically as a signed area under the curve $y = f(x)$. In particular, if $f(x) \ge 0$ on $[a, b]$, the integral is always non-negative and matches the actual area under the curve of $y = f(x)$.

Many numerical problems are specified on discrete grids of data points rather than on continuous intervals. You can see that such discrete grids (partitions) occur naturally in the definition of Riemann sums. If the grid points include both end points, we call the numerical approximation a *closed integration.* If one or both end points are not included in the discrete grid, we call the numerical approximation an *open integration.* We refer to the numerical approximation of the integral of $f(x)$ on $[a, b]$ as the *numerical integral* $Sf[a, b]$.

Problem 6.2 (Numerical Integration) Assume that the continuous function $y = f(x)$ is defined on the discrete grid of data points (x_1, y_1), (x_2, y_2), ..., (x_{n+1}, y_{n+1}), where $a = x_1 < x_2 < \ldots < x_{n+1} = b$. Find the numerical integral $Sf[a, b]$.

Rectangular integration rules

A general principle for developing numerical integration rules is to replace the function $f(x)$ with simpler functions coinciding with f at the selected grid points, such as polynomial interpolants, and then define the numerical integral as the exact integral of the interpolant function. For example, the simplest numerical integration can be developed by assuming that the interval $[a, b]$ is so small that no intermediate partition points are introduced, so that $x_1 = a$ and $x_2 = b$ are the only grid points. There are two versions of numerical integration based on a single evaluation of the function $f(x)$ at the grid points (x_1, y_1) and (x_2, y_2):

$$\text{(left-point)} \quad Sf[x_1, x_2] = y_1(x_2 - x_1), \tag{6.47}$$

$$\text{(right-point)} \quad Sf[x_1, x_2] = y_2(x_2 - x_1). \tag{6.48}$$

If we also know the value $y_{1.5} = f(x_{1.5})$ at the middle point $x_{1.5} = (x_1 + x_2)/2$, then we can use the following version of numerical integration based on a single evaluation of the function $f(x)$ at the grid point $(x_{1.5}, y_{1.5})$:

$$\text{(midpoint)} \quad Sf[x_1, x_2] = y_{1.5}(x_2 - x_1). \tag{6.49}$$

According to the numerical integrations (6.47)–(6.49), the signed area under the curve $y = f(x)$ on $[a, b]$ is replaced by the area of a rectangle with the height evaluated at the left-end, right-end, and middle points of the interval $[a, b]$, respectively. These integration formulas, called the *rectangular rules*, are equivalent to the integrals of the constant interpolation $y = P_0(x)$ through a single point, either the left-end point (x_1, y_1), or the right-end point (x_2, y_2), or the middle point $(x_{1.5}, y_{1.5})$.

Trapezoidal integration rule

The rectangular rules solve Problem 6.2 in terms of *open integration* because the function $f(x)$ is not evaluated on the *closed* discrete grid on $[a, b]$, including both end points. When $n = 1$, the only solution of the closed integration problem is possible if both data points (x_1, y_1) and (x_2, y_2) are connected by the linear interpolation (see Section 5.2):

$$y = p_1(x) = y_1 + \frac{y_2 - y_1}{x_2 - x_1}(x - x_1).$$

Integrating the interpolating polynomial $y = p_1(x)$ on $[x_1, x_2]$, we find a solution to Problem 6.2:

$$Sf[x_1, x_2] = \frac{1}{2}(y_1 + y_2)(x_2 - x_1). \tag{6.50}$$

This integration formula is called the *trapezoidal rule* because it recovers the area of a trapezoid with the heights at the points (x_1, y_1) and (x_2, y_2).

The numerical integration rules are illustrated for the integral

$$\int_0^a \sqrt{1-x^2}dx = \frac{1}{2}\left(\arcsin a + a\sqrt{1-a^2}\right). \tag{6.51}$$

The following MATLAB commands generate Figure 6.4, illustrating the midpoint rectangular rule (6.49) and the trapezoidal rule (6.50) applied to the integral (6.51) with $a = 0.75$.

```
>> x = linspace(0,0.75,101);
>> y = sqrt(1-x.^2);
>> x = [0,x,0.75,0]; y = [0,y,0,0];
>> fill(x,y,'c'); hold on;
>> x1 = [0,0.75]; y1 = sqrt(1-x1.^2);
>> y2 = sqrt(1-0.375^2)*ones(size(x1));
>> plot(x1,y1,'r',x1,y2,'b')
```

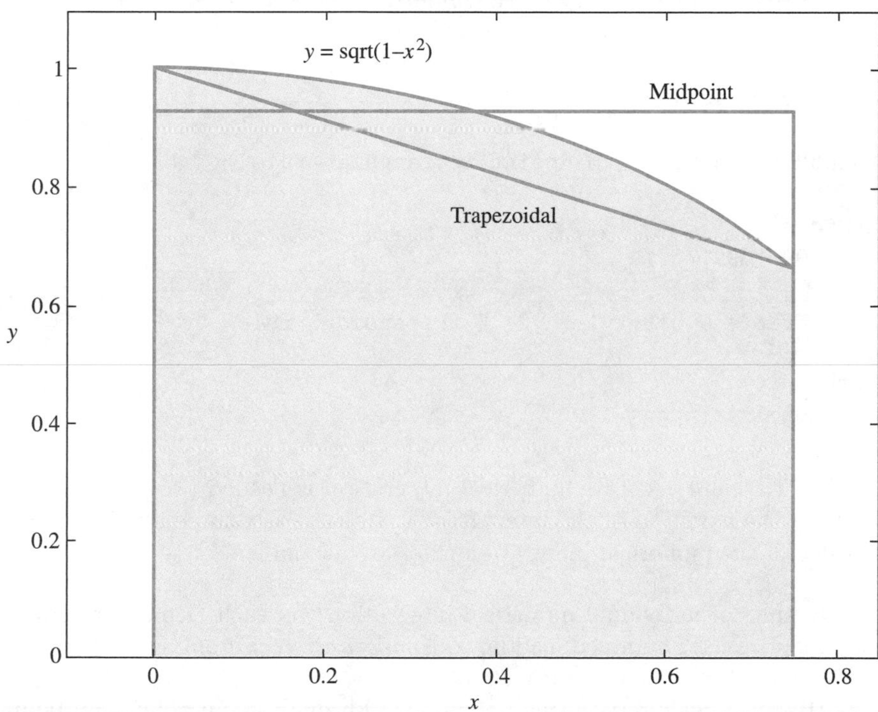

Figure 6.4 Simplest integration rules for the integral (6.51).

The following MATLAB commands illustrate the four numerical integration rules for the integral (6.51) with $a = 0.1$. You can see that the left-end and right-end rectangular rules are inaccurate: the left-end rectangle overshoots the actual area under the circle and the right-end rectangle undershoots the area under the circle. The midpoint and trapezoidal rules give better numerical values of the same accuracy.

```
>> a = 0.1; x1 = 0; x2 = a; x1_5 = a/2;
>> y1 = sqrt(1-x1^2); y2 = sqrt(1-x2^2); y1_5 = sqrt(1-x1_5^2);
>> IntExact = 0.5*(asin(x2)+x2*y2) % exact answer

IntExact =
   0.09983308243611

>> IntLeft = h*y1 % left rectangular rule

IntLeft =
   0.10000000000000

>> IntRight = h*y2 % right rectangular rule

IntRight =
   0.09949874371066

>> IntMid = h*y1_5 % midpoint rectangular rule

IntMid =
   0.09987492177719

>> IntTrap = 0.5*h*(y1 + y2) % trapezoidal rule

IntTrap =
   0.09974937185533
```

When a more accurate numerical integration is needed, we can use more points in the partition of the interval $[a, b]$. In this case, three possible modifications of the numerical integration rules are available:

1. Results of individual numerical integrations on each elementary subinterval can be summed leading to *composite integration rules.*

2. Higher-order interpolating polynomials through many points can be integrated leading to multipoint numerical integration rules known as the *Newton–Cotes closed integration formulas.*

3. A combination of methods (1) and (2) can be used to lead to recursive rules known as *Romberg integrations*.

Composite integration rules

In what follows, we use only the trapezoidal and midpoint rules because they have higher accuracy compared to the left-end and right-end rectangular rules. We consider $n+1$ partition points on the integration interval $[a, b]$, therefore dividing it into n subintervals. If the step size of each subinterval is constant, we say that the data grid is *uniform*. Alternatively, if the step size of each subinterval is adjusted depending on the rate of change of the function $y = f(x)$, we say that the data grid is *adaptive*.

A *composite trapezoidal rule* is based on the piecewise linear interpolation between each pair of two adjacent data points. In the case of uniform grids, we add all areas of individual trapezoids and obtain the summation formulas:

$$Sf[a,b] = \frac{h}{2}\left(y_1 + 2y_2 + 2y_3 + \ldots + 2y_n + y_{n+1}\right), \tag{6.52}$$

where h is the step size of the uniform grid. A *composite midpoint rule* is based on the piecewise constant interpolation through the midpoint of each interval. In the case of uniform grids, we sum all areas of individual rectangles and obtain the summation formulas:

$$Sf[a,b] = h\left(y_{1.5} + y_{2.5} + y_{3.5} + \ldots + y_{n.5}\right). \tag{6.53}$$

When the midpoint values are not available, we can replace them with the average of the given data points, in which case the composite midpoint rule (6.53) reduces to the composite trapezoidal rule (6.52).

The next example shows applications of the two composite integration rules to the integral

$$\int_0^1 \sqrt{1-x^2}dx = \frac{\pi}{4}. \tag{6.54}$$

The integral represents the quarter of the area of the unit circle. When $n = 100$, both integration rules give an accurate answer up to four decimal places.

```
>> a = 0; b = 1; h = 0.01;
>> x = a : h : b; n = length(x)-1;
>> y = sqrt(1-x.^2);
>> IntExact = pi/4 % exact answer

IntExact =
   0.78539816339745
```

```
>> yMid = sqrt(1-(x+0.5*h).^2);
>> IntMid = h*sum(yMid(1:n)) % composite midpoint rule

IntMid =
   0.78548421447500

>> IntTrap = 0.5*h*(y(1) + 2*sum(y(2:n)) + y(n+1))
                    % composite trapezoidal rule

IntTrap =
   0.78510425794476
```

A graphical illustration of the composite midpoint rule is shown in Figure 6.5, which is generated by the following MATLAB code:

```
>> x = linspace(0,1,101); y = sqrt(1-x.^2);
>> x1 = linspace(0,1,11)+0.05; y1 = sqrt(1-x1.^2);
>> plot(x,y); hold on; bar(x1,y1)
```

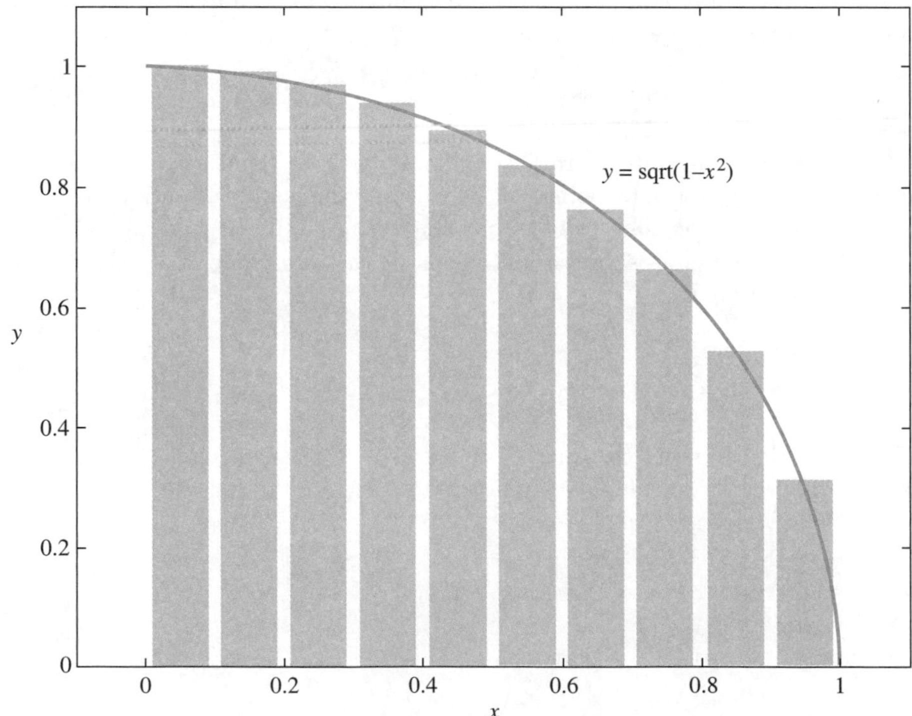

Figure 6.5 Composite midpoint integration rule for the integral (6.54).

Numerical integrals can be evaluated with the MATLAB function quad, which takes as input the integrand function and the limits of integration and outputs the value of the integral. An optional fourth input argument of the function quad can be used to set up a tolerance level. The next example computes the integral (6.54) with the default tolerance of 10^{-6} and with higher tolerance levels. Comparison with the exact answer confirms that the level of tolerance corresponds to the level of accuracy of the numerical approximation.

```
>> I1 = quad('sqrt(1-x.^2)',0,1)

I1 =
   0.78539448965158

>> I2 = quad('sqrt(1-x.^2)',0,1,10^(-8))

I2 =
   0.78539810614836

>> I3 = quad('sqrt(1-x.^2)',0,1,10^(-12))

I3 =
   0.78539816339244
```

Errors of integration rules

We can calculate the truncation errors of the numerical integration rules using the Taylor series (see Theorem 5.2). We distinguish between the *local truncation error* of an individual integration on the elementary subinterval $[x_i, x_{i+1}]$ and the *global truncation error* of the composite integration on the integration interval $[a, b]$. It is clear that the global error must be much larger than the local error. Given the integration interval $[a, b]$, the global error defines the actual error of numerical integration.

Using the Taylor series (5.10) with $n = 2$ at the point $x = x_1$ and the forward difference representation (6.5) for the first derivative $f'(x_1)$, we find the local truncation error of the trapezoidal rule (6.50):

$$\int_{x_1}^{x_2} f(x)dx - \frac{1}{2}\left(f(x_1) + f(x_2)\right)h = -\frac{1}{12}f''(\xi)h^3, \qquad \xi \in [x_1, x_2], \quad (6.55)$$

where $h = x_2 - x_1$. Similarly, using the Taylor series (5.10) with $n = 2$ at the point $x = x_{1.5}$, we find the local truncation error of the midpoint rule (6.49):

$$\int_{x_1}^{x_2} f(x)dx - f(x_{1.5})h = \frac{1}{24}f''(\xi)h^3, \qquad \xi \in [x_1, x_2], \quad (6.56)$$

where $h = x_2 - x_1$ and $x_{1.5} = (x_1 + x_2)/2$.

You can check the validity of error formulas (6.55) and (6.56) by performing a simple computation. Consider the function $f(x) = x^2$ on the interval

$[0,1]$, for which the exact integral is $\frac{1}{3}$. When you apply the trapezoidal rule between $x_1 = 0$ and $x_2 = 1$, the numerical integral is $\frac{1}{2}$, which gives the truncation error $-\frac{1}{6}$. When you apply the midpoint rule at $x_{1.5} = \frac{1}{2}$, the numerical integral is $\frac{1}{4}$, which gives the truncation error $\frac{1}{12}$. In both cases, the error formulas (6.55) and (6.56) recover the exact answer to these simple computations since $f''(\xi) = 2$ is independent of ξ.

The global truncation error of the composite rules accumulates after n individual integrations so that it is roughly n times larger than the local truncation error. Because $hn = b - a$, a general rule for uniform grids is that the global truncation error is one order larger than the local truncation error, and it grows linearly with the length of the integration interval. Using the Taylor series computations (6.55) and (6.56) and the triangular inequality, the following theorem summarizes the global truncation error of the composite trapezoidal and midpoint rules.

Theorem 6.3 *Let $f(x)$ be twice continuously differentiable on $[a,b]$. Let $[a,b]$ be divided into n equally spaced subintervals with the step size $h = (b-a)/n$. The truncation error of the composite trapezoidal rule (6.52) satisfies*

$$\left| \int_a^b f(x)dx - Sf[a,b] \right| \le \frac{M_2(b-a)}{12} h^2, \tag{6.57}$$

while the truncation error of the composite midpoint rule (6.53) satisfies

$$\left| \int_a^b f(x)dx - Sf[a,b] \right| \le \frac{M_2(b-a)}{24} h^2, \tag{6.58}$$

where $M_2 = \max_{x \in [a,b]} |f''(x)|$.

Let us illustrate the error formulas (6.57) and (6.58) with the same example of the function $f(x) = x^2$ on the interval $[0,1]$. We divide the integration interval $[0,1]$ into n subintervals of the step size $h = 1/n$ and use the summation formulas

$$\sum_{k=1}^{n} k = \frac{n(n+1)}{2}, \qquad \sum_{k=1}^{n} k^2 = \frac{n(n+1)(2n+1)}{6}. \tag{6.59}$$

As a result, the global truncation error of the composite trapezoidal rule is

$$\frac{1}{3} - \frac{1}{2n}\left(1 + 2\sum_{k=1}^{n-1} \frac{k^2}{n^2}\right) = \frac{1}{3} - \frac{(1+2n^2)}{6n^2} = -\frac{h^2}{6},$$

while the global truncation error of the composite midpoint rule is

$$\frac{1}{3} - \frac{1}{4n^3}\sum_{k=1}^{n-1}(1+2k)^2 = \frac{1}{3} - \frac{(4n^2-1)}{12n^2} = \frac{h^2}{12}.$$

Both answers match the exact values from the error formulas (6.57) and (6.58). Note that the global errors of composite trapezoidal and midpoint rules are comparable, which is also demonstrated in the previous numerical examples.

Exercise 6.8 Show numerically that the global truncation error of the composite trapezoidal rule grows linearly with the length L of the integration interval for the function $f(x) = \sin x$ on $[0, L]$.

Well-posedness of integration rules

If the step size h between two points becomes smaller, the truncation error of the summation rule decreases. For example, if the numerical computations are performed with double the number of subintervals (that is, the step size is halved), the global truncation errors of the trapezoidal and midpoint rules are reduced by the factor of 4, that is,

$$\frac{E(2n)}{E(n)} = \frac{E(h/2)}{E(h)} \approx 2^{-2} = \frac{1}{4}.$$

When the step size h reduces to zero, the truncation error converges to zero, but the limiting error could be nonzero as a result of the round-off error of numerical integration. By adding all round-off errors, we obtain the upper bound for the global round-off error of the composite trapezoidal rule:

$$\frac{h}{2}\left|e_1 + 2e_2 + \ldots + 2e_n + e_{n+1}\right| \le hn\varepsilon = \varepsilon(b-a),$$

where ε is the machine precision for the floating-point number system. The same upper bound is obtained for *all* composite integration rules. Because the global truncation error of the composite integration rule converges to zero and the global round-off error converges to a constant, the numerical integration is a *well-posed* numerical procedure. The numerical integral converges to the exact value as $h \to 0$, except for the constant round-off error.

The next example illustrates the behavior of the global truncation error for the composite trapezoidal and midpoint rules. You integrate the function $f(x) = \sqrt{1-x^2}$ on two intervals $[0, 0.5]$ and $[0, 1]$ (see (6.51) for the exact integrals). On the interval $[0, 0.5]$, the second derivative of $f(x)$ is bounded and the global error $E(h)$, in agreement with Theorem 6.3, is $O(h^2)$. Because the second derivative of $f(x)$ diverges as $x \to 1$, the global error $E(h)$ converges much slower as $h \to 0$. The numerical example shows that the global error $E(h)$ is $O(h^{3/2})$ in this case.

```
% global_truncation_error
figure(1); hold on;
  for j = 1 : 2
    if (j == 1)
        a = 0.5; n1 = 5; n2 = 500;
    else
        a = 1; n1 = 10; n2 = 1000;
    end
    IntExact = 0.5*(asin(a)+a*sqrt(1-a*a));
    k = 1;
    for n = n1 : 1 : n2
        h = a/n;
        x = linspace(0,a,n+1);
        y = sqrt(1-x.*x);
        yM = sqrt(1-(x+0.5*h).*(x+0.5*h));
        IntTrap = h*(y(1)+2*sum(y(2:n))+y(n+1))/2;
        IntMid = h*sum(yM(1:n));
        ErrTrap(k) = abs(IntTrap-IntExact);
        ErrMid(k) = abs(IntMid-IntExact);
        hst(k) = h;
        k = k+1;
    end
    aTrap = polyfit(log(hst),log(ErrTrap),1);
    powerTrap = aTrap(1)
    aMid = polyfit(log(hst),log(ErrMid),1);
    powerMid = aMid(1)
    loglog(hst,ErrTrap,'g',hst,ErrMid,'r');
end
```

The MATLAB script `global_truncation_error` produces the following numerical output and displays Figure 6.6.

```
>> global_truncation_error

powerTrap =
   1.99996570660879

powerMid =
   1.99994000376286

powerTrap =
   1.49975007686818

powerMid =
   1.49944822327719
```

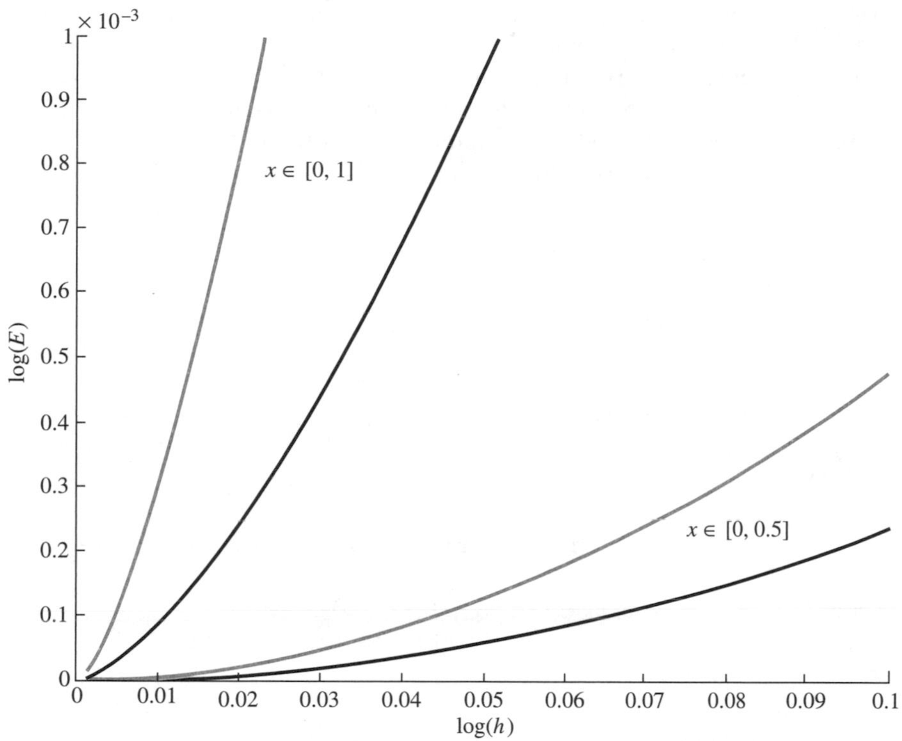

Figure 6.6 Numerical approximations of the integral (6.51) on $[0, 0.5]$ and $[0, 1]$ with the composite trapezoidal rule (bright curve) and the composite midpoint rule (dark curve).

Integration of improper integrals

Since the composite trapezoidal and midpoint rules have the same accuracy, their combination can be used for numerical integration of functions $f(x)$ with singularities. The integrals of functions with singularities on finite intervals are referred to as *improper integrals*. Depending on the character of the singularities, the improper integrals may diverge. For instance, the improper integral

$$\int_0^1 \frac{dx}{\sqrt{1-x^2}} = \frac{\pi}{2} \tag{6.60}$$

converges to an exact value, whereas the improper integral

$$\int_0^{\pi/2} \frac{\cos x}{\sin x} dx = \infty \tag{6.61}$$

diverges. The next example illustrates a combination of the composite trapezoidal rule in the interior points of the grids and the individual midpoint rule at the end point, where the singularity of the function $f(x)$ occurs. When the step size is decreased, the first integral converges to the exact value (6.60), whereas the second integral (6.61) diverges. The output of the MATLAB script `improper_integration` is shown in Figure 6.7.

```
% improper_integration
IntExact = pi/2; nn = 10 : 10 : 40000;
for k = 1:length(nn)
    n = nn(k);
    hh = 1./n;
    x = linspace(0,1,n+1);
    y1 = 1./sqrt(1-x(1:n).^2);
    y1M = 1/sqrt(1-(1-0.5*hh).^2);
    h(k) = hh;
    IntTrap1(k) = hh*(y1(1)+2*sum(y1(2:n-1))+y1(n))/2 + hh*y1M;
end
figure(1); % the improper integral (6.60)
semilogx(h,IntTrap1,'g',h,IntExact*ones(size(h)),':g')
for k = 1:length(nn)
    n = nn(k);
    hh = pi./(2*n);
    x = linspace(0,pi/2,n+1);
    y2 = cos(x(2:n+1))./sin(x(2:n+1));
    y2M = cos(0.5*hh)/sin(0.5*hh);
    h(k) = hh;
    IntTrap2(k) = hh*y2M + hh*(y2(1)+2*sum(y2(2:n-1))+y2(n))/2;
end
figure(2); % the improper integral (6.61)
semilogx(h,IntTrap2,'b')
```

This example shows that convergence of the numerical integration to a finite number becomes slow if there are singularities in the function $f(x)$ and its derivatives on the integration interval $[a, b]$. As another inconvenience, the function $f(x)$ may only be available in a tabular form and additional computations of data points could cause computational or experimental difficulties. For these reasons, it is often better to improve the accuracy of the trapezoidal and midpoint rules and to develop a multipoint integration rule of higher accuracy.

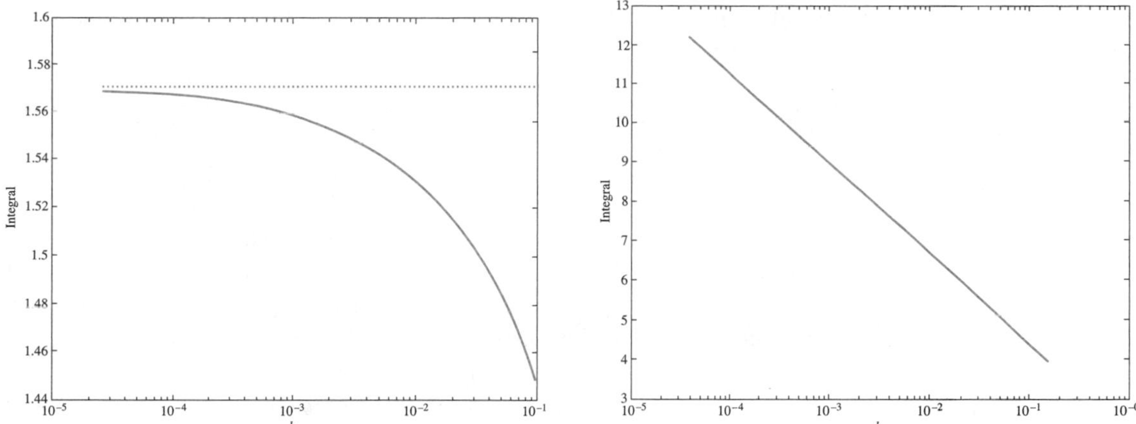

Figure 6.7 Numerical integration of improper integrals (6.60) and (6.61). Left: The numerical integration converges to the exact value (6.60) shown by a dotted curve. Right: The numerical integration of the integral (6.61) diverges.

6.6 Newton–Cotes Integration Rules

We consider again Problem 6.2 for $n \geq 1$, where the function $y = f(x)$ on the interval $[a, b]$ is defined by the set of data points $\{(x_k, y_k)\}_{k=1}^{n+1}$. The given set of data points can be connected by the interpolation polynomial $p_n(x)$ (see Section 5.2). The $(n+1)$-point numerical integration of the function $y = f(x)$ on $[a, b]$ is defined by the integral of the interpolating polynomial $y = p_n(x)$:

$$S_n f[a, b] = \int_a^b p_n(x) dx. \tag{6.62}$$

Simpson's integration rule

When $n = 1$, the linear interpolation between the two end points of the integration interval recovers the trapezoidal rule of Section 6.5 with $S_1 f[a, b] = S f[a, b]$. When $n = 2$, the quadratic polynomial $p_2(x)$ between two end points (x_1, y_1) and (x_3, y_3) and an intermediate point (x_2, y_2), $x_1 < x_2 < x_3$ can be written in the Newton form:

$$p_2(x) = y_1 + \left(\frac{y_2 - y_1}{x_2 - x_1}\right)(x - x_1) + \frac{1}{x_3 - x_1}\left(\frac{y_3 - y_2}{x_3 - x_2} - \frac{y_2 - y_1}{x_2 - x_1}\right)(x - x_1)(x - x_2).$$

Integration of $p_2(x)$ produces the general three-point numerical integration rule:

$$S_2 f[x_1, x_3] = \frac{x_3 - x_1}{6} \left(\frac{-2x_1 + 3x_2 - x_3}{x_2 - x_1} y_1 + \frac{-4x_1 + 3x_2 + x_3}{x_2 - x_1} y_2 + \frac{x_1 - 3x_2 + 2x_3}{x_3 - x_2} (y_3 - y_2) \right). \quad (6.63)$$

Looking at this complicated expression, it is hard to believe that the formula (6.63) matches the well-known *Simpson rule* in the case of the equally spaced grid. However, if x_2 is a midpoint on $x \in [x_1, x_3]$ so that $x_2 - x_1 = x_3 - x_2 = h$, the three-point rule (6.63) becomes:

$$S_2 f[x_1, x_3] = \frac{h}{3} (y_1 + 4y_2 + y_3) . \quad (6.64)$$

The following MATLAB commands generate Figure 6.8, which shows the interpolating polynomial $p_2(x)$ for the integral (6.51) with $a = 0.75$ at the equally spaced grid with $h = 0.375$:

```
>> x = linspace(0,0.75,101); y = sqrt(1-x.^2);
>> xE = [0,x,0.75,0]; yE = [0,y,0,0];
>> fill(xE,yE,'c'); hold on;
>> h = 0.375; x1 = [0,h,2*h]; y1 = sqrt(1-x1.^2);
>> yInt = y1(1)+(y1(2)-y1(1))*(x-x1(1))/h;
>> yInt = yInt+((y1(3)-y1(2))/h-(y1(2)-y1(1))/h)*(x-x1(1)).*(x-x1(2))/(2*h);
>> plot(x1,y1,'.b',x,yInt,'r');
```

Exercise 6.9 Prove that the Simpson rule $S(f)$ on the interval with end points x_1 and x_3 and midpoint x_2 is related to the trapezoidal rule $T(f)$ and the midpoint rule $M(f)$ by

$$S(f) = \frac{1}{3} T(f) + \frac{2}{3} M(f).$$

Errors of Simpson's rule

The truncation error is different between the three-point rule with variable step size (6.63) and the Simpson rule (6.64). According to the strategy of our book, we shall confirm numerically that the local truncation error of the general three-point integration rule (6.63) is

$$\int_{x_1}^{x_3} f(x)dx - S_2 f[x_1, x_3] = \frac{1}{144} f^{(3)}(\xi)(x_3 - x_1)^3 (2x_2 - x_1 - x_3), \quad (6.65)$$

where $\xi \in [x_1, x_3]$, while the local truncation error of the Simpson rule (6.64) is

$$\int_{x_1}^{x_3} f(x)dx - S_2 f[x_1, x_3] = -\frac{1}{90} f^{(4)}(\xi) h^5, \quad (6.66)$$

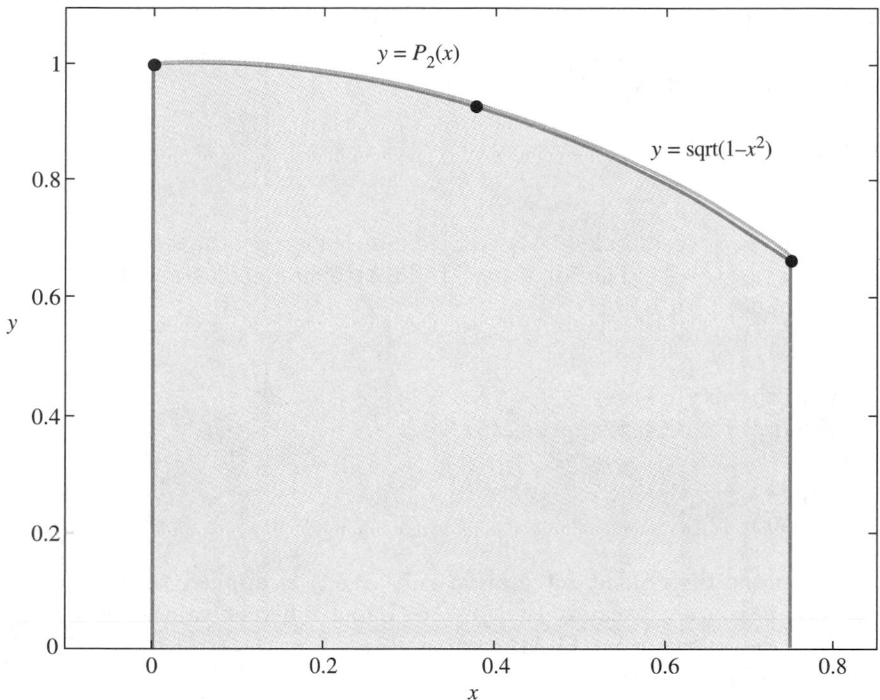

Figure 6.8 Simpson's integration rule for the integral (6.51).

where $\xi \in [x_1, x_3]$. Therefore, the Simpson rule for an equally spaced discrete grid admits a remarkable cancellation of the $O(h^4)$ term in the truncation error of the general three-point integration rule (6.65).

Exercise 6.10 Use Theorem 5.6 to prove analytically the error formulas (6.65) and (6.66).

To verify the error formulas (6.65) and (6.66) numerically, we study how the error of the integration rules depends on the derivatives of the function $f(x)$. Consider integrals of the power functions $f(x) = x^3$ and $f(x) = x^4$ on the interval $[1, 3]$, when

$$\int_1^3 x^3 dx = 20, \qquad \int_1^3 x^4 dx = \frac{242}{5}.$$

The following commands show that when the Simpson rule (6.64) is applied to $f(x) = x^3$, the numerical error is zero because $f^{(4)}(\xi) = 0$.

```
>> x = [1,2,3]; y = x.^3;
>> IntSimp = (y(1)+4*y(2) + y(3))/3;
>> ErrSimp = IntSimp-20

ErrSimp =
     0
```

When the Simpson rule (6.64) is applied to $f(x) = x^4$, the error is constant because $f^{(4)}(\xi) = 24$. The following MATLAB commands confirm the error formula (6.66) with $h = 1$:

```
>> y = x.^4;
>> IntSimp = (y(1)+4*y(2) + y(3))/3;
>> ErrSimp = 90*(IntSimp-242/5)/24

ErrSimp =
    1.0000
```

When the three-point integration rule (6.63) is applied to $f(x) = x^3$, the error is nonzero because $f^{(3)}(\xi) = 6$. Using different values for x_2, say, $x_2 = 1.5$, the following MATLAB commands confirm the error formula (6.65) for a single computation:

```
>> x = [1,1.5,3]; y = x.^3;
>> IntGen = (x(3)-x(1))*((-2*x(1)+3*x(2)-x(3))/(x(2)-x(1))*y(1)...
              + (-4*x(1)+3*x(2)+x(3))/(x(2)-x(1))*y(2)...
              + (x(1)-3*x(2)+2*x(3))/(x(3)-x(2))*(y(3)-y(2)))/6;
>> ErrGen = -144*(IntGen-20)/(6*8*(2*x(2)-x(1)-x(3)))

ErrGen =
    1.0000
```

Exercise 6.11 Plot the error of numerical integration for $f(x) = x^3$ with the three-point rule (6.63) as a function of x_2 for fixed $x_1 = 1$ and $x_3 = 3$ on $[1, 3]$ and show that the error is the linear function of x_2 according to the factor $(2x_2 - x_1 - x_3)$.

To study how the error of the integration rules depends on the step size h, consider the integral for the power function $f(x) = x^5$ on the interval $[1, 1 + 2h]$, where

$$\int_1^{1+2h} x^5 dx = \frac{1}{6}\left((1 + 2h)^6 - 1\right).$$

Let us set $x_1 = 1$, $x_3 = 1 + 2h$ and compute the three-point rule (6.63) with $x_2 = 1 + 0.5h$ and the Simpson rule (6.64) with $x_2 = 1 + h$. By using the least-square approximation for the power function $E(h) = ch^p$ (see Section 5.6), we find the power p that describes convergence of the truncation error in the limit $h \to 0$.

```
% error_integration_rules
n1 = 10; n2 = 1000; k = 1;
for n = n1 : 1 : n2
    h = 1/n; x = 1+[0,h,2*h]; y = x.^5;
    IntExact = ((1+2*h)^6-1)/6;
    IntSimp = h*(y(1)+4*y(2) + y(3))/3; % Simpson rule
    ErrSimp(k) = abs(IntSimp-IntExact);
    x = 1+[0,0.5*h,2*h]; y = x.^5;       % three-point rule
    IntGen = (x(3)-x(1))*((-2*x(1)+3*x(2)-x(3))/(x(2)-x(1))*y(1)...
                + (-4*x(1)+3*x(2)+x(3))/(x(2)-x(1))*y(2)...
                + (x(1)-3*x(2)+2*x(3))/(x(3)-x(2))*(y(3)-y(2)))/6;
    ErrGen(k) = abs(IntGen-IntExact);
    hst(k) = h; k = k+1;
end
aSimp = polyfit(log(hst),log(ErrSimp),1); powerSimp = aSimp(1)
aGen = polyfit(log(hst),log(ErrGen),1); powerGen = aGen(1)
```

The output of the MATLAB script `error_integration_rules` shows that $p = 5$ for the Simpson rule and $p = 4$ for the three-point rule, according to the error formulas (6.65) and (6.66).

```
>> error_integration_rules

powerSimp =
    5.0079

powerGen =
    4.0156
```

Errors of multipoint integrations

Let us develop a numerical algorithm to compute the $(n+1)$-point integration rule from the integral of the interpolating polynomial (6.62) with $n \geq 1$. Although the multipoint integration rule is valid for non-uniform grids with variable step size, explicit integration rules are available only for the uniform grid on the interval $[x_1, x_{n+1}]$:

$$x_k = x_1 + (k-1)h, \qquad k = 1, 2, \ldots, n+1, \tag{6.67}$$

where $h = (x_{n+1} - x_1)/n$ is the step size h and the interval $[x_1, x_{n+1}]$ is divided into n equally spaced subintervals. Explicit multipoint integration rules are obtained similarly to how multipoint numerical derivatives are obtained in Section 6.3. This algorithm is based on the following result from *Numerical Analysis*:

Theorem 6.4

1. *Let $n = 2m - 1$ for $m = 1, 2, \ldots$. Let $f(x)$ be a $(2m)$-times continuously differentiable function on the interval $[x_1, x_{2m}]$. Let the $(2m)$-point integration rule be defined by (6.62) on the equally spaced grid (6.67). There is a constant $C_m > 0$ such that*

$$\left| \int_{x_1}^{x_{2m}} f(x)dx - S_{2m-1} f[x_1, x_{2m}] \right| \leq C_m M_{2m} h^{2m+1}, \qquad (6.68)$$

where $M_{2m} = \max_{x \in [x_1, x_{2m}]} |f^{(2m)}(x)|$.

2. *Let $n = 2m$ for $m = 1, 2, \ldots$. Let $f(x)$ be a $(2m+2)$-times continuously differentiable function on the interval $[x_1, x_{2m+1}]$. Let the $(2m+1)$-point integration rule be defined by (6.62) on the equally spaced grid (6.67). There is a constant $C_m > 0$ such that*

$$\left| \int_{x_1}^{x_{2m+1}} f(x)dx - S_{2m} f[x_1, x_{2m+1}] \right| \leq C_m M_{2m+2} h^{2m+3}, \qquad (6.69)$$

where $M_{2m+2} = \max_{x \in [x_1, x_{2m+1}]} |f^{(2m+2)}(x)|$.

It follows from Theorem 6.4 that a cancellation of one power of h occurs in each even multipoint integration rule, similar to the cancellation of the power h^4 in the Simpson rule $S_2 f[x_1, x_3]$.

Exercise 6.12 Using the power functions $f(x) = x^k$ for $k = 5, 6, 7, 8, 9, 10$, show numerically that the $O(h^6)$, $O(h^8)$, and $O(h^{10})$ terms in the truncation error of the corresponding multipoint integration rule cancel.

We use Theorem 6.4 to develop a general explicit solution to Problem 6.2 on the uniform grid (6.67). This general solution is referred to as the *Newton–Cotes closed integration rule.* Let the general $(n+1)$-point integration rule be expressed in the form:

$$S_n f[x_1, x_{n+1}] = h \sum_{k=1}^{n+1} w_k f(x_k), \qquad (6.70)$$

where $w_1, w_2, \ldots, w_{n+1}$ are unknown weights. Following Theorem 6.4, we impose the conditionthat the integration rule (6.70) must be exact for the power functions $y = x^m$ with $m = 0, 1, \ldots, n$. This condition results in the inhomogeneous linear system:

$$h \sum_{k=1}^{n+1} w_k x_k^j = \frac{1}{j+1}, \qquad j = 0, 1, \ldots, n. \tag{6.71}$$

Because all x_k are distinct in the grid (6.67), a unique solution of the linear system (6.71) exists. Although the linear system (6.71) is generated by the ill-conditioned Vandermonde matrix (see Section 5.2), the practically important summation rules can be computed from this solution for small values of n.

```
>> format rat;
>> for n = 1 : 5
       x = linspace(0,1,n+1); h = 1/n;
       A = fliplr(vander(x))';
       b = 1./(1:n+1)';
       w = (A\b)'/h
   end

w =

     1/2          1/2

w =

     1/3          4/3          1/3

w =

     3/8          9/8          9/8          3/8

w =

    14/45        64/45        8/15        64/45       14/45

w =

    95/288     125/96      125/144     125/144     125/96       95/288
```

In addition to the trapezoidal and Simpson's rules (6.50) and (6.64), the next-order closed integration rules are called the *Simpson 3/8* and *Boole's* rules:

$$S_3 f[x_1, x_4] = \frac{3h}{8} \left(y_1 + 3y_2 + 3y_3 + y_4 \right), \tag{6.72}$$

$$S_4 f[x_1, x_5] = \frac{2h}{45} \left(7y_1 + 32y_2 + 12y_3 + 32y_4 + 7y_5 \right). \tag{6.73}$$

Composite Simpson's rule

Combining integration rules for $(n+1)$ data points with the composite integration rules on the fixed interval $[a, b]$, we can obtain composite Simpson's and other multipoint composite integration rules. You should realize, however, that if the data grid is uniform and if the $(n+1)$-point composite integration rule is used, the integration interval has to be divided into an n-multiple number of intervals. In this case, the piecewise polynomial interpolation is based on polynomial $p_n(x)$ between end points of the intervals that include $(n-1)$ interior equally spaced grid points.

When the number of subintervals is even, we apply the *composite Simpson rule*:

$$S_2 f[a,b] = \frac{h}{3}\left(y_1 + 4y_2 + 2y_3 + \ldots + 2y_{n-1} + 4y_n + y_{n+1}\right). \tag{6.74}$$

It follows from the error formula (6.66) that the composite Simpson rule (6.74) has the global truncation error bounded by

$$\left| \int_a^b f(x)dx - S_2 f[a,b] \right| \le \frac{M_4}{180} h^4 (b-a), \tag{6.75}$$

where $M_4 = \max_{x \in [a,b]} |f^{(4)}(x)|$. The next example shows numerical computations of the composite Simpson rule for the function $f(x) = \text{sech}^2 x$ on the interval $[0, 1]$, in comparison with the composite trapezoidal rule. The exact value of the integral is

$$\int_0^1 \text{sech}^2 x dx = \tanh 1.$$

The MATLAB script `composite_Simpson's_rule` contains the relevant computations.

```
% composite_Simpson's_rule
a = 1; IntExact = tanh(a);
k = 1; n1 = 10; n2 = 1000;
 for n = n1 : 2 : n2
    h = a/n; x = linspace(0,a,n+1); y = 1./(cosh(x).^2);
    IntTrap = h*(y(1)+2*sum(y(2:n))+y(n+1))/2;
    ErrTrap(k) = abs(IntTrap-IntExact); % composite trapezoidal rule
    IntSimp = h*(y(1)+4*sum(y(2:2:n))+2*sum(y(3:2:n-1))+y(n+1))/3;
    ErrSimp(k) = abs(IntSimp-IntExact); % composite Simpson's rule
    hst(k) = h; k = k+1;
 end
aTrap = polyfit(log(hst),log(ErrTrap),1); powerTrap = aTrap(1)
aSimp = polyfit(log(hst),log(ErrSimp),1); powerSimp = aSimp(1)
loglog(hst,ErrTrap,'g',hst,ErrSimp,'r');
```

When the MATLAB script `composite_Simpson's_rule` is executed, it shows that the global truncation error is $O(h^2)$ for the composite trapezoidal rule and is $O(h^4)$ for the composite Simpson rule. Figure 6.9 shows that the composite Simpson rule converges much faster than the composite trapezoidal rule does as $h \to 0$.

```
>> composite_Simpson's_rule

powerTrap =
    2.0000

powerSimp =
    3.9978
```

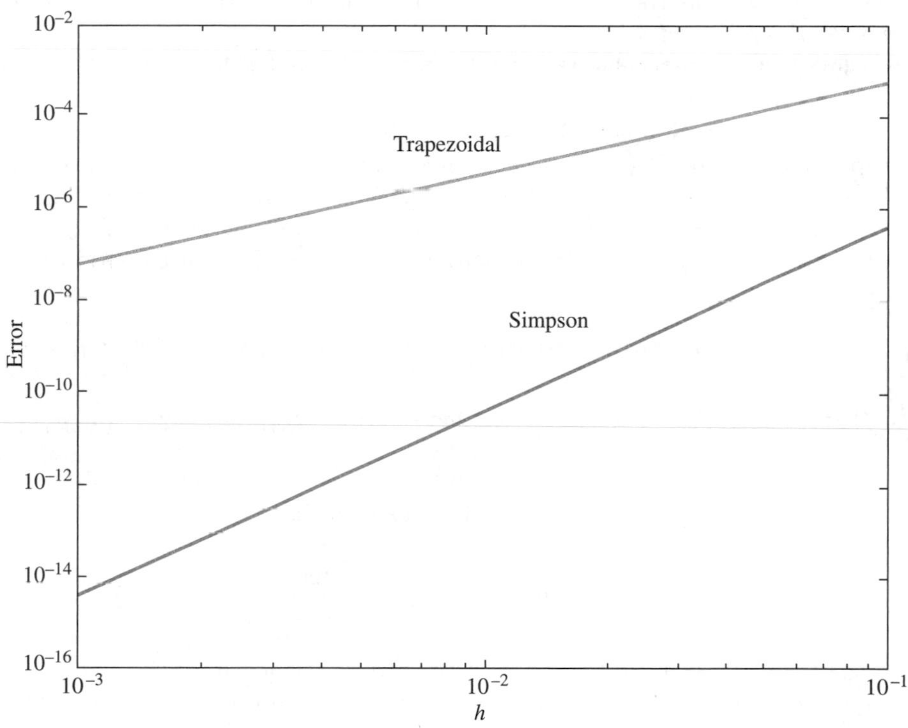

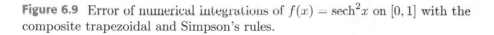

Figure 6.9 Error of numerical integrations of $f(x) = \text{sech}^2 x$ on $[0, 1]$ with the composite trapezoidal and Simpson's rules.

The multipoint Newton–Cotes integration rules are expressed by the weights $\{w_k\}_{k=1}^{n+1}$ in the summation formula (6.70), which are sign indefinite for $n \geq 10$. (You may check this claim by extending computations of the coefficients w_k for $n \geq 10$ in the previous MATLAB code.) As a result, numerical computations of the summation formula (6.70) become less efficient because of accumulating round-off errors. The Simpson rule and its composition remain the most important methods of numerical integration. The Simpson rule is used for many practical computations, including the Runge–Kutta initial-value ODE solvers (see Section 9.2).

6.7 Romberg Integration

The accuracy of numerical integration can be improved by using the Richardson extrapolation of Section 6.4. This extrapolation is applied to cancel the leading order of the truncation error in the elementary integration rules, such as composite trapezoidal and midpoint rules. In the context of numerical integration, the recursive algorithm based on Richardson extrapolation is called the *Romberg integration.*

First-order integration

For instance, consider the composite trapezoidal rule for the function $f(x)$ integrated on the interval $[a, b]$ with the uniform grid consisting of $(n+1)$ data points with spacing $h = (b-a)/n$. Let us denote the composite trapezoidal rule by $R_1(h) = Sf[a, b]$. According to the error formula (6.57), the global truncation error of the composite trapezoidal is $O(h^2)$ if $f(x)$ is twice differentiable on $[a, b]$. If we take n to be even, we can perform two computations of the composite trapezoidal rule with two approximations:

$$\int_a^b f(x)dx = R_1(h) + E(h), \quad R_1(h) = \frac{h}{2}\left(y_1 + 2y_2 + 2y_3 + \ldots + 2y_n + y_{n+1}\right),$$

$$\int_a^b f(x)dx = R_1(2h) + \tilde{E}(2h), \qquad R_1(2h) = h\left(y_1 + 2y_3 + \ldots + 2y_{n-1} + y_{n+1}\right).$$

Theorem 6.3 implies that there exists an α_2 such that

$$\alpha_2 = \lim_{h\to 0}\frac{E(h)}{h^2} = \lim_{h\to 0}\frac{\tilde{E}(2h)}{4h^2},$$

so that a modified integration rule $R_2(h)$ can be defined from the composite trapezoidal rule $R_1(h)$ by canceling the term $O(h^2)$ in the truncation error:

$$\begin{aligned} R_2(h) &= \frac{1}{3}\left[4R_1(h) - R_1(2h)\right] \\ &= \frac{h}{3}\left(y_1 + 4y_2 + 2y_3 + \ldots + 2y_{n-1} + 4y_n + y_{n+1}\right). \end{aligned}$$

The integration rules $R_1(h)$ and $R_2(h)$ are enumerated according to iterations of the Romberg integration algorithm. You can see that the second Romberg integration rule $R_2(h)$ recovers the composite Simpson rule $S_2 f[a,b]$, whose global truncation error is $O(h^4)$ according to the error formula (6.75). Therefore, we can continue the algorithm by canceling the $O(h^4)$ term in the truncation error of the composite Simpson rule $R_2(h) = S_2 f[a,b]$:

$$\begin{aligned} R_3(h) &= \frac{1}{15}\left(16R_2(h) - R_2(2h)\right) \\ &= \frac{2h}{45}\left(7y_1 + 32y_2 + 12y_3 + 32y_4 + 14y_5 + \dots \right. \\ &\qquad \left. + 14y_{n-3} + 32y_{n-2} + 12y_{n-1} + 32y_n + 7y_{n+1}\right). \end{aligned}$$

This formula assumes that n is a multiple of 4. You can see that the third Romberg integration rule $R_3(h)$ recovers the composite Boole rule $S_4 f[a,b]$. To proceed with further iterations, observe that the global truncation error of the Boole rule is $O(h^6)$ and, in general, the global truncation error of the Romberg integration $R_m(h)$ is $O(h^{2m})$. Although Theorem 6.4 can be used for the Boole rule ($n = 5$), it cannot be applied to the next Romberg integration rules because the data points in the uniform grid (6.67) are collected differently between the Romberg integration rule $R_m(h)$ for $m > 3$ and the multipoint Newton–Cotes integration rule. We use here a different result from *Numerical Analysis* (see [23] for the proof):

Theorem 6.5 *Let f be an infinitely smooth function on the interval $[a,b]$. Let $[a,b]$ be divided into n equally spaced subintervals with the step size $h = (b-a)/n$. The global truncation error of the composite trapezoidal rule (6.52) can be extended into the power series in even powers of h up to any order $N \geq 1$:*

$$\int_a^b f(x)dx - Sf[a,b] = \alpha_1 h^2 + \alpha_2 h^4 + \dots + \alpha_N h^{2N} + \dots, \qquad (6.76)$$

where α_k, $k = 1, 2, \dots, N, \dots$ are some constants, which are h-independent.

Recursive integrations

To develop a recursive algorithm of Romberg integrations, we introduce the uniform grid on $[a,b]$ with n subintervals, such that the number n is a multiple of 2^m, where m is the order of the Romberg integration rule. It follows from Theorem 6.5 that the exact integral is approximated by $R_m(h)$ with a truncation error that is $O(h^{2m})$:

$$\int_a^b f(x)dx = R_m(h) + \alpha_m h^{2m}, \qquad m \geq 1. \qquad (6.77)$$

By eliminating the truncation error with two computations of $R_m(h)$ and $R_m(2h)$, we obtain the recursive formula:

$$R_{m+1}(h) = \frac{4^m R_m(h) - R_m(2h)}{4^m - 1} = R_m(h) + \frac{R_m(h) - R_m(2h)}{4^m - 1}. \qquad (6.78)$$

The recursive Romberg integration rules can be organized in the table, similar to the table of Richardson extrapolations for numerical derivatives.

Step Size	Trapezoidal Rule	Simpson's Rule	Boole's Rule	4th Romberg Integration
h	$R_1(h)$	$R_2(h)$	$R_3(h)$	$R_4(h)$
$2h$	$R_1(2h)$	$R_2(2h)$	$R_3(2h)$	
$4h$	$R_1(3h)$	$R_2(3h)$		
$8h$	$R_1(4h)$			

The computations of the recursive Romberg integrations are coded in the MATLAB function `Romberg`:

```
function Rnum = Romberg(f,a,b,maxRomb)
% computes recursive Romberg integrations
% f - string for the name of function f(x)
% a,b - lower and upper limits of integration
% maxRomb - the number of recursive Romberg integrations
% Rnum - vector of lowest-error Romberg integrations

R = ones(maxRomb,maxRomb);
hmin = (b-a)/2^(maxRomb-1); % minimal step size
for k = 1 : maxRomb % trapezoidal rule for the first iteration
    h = 2^(k-1)*hmin;
    x = a : h : b;
    y = eval(f);
    lenY = length(y);
    R(k,1) = 0.5*h*(y(1) + 2*sum(y(2:lenY-1)) + y(lenY));
end
for k = 2 : maxRomb % Romberg integrations
    for kk = 1 : (maxRomb-k+1)
        R(kk,k) = R(kk,k-1)+(R(kk,k-1) - R(kk+1,k-1))/(4^(k-1)-1);
    end
end
Rnum = R(1,:);
```

This function is used to evaluate numerically the integral:

$$\int_0^{2\pi} e^{-x} \sin x dx = \frac{1 - e^{-2\pi}}{2}. \tag{6.79}$$

```
>> f = 'exp(-x).*sin(x)';
>> Rnum = Romberg(f,0,2*pi,20);
>> format long; R = Rnum(1:4)

R =
   0.49906627862218 0.49906627863411 0.49906627863410 0.49906627863410

>> IntExact = (1-exp(-2*pi))/2

IntExact =
   0.49906627863415

>> Error = abs(IntExact - Rnum(1:4))

Error =
  1.0e-010 *
   0.11968537272367   0.00036637359813   0.00041078251911   0.00042243986087
```

The algorithm converges quickly to the exact value up to round-off error, which is invariant in all higher-order numerical integration rules and is defined by the machine precision ε multiplied by the integration interval. Because no further increase in accuracy of numerical integration can be obtained after few iterations, the algorithm can be terminated when the relative error falls below an error tolerance. In the previous example, the error of the Romberg integrations converges to the value of the order of 10^{-14}, which is at the level of round-off errors.

The Romberg integration algorithm allows you to obtain accurate numerical approximations of the integral in a few iterations. No explicit computations of weighted summation formulas are required except for the computations of the composite trapezoidal rules in the first column of the table of Romberg integrations. The algorithm converges quickly if the function $f(x)$ can be differentiated infinitely many times on the integration interval. The convergence becomes slower if singularities may occur in the function or its derivatives on the integration interval. A combination of composite trapezoidal and midpoint rules can be useful to avoid integrations of functions with singularities.

Exercise 6.13 Compute the table of Romberg integration rules starting with the midpoint rule for the function $f(x) = \frac{1}{\sqrt{1-x^2}}$ on the interval $[-1, 1]$.

A more practical form of the Romberg integration algorithm can be achieved if the number of points at each new iteration is doubled dynamically compared to the previous iteration. As a result, the step size h becomes smaller and smaller until it reaches the level of maximal precision or until the algorithm is terminated if the required accuracy of the numerical integral is reached. In this case, the value of $R_m(h)$ stands for the numerical integral computed at the new (smaller) step size, while the value of $R_m(2h)$ stands for the numerical integral computed at the previous (larger) step size. See [25] for a discussion of this algorithm.

Exercise 6.14 Write a MATLAB script for the Romberg integration algorithm that reduces the step size h in obtaining Romberg integration rules of higher accuracy. Compute numerically the integral of $f(x) = e^{-x} \sin x$ on the interval $[0, 2\pi]$.

Other methods of numerical integration rely on adaptive integration, either by estimating the errors of uniform integration or by using nodal points from zeros of orthogonal polynomials. The first method, called the *adaptive integration*, is described in Section 9.3 in the context of initial-value ODE solvers. The second method, called *Gaussian quadratures*, is described in Section 6.8. The main idea of these methods is that if the interpolation points are not equally spaced, better accuracy in the numerical approximation of the integral $\int_a^b f(x)dx$ can be achieved with the same number of evaluations of the integrand $f(x)$ compared to the same integration rule with the equally spaced interpolation points.

6.8 Gaussian Quadrature Rules

We consider again the solution to Problem 6.2 by using the interpolating polynomial $p_n(x)$ through $(n+1)$ data points and working with the integral (6.62). However, we do not assume now the equally spaced distribution of grid points $\{x_k\}_{k=1}^{n+1}$ that results in the Newton–Cotes summation rules (6.70). Instead, let us reformulate Problem 6.2 in a different form.

Problem 6.3 (Gaussian Open Quadrature) Let the numerical integral of $f(x)$ on $[a, b]$ be expressed by the finite sum

$$S_n f[a,b] = \frac{b-a}{n+1} \sum_{k=1}^{n+1} w_k f(x_k) \tag{6.80}$$

on the grid points $\{x_k\}_{k=1}^{n+1}$, such that $a < x_1 < x_2 < \ldots < x_n < x_{n+1} < b$. Find the distribution of grid points so that the summation formula (6.80) becomes exact for any polynomial of degree d with $d \geq n$.

It is obvious that Problem 6.3 with $d = n$ has a solution on any set of grid points because the truncation error of the polynomial interpolant $p_n(x)$ is identically zero if $f(x)$ is a polynomial of degree $d = n$. It is also clear that Problem 6.3 with $d \geq 2n + 2$ has no solution. Indeed, by Theorem 5.6, the error of the polynomial interpolation is

$$f(x) - p_n(x) = \frac{1}{(n+1)!} f^{(n+1)}(\xi)\omega_{n+1}(x), \qquad (6.81)$$

where

$$\omega_{n+1}(x) = (x - x_1)(x - x_2)\dots(x - x_{n+1}).$$

The numerical integral $Sf[a, b]$ is *exact* if and only if

$$\forall f(x) \in \mathcal{P} \qquad \int_a^b f^{(n+1)}(x)\omega_{n+1}(x)\,dx = 0. \qquad (6.82)$$

If $f(x)$ is a polynomial of degree $d \geq 2n + 2$, then $f^{(n+1)}(x)$ is a polynomial of degree $d - n - 1 \geq n + 1$. Because $\omega_{n+1}(x)$ is a polynomial of degree $n + 1$, the integral is definitely nonzero for at least $f^{(n+1)}(x) = \omega_{n+1}(x)$. Therefore, the numerical integral $S_n f[a, b]$ is not exact for the corresponding polynomial $f(x)$.

We show that Problem 6.3 has a solution for $d = 2n + 1$, and the corresponding summation formula (6.80) is optimal in the sense that it gives the exact integral $\int_a^b f(x)dx$ for the *largest* class of polynomial functions. The corresponding numerical integral $S_n f[a, b]$ is called the *Gaussian open quadrature rule.*

Exercise 6.15 Show that the Gaussian open quadrature of Problem 6.3 with $n = 0$ is equivalent to the midpoint rule at $x_1 = (a + b)/2$.

Two-point Gaussian quadrature

Before developing a general solution for Problem 6.3 with $d = 2n+1$, let us give an explicit example for $n = 1$ on $[-1, 1]$, that is, for two data points x_1, x_2, such that $-1 < x_1 < x_2 < 1$. We require that the summation formulas be exact for a polynomial of degree $d = 3$, that is, for the set of four power functions $\{1, x, x^2, x^3\}$. By using the summation formula (6.80), we set the system of four equations for two weights w_1, w_2 and two grid points x_1, x_2:

$$\int_{-1}^{1} dx = 2 = w_1 + w_2, \qquad \int_{-1}^{1} x dx = 0 = w_1 x_1 + w_2 x_2,$$

$$\int_{-1}^{1} x^2 dx = \frac{2}{3} = w_1 x_1^2 + w_2 x_2^2, \qquad \int_{-1}^{1} x^3 dx = 0 = w_1 x_1^3 + w_2 x_2^3.$$

It follows from the second and fourth equations that $x_1^2 = x_2^2$ for $w_1, w_2 \neq 0$, and therefore, $x_1 = -x_2$ and $w_1 = w_2$. Then, the first and third equations show that $x_2 = \frac{1}{\sqrt{3}} = -x_1$, after which the solution for w_1, w_2 is $w_1 = w_2 = 1$. Thus, the Gaussian quadrature with $n = 1$ and $d = 3$ on $[-1, 1]$ is the summation rule

$$Sf[-1,1] = f\left(-\frac{1}{\sqrt{3}}\right) + f\left(\frac{1}{\sqrt{3}}\right).$$

Using a linear transformation

$$x = \frac{a+b}{2} + z\frac{b-a}{2}, \qquad -1 < z < 1, \tag{6.83}$$

we can obtain the equivalent Gaussian quadrature on the interval $[a, b]$:

$$Sf[a,b] = \frac{b-a}{2}\left[f\left(\frac{a+b}{2} - \frac{b-a}{2\sqrt{3}}\right) + f\left(\frac{a+b}{2} + \frac{b-a}{2\sqrt{3}}\right)\right], \tag{6.84}$$

where the prefactor $\frac{b-a}{2}$ is caused by the transformation of the integral variable

$$\int_a^b f(x)dx = \frac{b-a}{2}\int_{-1}^{1} f(z)dz.$$

Furthermore, if we split the interval $[a, b]$ into n elementary subintervals with the equal step size $h = (b-a)/n$, we can apply the composite Gaussian quadrature rule in the following form:

$$\begin{aligned} Sf[a,b] = &\frac{h}{2}\sum_{k=1}^{n} f\left(a + h(k-1) + \frac{h}{2}\left(1 - \frac{1}{\sqrt{3}}\right)\right) \\ &+\frac{h}{2}\sum_{k=1}^{n} f\left(a + h(k-1) + \frac{h}{2}\left(1 + \frac{1}{\sqrt{3}}\right)\right). \end{aligned} \tag{6.85}$$

The MATLAB script `Gaussian_quadrature` illustrates the composite Gaussian quadrature rule (6.85) for the function $f(x) = \sqrt{1-x^2}$ on $[0, a]$ with $a = 0.5$ and $a = 1$.

```
% Gaussian_quadrature
for j = 1 : 2     % two computations
    if (j == 1)
        a = 0.5; n1 = 5; n2 = 500;
    else
        a = 1; n1 = 10; n2 = 1000;
    end
    IntExact = 0.5*(asin(a)+a*sqrt(1-a*a)); k = 1;
    for n = n1 : 1 : n2   % two-point Gaussian quadrature
        h = a/n;
        x1 = linspace(0,a,n+1)+h/2*(1-1/sqrt(3));
```

```
        x2 = linspace(0,a,n+1)+h/2*(1+1/sqrt(3));
        y1 = sqrt(1-x1.*x1); y2 = sqrt(1-x2.*x2);
        IntGauss = h*(sum(y1(1:n))+sum(y2(1:n)))/2;
        ErrGauss(k) = abs(IntGauss-IntExact);
        hst(k) = h; k = k+1;
    end
    aGauss = polyfit(log(hst),log(ErrGauss),1);
    powerGauss = aGauss(1)
end
```

The outcome of the MATLAB script `Gaussian_quadrature` shows that the global truncation error is $O(h^4)$ on $[0, 0.5]$ when the function $f(x) = \sqrt{1-x^2}$ is infinitely smooth. The convergence is only $O(h^{3/2})$ on $[0, 1]$ because the derivative of $f(x)$ is unbounded as $x \to 1$. Comparison with the composite trapezoidal and midpoint rules in Section 6.5 shows that the global error of the Gaussian open quadrature rule is two orders smaller on $[0, 0.5]$ and is of the same order on $[0, 1]$.

```
>> Gaussian_quadrature

powerGauss =
    4.0150

powerGauss =
    1.5004
```

Exercise 6.16 Compute the grid points x_1, x_2, x_3 and weights w_1, w_2, w_3 from the system of equations that solves Problem 6.3 with $n = 2$ on $[-1, 1]$ and obtain that

$$x_1 = -\frac{\sqrt{3}}{\sqrt{5}}, \; x_2 = 0, \; x_3 = \frac{\sqrt{3}}{\sqrt{5}}, \qquad w_1 = \frac{5}{6}, \; w_2 = \frac{4}{3}, \; w_3 = \frac{5}{6}.$$

Apply the three-point Gaussian quadrature to the integral of the function $f(x) = \sqrt{1-x^2}$ on $[0, a]$ with $a = 0.5$ and $a = 1$ and show that the global error is $O(h^6)$ for $a = 0.5$ and $O(h^{3/2})$ for $a = 1$.

n-point Gaussian quadratures

We shall now construct a general solution to Problem 6.3 with $d = 2n+1$ and $n \geq 1$ on the interval $[-1, 1]$. A more general solution on the interval $[a, b]$ is immediately given by the transformation (6.83). It is clear intuitively that the general solution does indeed exist for $d = 2n + 1$ because the system of $(2n + 2)$ equations from the formulation of Problem 6.3 is set for $(n + 1)$ weights $\{w_k\}_{k=1}^{n+1}$ and for $(n+1)$ grid points $\{x_k\}_{k=1}^{n+1}$. However, this system is nonlinear and it does not necessarily have a solution. We shall use orthogonal

Legendre polynomials $P_n(x)$ from Section 5.7 to identify a better solution to Problem 6.3 with $d = 2n + 1$.

Because $\omega_{n+1}(x)$ in the error formula (6.81) must be orthogonal to a general polynomial $f^{(n+1)}(x)$ of degree n according to the condition (6.82), the polynomial $\omega_{n+1}(x)$ of degree $n+1$ must be a member of a sequence of orthogonal polynomials $\{\pi_k(x)\}_{k\in\mathbb{N}}$, where $\pi_k(x)$ is a polynomial of degree k and

$$\langle \pi_k, \pi_n \rangle = \int_{-1}^{1} \pi_k(x)\pi_n(x)dx = 0 \qquad \forall k \neq n.$$

The set of orthogonal polynomials $\{\pi_k(x)\}_{k\in\mathbb{N}}$ is proportional to the set of Legendre polynomials $\{P_k(x)\}_{k\in\mathbb{N}}$, which is introduced in Section 5.7, that is,

$$P_0 = 1, \quad P_1 = x, \quad P_2 = \frac{1}{2}(3x^2 - 1), \quad P_3 = \frac{1}{2}x(5x^2 - 3),$$

and general formula given by

$$P_n(x) = \frac{1}{n!2^n}\frac{d^n}{dx^n}(x^2 - 1)^n.$$

Therefore, the set of discrete grid points $\{x_k\}_{k=1}^{n+1}$ for the Gaussian quadrature rules on $[-1, 1]$ is defined by the roots of the Legendre polynomial $P_n(x)$.

We will show that roots of $P_n(x)$ for any $n \in \mathbb{N}$ are all simple and located inside the interval $[-1, 1]$. Indeed, if $P_n(x)$ has a multiple root $x = x_0$, then $\phi(x) = P_n(x)/(x - x_0)^2$ is a polynomial of degree $n - 2$, which is thus orthogonal to the polynomial $P_n(x)$. However, the orthogonality of $\phi(x)$ and $P_n(x)$ is impossible because

$$\int_{-1}^{1} \frac{P_n^2(x)dx}{(x - x_0)^2} > 0.$$

Therefore, all roots of $P_n(x)$ are simple. If $P_n(x)$ has only m simple roots $\{x_k\}_{k=1}^{m}$ inside $[-1, 1]$ for $m < n$, then $P_n(x)$ is orthogonal to the product $\psi(x) = (x - x_1)(x - x_2)\dots(x - x_m)$ and the function $P_n(x)/\psi(x)$ is either positive or negative definite on $[-1, 1]$. Then, the orthogonality between $P_n(x)$ and $\psi(x)$ is impossible because the integrand

$$\int_{-1}^{1} P_n^2(x)\frac{\psi(x)}{P_n(x)}dx$$

is also either positive or negative definite on $[-1, 1]$. Moreover, the roots can not occur on the end points of the interval $[-1, 1]$ because $P_n(1) = 1$ and $P_n(-1) = (-1)^n$ as a result of the normalization and symmetry of Legendre polynomials. Therefore, all roots of $P_n(x)$ belong to the open interval $(-1, 1)$. For example,

$$n = 1: \ \{0\}, \quad n = 2: \left\{-\frac{1}{\sqrt{3}}, \frac{1}{\sqrt{3}}\right\}, \quad n = 3: \left\{-\frac{\sqrt{3}}{\sqrt{5}}, 0, \frac{\sqrt{3}}{\sqrt{5}}\right\}.$$

The roots of $P_n(x)$ for $n = 1, 2, 3$ are of course the same as those obtained previously. Zeros of orthogonal polynomials can be computed from eigenvalues of the Jacobi matrix related to the three-term recurrence relation [24]. The numerical values of these zeros for small k can be computed from the MATLAB function `root` as in the following example:

```
>> for k = 1 : 5
        c = poly([ones(1,k),-ones(1,k)]);
        for m = 1 : k
            c = polyder(c);
        end
        x = roots(c)'
    end

x =
     0

x =
    0.5774   -0.5774

x =
         0    0.7746   -0.7746

x =
    0.8611   -0.8611    0.3400   -0.3400

x =
         0    0.9062    0.5385   -0.9062   -0.5385
```

The first three rows of roots are truncations of the irrational numbers for the roots of $P_n(x)$ with $n = 1, 2, 3$.

When the grid points $\{x_k\}_{k=1}^{n+1}$ are found from roots of the Legendre polynomials, the weights $\{w_k\}_{k=1}^{n+1}$ can be found from the linear system of the first $n + 1$ equations for exact integrations of the power functions $\{1, x, \ldots, x^n\}$ on $[-1, 1]$. This algorithm is similar to the computations of the weights in the Newton–Cotes integration rules in Section 6.6. Continuing the previous example, you can compute the weights $\{w_k\}_{k=1}^{n+1}$ of the Gaussian quadrature rules on $[-1, 1]$ with $1 \leq n \leq 5$:

```
>> format rat;
>> for n = 1 : 5
       A = fliplr(vander(x))';
       b = n./(1:n)'; b(2:2:n) = 0;
       w = (A\b)'
   end
```

```
w =
     1

w =
     1           1

w =
    4/3         5/6         5/6

w =
  1054/1515   1054/1515    973/746     973/746

w =
   64/45       478/807    1047/875     478/807     1047/875
```

The first three rows coincide with the coefficients $\{w_k\}_{k=1}^{n+1}$ for $n = 0, 1, 2$. It is obvious that the weights $\{w_k\}_{k=1}^{n+1}$ can be obtained in the implicit form if the set of Lagrange interpolating polynomials $\{L_{k,n}(x)\}_{k=1}^{n+1}$ from Section 5.3 is used, such that the function $f(x)$ is interpolated by the polynomial $p_n(x) = \sum_{k=1}^{n+1} f(x_k)L_{k,n}(x)$. By integrating the polynomial interpolant, we immediately obtain that

$$w_k = \frac{(n+1)}{2}\int_{-1}^{1} L_{k,n}(x)dx, \qquad k = 1, 2, \dots, n+1. \tag{6.86}$$

Of course, this formula is computationally inefficient because it is difficult to evaluate the integrals analytically for any k and n. You can just check for $n = 0$ and $x_1 = 0$ that $w_1 = 1$ and for $n = 1$ and $x_2 = -x_1 = \frac{1}{\sqrt{3}}$ that

$$w_{1,2} = \pm\frac{\sqrt{3}}{2}\int_{-1}^{1}\left(x \pm \frac{1}{\sqrt{3}}\right)dx = \pm\frac{\sqrt{3}}{4}\left(\left(1 \pm \frac{1}{\sqrt{3}}\right)^2 - \left(1 \mp \frac{1}{\sqrt{3}}\right)^2\right) = 1.$$

Exercise 6.17 Compute the weights $w_1 = w_3 = \frac{5}{6}$ and $w_2 = \frac{4}{3}$ from the exact integration formula (6.86) with $n = 2$.

Error of Gaussian quadratures

The error of the Gaussian integration rule follows from the Hermite interpolation polynomials (see Section 12.2). The main result is given by the following theorem (see [7, 24] for the proofs):

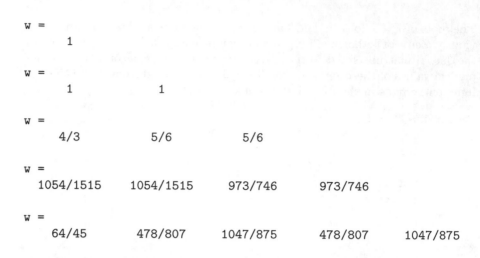

Theorem 6.6 *Let $f(x)$ be a $(2n+2)$-times continuously differentiable function on $[-1, 1]$. Let $S_n f[-1, 1]$ be the Gaussian quadrature of the function*

$f(x)$ with the nodes $\{x_k\}_{k=1}^{n+1}$ from zeros of the Legendre polynomial $P_n(x)$ and weights $\{w_k\}_{k=1}^{n+1}$ from the Lagrange polynomials (6.86). Then,

$$\left| \int_{-1}^{1} f(x)dx - S_n f[-1,1] \right| \leq \frac{2^{2n+3}[(n+1)!]^4}{(2n+3)![(2n+2)!]^2} M_{2n+2}, \tag{6.87}$$

where $M_{2n+2} = \max_{x \in [-1,1]} |f^{(2n+2)}(x)|$.

The upper bound (6.87) becomes an exact equality in the case if $f(x)$ is a polynomial of degree $d = 2n+2$. The MATLAB script `error_Gaussian_quadrature` gives a numerical proof of the error formula (6.87) for $f(x) = x^{2n+2}$ and $n = 0, 1, \ldots, 5$.

```
% error_Gaussian_quadrature
for k = 1 : 5
    c = poly([ones(1,k),-ones(1,k)]);
    for m = 1 : k
        c = polyder(c);
    end
    x = roots(c)'; % zeros of Legendre polynomials
    A = fliplr(vander(x))';
    b = k./(1:k)'; b(2:2:k) = 0;
    w = (A\b)';     % weights of the Gaussian quadrature
    C = 2^(2*k+1)*factorial(k)^4/(factorial(2*k+1)*factorial(2*k));
    E = (2/(2*k+1)-2*w*(x'.^(2*k))/k)/C
end
```

The output shows that $E = 1$ for $n = 0, 1, \ldots, 5$, which indicates that the error formula (6.87) is an exact equality for $f(x) = x^{2n+2}$.

```
>> error_Gaussian_quadrature

E =

     1


E =

    1.0000


E =

    1.0000


E =

    1.0000
```

```
E =
    1.0000
```

Similar to the previous case, the transformation (6.83) allows us to extend all Gaussian quadrature rules from $[-1, 1]$ to $[a, b]$. The numerical integrals of higher accuracy for fixed values of n are obtained by dividing the interval $[a, b]$ into equally spaced intervals. The composite Gaussian quadrature rules are employed in the limit when the step size h becomes small. It follows from Theorem 6.6 that the global truncation error of the composite Gaussian quadrature rule converges as $h \to 0$ according to the law

$$\left| \int_a^b f(x)dx - S_n f[a, b] \right| \leq C_n M_{2n+2} h^{2n+2}, \tag{6.88}$$

where M_{2n+2} are the same as earlier, C_n are some constants, and h is the step size on $[a, b]$. The following MATLAB script `error_composite_Gauss_quadrature` illustrates convergence of the composite Gaussian quadrature rule with $n = 2$ for the integral $\int_0^1 \operatorname{sech}^2 x dx = \tanh 1$. The weights and grid points are taken from Exercise 6.16.

```
% error_composite_Gauss_quadrature
a = 1; IntExact = tanh(a); k = 1; n1 = 10; n2 = 20;
for n = n1 : 2 : n2 % three-point Gauss quadrature
      h = a/n;
      x1 = linspace(0,a,n+1)+h/2*(1-sqrt(3)/sqrt(5));
      x2 = linspace(0,a,n+1)+h/2;
      x3 = linspace(0,a,n+1)+h/2*(1+sqrt(3)/sqrt(5));
      y1 = 1./(cosh(x1).^2);
      y2 = 1./(cosh(x2).^2);
      y3 = 1./(cosh(x3).^2);
      IntGauss = h*(5*sum(y1(1:n))+8*sum(y2(1:n))+5*sum(y3(1:n)))/18;
      ErrGauss(k) = abs(IntGauss-IntExact);
      hst(k) = h; k = k+1;
   end
aGauss = polyfit(log(hst),log(ErrGauss),1);
powerGauss = aGauss(1)
```

The output of the script shows that the error of the composite Gaussian quadrature rule with $n = 2$ is $O(h^6)$, where h is the step size for the equally spaced partition on $[a, b]$. This result can be compared with the $O(h^4)$ error for the composite Gaussian quadrature with $n = 1$.

```
>> error_composite_Gauss_quadrature

powerGauss =
    6.0083
```

Gaussian quadrature rules can be associated with more general integrals of the function $f(x)$ that involve any prescribed weights $w(x)$ such that $\int_a^b w(x)f(x)dx$. The function $w(x)$ may have some weak singularities at the end points of the integration interval, for example, for $w(x) = (1 - x^2)^{1/2}$ on $[-1, 1]$. More general orthogonal polynomials with weighted inner products are required in the construction of the Gaussian quadrature rule. General summation formulas exist in specialized literature on numerical integration. Also, the Gaussian quadrature rules can be extended for closed numerical integration.

6.9 Summary and Notes

In this chapter, we construct numerical approximations for the first-order and higher-order derivatives of continuously differentiable functions and for integrals of continuous functions:

- **Section 6.1:** Three main methods for numerical derivatives, namely, the forward, backward, and central differences, solve the problem of numerical differentiation (Problem 6.1). Truncation errors of the numerical approximations are defined by the Taylor series, and their convergence rates are determined by powers of the step size of the discrete grid. We show that the problem of numerical differentiation is ill-posed in the sense that the round-off error starts to increase when the step size becomes too small.

- **Section 6.2:** Higher-order numerical derivatives are defined from higher-order derivatives of the interpolating polynomials. Hierarchies of forward-difference, backward-difference, and central-difference numerical approximations for higher-order derivatives are constructed using repeated matrix multiplications. Their truncation errors are classified from the truncation errors of the polynomial interpolation (Theorem 6.1).

- **Section 6.3:** It follows from the convergence rate of the truncation error of numerical derivatives (Theorem 6.6.2) that the accuracy of numerical derivatives can be increased if the first-order derivative is approximated by the derivative of the interpolating polynomial of a higher degree. A hierarchy of multipoint first-order numerical derivatives is constructed using a numerical algorithm based on computations of derivatives of power functions.

- **Section 6.4:** The numerical algorithm of Richardson extrapolations computes multipoint numerical derivatives by recursive cancellation of their truncation errors. This algorithm is well-posed numerically, but the multipoint numerical derivatives differ from those obtained directly in Section 6.3.

- **Section 6.5:** Two main approximations of the numerical integrals, namely, the rectangular and trapezoidal rules, solve the problem of numerical integration (Problem 6.2). Compositions of these integration rules can be applied for continuous functions and functions with point singularities when the integration interval is divided into a number of elementary subintervals. Truncation errors of the composite integration rules are categorized from the Taylor series (Theorem 6.3), while the round-off error is uniformly bounded by the machine precision multiplied by the length of the integration interval. This makes the numerical integration a well-posed numerical procedure.

- **Section 6.6:** Using polynomial interpolations of higher degree, higher-order integration rules can be obtained, such as Simpson's and Boole's rules as well as their compositions. Using the analysis of the error of the numerical integration (Theorem 6.4, a hierarchy of Newton–Cotes closed integration rules is constructed using a numerical algorithm based on computations of integrals of power functions.

- **Section 6.7:** Using Richardson extrapolations and the global truncation error of the trapezoidal rule (Theorem 6.5), higher-order numerical integrals can be computed by a recursive algorithm of Romberg integrations.

- **Section 6.8:** Using polynomial interpolations on grid points from zeros of orthogonal Legendre polynomials, a hierarchy of Gaussian open quadrature rules that define numerical integrals of higher accuracy can be built (Problem 6.3). The accuracy of the Gaussian quadrature is optimal in the sense that polynomials of the largest possible degree can be integrated exactly by using summation formulas. The errors of the Gaussian open quadrature rules are categorized in Theorem 6.6.

Definitions, properties, and computational techniques for derivatives and integrals are studied in *Calculus* (the suggested text is [26]). Algorithms and truncation errors of the numerical derivatives and integrals are studied in *Numerical Analysis* (the suggested text is [7]). Advanced integration methods that involve Gaussian quadratures and adaptive integration rules are covered in advanced courses of *Numerical Analysis* (the suggested text is [24]).

6.10 Exercises

1. Use Taylor's theorem and find coefficients α, β, γ, and δ such that

$$f'(x_0) = \alpha f(x_0) + \beta f(x_0 + h) + \gamma f(x_0 + 2h) + O(h^2)$$
$$f'(x_0) = \alpha f(x_0) + \beta f(x_0 + h) + \gamma f(x_0 + 2h) + \delta f(x_0 + 3h) + O(h^3)$$

for an infinitely smooth $f(x)$ near $x = x_0$.

2. Approximate the derivative of $f(x) = \log(1+x^2)$ at $x = 1$ numerically by using forward, backward, and central differences. Compute the distance between the numerical approximations and exact derivatives and plot the computational error versus the step size. Confirm that the error of the forward and backward differences scales as $O(h)$ and the error of the central differences scales as $O(h^2)$ for small values of h.

3. Compute the set of data points from the function $f(x) = \exp(-x^2)$ at the equally spaced grid on $x \in [0, 2]$ with $h = 0.5$. Find closed finite difference approximations of the first-order and second-order numerical derivatives for all points of the grid with the errors $O(h)$ and $O(h^2)$.

4. Compute the set of data points from the function $f(x) = \sin(x^3)$ on $[0, 1]$. Find the interpolating polynomial between the equally spaced grid points with step size h, and compute first-order derivatives from the derivative of the interpolating polynomials. Plot the computational error of the derivative at $x = 1$ as a function of h.

5. Consider the function $f(x) = x\sin(1/x)$, which is continuous at $x = 0$ but has no bounded derivative as $x \to 0$. Compute the central difference approximation for the first derivative and show that the central differences diverge in the limit $h \to 0$.

6. Consider the function $f(x) = x(1 - x^2 + 2x^4 - x^6)$ and approximate the derivative at $x = 1$ with a hierarchy of central difference approximations of different orders. At what order does the central difference approximation recover the exact answer? Repeat the same example with the Richardson extrapolation algorithm and check if the algorithm is truncated at a finite number of iterations.

7. Apply the Richardson extrapolation algorithm to evaluate the second derivative of $f(x) = \text{sech}^2(x)$ at $x = 1$. Terminate iterations when the total distance between two successive approximations becomes smaller than 10^{-8}. Plot the absolute error versus the number of iterations.

8. Consider the integrals

$$\int_0^{\pi/2} \cos x dx, \qquad \int_1^2 \log x \, dx.$$

Estimate the number of subintervals n such that the global error of the composite trapezoidal rule is smaller than 10^{-8}. Apply the composite trapezoidal rule and compute its global error. Confirm the correctness of the estimate.

9. Use the Simpson rule and obtain the exact value of the integral

$$\int_0^1 x(1-x^2)dx$$

using only three equally spaced points on $[0,1]$.

10. Use the composite trapezoidal, Simpson's and Boole's rules to approximate numerically the integral

$$\int_0^1 \frac{dx}{1+x^2}.$$

Plot the total error E versus the step size h, and fit the dependence with the least-square power fit $E = Ch^p$. Confirm that the error of the composite trapezoidal rule has $p = 2$, the error of the composite Simpson rule has $p = 4$, and the error of the composite Boole rule has $p = 6$.

11. Use the composite trapezoidal and midpoint rules to approximate numerically the integrals

$$\int_0^1 x^{3/2} dx, \qquad \int_0^1 x^{5/2} dx.$$

Plot the global error versus the step size and find the least-square power fits $E = Ch^p$. Explain why $p \neq 2$ for some computations. Repeat the exercise for the Simpson rule and explain why $p \neq 4$ for all computations.

12. Use the MATLAB function `quad` and compute the antiderivative $g(x)$ of the function $f(x)$ at the equally spaced grid points on $[0,1]$ with the integration constant such that $g(0) = 0$. Evaluate the antiderivative $g(x)$ analytically and numerically and plot the computational error for the function $f(x) = \cos(10\pi x)$.

13. By using a hierarchy of Newton–Cotes closed integration rules from the trapezoidal rule to the 10th-order integration rule, evaluate the following integral numerically and analytically and compute the corresponding error:

$$\int_0^1 \frac{dx}{1+x}.$$

14. Consider the improper integral on the semi-infinite line

$$\int_0^\infty \frac{dx}{\cosh^6 x} = \frac{8}{15}.$$

Use the composite midpoint rule on the uniform grid $x_k = h(k-1)$ for $k = 1, 2, \ldots, n+1$, where $L = hn$ is large. Plot the global truncation error E versus h for a given $L = 10$ and show that the error does not vanish in the limit of small h. Confirm that the remaining error depends on L and that it becomes small when L gets larger.

15. Consider the improper integral on the infinite line

$$\int_{-\infty}^\infty \frac{\sinh^2 x}{\cosh^4 x} dx = \frac{2}{3}.$$

Rewrite the integral in variable $z = \tanh x$ on the finite interval $[-1, 1]$. Use the composite trapezoidal rule on an equally spaced grid and plot the global truncation error E versus the step size h. Show that the error converges to a numerical zero in the limit of small h.

16. The length of the curve $y = g(x)$ on $[a, b]$ is defined by the integral

$$L[g] = \int_a^b \sqrt{1 + (g'(x))^2} dx.$$

Use the trapezoidal rule and approximate the length of the curve $g(x) = \sin(x)$ on $[0, \pi/2]$ with $h = \pi/m$, $m \geq 2$. Plot the total error versus m.

17. Write a MATLAB function that computes a vector of weights in the composite Boole rule based on the formula (6.73) from the number of breaking points. Add a condition that checks that the number of breaking points agrees with the requirements of the composite Boole rule.

18. Apply the Romberg integration algorithm to evaluate the integral

$$\int_0^\pi \sin^3 x dx.$$

Terminate iterations when the total distance between two successive approximations becomes smaller than 10^{-12}. Plot the absolute error versus the number of iterations.

19. Apply the trapezoidal rule to compute the Bessel function $J_0(x)$ defined by the integral

$$J_0(x) = \frac{1}{\pi} \int_0^\pi \cos(x \sin t) dt.$$

Plot $J_0(x)$ on $[0, 10]$ for $h = 0.1$ and $h = 0.01$.

20. Consider the gamma function defined as the improper integral on the semi-infinite line

$$\Gamma(x) = \int_0^\infty e^{-t} t^{x-1} dt.$$

Confirm the values $\Gamma(n+1) = n!$ for positive integer n by approximating the integral numerically at $x = n$ for $n = 1, 2, 3, 4, 5$ on $[0, L]$ in the limit $h \to 0$ and $L \to \infty$.

21. Consider the integral

$$\int_0^1 x^{20} dx = \frac{1}{21}$$

and apply the hierarchy of Gaussian quadrature rules with $a = 0$, $b = 1$, and n ranging from $n = 2$ to $n = 10$. Show that the error is nonzero for $n < 10$ but it vanishes for $n = 10$.

22. Use the Gaussian quadratures with $n = 3, 4, 5$ and approximate the integral

$$\log(x) = \int_1^x \frac{dt}{t}$$

on equally spaced discrete grids on $[1, 3]$. Plot the numerical results and the function $\log(x)$ on $[1, 3]$. Find the best power fit for the convergence rates $O(h^p)$ of the composite Gaussian quadrature rules with $n = 3, 4, 5$ as $h \to 0$, where h is the step size on the interval $[1, 3]$.

CHAPTER 7

Vector Calculus

THE CONCEPTS AND TECHNIQUES related to differentiation and integration introduced in Chapter 6 applied to functions of the form $f : \mathbb{R} \to \mathbb{R}$. In this chapter, we generalize them to functions from $\mathbb{R}^n$ to $\mathbb{R}^m$, where n and m are any integers. Because $\mathbb{R}^n$ and $\mathbb{R}^m$ are prototype examples of vector spaces, such functions provide an explicit link between *Calculus* and *Linear Algebra*, which is often neglected in books that treat the subjects separately. One reason for neglecting this connection with *Linear Algebra* is the perception that the several special cases depending on specific values for n and m give rise to seemingly unrelated concepts. For instance, scalar functions of several variables of the form $f : \mathbb{R}^n \to \mathbb{R}$ appear to be very different from a vector–valued function of one variable of the form $\mathbf{s} : \mathbb{R} \to \mathbb{R}^m$. In what follows, we acknowledge these differences by presenting the relevant concepts to each special case in different sections, while emphasizing the role of some unifying ideas, such as the concept of a derivative as a linear map. Along the way, we make full use of MATLAB's graphical capabilities as tools for understanding and handling the new concepts.

The first three sections refer to scalar functions of several variables. Section 7.1 sets up the basic concepts of domains, graphs, level curves, and level surfaces, all properly illustrated by MATLAB examples. Section 7.2 provides a first encounter with differentiation in more than one dimension through the idea of partial derivatives. In Section 7.3, the differentiability of a function at a point is characterized through the existence of a linear approximation to the function near the point. Such approximation is therefore a linear map from $\mathbb{R}^n$ to $\mathbb{R}$, which according to Chapter 2 corresponds to an $n \times 1$ row vector, called the gradient vector.

Section 7.4 turns the discussion around to functions of the form $\mathbf{s} : \mathbb{R} \to \mathbb{R}^m$, whose images trace curves in $\mathbb{R}^m$ as the single variable varies in $\mathbb{R}$. For historical reasons stemming from applications in *Physics*, such functions are called paths. Accordingly, the linearization of a path at a point is called the tangent to the path. Because this gives rise to a linear map from $\mathbb{R}$ to $\mathbb{R}^m$, it must be associated with a $1 \times m$ column vector, called the tangent vector.

After these two important special cases, Section 7.5 treats the general case of functions of the form $\mathbf{F} : \mathbb{R}^n \to \mathbb{R}^m$, which are called vector fields, again for historical reasons. The linearization of such a function at a given point is then a linear map from $\mathbb{R}^n \to \mathbb{R}^m$, which is therefore associated with an $n \times m$ matrix, called the Jacobian matrix. This completes the characterization of derivatives as linear maps, with the gradient and tangent vectors being obtained in the special cases $n = 1$ and $m = 1$, respectively. The related concepts of divergence and curl are also introduced in this section, together with their physical and geometrical interpretation with the help of MATLAB graphics.

Having dealt with differentiation, the next two sections focus on integration. Section 7.6 introduces the concept of line integrals of both scalar functions and vector fields along paths. Section 7.7 presents the corresponding concept of surface integrals of scalar functions and vector fields across surfaces. Finally, Section 7.8 relates differentiation and integration through the theorems of Green, Stokes, and Gauss.

7.1 Scalar Functions of Several Variables

A function of the form $f : \mathbb{R}^n \to \mathbb{R}$ is called a *scalar function*, because it maps the vector $\mathbf{x} = [x_1, \dots, x_n]$ into the scalar $f(\mathbf{x})$. Alternatively, we can say that f is a real-valued function of n variables, stressing the dependence of the real number $f(x_1, \dots, x_n)$ on the variables $x_1, \dots, x_n$. Examples of these functions abound in applications of mathematics. For instance, your body mass index can be expressed as follows:

$$b(w, h) = \frac{w}{h^2},$$

where w is your weight in kilograms and h is your height in meters.

Domains and graphs

The best strategy when approaching multivariate functions for the first time is to try to generalize the familiar concepts from one-variable *Calculus*, one by one, and realize the differences and similarities as they come. For example, the *domain* for a function of several variables is the subset $D(f) \subset \mathbb{R}^n$ where we want it to be defined. When not specified, the domain is understood to be the largest subset of $\mathbb{R}^n$ on which f makes sense. In the body mass example, it is clear that the variables (w, h) should be positive numbers, so that its domain is $D = \mathbb{R}^2_+$, even though the formula makes sense for negative values as well. On the other hand, in the absence of further specifications, the domain for the function $f(x, y, z) = \sqrt{16 - x^2 - y^2 - z^2}$ is implicitly understood to be the set

$$D(f) = \{(x, y, z) \in \mathbb{R}^3 : x^2 + y^2 + z^2 \leq 16\}.$$

Next, consider the graphic visualization of scalar functions of several variables. In the simple case of $f : \mathbb{R} \to \mathbb{R}$, you can calculate the value $y = f(x)$ corresponding to each value of x in the domain of f and depict points of the form $(x, y) = (x, f(x))$ on a two-dimensional plane. As you recall, this is exactly the syntax for the MATLAB command `plot`, as in

```
>> x = [-1:0.01:1];
>> y = (x.^2).^(1/3)-x.^2;
>> plot(x,y);
```

which results in Figure 7.1. The *graph* of $f(x) = x^{2/3} - x^2$ in this case is naturally defined as the set

$$\Gamma(f) = \{(x, f(x)) \in \mathbb{R}^2, x \in D(f)\}. \tag{7.1}$$

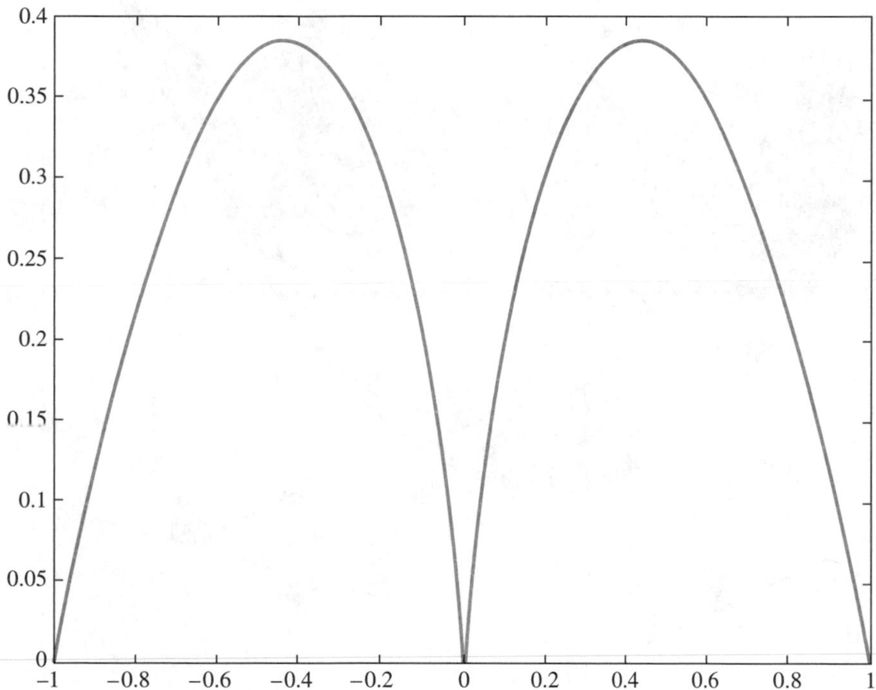

Figure 7.1 Graph for the function $f(x) = x^{2/3} - x^2$.

Surfaces and level curves

For functions of two variables, we can calculate the values $z = f(x, y)$ and depict the points $(x, y, z) = (x, y, f(x, y))$ in a three-dimensional space so that its graph is defined as the set

$$\Gamma(f) = \{(x, y, f(x, y)) \in \mathbb{R}^3, (x, y) \in D(f)\}. \tag{7.2}$$

As explained in Section 1.6, such a graph can be obtained with the MATLAB command `surf`, used in conjunction with the auxiliary command `meshgrid`:

```
>> x = [-4:0.5:4]; y = x;
>> [X,Y] = meshgrid(x,y); % creates 2-D arrays
>> z = X.^2+Y.^2;    % uses the 2-D arrays for component-wise operations
>> surf(x,y,z);
```

which produces the *paraboloid* $z = x^2 + y^2$ depicted in Figure 7.2.

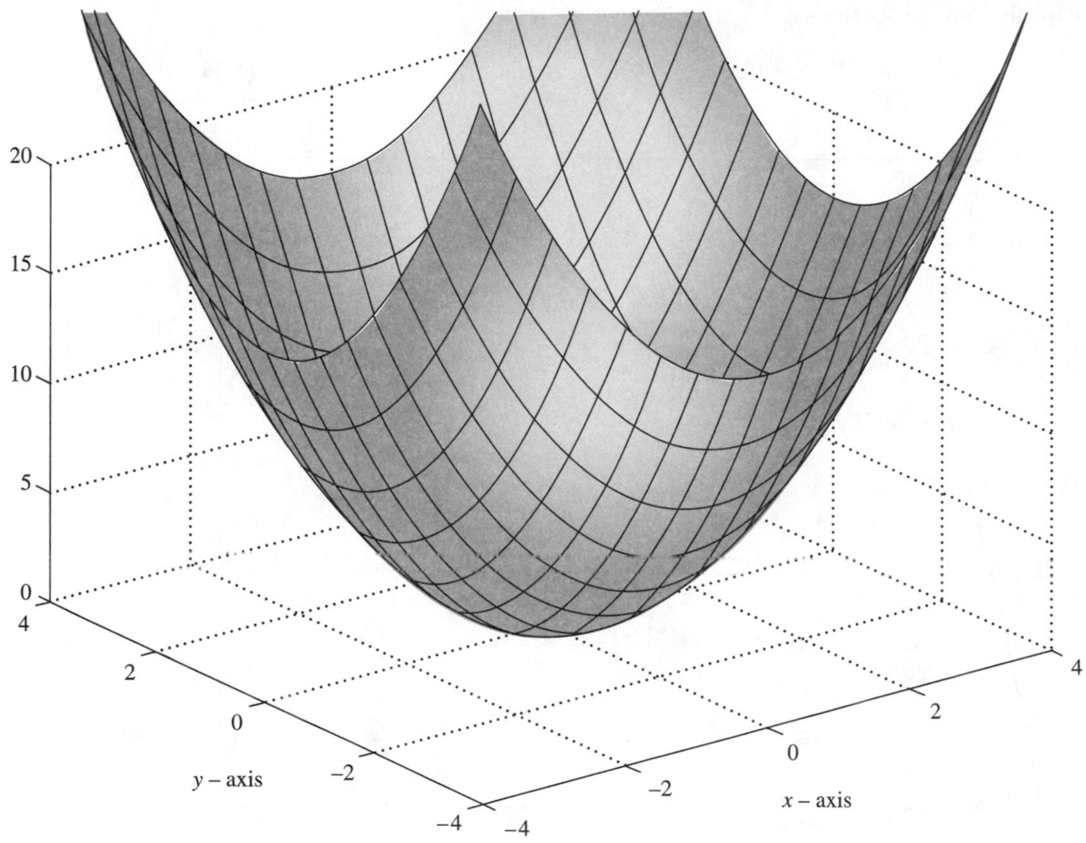

Figure 7.2 Graph for the function $f(x, y) = x^2 + y^2$.

The capability of visualizing three-dimensional graphs like these is one of the most appealing advantages of using a computer while trying to understand functions of several variables, and you should make it a personal rule to always plot such graphs when studying the properties of a novel function. Given that visual aids like these are relatively recent, many other related concepts were created to help us gain graphical intuition about a function without actually having to draw a three-dimensional plot, and many of them remain useful even after the popularization of computers.

For instance, given a constant c, the *level curve* of value c for a function $f : \mathbb{R}^2 \to \mathbb{R}$ is defined as the set of points (x, y) in the domain of f such that $f(x, y) = c$. In other words, the points on a level curve of value c tells us where the graph of f has height c. Put differently, a level curve of value c is just the intersection of the graph of f with a horizontal plane $z = c$ projected down to the xy-plane. Conversely, once we draw enough level curves for a given function f on the same xy-plane, we can then reconstruct the graph of f in our minds by imagining lifting the level curves up to their corresponding value.

In MATLAB, the function `contour` can be used to draw level curves in a two-dimensional plot, using a color scale to represent their heights (the value c earlier), ranging from blue (smallest) to red (highest). You can also use the function `clabel` to label each level curve explicitly with its corresponding height. Alternatively, you can use the function `surfc` to plot the graph and the level curves of a function of two variables on the same three-dimensional plot. For the paraboloid of the previous example, these plots are obtained from the following commands:

```
>> x = [-4:0.5:4]; y = x; [X,Y] = meshgrid(x,y);z = X.^2+Y.^2;
>> subplot(1,2,1); C = contour(x,y,z); clabel(C);
>> subplot(1,2,2); surfc(x,y,z);
```

which produce Figure 7.3.

In some applications, level curves are important in their own right, not just as tools for visualization. For instance, the Cobb–Douglas production function is a simplified model for the total output of an economic activity for given amounts of labor-hours L and invested capital K. Its general functional form is as follows:

$$P(L, K) = aL^{\alpha}K^{\beta}, \tag{7.3}$$

for constants a, α, and β, depending on the type of technology employed. The graph and level curves for the parameters $a = 1, \alpha = 0.4, \beta = 0.6$ are obtained from the commands:

```
>> x = [0:10:200]; y = x;
>> [X,Y] = meshgrid(x,y); z = X.^(0.4).*Y.^(0.6);
>> subplot(1,2,1); C = contour(x,y,z); clabel(C);
>> subplot(1,2,2); surf(x,y,z);
```

For a given output level, a typical question you may want to address in such production models is how many labor-hours can be saved by increasing investment (or the opposite, in perverse places with cheap labor). The level curves shown in Figure 7.4 provide much clearer answers to these questions than the graph for the function does.

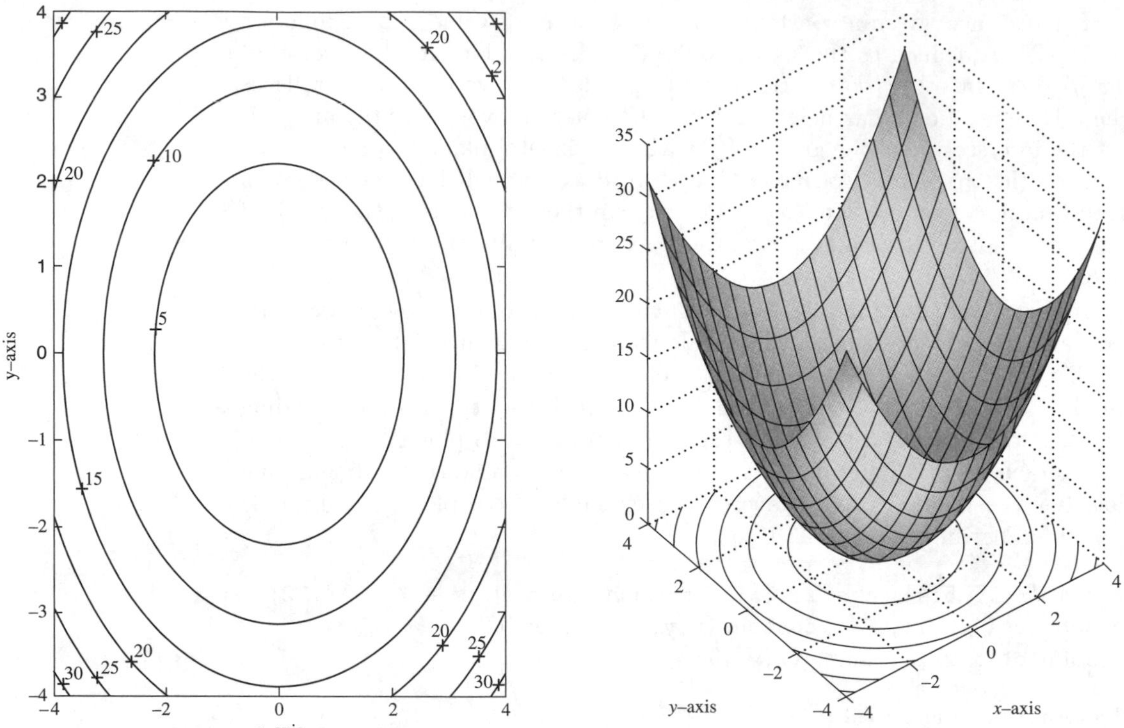

Figure 7.3 Level curves for the function $f(x, y) = x^2 + y^2$.

Level surfaces

Moving on to functions of three variables, we are faced with the fact that their graphs are subsets of a four-dimensional space so that direct visualization becomes an insoluble problem. All we can do in this case is resort to indirect visualization methods, where MATLAB proves to be particularly effective. For instance, we can define a *level surface* of value c for a function $f : \mathbb{R}^3 \to \mathbb{R}$ as the set of points $(x, y, z) \in \mathbb{R}^3$ such that $f(x, y, z) = c$. For example, it is clear that the level surfaces for the function $f(x, y, z) = x^2 + y^2 + z^2$ are spheres centered at the origin with radius equal to $\sqrt{c}$.

For more complicated functions, you can find level surfaces with the MATLAB command `isosurface` and then plot them using the command `patch`. The following commands show how to do this for the function $f(x, y, z) = x^2 + y^2 - z^2$, adjusting the settings for `patch` so that the level surfaces are plotted in superposition, as shown in Figure 7.5.

```
>> x = [-2:0.2:2]; y = x; z = x;
>> [X,Y,Z] = meshgrid(x,y,z); %  creates 3-D arrays
>> v = X.^2+Y.^2-Z.^2; % uses 3-D arrays for component-wise operations
>> s1 = isosurface(x,y,z,v,-1);  % surface level of value -1
```

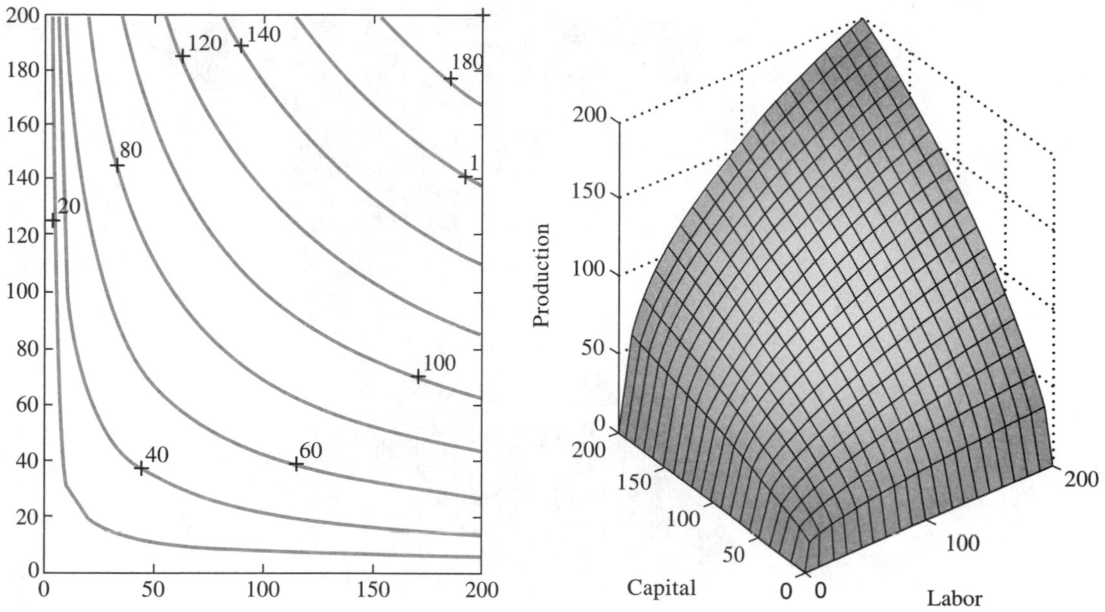

Figure 7.4 Level curves and graph for a Cobb–Douglas production function $P = aL^{\alpha}K^{\beta}$ with $a = 1$, $\alpha = 0.4$, and $\beta = 0.6$.

```
>> s2 = isosurface(x,y,z,v,0);     % surface level of value 0
>> s3 = isosurface(x,y,z,v,1);     % surface level of value 1
>> p1 = patch(s1); p2 = patch(s2); p3 = patch(s3); view(3);
>> set(p1,'FaceColor','blue','EdgeColor','none');
>> set(p2,'FaceColor','green','FaceAlpha',0.5,'EdgeColor','none');
>> set(p3,'FaceColor','red','FaceAlpha',0.7,'EdgeColor','none');
```

You can observe that when c is positive, the level surfaces are single-sheeted hyperboloids, whereas for negative c, you obtain hyperboloids of two sheets. When $c = 0$, these degenerate in a double cone centered on the z-axis. Although you can reach all these conclusions analytically, it is clear that being able to determine and plot the level surfaces of much more complicated functions is a powerful visualization tool.

As useful as they can be, level surfaces provide little information on the actual changes in value for the function as you move from one point in the three-dimensional space to another. Alternatively, you can assign colors to each point (x, y, z) depending on the value $f(x, y, z)$, effectively using them as an "extra dimension." Notice that this is what MATLAB regularly does when plotting surfaces such as those shown in Figures 7.2 and 7.3, although

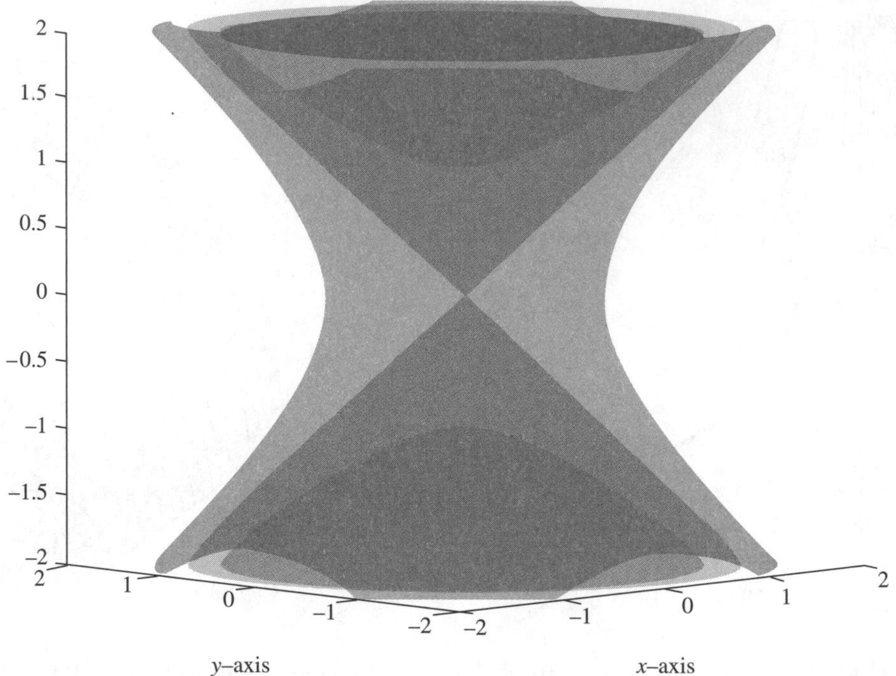

Figure 7.5 Level surfaces for the function $f(x, y, z) = x^2 + y^2 - z^2$.

in such cases the coloring is redundant since the height of the graph already indicates the value of a function of two variables.

With the help of this extra dimension, you can probe functions of three variables by looking at how the colors vary on specific regions of the three-dimensional space. In MATLAB, this is achieved with the command `slice`, which is applied to the function $f(x, y, z) = x^2 + y^2 - z^2$:

```
>> x = [-2:0.2:2]; y = x; z = x; [X,Y,Z] = meshgrid(x,y,z);
>> v = X.^2+Y.^2-Z.^2;
>> subplot(1,2,1);
>> slice(x,y,z,v,2,2,-2); % slices through the planes x=2,y=2,z=-2
>> subplot(1,2,2); [xs,ys] = meshgrid(x,y);
>> zs = -xs.^2+2; % another surface of interest
>> slice(x,y,z,v,xs,ys,zs);  % slices through the surface xs,ys,zs
```

In the left panel of Figure 7.6, you can see how the function behaves along selected planes. For instance, you can see that on the plane $z = -2$, the function has a low value (blue color), which slowly increases as we move away from the center point $(0, 0, -2)$, whereas on the planes $x = 2$ and $y = 2$,

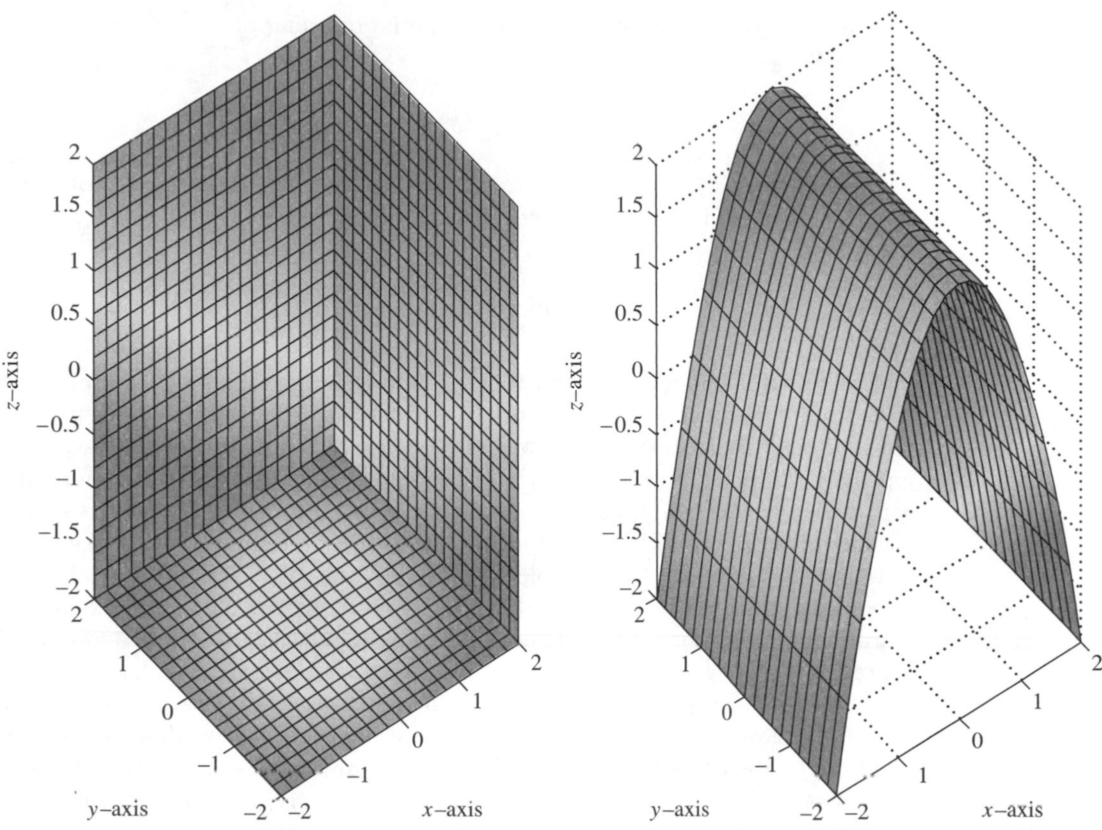

Figure 7.6 Values for the function $f(x, y, z) = x^2 + y^2 - z^2$ along the planes $x = 2$, $y = 2$, and $z = -2$ (left panel) and along the surface $z = -x^2 + 2$ (right panel). Color scale in MATLAB ranges from blue (lowest) to red (highest), shown as different shades of gray in this figure.

you can recognize the contours of the level surfaces depicted in Figure 7.5, as well as have an idea of how fast the function varies along different directions. But you can also probe the function along any other surface of interest. For instance, if this function were used to model the temperature on a rooftop shaped according to $z = -x^2 + 2$, then the plot on the right panel of Figure 7.6 helps you find its hot and cold spots.

Limits and continuity

At this point, courses in *Vector Calculus* introduce the concept of *limit* for scalar functions of several variables, either by generalizing the "epsilons and deltas" definition used in *Calculus* or by specializing the notions of open sets and limit points from abstract courses in *Topology*. Whichever the definition, we assume that you know or can consult the meaning of expressions such as follows:

$$\lim_{\mathbf{x}\to\mathbf{x}_0} f(\mathbf{x}) = L,$$

where $\mathbf{x}, \mathbf{x}_0 \in \mathbb{R}^n$, $f : \mathbb{R}^n \to \mathbb{R}$, and $L \in \mathbb{R}$. Armed with the concept of limit, we say that a function of several variables is *continuous* at a point $\mathbf{x}_0$ if it satisfies

$$\lim_{\mathbf{x}\to\mathbf{x}_0} f(\mathbf{x}) = f(\mathbf{x}_0).$$

The most important aspect to keep in mind when dealing with limits and continuity of functions of several variables is that the limit at a given point often fails to exist because the function approaches different values as we approach the point from different directions. A classic example is the function $f(x, y) = x^2/(x^2 + y^2)$, which does not have a limit at $(0, 0)$ because it equals 1 along the x–axis (when $y = 0$) but vanishes along the y–axis (when $x = 0$). Although it is perfectly possible to plot this kind of function using `surf`, it requires additional care for grid points around the discontinuity, as well as around points where the function is not defined. The MATLAB command `ezsurf`, however, does the filtering of discontinuities and singularities automatically. As suggested by the prefix in its name, `ezsurf` is part of a suite of higher-level commands designed to generate plots easily, that is, without having to specify data points explicitly. As is always the case with computer packages, friendliness comes at the price of less control by the user, and so we recommend the use of `ez` commands with moderation.

Continuing the previous example, you obtain the graph of the function $f(x, y) = x^2/(x^2 + y^2)$ with the single line

```
>> ezsurf('x^2/(x^2+y^2)')
```

resulting in Figure 7.7.

7.2 Partial Derivatives and Differentiability

Recall from *Vector Calculus* that the *partial derivative* of the function $f : \mathbb{R}^n \to \mathbb{R}$ with respect to its kth variable at a point $\mathbf{x}$ is given by

$$\frac{\partial f}{\partial x_k}(\mathbf{x}) = \lim_{h\to 0} \frac{f(\mathbf{x} + h\mathbf{e}_k) - f(\mathbf{x})}{h}, \tag{7.4}$$

where $\mathbf{e}_k$ is the kth vector in the canonical basis for $\mathbb{R}^n$ defined in Section 2.1. In other words, you calculate $\frac{\partial f}{\partial x_k}$ by taking the derivative of the function of a single variable obtained from f by keeping all variables but x_k constant.

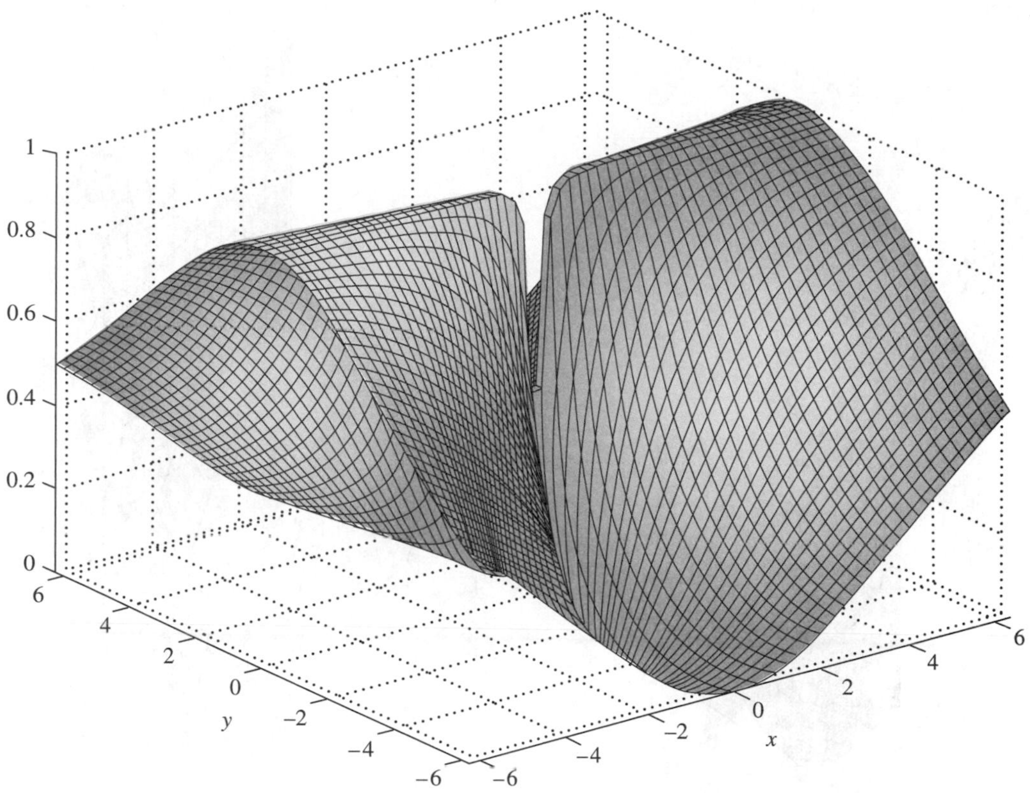

Figure 7.7 Graph for the function $f(x, y) = x^2/(x^2 + y^2)$, which does not have a limit at $(0, 0)$.

Therefore, from a numerical point of view, calculating partial derivatives essentially reduces to the techniques developed for Problem 6.1, so that we do not formulate a separate problem for them. Instead, you can use MATLAB to gain a geometrical understanding of partial derivatives.

For example, consider the function $f(x, y) = x^2 - xy - y^2$, whose partial derivatives are

$$\frac{\partial f}{\partial x} = 2x - y, \qquad \frac{\partial f}{\partial y} = -x - 2y.$$

In the left panel of Figure 7.8, you can see the intersection of the function f with the plane $y = -1$, corresponding to the function of one variable $f(x, -1) = x^2 + x - 1$, while in the right panel you see the intersection with the plane $x = 1$, corresponding to $f(1, y) = 1 - y - y^2$. From the definition

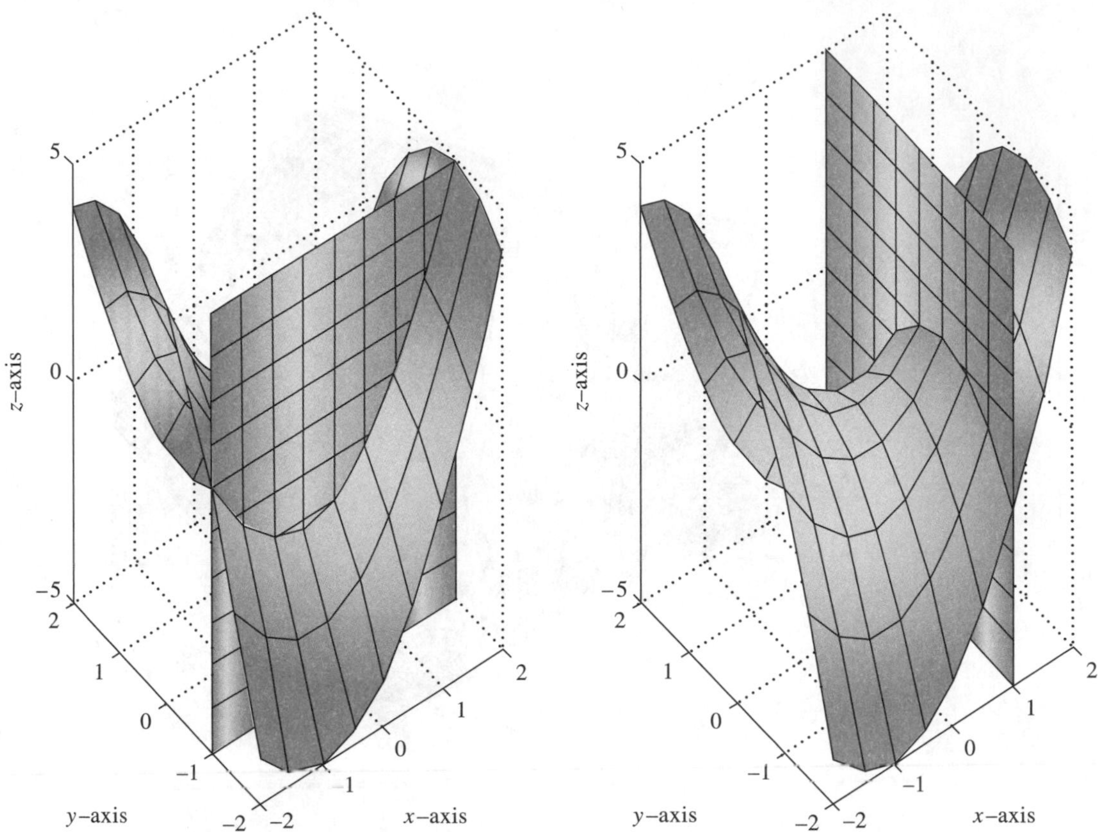

Figure 7.8 Intersections of the function $f(x, y) = x^2 - xy - y^2$ with the planes $y = -1$ (left) and $x = 1$ (right).

(7.4), the slopes of the tangents to these intersections are then given by the partial derivatives. For instance, at the point $(1, -1)$, you have

$$\frac{\partial f}{\partial x}(1, -1) = 3 > 0, \qquad \frac{\partial f}{\partial y}(1, -1) = 1 > 0,$$

corresponding to upward-sloping tangents at this point, as can be readily verified in Figure 7.8. The MATLAB commands for generating this figure follow, and you should in particular notice how to plot planes of the form $x = a$ or $y = b$ using the function `surf`.

```
>> x = [-2:0.5:2]; y = x;
>> [X,Y] = meshgrid(x,y); z = X.^2-X.*Y-Y.^2;
>> subplot(1,2,1); surf(x,y,z); hold on;
>> z1 = [-5:5]; % the range for the z-axis
```

```
>> [X1,Z1] = meshgrid(x,z1); % 2-D arrays for plotting the plane y=-1
>> Y1 = -1*ones(size(X1)); % 2-D array of the same size as above
>> surf(X1,Y1,Z1);
>> subplot(1,2,2); surf(x,y,z); hold on;
>> [Y2,Z2] = meshgrid(y,z1); % 2-D arrays for plotting the plane x=1
>> X2 = ones(size(Y2));  % 2-D  array of the same size as above
>> surf(X2,Y2,Z2);
```

Higher-order partial derivatives are defined in a recursive manner, given that $\frac{\partial f}{\partial x_k}$ is itself a function of several variables. For example, the first- and second-order partial derivatives for $f(x,y) = \sin(\pi x)y^2$ are

$$\frac{\partial f}{\partial x} = \pi\cos(\pi x)y^2, \qquad \frac{\partial f}{\partial y} = 2\sin(\pi x)y$$
$$\frac{\partial^2 f}{\partial x^2} = -\pi^2\sin(\pi x)y^2, \qquad \frac{\partial^2 f}{\partial y^2} = 2\sin(\pi x)$$
$$\frac{\partial^2 f}{\partial x\partial y} = \frac{\partial^2 f}{\partial y\partial x} = 2\pi\cos(\pi x)y.$$

The information obtained from partial derivatives in terms of intervals of increase, decrease, and concavity is the same as in one-variable *Calculus*, as you can verify in Figure 7.9, generated by the following commands:

```
>> x = [-1:0.2:1]; y = x; [X,Y] = meshgrid(x,y);
>> f = sin(pi*X).*Y.^2; subplot(2,3,1); surf(x,y,f);
>> fx = pi*cos(pi*X).*Y.^2; subplot(2,3,2); surf(x,y,fx);
>> fy = 2*sin(pi*X).*Y; subplot(2,3,3); surf(x,y,fy);
>> fxx = -pi^2*sin(pi*X).*Y.^2; subplot(2,3,4); surf(x,y,fxx);
>> fyy = 2*sin(pi*X); subplot(2,3,5); surf(x,y,fyy);
>> fxy = 2*pi*cos(pi*X).*Y; subplot(2,3,6); surf(x,y,fxy);
```

Tangent plane

With the help of partial derivatives, we can define the *tangent plane* to the graph $f : \mathbb{R}^2 \to \mathbb{R}$ at the point (x_0, y_0) as

$$t(x,y) = f(x_0,y_0) + \frac{\partial f}{\partial x}(x_0,y_0)(x-x_0) + \frac{\partial f}{\partial y}(x_0,y_0)(y-y_0). \tag{7.5}$$

Observe that this plane gives a linear approximation for the function f around the point (x_0, y_0). We then say that the function is *differentiable* at (x_0, y_0) if this is a *good* approximation, in the sense that

$$\frac{f(x,y)-t(x,y)}{\|[x,y]-[x_0,y_0]\|} \to 0 \quad \text{as} \quad [x,y] \to [x_0,y_0], \tag{7.6}$$

where $\|[x,y]\|$ is the Euclidean vector norm for the two-dimensional vector $[x,y]$ (see Section 2.4). The definitions of tangent plane and differentiability

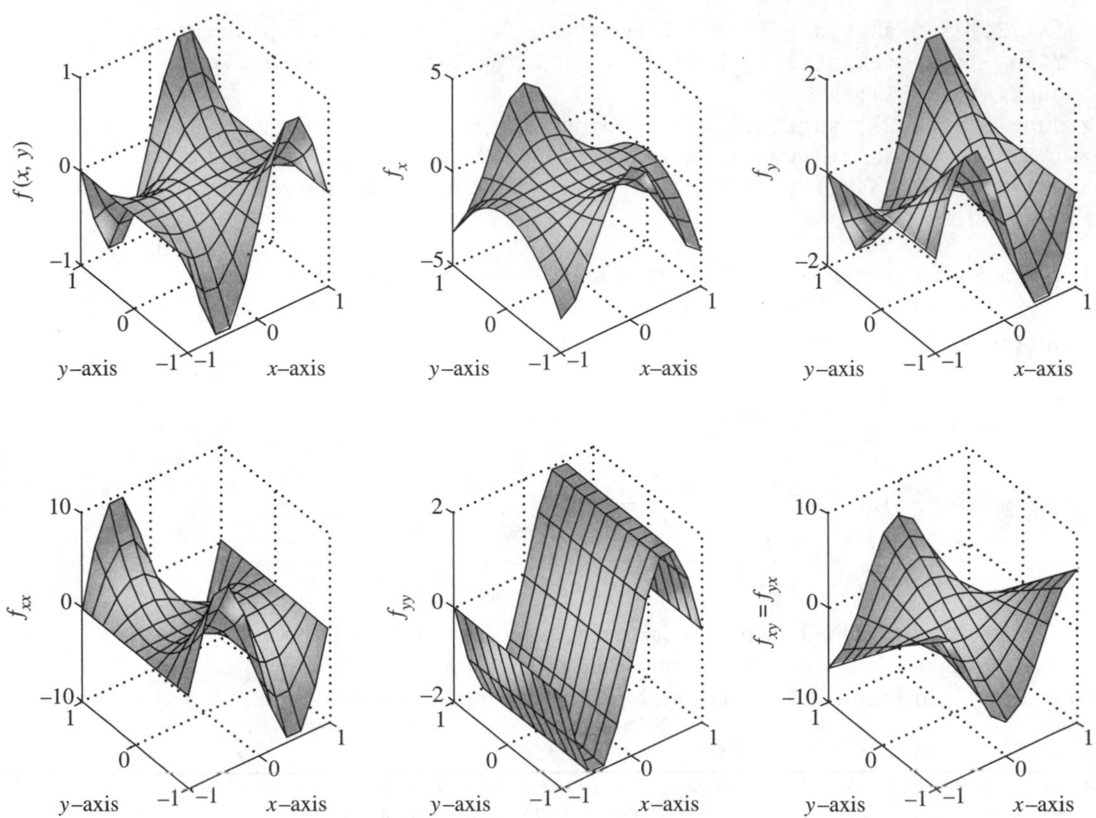

Figure 7.9 Graphs for the function $f(x, y) = \sin(\pi x)y^2$ and its first-order and second-order partial derivatives.

for scalar functions of more than two variables are immediate generalizations of (7.5) and (7.6). Given these definitions, it is an easy task to establish that if $f : \mathbb{R}^n \to \mathbb{R}$ is differentiable at $\mathbf{x}_0$, then it is also continuous at $\mathbf{x}_0$.

Notice that the existence of partial derivatives is a necessary condition for a function of several variables to be differentiable because they are used in the definition for the tangent plane. However, having well-defined partial derivatives does not guarantee differentiability. For a canonical example, consider the function

$$f(x, y) = \begin{cases} \dfrac{xy}{\sqrt{x^2 + y^2}} & \text{if } (x, y) \neq (0, 0) \\ 0 & \text{if } (x, y) = (0, 0) \end{cases},$$

which is continuous at $(0,0)$ and has partial derivatives equal to

$$\frac{\partial f}{\partial x} = \begin{cases} \dfrac{y^3}{(x^2+y^2)^{3/2}} & \text{if } (x,y) \neq (0,0) \\ 0 & \text{if } (x,y) = (0,0) \end{cases}$$

$$\frac{\partial f}{\partial y} = \begin{cases} \dfrac{x^3}{(x^2+y^2)^{3/2}} & \text{if } (x,y) \neq (0,0) \\ 0 & \text{if } (x,y) = (0,0). \end{cases}$$

Even though the partial derivatives are defined everywhere, you can easily show in the next exercise that this function is *not* differentiable at $(0,0)$.

Exercise 7.1 Show that for the function defined in the previous paragraph, the limit in (7.6) does not exist when (x,y) approaches $(0,0)$ along the line $x=y$. Moreover, show that its partial derivatives are *not* continuous at $(0,0)$.

The issues raised in the previous exercise are illustrated in Figure 7.10, obtained from the following commands, which shows the crinkle in the graph of $f(x,y)$ and the discontinuity in the graph of f_y.

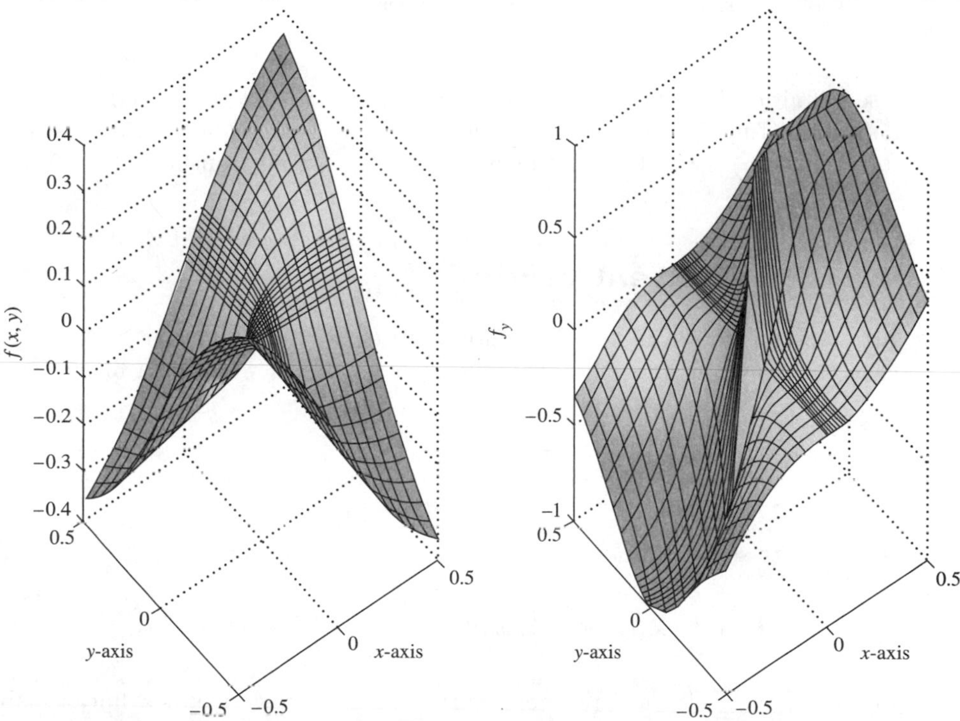

Figure 7.10 Graphs for the function $f(x,y) = xy/\sqrt{x^2+y^2}$ and its partial derivative dy.

```
>> x = [-.5:0.05:-0.1 0.09:0.01:-0.01 0.01:0.01:0.09 0.1:0.05:0.5];
>> y = x; [X,Y] = meshgrid(x,y);
>> f = X.*Y./sqrt(X.^2+Y.^2);
>> subplot(1,2,1); surf(x,y,f);
>> fy = X.^3./sqrt(X.^2+Y.^2).^3;
>> subplot(1,2,2); surf(x,y,fy);
```

Fortunately, when a function has continuous partial derivatives, the following theorem states that it must be differentiable.

Theorem 7.1 *If a function $f : \mathbb{R}^n \to \mathbb{R}$ has first-order partial derivatives that are continuous at a point $\mathbf{x}_0$, then f is differentiable at $\mathbf{x}_0$.*

The converse to this theorem, however, is not true. For example, the following exercise exhibits a differentiable function with discontinuous partial derivatives.

Exercise 7.2 Consider the function

$$f(x,y) = \begin{cases} (x^2+y^2)\sin\left(\frac{1}{\sqrt{x^2+y^2}}\right) & \text{if } (x,y) \neq (0,0) \\ 0 & \text{if } (x,y) = (0,0). \end{cases}$$

Use definition (7.4) to calculate its partial derivatives at $(0,0)$, and then use definition (7.6) to show that f is differentiable at $(0,0)$. Obtain the derivatives for points $(x,y) \neq (0,0)$ and conclude that they are not continuous at $(0,0)$.

7.3 The Gradient Vector

An alternative definition of differentiability for a scalar function $f : \mathbb{R}^2 \to \mathbb{R}$ is to say that f is differentiable at $\mathbf{x}_0 = (x_0, y_0)$ if there exists a linear map $\mathbf{T}_{\mathbf{x}_0} : \mathbb{R}^2 \to \mathbb{R}$ such that

$$\frac{f(\mathbf{x}) - f(\mathbf{x}_0) - \mathbf{T}_{\mathbf{x}_0}(\mathbf{x}-\mathbf{x}_0)}{\|\mathbf{x}-\mathbf{x}_0\|} \to 0 \qquad \text{as } \mathbf{x} \to \mathbf{x}_0. \tag{7.7}$$

It is easy to see that when (7.6) holds, then the linear map

$$\mathbf{T}_{\mathbf{x}_0}(\mathbf{x}-\mathbf{x}_0) = \frac{\partial f}{\partial x}(x_0,y_0)(x-x_0) + \frac{\partial f}{\partial y}(x_0,y_0)(y-y_0) \tag{7.8}$$

satisfies (7.7). With a little more work, one can show that if a linear map satisfying (7.7) exists, then it must be unique, which implies that the two definitions of differentiability are indeed equivalent. As before, both (7.7) and

(7.8) can be generalized immediately to scalar functions of more than two variables.

Gradient and differentials

From the discussion in Section 2.2 on the relation between linear maps and matrices, you know that each choice of basis for $\mathbb{R}^n$ uniquely associates the linear map $\mathbf{T}_{\mathbf{x}_0} : \mathbb{R}^n \to \mathbb{R}$ with a $1 \times n$ matrix. It is clear from (7.8) that, for the canonical basis in $\mathbb{R}^n$, this matrix is given by the row-vector

$$\nabla f(\mathbf{x}_0) = \left[\frac{\partial f}{\partial x_1}(\mathbf{x}_0), \dots, \frac{\partial f}{\partial x_n}(\mathbf{x}_0)\right], \tag{7.9}$$

called the *gradient* of f and alternatively denoted by ∇f or $\mathbf{grad} f$. A mnemonic way to help you to remember the gradient vector is to consider the "row-vector"

$$\nabla = \left[\frac{\partial}{\partial x_1}, \dots, \frac{\partial}{\partial x_n}\right] \tag{7.10}$$

and interpret ∇f as "matrix multiplication" of ∇ by the "scalar" f.

Using the gradient vector, we can rewrite (7.8) in matrix form as

$$\mathbf{T}_{\mathbf{x}_0}(\mathbf{x} - \mathbf{x}_0) = \nabla f(\mathbf{x}_0) \cdot (\mathbf{x} - \mathbf{x}_0) = \left[\frac{\partial f}{\partial x_1}(\mathbf{x}_0), \dots, \frac{\partial f}{\partial x_n}(\mathbf{x}_0)\right] \begin{bmatrix} dx_1 \\ \vdots \\ dx_n \end{bmatrix},$$

where we have introduced the components of the difference vector $\mathbf{dx} = (\mathbf{x} - \mathbf{x}_0)$. If we further denote the scalar $\mathbf{T}_{\mathbf{x}_0}(\mathbf{x} - \mathbf{x}_0)$ by df, then this last equation can be expressed as

$$df = \frac{\partial f}{\partial x_1} dx_1 + \cdots \frac{\partial f}{\partial x_n} dx_n. \tag{7.11}$$

In *Calculus* texbooks, such df is called the *total differential* of f and is often given completely mysterious meanings such as "the infinitesimal change in f" resulting from "infinitesimal changes dx_k." We hope that our discussion helps you understand the clear linear algebra concepts underlying expressions such as (7.11).

For a geometrical understanding of the gradient vector, observe that the gradient of a function of two variables $f : \mathbb{R}^2 \to \mathbb{R}$ is a two-dimensional vector whose components depend on the variables (x, y). That is, the gradient itself is a function $\nabla f : \mathbb{R}^2 \to \mathbb{R}^2$ that associates a two-dimensional vector to each point (x, y) in the domain of f, and can therefore be visualized using the standard association between vectors and arrows in the plane. For example, the gradient for the function $f(x, y) = x^2 - y^2$ is the vector

$$\nabla f(x, y) = [2x, -2y],$$

so that you have, for instance, $\nabla f(1,2) = [2,-4]$, which you can depict as an arrow with components $[2,-4]$ based at the point $(1,2)$.

In MATLAB, gradient vectors are obtained by using the function `gradient`, which uses the finite differences introduced in Section 6.1 to calculate *numerical derivatives*. As you recall from Problem 6.1, numerical derivatives for functions of one variable require specifying the function on a one-dimensional array of discrete points. Therefore, to calculate the gradient of a function of two variables, we need to specify the values of f on a two-dimensional array of discrete points (one dimension for each partial derivative). Fortunately, this is exactly what you have been doing with the command `meshgrid` when plotting surfaces using `surf`, and so the procedure should be familiar to you by now. After the numerical gradient has been calculated, you use the command `quiver` to plot the resulting vectors, as shown in Figure 7.11 for the function $f(x,y) = x^2 - y^2$:

```
>> x = [-4:0.5:4]; y = x;
>> [X,Y] = meshgrid(x,y); f = X.^2-Y.^2;
>> [Dx,Dy] = gradient(f); quiver(x,y,Dx,Dy);
```

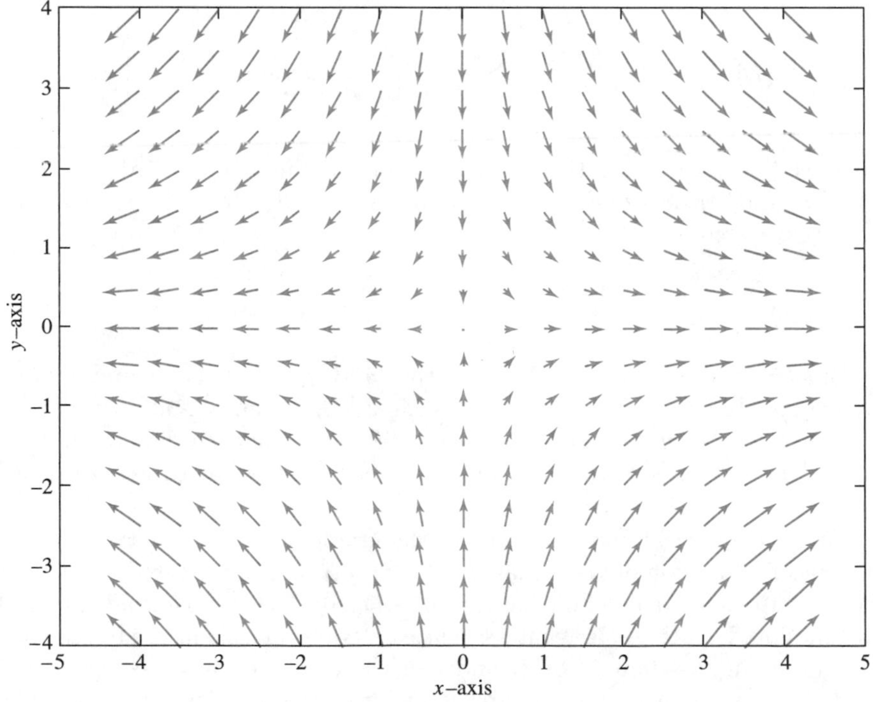

Figure 7.11 Gradient vectors for the function $f(x,y) = x^2 - y^2$.

Directional derivative

If you look back at the definition of partial derivatives, you find the ubiquitous basis vector $\mathbf{e}_k$ making its appearance in (7.4), meaning that the partial derivative $\frac{\partial f}{\partial x_k}$ can be interpreted as the rate of change of the function f as we move in the direction of $\mathbf{e}_k$. Put in this way, you immediately realize that there should be nothing special about these particular directions and that we ought to be able to calculate the rate of change of a scalar function of several variables in whichever direction we please. This leads to the definition of the *directional derivative* of $f : \mathbb{R}^n \to \mathbb{R}$ at $\mathbf{x}_0$ in the direction of a unit vector $\mathbf{v} \in \mathbb{R}^n$ as

$$D_{\mathbf{v}} f(\mathbf{x}_0) = \lim_{h \to 0} \frac{f(\mathbf{x} + h\mathbf{v}) - f(\mathbf{x})}{h}. \tag{7.12}$$

It can then be shown, using the Chain Rule for functions of several variables, that directional derivatives and the gradient vector are related through

$$D_{\mathbf{v}} f(\mathbf{x}_0) = \nabla f(\mathbf{x}_0) \cdot \mathbf{v}. \tag{7.13}$$

Besides being of practical relevance for calculating directional derivatives, expression (7.13) provides a geometrical interpretation of the gradient vector as the direction of fastest increase for the function f, as shown in the next exercise.

Exercise 7.3 Use the Cauchy–Schwarz inequality $|\mathbf{u} \cdot \mathbf{v}| \leq \|\mathbf{u}\| \|\mathbf{v}\|$ and the relation (7.13) to show that the maximum value for the directional derivative $D_{\mathbf{v}} f(\mathbf{x}_0)$ is $\|\nabla f(\mathbf{x}_0)\|$ and is achieved when $\mathbf{v}$ has the same direction as the gradient vector $\nabla f(\mathbf{x}_0)$.

Another consequence of (7.13) is that if the rate of change of f in the direction of $\mathbf{v}$ vanishes at a point $\mathbf{x}_0$, then the vector $\mathbf{v}$ must be orthogonal to $\nabla f(\mathbf{x}_0)$. Because level curves and level surfaces were defined as sets where the function is constant, this observation provides the basis for the following result:

Theorem 7.2 *Let $f : \mathbb{R}^n \to \mathbb{R}$ be a differentiable function with continuous partial derivatives. Then the gradient vector is orthogonal to the level surfaces of f in the sense that, if $\mathbf{v} \in \mathbb{R}^n$ is the tangent vector at a point $\mathbf{x}_0$ to a curve on the level surface, we must have $\nabla f(\mathbf{x}_0) \cdot \mathbf{v} = 0$.*

Figure 7.12 provides an illustration for this theorem applied to the function $f(x, y) = x^2 - y^2$ of the previous example by superimposing its level curves and gradient vector on the same plot:

```
>> quiver(x,y,Dx,Dy); hold on; surfc(x,y,f);
```

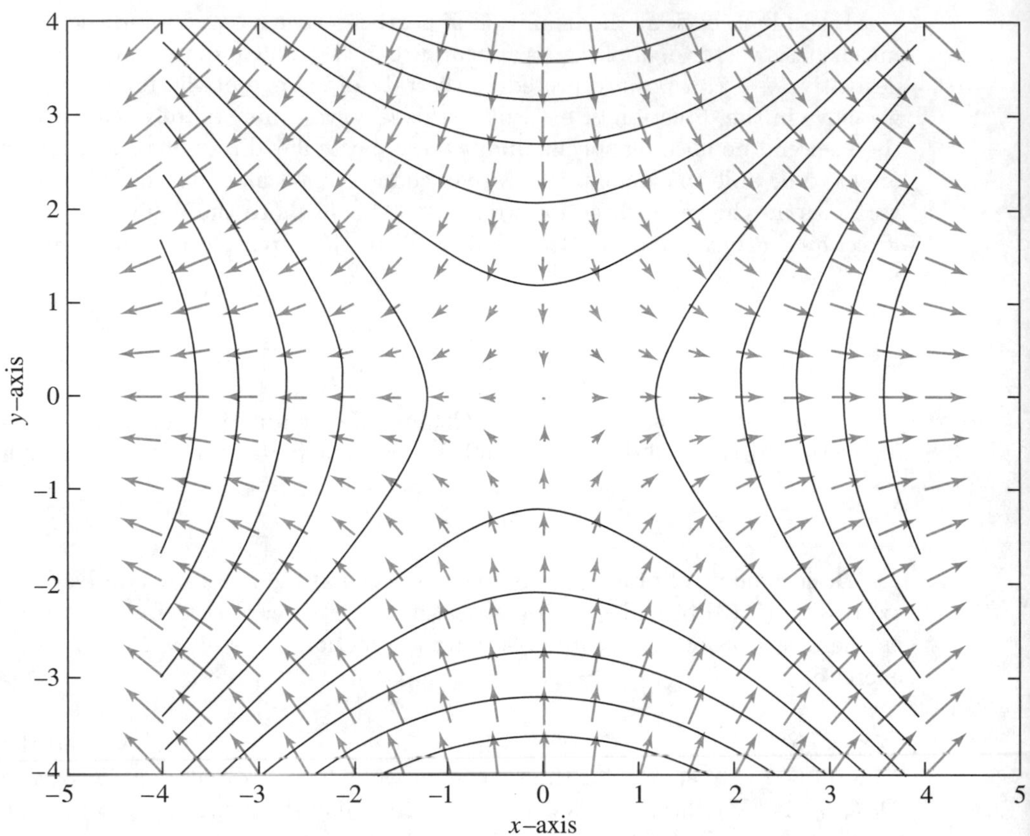

Figure 7.12 Gradient vectors and level curves for the function $f(x, y) = x^2 - y^2$.

For an example with three variables, consider the function $f(x, y, z) = x^2 + y^2 + z^2$, whose gradient is

$$\nabla f(x, y, z) = [2x, 2y, 2z] .$$

By recalling that the level surfaces for this function are spheres centered at the origin, you can see that the gradient vector at (x, y, z) is orthogonal to the tangent line of any curve on the sphere, as shown in Figure 7.13, obtained from the following commands:

```
>> x = [-2:0.2:2]; y = x; z = x;
>> [X,Y,Z] = meshgrid(x,y,z); f = X.^2+Y.^2+Z.^2;
>> s1 = isosurface(x,y,z,f,4); s2=isosurface(x,y,z,f,2);
```

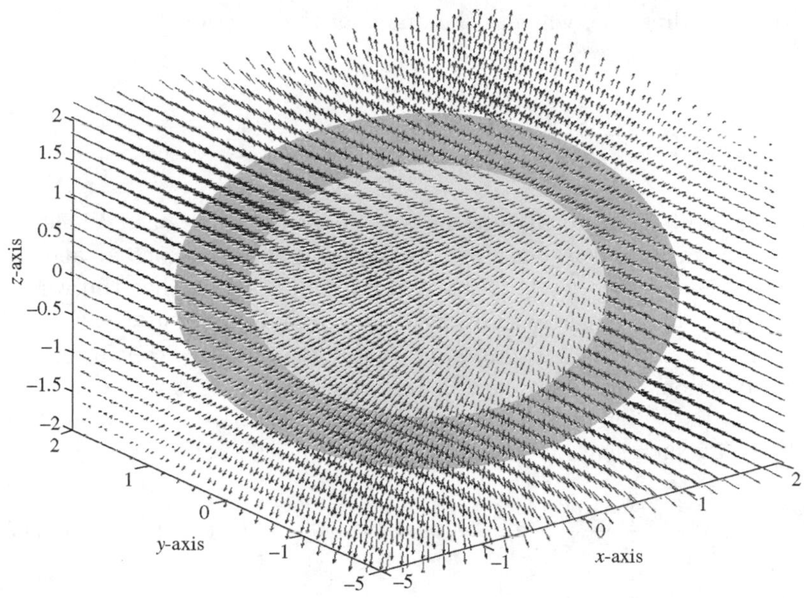

Figure 7.13 Gradient vector and level surfaces for the function $f(x, y) = x^2 + y^2 + z^2$.

```
>> p1 = patch(s1); view(3);
>> set(p1,'FaceColor','red','FaceAlpha',0.4,'EdgeColor','none');
>> p2 = patch(s2); set(p2,'FaceColor','yellow','EdgeColor','none');
>> [Dx,Dz,Dy] = gradient(f); hold on; quiver3(X,Y,Z,Dx,Dy,Dz);
```

7.4 Paths

We now consider maps of the form $\mathbf{s} : \mathbb{R} \to \mathbb{R}^m$. When the domain of such a map consists of an interval $[a, b]$ of the real line, we denote it by $\mathbf{s} : [a, b] \to \mathbb{R}^m$ and call it a *path.* As the variable t varies in the interval $[a, b]$, the image $\mathbf{s}(t)$ traces a curve in $\mathbb{R}^m$. You should be careful, however, not to call this curve a *graph* because, following the definitions in (7.1) and (7.2), the graph of a function $\mathbf{s} : \mathbb{R} \to \mathbb{R}^m$ consists of points of the form $(t, \mathbf{s}(t))$, which is a subset of $\mathbb{R}^{m+1}$. In other words, the curve traced by $\mathbf{s}(t)$, which we can see in the cases $m = 2$ and $m = 3$, shows only the *image* of the function $\mathbf{s}$, with the variable t running in the back of your mind. For example, the image for the function

$$\mathbf{s}_1(t) = \begin{bmatrix} \cos t \\ \sin t \end{bmatrix}, \qquad 0 \leq t \leq 2\pi \tag{7.14}$$

is a circle of radius one, whereas the image of the function

$$\mathbf{s}_2(t) = \begin{bmatrix} \cos t \\ \sin t \\ t \end{bmatrix}, \quad 0 \leq t \leq 4\pi \tag{7.15}$$

consists of two laps of a helix around a cylinder of radius one. When we are interested in the curves themselves, we say that the functions $\mathbf{s}_1$ and $\mathbf{s}_2$ are *parametrizations* of the circle and the helix, emphasizing that the variable t is just a *parameter* that is not seen in their graphic representations.

Tangent vector

Before plotting some examples, let us examine the concept of derivative for such functions. Similar to the definition in (7.7), we say that a function $\mathbf{s} : \mathbb{R} \to \mathbb{R}^m$ is differentiable at a point $t_0 \in \mathbb{R}$ if there exists a linear map $\mathbf{T}_{t_0} : \mathbb{R} \to \mathbb{R}^m$ such that

$$\frac{\|\mathbf{s}(t) - \mathbf{s}(t_0) - \mathbf{T}_{t_0}(t - t_0)\|}{|t - t_0|} \to 0 \qquad \text{as } t \to t_0. \tag{7.16}$$

Now observe that the scalar factor $1/|t-t_0|$ can be put inside the vector norm appearing in the preceding limit. Moreover, using the rules of multiplication of a vector by a scalar, we can multiply this factor by each component of the vector appearing inside the norm. With one more ingredient (continuity of the norm), we see that the same limit (7.16) needs to hold for each component of the vector $\mathbf{s}(t)$. Putting it all together means that, for the canonical basis in $\mathbb{R}^m$, the linear map $\mathbf{T}_{t_0}$ must be given by

$$\mathbf{T}_{t_0} = \begin{bmatrix} \dot{s}_1(t_0) \\ \vdots \\ \dot{s}_m(t_0) \end{bmatrix}, \tag{7.17}$$

where $\dot{s}_k(t)$ is the standard notation for the derivative $\frac{ds_k}{dt}(t)$. When this map $\mathbf{T}_{t_0}$ exists, we call it the derivative of the function $\mathbf{s}(t)$ and denote it by $\dot{\mathbf{s}}(t)$ to remind us that it is simply a column vector whose components are the derivatives of each component $s_k(t)$, $k = 1, \dots, m$. For example, the derivatives for the preceding functions are

$$\dot{\mathbf{s}}_1(t) = \begin{bmatrix} -\sin t \\ \cos t \end{bmatrix}, \qquad \dot{\mathbf{s}}_2(t) = \begin{bmatrix} -\sin t \\ \cos t \\ 1 \end{bmatrix}.$$

Using the derivative of a path $\mathbf{s}$, we define its *tangent vector* at a point $\mathbf{s}(t_0)$ as the vector $\dot{\mathbf{s}}(t_0)$. In Figure 7.14, you can see the curves and tangent vectors for the preceding functions, obtained as a different application of the commands quiver and quiver3:

```
>> t = [0:0.1:2*pi]; subplot(1,2,1); plot(cos(t),sin(t)); hold on;
>> t = [0:0.5:2*pi]; quiver(cos(t),sin(t),-sin(t),cos(t));
>> t = [0:0.1:4*pi]; subplot(1,2,2); plot3(cos(t),sin(t),t); hold on;
>> t = [0:0.5:4*pi]; quiver3(cos(t),sin(t),t,-sin(t),cos(t),1);
```

Arc length

If $\mathbf{s} : [a, b] \rightarrow \mathbb{R}^3$ represents a trajectory for a moving body in three-dimensional vector space, then we should be able to determine the total distance traveled between times $t = a$ and $t = b$. For rectilinear trajectories, this is simply given by the Euclidean distance $\|\mathbf{s}(a) - \mathbf{s}(b)\|$ between the points $\mathbf{s}(a)$ and $\mathbf{s}(b)$. We can then approximate a curved trajectory by several straight-line segments connecting intermediate points $\mathbf{s}(t_i)$, for $a = t_0 < \ldots < t_i < \ldots < t_n = b$. The Euclidean distance between each pair of intermediate points is then

$$\|\mathbf{s}(t_i) - \mathbf{s}(t_{i-1})\| = \sqrt{(s_1(t_i) - s_1(t_{i-1}))^2 + (s_2(t_i) - s_2(t_{i-1}))^2 + (s_3(t_i) - s_3(t_{i-1}))^2}.$$

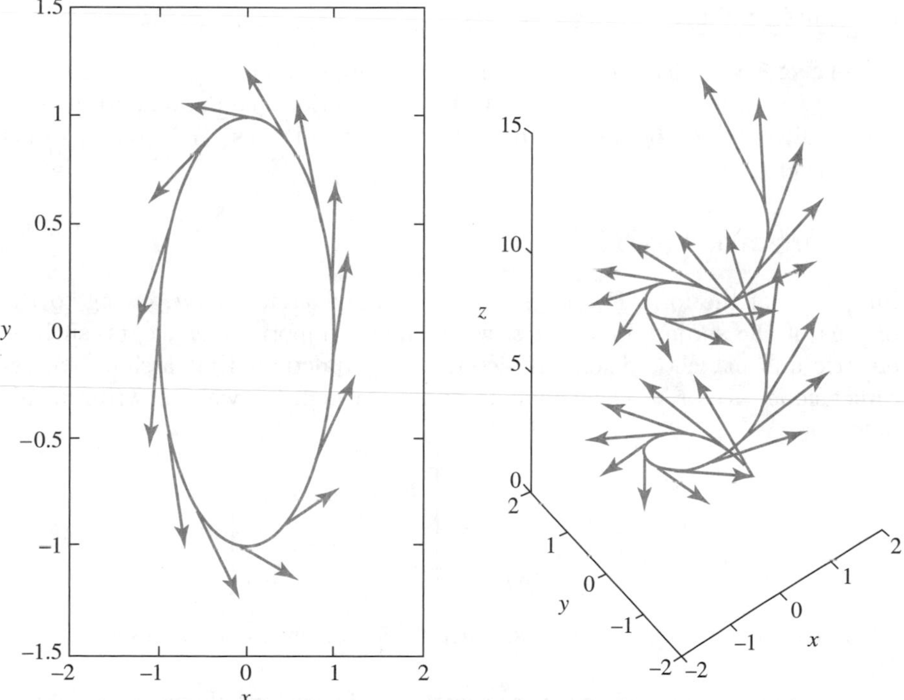

Figure 7.14 Plots for curves traced by the vector functions $\mathbf{s}_1(t) = [\cos t, \sin t]$ and $\mathbf{s}_2(t) = [\cos t, \sin t, t]$ and their tangent vectors.

Assuming that the component functions $s_1(t), s_2(t), s_3(t)$ are differentiable, we can use the *Mean Value Theorem* to rewrite this as

$$\|\mathbf{s}(t_i) - \mathbf{s}(t_{i-1})\| = \sqrt{\left(\frac{ds_1}{dt}(\xi_i)\right)^2 + \left(\frac{ds_2}{dt}(\eta_i)\right)^2 + \left(\frac{ds_3}{dt}(\nu_i)\right)^2}\,(t_i - t_{i-1}),$$

where ξ_i, η_i, ν_i are points in the interval (t_{i-1}, t_i). Adding all these distances together, we clearly obtain a Riemann sum, and taking the limit as the number of intermediate points t_i goes to infinity results in an integral. This leads to defining the *arc length* of a continuously differentiable path $\mathbf{s} : [a, b] \to \mathbb{R}^m$ as

$$L(\mathbf{s}) = \int_a^b \|\dot{\mathbf{s}}(t)\| dt, \tag{7.18}$$

where $\dot{\mathbf{s}}$ denotes the derivative of $\mathbf{s}(t)$ as in (7.17).

Back to the example where $\mathbf{s}(t)$ is a trajectory in space, observe that the tangent vectors $\dot{\mathbf{s}}(t_0)$ correspond to the *velocity* of the moving body at time t_0, with $\|\dot{\mathbf{s}}(t_0)\|$ being its *speed*. Therefore, the integral formula in (7.18) is simply an elaborate version of the familiar rule to calculate distance as speed multiplied by time.

Exercise 7.4 Calculate the length of the ellipse $\mathbf{s}(t) = [2\cos t, 3\sin t]$ for $0 \leq t \leq 2\pi$ using (7.18) and the MATLAB function `quad`. Compare it with the result obtained by adding the elementary lengths $\|\mathbf{s}(t_i) - \mathbf{s}(t_{i-1})\|$ with $(t_i - t_{i-1}) = 2\pi/100$.

7.5 Vector Fields

Functions of the form $\mathbf{F} : \mathbb{R}^n \to \mathbb{R}^m$ are called *vector fields*, owing to the origins of the subject in *Physics*, where many important concepts, such as electric and magnetic fields, are described by functions that assign a three-dimensional vector to each point in space. In general, we can write vector fields as

$$\mathbf{F}(\mathbf{x}) = \begin{bmatrix} F_1(\mathbf{x}) \\ \vdots \\ F_m(\mathbf{x}) \end{bmatrix} = \begin{bmatrix} F_1(x_1, \dots, x_n) \\ \vdots \\ F_m(x_1, \dots, x_n) \end{bmatrix}, \tag{7.19}$$

where each component is itself a scalar function of several variables $F_k : \mathbb{R}^n \to \mathbb{R}$, $k = 1, \dots, m$.

You have seen already a concrete example of a vector field when we defined the gradient of scalar functions of several variables in Section 7.3. Figures 7.11 and 7.12 show two-dimensional vectors assigned to each point in the plane, therefore providing a visualization technique for vector fields of the form $\mathbf{F} : \mathbb{R}^2 \to \mathbb{R}^2$. Similarly, Figure 7.13 is an attempt to visualize a vector field of the form $\mathbf{F} : \mathbb{R}^3 \to \mathbb{R}^3$ by assigning a three-dimensional vector to each point in space. As in the previous section, notice that we have avoided using the word *graph* in this context, since the graph of a function $\mathbf{F} : \mathbb{R}^n \to \mathbb{R}^m$ must be a subset of $\mathbb{R}^{n+m}$, which is obviously impossible to visualize when $(n+m) > 3$.

Jacobian matrix

Similar to (7.7) and (7.16), we say that a vector field $\mathbf{F} : \mathbb{R}^n \to \mathbb{R}^m$ is *differentiable* at $\mathbf{x}_0$ if there exists a linear map $\mathbf{T}_{\mathbf{x}_0} : \mathbb{R}^n \to \mathbb{R}^m$ such that

$$\frac{\|\mathbf{F}(\mathbf{x}) - \mathbf{F}(\mathbf{x}_0) - \mathbf{T}_{\mathbf{x}_0}(\mathbf{x} - \mathbf{x}_0)\|}{\|\mathbf{x} - \mathbf{x}_0\|} \to 0, \qquad \text{as } \mathbf{x} \to \mathbf{x}_0. \tag{7.20}$$

By treating each component function $F_k : \mathbb{R}^n \to \mathbb{R}$, $k = 1, \ldots, m$ separately, we see that this linear map must be given by the $m \times n$ matrix

$$\mathbf{T}_{\mathbf{x}_0} = \begin{bmatrix} \dfrac{\partial F_1}{\partial x_1} & \cdots & \dfrac{\partial F_1}{\partial x_n} \\ \vdots & & \vdots \\ \dfrac{\partial F_m}{\partial x_1} & \cdots & \dfrac{\partial F_m}{\partial x_n} \end{bmatrix}. \tag{7.21}$$

The map $\mathbf{T}_{\mathbf{x}_0}$, called the *Jacobian matrix*, is then said to be the derivative for the vector field $\mathbf{F}$ at the point $\mathbf{x}_0$ and is denoted by $\mathbf{DF}(\mathbf{x}_0)$. For the special case $m = 1$ and $n > 1$, you can see that it reduces to the gradient vector ∇F in (7.9), whereas for the case $n = 1$ and $m > 1$ you obtain the derivative vector $\dot{\mathbf{s}}$ in (7.17). Clearly, when $m = n = 1$, you recover the usual derivative of a scalar function of one variable.

Integral curves

An important example of a vector field arises in connection with tangent vectors to curves in space, but requires a bit of care in its definition. By recalling the concepts introduced in the previous section, you can see that, for a given function $\mathbf{s}(t) = [s_1(t), \ldots, s_m(t)]$, the function $\dot{\mathbf{s}}(t)$ is not a vector field because it only depends on the single variable t. However, you might be tempted to think of the tangent vector $\dot{\mathbf{s}}(t_0)$ associated with each point $(s_1(t_0), \ldots, s_m(t_0))$ as being part of a vector field in $\mathbb{R}^m$, as suggested by Figure 7.14. The problem with this is that such tangent vectors are only assigned to points on the image of the function $\mathbf{s} = \mathbf{s}(t)$, and you have effectively no idea about the putative vector field for points outside this curve. As is often the case in mathematics, the way around this problem is to turn it inside out: start with a well-defined vector field and try to find the curves to which each vector is tangent. In other words, given a vector field $\mathbf{F}$, we seek a vector function $\mathbf{s}$ such that

$$\dot{\mathbf{s}}(t) = \mathbf{F}(\mathbf{s}(t)). \tag{7.22}$$

These are then called the *integral curves* for the vector field because the process for finding them must somehow be the inverse of taking derivatives.

As a matter of fact, finding the integral curves of a vector field is quite an involved process because obtaining the function $\mathbf{s}$ in (7.22) requires solving a system of *ordinary differential equations*, which are covered in Chapter 9. However, verifying that certain curves have a given vector field as their tangent vector is usually straightforward, as you can see in the next exercise.

Exercise 7.5 Verify that the vector field $\mathbf{F}(x, y) = [-y, x]$ has a family of integral curves of the form $\mathbf{s}(t) = [r\cos(t+\varphi), r\sin(t+\varphi)]$ for parameters $r \geq 0$ and $0 \leq \varphi \leq 2\pi$. Find the particular curve in this family that has tangent vector $[-y_0, x_0]$ at the point (x_0, y_0).

The integral curves and some of the vectors for the vector field in Exercise 7.5 are shown in Figure 7.15, obtained from the following commands:

```
>> x = [-4:0.5:4]; y = x; [X,Y] = meshgrid(x,y);
>> quiver(X,Y,-Y,X); hold on;
>> t = [0:0.1:2*pi]; plot(cos(t),sin(t)); plot(2*cos(t),2*sin(t),'r');
>> plot(3*cos(t),3*sin(t),'g'); plot(4*cos(t),4*sin(t),'k');
```

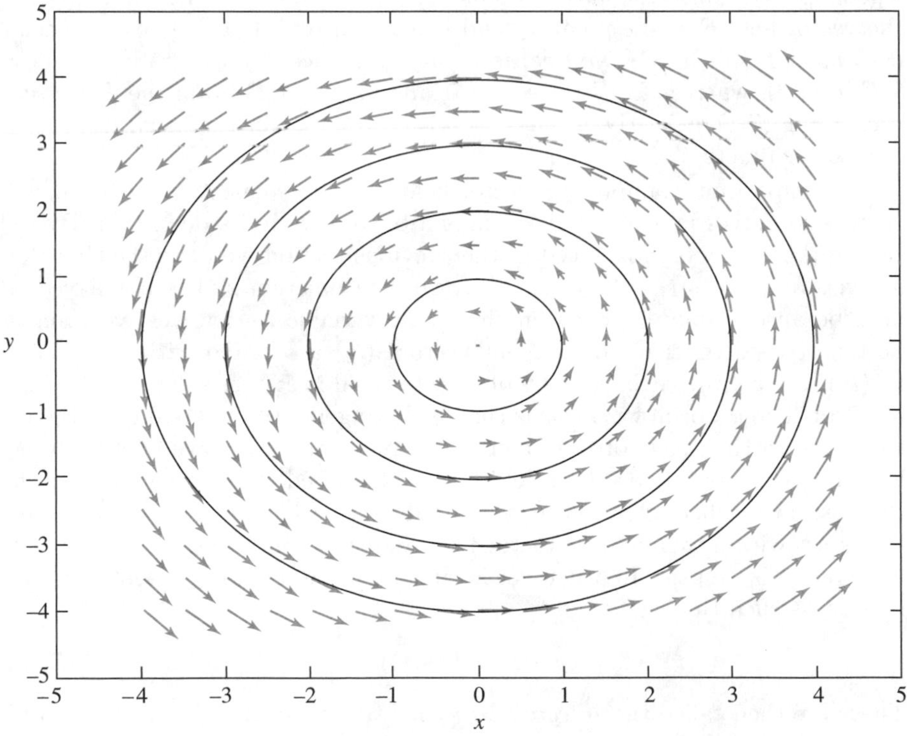

Figure 7.15 The vector field $\mathbf{F}(x, y) = [-y, x]$ and some of its integral curves.

Historically, the most important examples of curves in space were those representing the trajectories of moving bodies, with their tangent vectors corresponding to the velocity at a given point on the trajectory, as discussed in the previous section. Accordingly, many of the concepts developed in *Vector Calculus* owe their name to having a *velocity field* as motivation. For instance, in the context of the velocity field of a moving fluid, the integral curves (7.22) were appropriately called *flow lines*, a term that then became synonymous with integral curves for vector fields in general.

Divergence

Another concept stemming from the study of moving fluids is the *divergence* of a vector field $\mathbf{F} : \mathbb{R}^n \to \mathbb{R}^n$, defined as the scalar function

$$\text{div } \mathbf{F}(x_1, \ldots, x_n) = \frac{\partial F_1}{\partial x_1} + \cdots + \frac{\partial F_n}{\partial x_n}. \tag{7.23}$$

Using the operator ∇ defined in (7.10), you can symbolically write the divergence of a vector field as the "scalar product" $\nabla \cdot \mathbf{F}$. Its physical interpretation is that if $\mathbf{v}$ represents the velocity field of fluid, than $\nabla \cdot \mathbf{v}(\mathbf{x_0})$ gives the rate of *outward flux* per unit volume at the point $\mathbf{x}_0$. For instance, the divergence for the vector field $\mathbf{v}(x, y) = [xy, x^2]$ is

$$\nabla \cdot \mathbf{v} = \frac{\partial v_1}{\partial x} + \frac{\partial v_2}{\partial y} = y,$$

which means that the fluid must be flowing outwardly (or diverging) on regions where $y > 0$ and inwardly (or converging) on regions where $y < 0$. By contrast, a fluid whose velocity field is $\mathbf{v}(x, y) = [-x, -y]$ converges everywhere because its divergence is constant and equal to -2. The mathematical basis for these interpretations is *Gauss' Theorem* introduced in Section 7.8. For now, you can verify these statements intuitively in Figure 7.16, while at the same time getting to know the MATLAB command `divergence`, which computes the scalar function $\nabla \cdot \mathbf{F}$ from the vector field $\mathbf{F}$:

```
>> x = [-4:0.5:4]; y = x; [X,Y] = meshgrid(x,y);
>> div1 = divergence(X,Y,X.*Y,X.^2);
>> subplot(1,2,1); surf(x,y,div1);
>> hold on; quiver(X,Y,X.*Y,X.^2);
>> div2 = divergence(X,Y,-X,-Y);
>> subplot(1,2,2); surf(X,Y,div2);
>> hold on; quiver(X,Y,-X,-Y);
```

Curl

Given the preceding interpretation for the divergence function, a vector field is said to be *incompressible* if its divergence is zero. For example, the vector field in Figure 7.15 is incompressible, meaning that a fluid having $\mathbf{F}$ as its velocity field simply rotates around without either expanding or contracting.

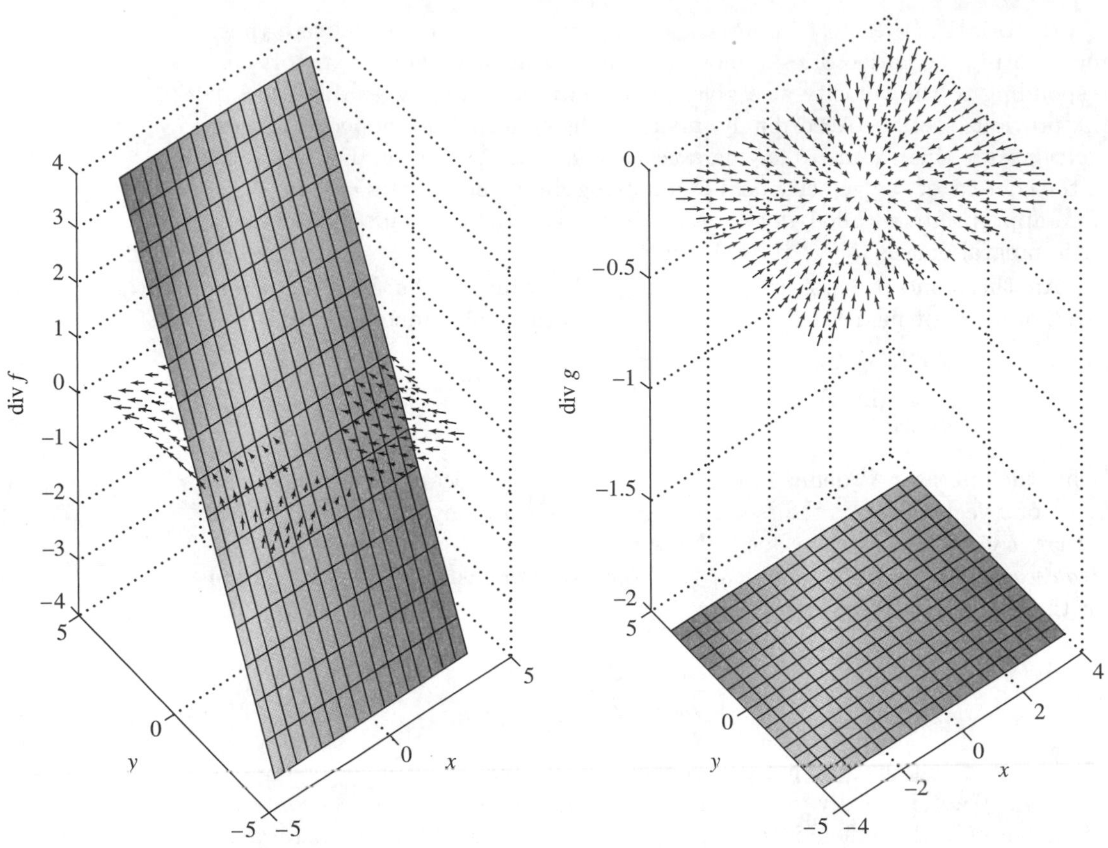

Figure 7.16 The vector fields $\mathbf{F}(x, y) = [xy, x^2]$ and $\mathbf{G}(x, y) = [-x, -y]$ and their divergences.

Rotations like these lead us to the concept of the *curl* of a vector field $\mathbf{F} : \mathbb{R}^3 \to \mathbb{R}^3$, defined as the vector

$$\operatorname{curl} \mathbf{F} = \begin{bmatrix} \dfrac{\partial F_3}{\partial x_2} - \dfrac{\partial F_2}{\partial x_3} \\ \\ \dfrac{\partial F_1}{\partial x_3} - \dfrac{\partial F_3}{\partial x_1} \\ \\ \dfrac{\partial F_2}{\partial x_1} - \dfrac{\partial F_1}{\partial x_2} \end{bmatrix}. \tag{7.24}$$

Whereas the use of the operator ∇ in the definition of gradient vector and divergence was a mere rewriting of easy formulas, it fully justifies its appearance now because the complicated expression (7.24) can be conveniently rewritten as

$$\text{curl } \mathbf{F} = \nabla \times \mathbf{F} = \left[\frac{\partial}{\partial x_1}, \frac{\partial}{\partial x_2}, \frac{\partial}{\partial x_3}\right] \times \begin{bmatrix} F_1(\mathbf{x}) \\ F_2(\mathbf{x}) \\ F_3(\mathbf{x}) \end{bmatrix}, \tag{7.25}$$

where $\times$ denotes the cross product in $\mathbb{R}^3$ defined in Section 1.3.

An example of cross product is the expression

$$\mathbf{v} = \boldsymbol{\omega} \times \mathbf{r},$$

which relates the *angular velocity* $\boldsymbol{\omega}$ of a particle moving on a circular trajectory of radius r to its velocity field $\mathbf{v}$. In the next exercise, you can use this example to provide a physical interpretation for the curl of a vector field.

Exercise 7.6 Let $\mathbf{r} = [x, y, z]$ be the position of particles of a rigid body rotating around an axis l with velocity field $\mathbf{v}$. Show that $\nabla \times \mathbf{v} = 2\boldsymbol{\omega}$.

For example, the curl of the vector field $\mathbf{F}(x, y) = [-y, x, 0]$ is

$$\nabla \times \mathbf{F} = \begin{bmatrix} 0 \\ 0 \\ 2 \end{bmatrix},$$

meaning that the constant vector $\boldsymbol{\omega} = [0, 0, 1]$ is the angular velocity of a rigid body rotating around the z-axis with velocity $\mathbf{F}$. In particular, the trajectories of each particle are the circular integral lines of $\mathbf{F}$ shown in Figure 7.15. In MATLAB, the curl of a vector field is calculated by using the command `curl`, which we use in the following code to visualize $\nabla \times \mathbf{F}$ on Figure 7.17:

```
>> x = [-4:4]; y = x; z = x; [X,Y,Z] = meshgrid(x,y,z);
>> [cx,cy,cz] = curl(X,Y,Z,-Y,X,zeros(size(x)));
>> quiver3(X,Y,zeros(size(x)),cx,cy,cz,0);
>> hold on; quiver(X,Y,-Y,X);
```

Observe that the interpretation provided by Exercise 7.6 is only valid for rotating rigid bodies, for which the angular velocity of a particle is well defined. Clearly, this limits the choices of possible vector fields because they must satisfy $\mathbf{v} = \boldsymbol{\omega} \times \mathbf{r}$. To obtain a physical interpretation for the curl of more general vector fields, we need to consider the velocity field $\mathbf{v}$ of a moving fluid. In this context, $\mathbf{w} = \nabla \times \mathbf{v}$ is called the *vorticity* of the flow and is intimately related to the idea of *circulation*, which we introduce in Section 7.8 in connection with *Stokes's Theorem.* Loosely speaking, the vorticity measures the intensity and direction of vortices (i.e., whirlpools) along the flow. That

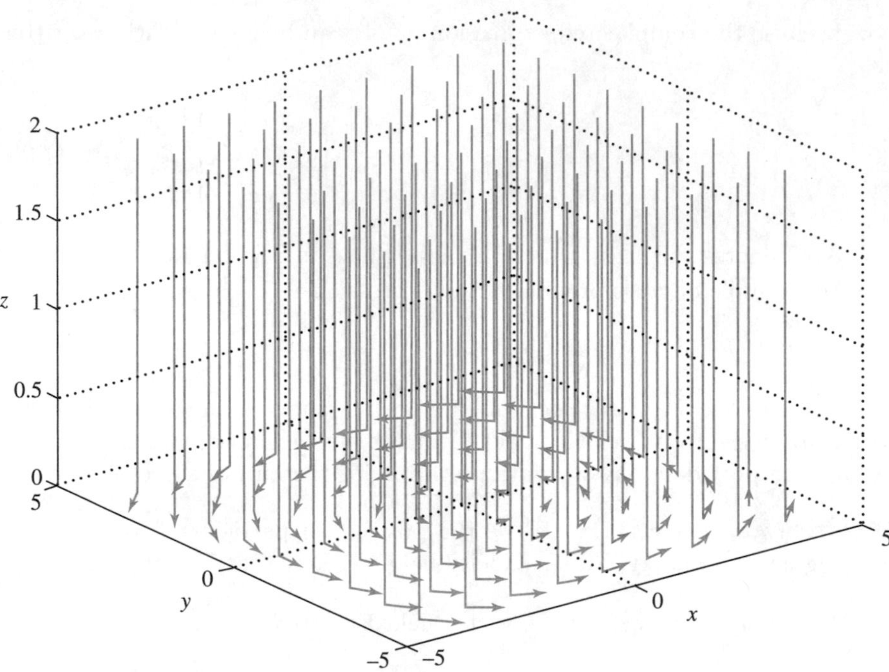

Figure 7.17 The vector field $\mathbf{F}(x, y) = [-y, x, 0]$ and its curl.

is, if a small pebble is placed on a point where the vorticity of the flow is nonzero, then the fluid will cause it to rotate around an internal axis pointing in the direction of $\mathbf{w}$ with a speed of rotation proportional to $\|\mathbf{w}\|$.

Based on this interpretation, a vector field $\mathbf{F}$ is said to be *irrotational* if its curl is zero. Observe, however, that this definition has nothing to do with the shape of the flow lines themselves. For example, you can verify with an explicit calculation that the vector field

$$\mathbf{F} = \begin{bmatrix} \dfrac{y}{x^2 + y^2} \\ -\dfrac{x}{x^2 + y^2} \\ 0 \end{bmatrix}, \qquad (x, y) \neq (0, 0) \tag{7.26}$$

has a vanishing curl, being therefore irrotational. However, its flow lines clearly rotate around the origin, as suggested by Figure 7.18, obtained from:

```
>> x = [-2:0.2:-0.2 0.2:0.2:2]; y = x; [X,Y] = meshgrid(x,y);
>> quiver(X,Y,Y./(X.^2+Y.^2),-X./(X.^2+Y.^2));
```

The correct interpretation of Figure 7.18 is that if you place a small pebble in a fluid with that velocity field, then the pebble will *not* rotate around any

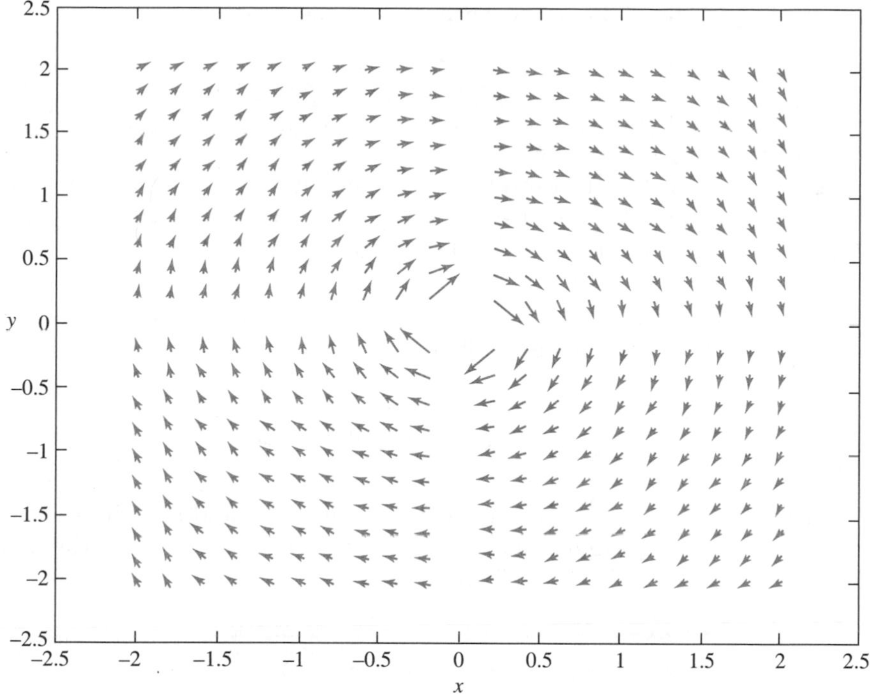

Figure 7.18 The irrotational vector field $\mathbf{F}(x, y) = \left[\frac{y}{x^2+y^2}, -\frac{x}{x^2+y^2}, 0\right]$.

of its own axes, but will nevertheless rotate around the origin according to the flow lines of the fluid. By contrast, if the velocity field for the fluid were the vector field of Figure 7.17, then the pebble would not only rotate around the origin but would also spin around its own vertical axis.

7.6 Line Integrals

Suppose that you are interested in calculating the total mass of a thin wire whose location in space is given by the path $\mathbf{s} : [a, b] \to \mathbb{R}^3$, for example, as in (7.15). Suppose further that the density of the wire can vary from point to point and therefore should be represented by a scalar function $f : \mathbb{R}^3 \to \mathbb{R}$, whose domain contains the image of $\mathbf{s}(t)$. By following the same strategy used to define arc lengths in Section 7.4, you can consider intermediate points $a = t_0 < t_1 \ldots < t_{n-1} < t_n = b$, take the density to be constant between the points $\mathbf{s}(t_{i-1})$ and $\mathbf{s}(t_i)$ (say, equal to $f(\mathbf{s}(t_{i-1}))$, and approximate the mass of the segment of wire connecting these points by

$$f(\mathbf{s}(t_{i-1}))\|\mathbf{s}(t_i) - \mathbf{s}(t_{i-1})\|. \qquad (7.27)$$

With the help of the Mean Value Theorem, this becomes

$$f(\mathbf{s}(t_{i-1}))\sqrt{\left(\frac{ds_1}{dt}(\xi_i)\right)^2+\left(\frac{ds_2}{dt}(\eta_i)\right)^2+\left(\frac{ds_3}{dt}(\nu_i)\right)^2}\,(t_i-t_{i-1}),$$

for some ξ_i, η_i, ν_i in the interval (t_{i-1}, t_i). By adding these elementary masses and taking the limit as the number of intermediate points goes to infinity, you obtain the total mass as an integral of the scalar function $f(x, y, z)$ along the image of the vector function $\mathbf{s}(t)$.

Line integral of a scalar function

This procedure leads to defining the *line integral* of a continuous function $f : \mathbb{R}^m \to \mathbb{R}$ along a continuously differentiable path $\mathbf{s} : [a, b] \to \mathbb{R}^m$ as

$$\int_{\mathbf{s}} f ds = \int_a^b f(\mathbf{s}(t))\|\dot{\mathbf{s}}(t)\|dt. \tag{7.28}$$

For two-dimensional vector functions $\mathbf{s} : [a, b] \to \mathbb{R}^2$ and a positive scalar function $f : \mathbb{R}^2 \to \mathbb{R}$, the line integral (7.28) can be interpreted as the area of a fence with height $f(\mathbf{s}(t))$ built on the image of $\mathbf{s}(t)$, as shown in Figure 7.19, which is generated by the following commands:

```
>> t = [0:0.05:3*pi]; a = zeros(1,length(t));
>> quiver3(t,cos(t),a,a,a,4+2*sin(t).*cos(t),0);
```

For an example of a line integral with applications in *Differential Geometry*, let $\mathbf{s}(t) : [a, b] \to \mathbb{R}^3$ be a path with $\dot{\mathbf{s}}(t) \neq \mathbf{0}$ for all $t \in [a, b]$, and define the *curvature* of its image at the point $t = t_0$ as

$$k(\mathbf{s}(t_0)) = \frac{\|\dot{\mathbf{s}}(t_0) \times \ddot{\mathbf{s}}(t_0)\|}{\|\dot{\mathbf{s}}(t_0)\|^3}. \tag{7.29}$$

Then, the *total curvature* of the image of $\mathbf{s}(t)$ is given by

$$\int_{\mathbf{s}} k ds = \int_a^b k(\mathbf{s}(t))\|\dot{\mathbf{s}}(t)\|dt = \int_a^b \frac{\|\dot{\mathbf{s}}(t) \times \ddot{\mathbf{s}}(t)\|}{\|\dot{\mathbf{s}}(t)\|^2}dt. \tag{7.30}$$

For instance, it is easy to see that the curvature for the circle $\mathbf{s} = [r \sin t, r \cos t]$ is constant and equal to $1/r$, implying that its total curvature is equal to 2π. Less obvious, but still intuitive enough, is the fact that the total curvature of any other closed curve is greater than 2π. It is also clear that by introducing *knots* along the image of $\mathbf{s}(t)$ you can increase its total curvature (think of how many times you need to bend a string to make a knot). A far less obvious result is that there should be a lower bound for the total curvature of knotted curves. This bound is found in the celebrated *Fary–Milnor Theorem*, which states that if the total curvature of a three-dimensional curve is less than or equal to 4π, then the curve can be unknotted.

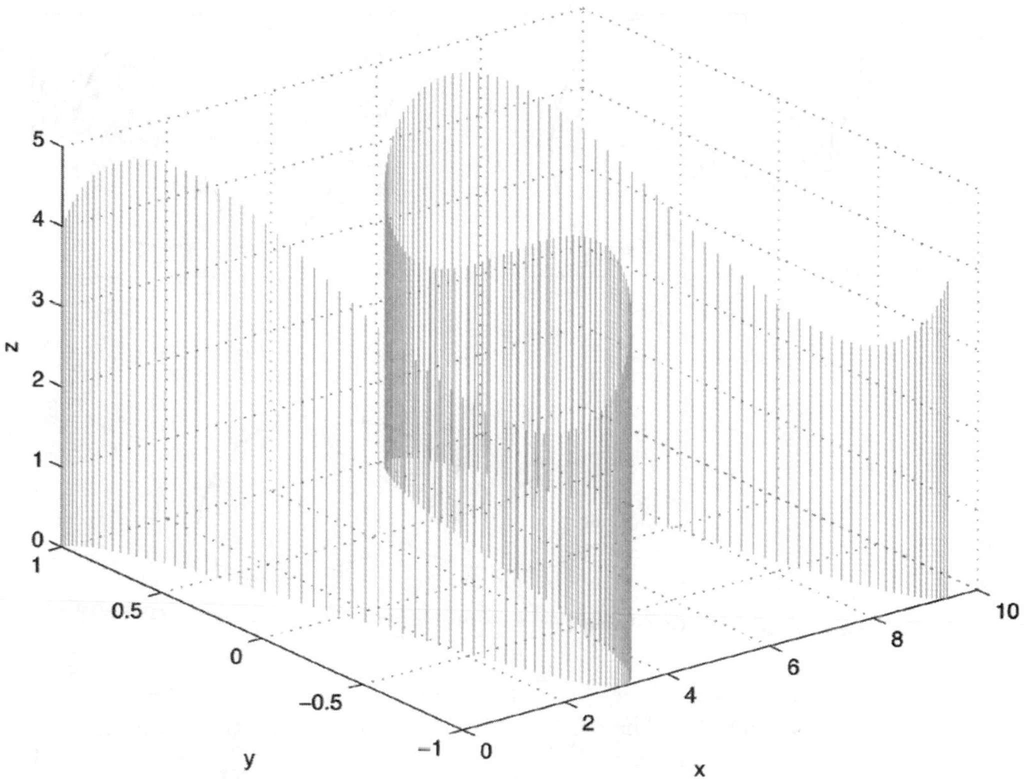

Figure 7.19 Line integral of the scalar function $f(x, y) = 4 + 2y\sin(x)$ along the path $\mathbf{s}(t) = [t, \cos t]$.

These possibilities are illustrated by the curves shown in Figure 7.20, which is obtained by using the following commands:

```
>> subplot(2,2,1); % cardioid
>> plot((sin(t)-1).*cos(t),(sin(t)-1).*sin(t),'r');
>> subplot(2,2,2); % limacon
>> plot((3*sin(t)+2).*cos(t),(3*sin(t)+2).*sin(t));
>> subplot(2,2,3:4); %trefoil knot
>> plot3(4*cos(2*t)+2*cos(t),4*sin(2*t)-2*sin(t),sin(3*t),'g');
```

The first curve is called a *cardioid*, because it resembles a heart, and is given by

$$\mathbf{s}_1(t) = \begin{bmatrix} (\sin t - 1)\cos t \\ (\sin t - 1)\sin t \end{bmatrix}. \tag{7.31}$$

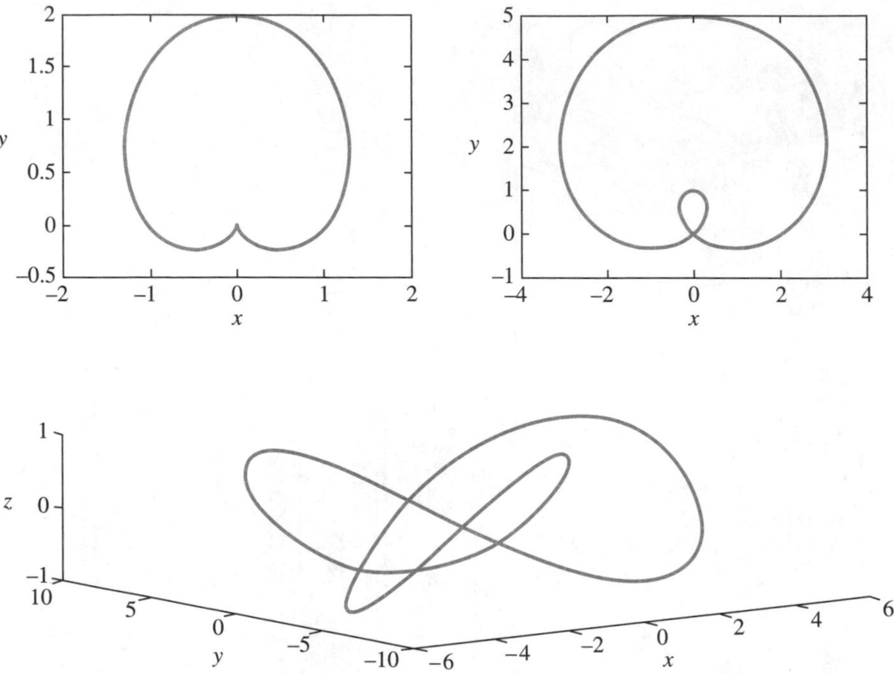

Figure 7.20 The cardioid, the limaçon, and the trefoil knot.

An explicit integration shows that its total curvature is 3π. The second curve is a *limaçon*, which is French for snail, and is given by

$$\mathbf{s}_2(t) = \begin{bmatrix} (3\sin t + 2)\cos t \\ (3\sin t + 2)\sin t \end{bmatrix}. \tag{7.32}$$

Notice that it contains an extra lace when compared with the cardioid, and so you can expect its total curvature to be larger. Indeed, another explicit integration (using the $\tan(t/2)$ substitution technique) gives its total curvature as equal to 4π. Notice, however, that the Fary–Milnor Theorem tells you that it can be unknotted, as is indeed the case when you think of it as embedded in three-dimensional spaces (having the extra dimension to undo the lace).

The third curve, on the other hand, given by

$$\mathbf{s}_3(t) = \begin{bmatrix} 4\cos 2t + 2\cos t \\ 4\sin 2t - 2\sin t \\ \sin 3t \end{bmatrix}, \tag{7.33}$$

is the *trefoil knot*, which cannot be unknotted. The Fary–Milnor Theorem then says that its total curvature must be strictly greater than 4π.

Exercise 7.7 Calculate the total curvature for the cardioid (7.31) and the limaçon (7.32) by explicit integration. Then, use the MATLAB integration function `quad` to find the total curvature for the trefoil knot (7.33).

Line integral of vector fields

To motivate our next definition, recall from elementary *Physics* that the work done by a constant force $\mathbf{F}$ when moving a particle along a straight-line segment between $\mathbf{x}$ and $\mathbf{x}_0$ is given by the scalar $\mathbf{F}\cdot(\mathbf{x}-\mathbf{x}_0)$. Now suppose that you want to calculate the work done by the force vector field $\mathbf{F} : \mathbb{R}^3 \to \mathbb{R}^3$ when moving an object along a general trajectory $\mathbf{s}(t)$. You can then follow the familiar strategy setting the force to be constant $\mathbf{F}(\mathbf{s}(t_{i-1}))$ on the segment connecting the points $\mathbf{s}(t_{i-1})$ and $\mathbf{s}(t_i)$ and approximate the work done on this segment by

$$\mathbf{F}(\mathbf{s}(t_{i-1}))\cdot(\mathbf{s}(t_i)-\mathbf{s}(t_{i-1})).$$

By reproducing the same arguments as before, we are ready to define the *line integral* of a continuous vector field $\mathbf{F} : \mathbb{R}^m \to \mathbb{R}^m$ along a continuously differentiable path $\mathbf{s} : \mathbb{R} \to \mathbb{R}^m$ as

$$\int_{\mathbf{s}} \mathbf{F}\cdot d\mathbf{s} = \int_a^b \mathbf{F}(\mathbf{s}(t))\cdot\dot{\mathbf{s}}(t)dt. \tag{7.34}$$

That is, since $\dot{\mathbf{s}}(t)$ has the direction of the tangent to the path at the point $\mathbf{s}(t)$, you can see that the line integral can be interpreted as a sum of the tangential components of the vector field $\mathbf{F}$ along the path $\mathbf{s}$.

In the special case where $\mathbf{F} = \nabla f$ for some scalar function f, we can use the chain rule and rewrite the preceding integrand in the form

$$\nabla f(\mathbf{s}(t))\cdot\dot{\mathbf{s}}(t) = \frac{df(\mathbf{s}(t))}{dt}.$$

By using the Fundamental Theorem of Calculus, we conclude that

$$\int_{\mathbf{s}} \nabla f\cdot d\mathbf{s} = f(\mathbf{s}(b)) - f(\mathbf{s}(a)). \tag{7.35}$$

In particular, the line integral of a gradient vector field depends only on the end points $\mathbf{s}(a)$ and $\mathbf{s}(b)$ but not on the path connecting them.

Exercise 7.8 Use MATLAB to compute the line integral of the vector field $\mathbf{F} = [yz, xz, xy]$ along the path $\mathbf{s} = [\cos t, \sin t, t]$ for $0 \le t \le 4\pi$. Compare it with the line integral along the straight line connecting the points $(1, 0, 0)$ and $(1, 0, 4\pi)$.

7.7 Surface Integrals

We now consider the special case of functions of the form $\mathbf{S}: \mathbb{R}^2 \to \mathbb{R}^3$. Instead of treating them as vector fields as in Section 7.5, we interpret the image $\mathbf{S}(u,v)$ as tracing a surface in $\mathbb{R}^3$ as (u,v) vary on a domain $D \in \mathbb{R}^2$, which in most cases is just the rectangle $D = [a,b] \times [c,d]$. By doing so, we obtain a way to describe surfaces that is much different from the graphs of scalar functions of two variables treated in Section 7.1. Namely, although both are subsets of $\mathbb{R}^3$, the *graph* of a scalar function $f: \mathbb{R}^2 \to \mathbb{R}$ consists of points of the form $(x, y, f(x,y))$, whereas the *image* of a function $\mathbf{S}: [a,b] \times [c,d] \to \mathbb{R}^3$ consists of points of the form $(S_1(u,v), S_2(u,v), S_3(u,v))$. That is, in one case we see both the image $f(x,y)$ and the variables (x,y) in the same plot, whereas in the other we see only the image $\mathbf{S}(u,v)$ and have to keep the variables (u,v) in the darkness of our minds. By analogy to the role played by the variable t in Section 7.4, we treat the variables (u,v) as parameters and call $\mathbf{S}(u,v)$ a *parametric surface.*

Clearly, the graph of any scalar function $f: \mathbb{R}^2 \to \mathbb{R}$ can be represented as a parametric surface simply by setting $S_1(u,v) = u$, $S_2(u,v) = v$, and $S_3(u,v) = f(u,v)$. However, the advantage of parametric surfaces is that they have a much larger scope. For example, the *torus* shown in Figure 7.21 is clearly not the graph of any function (can you see why?), but can still be described by the parametric surface

$$\mathbf{S}(u,v) = \begin{bmatrix} (5 + 2\sin u)\cos v \\ (5 + 2\sin u)\sin v \\ 2\cos u \end{bmatrix}, \tag{7.36}$$

for $0 \le u \le 2\pi$ and $0 \le v \le 2\pi$.

Plotting parametric surfaces in MATLAB requires only a minor modification of the procedure used so far for plotting graphs of scalar functions of two variables. The following example shows the commands used to plot the torus (7.36) on Figure 7.21:

```
>> u = linspace(0,2*pi,20); v = u; [U,V] = meshgrid(u,v);
>> X = (5+2*sin(U)).*cos(V);
>> Y = (5+2*sin(U)).*sin(V);
>> Z = 2*cos(U);
>> surf(X,Y,Z);
```

If we now fix the parameter $v = v_0$, then $\mathbf{S}(u, v_0)$ becomes a path depending only on the single variable u. Similarly, fixing $u = u_0$ leads to a path $\mathbf{S}(u_0, v)$ depending only on the variable v. We can then take the derivative of

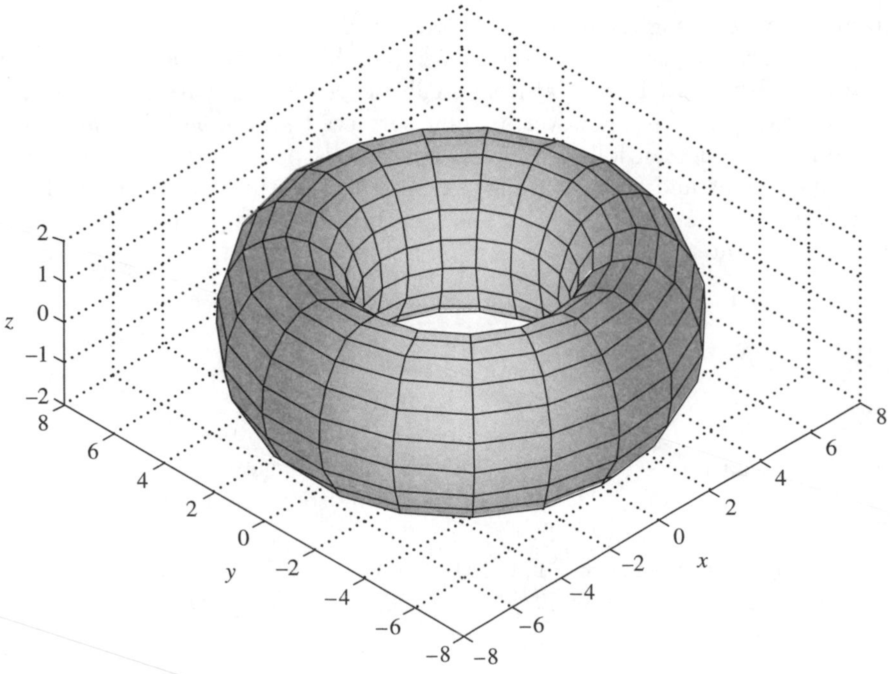

Figure 7.21 A torus.

each of these paths and define their tangent vectors at the point $\mathbf{S}(u_0, v_0)$ as

$$\mathbf{t}_u(u_0, v_0) = \left.\frac{\partial \mathbf{S}(u, v_0)}{\partial u}\right|_{u=u_0}, \tag{7.37}$$

$$\mathbf{t}_v(u_0, v_0) = \left.\frac{\partial \mathbf{S}(u_0, v)}{\partial v}\right|_{v=v_0}. \tag{7.38}$$

We say that a surface is *regular* at the point $\mathbf{S}(u_0, v_0)$ whenever

Regular surfaces

$$\mathbf{t}_u(u_0, v_0) \times \mathbf{t}_v(u_0, v_0) \neq \mathbf{0},$$

because we can then define the *tangent plane* as the plane determined by the tangent vectors. In this case, we call the vector

$$\mathbf{n}(u_0, v_0) = \frac{\mathbf{t}_u(u_0, v_0) \times \mathbf{t}_v(u_0, v_0)}{\|\mathbf{t}_u(u_0, v_0) \times \mathbf{t}_v(u_0, v_0)\|} \tag{7.39}$$

a *unit normal* to the surface at the point $\mathbf{S}(u_0, v_0)$. Observe that the vector

$$\tilde{\mathbf{n}}(u_0, v_0) = \frac{\mathbf{t}_v(u_0, v_0) \times \mathbf{t}_u(u_0, v_0)}{\|\mathbf{t}_u(u_0, v_0) \times \mathbf{t}_v(u_0, v_0)\|} = -\mathbf{n}(u_0, v_0),$$

obtained by reversing the order in the cross product, is also a unit normal to the surface, which corresponds to the fact that there are two sides to a surface at any regular point. That is, at least locally, every regular parametric surface is two sided. There are, however, regular surfaces that are *globally one sided*, meaning that one can follow a closed continuous path starting on one side of a given point and ending at the opposite side of the same point without ever changing sides along the way. The archetypical example of this seemingly paradoxical statement is a *Möbius strip*, parameterized by

$$\mathbf{M}(u,v) = \begin{bmatrix} (2+v\cos u/2)\cos u \\ (2+v\cos u/2)\sin u \\ v\sin u/2 \end{bmatrix}, \tag{7.40}$$

for $0 \le u \le 2\pi$ and $-1 \le v \le +1$. You can observe the one-sidedness of the Möbius strip in Figure 7.22, which is created by the following commands:

```
>> u = linspace(0,2*pi,20); v = linspace(-1,1,20); [U,V] = meshgrid(u,v);
>> surf((2+V.*cos(U/2)).*cos(U),(2+V.*cos(U/2)).*sin(U),V.*sin(U/2))
```

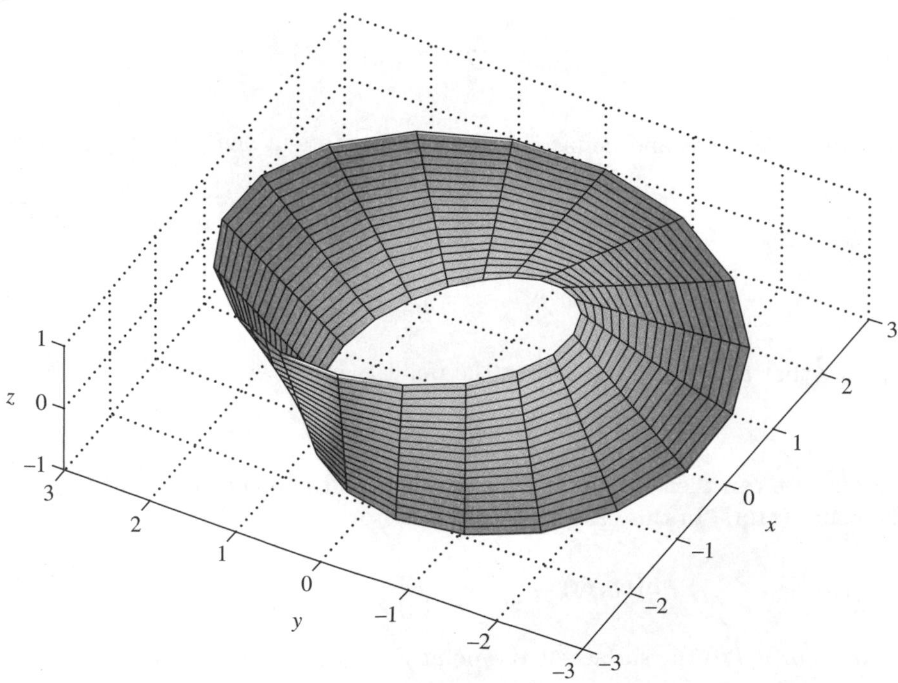

Figure 7.22 A Möbius strip.

As you can verify in the next exercise, the one-sidedness of a Möbius strip reflects itself in a discontinuity of the unit normal vector along a closed path on the surface. We say that a regular parametric surface $\mathbf{S} : \mathbb{R}^2 \to \mathbb{R}^3$ is *oriented* if we can continuously assign a unit normal to all of its points, which is in turn equivalent to being globally two-sided.

Exercise 7.9 Show that the vector field $\mathbf{n}$ in (7.39) has a discontinuity at $u = 0$ along the path $\mathbf{M}(u, 0)$, $0 \leq u \leq 2\pi$, $v = 0$ on the Möbius strip (7.40).

Surface area

We now want to provide a meaningful definition for the area of a parametric surface. The idea is to follow the same strategy used for the arc length of a path in Section 7.4, by considering a partition $a < \ldots < u_i < \ldots < b$ and $c < \ldots < v_j < \ldots < d$ and looking at the surface area between each four corners $\mathbf{S}(u_{i-1}, v_{j-1})$, $\mathbf{S}(u_{i-1}, v_j)$, $\mathbf{S}(u_i, v_{j-1})$, and $\mathbf{S}(u_i, v_j)$. Each of these areas can be approximated by the area of a parallelogram with sides $\Delta\mathbf{S}_i = \mathbf{S}(u_i, v_j) - \mathbf{S}(u_{i-1}, v_j)$ and $\Delta\mathbf{S}_j = \mathbf{S}(u_i, v_j) - \mathbf{S}(u_i, v_{j-1})$, which in turn is given by the norm $\|\Delta\mathbf{S}_i \times \Delta\mathbf{S}_j\|$. By using the Mean Value Theorem yet one more time, we see that

$$\|\Delta\mathbf{S}_i \times \Delta\mathbf{S}_j\| \approx \|\mathbf{t}_u(u_i, v_j) \times \mathbf{t}_v(u_i, v_j)\|(u_i - u_{i-1})(v_j - v_{j-1}),$$

which leads to a double integral when we add all areas together and take the limit as the number of points in the two partitions goes to infinity.

Motivated by these approximations, we define the *area* of a continuously differentiable regular surface $\mathbf{S} : [a, b] \times [c, d] \to \mathbb{R}^3$ as

$$A(\mathbf{S}) = \int_c^d \int_a^b \|\mathbf{t}_u \times \mathbf{t}_v\| du\, dv. \tag{7.41}$$

For example, for a sphere of radius r parameterized by

$$\mathbf{S}(u, v) = \begin{bmatrix} r \sin u \cos v \\ r \sin u \sin v \\ r \cos u \end{bmatrix}, \tag{7.42}$$

with $0 \leq u \leq \pi$, $0 \leq v \leq 2\pi$, you have

$$\mathbf{t}_u = \begin{bmatrix} r \cos u \cos v \\ r \cos u \sin v \\ -r \sin u \end{bmatrix}, \quad \mathbf{t}_v = \begin{bmatrix} -r \sin u \sin v \\ r \sin u \cos v \\ 0 \end{bmatrix}, \quad \mathbf{t}_u \times \mathbf{t}_v = \begin{bmatrix} r^2 \sin^2 u \cos v \\ r^2 \sin^2 u \sin v \\ r^2 \sin u \cos u \end{bmatrix},$$

leading to

$$A(\mathbf{S}) = \int_0^{2\pi} \int_0^{\pi} \|\mathbf{t}_u \times \mathbf{t}_v\| du dv = \int_0^{2\pi} \int_0^{\pi} r^2 \sin u du dv = 4\pi r^2.$$

Observe that the preceding integral came out easily as a result of several lucky simplifications occurring in the calculation of $\mathbf{t}_u \times \mathbf{t}_v$ and its norm. For more complicated functions, you almost certainly will have to resort to numerical integration of (7.41), after having obtained an expression for $\|\mathbf{t}_u \times \mathbf{t}_v\|$ in terms of u and v. We illustrate the procedure with the MATLAB command `dblquad`, using the sphere of radius $r = 2$ as an example. We first need to create a function to use as an integrand. This is done by creating the file `areasphere.m`:

```
function z=asphere(u,v)
% computes tangent vectors, normal vector and elementary areas
ell = length(u);
tu = [2*cos(u)*cos(v);2*cos(u)*sin(v);-2*sin(u)];
tv = [-2*sin(u)*sin(v);2*sin(u)*cos(v);zeros(1,ell)];
n = cross(tu,tv); % normal vector
for i=1:ell
    z(i)=norm(n(:,i));  % elementary areas
end
```

Next, we pass this function as an argument for `dblquad`, together with the appropriate integration limits, and obtain the desired area:

```
>> area = dblquad('asphere',0,pi,0,2*pi)

area =
    50.2655
```

Exercise 7.10 Calculate the area of the torus (7.36) using the formula (7.41) and the MATLAB command `dblquad`.

Surface integrals

Pursuing the same example as in Section 7.6, suppose that you are interested in the total mass of a continuously differentiable regular surface whose density is given by a continuous scalar function $f : \mathbb{R}^3 \to \mathbb{R}$. Following the same steps as before, you are led to conclude that such total mass is obtained as the *surface integral* of f over the surface $\mathbf{S} : [a, b] \times [c, d] \to \mathbb{R}^3$, which is defined as

$$\int\int_{\mathbf{S}} f dS = \int_c^d \int_a^b f(\mathbf{S}(u, v)) \|\mathbf{t}_u \times \mathbf{t}_v\| du dv. \tag{7.43}$$

For example, if the density of the sphere (7.42) is

$$f(\mathbf{S}(u, v))) = (S_3(u, v))^2 = r^2 \cos^2 u,$$

an explicit integration using this formula gives a total mass equal to $4\pi r^4/3$. We can verify this numerically (with $r = 2$) by first creating the integrand

```
function m=msphere(u,v)
% computes tangent vectors, normal vector and elementary masses
ell = length(u);
tu = [2*cos(u)*cos(v);2*cos(u)*sin(v);-2*sin(u)];
tv = [-2*sin(u)*sin(v);2*sin(u)*cos(v);zeros(1,ell)];
n = cross(tu,tv); % normal vector
for i=1:ell
    z(i)=norm(n(:,i));
end
m = 4*(cos(u).^2).*z; % elementary masses
```

and then using the command `dblquad` to evaluate the integral:

```
>> mass = dblquad('msphere',0,pi,0,2*pi)

mass =
    67.0206
```

We now want to introduce the concept of surface integral of a vector field by using yet another example from *Physics* as motivation. Suppose that a fluid with uniform density ρ moves with constant velocity $\mathbf{v}$ across the surface of a parallelogram with sides $\mathbf{x}$ and $\mathbf{y}$. Then, the amount of fluid that crosses this surface during a time Δt must fill the parallelepiped with sides $\mathbf{x}$, $\mathbf{y}$, and $\mathbf{v}\Delta t$ in space, whose volume is given by $(\mathbf{x} \times \mathbf{y}) \cdot \mathbf{v}\Delta t$. Therefore, because the mass of fluid in this parallelepiped is $\rho(\mathbf{x} \times \mathbf{y}) \cdot \mathbf{v}\Delta t$, if we denote $\mathbf{F} = \rho\mathbf{v}$, we find that the *rate* of fluid flow across the surface per unit of time is simply given by $\mathbf{F} \cdot (\mathbf{x} \times \mathbf{y})$.

We are then led to define the *surface integral* of a continuous vector field $\mathbf{F} : \mathbb{R}^3 \to \mathbb{R}^3$ over an oriented surface $\mathbf{S} : [a, b] \times [c, d] \to \mathbb{R}^3$ as

$$\iint_{\mathbf{S}} \mathbf{F} \cdot d\mathbf{S} = \int_c^d \int_a^b \mathbf{F}(\mathbf{S}(u, v)) \cdot (\mathbf{t}_u \times \mathbf{t}_v) du dv. \tag{7.44}$$

We interpret this as the *flux* of vector field $\mathbf{F}$ across the surface $\mathbf{S}$. Observe that the term $\mathbf{F}(\mathbf{S}(u, v)) \cdot (\mathbf{t}_u \times \mathbf{t}_v)$ corresponds to the *normal* component of the vector field $\mathbf{F}$ across the surface $\mathbf{S}$, and therefore depends on the orientation of the surface (that is, the order in which we take the cross product in (7.44)).

For example, let

$$\mathbf{E}(\mathbf{x}) = \frac{q\mathbf{x}}{4\pi\varepsilon_0 \|\mathbf{x}\|^3}, \quad \mathbf{x} \neq \mathbf{0} \tag{7.45}$$

denote the electric field created by a point charge q located at the origin, where ε_0 is a constant that depends on the units adopted (for instance, in the International System of Units, $\varepsilon_0 \approx 8.8542 \times 10^{-12}$ $\mathrm{C}^2/\mathrm{N} \cdot \mathrm{m}^2$). Suppose that you want to calculate the flux of this electric field across the sphere (7.42) when $q = 10$ and $r = 2$ in units such that $\varepsilon_0 = 1$. Start by preparing the integrand:

```
function e=electric(u,v)
% computes tangent vectors, normal vector and normal components
 tu = [2*cos(u)*cos(v);2*cos(u).*sin(v);-2*sin(u)];
 tv = [-2*sin(u)*sin(v);2*sin(u)*cos(v);zeros(1,length(u))];
 n = cross(tu,tv); % normal vector
 field = (10/(16*pi))*[sin(u)*cos(v);sin(u)*sin(v);cos(u)];
 for i=1:length(u)
    e(i)=field(:,i)'*n(:,i);  % normal components
 end
```

and then use `dblquad` to evaluate the integral:

```
>> flux=dblquad('electric',0,pi,0,2*pi)

flux =
   10.0000
```

The fact that this flux equals q/ε_0 is a special case of *Gauss's law*, which states that the flux of an electric field across any closed surface is a constant multiple of the net electric charge inside it.

Exercise 7.11 Use MATLAB to verify Gauss's law for the flux of the electric field (7.45) for a charge $q = 10$ across a cube centered at the origin with sides of length $\ell = 1$.

7.8 Integral Theorems

We review in this section the important theorems of Green's, Stokes's, and Gauss's, relating differential and integral properties of vector fields. Both from a theoretical point of view and regarding their applications, these theorems correspond to the pinnacle of vector calculus, and being well versed in them brings tremendous advantages to a student in any mathematical field.

Simple regions

To state the first theorem, we define a *y-simple* region D in $\mathbb{R}^2$ as a set of points (x, y) of the form

$$D = \{a \leq x \leq b,\ \ f_1(x) \leq y \leq f_2(x)\},$$

for some continuous functions f_1 and f_2. Similarly, we define an *x-simple* region D as a set of points (x, y) of the form

$$D = \{g_1(y) \leq x \leq g_2(y),\ \ c \leq y \leq d\},$$

for some continuous functions g_1 and g_2. Clearly, both types of regions are very special cases of parametric surfaces defined in Section 7.7. To see this, all you have to do is take the region D as the *domain* for the parameters

(u, v) and observe that it coincides with the image of $\mathbf{S}(u, v) = (u, v, 0)$. To convince yourself of this statement, try the following MATLAB commands, which make explicit use of `surf` to illustrate the point, resulting in Figure 7.23:

```
>> u = linspace(0,2*pi,20); [U,A] = meshgrid(u,u);
>> for i=1:20 % creates the domain for a y-simple region
V(:,i)=linspace(-sin(u(i)),sin(u(i)),20)';
end
>> subplot(1,2,1); surf(U,V,zeros(20,20));
>> v = linspace(-1,1,20); [B,V] = meshgrid(v,v);
>> for i=1:20 % creates the domain for an x-simple region
U(i,:)=linspace(0,v(i)^3-v(i)+1,20);
end
>> subplot(1,2,2); surf(U,V,zeros(20,20));
```

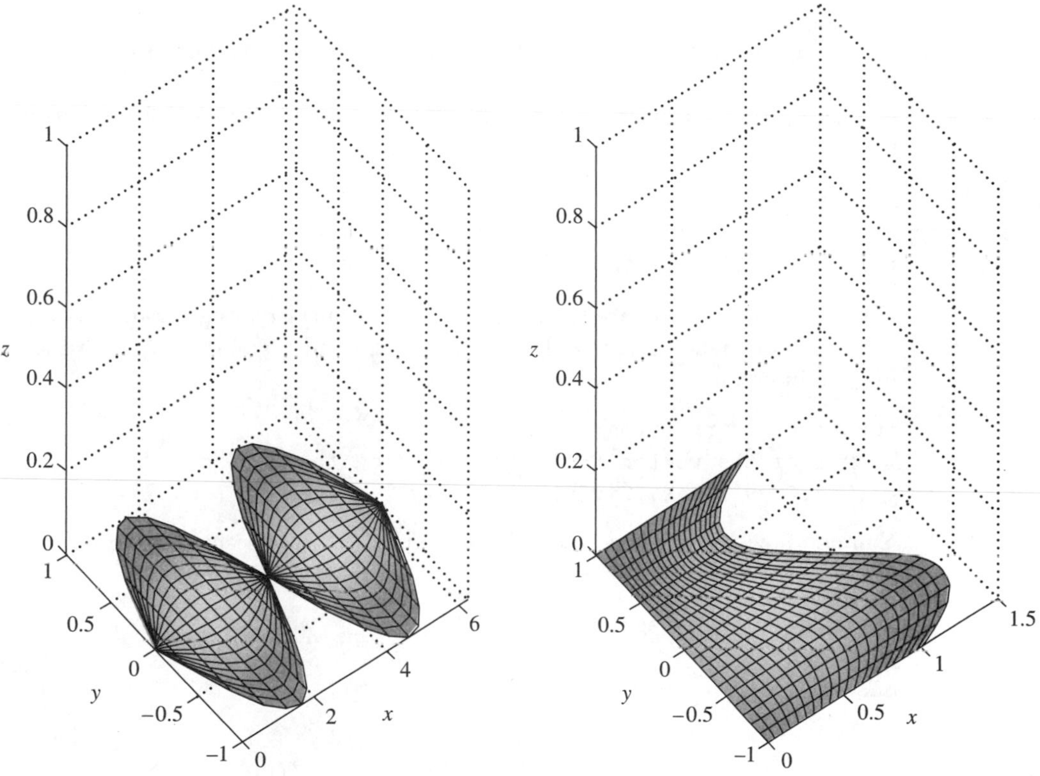

Figure 7.23 The y-simple region $D_1 = \{0 \leq x \leq 2\pi,\ -\sin x \leq y \leq \sin x\}$ and the x-simple region $D_2 = \{0 \leq x \leq y^3 - y + 1,\ -1 \leq y \leq 1\}$

Green's theorem

As a final piece of preparation, we shall say that the region is *simple* if it is simultaneously an x- and y-simple region. Observe that the *boundary* of a simple region traces a closed curve in $\mathbb{R}^2$, which we denote by ∂D. If we can then find a one-to-one path $\mathbf{s} : [t_0, t_1] \to \mathbb{R}^2$, with $\mathbf{s}(t_0) = \mathbf{s}(t_1)$, whose image coincides with ∂D, we say that it is a parametrization for the boundary of D. Moreover, we say that the path $\mathbf{s}$ has *positive orientation* if it traces ∂D in a counterclockwise manner as the parameter t varies in $[t_0, t_1]$. We now have all the necessary ingredients for *Green's theorem*:

Theorem 7.3 (Green) *Let $\mathbf{F} : \mathbb{R}^2 \to \mathbb{R}^2$ be a continuously differentiable vector field, and let $\mathbf{s} : [t_0, t_1] \to \mathbb{R}^2$ be a positively oriented, continuously differentiable parametrization for the boundary ∂D of a simple region D in $\mathbb{R}^2$. Then,*

$$\int\int_D \left(\frac{\partial F_2}{\partial x_1} - \frac{\partial F_1}{\partial x_2} \right) dx\, dx2 = \int_{\partial D} \mathbf{F} \cdot d\mathbf{s}. \tag{7.46}$$

As a verification of Green's theorem, you can use MATLAB to calculate both sides of Equation (7.46) for the vector field $\mathbf{F} = [x^2y^5, x^3y^2]$ when ∂D is a unit circle parametrized as $\mathbf{s}(t) = [\cos t, \sin t]$, for $0 \le t \le 2\pi$. As usual, first you need to prepare the integrand for the line integral:

```
function f=line_integrand(t)
% tangential component of the vector field evaluated on the path:
f=(cos(t).^2.*sin(t).^5).*(-sin(t))+(cos(t).^3.*sin(t).^2).*cos(t);
```

Then, you need to prepare the integrand for the surface integral, and since the command `dblquad` accepts only rectangular domains, you do this by truncating the integrand to zero outside the circle:

```
function g=surf_integrand(x,y)
% curl of the vector field evaluated on the surface:
g=(3*x.^2*y^2-5*x.^2*y^4).*((x.^2+y^2)<=1);
```

You can now perform the integrals and verify (7.46) subject to the numerical error:

```
>> quad('line_integrand',0,2*pi)

ans =
    0.1473

>> dblquad('surf_integrand',-1,1,-1,1)

ans =
    0.1475
```

A generalization of Green's theorem for vector fields and surfaces in $\mathbb{R}^3$ is provided by *Stokes's theorem*. Before stating it, let us define the boundary $\partial \mathbf{S}$ of a one-to-one surface $\mathbf{S} : \mathbb{R}^2 \to \mathbb{R}^3$ with domain $D \subset \mathbb{R}^2$ as the image of ∂D under $\mathbf{S}$. That is, if ∂D is a curve parametrized by $t \to [u(t), v(t)]$, then $\partial \mathbf{S}$ is a curve parametrized by $t \to \mathbf{S}(u(t), v(t))$. For oriented surfaces, we say that this boundary is *positively oriented* if it is traced in a counterclockwise manner with respect to the unit normal for the surface.

Stokes's theorem

Theorem 7.4 (Stokes) *Let $\mathbf{F} : \mathbb{R}^3 \to \mathbb{R}^3$ be a continuously differentiable vector field, and let $\mathbf{S} : \mathbb{R}^2 \to \mathbb{R}^3$ be a one-to-one oriented differentiable surface. Assume further that the domain of $\mathbf{S}$ is a simple region D in $\mathbb{R}^2$ and that $\partial \mathbf{S} = \mathbf{S}(\partial D)$ is positively oriented. Then,*

$$\iint_{\mathbf{S}} \nabla \times \mathbf{F} \cdot d\mathbf{S} = \int_{\partial \mathbf{S}} \mathbf{F} \cdot ds. \tag{7.47}$$

You can illustrate Stokes's theorem with

$$\mathbf{F}(x, y, z) = \begin{bmatrix} e^{xy} \\ yz^2 \\ e^{x^2} y \end{bmatrix} \quad \Rightarrow \quad \nabla \times \mathbf{F} = \begin{bmatrix} e^{x^2} - 2yz \\ -2xye^{x^2} \\ -xe^{xy} \end{bmatrix},$$

over the hemisphere obtained from (7.42) with $r = 2$, $0 \le u \le \pi$, and $-\pi/2 \le v \le \pi/2$. The hemisphere is shown in Figure 7.24, which is obtained from the following commands:

```
>> u = linspace(0,pi,20); v = linspace(-pi/2,pi/2,20);
>> [U,V] = meshgrid(u,v);
>> surf(2*sin(U).*cos(V),2*sin(U).*sin(V),2*cos(U));
```

Here, $D = [0, \pi] \times [-\pi/2, \pi/2]$ and the image of the boundary of this rectangle under the map $\mathbf{S}$ is the circle $\partial \mathbf{S}(t) = [0, 2\sin t, 2\cos t]$, for $0 \le t \le 2\pi$. Observe that this circle is traced in a counterclockwise manner with respect to the orientation of the surface provided by the unit normal at the point $(2, 0, 0)$ in the negative x direction, and you need to make sure that the order in which you take the cross product $\mathbf{t}_u \times \mathbf{t}_v$ in the surface integral is consistent with this choice of orientation.

Again, start by preparing the integrand for the line integral:

```
function f=line_integrand_2(t)
% tangential component of the vector field along the path
f=16*cos(t).^3.*sin(t)-4*sin(t).^2;
```

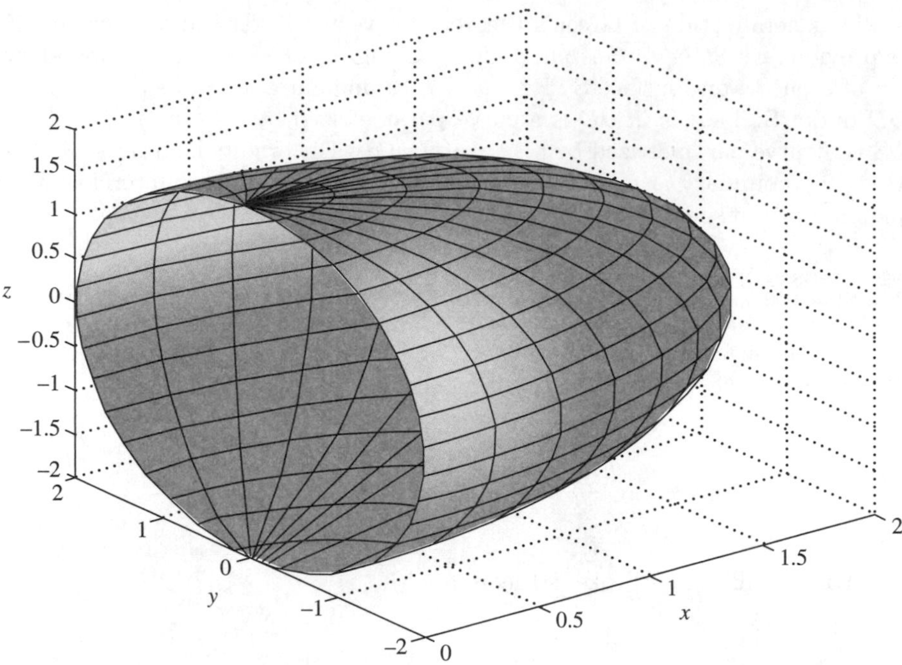

Figure 7.24 A hemisphere used as an example of Stokes's theorem. Notice that the boundary circle is positively oriented (counterclockwise) when the unit normal to the sphere points in the *inward* direction.

Next, prepare the integrand for the surface integral:

```
function g=surf_integrand_2(u,v)
% tangent vectors:
tu=[2*cos(u)*cos(v);2*cos(u)*sin(v);-2*sin(u)];
tv=[-2*sin(u)*sin(v);2*sin(u)*cos(v);zeros(1,length(u))];
% normal vector with the correct orientation:
N=cross(tv,tu);
% curl of the vector field evaluated on the surface:
curlf=[exp((2*sin(u)*cos(v))^2)-8*sin(u)*sin(v).*cos(u);
    -8*sin(u).^2*cos(v)*sin(v).*exp((2*sin(u)*cos(v)).^2);
    2*sin(u)*cos(v).*exp(8*sin(u).^2*cos(v)*sin(v))];
% normal component
for i=1:length(u)
    g(i)=curlf(:,i)'*N(:,i);
end
```

Finally, evaluate the integrals, thereby verifying (7.47):

```
>> quad('line_integrand_2',0,2*pi)

ans =
  -12.5664

>> dblquad('surf_integrand_2',0,pi,-pi/2,pi/2)

ans =
  -12.5664
```

The line integral appearing in (7.47) gives the total tangential component of the vector field along the boundary curve and is called the *circulation* of $\mathbf{F}$ around the $\partial\mathbf{S}$. For instance, if $\mathbf{F}$ is a force field acting on a body, then its circulation measures the total work necessary to move the body along the closed curve $\partial\mathbf{S}$. Alternatively, if $\mathbf{v}$ is the velocity vector field of a fluid, then its circulation $\int_{\partial\mathbf{S}} \mathbf{v}\cdot d\mathbf{s}$ measures the tendency for the fluid to move clockwise along $\partial\mathbf{S}$. You, therefore, see from (7.47) that the curl at a given point measures the fluid's circulation per unit area, justifying the interpretation introduced in Section 7.5.

We must mention that Stokes's theorem requires the surface to be oriented. If this hypothesis is removed, the theorem is no longer valid, as you can verify in the next exercise.

Exercise 7.12 Explicitly calculate both sides of Equation (7.47) for the irrotational vector field given in (7.26) and the Möbius strip defined in (7.40).

Gauss's theorem

The final theorem of this chapter relates a double integral over a closed surface to a triple integral over the volume enclosed by it. We are interested in a *simple volume* $V \subset \mathbb{R}^3$, implying that the boundaries ∂V can be described as the union of at most six *faces*. The surface integral of a vector field over the surface ∂V is then understood as the sum of the surface integral over each of these faces, with the convention that their unit normal vectors point outward from the volume V.

Theorem 7.5 (Gauss) *Let $\mathbf{F} : \mathbb{R}^3 \to \mathbb{R}^3$ be a continuously differentiable vector field, and let V be a simple volume with an oriented boundary $\partial V = \mathbf{S}$. Then,*

$$\iiint_V (\nabla \cdot \mathbf{F})\, dV = \iint_{\mathbf{S}} \mathbf{F} \cdot d\mathbf{S}. \tag{7.48}$$

As an illustration of Gauss's theorem, consider the vector field

$$\mathbf{F} = \begin{bmatrix} x^4 e^y \\ e^x \cos y \\ \sin x + y^2 z \end{bmatrix}$$

over the box bounded by the planes $x = 0, x = 1$, $y = 0, y = 2$, and $z = -1, z = 1$. We have that

$$\nabla \cdot \mathbf{F} = 4x^3 e^y - e^x \sin y + y^2,$$

and so the triple integral can be calculated by using the command **triquad**:

```
>> divF = '4*x.^3.*exp(y)-exp(x).*sin(y)+y.^2';
>> triplequad(divF,0,1,0,2,-1,1)

ans =
   13.2448
```

Next, the total surface integral of **F** across all the faces of the box is as follows:

```
>> F1 = '-1*(sin(x)+y^2*(-1))'; s(1) = dblquad(F1,0,1,0,2); % plane z=-1
>> F2 = 'sin(x)+y^2'; s(2) = dblquad(F2,0,1,0,2); % plane z=1
>> F3 = '-1*(exp(x)*cos(0))'; s(3) = dblquad(F3,0,1,-1,1); % plane y=0
>> F4 = 'exp(x)*cos(2)'; s(4) = dblquad(F4,0,1,-1,1); % plane y=2
>> F5 = '-1*(0*exp(y))'; s(5) = dblquad(F5,0,2,-1,1); % plane x=0
>> F6 = 'exp(y)'; s(6) = dblquad(F6,0,2,-1,1); % plane x=1
>> sum(s)

ans =
   13.2448
```

which confirms (7.48).

Suppose now that $\mathbf{F} = \rho\mathbf{v}$, where $\mathbf{v}$ is the velocity field of a fluid with density ρ. Then, according to the discussion in Section 7.7, you can see that the left-hand side of (7.48) gives the total rate in which fluid mass flows outward in the region V. You can then conclude from Gauss's theorem that the divergence of the velocity field at a given point measures the fluid's outward flux per unit of volume, justifying the interpretation given in Section 7.5.

7.9 Summary and Notes

This chapter generalizes the concepts of *Calculus* to functions of the form $\mathbf{F} : \mathbb{R}^n \to \mathbb{R}^m$ and explores them using the commands and graphical tools of MATLAB.

- **Section 7.1:** A *scalar function* of several variables is a function of the form $f : \mathbb{R}^n \to \mathbb{R}$. Its *domain* $D(f)$ is the largest subset of $\mathbb{R}^n$ where the function can be defined, while its *graph* $\Gamma(f)$ is the subset of $\mathbb{R}^{n+1}$ consisting of points of the form $(\mathbf{x}, f(\mathbf{x}))$. A *level curve* of value c for a function $f : \mathbb{R}^2 \to \mathbb{R}$ is the set of points (x, y) satisfying $f(x, y) = c$. Similarly, a *level surface* of value c for a function $f : \mathbb{R}^3 \to \mathbb{R}$ is the set of points (x, y, z) satisfying $f(x, y, z) = c$.

- **Section 7.2:** *Partial derivatives* give the rate of change of a function $f : \mathbb{R}^n \to \mathbb{R}$ with respect to one of its variables when all the others are kept fixed, whereas the *tangent plane* gives a linear approximation of f near a point $\mathbf{x}_0$. When this approximation is good, the function is said to be *differentiable*.

- **Section 7.3:** The *gradient* of a function $f : \mathbb{R}^n \to \mathbb{R}$ is an n-dimensional row vector with components $\frac{\partial f}{\partial x_i}$, $i = 1, \dots, n$. Using the *nabla* operator $\nabla = \left[\frac{\partial}{\partial x_1}, \dots, \frac{\partial}{\partial x_n}\right]$, the gradient can be written as ∇f. The *directional derivative* measures the rate of change of a function $f : \mathbb{R}^n \to \mathbb{R}$ in the direction of a vector $\mathbf{v}$ and can be obtained with the help of the gradient vector as $D_{\mathbf{v}} f(\mathbf{x}_0) = \nabla f(\mathbf{x}_0) \cdot \mathbf{v}$. Theorem 7.2 gives further geometrical interpretation of the gradient as a vector that is perpendicular to the level surfaces of f.

- **Section 7.4:** A *path* is a function of the form $\mathbf{s} : [a, b] \to \mathbb{R}^m$. Its *tangent vector* at $\mathbf{s}(t_0)$ is given by $\dot{\mathbf{s}}(t_0)$. The arc length for the path $\mathbf{s}$ between the points a and b is $L(\mathbf{s}) = \int_a^b \|\dot{\mathbf{s}}(t)\| dt$.

- **Section 7.5:** A *vector field* is a function of the form $\mathbf{F} : \mathbb{R}^n \to \mathbb{R}^m$. The analogue of its derivative is the *Jacobian matrix*, defined as the $m \times n$ matrix $\mathbf{DF}(\mathbf{x}_0)$ with components $\frac{\partial F_i}{\partial x_j}$, $i = 1, \dots, m$, $j = 1, \dots, n$. The *divergence* of a vector field $\mathbf{F} : \mathbb{R}^n \to \mathbb{R}^n$ is the scalar $\nabla \cdot \mathbf{F}$ and measures the outward flux of the vector field per unit of volume. The vector field is called *incompressible* if its divergence is zero. The *curl* of a vector field $\mathbf{F} : \mathbb{R}^3 \to \mathbb{R}^3$ is the vector $\nabla \times \mathbf{F}$ and measures the intensity and direction of vortices of the vector field. The vector field is called *irrotational* if its curl is zero.

- **Section 7.6:** The *line integral* of a scalar function $f : \mathbb{R}^m \to \mathbb{R}$ along a path $\mathbf{s} : [a, b] \to \mathbb{R}^m$ is defined as $\int_{\mathbf{s}} f ds = \int_a^b f(\mathbf{s}(t)) \|\dot{\mathbf{s}}\| dt$. The line integral of a vector field $\mathbf{F} : \mathbb{R}^m \to \mathbb{R}^m$ along a path $\mathbf{s} : \mathbb{R} \to \mathbb{R}^m$ is defined as $\int_{\mathbf{s}} \mathbf{F} \cdot d\mathbf{s} = \int_a^b \mathbf{F}(\mathbf{s}(t)) \cdot \dot{\mathbf{s}}(t) dt$. In the special case $\mathbf{F} = \nabla f$, the line integral reduces to $\int_{\mathbf{s}} \nabla f \cdot d\mathbf{s} = f(\mathbf{s}(b)) - f(\mathbf{s}(a))$, which depends only on the end points but not on the path itself.

- **Section 7.7:** A *parametric surface* is a map of the form $\mathbf{S} : \mathbb{R}^2 \to \mathbb{R}^3$. Its tangent vectors are defined as $\mathbf{t}_u = \frac{\partial \mathbf{S}}{\partial u}$ and $\mathbf{t}_v = \frac{\partial \mathbf{S}}{\partial v}$, and the surface is said to be *regular* at a point $\mathbf{S}(u_0, v_0)$ if $\mathbf{t}_u(u_0, v_0) \times \mathbf{t}_v(u_0, v_0) \neq \mathbf{0}$. The *area* of a regular surface is given by $A(\mathbf{S}) = \int_c^d \int_a^b \|\mathbf{t}_u \times \mathbf{t}_v\| du dv$, whereas the *surface integral* of a scalar function $f : \mathbb{R}^3 \to \mathbb{R}$ over $\mathbf{S}$ is $\int\int_{\mathbf{S}} f dS = \int_c^d \int_a^b f(\mathbf{S}(u, v)) \|\mathbf{t}_u \times \mathbf{t}_v\| du dv$. Similarly, the surface integral of a vector field $\mathbf{F} : \mathbb{R}^3 \to \mathbb{R}^3$ over $\mathbf{S}$ is defined as $\int\int_{\mathbf{S}} \mathbf{F} \cdot d\mathbf{S} = \int_c^d \int_a^b \mathbf{F}(\mathbf{S}(u, v)) \cdot (\mathbf{t}_u \times \mathbf{t}_v) du dv$.

- **Section 7.8:** *Green's theorem* states that the surface integral of the curl of a vector field $\mathbf{F} : \mathbb{R}^2 \to \mathbb{R}^2$ over a two-dimensional region D equals the line integral of $\mathbf{F}$ along the boundary of D, that is, $\int\int_D \nabla \times \mathbf{F} \cdot d\mathbf{S} = \int_{\partial D} \mathbf{F} \cdot d\mathbf{s}$. Its generalization to three-dimensional vector fields $\mathbf{F} : \mathbb{R}^3 \to \mathbb{R}^3$ is given by *Stokes's theorem*, where the surface integral is performed over a surface $\mathbf{S}$ and the line integral along its boundary $\partial \mathbf{S}$, that is, $\int\int_{\mathbf{S}} \nabla \times \mathbf{F} \cdot d\mathbf{S} = \int_{\partial \mathbf{S}} \mathbf{F} \cdot d\mathbf{s}$. Finally, *Gauss's theorem* states that the integral of the divergence of a vector field $\mathbf{F} : \mathbb{R}^3 \to \mathbb{R}^3$ over a volume V equals the surface integral of $\mathbf{F}$ over the surface enclosing V, that is, $\int\int\int_V (\nabla \cdot \mathbf{F})\, dV = \int\int_{\mathbf{S}} \mathbf{F} \cdot d\mathbf{S}$.

The material for this chapter is treated at an introductory level in [26]. For a more thorough exposition, including a wealth of examples and historical notes, we recommend [18]. The use of MATLAB commands in relation to vector calculus is well described in [11].

7.10 Exercises

1. Use the commands `contour` and `clabel` to find level curves for the following functions:

   (a) $f(x, y) = \sqrt{4 - x^2 - y^2}$
   (b) $f(x, y) = \sqrt{x^2 + y^2}$
   (c) $f(x, y) = x^2 - y^2$
   (d) $f(x, y) = |xy|$
   (e) $f(x, y) = x^2 + 4y^2$
   (f) $f(x, y) = \sqrt{1 - \frac{x^2}{4} - \frac{y^2}{9}}$.

2. Use the commands `surfc` and `surf` to plot the graphs of the functions in the previous exercise with and without their level curves.

3. Compare the level curves in Figure 7.4 with those of a Cobb–Douglas production function with parameters $a = 1$, $\alpha = 0.25$, and $\beta = 0.75$,

and decide in which case a change in the investment capital K has a greater impact on the number of labor hours L necessary to achieve a fixed level of production.

4. (a) Use the command `isosurface` to obtain the level surfaces for the function $f(x, y, z) = 4x^2 + y^2 + 9z^2$ corresponding to the levels $c = 1, 4, 9$.
   (b) Use the command `slice` to see how this function varies along the planes $x = -2$, $y = -2$, and $z = 2$.
   (c) Use the command `slice` to see how this function varies on the surface $z = x^2 - y^2$.

5. Use the command `ezsurf` to obtain the graphs of the following functions. In each case, either find $\lim_{(x,y)\to(0,0)} f(x, y)$ if it exists, or explain why it does not exist.
   (a) $f(x, y) = \dfrac{x^2 y}{x^2 + y^2}$.
   (b) $f(x, y) = \dfrac{x^2 y}{x^4 + y^2}$.
   (c) $f(x, y) = \dfrac{\sin(x^2 + y^2)}{x^2 + y^2}$.
   (d) $f(x, y) = \dfrac{\sin(xy)}{x^2}$.
   (e) $f(x, y) = \dfrac{\cos(xy) - 1}{x}$.

6. (a) Find the partial derivatives of $f(x, y) = \dfrac{xy}{x^2 + y^2}$ and verify that they are well defined at $(x, y) = (0, 0)$.
   (b) Use the command `surf` to obtain the graphs of f and its partial derivatives.
   (c) Show that $\frac{\partial f}{\partial x}$ and $\frac{\partial f}{\partial y}$ are not continuous at $(x, y) = (0, 0)$.
   (d) Show that f is not differentiable at $(x, y) = (0, 0)$.

7. (a) Find the partial derivatives of $f(x, y) = \sqrt{1 - \frac{x^2}{4} - \frac{y^2}{9}}$.
   (b) Obtain the expression for its tangent plane at the point $(x_0, y_0) = (1, 1)$.
   (c) Use the command `surf` to plot the graph of f and its tangent plane at $(x_0, y_0) = (1, 1)$ on the same figure.

8. Let $f(x, y) = 10 - 2x^2 - 3y^2$ represent the height of a mountain at points with coordinates (x, y).

(a) Use the command `surf` to plot the graph of f.

(b) Find the analytic expression for the gradient $\nabla f(x, y)$.

(c) Find the direction of fastest increase of f at the point $(x, y) = (2, 0)$.

(d) If a river had its source at the point $(x, y) = (-1, 1)$, in which direction would it begin to flow?

(e) Compute the gradient of f numerically using the command `gradient`.

(f) Use the commands `quiver` and `countour` to plot level curves and the gradient of f on the same figure.

9. Use the command `plot` or `plot3` to plot the image of the following paths:

(a) $\mathbf{s}(t) = [2\sin t, 3\cos t]$, $0 \le t \le 2\pi$.

(b) $\mathbf{s}(t) = [t - \sin t, 1 - \cos t]$, $0 \le t \le 6\pi$.

(c) $\mathbf{s}(t) = [t^3/6, 2t, t^2]$, $-6 \le t \le 4$.

(d) $\mathbf{s}(t) = [\sin t, \cos t, \sin(4t)]$, $0 \le t \le 2\pi$.

(e) $\mathbf{s}(t) = [(5 + \sin 50t)\cos t, (5 + \sin 50t)\sin t, \cos 50t]$, $0 \le t \le 2\pi$.

10. Use the command `quiver` or `quiver3` to plot some of the tangent vectors for the paths in the previous exercise.

11. Find the lengths of the paths in Exercise 9 using the command `quad` to evaluate integrals in MATLAB.

12. The DNA molecule, discovered by James Watson, Francis Crick, Maurice Wilkins, and Rosalind Franklin, has the shape of two parallel helices, each given by $\mathbf{s}(t) = [a\cos t, a\sin t, bt/2\pi]$, where $a = 10^{-8}$cm is the radius of the helix and $b = 3.4 \times 10^{-7}$cm is the height the helix rises during a complete turn. Given that a DNA molecule has about 2.9×10^8 turns, estimate the total length of each helix.

13. Consider the vector field

$$\mathbf{F}(x, y) = \begin{bmatrix} xy - 1 \\ y^2 - x - 2 \end{bmatrix}.$$

(a) Find the Jacobian matrix $\mathbf{DF}(x, y)$.

(b) Obtain an expression for $\mathbf{DF}(x, y)^{-1}$ and determine the values of (x, y) for which the Jacobian matrix is invertible.

(c) Verify your answer to the previous item using MATLAB.

14. Consider the vector field

$$\mathbf{F}(r,\phi,\theta) = \begin{bmatrix} r\sin\phi\cos\theta \\ r\sin\phi\sin\theta \\ r\cos\phi \end{bmatrix}.$$

(a) Find the Jacobian matrix $\mathbf{DF}(r,\phi,\theta)$.

(b) Obtain an expression for the determinant of $\mathbf{DF}(r,\phi,\theta)$ and determine the values of (r,ϕ,θ) for which the Jacobian matrix is invertible.

(c) Verify your answer to the previous item using MATLAB.

15. Consider the electric field

$$\mathbf{E}(\mathbf{x}) = \frac{q\mathbf{x}}{4\pi\varepsilon_0\|\mathbf{x}\|^3}.$$

(a) Obtain the integral curves of $\mathbf{E}$.

(b) Calculate the divergence div $\mathbf{E}$ and verify your answer in MATLAB by using the command `divergence`.

(c) Use the commands `quiver3` and `surf` to plot the vector field and its divergence on the same figure.

(d) Find a scalar function $f:\mathbb{R}^3\to\mathbb{R}$ satisfying $\mathbf{E}=\nabla f$.

(e) Use the command `quad` to calculate the line integral of $\mathbf{E}$ along the line segment connecting the points $(1,0,0)$ and $(0,0,1)$.

(f) Use the command `quad` to calculate the line integral of $\mathbf{E}$ along the arc of the circle connecting the points $(1,0,0)$ and $(0,0,1)$ over a sphere of radius one centered at the origin.

(g) Compare the results of the two previous items with the difference $f(0,0,1)-f(1,0,0)$.

16. Consider the vector field

$$\mathbf{F} = \begin{bmatrix} \dfrac{y}{x^2+y^2} \\ \\ -\dfrac{x}{x^2+y^2} \\ 0 \end{bmatrix}, \qquad (x,y)\neq(0,0).$$

(a) Verify that $\nabla\times\mathbf{F}=\mathbf{0}$.

(b) Use the command `quad` to compute the line integral of $\mathbf{F}$ along the boundary of a square of an area equal to one centered at the point $(x,y,z)=(0,0,0)$.

(c) Use the result of the previous item to conclude that there is no scalar function $f : \mathbb{R}^3 \to \mathbb{R}$ satisfying $\mathbf{F} = \nabla f$.

(d) Calculate the surface integral of $\nabla \times \mathbf{F}$ across the same square and explain how the result does not contradict Stokes's theorem.

17. Use the commands `dblquad` and `triquad` to verify Gauss's theorem for the vector field $\mathbf{F} = [2x^2, y^3z^2, xz^3]$ and a sphere $\mathbf{S}$ of radius one centered at the origin.

CHAPTER **8**

Zeros and Extrema of Functions

THE QUEST FOR SOLUTIONS of algebraic equations is almost as old as mathematics itself. Ancient Egyptians and Babylonians already knew methods for solving some practical problems that are described by quadratic equations. General quadratic equations were subsequently treated by Hindu and Persian mathematicians, and by the 12th century the method of *completing squares* was essentially known. Around the 16th century, great emphases were put in using letters to represent the coefficients of general equations, with the corresponding solutions sought as exact formulas in terms of these coefficients. This led to the *quadratic formula*

$$x = \frac{-b \pm \sqrt{b^2 - 4ac}}{2a},$$

summarizing previous knowledge regarding the solutions for the quadratic equation $ax^2 + bx + c = 0$, and the discovery of new formulas for the solutions of cubic and quartic equations. As physicist Richard Feynman put it, these truly remarkable developments helped "freeing man from the intimidation of ancients" during the Renaissance. They also propelled mathematicians to search for closed formulas (in terms of a finite number of arithmetic operations) for solutions of algebraic equations of higher degrees. The search stopped in 1824, when Abel proved that such closed-form solutions cannot be found in general for polynomial equations of degree greater than four. We have already discussed this result in Section 4.1, in connection to eigenvalues and the characteristic polynomial of a matrix. The conclusion we reached there, namely, that eigenvalues for matrices of dimension larger than four must be found through iterative numerical algorithms, is equally valid for finding solutions of higher degree polynomial equations, and this chapter is devoted to such algorithms.

More generally, we seek to solve nonlinear equations involving more complicated functions, such as trigonometric or transcendental ones. In general, a nonlinear equation in one unknown can be expressed as $f(x) = 0$, where $f : \mathbb{R} \to \mathbb{R}$ is a nonlinear scalar function of one variable. In Section 8.1, we consider numerical methods for solving this type of equation. Such methods can be generalized to deal with nonlinear functions of the form $\mathbf{F} : \mathbb{R}^n \to \mathbb{R}^n$, which are the subject of Section 8.2. Finding the solution to the vector equation $\mathbf{F}(\mathbf{x}) = \mathbf{0}$ is then equivalent to solving a system of nonlinear equations.

The last two sections of this chapter are devoted to the closely related problem of finding the points where a nonlinear function has an extreme value, either a minimum or a maximum. When a function $f : \mathbb{R} \to \mathbb{R}$ is differentiable, such extreme values occur at points satisfying $f'(x) = 0$, and this equation can be solved numerically using the methods of Section 8.1. We explore this in Section 8.3, but also consider algorithms for finding extreme points of functions that are not differentiable or whose derivatives have too complicated expressions. Similarly, the extreme values of a multivariate differentiable function $f : \mathbb{R}^n \to \mathbb{R}$ occur at points satisfying $\nabla f(\mathbf{x}) = 0$. Because the gradient ∇f is a special case of a nonlinear function from $\mathbb{R}^n$ to $\mathbb{R}^n$, we can search for extreme points for f by solving the equation $\nabla f(\mathbf{x}) = 0$ using the methods of Section 8.2. We discuss this in Section 8.4, particularly in connection to a multidimensional version of the second derivative test, as well as methods for finding extreme values without relying on the roots of the gradient vector field.

8.1 One-Dimensional Root Finding

Suppose you want to find the value for x such that

$$2 \log x = 2 - \sqrt{x}. \tag{8.1}$$

As usual, it is always a good idea to start with a graphical visualization, so you can use the following MATLAB commands to plot the functions on both sides of this equation:

```
>> x = 1 : 0.01 : 10;
>> f1 = 2*log(x); f2 = 2-x.^(1/2);
>> plot(x,f1,x,f2);
```

As you can see in Figure 8.1, the nonlinear equation (8.1) has a solution x on the interval $[1, 2]$. In what follows we devise numerical algorithms that find this number with some accuracy. Because we can always rewrite equations like (8.1) as

$$2 \log x + \sqrt{x} - 2 = 0,$$

you can see that finding their solutions is equivalent to finding the *zeros*, or *roots*, of nonlinear functions such as $f(x) = 2 \log x + \sqrt{x} - 2$.

Problem 8.1 (Root Finding) Given a continuous function $f : \mathbb{R} \to \mathbb{R}$, find all the solutions of the equation $f(x) = 0$ for x in the interval $[a, b]$.

Roots of polynomials

If the function is a polynomial $p_n(x)$ of degree n, there exist exactly n roots (up to multiplicity), according to the *Fundamental Theorem of Algebra* (Theorem 5.1). However, some roots could be complex valued, whereas some of

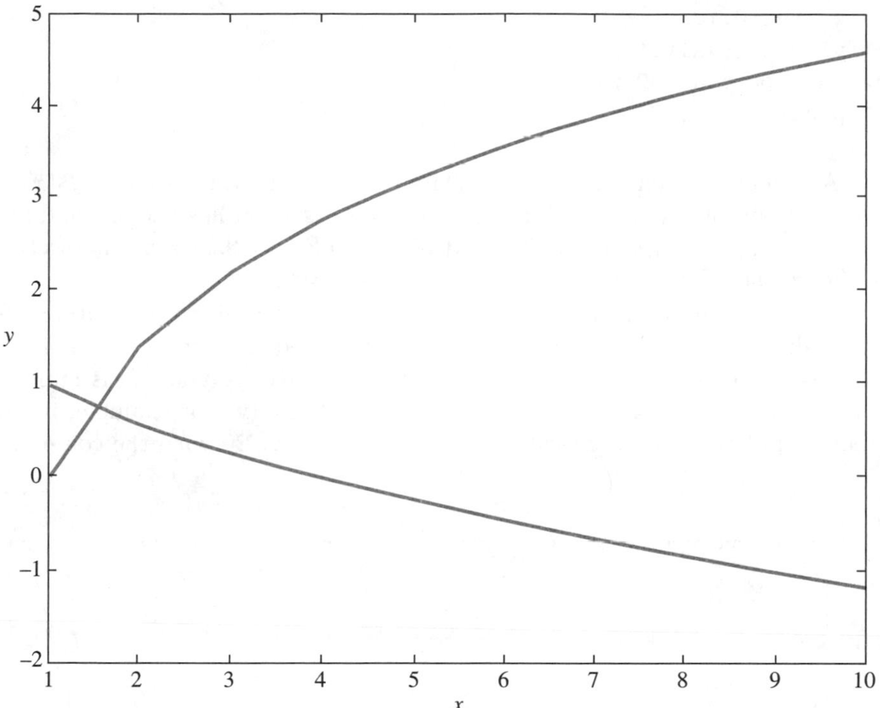

Figure 8.1 The graphical solution for $2 \log x - 2 - \sqrt{x}$.

the real-valued roots may not belong to the given interval $[a, b]$. The following example computes the roots of the cubic polynomials

$$f_1(x) = x^3 - 2x^2 - 3x + 2, \qquad f_2(x) = x^3 - 2x^2 - 3x - 2,$$

using the MATLAB function `roots` and plots their corresponding graphs:

```
>> c1 = [ 1,-2,-3,2];
>> p1 = roots(c1)'

p1 =
    2.8136   -1.3429    0.5293

>> c2 = [ 1,-2,-3,-2];
>> p2 = roots(c2)'

p2 =
   3.1528             -0.5764 - 0.5497i  -0.5764 + 0.5497i
```

```
>> x = -2:0.1:4;
>> y1 = polyval(c1,x);
>> y2 = polyval(c2,x);
>> plot(x,y1,x,y2)
```

Although the function $f_2(x)$ is related to $f_1(x)$ by a translation $f_2(x) = f_1(x) - 4$, all three roots for $f_1(x)$ are real, whereas $f_2(x)$ has one real and two complex roots. The real roots can be identified in Figure 8.2 as the intersection of the graphs of $f_1(x)$ and $f_2(x)$ with the line $y = 0$.

The root-finding algorithm used for `roots` is based on computing the eigenvalues for the companion matrix as mentioned in Section 4.1. This algorithm could be very inaccurate if the degree of the polynomial is large or if the real-valued roots have high algebraic multiplicity. For example, if you create a polynomial $p_{20}(x)$ with roots at $x = 1, 2, \ldots, 20$ with the command

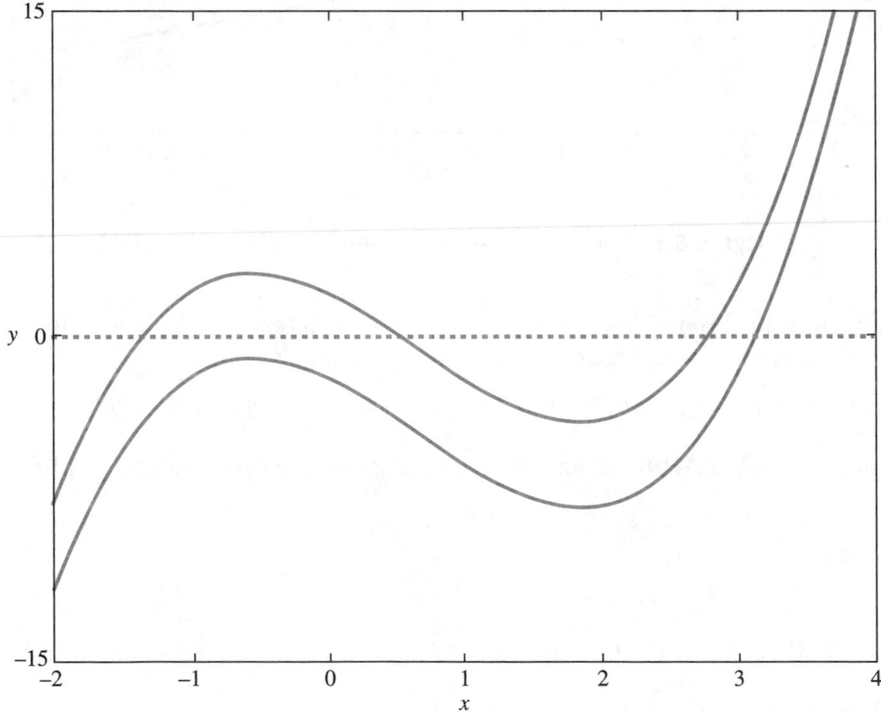

Figure 8.2 The three real roots of $f_1(x) = x^3 - 2x^2 - 3x + 2$ and the one real root of $f_2(x) = x^3 - 2x^2 - 3x - 2$.

poly, and then compute its roots from its coefficients, you may observe that the roots are found with a numerical error in the second significant digit:

```
>> r1 = roots(poly(1:20))'

r1 =
  Columns 1 through 7
   20.0003    18.9970    18.0118    16.9695    16.0509    14.9319    14.0684

  Columns 8 through 14
   12.9472    12.0345    10.9836    10.0063     8.9983     8.0003     7.0000

  Columns 15 through 20
   6.0000     5.0000     4.0000     3.0000     2.0000     1.0000
```

As a second example, consider the root $x = 1$ of multiplicity six for the polynomial $P_6(x) = (x-1)^6$. The numerical algorithm completely misses this root and instead returns complex-valued numerical roots:

```
>> r2 = roots(poly([1 1 1 1 1 1]))'

r2 =
  Columns 1 through 3
   1.0042 - 0.0025i   1.0042 + 0.0025i   1.0000 - 0.0049i

  Columns 4 through 6
   1.0000 + 0.0049i   0.9958 - 0.0024i   0.9958 + 0.0024i
```

Exercise 8.1 Use the MATLAB function **roots** to compute the roots for the Hermite polynomials defined by the recursion formula

$$H_{n+1}(x) - 2xH_n(x) + 2nH_{n-1}(x) = 0,$$

for $n = 1, 2, \ldots, 100$, starting with the first two polynomials $H_0(x) = 1$ and $H_1(x) = 2x$. Plot all roots versus n.

Zeros of general nonlinear functions, including polynomials, can be obtained by using the MATLAB function fzero, which takes the function and the value for a starting approximation as arguments and returns the value of the nearest zero using a combination of some of the iterative algorithms discussed in this section. The next example shows that it can be used to find all three roots of the cubic polynomial $f_1(x) = x^3 - 2x^2 - 3x + 2$ from the starting approximations at $x = -1$, $x = 0$, and $x = 2$:

```
>> z1 = fzero('x.^3-2*x.^2-3*x+2',2)

z1 =
    2.8136

>> z2 = fzero('x.^3-2*x.^2-3*x+2',-1)

z2 =
   -1.3429

>> z3 = fzero('x.^3-2*x.^2-3*x+2',0)

z3 =
    0.5293
```

In the case of the cubic polynomial $f_2(x) = x^3 - 2x^2 - 3x - 2$, the function `fzero` finds only one real root from any starting approximation, such as from $x = -2$ and $x = 2$:

```
>> z4 = fzero('x.^3-2*x.^2-3*x-2',2)

z4 =
    3.1528

>> z5 = fzero('x.^3-2*x.^2-3*x-2',-2)

z5 =
    3.1528
```

The search fails when no real roots of the nonlinear equation $f(x) = 0$ exist, as in the case of the function $f(x) = 1 + x^2$:

```
>> z6 = fzero('1 + x^2',1)

Exiting fzero: aborting search for an interval containing a sign
change because NaN or Inf function value encountered during search
(Function value at -1.716199e+154 is Inf) Check function or try
again with a different starting value.

z6 =
   NaN
```

Sensitivity

Before introducing specific root-finding algorithms, it is instructive to consider the sensitivity of Problem 8.1. Recall from Section 1.8 that the condition number of a numerical problem is defined as the ratio between the relative output error and the relative input error. In particular, according to Exercise 1.2, the condition number for finding x satisfying $y = f(x)$ for a given y is approximately

$$\left|\frac{f(x)}{xf'(x)}\right|. \tag{8.2}$$

Because a root-finding problem is a special case of this when $y = 0$, you might want to use (8.2) to quantify its sensitivity. But this would lead to an identically zero condition number, regardless of the function f because $f(x) = 0$ by definition in the root-finding problem. The way around this difficulty is to consider an alternative definition of a condition number as the ratio of absolute errors, instead of relative ones. This still serves as a measure of sensitivity, with a large condition number corresponding to an ill-conditioned problem. When f is differentiable, the same reasoning used in Exercise 1.2 leads us to conclude that such an absolute condition number for the root-finding problem is approximately equal to

$$\frac{1}{|f'(x)|}. \tag{8.3}$$

In other words, the root-finding problem is well conditioned if the function f has a large derivative near the root. This is intuitively clear because a large derivative means that the graph of the function is close to vertical near the root, making it easy to identify where it crosses the x-axis. On the other hand, if the function has a very small derivative near the root, then its graph is close to horizontal near the root, making it hard to locate where it crosses the x-axis and leading to an ill-conditioned root-finding problem. Such a situation arises, for example, when x_* is a *multiple root* for f, in which case $f'(x_*) = 0$ and the absolute condition number (8.3) is infinite.

Bisection method

The first algorithm we present for the solution of Problem 8.1 is called the *bisection method* and consists of locating the zeros of a continuous function by identifying its sign changes over an interval containing it. It relies on the following theorem from *Calculus*:

Theorem 8.1 (Intermediate Value) *Let $f(x)$ be a continuous function on $[a, b]$ with $f(a)f(b) < 0$. Then, there exists at least one point $x_* \in (a, b)$, such that $f(x_*) = 0$.*

In other words, if $f(a)$ and $f(b)$ are of different signs, then the function f has at least one root in this interval. We then divide the original interval into two subintervals and take the midpoint to be the first approximation x_1 for the true root x. We iterate the algorithm by inspecting the value of $f(x_1)$ and selecting the subinterval with a sign change at the end points. We then divide this new interval in half and select its midpoint as the new approximation.

This numerical procedure allow us to approximate the roots to any degree of accuracy above the level of machine precision. When the first solution for $f(x) = 0$ is found, we can search for more roots by splitting the initial interval into a number of smaller subintervals and looking for more sign changes in the function between two adjacent end points in each subinterval.

Let $e_k = |x_k - x_*|$ be the error in the kth approximation x_k for the exact root x_*. Since the length of the searching interval in the bisection method gets cut in half with each iteration, the error converges to zero according to the power law:

$$e_k \leq \left(\frac{1}{2}\right)^k (b-a), \qquad k \geq 0. \tag{8.4}$$

By using the definition of convergence rates in Section 1.8, you can see that the bisection method converges *linearly*. The next example illustrates the bisection method applied to the transcendental function

$$f(x) = 2\log x + \sqrt{x} - 2, \tag{8.5}$$

which, according to Figure 8.1, has only one root in the interval $[1, 2]$. The MATLAB script `bisection_method` executes iterations of the bisection method for this function. The iterations stop when the desired tolerance is reached.

```
% bisection_method
a = 1; b = 2; tolerance = 0.0000001; k = 1;
ya = 2*log(a)+sqrt(a)-2; yb = 2*log(b)+sqrt(b)-2; %
while (abs(b-a) > tolerance)  % iterations of the bisection method
    x(k) = (a+b)/2;
    y(k) = 2*log(x(k))+sqrt(x(k))-2;
    if (y(k) == 0)
        break;
    elseif (sign(y(k)) == sign(ya))
        a = x(k); ya = y(k);
    else
        b = x(k); yb = y(k);
    end
    k=k+1;
end
fprintf('Root x = %12.10f found after %2.0f iterations \n',x(k-1),k-1);
```

You can compare the approximation obtained from this script with the root found by the MATLAB function `fzero`:

```
>> bisection_method
Root x = 1.4796387553 found after 24 iterations
```

```
>> xZeroMATLAB = fzero('2*log(x)+sqrt(x)-2',1.5)

xZeroMATLAB =
   1.47963877016712
```

To confirm the validity of the error upper bound (8.4), assume that e_k obeys the power law $e_k = C^k e_0$ and find the factor C from the linear regression (see Section 5.6) of the logarithmic dependence

$$\log e_k = k \log C + \log e_0.$$

The following commands implement this linear regression and plot the result in Figure 8.3, together with the graph of $\log e_k$ versus k:

```
>> K = [1:k-1]; Error=abs(x-xZeroMATLAB);
>> a = polyfit(K,log(Error),1);
>> ErrorLinear = exp(a(2)+a(1)*K);
>> semilogy(K,Error,'.',K,ErrorLinear,':r')
```

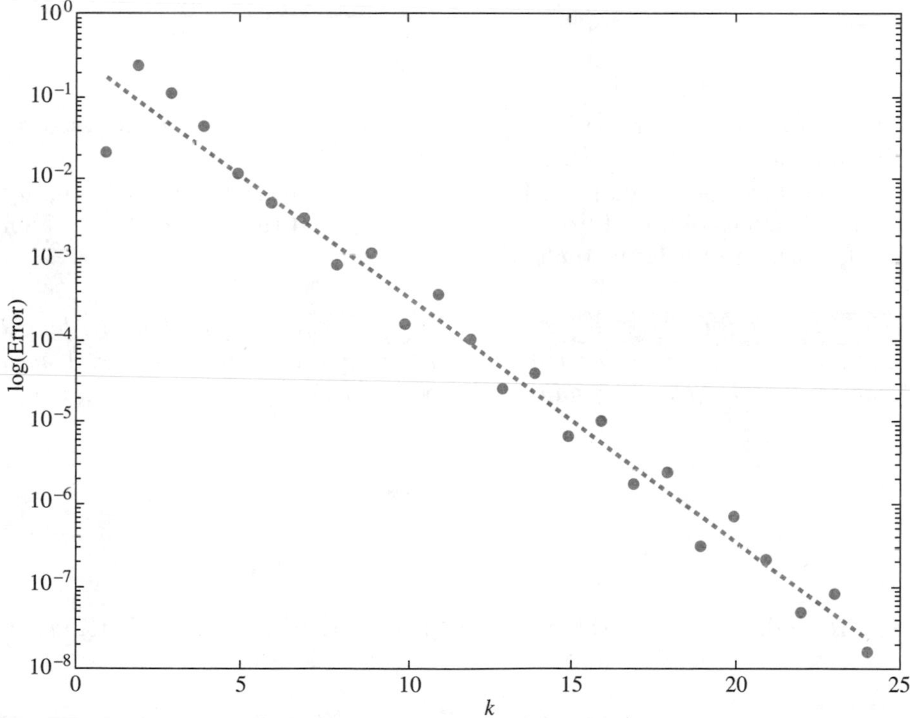

Figure 8.3 Iterations of the bisection method.

```
>> C = exp(a(1))

C =
   0.50054722509846
```

The numerical value of C is close to the value $\frac{1}{2}$ in (8.4).

Fixed-point iterations

Because the bisection method converges slowly (linearly) near the solution of the nonlinear equation $f(x) = 0$, we try to solve Problem 8.1 with faster iterative methods. In particular, we consider a sequence of numerical approximations x_k obtained from a starting approximation x_0 by the *fixed-point iterations*:

$$x_{k+1} = g(x_k). \tag{8.6}$$

The function g is assumed to be continuously differentiable on the interval $[a, b]$ and is chosen in such a way that the nonlinear equation $f(x) = 0$ is equivalent to $x = g(x)$. In other words, we choose g so that a root for the function f is a *fixed point* for the function g. Observe that for a given function f, there are many functions g that satisfy this criterion. For example, both

$$g_1(x) = x - 2\log x - \sqrt{x} + 2$$
$$g_2(x) = (2 - 2\log x)^2$$

are possible choices for the fixed-point function g when you are interested in the roots for thc function $f(x) = 2\log x + \sqrt{x} - 2$.

The choice of a fixed-point function affects the convergence and stability of the iterations obtained from it through (8.6), and these depend in turn on the following result from *Analysis*:

Theorem 8.2 (Contraction-Mapping Principle) *Suppose that, (i) for all $x \in [a, b]$, (i) $g(x) \in [a, b]$, and (ii) there exists a constant $0 \le C < 1$ such that $|g'(x)| \le C$. Then, for any starting point x_0 in $[a, b]$, the sequence $x_{k+1} = g(x_k)$ satisfies*

$$\lim_{k\to\infty} x_k = x_*,$$

where $x_* = g(x_*)$ is the unique fixed point for g in $[a, b]$.

To understand why this powerful theorem is true, observe that if $x_* = g(x_*)$ is a fixed point for g, then after $k + 1$ iterations in (8.6) we have

$$|x_{k+1} - x_*| = |g(x_k) - g(x_*)| = |g'(\xi_k)||x_k - x_*|,$$

for some ξ_k between x_* and x_k given by the Mean Value Theorem of *Calculus*. But regardless of what ξ_k is, we know that $|g'(\xi_k)| \leq C$, and so repeating this argument $(k+1)$ times leads to

$$|x_{k+1} - x_*| \leq C^{k+1}|x_0 - x_*|, \tag{8.7}$$

which converges to zero as $k \to \infty$, since $C < 1$. We also see from this argument that $e_{k+1} \leq Ce_k$, which means that the rate of convergence for general fixed-point iterations is at least *linear*.

Unfortunately, if $|g'(x_*)| > 1$, then any contraction obtained at early steps is lost when we get closer to x_*. In this case, the fixed point is said to be *unstable* and the iterations in (8.6) fail to converge even if the fixed point exists. The divergence occurs no matter how close the initial approximation x_0 is chosen from the fixed point x_*, as long as $x_0 \neq x_*$.

Direct method

A particular type of fixed-point iterations, called *direct method*, is obtained when we set

$$g(x) = x + \alpha f(x) \tag{8.8}$$

for $\alpha \neq 0$. Since in this case $g'(x) = 1 + \alpha f'(x)$, you can see that the stability condition $|g'(x)| < 1$ translates into $-2 < \alpha f'(x) < 0$, which can be controlled by choosing an appropriate value for $\alpha \neq 0$. Moreover, we can improve convergence dramatically by trying to make the value $|g'(x_*)|$ at a fixed point x_* as small as possible. In particular, if $f'(x_*) \neq 0$ and we choose $\alpha = -1/f'(x_*)$, then $g'(x_*) = 0$. In this case, if $g(x)$ is twice continuously differentiable near $x = x_*$, the Taylor series expansion leads to

$$g(x_k) - g(x_*) = \frac{1}{2}g''(\xi_k)(x_k - x_*)^2,$$

for some ξ_k between x_* and x_k. Therefore, provided g'' is uniformly bounded near the fixed point x_*, we see that

$$e_{k+1} \leq De_k^2, \tag{8.9}$$

for some constant $0 < D < \infty$, which implies that we can achieve a *quadratic* convergence rate in this case. Recall that algorithms with quadratic convergence are much more efficient because the number of accurate digits in the numerical approximations is doubled after each iteration.

The next example illustrates iterations of the direct method for the same function (8.5) starting with $x_0 = 1$ when $\alpha = -1$ and $\alpha = -1/f'(x_*) \approx -0.5673$.

```
% direct_method
xZeroMATLAB = fzero('2*log(x)+sqrt(x)-2',1.5); % 'exact' value of zero
epsilon = 1; tol = 0.0000001; total = 100; alpha = -1; k = 1;
x(1)= 1; Error1(1) = abs(x(1)-xZeroMATLAB);
while ((epsilon > tol) & (k < total))   % the direct method
    x(k+1) = x(k) + alpha*(2*log(x(k))+sqrt(x(k))-2);
    epsilon = abs(x(k+1)-x(k));
    Error1(k+1) = abs(x(k+1) - xZeroMATLAB);
    k = k+1;
end
fprintf('Root x = %12.10f is found after %2.0f iterations\n',x(k),k-1);
fDer = 2/xZeroMATLAB + 1/(2*sqrt(xZeroMATLAB));
alpha = -1/fDer; % improved value for alpha
clear x, x(1)=1; j = 1; epsilon = 1;
Error2(1)=abs(x(1)-xZeroMATLAB);
while ((epsilon > tol) & (j < total))  % the direct method
    x(j+1) = x(j) + alpha*(2*log(x(j))+sqrt(x(j))-2);
    epsilon = abs(x(j+1)-x(j));
    Error2(j+1) = abs(x(j+1) - xZeroMATLAB);
    j = j+1;
end
fprintf('Root x = %12.10f is found after %2.0f iterations\n',x(j),j-1);
```

You can now compare the performance of the iterations for the two different values of α:

```
>> direct_method

Root x = 1.4796388049 is found after 61 iterations
Root x = 1.4796387702 is found after  5 iterations
```

You can see that for $\alpha = -1$ the rate of convergence is slower than that of the bisection method because more iterations are needed for the same tolerance level, but the rate improves for $\alpha = -0.5673$, corresponding to the case when $g'(x_*)$ is close to zero. The following commands implement a linear regression on the error function $e_k = C^k e_0$ and display the results in Figure 8.4, where you can observe that for the second value of α the error drops below the level of machine precision in a few iterations:

```
>> K = [0:k-1]; a1 = polyfit(K,log(Error1),1);
>> ErrorLinear1 = exp(a1(2)+a1(1)*K);
>> semilogy(K-1,Error1,'.b',K,ErrorLinear1,':r');
>> J = [0:j-1]; a2 = polyfit(J,log(Error2),1);
>> ErrorLinear2 = exp(a2(2)+a2(1)*J);
>> semilogy(J,Error2,'.g',J,ErrorLinear2,':m');
>> C1=exp(a1(1)), C2 = exp(a2(1))
```

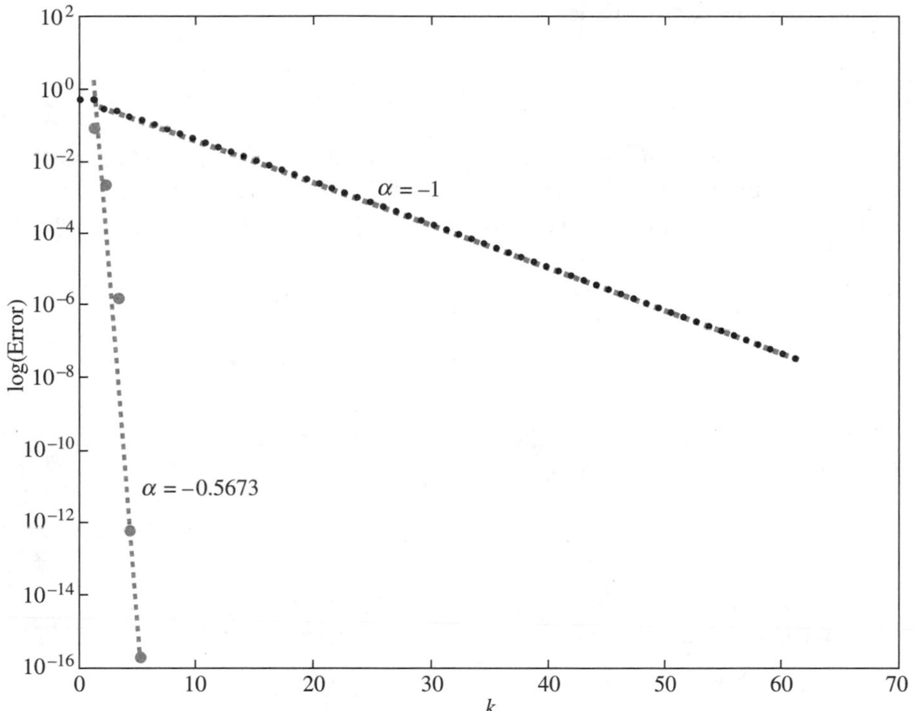

Figure 8.4 Iterations of the direct method for two values of α.

```
C1 =
    0.76247934812923

C2 =
     5.801295562812796e-04
```

Newton–Raphson method

Another special type of fixed-point iteration is the root finding algorithm known as the *Newton–Raphson method*, which is based on the tangent approximation to a twice differentiable function $f(x)$ at the given point x_k,

$$f(x) = f(x_k) + f'(x_k)(x - x_k) + O(x - x_k)^2. \tag{8.10}$$

Setting $f(x) = 0$ in (8.10), you can approximate the root x_* for the nonlinear function f by the zero of the tangent line, that is,

$$x_{k+1} = x_k - \frac{f(x_k)}{f'(x_k)}. \tag{8.11}$$

Comparing this with (8.6) leads to

$$g(x) = x - \frac{f(x)}{f'(x)}, \qquad g'(x) = \frac{f(x)f''(x)}{(f'(x))^2}.$$

If x_* is a *simple root* for f, that is $f(x_*) = 0$ and $f'(x_*) \neq 0$, then the preceding function g satisfies $g'(x_*) = 0$ and the Newton–Raphson method converges quadratically by the same argument leading to (8.9), provided g'' is uniformly bounded near x_*.

On the other hand, when x_* is a *multiple root*, that is $f(x_*) = f'(x_*) = 0$, we can always factorize the function f near the root as

$$f(x) = (x - x_*)^m \tilde{f}(x),$$

where $\tilde{f}(x_*) \neq 0$ and $m > 1$ is the multiplicity for the root. Inserting this into the expression for $g'(x)$ gives

$$g'(x_*) = \frac{m-1}{m} < 1,$$

which leads to a linear convergence as in (8.7) with $C = 1 - 1/m$.

The next example illustrates iterations of the Newton–Raphson method for the functions $f(x) = 2\log x + \sqrt{x} - 2$ and $f(x) = (x-1)^6$.

```
% Newton_method
% function f(x) = 2*log(x)+sqrt(x)-2
xZeroMATLAB = fzero('2*log(x)+sqrt(x)-2',1.5); % 'exact' value
epsilon = 1; tol = 0.0000001; total = 100; k = 1; x(1) = 1;
Error(1) = abs(x(1)-xZeroMATLAB);
while ((epsilon > tol) & (k < total))  % the Newton method
    f = 2*log(x(k))+sqrt(x(k))-2;
    fDer = 2/x(k)+1/(2*sqrt(x(k)));
    x(k+1) = x(k)-f/fDer;
    epsilon = abs(x(k+1)-x(k));
    Error1(k+1) = abs(x(k+1) - xZeroMATLAB);
    k = k+1;
end
fprintf('Root x = %12.10f found after %2.0f iterations\n',x(k),k-1);
    % function f(x) = (x-1)^6
epsilon = 1; tol = 0.0000001; total = 100; x(1) = 0; j = 1;
Error2(1) = abs(x(1)-1);
while ((epsilon > tol) & (k < total))  % the Newton method
    f = (x(j)-1)^6;
    fDer = 6*(x(j)-1)^5;
    x(j+1) = x(j)-f/fDer;
    epsilon = abs(x(j+1)-x(j));
```

```
    Error2(j+1) = abs(x(j+1) - 1);
    j = j+1;
  end
fprintf('Root x = %12.10f found after %2.0f iterations \n',x(j),j-1);
```

When Newton_method is executed, it shows a faster (quadratic) rate of convergence for the simple zero of $f(x) = 2\log x + \sqrt{x} - 2$ and slower (linear) rate of convergence for the multiple zero $x = 1$ of $f(x) = (x-1)^6$:

```
>> Newton_method
Root x = 1.4796387702 found after  5 iterations
Root x = 0.9999995371 found after 80 iterations
```

The errors for both functions are shown in Figure 8.5. Moreover, you can verify that the constant C in the linear rate (8.7) is close to the value $1 - 1/m = 5/6$:

```
>> K=[0:k-1]; J = [0:j-1];
>> a = polyfit(J,log(Error2),1);
```

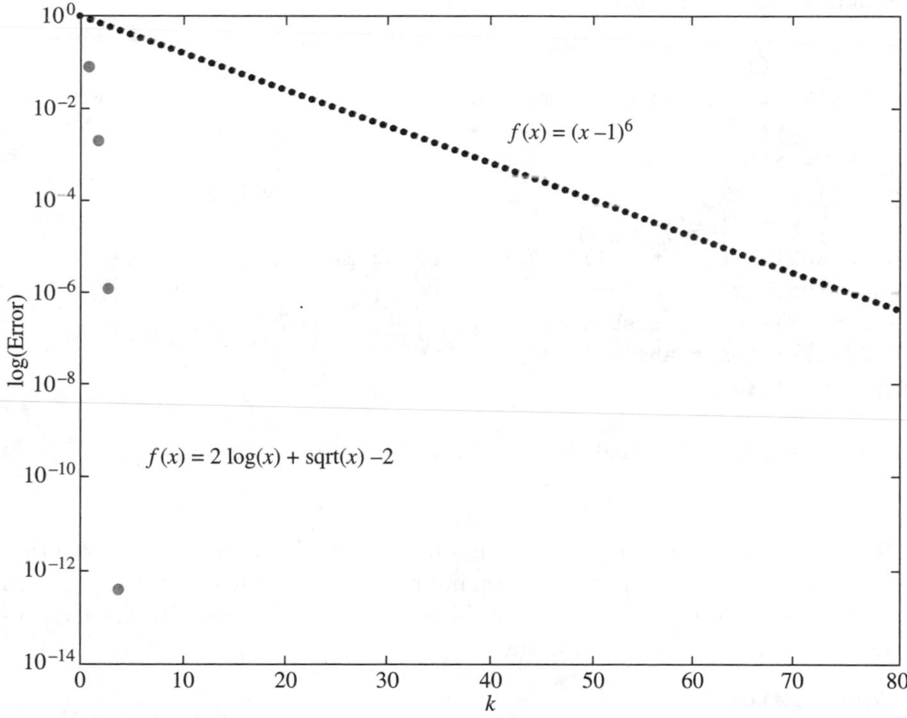

Figure 8.5 Iterations of the Newton–Raphson method for a simple root of $f(x) = 2\log x + \sqrt{x} - 2$ and a multiple root of $f(x) = (x-1)^6$.

```
>> ErrorLinear = exp(a(2)+a(1)*J);
>> semilogy(J,Error2,'.b',J,ErrorLinear,':r',K,Error1,'.g');
>> C = exp(a(1))

C =
   0.83333333333305
```

We remark that the convergence results for the Newton–Raphson method are only local. If the initial approximation x_0 is chosen far from the root x_*, the iterations may converge to a different root, or they may diverge to plus or minus infinity. One way to avoid this is to use the bisection method first for finding a fairly small interval with different signs of $f(x)$ at the end points, and then switch to the Newton–Raphson method for a fast convergence.

The next script illustrates a combination of bisection and Newton–Raphson methods in an algorithmic search for roots of the nonlinear equation $f(x) = \cos x \cosh x + 1$ on the interval $[0, 20]$:

```
% many_zeros
xInt = 0 : 0.01 : 20;  % a dense grid
yInt = cos(xInt).*cosh(xInt) + 1; plot(xInt,yInt);
tol = 10^(-14);
total = 100; a = sign(yInt(1)); j = 1;
while (j < length(xInt))  % the bisection method
    j = j + 1;
    if (sign(yInt(j)) ~= a)
        a = sign(yInt(j));
        k = 0; eps = 1; x = xInt(j-1);
        while ((eps > tol) & (k < total))  % the Newton method
            f = cos(x)*cosh(x)+1;
            fDer = sinh(x)*cos(x)-cosh(x)*sin(x);
            xx = x-f/fDer; eps = abs(xx-x);
            x = xx; k = k+1;
        end
        fprintf('Root x = %12.10f found after %2.0f iterations \n',x,k);
    end
end
```

You can see from the graph in Figure 8.6 that six solutions of the equation $\cos x \cosh x + 1 = 0$ exist on the given interval, and the output for `many_zeros` shows that with the combined use of the bisection and Newton–Raphson methods you can detect all six of them:

```
>>many_zeros

Root x = 1.8751040687 found after  4 iterations
Root x = 4.6940911330 found after  4 iterations
```

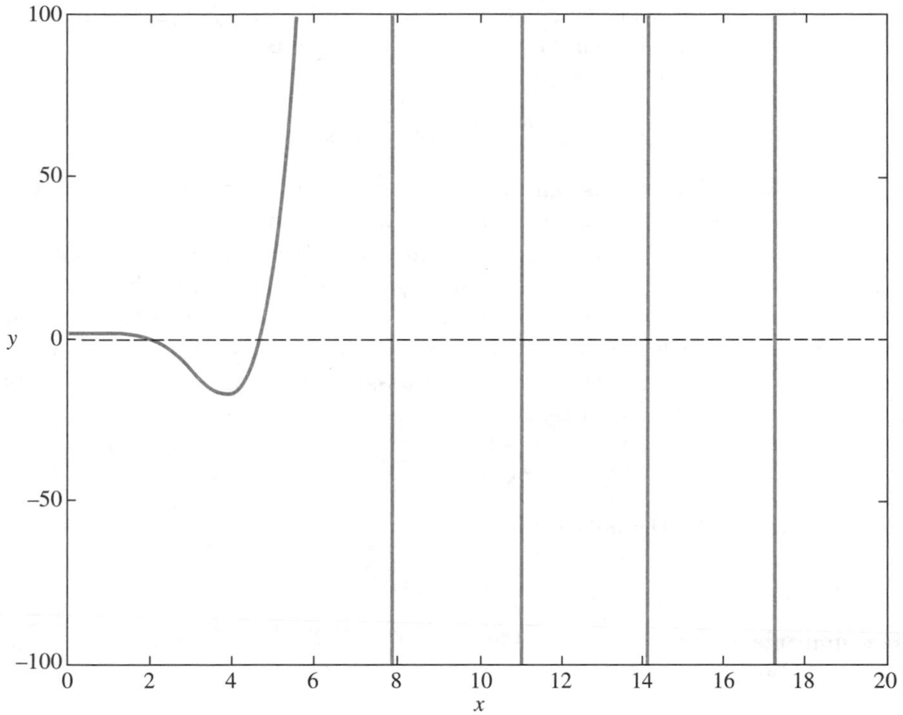

Figure 8.6 The graph of the function $f(x) = \cos x \cosh x + 1$.

```
Root x = 7.8547574382 found after  4 iterations
Root x = 10.9955407349 found after  4 iterations
Root x = 14.1371683910 found after  4 iterations
Root x = 17.2787595321 found after  4 iterations
```

Secant method

The Newton–Raphson method requires the evaluation of $f'(x_k)$ for each of the iterations (8.11). If the function f is not available in analytical form (say, it is only defined numerically as in the shooting method for the boundary-value problems of Section 10.2), then the derivative has to be estimated numerically. Even if f is expressed analytically, the derivative f' might have such a complex expression that its computations could result in a large round-off error. In all these cases, it is better to approximate $f'(x_k)$ by the first-order backward numerical derivative of Section 6.1:

$$Df[x_k] = \frac{f(x_k) - f(x_{k-1})}{x_k - x_{k-1}}. \tag{8.12}$$

This modification of the Newton–Raphson method is called the *secant method* because the backward numerical derivative in (8.12) corresponds to the slope of the secant line between the points $(x_k, f(x_k))$ and $(x_{k-1}, f(x_{k-1}))$. Inserting (8.12) into (8.11) as an approximation for $f'(x_k)$ leads to

$$x_{k+1} = x_k - \frac{f(x_k)(x_k - x_{k-1})}{f(x_k) - f(x_{k-1})}. \tag{8.13}$$

Unlike (8.11), which needs only x_k to produce the next iterate x_{k+1}, the iterations (8.13) use both x_k and x_{k-1} at each step, therefore requiring two initial approximations x_0 and x_1 at the first step. It is common that the two initial points are chosen arbitrarily, but close to each other. Like the Newton–Raphson method, the secant method is always a contraction near the solution x_* of the nonlinear equation $f(x) = 0$, but the rate of convergence is slower. For a simple root, its convergence is *superlinear* in the sense that there exists a constant $0 < C < \infty$ such that

$$e_{k+1} \leq C e_k^p, \tag{8.14}$$

where p is given by the *golden ratio*

$$p = (1 + \sqrt{5})/2 \approx 1.618.$$

For multiple roots, the secant method converges linearly, similarly to the Newton–Raphson method.

Exercise 8.2 Verify numerically that the secant method converges linearly to the triple root $x_* = 1$ for the function $f(x) = (x-1)^3$.

The next example compares iterations of the Newton–Raphson and secant methods for the function

$$f(x) = x\sin(1/x) - \frac{1}{2}e^{-x}.$$

You can start by plotting the graphs of the functions $y = x\sin(1/x)$ and $y = 0.5e^{-x}$ in Figure 8.7:

```
>> xInt = 0.01 : 0.01 : 1;
>> yInt1 = xInt.*sin(1./xInt);
>> yInt2 = 0.5*exp(-xInt);
>> plot(xInt,yInt1,'b',xInt,yInt2,'g');
```

The graph shows that there exists a unique solution for $f(x) = 0$ on the interval $[0, 1]$. The following script implements both the secant and Newton–Raphson methods for this function:

```
% secant_method
xZeroMATLAB = fzero('x*sin(1/x)-0.5*exp(-x)',0.5); tol = 10^(-14);
total = 100; eps = 1; k = 0; x = 0.5;
    disp('Iterations of the Newton--Raphson method');
```

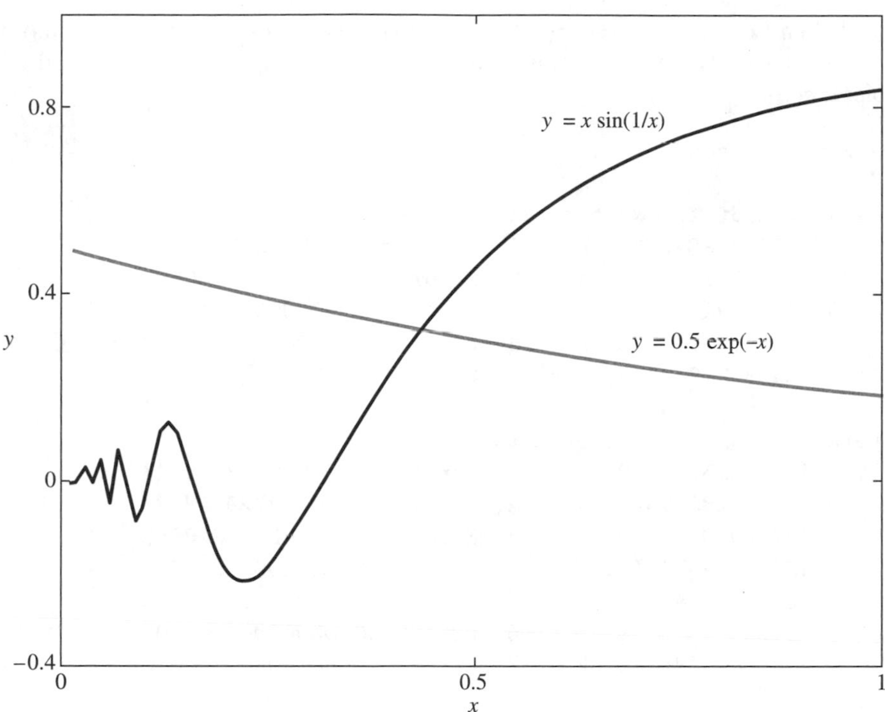

Figure 8.7 The graphs of the functions $y = x\sin(1/x)$ and $y = 0.5e^{-x}$.

```
while ((eps > tol) & (k < total))  % the Newton method
    f = x*sin(1/x)-0.5*exp(-x);
    f1 = sin(1/x)-cos(1/x)/x+0.5*exp(-x);
    xx = x-f/f1; eps = abs(xx-x);
    x = xx; k = k+1;
    Error = abs(x - xZeroMATLAB);
    fprintf('x = %12.14f, k = %2.0f, Error = %12.14f\n',x,k,Error);
end
    disp('Iterations of the Secant method');
eps = 1; k = 0; x = 0.5; f = x*sin(1/x)-0.5*exp(-x);
x0 = x; f0 = f; x = x + 0.01;
while ((eps > tol) & (k < total))  % the secant method
    f = x*sin(1/x)-0.5*exp(-x);
    xx = x-f*(x-x0)/(f-f0); eps = abs(xx-x);
    x0 = x; f0 = f; x = xx; k = k+1;
    Error = abs(x - xZeroMATLAB);
    fprintf('x = %12.14f, k = %2.0f, Error = %12.14f\n',x,k,Error);
end
```

When executed, this script shows that the secant method takes two more iterations than the Newton–Raphson method does to get within the required tolerance:

```
>>secant_method

Iterations of the Newton--Raphson method
x = 0.42596869816768, k =  1, Error = 0.00861529127990
x = 0.43445261944470, k =  2, Error = 0.00013137000289
x = 0.43458395830151, k =  3, Error = 0.00000003114608
x = 0.43458398944758, k =  4, Error = 0.00000000000000
x = 0.43458398944758, k =  5, Error = 0.00000000000000

Iterations of the Secant method
x = 0.42459025326612, k =  1, Error = 0.00999373618147
x = 0.43602128357945, k =  2, Error = 0.00143729413187
x = 0.43460966745010, k =  3, Error = 0.00002567800252
x = 0.43458392271514, k =  4, Error = 0.00000006673244
x = 0.43458398945068, k =  5, Error = 0.00000000000309
x = 0.43458398944758, k =  6, Error = 0.00000000000000
x = 0.43458398944758, k =  7, Error = 0.00000000000000
```

In practical examples, the lower *flop* count per iteration for the secant method compensates its slower convergence, resulting in a smaller computational cost than for the Newton–Raphson method. As with the Newton–Raphson method, the convergence of the secant method is guaranteed only near the root so that it should be used in combination with a global root-finding algorithm such as the bisection method.

8.2 Multidimensional Root Finding

Systems of linear equations can be solved with the direct and iterative algorithms discussed in Sections 2.5 and 2.6. When expressed in the matrix form $\mathbf{Ax} = \mathbf{b}$, Theorem 2.3 tell us that square linear systems may have (1) no solutions at all, (2) a unique solution, or (3) infinitely many solutions, depending on rather simple properties of the range and null spaces of $\mathbf{A}$. For systems of nonlinear equations, the situation can be more complicated. For instance, suppose you want to find all the solutions for the system

$$\begin{cases} xy & = 1 \\ y^2 - x & = 2 \end{cases}. \tag{8.15}$$

Since the solutions to this system are the intersections of the curves $xy = 1$ and $y^2 - x = 2$, you can use MATLAB to investigate their location before calculating them. For example, you can use the MATLAB function `contour` as follows:

```
>> x = [-3:0.01:3]; y = x; [X,Y] = meshgrid(x,y);
>> contour(X,Y,X.*Y,[1 1]) % plot for the curve xy=1
>> contour(X,Y,Y.^2-X,[2 2]) % plot for the curve y^2-x=2
```

As you can see in Figure 8.8, there are three distinct solutions for the system (8.15) within the square $[-3,3] \times [-3,3]$. These solutions can be obtained analytically by substituting $y = 1/x$ into the second equation in (8.15), leading to the cubic equation

$$x^3 + 2x^2 - 1 = 0.$$

Because this polynomial can be factorized as $x^3+2x^2-1 = (x+1)(x^2+x-1)$, we see that the three solutions of the system (8.15) are the following (x, y) pairs:

$$(-1,-1), \qquad \left(\frac{-1+\sqrt{5}}{2}, \frac{1+\sqrt{5}}{2}\right), \qquad \left(\frac{-1-\sqrt{5}}{2}, \frac{1-\sqrt{5}}{2}\right). \qquad (8.16)$$

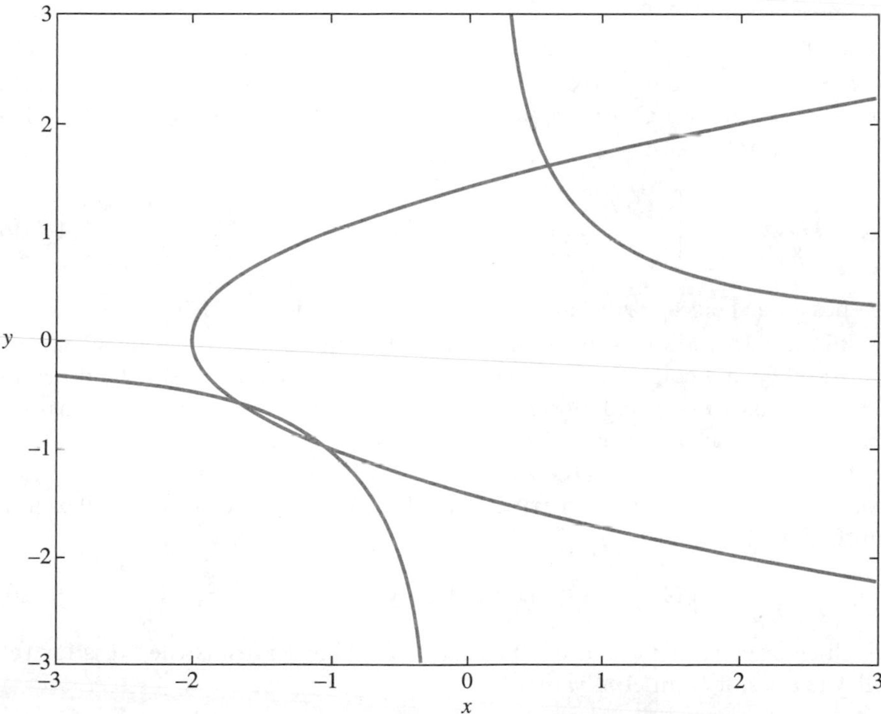

Figure 8.8 The curves $xy = 1$ and $y^2 - x = 2$ on the region $[-3,3] \times [-3,3]$.

In practice, this kind of analytic solution is rarely available, leading us to the task of devising general numerical methods for solving nonlinear systems. Returning to the example, observe that finding the solutions for the nonlinear system (8.15) is equivalent to finding the zeros of the nonlinear vector function

$$\mathbf{F}(x, y) = \begin{bmatrix} xy - 1 \\ y^2 - x - 2 \end{bmatrix}. \tag{8.17}$$

In general, the solution of a nonlinear system of n equations with n variables is defined by the following generalized version of Problem 8.1:

Problem 8.2 (Multidimensional Root Finding) Given a continuous vector function $\mathbf{F} : \mathbb{R}^n \to \mathbb{R}^n$, find all the solutions to the equation $\mathbf{F}(\mathbf{x}) = \mathbf{0}$ for $\mathbf{x}$ in the region $[a_1, b_1] \times \cdots \times [a_n, b_n]$.

Generalized fixed-point iterations

Unfortunately, generalizing the bisection method to this multidimensional setting is not straightforward. Instead, we rely on the following multidimensional version of fixed-point iterations:

$$\mathbf{x}_{k+1} = \mathbf{G}(\mathbf{x}_k), \tag{8.18}$$

where $\mathbf{G} : \mathbb{R}^n \to \mathbb{R}^n$ is a continuous function chosen in such a way that $\mathbf{F}(\mathbf{x}) = \mathbf{0}$ is equivalent to $\mathbf{x} = \mathbf{G}(\mathbf{x})$. As before, many different choices for a fixed-point function $\mathbf{G}$ are available for finding the zeros of the same function $\mathbf{F}$. For example, both

$$\mathbf{G}_1(x, y) = \begin{bmatrix} 1/y \\ -\sqrt{x+2} \end{bmatrix} \quad \text{and} \quad \mathbf{G}_2(x, y) = \begin{bmatrix} 1/y \\ (x+2)/y \end{bmatrix} \tag{8.19}$$

are possible choices corresponding to the function $\mathbf{F}$ in (8.17).

Similar to the one-dimensional case, the choice of a function $\mathbf{G}$ affects the stability and convergence properties of the iterations (8.18). The previous criterion was to seek a function g such that $g'(x)$ was as small as possible near a fixed point. In the multidimensional case, an analogous role is played by the spectral radius $\rho(\mathbf{DG}(\mathbf{x}))$ of the Jacobian matrix $\mathbf{DG}(\mathbf{x})$ computed from derivatives of $\mathbf{G}(\mathbf{x})$. In particular, if $\rho(\mathbf{DG}(\mathbf{x})) \leq C < 1$ for all $\mathbf{x}$ in a subset $S \subset \mathbb{R}^n$, then

$$\|\mathbf{G}(\mathbf{x}) - \mathbf{G}(\mathbf{y})\| < C\|\mathbf{x} - \mathbf{y}\| \qquad \forall \mathbf{x}, \mathbf{y} \in S, \tag{8.20}$$

which characterizes $\mathbf{G}$ as a *contraction* on S. This serves as the basis for the following result from *Analysis*:

Theorem 8.3 (Contraction Mapping Principle) *Suppose that $\mathbf{G} : \mathbb{R}^n \to \mathbb{R}^n$ satisfies (8.20) on $S \subset \mathbb{R}^n$ and that $\mathbf{G}(S) \subset S$. Then for any starting* *point* $\mathbf{x}_0 \in S$, *the sequence* $\mathbf{x}_{k+1} = \mathbf{G}(\mathbf{x}_k)$ *satisfies*

$$\lim_{k\to\infty} \mathbf{x}_k = \mathbf{x}_*,$$

where $\mathbf{x}_* = \mathbf{G}(\mathbf{x}_*)$ *is the unique fixed point of* $\mathbf{G}$ *in* S.

For example, the function $\mathbf{G}_1$ proposed in (8.19) is a candidate for fixed-point iterations for the roots $(-1,-1)$ and $(-1.6180, -0.6180)$, although obviously not suited for $(0.6180, 1.6180)$ because the second component of $\mathbf{G}_1(x, y)$ is always negative. You then have

$$\mathbf{DG}_1(x,y) = \begin{bmatrix} 0 & -1/y^2 \\ -1/2\sqrt{x+2} & 0 \end{bmatrix}, \tag{8.21}$$

whose spectral radius near the relevant roots can be calculated as follows:

```
>> [x,y]=[-1,-1];
>> rho1=max(abs(eig([0 -1/y^2;-1/(2*sqrt(x+2)) 0])))

rho1 =
    0.7071

>> [x,y]=[(-1-√5)/2, (1-√5/2]
>> rho3=max(abs(eig([0 -1/y^2;-1/(2*sqrt(x+2)) 0])))

rho3 =
    1.4553
```

This indicates that fixed-point iterations using the function $\mathbf{G}_1(x, y)$ will converge when started sufficiently close to the solutions $(-1,-1)$, but will fail to converge to the solution $(-1.6180, -0.6180)$. You can verify this using the following script:

```
% multi_fixed_point
tol = 0.0000001; total = 100; x = [-1.6;-0.6]; k = 0; epsilon=1;
while ((epsilon > tol) & (k < total)) % fixed point iterations
    xnew = [1/x(2); -sqrt(x(1)+2)];
    epsilon = norm(xnew-x);
    x=xnew; k = k+1;
end
fprintf('x = %12.10f, y=%12.10f found for k= %2.0f \n',x(1),x(2),k);
```

When executed, this script show convergence (albeit slow) to the root $(-1,-1)$, where the spectral radius of the Jacobian is less than one, even when the initial approximation is already very close to the root $(-1.6180, -0.6180)$, where the spectral radius of the Jacobian is greater than one.

```
>> multi_fixed_point
x = -1.0000000293, y=-1.0000000135 found after k= 58 iterations
```

Exercise 8.3 Show that the spectral radius of the Jacobian for the function $\mathbf{G}_2$ in (8.19) is less than one near the root $(0.6180, 1.6180)$ but greater than one near the other two roots. Verify that fixed-point iterations using $\mathbf{G}_2$ converge to $(0.6180, 1.6180)$. Obtain a new fixed-point function $\mathbf{G}_3$ that can be used for finding the root $(-1.6180, -0.6180)$.

Generalized Newton–Raphson method

As a way of improving the convergence rate for fixed-point iterations, we now turn to a generalized version of the Newton–Raphson method, based on a linear approximation for $\mathbf{F}$ near a point $\mathbf{x}_k$. In other words, we consider the first-order Taylor series expansion

$$\mathbf{F}(\mathbf{x}) = \mathbf{F}(\mathbf{x}_k) + \mathbf{DF}(\mathbf{x}_k)(\mathbf{x} - \mathbf{x}_k) + O(\|\mathbf{x} - \mathbf{x}_k\|^2), \tag{8.22}$$

which you may readily recognize as the multidimensional analogue of the expansion (8.10), with the Jacobian matrix $\mathbf{DF}(\mathbf{x}_k)$ playing the role of the derivative of $\mathbf{F}$ at the point $\mathbf{x}_k$. Setting $\mathbf{F}(\mathbf{x}) = \mathbf{0}$, we approximate the root $\mathbf{x}_*$ by $\mathbf{x}_{k+1}$ satisfying the linear system

$$\mathbf{DF}(\mathbf{x}_k)(\mathbf{x}_{k+1} - \mathbf{x}_k) = -\mathbf{F}(\mathbf{x}_k), \tag{8.23}$$

which is the multidimensional analogue of Equation (8.11). Therefore, iterations for the generalized Newton–Raphson are well defined provided $\mathbf{DF}(\mathbf{x}_k)$ is a nonsingular matrix, in analogy to the condition $f'(x_k) \neq 0$. Rewriting (8.23) in terms of the fixed-point iterations (8.18) leads to the function

$$\mathbf{G}(\mathbf{x}) = \mathbf{x} - \mathbf{DF}^{-1}(\mathbf{x})\mathbf{F}(\mathbf{x}). \tag{8.24}$$

By using Theorem 8.3, you can see that the generalized Newton–Raphson method converges provided the spectral radius for the Jacobian of $\mathbf{G}$ is less than one near the root $\mathbf{x}_*$. By taking derivatives of expression (8.24), you can verify that the matrix $\mathbf{DG}(\mathbf{x}_*)$ is identically zero whenever the matrix $\mathbf{DF}(\mathbf{x}_*)$ is nonsingular, which is analogous to the fact that $g'(x_*) = 0$ whenever $f'(x_*) \neq 0$ in the one-dimensional case, leading to a *quadratic* convergence rate.

For example, the Jacobian for the function $\mathbf{F}$ in (8.17) is

$$\mathbf{DF}(\mathbf{x}) = \begin{bmatrix} y & x \\ -1 & 2y \end{bmatrix}. \tag{8.25}$$

Even though it is not hard to compute the inverse of this simple 2×2 matrix, we will proceed by solving the linear system (8.23) for each iteration instead, since this should be the general approach when dealing with more higher-dimensional functions. We choose $(-1.5, -1.5)$, $(1, 2)$, and $(-1.5, -0.5)$ as starting approximations for the roots indicated in Figure 8.8 and implement the generalized Newton–Raphson method in the following script:

```
% multi_Newton
tol = 0.0000001; total = 100; k = 0; epsilon = 1;
xExact = [-1;-1]; x = [-1.5;-1.5]; % first root
while ((epsilon > tol) & (k < total)) % Newton method
    F = [x(1)*x(2)-1;x(2)^2-x(1)-2];
    DF = [x(2) x(1);-1 2*x(2)];
    xnew = x - DF\F; epsilon = norm(xnew-x);
    Error = norm(xnew-xExact);
    x=xnew; k = k+1;
    fprintf('x = %12.10f, y=%12.10f, k= %2.0f, Error= %12.10f \n', ...
        x(1),x(2),k,Error);
end
k = 0; epsilon = 1;
xExact = [(-1+5^(1/2))/2;(1+5^(1/2))/2]; x = [1;2]; % second root
while ((epsilon > tol) & (k < total)) % Newton method
    F = [x(1)*x(2)-1;x(2)^2-x(1)-2];
    DF = [x(2) x(1);-1 2*x(2)];
    xnew = x-DF\F; epsilon = norm(xnew-x);
    Error = norm(xnew-xExact);
    x = xnew; k = k+1;
    fprintf('x = %12.10f, y=%12.10f, k= %2.0f, Error= %12.10f \n', ...
        x(1),x(2),k,Error);
end
k = 0; epsilon = 1;
xExact = [(-1-5^(1/2))/2;(1-5^(1/2))/2]; x = [-1.5;-0.5]; % third root
while ((epsilon > tol) & (k < total)) % Newton method
    F = [x(1)*x(2)-1;x(2)^2-x(1)-2];
    DF = [x(2) x(1);-1 2*x(2)];
    xnew = x-DF\F; epsilon = norm(xnew-x);
    Error = norm(xnew-xExact);
    x=xnew; k = k+1;
    fprintf('x = %12.10f, y=%12.10f, k= %2.0f, Error= %12.10f \n', ...
        x(1),x(2),k,Error);
end
```

When executed, this script confirms the quadratic convergence of the generalized Newton–Raphson method, as observed in the number of significant digits in the approximations for the roots, which doubles with each iteration:

```
>> multi_Newton

x = -1.1250000000, y=-1.0416666667, k=  1, Error= 0.1317615692
x = -1.0085132890, y=-0.9967469546, k=  2, Error= 0.0091136378
x = -0.9999326717, y=-1.0000390824, k=  3, Error= 0.0000778494
```

```
x = -0.9999999932, y=-1.0000000042, k=  4, Error= 0.0000000080
x = -1.0000000000, y=-1.0000000000, k=  5, Error= 0.0000000000

x = 0.6666666667, y=1.6666666667, k=  1, Error= 0.0687769927
x = 0.6190476190, y=1.6190476190, k=  2, Error= 0.0014334897
x = 0.6180344478, y=1.6180344478, k=  3, Error= 0.0000006492
x = 0.6180339887, y=1.6180339887, k=  4, Error= 0.0000000000
x = 0.6180339887, y=1.6180339887, k=  5, Error= 0.0000000000

x = -1.6250000000, y=-0.6250000000, k=  1, Error= 0.0098514276
x = -1.6180555556, y=-0.6180555556, k=  2, Error= 0.0000305001
x = -1.6180339890, y=-0.6180339890, k=  3, Error= 0.0000000003
x = -1.6180339887, y=-0.6180339887, k=  4, Error= 0.0000000000
```

Despite the fast convergence, the Newton–Raphson method can become rather expensive in higher dimensions because of the number of *flops* necessary to solve the system (8.23) for each iteration. In these circumstances, direct application of fixed-point iterations, even with just linear convergence, might become competitive.

Secant methods

Similar to its one–dimensional counterpart, the Newton–Raphson method requires the evaluation of the Jacobian matrix $\mathbf{DF}(\mathbf{x}_k)$ at each iteration. When an analytic expression for $\mathbf{F}$ is either not available or produces partial derivatives that are too complicated for efficient computations, it becomes necessary to approximate the Jacobian matrix numerically. But instead of numerically approximating each partial derivative in the Jacobian, we look for an approximation $\mathbf{D}_k$ satisfying the *secant equation*

$$\mathbf{D}_k(\mathbf{x}_k - \mathbf{x}_{k-1}) = \mathbf{F}(\mathbf{x}_k) - \mathbf{F}(\mathbf{x}_{k-1}), \tag{8.26}$$

which you may recognize as the multidimensional analogue of Equation (8.12). There are several ways of choosing a Jacobian approximation satisfying the secant Equation (8.26). One such algorithm is called *Broyden's method* and consists of specifying initial approximations $\mathbf{x}_0$ and $\mathbf{D}_0$ and proceeding with iterations that compute a new approximation for the root by solving the linear system

$$\mathbf{D}_k(\mathbf{x}_{k+1} - \mathbf{x}_k) = -\mathbf{F}(\mathbf{x}_k), \tag{8.27}$$

followed by an update in the Jacobian according to

$$\mathbf{D}_{k+1} = \mathbf{D}_k + \frac{\mathbf{F}(\mathbf{x}_{k+1}) * (\mathbf{x}_{k+1} - \mathbf{x}_k)'}{(\mathbf{x}_{k+1} - \mathbf{x}_k)' * (\mathbf{x}_{k+1} - \mathbf{x}_k)}. \tag{8.28}$$

In the next script we implement the iterations for Broyden's method applied to the third root of the nonlinear function $\mathbf{F}$ from (8.17), taking $\mathbf{x}_0 = [-1.5, -0.5]$ and $\mathbf{D}_0 = \mathbf{1}_2$ as initial approximations:

```
% broyden
tol = 0.0000001; total = 100;  k = 0; epsilon = 1;
xExact = [(-1-5^(1/2))/2;(1-5^(1/2))/2]; x = [-1.5;-0.5]; % third root
D = [1 0;0 1];
while ((epsilon > tol) & (k < total)) % Newton method
    F = [x(1)*x(2)-1;x(2)^2-x(1)-2];
    xnew = x-D\F; epsilon = norm(xnew-x);
    Error = norm(xnew-sol(3,:)');
    D = D+[xnew(1)*xnew(2)-1;xnew(2)^2-xnew(1)-2]*(xnew-x)'/((xnew-x)'*(xnew-x));
    x = xnew; k = k+1;
    fprintf('x = %12.10f, y=%12.10f, k= %2.0f, Error= %12.10f \n', ...
        x(1),x(2),k,Error);
end
```

When executed, this script shows a slower convergence to the root $(-1.6180, -0.6180)$ than the Newton–Raphson method does:

```
>> broyden
x = -1.2500000000, y=-0.2500000000, k=  1, Error= 0.5204786583
x = -1.6428571429, y=-0.6428571429, k=  2, Error= 0.0351052412
x = -1.6132075472, y=-0.6132075472, k=  3, Error= 0.0068256191
x = -1.6179808841, y=-0.6179808841, k=  4, Error= 0.0000751013
x = -1.6180341036, y=-0.6180341036, k=  5, Error= 0.0000001625
x = -1.6180339887, y=-0.6180339887, k=  6, Error= 0.0000000000
x = -1.6180339887, y=-0.6180339887, k=  7, Error= 0.0000000000
```

In general, secant methods have only a superlinear, not quadratic, convergence rate, which might be compensated by savings in *flop* counting when compared to the evaluation of partial derivatives in the Newton–Raphson method.

The MATLAB function `fsolve` contains robust implementations of the algorithms discussed in this section, as well as more sophisticated numerical routines. It allows the user to specify a variety of options, such as which algorithm is to be used, the maximum number of iterations, and the tolerance level. Among its outputs, you can find the value obtained for the root, the value for the nonlinear function at such root (to check if it is actually close to zero), the number of iterations used for that root, and the reason for terminating the algorithm.

Exercise 8.4 Use the MATLAB function `fsolve` to obtain three distinct solutions for the system (8.15).

Before leaving this section, let us mention that the absolute condition number of the multidimensional root-finding problem for a differentiable function $\mathbf{F}$ is approximately equal to

$$\|\mathbf{DF}^{-1}(\mathbf{x})\|, \tag{8.29}$$

in direct analogy with (8.3). In particular, multidimensional root finding is ill–conditioned whenever the Jacobian $\mathbf{DF}$ is close to singular near the root.

8.3 One-Dimensional Minimization

Recall from *Calculus* that a function $f : \mathbb{R} \to \mathbb{R}$ has a *local minimum* at a point c if $f(c) \leq f(x)$ for all x in an open interval around c. If $f(c) \leq f(x)$ holds for all x in the domain of f, then the function is said to have a *global minimum* at c. Global and local maxima are defined analogously, using the reversed inequality $f(c) \geq f(x)$. You can immediately see that a function f has a local (respectively, global) maximum at c if and only if the function $-f$ has a local (respectively, global) minimum at the same point. Therefore, we can formulate all the statements of this section in terms of minimum values since the corresponding statement for maximum values can be immediately obtained by replacing f by $-f$. Accordingly, we are interested in the following problem:

Problem 8.3 (Minimization) Find the global minimum and all local minima of a continuous function $f : \mathbb{R} \to \mathbb{R}$ on a closed interval $[a, b]$.

Solutions for this problem are based on the following result from *Calculus*:

Theorem 8.4 *If $f : \mathbb{R} \to \mathbb{R}$ is a continuous function on a closed interval $[a, b]$, then f has a global minimum at a point $c \in [a, b]$. Moreover, if f has a local minimum at a point c where f is differentiable, then $f'(c) = 0$.*

Observe that the point c where a continuous function attains its global minimum on a closed interval $[a, b]$ can be located at the endpoints of the interval, in which case it is not a local minimum (because there is no open interval around it) and $f'(c)$ does not need to vanish. On the other hand, a point c where a local minimum occurs (therefore contained in the interior of $[a, b]$) might or might not be the global minimum, depending on the value of f at the other local minima and the endpoints. For example, consider the function $f(x) = x \sin(1/x)$ on the interval $[0.1, 0.5]$, plotted together with its derivative using the following commands:

```
>> x = 0.1:0.001:0.5;
>> y = x.*sin(1./x); yprime = sin(1./x)-cos(1./x)./x;
>> subplot(2,1,1); plot(x,y);
>> subplot(2,1,2); plot(x,yprime);
```

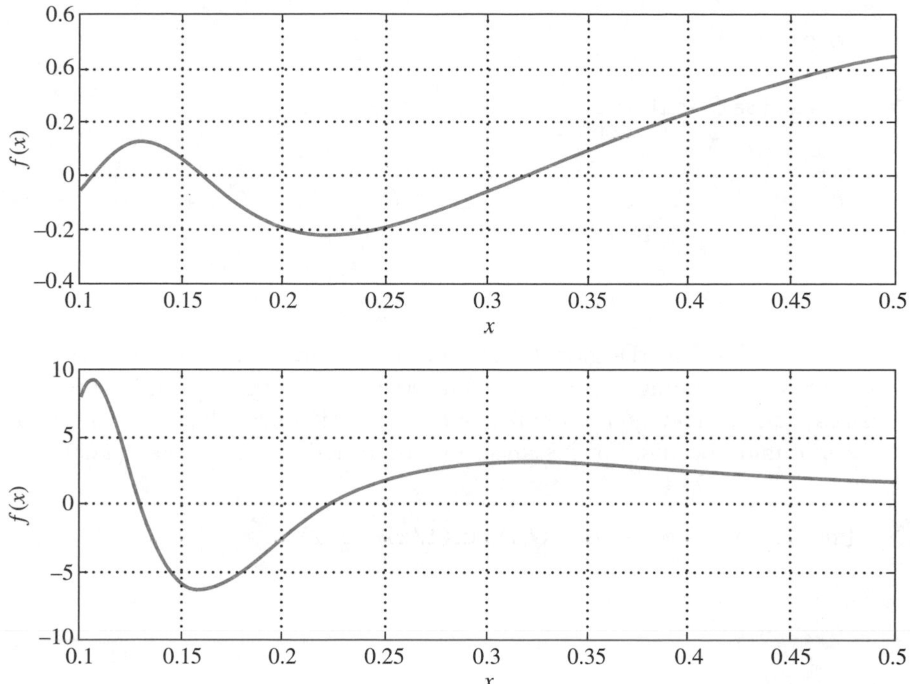

Figure 8.9 The graphs for the function $f(x) = x\sin(1/x)$ and its derivative.

You can see from Figure 8.9 that the global maximum for this function on the interval $[0.1, 0.5]$ occurs at the endpoint $b = 0.5$, despite the fact that $f'(0.5) \neq 0$. You can also observe that the function has a local maximum somewhere between the points $x = 0.1$ and $x = 0.15$ and that its derivative vanishes at this point. Moreover, the function has a local minimum somewhere between $x = 0.2$ and $x = 0.25$, where its derivative vanishes, which also happens to be a global minimum over the interval $[0.1, 0.5]$.

The distinctions between global and local minima are reflected in the numerical algorithms used to find them. For example, the MATLAB function **fminbnd**, based on the *golden section search* and *successive parabolic interpolation* described later in this section, returns the global minimum of a scalar function of one variable over a closed interval $[a, b]$. For instance, for the function $f(x) = x\sin(1/x)$ over the interval $[0.1, 0.5]$, you have:

```
>> [xmin,fmin]=fminbnd('x*sin(1/x)',0.1,0.5)

xmin =
    0.2225
```

```
fmin =
   -0.2172

>> [xmax,fmax]=fminbnd('-x*sin(1/x)',0.1,0.5)

xmax =
    0.5000

fmax =
   -0.4546
```

which correspond to the global minimum and global maximum observed in Figure 8.9. For finding a local minimum, you can use the MATLAB function `fminsearch`, which requires an initial approximation near the point where the local minimum occurs. For instance, for the function $f(x) = x\sin(1/x)$, you have

```
>> [xmin,fmin]=fminsearch('x*sin(1/x)',0.2)

xmin =
    0.2226

fmin =
   -0.2172

>> [xmax,fmax]=fminsearch('-x*sin(1/x)',0.15)

xmax =
    0.1294

fmax =
   -0.1284
```

corresponding to the local minimum and local maximum shown in Figure 8.9.

As you can see from both Theorem 8.4 and Figure 8.9, there exists a close relationship between Problems 8.1 and 8.3 because the roots for the nonlinear function f' are the candidates for local minima for the function f. That is, you can use one of the algorithms developed for Problem 8.1 to find the roots of f' as a first step, and then identify the local minima (or maxima) for f among them.

Second derivative test

This second step is necessary because $f'(c) = 0$ is not a sufficient condition for f to have a local minimum (or a maximum) at c. When the function is twice-differentiable, then a sufficient condition for it to have a local minimum at point c satisfying $f'(c) = 0$ is that $f''(c) > 0$. Similarly, if $f'(c) = 0$ and $f''(c) < 0$, then f has a local maximum at $x = c$. This is called the *second derivative test* for critical points and is based on the fact that the second derivative determines the *concavity* of a function around a point. Notice that the second derivative test cannot be used conclusively when $f''(c) = 0$. For example, for the function $f(x) = x^3$, you have that $f'(0) = f''(0) = 0$, but f does not have either a maximum or a minimum at $x = 0$. On the hand, for the function $f(x) = x^4$, you also have $f'(0) = f''(0) = 0$ and f has a minimum at $x = 0$.

An obvious limitation to finding the minima of a function as the roots of its derivative is that this requires the function to be differentiable at the minimum point. As formulated in Problem 8.3, you will often encounter practical situations where you want to minimize a function that is continuous but not differentiable at the minimum, such as $f(x) = |x|$. More generally, even when the function is differentiable, its derivative might not be available in closed form, or can be too complicated for efficient computations. For all of these reasons, it is important to develop numerical methods specifically designed for solving a minimization problem without the need to compute derivatives.

Golden section search

We start with a method that uses the increase and decrease properties of a function near a minimum similarly to the way the bisection method uses a sign change near a root. For this, let us say that a function f is *unimodal* on an interval $[a_0, b_0]$ if it has a unique minimum and no local maximum in this interval (which you can always ensure by making the interval around a local minimum sufficiently small). In this case, we select two points $a_1 < b_1$ inside the interval $[a_0, b_0]$ and observe that, if $f(a_1) < f(b_1)$, the minimum cannot be achieved in the subinterval $(b_1, b_0]$, whereas if $f(a_1) > f(b_1)$, the minimum cannot be achieved in the subinterval $[a_0, a_1)$. We can therefore concentrate the search either on $[a_0, b_1]$ or on $[a_1, b_0]$, which are smaller than the initial interval $[a_0, b_0]$. In analogy with the bisection method, we would like to reduce the length of the intervals by a fixed fraction at each iteration, if only to have a definite control of the error at each step. This can be achieved if we set

$$b_1 - a_0 = s(b_0 - a_0) \tag{8.30}$$

$$b_0 - a_1 = s(b_0 - a_0), \tag{8.31}$$

so that both $[a_0, b_1]$ and $[a_1, b_0]$ have length equal to $s(b_0 - a_0)$. Moreover, because a_1 lies inside the interval $[a_0, b_1]$ and b_1 lies inside the interval $[a_1, b_0]$, we already know the value for the function at one point inside whichever interval is retained. That is, if we reuse this point, all we need to do to continue the iterations is to choose one more point inside the new subinterval. But to obtain consistent lengths we must have

$$b_1 - a_1 = (1 - s)(b_0 - a_1)$$

$$a_1 - a_0 = s(b_1 - a_0),$$

which can only be satisfied simultaneously with (8.30) and (8.31) if $s^2 = 1 - s$. Because the positive solution to this equation is $s = (\sqrt{5} - 1)/2 \approx 0.618$, this method is known as the *golden section search*, in reference to the golden ratio $(\sqrt{5}+1)/2 \approx 1.618$.

In the next script, we implement the golden section search for the minimum of $f(x) = x\sin(1/x)$ over the interval $[0.15, 0.5]$, where the unimodal condition is satisfied:

```
% golden_section
tol = 0.000001; k = 0;
options = optimset('TolX',tol); % sets the tolerance for fminbnd
[xmin,fmin] = fminbnd('x*sin(1/x)',0.15,0.5,options);
a = 0.15; b = 0.5;     % initial interval
s = (sqrt(5)-1)/2;  % golden section
a1 = a+(1-s)*(b-a); % left intermediate point
fa1 = a1*sin(1/a1);
b1 = a+s*(b-a);     % right intermediate point
fb1 = b1*sin(1/b1);
while (b-a > tol) % the golden section search
 if fa1>fb1      % minimum must be to the right
    a=a1;      % left end point
    a1=b1;     % left intermediate point
    fa1=fb1;
    b1=a+s*(b-a); % right intermediate point
    fb1=b1*sin(1/b1);
 else          % minimum must be to the left
    b=b1;      % right end point
    b1=a1;     % right intermediate point
    fb1=fa1;
    a1=a+(1-s)*(b-a); % left intermediate point
    fa1=a1*sin(1/a1);
 end
 x = b1; Error = abs(x-xmin);k=k+1;
end
fprintf('x = %12.10f, Error= %12.10f, k= %2.0f \n',x,Error,k);
```

Upon execution, you can observe the convergence of the golden section search to the same minimum found by `fminbnd`:

```
>> golden_section

x = 0.2225483432, Error= 0.0000002015, k= 27
```

From the same argument used for the bisection method, we conclude that the golden section search has a linear convergence with constant $C \approx 0.618$, which explains the large number of iterations in the preceding example.

Successive parabolic interpolation

Observe that only the relative size of the function values is used in the golden section search, much in the same way that only the sign of the function is used in the bisection method. To obtain a better convergence rate, you might want to use more information from the function values. The idea for the next method, known as *successive parabolic interpolation*, is to approximate a given function $f(x)$ by a parabola $y = c_1x^2 + c_2x + c_3$ and take its minimum at $x_{\min} = -\frac{c_2}{2c_1}$ as an approximation for the minimum of f. For this, let $x_1 < x_2 < x_3$ be three arbitrary points in the interval $[a, b]$ and denote $y_1 = f(x_1), y_2 = f(x_2), y_3 = f(x_3)$. Then, a little bit of algebra shows that the minimum for the unique parabola determined by these three points is achieved at

$$x_{min} = \frac{1}{2}\frac{(y_1 - y_2)(x_3 - x_2)^2 - (y_3 - y_2)(x_2 - x_1)^2}{(y_3 - y_2)(x_2 - x_1) + (y_1 - y_2)(x_3 - x_2)}. \tag{8.32}$$

Exercise 8.5 Derive Expression (8.32).

We then choose $x_{\min}$ and the two points out of x_1, x_2, x_3 that provide the best parabola to approximate the function $f(x)$ in the next iteration. In the next script, we implement these iterations for the function $f(x) = x\sin(1/x)$ on the interval $[0.15, 0.5]$:

```
% successive_parabolic_interpolation
tol = 0.000001; k = 0;
options = optimset('TolX',tol); % sets the tolerance for fminbnd
[xmin,fmin] - fminbnd('x*sin(1/x)',0.1,0.5,options);
a = 0.15; b = 0.5;        % initial interval
x1 = a; x2 = (a+b)/2; x3 = b;   % initial three points
while x3-x1 > tol    % the successive parabolic interpolation

    y1 = x1*sin(1/x1); y2 = x2*sin(1/x2); y3 = x3*sin(1/x3);

    x = ((y2-y1)*(x3^2-x2^2)-(y3-y2)*(x2^2-x1^2))/...
        (2*((y3-y2)*(x1-x2)+(y2-y1)*(x3-x2)));
    if (x2-x)>0          % minimum must be to the left
        x3=x2;
    else                 % minimum must be to the right
        x1=x2;
    end
    x2=x;
    Error=abs(x-xmin);k=k+1;
end
fprintf('x = %12.10f, Error= %12.10f, k= %2.0f \n',x,Error,k);
```

You can then observe a faster convergence for this method when compared to the golden section search:

```
>> parabolic

x = 0.2225481587, Error= 0.0000000239, k= 18
```

In general, the method of successive parabolic interpolation converges with a superlinear rate when started sufficiently close to the minimum in an interval where the function is unimodal. To obtain a faster convergence rate, we must use higher-order methods, that is, methods based on evaluations of the derivative as well as the function itself. Because this turns out to be equivalent to using root-finding algorithms for solving $f'(x) = 0$, we leave it as an exercise.

Exercise 8.6 Consider the function $f(x) = x\sin(1/x)$ over the interval $[0.1, 0.2]$. Calculate its local maximum using the golden section search, successive parabolic interpolation, and the Newton–Raphson method for $f'(x) = 0$.

Sensitivity

The sensitivity of Problem 8.3 varies depending on the type of method we consider. For methods relying on finding the root for the equation $f'(x) = 0$, we can use the results from Section 8.1 to conclude that the absolute condition number for the minimization problem is approximately equal to

$$\frac{1}{|f''(x)|}. \tag{8.33}$$

Therefore, the minimization problem is ill-conditioned whenever the minimum occurs at a point where the second derivative of f vanishes. In particular, if $f''(x) = 0$ near the minimum, then the Newton–Raphson method is a poor choice for solving $f'(x) = 0$.

Consider now methods that do not rely on evaluations of the first derivative, but rather on the function f itself, such as the golden section search. In this case, using the fact that $f'(x_{min}) = 0$, a Taylor expansion near the minimum x_{min} gives

$$f(x) = f(x_{min}) + \frac{1}{2}f''(\xi)(x - x_{min})^2,$$

for some $x_{min} \leq \xi \leq x$. Therefore, by treating $|f(x) - f(x_{min})|$ as an input error and $|x - x_{min}|$ as an output error, we see that the absolute condition number in this case is approximately equal to

$$\sqrt{\frac{2}{|f''(x)|}}, \tag{8.34}$$

which can be much smaller than (8.33) when the second derivative is small.

8.4 Multidimensional Minimization

We now consider the extreme points of scalar functions of the form $f : \mathbb{R}^n \to \mathbb{R}$ with a domain $D \subset \mathbb{R}^n$. In analogy with the one-dimensional case, we say that f has a *local minimum* at a point $\mathbf{c} \in \mathbb{R}^n$ if $f(\mathbf{c}) \leq f(\mathbf{x})$ for all $\mathbf{x}$ in a *neighborhood* of $\mathbf{c}$. In this context, a neighborhood replaces the concept of an open interval around the local minimum, that is, it consists of points $\mathbf{x}$ such that $\|\mathbf{x} - \mathbf{c}\| < r$ for some positive constant r. If $f(\mathbf{c}) \leq f(\mathbf{x})$ holds for all $\mathbf{x}$ in the domain of f, then $\mathbf{c}$ is said to be a *global minimum*. The concepts of *local maximum* and *global maximum* are defined similarly, using the reverse inequality $f(\mathbf{c}) \geq f(\mathbf{x})$. As in the one-dimensional case, we formulate all the statements of this section in terms of minimum values since the corresponding statements for maximum values can be immediately obtained by replacing f with $-f$.

Problem 8.4 (Minimization) Find the global minimum and all the local minima of a continuous function $f : \mathbb{R}^n \to \mathbb{R}$ on the region $[a_1, b_1] \times \cdots \times [a_n, b_n]$.

The existence of a solution to Problem 8.4 rests on the following result from *Analysis*, which makes use of the concepts of differentiability and gradient of a function of several variables defined in Section 7.3.

Theorem 8.5 *If $f : \mathbb{R}^n \to \mathbb{R}$ is a continuous function on $D = [a_1, b_1] \times \cdots \times [a_n, b_n]$, then f has a global minimum at a point $\mathbf{c} \in D$. Moreover, if f has a local minimum at a point $\mathbf{c}$ where it is differentiable, then $\nabla f(\mathbf{c}) = \mathbf{0}$.*

Analogously to the one-dimensional case, there is a close relationship between Problems 8.2 and 8.4 in the sense that the solutions to the nonlinear system $\nabla f(\mathbf{x}) = \mathbf{0}$ are the candidates for local minima for the function $f : \mathbb{R}^n \to \mathbb{R}$. In other words, you can use the algorithms described in Section 8.2 to find the solutions to $\nabla f(\mathbf{x}) = \mathbf{0}$, which are called *critical points*, and then verify if they correspond to points where f has a minimum (or a maximum).

Hessian test

To verify if a function has a minimum or a maximum at a point $\mathbf{c}$ satisfying $\nabla f(\mathbf{c}) = \mathbf{0}$, you can use a multidimensional analogue of the second derivative test. Just as the second derivative of a scalar function of one variable is defined as the derivative of its first derivative, we obtain the second derivative of a scalar function of several variables by taking the derivative of its gradient. Since ∇f maps $\mathbb{R}^n$ to $\mathbb{R}^n$, you know from Section 7.5 that its derivative at a point $\mathbf{x}_0$ is given by its $n \times n$ Jacobian matrix. Because each component of the gradient vector is of the form $\frac{\partial f}{\partial x_i}$, the components of the Jacobian of the gradient are given in terms of the second partial derivatives $\frac{\partial^2 f}{\partial x_j \partial x_i}$. This defines the *Hessian matrix* of a function $f : \mathbb{R}^n \to \mathbb{R}$ as

$$\mathbf{H}_f(\mathbf{x}) = \begin{bmatrix} \frac{\partial^2 f(\mathbf{x})}{\partial x_1^2} & \frac{\partial^2 f(\mathbf{x})}{\partial x_2 \partial x_1} & \cdots & \frac{\partial^2 f(\mathbf{x})}{\partial x_n \partial x_1} \\ \frac{\partial^2 f(\mathbf{x})}{\partial x_1 \partial x_2} & \frac{\partial^2 f(\mathbf{x})}{\partial x_2^2} & \cdots & \frac{\partial^2 f(\mathbf{x})}{\partial x_n \partial x_2} \\ \vdots & \vdots & \ddots & \vdots \\ \frac{\partial^2 f(\mathbf{x})}{\partial x_1 \partial x_n} & \frac{\partial^2 f(\mathbf{x})}{\partial x_2 \partial x_n} & \cdots & \frac{\partial^2 f(\mathbf{x})}{\partial x_n^2} \end{bmatrix}. \tag{8.35}$$

As with scalar functions of a single variable, you can express the function $f : \mathbb{R}^n \to \mathbb{R}$ near a point $\mathbf{x}_0$ in terms of its Taylor expansion as

$$f(\mathbf{x}) = f(\mathbf{x}_0) + \nabla f(\mathbf{x}_0)' * (\mathbf{x} - \mathbf{x}_0) + \frac{1}{2}(\mathbf{x} - \mathbf{x}_0)' * \mathbf{H}_f(\xi) * (\mathbf{x} - \mathbf{x}_0), \tag{8.36}$$

where $\boldsymbol{\xi} = \alpha\mathbf{x}_0 + (1 - \alpha)\mathbf{x}$ for some $\alpha \in [0, 1]$. If $\mathbf{x}_0 = \mathbf{c}$ is a point satisfying $\nabla f(\mathbf{c}) = \mathbf{0}$, this expression reduces to

$$f(\mathbf{x}) = f(\mathbf{c}) + \frac{1}{2}(\mathbf{x} - \mathbf{c})' * \mathbf{H}_f(\boldsymbol{\xi}) * (\mathbf{x} - \mathbf{c}). \tag{8.37}$$

If the second partial derivatives of f are continuous, then the mixed derivatives are equal, which implies that the Hessian matrix in (8.35) is symmetric. Moreover, if the Hessian matrix $\mathbf{H}_f$ is positive definite at the point $\mathbf{c}$, then it is positive definite at the nearby point $\boldsymbol{\xi}$, also by continuity. In this case, using the definition in Section 2.7, you can see that the last term in (8.37) is a positive number, meaning that $f(\mathbf{x}) > f(\mathbf{c})$ for all $\mathbf{x}$ in a neighborhood of $\mathbf{c}$. In other words, the function f has a local minimum at $\mathbf{c}$. Similarly, if $-\mathbf{H}_f(\mathbf{c})$ is positive definite, then the last term in (8.37) is negative, implying that $f(\mathbf{x}) < f(\mathbf{c})$ for all $\mathbf{x}$ in a neighborhood of $\mathbf{c}$, meaning that f has a local maximum at $\mathbf{c}$. Finally, if the matrix $\mathbf{H}_f(\mathbf{c})$ is indefinite, then $\mathbf{c}$ is called a *saddle point*.

Let us examine these possibilities using the function

$$f(x, y) = 3x - x^3 - 2y^2 + y^4. \tag{8.38}$$

The candidates for extreme points are the solutions to the nonlinear system

$$\nabla f(x, y) = [3 - 3x^2, -4y + 4y^3] = [0, 0]. \tag{8.39}$$

In general, you need to use numerical root-finding algorithms to find the points $\mathbf{c}$ satisfying $\nabla f(\mathbf{c}) = \mathbf{0}$. For this simple function it is easy to see that the solutions to (8.39) are the points (x, y) with $x = \pm 1$ and $y = 0$ or $y = \pm 1$. In order to classify these critical points, observe that the Hessian for this function is

$$\mathbf{H}_f(x, y) = \begin{bmatrix} -6x & 0 \\ 0 & -4 + 12y^2 \end{bmatrix}.$$

For this example, the Hessian matrix turns out to be diagonal, which means that its eigenvalues can be easily calculated. By recalling from Section 4.4 that a symmetric matrix is positive definite if and only if all of its eigenvalues are positive, you can conclude that the critical points $(-1,1)$ and $(-1,-1)$ correspond to local minima for f because the eigenvalues of the Hessian matrix at these points are $\lambda_1 = 6$ and $\lambda_2 = 8$. Similarly, the critical point $(1,0)$ corresponds to a local maximum because the Hessian matrix at this point has eigenvalues $\lambda_1 = -6$ and $\lambda_2 = -4$, being therefore negative definite. Finally, the critical points $(1,1)$, $(1,-1)$, and $(-1,0)$ are saddle points because the Hessian matrix at these points has both a positive and a negative eigenvalue, being therefore indefinite. These extreme points are shown in Figure 8.10, obtained from the following:

```
>> x=[-2:0.1:2];y=x;[X,Y]=meshgrid(x,y);
>> z=3.*X-X.^3-2.*Y.^2+Y.^4;
>> surf(x,y,z)
```

In general, calculating all the eigenvalues of a matrix is an expensive way to check for positivity. A much better alternative is to try to perform a

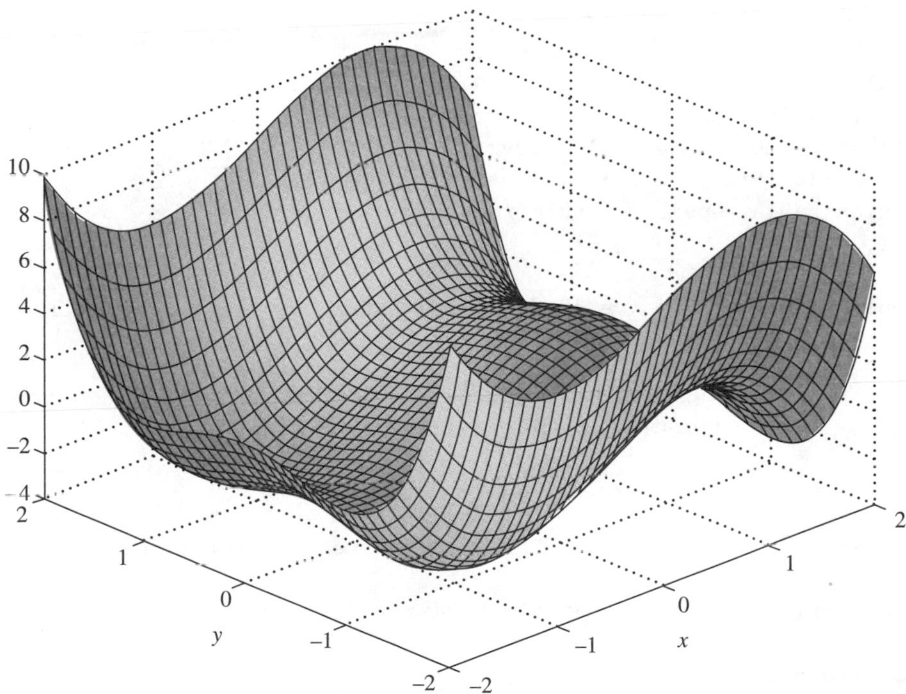

Figure 8.10 Extreme points for the function $f(x, y) = 3x - x^3 - 2y^2 + y^4$.

Cholesky factorization, since this can be achieved if and only if the matrix is positive definite (see Section 2.7).

Newton's method

For an example in which the critical point must be found numerically, consider the function

$$f(x, y) = 10x^2y - 5x^2 - 4y^2 - x^4 - 2y^4, \tag{8.40}$$

whose gradient and Hessian are

$$\nabla f(x, y) = [20xy - 10x - 4x^3, 10x^2 - 8y - 8y^3],$$

and

$$\mathbf{H}_f(x, y) = \begin{bmatrix} 20y - 10 - 12x^2 & 20x \\ 20x & -8 - 24y^2 \end{bmatrix}.$$

Using the fact that the Hessian matrix is the Jacobian of the gradient, the following script implements Newton's method to find one solution for the nonlinear equation $\nabla f = \mathbf{0}$:

```
% newton_min
tol = 0.000001; total = 100; k = 0; epsilon = 1;
x = [1;1]; % initial guess
while ((epsilon > tol) & (k < total)) % Newton method
    nabla = [20*x(1)*x(2)-10*x(1)-4*x(1)^3;10*x(1)^2-8*x(2)-8*x(2)^3];
    Hessian = [20*x(2)-10-12*x(1)^2 20*x(1);20*x(1) -8-24*x(2)^2];
    xnew = x - Hessian\nabla; epsilon = norm(xnew-x);
    x=xnew; k = k+1;
end
fprintf('Critical point x = %12.10f, y=%12.10f found after %2.0f iterations \n',...
    x(1),x(2),k);
```

You can now execute this script and use the Cholesky factorization technique to test for positivity of the Hessian at the critical point:

```
>> newton_min

Critical point x = 0.8566567332, y=0.6467721517 found after 10 iterations

>> chol([20*x(2)-10-12*x(1)^2 20*x(1);20*x(1) -8-24*x(2)^2])
??? Error using ==> chol
Matrix must be positive definite.

>> chol(-1*[20*x(2)-10-12*x(1)^2 20*x(1);20*x(1) -8-24*x(2)^2])
??? Error using ==> chol
Matrix must be positive definite.
```

Because the Cholesky factorization fails for both $\mathbf{H}_f$ and $-\mathbf{H}_f$, you can conclude that the Hessian matrix is indefinite at the critical point $[0.8566, 0.6468]$, implying that it is a saddle point for f. As usual, Newton's method converges quadratically to a critical point provided the iterations start sufficiently close to it. A drawback of the method is the requirement to evaluate the Hessian matrix at each step, which might not be possible for functions whose second derivatives have complicated expressions. This can be overcome by approximating the Hessian matrix in a variety of ways, such as by using the secant methods discussed in Section 8.2.

Exercise 8.7 Use Newton's method with different initial guesses to find the other four critical points for the function $f(x, y) = 10x^2y - 5x^2 - 4y^2 - x^4 - 2y^4$ and classify them according to the positivity of the corresponding Hessian matrices. Plot the graph of the function and locate its extreme points.

Steepest descent

Even after replacing the Hessian matrix with a suitable approximation $\mathbf{H}_k$, Newton's method and its variants still require the solution of a linear system of the form $\mathbf{H}_k\mathbf{s} = -\nabla f(\mathbf{x}_k)$ at each iteration. The *steepest descent* method is an alternative that avoids this requirement. Given a point $\mathbf{x}_0$ such that $\nabla f(\mathbf{x_0}) \neq \mathbf{0}$, you know from Exercise 7.3 that the function f has its largest rate of *decrease* in the direction $-\nabla f(\mathbf{x}_0)$. This means that $-\nabla f(\mathbf{x}_0)$ is a good candidate for a *search direction* for the minimum of f. In other words, we look for the minimum of the one-variable function

$$g(t) = f(\mathbf{x}_0 - t\nabla f(\mathbf{x}_0))$$

using any of the methods from Section 8.3. Suppose that this minimum is achieved at the point t_0. We can test to see if $\mathbf{x}_1 = \mathbf{x}_0 - t\nabla f(\mathbf{x}_0)$ is sufficiently close to a local minimum for f by checking if $\nabla f(\mathbf{x}_1)$ is sufficiently close to zero. That is, if $\nabla f(\mathbf{x}_1)$ is smaller than a specified tolerance, we deem $\mathbf{x}_1$ to correspond to a local minimum for f. Otherwise, we take it as a new starting point and repeat the procedure.

As an illustration, consider the function $f(x, y) = xye^{-(x^2+y^2)}$, whose gradient is

$$\nabla f = [(1 - 2x^2)ye^{-(x^2+y^2)}, (1 - 2y^2)xe^{-(x^2+y^2)}].$$

The following script implements the steepest descent method using `fminsearch` for the one-dimensional minimization step.

```
% steepest
tol = 0.000001; total = 100; k = 0; epsilon = 1;
x = [0;-1]; % initial guess
% gradient calculated at the initial guess
s=[(1-2*x(1)^2)*x(2)*exp(-x(1)^2-x(2)^2);(1-2*x(2)^2)*x(1)*exp(-x(1)^2-x(2)^2)];
while ((epsilon > tol) & (k < total)) % steepest descent method
```

```
    tmin=fminsearch(@(t)(x(1)-t*s(1))*(x(2)-t*s(2))...
            *exp(-(x(1)-t*s(1))^2-(x(2)-t*s(2))^2),1);
    xnew = x-tmin*s;
    x=xnew;
    s=[(1-2*x(1)^2)*x(2)*exp(-x(1)^2-x(2)^2);
    (1-2*x(2)^2)*x(1)*exp(-x(1)^2-x(2)^2)];
    epsilon=norm(s);
    k = k+1;
end
fprintf('Minimum found at x = %12.10f, y=%12.10f after %2.0f iterations \n',...
    x(1),x(2),k);
```

When executed, this script produces the following outcome:

```
>> steepest

Minimum found at x = 0.7071067811, y=-0.7071067811 after  3 iterations
```

It is easy to see that the same script, when started at the initial guess $\mathbf{x}_0 = [-1, 0]$, leads to the other local minimum for the function $f(x, y) = xye^{-(x^2+y^2)}$.

The convergence of the steepest descent method depends on the shape of the level curves of f. Generally, oblong level curves lead to slow convergence, whereas circular level curves lead to fast convergence, as can be verified in the next exercise.

Exercise 8.8 Implement the steepest descent method for finding the minimum of the function $f(x, y) = x^2 + 5y^2$ starting at the point $\mathbf{x}_0 = [5, 1]$ with a tolerance of 10^{-6}. Plot the 10 first points $\mathbf{x}_k$ and verify that the lines connecting any 3 consecutive points are perpendicular. In the same graph, plot the level curves of f passing through these points.

The same arguments at the end of Section 8.3 apply to the sensitivity of multidimensional minimization problems. In particular, the absolute condition number is generally inversely proportional to $\|\mathbf{H}_f^{-1}(\mathbf{x})\|$. In other words, multidimensional minimization is ill-conditioned whenever the Hessian matrix is close to singular near the minimum.

8.5 Summary and Notes

In this chapter we discuss methods for finding roots and extreme values of nonlinear functions.

- **Section 8.1:** A solution of the nonlinear equation $f(x) = 0$ is called a *root* for the nonlinear function $f : \mathbb{R} \to \mathbb{R}$. Problem 8.1 consists of finding the roots of a continuous function on a closed interval $[a, b]$. The *bisection* method is based on the fact that a root must exist whenever the function changes sign on an interval, as a result of Theorem 8.1. The *fixed-point iterations* method is based on the fact that the solution to $f(x) = 0$ is equivalent to the solution to $g(x) = x$ for appropriate fixed-point functions g. When g is a *contraction*, the point satisfying $g(x_*) = x_*$ is the limit of the iterations $x_{k+1} = g(x_k)$, as a result of Theorem 8.2. A special case of fixed-point iterations is the *Newton–Raphson* method, based on a tangent approximation for f. An alternative to the Newton–Raphson method that avoids the need to evaluate the derivative of f is the *secant* method, based on a secant approximation for the derivative.

- **Section 8.2:** Problem 8.2 consists of finding the roots for the nonlinear function $\mathbf{F} : \mathbb{R}^n \to \mathbb{R}^n$ on region $[a_1, b_1] \times \cdots \times [a_n, b_n]$, or equivalently, the solutions to the nonlinear system of equations $\mathbf{F}(\mathbf{x}) = \mathbf{0}$. This is solved using generalized *fixed-point iterations* of the form $\mathbf{x}_{k+1} = \mathbf{G}(\mathbf{x}_k)$, whose convergence depend on the spectral radius of the Jacobian matrix $\mathbf{DG}$. Special cases of such iterations are the generalized *Newton–Raphson* and *secant* methods.
- **Section 8.3:** Problem 8.3 consists of finding all the local minima for a function $f : \mathbb{R} \to \mathbb{R}$ over a closed interval $[a, b]$. Because a maximum for f is a minimum for $-f$, a trivial modification of this problem encompasses all the extreme values for f. When f is differentiable, these extreme values occur at points satisfying $f'(x) = 0$, called *critical points*. The minima for f can be obtained by first searching for the roots of its derivative, for instance, using the Newton–Raphson method. After the critical points are found, minima and maxima can be detected using the *second derivative test*. Alternatively, a minimum for f can be searched for without using the values of its derivative, for example, using the *golden section search* or *successive parabolic interpolations*.
- **Section 8.4:** Problem 8.4 considers the search of extreme values for a function of the form $f : \mathbb{R}^n \to \mathbb{R}$. When f is differentiable, such values occur at points satisfying $\nabla f(\mathbf{x}) = \mathbf{0}$ so that the minima for f can be obtained by first searching for the roots of its gradient, for instance, using the generalized Newton–Raphson method. After these critical points are found, minima and maxima can be detected using the *Hessian test*. An alternative to the Newton–Raphson method is to search for minima for the function f using the *steepest descent method*, which does not require the evaluation of the Hessian matrix.

Additional algorithms for all the problems in this chapter can be found in [12]. Taylor's theorem for many variables and the Hessian test are presented in [18]. The methods described in Section 8.4 pertain to what is called *unconstrained optimization*, in the sense that the function to be minimized is not subject to any additional constraints. For an introduction to constrained optimization, we recommend [12].

8.6 Exercises

1. (a) Based on Figure 8.2, obtain suitable initial intervals for finding the roots of $f(x) = x^3 - 2x^2 - 3x + 2$ using the bisection method.
   (b) For each of the roots, predict how many iterations of the bisection method are sufficient to obtain an approximation error smaller than 10^{-5}.
   (c) Obtain an estimate for the absolute condition number for the problem of finding each of these roots.
   (d) Write a MATLAB script implementing the bisection method for the preceding function and confirm the number of iterations predicted in the previous item.

2. Consider the function $f(x) = x + \log x$.
   (a) Plot the functions $y = x$ and $y = -\log x$ over the interval $(0, 2]$ on the same axes.
   (b) Using the intermediate value theorem, explain why there is exactly one root for f on $[0, 1]$.
   (c) Write a MATLAB script implementing the fixed-point iterations $x_{k+1} - g(x_k)$ using the function $g(x) = -\log x$.
   (d) Calculate the first 10 iterations starting from $x_0 = 1/2$ and from $x_0 = 3/4$. Decide if these iterations converge and, if they do, describe the type of convergence.
   (e) Observe that $g(x) = -\log x$ corresponds to the direct method with $\alpha = -1$. Repeat the last two items using $g(x) = x + \alpha f(x)$, with $\alpha = -1/f'(\widehat{x})$, where $\widehat{x}$ is an approximation for the root of $f(x)$ obtained in the previous items.

3. Decide whether the iterative sequence

$$x_{k+1} = \cos(x_k), \quad x_0 = 1$$

   converges to the solution of $x - \cos x = 0$ and at what rate.
   (a) Write a MATLAB script implementing the preceding iterations stopping either at a maximum number I_{max} or when the distance between two consecutive approximations is smaller than 10^{-6}.
   (b) Describe the Newton–Raphson method for solving the equation $x - \cos x = 0$ and write a MATLAB script implementing it with the same stopping criteria used in the previous item.

4. Consider the problem of solving the nonlinear equation $e^x = 3x^2$.

   (a) Plot the graphs of $y = e^x$ and $y = 3x^2$ on the same axes.

   (b) Use the bisection method to find an approximate solution x_0 with an error smaller than $1/2$.

   (c) Prove that the Newton–Raphson method for solving $e^x = 3x^2$ converges using this value of x_0 and describe the type of convergence.

   (d) Write a MATLAB script implementing the iterations for the Newton–Raphson method for this equation starting from x_0 from the previous items. Stop the iterations when the distance between two consecutive approximations is smaller than 10^{-9}.

   (e) Explain how to use the secant method to solve $e^x = 3x^2$ and write a MATLAB script implementing it, starting from x_0. Stop the iterations when $e_k = |x_k - \widehat{x}|$ is smaller than 10^{-6}, where $\widehat{x}$ is the approximation obtained using the Newton–Raphson method in the previous item.

   (f) Describe the convergence for the secant method in the previous item.

5. Consider the nonlinear equations

$$\begin{aligned} x &= xy + y^3 + 0.1 \\ y &= x^2 + y^2. \end{aligned}$$

   (a) Determine if the transformation $\mathbf{G} : \mathbb{R}^2 \to \mathbb{R}^2$ defined by the right-hand side of these equations maps the domain $D = \{0.09 \leq x \leq 0.11, 0 \leq y \leq 0.02\}$ into itself.

   (b) Find the spectral radius of the Jacobian of $\mathbf{G}$ at $[x^{(0)}, y^{(0)}] = [0.1, 0]$.

   (c) Write a MATLAB script implementing fixed-point iterations for solving these nonlinear equations starting from $[x^{(0)}, y^{(0)}] = [0.1, 0]$.

6. Consider the nonlinear equations

$$\begin{aligned} x^3 + 2xy + 2y^2 - 1 &= 0 \\ 3x^2 + y &= 0. \end{aligned}$$

   (a) Estimate $\|\mathbf{DF}^{-1}\|$ at the point $[x^{(0)}, y^{(0)}] = [1, 1]$, where $\mathbf{F} : \mathbb{R}^2 \to \mathbb{R}^2$ is the transformation defined by the left-hand side of the preceding equations.

   (b) Write a MATLAB script implementing the Newton–Raphson method to find a solution for this nonlinear system starting from $[x^{(0)}, y^{(0)}] = [1, 1]$.

(c) Write a MATLAB script implementing Broyden's method for this nonlinear system starting from $[x^{(0)}, y^{(0)}] = [1, 1]$ and using $\mathbf{D}_0 = \mathbf{1}$ as the initial approximation for the Jacobian matrix.

7. (a) Use the command `plot` to decide whether the function $f(x) = x^2 + \frac{1}{x}$ is unimodal on the interval $(0, 2]$.

   (b) Determine the local minimum of f on $[0, 2]$ by solving the equation $f'(x) = 0$ using the Newton–Raphson method.

   (c) Confirm that the solution obtained in the previous item corresponds to a local minimum using the second derivative test.

8. (a) Use the command `plot` to decide whether the function $f(x) = \frac{x^2}{2} - \sin x$ is unimodal on the interval $[-1, 3]$.

   (b) Write a MATLAB script implementing the golden search method for finding the minimum of $f(x)$ on $[-1, 3]$ with a 10^{-6} tolerance level.

   (c) Use Newton–Raphson to find the solution to $f'(x) = 0$ with the same tolerance, and compare the two methods in terms of convergence and difficulty of implementation.

9. (a) Write a MATLAB script implementing the method of successive parabolic approximations to find the local minimum of $f(x) = e^x - 3x^2$ on the interval $[-1, 3.5]$. Stop the iterations when the distance between two successive approximations is smaller than 10^{-6}.

   (b) Modify the script to find the local maximum of $f(x)$ on the same interval.

   (c) Use the Newton–Raphson method to find solutions of $f'(x) = 0$ and identify the local maximum and local minimum using the second derivative test.

10. Use the command `surf` to plot the graphs of the following functions. Find their critical points by solving the system $\nabla f(\mathbf{x}) = \mathbf{0}$ with the Newton–Raphson method and classify them as local maxima, local minima, or saddle points according to the Hessian test.

    (a) $f(x, y) = xye^{-x^2-y^2}$

    (b) $f(x, y) = 3x + 6x^2 + y^3 + 2xy^2 - x^3 - y^4$

    (c) $f(x, y) = e^{x/2} + 2y^4 - x^3 + 6\cos y$.

11. Use the commands `contour` and `surfc` to obtain the level curves and graphs of the following functions. Find their local minima using the

steepest descent method and the method of your choice for the one-dimensional minimization steps.

(a) $f(x,y) = -xye^{-x^2-y^2}$

(b) $f(x,y) = 3x^2 + 2y^2 - xy$

(c) $f(x,y) = x^2 + y^2 + \frac{1}{(x^2+1)(y^2+1)}$.

12. Use the commands `contour` and `surfc` to obtain the level curves and graph of the *Rosenbrock function* $f(x,y) = 100(y-x^2)^2 + (1-x)^2$. Find its minimum using the numerical method of your choice.

CHAPTER 9

Initial-Value Problems for ODEs

DIFFERENTIAL EQUATIONS DESCRIBE the evolution of various quantities in many physical, chemical, biological, and economical problems. When the quantities depend on one variable, such as time, and do not depend on other variables, such as spatial coordinates, they may satisfy certain relations between the function and its first- and higher-order derivatives called *ordinary differential equations*, or simply *ODEs*. Depending on the applied context of the problem, the only independent variable can have different meanings, for example, time, a space coordinate, a parameter. When the ODEs are used to model the evolution dynamics of a physical system starting with a given initial state, it makes sense to call the independent variable the *time variable*, as we do in this chapter.

Two mathematical problems are associated with ordinary differential equations depending on whether the system of ODEs is supplemented by an *initial* value of the function at a single instance of the independent variable or its *boundary* values at distinct instances of the independent variable.

This chapter covers existence and properties of solutions of the initial-value problems for ODEs, algorithms and errors of their numerical approximations, and an interplay between the numerical algorithms and their convergence, stability, and robustness.

9.1 Approximations of Solutions

We consider an ODE system for a vector function $\mathbf{y} = \mathbf{y}(t)$ of a scalar variable t called the *time variable*. The initial-value problem consists of finding $\mathbf{y}(t)$ given its derivative $\mathbf{y}'(t)$ and the value of $\mathbf{y}(t)$ at a specific point, say, at $t = 0$:

Problem 9.1 (Initial-Value ODE Problem) Let $\mathbf{y} : \mathbb{R}_+ \to \mathbb{R}^d$ be a vector function of the time variable $t \in \mathbb{R}_+$, such that

$$\frac{d\mathbf{y}}{dt} = \mathbf{f}(t, \mathbf{y}), \qquad t \geq 0 \tag{9.1}$$

and $\mathbf{y}(0) = \mathbf{y}_0$, where $\mathbf{f} : \mathbb{R}_+ \times \mathbb{R}^d \mapsto \mathbb{R}^d$ is a given vector field and $\mathbf{y}_0 \in \mathbb{R}^d$ is the initial value. Find a continuously differentiable function $\mathbf{y}(t)$ satisfying (9.1) for $t \in [0, T)$, where $0 \leq T \leq \infty$ is the maximal existence interval, which may depend on $\mathbf{f}$ and $\mathbf{y}_0$.

Well-posed ODE problems

The ODE system (9.1) makes sense only if the solution $\mathbf{y}(t)$ exists for $T > 0$. Otherwise, if $T = 0$, the ODE system (9.1) cannot be used for modeling of any applied problem. For instance, when the initial data $\mathbf{y}_0$ is taken at a singularity of the vector field $\mathbf{f}(t, \mathbf{y})$, it may result in $T = 0$. In this case, we have to correct the vector field by some other physical factors to avoid singular values. It is in fact easy to distinguish consistent (*well-posed*) ODE systems (with $T > 0$) from inconsistent (*ill-posed*) ODE systems (with $T = 0$).

Theorem 9.1 (Initial-Value ODE Problem) *If $\mathbf{f}(t, \mathbf{y})$ is continuously differentiable in $\mathbf{y} \in D \subset \mathbb{R}^d$ and continuous in $t \in [0, T]$, then there exists a unique solution $\mathbf{y}(t)$ of Problem 9.1 for each $\mathbf{y}_0 \in D$ as long as the solution $\mathbf{y}(t)$ remains in D for $t \in [0, T]$.*

We construct four explicit examples of scalar ODEs ($d = 1$) that illustrate Theorem 9.1 with four generic types of behavior of the solution $y(t)$. The initial-value problem

$$\frac{dy}{dt} = -2ty, \qquad y(0) = 1 \tag{9.2}$$

has a unique solution $y(t) = e^{-t^2}$ for $t \geq 0$ ($T = \infty$) because the vector field $f(t, y) = -2ty$ is smooth in $y \in \mathbb{R}$ and $t \in \mathbb{R}$ and the solution $y(t)$ remains finite for all values of t. The initial-value problem

$$\frac{dy}{dt} = y^2, \qquad y(0) = 1 \tag{9.3}$$

has a unique solution $y(t) = 1/(1-t)$ for $0 \leq t < 1$ ($T = 1$) because the vector field $f(y) = y^2$ is smooth in $y \in \mathbb{R}$, but the solution $y(t)$ becomes unbounded as $t \to 1$ (indicating a finite time blow-up). The initial-value problem

$$\frac{dy}{dt} = \sqrt{y-1}, \qquad y(0) = 1 \tag{9.4}$$

has two solutions $y(t) = 1$ and $y(t) = 1 + \frac{1}{4}t^2$. Both solutions are valid for $t \geq 0$ ($T = \infty$), and they match the specific initial value $y(0) = 1$. Therefore, uniqueness of solution to the ODE problem (9.4) is lost. In agreement with Theorem 9.1, this happens because of divergence of the first derivative of $f(y) = \sqrt{y-1}$ at $y = 1$. Finally, the initial-value problem

$$\frac{dy}{dt} = \frac{1}{y-1}, \qquad y(0) = 1 \tag{9.5}$$

illustrates the situation when the vector field $f(y) = 1/(y-1)$ diverges at the initial value $y = 1$. Although Theorem 9.1 does not guarantee the existence of any solution to the initial-value problem (9.5), the solution still exists! The unique continuous solution is given by $y(t) = 1 + \sqrt{2t}$, and it is valid for $t \geq 0$ ($T = \infty$). Therefore, even if the ODE problem does not satisfy the sufficient condition of Theorem 9.1, Problem 9.1 may still possess a solution in a weaker sense, for instance, with an infinite derivative of $y(t)$ at $t = 0$.

We shall develop algorithms for a numerical solution of Problem 9.1 in the regular case when a unique solution is guaranteed by the conditions of Theorem 9.1. The interval $[0, T]$ where a solution $\mathbf{y}(t)$ exists can be represented numerically by a discrete partition at selected time points

$$0 = t_0 < t_1 < t_2 < \ldots < t_n = T. \tag{9.6}$$

Because the solution $\mathbf{y}(t)$ is known at least at $t_0 = 0$, we can develop iterative methods to find numerical approximations of the solution $\mathbf{y}(t)$ at t_{k+1} from numerical approximations of the solution $\mathbf{y}(t)$ at t_0, t_1, ..., t_k. Such methods are classified into *single-step* ODE solvers, where the approximation for $y(t_{k+1})$ is obtained using the numerical approximation at t_k only, and *multi-step* ODE solvers, where the approximation for $y(t_{k+1})$ is obtained using numerical approximations at the earlier times t_0, t_1, ..., t_k.

Let $\mathbf{y}_k$ denote the numerical approximation of the solution $\mathbf{y}(t)$ at t_k. The distance between $\mathbf{y}_1$ and the exact solution $\mathbf{y}(t_1)$ after performing a single step of the iterative algorithm is called the *local truncation error* of the numerical method. Given a fixed maximal time of computations T, the distance between $\mathbf{y}_n$ and $\mathbf{y}(T)$ is referred to as the *global truncation error* of the numerical method. The global error can be thought of as the accumulated local error after n consecutive steps of the iterative algorithm.

Euler's method

The simplest single-step ODE solver is called *Euler's* method, or the *slope approximation.* The slope approximation follows immediately from the forward difference approximations of the numerical derivative, which is explained in Section 6.1. Indeed, on one hand, the ODE system (9.1) at t_0 implies that

$$\mathbf{y}'(t_0) = \mathbf{f}(t_0, \mathbf{y}_0),$$

where $\mathbf{y}_0$ is known. On the other hand, the numerical derivative for $\mathbf{y}'(t_0)$ is

$$D_1[\mathbf{y}_0] = \frac{\mathbf{y}_1 - \mathbf{y}_0}{t_1 - t_0},$$

where $\mathbf{y}_1$ is to be found. Introducing the time step $h = t_1 - t_0$ and using the numerical derivative instead of the exact derivative, we obtain the first step of Euler's method:

$$\mathbf{y}_1 = \mathbf{y}_0 + h\mathbf{f}(t_0, \mathbf{y}_0).$$

The single step of Euler's method has a simple graphical illustration: the numerical approximation $\mathbf{y}_1$ is evaluated at time t_1 along the tangent line to the point $\mathbf{y}_0$ at t_0 with the known slope $\mathbf{y}'(t_0) = \mathbf{f}(t_0, \mathbf{y}_0)$. The error is small if the time step $h = t_1 - t_0$ is small. More precisely, the local truncation error is measured by the Taylor series approximation in Theorem 5.2. Assuming that each component of the solution $\mathbf{y}(t) = [y_1(t), \ldots, y_d(t)]$ is twice continuously differentiable on the interval $[t_0, t_1]$, the distance between the numerical approximation $\mathbf{y}_1$ and the exact value $\mathbf{y}(t_1)$ is

$$\mathbf{y}(t_1) - \mathbf{y}_1 = \mathbf{y}(t_0) + h\mathbf{y}'(t_0) + \frac{1}{2}\mathbf{a}h^2 - \mathbf{y}_0 - h\mathbf{f}(t_0, \mathbf{y}_0) = \frac{1}{2}\mathbf{a}h^2, \tag{9.7}$$

where $h = t_1 - t_0$, $\mathbf{y}(t_0) = \mathbf{y}_0$, and $\mathbf{a}_j = \mathbf{y}_j''(\tau_j)$, for some $t_0 \leq \tau_j \leq t_1$. According to the definition of the order of the truncation error in Section 6.1, you can see that the truncation error of Euler's method is $O(h^2)$.

Convergence of Euler's method

When we perform further iterations, we have to take into account that the values $\mathbf{y}_1$, $\mathbf{y}_2$, and generally $\mathbf{y}_k$ are no longer given exactly but approximated by iterations of the Euler method with accumulated truncation errors. If we still try to fit the solution curve through the point $\mathbf{y}_k$ at t_k and use the slope from the ODE $\mathbf{y}'(t_k) = \mathbf{f}(t_k, \mathbf{y}_k)$, then the repeated use of Euler's method leads to the first-order difference equation:

$$\mathbf{y}_{k+1} = \mathbf{y}_k + h\mathbf{f}(t_k, \mathbf{y}_k), \qquad t_{k+1} = t_k + h. \tag{9.8}$$

For simplicity, we have used here the constant time step $h = t_{k+1} - t_k$ to perform iterations. More general methods with variable time step during iterations are referred to as *adaptive* methods (see Section 9.3).

The following theorem gives the upper bound for the global truncation error of the Euler method in the scalar case $d = 1$ (see [7] for its proof), using n subintervals of length h for the interval $[0, T]$ and assuming that the solution $y(t)$ remains twice continuously differentiable on the entire interval $[0, T]$.

Theorem 9.2 *Let $y(t)$ be a twice continuously differentiable solution of Problem 9.1 with $d = 1$ and $f(t, y)$ be continuously differentiable in y and continuous in t for $y = y(t)$ and $t \in [0, T]$. Then,*

$$|y(T) - y_n| \leq \frac{1}{2}CTM_2h, \tag{9.9}$$

where $M_2 = \max_{0 \leq t \leq T} |y''(t)|$,

$$C = \frac{(1 + hM_1)^n - 1}{TM_1} \leq \frac{e^{TM_1} - 1}{TM_1},$$

and the constant M_1 is used in the Lipschitz estimate $|f(t, y_1) - f(t, y_2)| \leq M_1|y_1 - y_2|$ for any y_1 and y_2 on $y = y(t)$ for $t \in [0, T]$.

Exercise 9.1 Prove from Taylor's series that

$$(1+x)^n \leq e^{nx},$$

for any $n > 0$ and $x > -1$.

It follows from Theorem 9.2 that the global truncation error is $O(h)$, which is the reason why the Euler method is called a *first-order* ODE solver. It is a general rule of all ODE solvers that the global truncation error is *one order larger* than the local truncation error is. The same rule applies to the numerical integration formulas described in Section 6.5.

Iterations of Euler's method are illustrated in the MATLAB script `Euler_method`, where numerical solutions of the ODEs (9.2) and (9.3) are computed.

```
% Euler_method
h = 0.1; y0 = 1; T = 3; n = T/h;
t(1) = 0; y(1) = y0;
for k = 1 : n                        % ODE (9.2)
    t(k+1) = t(k) + h;
    y(k+1) = y(k) + h*(-2*t(k)*y(k));
end
yExact = exp(-t.^2); eLocal = abs(y-yExact);
plot(t,yExact,'r',t,y,'.b',t,eLocal,':g');
 h = 0.02; T = 0.94; n = T/h;
t(1) = 0; y(1) = y0;
for k = 1 : n                        % ODE (9.3)
    t(k+1) = t(k) + h;
    y(k+1) = y(k) + h*y(k)^2;
end
yExact = 1./(1-t); eLocal = abs(y-yExact);
plot(t,yExact,'r',t,y,'.b',t,eLocal,':g');
```

The MATLAB script `Euler_method` plots the numerical approximation (dots), the exact solution (solid curve), and the computational error (dotted curve) in Figure 9.1. The left figure shows the numerical solution of the ODE (9.2) with the constant time step $h = 0.1$. The error grows at the initial stage but reduces to zero at the later stage. This behavior follows the exact solution $y(t) = e^{-t^2}$, which decays to zero. The right figure shows the numerical solution of the ODE (9.3) with the constant time step $h = 0.02$. The error grows, similarly to the exact solution $y(t) = 1/(1-t)$, which becomes infinite at the finite time $t = 1$.

The upper bound (9.9) explains the initial growth of the global error with the length of the interval $[0, T]$. Subsequent behavior of the global error depends on the behavior of the solution. You can see in Figure 9.1 that the

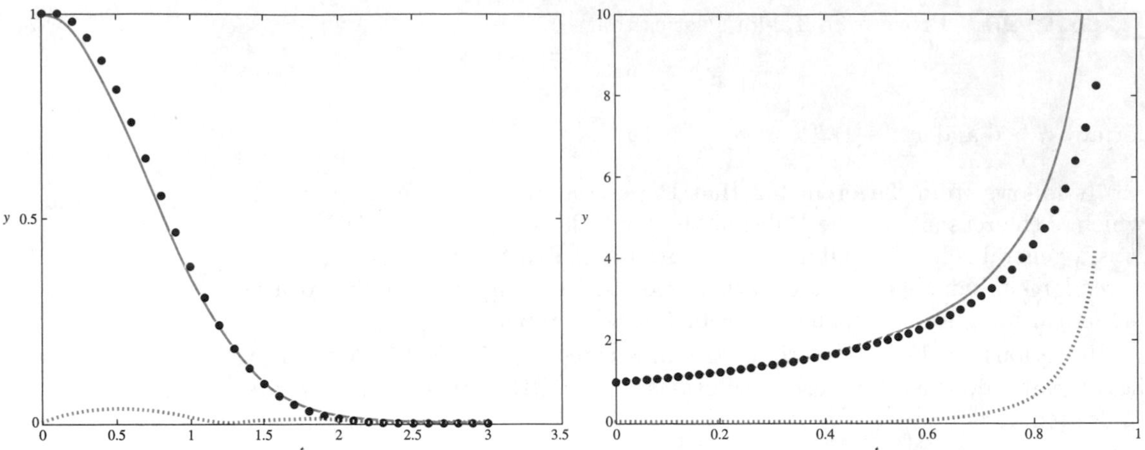

Figure 9.1 Numerical approximation of the Euler method (dots), exact solution (solid curve), and the truncation error (dotted curve) for the ODE (9.2) (*left*) and the ODE (9.3) (*right*).

global error reduces if the solution decays to zero and grows if the solution blows to infinity. When the time step h becomes small, the approximations given by Euler's method converge to the exact solution of the ODE, according to the error formula (9.9).

Let us illustrate the convergence of Euler's method for the ODE (9.4). To obtain a nonconstant solution, we use a perturbed initial condition $t_0 > 0$ and $y_0 = 1 + \frac{1}{4}t_0^2$ for small t_0, which coincides with the exact nonconstant solution $y(t) = 1 + \frac{1}{4}t^2$ of the initial-value problem (9.4). The MATLAB script `error_Euler` computes the numerical approximation and the local and global errors after one step and n steps on the interval $[t_0, t_0 + 4]$, where $t_0 = 10^{-4}$.

```
% error_Euler
e = 0.0001; t(1) = e; y(1) = 1+0.25*t(1)^2; T = 4+e;
nArray = 10 : 10 : 10000; % array for different values of h
for j = 1 : length(nArray)
    n = nArray(j); h(j) = T/n;
    for k = 1 : n            % Euler's method
            t(k+1) = t(k) + h(j);
            y(k+1) = y(k) + h(j)*sqrt(y(k)-1);
    end
    yExact = 1+0.25*t.^2;
    Eloc(j) = abs(y(2)-yExact(2));
    Eglob(j) = abs(y(n+1)-yExact(n+1));
```

```
end
aLoc = polyfit(log(h),log(Eloc),1); % power fit of local error
power1=aLoc(1), C1=exp(aLoc(2)), Eapr1 = C1*h.^power1;
aGlob = polyfit(log(h),log(Eglob),1); % power fit of global error
power2=aGlob(1), C2=exp(aGlob(2))/T, Eapr2 = T*C2*h.^power2;
loglog(h,Eloc,'.b',h,Eapr1,':r',h,Eglob,'.b',h,Eapr2,':r')
```

The output of the MATLAB script `error_Euler` shows parameters of the power fit $E(h) = Ch^p$ for the local and global errors. You can see that the values $p = 2$ and $C = \frac{1}{4}$ for the power fit of the local error coincide with the error formula (9.7) since the exact solution $y(t) = 1 + \frac{1}{4}t^2$ produces the constant second derivative $y''(t) = \frac{1}{2}$. The power fit for the global error gives the value p close to $p = 1$ and the value C different from $C = \frac{1}{4}$. Therefore, even in the case of solutions with constant second derivatives, the accumulated error is not the sum of individual local errors. Figure 9.2 shows the graphs of the local and global errors together with their power fits. It is obvious that the

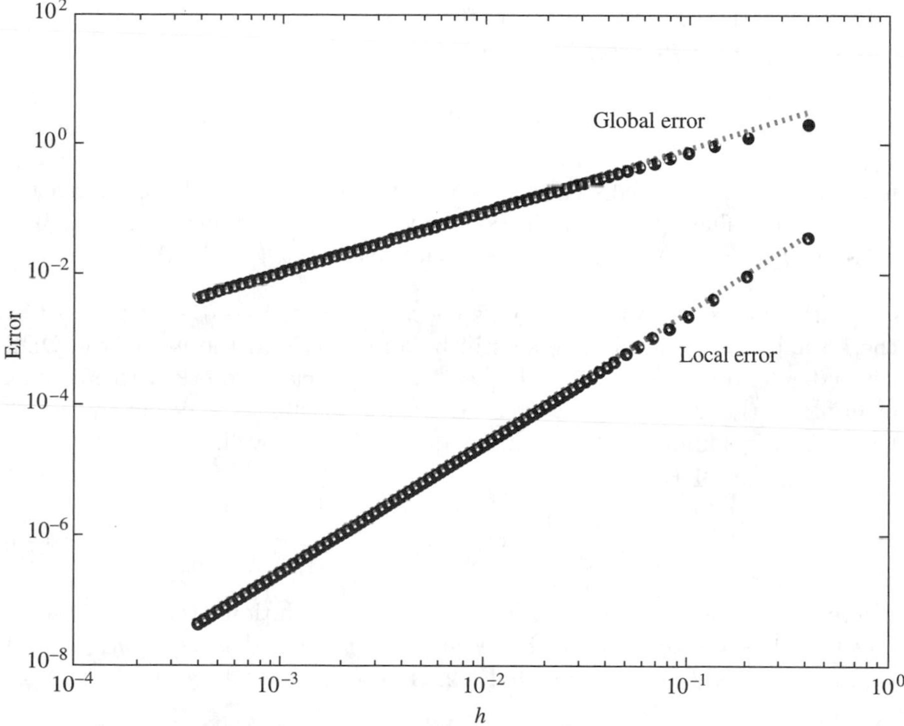

Figure 9.2 Local and global truncation errors for the explicit Euler method for the ODE (9.4).

global error is larger than the local error is because the former has first-order convergence whereas the latter has second-order convergence.

```
>> error_Euler

power1 =
        2.0000

C1 =
        0.2500

power2 =
        0.9595

C2 =
        1.9327
```

Exercise 9.2 Compute a numerical solution of the ODE problem $y' = y$, $y(0) = 1$ with the Euler method. By using the exact solution $y(t) = e^t$, compute and plot the error function

$$D_k = \frac{2|y(t_k) - y_k|}{t_k h}$$

in k, where h is the time step, $t_k = kh$, and y_k is a numerical approximation of the exact value $y(t_k)$. Plot the constant CM_2 in the upper bound (9.9) in k (when $T = t_k$) and estimate the distance between D_k and CM_2.

Stability of Euler's method

Although convergence of Euler's method (9.8) as $h \to 0$ is guaranteed by the error bound (9.9), there may still be an obstacle in the use of the ODE solver for finite values of h. To explain this obstacle we work with a simple example of the first-order linear ODE with a constant coefficient, although the difficulty is fundamental for many other ODEs as well.

The linear ODE

$$\frac{dy}{dt} = -\lambda y, \qquad y(0) = y_0, \tag{9.10}$$

where λ is a constant parameter, has the exact solution $y(t) = e^{-\lambda t} y_0$ for any $t \geq 0$. An exact relation between $y_k = y(t_k)$ and $y_{k+1} = y(t_{k+1})$ for $t_{k+1} = t_k + h$ follows from the ODE (9.10) after a single step:

$$y_{k+1} = e^{-\lambda h} y_k = \left(1 - \lambda h + \frac{1}{2!}\lambda^2 h^2 - \frac{1}{3!}\lambda^3 h^3 + \frac{1}{4!}\lambda^4 h^4 + \mathrm{O}(h^5)\right) y_k, \tag{9.11}$$

where we used the Taylor series approximation of the exponential function $e^{-\lambda h}$. When the first-order Euler method (9.8) with $f(y) = -\lambda y$ is used instead of the exact ODE (9.10), we obtain the first-order difference equation with constant coefficient

$$y_{k+1} = (1 - \lambda h) y_k. \tag{9.12}$$

Comparison of (9.11) and (9.12) shows that the local truncation error is $O(h^2)$, in agreement with the error formula (9.7). Let y_k be a numerical approximation of $y(t_k)$ and apply iterations of the first-order difference equation (9.12) on the equally spaced discrete grid $t_k = kh$, $k \in \mathbb{N}$. Then,

$$y_k = (1 - \lambda h)^k y_0 = e^{-\Lambda t_k} y_0, \qquad k \in \mathbb{N}, \tag{9.13}$$

where

$$\Lambda = -\frac{1}{h} \log(1 - \lambda h) = \lambda + \frac{1}{2}\lambda^2 h + O(h^2),$$

where we used the Taylor series approximation of the logarithmic function $\log(1 - \lambda h)$. The global truncation error (9.13) is $O(h)$, in agreement with the error bound (9.9). When $h \to 0$, both the local and global truncation errors of Euler's method converge to zero and the approximation y_k converges to the exact solution $y(t_k)$.

Now, consider iterations of the Euler method for finite values of h. If $\lambda > 0$, the exact solution $y(t) = e^{-\lambda t} y_0$ of the ODE (9.10) decays exponentially to zero as $t \to \infty$. However, the numerical approximation y_k of the first-order difference equation (9.12) decays to zero only if $\lambda h < 2$. Moreover, the decay is monotonic only if $\lambda h < 1$, while the decay occurs in a single iteration for $\lambda h = 1$ and alternates in sign for $1 < \lambda h < 2$. The sequence is nondecaying and sign-alternating for $\lambda h = 2$, while the sequence diverges to infinity for $\lambda h > 2$. According to this analysis, we say that Euler's method is *stable* for the linear ODE (9.10) only if the time step h satisfies the bound $h \leq 2/\lambda$. Thus, both convergence and stability of iterations of the Euler method can be controlled only if the time step h is chosen to be sufficiently small.

Exercise 9.3 Find numerical approximations of the solution of the ODE problem (9.2) by using Euler's method with $h = 0.6$ and observe the development of numerical instabilities of iterations. Do instabilities develop for $h = 0.5$ and $h = 0.4$?

When the parameter λ of the exponential decay (9.10) is sufficiently large, the time step h has to be sufficiently small so that the numerical solution would follow the rapidly decaying curve of the exact solution $y(t) = e^{-\lambda t} y_0$. The adjustment of the time step h to maintain proper stability and accuracy of the ODE solver is developed in the adaptive methods (see Section 9.3). However, even adaptive methods fail to provide adequate approximations to stiff systems of first-order differential equations (see Section 9.5).

9.2 Single-Step Runge–Kutta Solvers

We now generalize the idea of the Euler method and derive a family of single-step Runge–Kutta ODE solvers. The family of Runge–Kutta methods originates from the hierarchy of Newton–Cotes integration formulas, which we describe in Section 6.6. The Runge–Kutta methods provide a better control on convergence and stability of numerical approximations of solutions of differential equations. To emphasize the analogy between numerical ODE solvers and numerical integration rules, we rewrite the ODE (9.1) on the particular interval $[t_k, t_{k+1}]$ in the equivalent integral form

$$\mathbf{y}(t_{k+1}) - \mathbf{y}(t_k) = \int_{t_k}^{t_{k+1}} \mathbf{f}(t, \mathbf{y}(t))dt. \tag{9.14}$$

When the vector field $\mathbf{f} = \mathbf{f}(t)$ is independent on the unknown solution $\mathbf{y}(t)$, the right-hand side of the integral equation (9.14) is an integral of the explicit function of t and the numerical approximation of the integral on an elementary time interval $[t_k, t_{k+1}]$ can be developed with various integration rules, such as the trapezoidal and midpoint rules discussed in Section 6.5.

The Newton–Cotes integration rules in Section 6.6 differ in the order of the local truncation error. A similar characterization can be used for the Runge–Kutta ODE solvers. The Euler method described in Section 9.1 corresponds to the left-point rectangular rule (6.47) applied to the integral equation (9.14). This primitive integration rule has an $\mathrm{O}(h^2)$ local truncation error. The trapezoidal and midpoint integration rules have improved accuracy because their local truncation error is $\mathrm{O}(h^3)$. These integration rules can now be applied to the integral equation (9.14) and they become single-step ODE solvers.

Heun's and modified Euler's methods

Let $\mathbf{y}_k$ be a numerical approximation of the solution $\mathbf{y}(t)$ at t_k. Similarly, define numerical approximations $\mathbf{y}_{k+1/2}$ and $\mathbf{y}_{k+1}$ at times $t_{k+1/2} = t_k + h/2$ and $t_{k+1} = t_k + h$. The trapezoidal integration rule (6.50) for the integral equation (9.14) leads to the *Heun* method:

$$\mathbf{y}_{k+1} - \mathbf{y}_k = \frac{1}{2}h\left(\mathbf{f}(t_k, \mathbf{y}_k) + \mathbf{f}(t_{k+1}, \mathbf{y}_{k+1})\right), \tag{9.15}$$

while the mid-point integration rule (6.49) leads to the *modified Euler* method:

$$\mathbf{y}_{k+1} - \mathbf{y}_k = h\mathbf{f}(t_{k+1/2}, \mathbf{y}_{k+1/2}). \tag{9.16}$$

Both ODE solvers (9.15) and (9.16) are implicit in the sense that the unknown values $\mathbf{y}_{k+1}$ and $\mathbf{y}_{k+1/2}$ appear in the right-hand side of the difference equations. When the vector field $\mathbf{f}(t, \mathbf{y})$ is linear in $\mathbf{y}$, the difference equation (9.15) can be rewritten as an explicit map $\mathbf{y}_k \mapsto \mathbf{y}_{k+1}$ (see Section 9.5). However, when the vector field $\mathbf{f}(t, \mathbf{y})$ is nonlinear, we need a better strategy for explicit iterations of the difference equation (9.15) unless we are prepared to solve nonlinear equations at each elementary time step by a root finding algorithm (see Section 8.1). Similar questions rise in connection to the modified Euler method because the difference equation (9.16) is not closed for $\mathbf{y}_{k+1}$ and a numerical strategy to approximate $\mathbf{y}_{k+1/2}$ is needed when an elementary time step is performed for the map $\mathbf{y}_k \mapsto \mathbf{y}_{k+1}$. These simplest integration rules pose a new numerical problem, which can be formulated in a general form.

Problem 9.2 (Explicit nth Order Runge–Kutta Method) Let the interval $[t_k, t_{k+1}]$ be discretized on the grid of $(n+1)$ points $t_k^{(j)}$, $j = 0, 1, \ldots, n$, such that

$$t_k = t_k^{(0)} \le t_k^{(1)} \le \ldots \le t_k^{(n)} = t_{k+1}. \qquad (9.17)$$

Let $\mathbf{p}_k^{(j)}$, $j = 1, 2, \ldots, n$ be successive approximations of the solution $\mathbf{y}(t)$ at the times $t_k^{(j)}$ defined by the summation rule:

$$\mathbf{p}_k^{(j)} = \mathbf{y}_k + h \sum_{i=0}^{j-1} \alpha_{j-1,i} \mathbf{f}(t_k^{(i)}, \mathbf{p}_k^{(i)}), \qquad j = 1, 2, \ldots, n, \qquad (9.18)$$

where $\mathbf{p}_k^{(0)} = \mathbf{y}_k$ and $\mathbf{p}_k^{(n)} = \mathbf{y}_{k+1}$. Let $\mathbf{y}(t)$ be an exact solution of the integral equation (9.14) and let $\mathbf{y}_k = \mathbf{y}(t_k)$. Find the set of coefficients $\{\alpha_{j-1,i}\}_{0 \le i < j \le n}$, such that the local truncation error for the numerical approximation $\mathbf{y}_{k+1}$ is $O(h^{n+1})$.

Runge–Kutta ODE solvers in Problem 9.2 represent single-step methods because after performing a number of computations for $\mathbf{p}_k^{(j)}$, $j = 1, 2, \ldots, n-1$, the final computation for $\mathbf{p}_k^{(n)}$ provides an explicit map from $\mathbf{y}_k \mapsto \mathbf{y}_{k+1}$. Because of the relationship between the ODE (9.1) and the integral equation (9.14), the summation formula for $\mathbf{y}_{k+1}$ must be equivalent to a Newton–Cotes integration rule (6.70) if $\mathbf{f} = \mathbf{f}(t)$. The order of the local truncation error of the Newton–Cotes integration rules is defined by Theorem 6.4 and it may exceed the order of the local truncation error of the corresponding Runge–Kutta ODE solver. For instance, Simpson's rule (the Newton–Cotes integration rule with $m = 2$) has a local truncation error that is $O(h^5)$, while the Runge–Kutta ODE solver with $n = 3$ may have an error that is $O(h^4)$. Because the global truncation error is one order larger than the local truncation error, the Runge–Kutta ODE solver in Problem 9.2 has a global truncation error that is $O(h^n)$.

First- and second-order Runge–Kutta methods

When $n = 1$, the Runge–Kutta ODE solver (9.18) recovers the explicit Euler method (9.8) with $\alpha_{0,0} = 1$. It is easy to see that this is the only solution of Problem 9.2 in the case $n = 1$. When $n = 2$, the Runge–Kutta ODE solver (9.18) can be rewritten with four parameters $\alpha_{0,0}$, $\alpha_{0,1}$, $\alpha_{1,1}$, and α_0:

$$\begin{aligned} \mathbf{p}_k^{(1)} &= \mathbf{y}_k + \alpha_{0,0} h \mathbf{f}(t_k, \mathbf{y}_k), \\ \mathbf{y}_{k+1} &= \mathbf{y}_k + h\left(\alpha_{1,0}\mathbf{f}(t_k, \mathbf{y}_k) + \alpha_{1,1}\mathbf{f}(t_k + \alpha_0 h, \mathbf{p}_k^{(1)})\right), \end{aligned}$$

where $0 \le \alpha_0 \le 1$. The coefficients $\alpha_{0,0}$, $\alpha_{1,0}$, $\alpha_{1,1}$, and α_0 must be found from the condition that the local truncation error is $O(h^3)$ provided $\mathbf{y}_k$ is known exactly as $\mathbf{y}_k = \mathbf{y}(t_k)$. The distance between the exact value $\mathbf{y}(t_{k+1})$ and the numerical value $\mathbf{y}_{k+1}$ is computed as follows:

$$\begin{aligned} \mathbf{y}(t_{k+1}) &= \mathbf{y}(t_k) + h\mathbf{y}'(t_k) + \frac{1}{2}h^2\mathbf{y}''(t_k) + O(h^3) \\ &= \mathbf{y}_k + h\mathbf{f}(t_k, \mathbf{y}_k) + \frac{1}{2}h^2\left[\frac{\partial}{\partial t}\mathbf{f}(t_k, \mathbf{y}_k) + (\mathbf{f}\cdot\nabla)\,\mathbf{f}(t_k, \mathbf{y}_k)\right] + O(h^3) \end{aligned}$$

and

$$\begin{aligned} \mathbf{y}_{k+1} &= \mathbf{y}_k + h\left[\alpha_{1,0}\mathbf{f}(t_k, \mathbf{y}_k) + \alpha_{1,1}\mathbf{f}(t_k + \alpha_0 h, \mathbf{y}_k + \alpha_{0,0}h\mathbf{f}(t_k, \mathbf{y}_k))\right] \\ &= \mathbf{y}_k + h\left[(\alpha_{1,0} + \alpha_{1,1})\mathbf{f}(t_k, \mathbf{y}_k) + \alpha_{1,1}\alpha_0 h\frac{\partial}{\partial t}\mathbf{f}(t_k, \mathbf{y}_k) + \alpha_{0,0}\alpha_{1,1}h\,(\mathbf{f}\cdot\nabla)\,\mathbf{f}(t_k, \mathbf{y}_k) + O(h^2)\right]. \end{aligned}$$

The two expansions match up to $O(h^2)$ terms if

$$\alpha_{1,0} + \alpha_{1,1} = 1, \qquad \alpha_{1,1}\alpha_0 = \alpha_{1,1}\alpha_{0,0} = \frac{1}{2}.$$

The general solution of the system of equations has one free parameter:

$$\alpha_{0,0} = \alpha_0, \qquad \alpha_{1,1} = \frac{1}{2\alpha_0}, \qquad \alpha_{1,0} = \frac{2\alpha_0 - 1}{2\alpha_0}, \tag{9.19}$$

where $0 < \alpha_0 \le 1$. The explicit Heun method that breaks the implicit character of the trapezoidal rule (9.15) corresponds to the choice $\alpha_0 = 1$ in the solution (9.19):

$$\mathbf{p}_{k+1} = \mathbf{y}_k + h\mathbf{f}(t_k, \mathbf{y}_k), \tag{9.20}$$

$$\mathbf{y}_{k+1} = \mathbf{y}_k + \frac{1}{2}h\left(\mathbf{f}(t_k, \mathbf{y}_k) + \mathbf{f}(t_{k+1}, \mathbf{p}_{k+1})\right), \tag{9.21}$$

where the value $\mathbf{p}_{k+1} \equiv \mathbf{p}_k^{(1)}$ is a prediction for the solution $\mathbf{y}(t_{k+1})$ obtained with the explicit Euler method (9.8), while the value $\mathbf{y}_{k+1}$ is a correction for the solution $\mathbf{y}(t_{k+1})$ obtained with the explicit version of the Heun method

(9.15). According to the preceding general solution, the explicit Heun method preserves the local truncation error of the trapezoidal rule that is given by the error formula (6.55). The remainder term of order $O(h^3)$ can be written explicitly from the error formula:

$$\mathbf{y}(t_{k+1}) - \mathbf{p}_{k+1} = \frac{1}{2}\mathbf{b}h^2,$$

where $\mathbf{b}_j = \mathbf{y}_j''(\tau_j)$, for some $\tau_j \in [t_k, t_{k+1}]$, and

$$\mathbf{y}(t_{k+1}) - \mathbf{y}_{k+1} = \frac{1}{2}h\left(\mathbf{f}\left(t_k, \mathbf{p}_{k+1} + \frac{1}{2}\mathbf{b}h^2\right) - \mathbf{f}(t_{k+1}, \mathbf{p}_{k+1})\right) - \frac{1}{12}\mathbf{c}h^3, \tag{9.22}$$

where $\mathbf{c}_j = \mathbf{y}_j^{(3)}(\tilde{\tau}_j)$ for some $\tilde{\tau}_j \in [t_k, t_{k+1}]$. If the vector field $\mathbf{f}(t, \mathbf{y})$ is continuously differentiable in $\mathbf{y}$ (similarly to the condition of Theorem 9.1), then the approximation $\mathbf{y}_{k+1}$ for $\mathbf{y}(t_{k+1})$ has a local truncation error that is $O(h^3)$. Therefore, the global truncation error for the explicit Heun method (9.20)–(9.21) is $O(h^2)$.

The explicit modified Euler method that breaks the implicit character of the midpoint rule (9.16) corresponds to the choice $\alpha_0 = \frac{1}{2}$ in the solution (9.19):

$$\mathbf{p}_{k+1/2} = \mathbf{y}_k + \frac{1}{2}h\mathbf{f}(t_k, \mathbf{y}_k), \tag{9.23}$$

$$\mathbf{y}_{k+1} = \mathbf{y}_k + h\mathbf{f}(t_{k+1/2}, \mathbf{p}_{k+1/2}), \tag{9.24}$$

where the value $\mathbf{p}_{k+1/2} \equiv \mathbf{p}_k^{(1)}$ is a prediction for the solution $\mathbf{y}(t_{k+1/2})$ obtained with the explicit Euler method (9.8), while the value $\mathbf{y}_{k+1}$ is a correction for the solution $\mathbf{y}(t_{k+1})$ obtained with the explicit version of the modified Euler method (9.16). According to the preceding general solution, the explicit modified Euler method preserves the local truncation error of the midpoint rule that is given by the error formula (6.56). The remainder term of order $O(h^3)$ can be written explicitly from the error formula:

$$\mathbf{y}(t_{k+1/2}) - \mathbf{p}_{k+1/2} = \frac{1}{8}\mathbf{b}h^2,$$

where $\mathbf{b}_j = \mathbf{y}_j''(\tau_j)$ for some $\tau_j \in [t_k, t_{k+1}]$, and

$$\mathbf{y}(t_{k+1}) - \mathbf{y}_{k+1} = h\left(\mathbf{f}\left(t_{k+1/2}, \mathbf{p}_{k+1/2} + \frac{1}{8}\mathbf{b}h^2\right) - \mathbf{f}(t_{k+1/2}, \mathbf{p}_{k+1/2})\right) + \frac{1}{24}\mathbf{c}h^3, \tag{9.25}$$

where $\mathbf{c}_j = \mathbf{y}_j^{(3)}(\tilde{\tau}_j)$, for some $\tilde{\tau}_j \in [t_k, t_{k+1}]$. If the vector field $\mathbf{f}(t, \mathbf{y})$ is continuously differentiable in $\mathbf{y}$, then the distance between $\mathbf{y}(t_{k+1})$ and $\mathbf{y}_{k+1}$

has a local error that is $O(h^3)$. Therefore, the global truncation error for the modified Euler method (9.23)–(9.24) is $O(h^2)$.

The MATLAB script improved_Euler_methods illustrates the iterations and errors of the explicit Euler, Heun, and modified Euler methods for the ODE (9.2).

```
% improved_Euler_methods
h = 0.05; y0 = 1; T = 3; n = T/h; t(1) = 0;
yEul(1) = y0; yHeun(1) = y0; yMod(1) = y0;      % initial values
for k = 1 : n
    t(k+1) = t(k) + h;
    yEul(k+1) = yEul(k) + h*(-2*t(k)*yEul(k));  % Euler's method
    p = yHeun(k) + h*(-2*t(k)*yHeun(k));        % Heun's method
    yHeun(k+1) = yHeun(k) + 0.5*h*(-2*t(k)*yHeun(k)-2*t(k+1)*p);
    p = yMod(k) + 0.5*h*(-2*t(k)*yMod(k));      % modified Euler's method
    yMod(k+1) = yMod(k) + h*(-2*(t(k)+0.5*h)*p);
end
yExact = exp(-t.^2); % errors of the three ODE solvers
eEul = abs(yEul-yExact); normEul = max(eEul)
eHeun = abs(yHeun-yExact); normHeun = max(eHeun)
eMod = abs(yMod-yExact); normMod = max(eMod)
plot(t,eHeun,'.b',t,eMod,'.g');
```

The output of the MATLAB script improved_Euler_methods shows that both the Heun and the modified Euler methods give an efficient improvement in the reduction of the truncation error compared to the Euler method. This is because of the second-order convergence of the improved Euler method compared to the first-order convergence of the standard Euler method. Figure 9.3 shows the behavior of the distance between exact and numerical solutions as the time evolves and the exact solution of the ODE (9.2) decays to zero. The L^∞ norm for the maximal error and the graph of the pointwise error show that the modified Euler method based on the midpoint rule gives an even smaller error compared to the Heun method based on the trapezoidal rule.

```
>> improved_Euler_methods

normEul =
    0.0169

normHeun =
    4.7972e-004

normMod =
    2.6563e-004
```

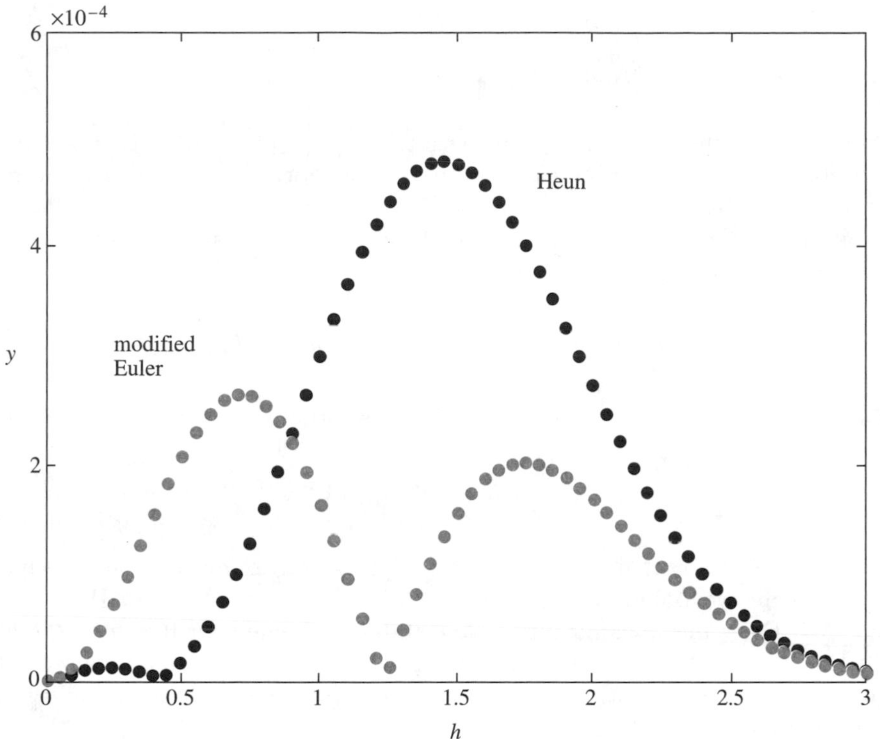

Figure 9.3 Global errors for the explicit Heun (9.20)–(9.21) and modified Euler (9.23)–(9.24) methods for the ODE (9.2).

Exercise 9.4 Consider convergence of the explicit Heun and modified Euler methods for the ODE (9.4) as the time step h is reduced. Use a power fit to show that the local truncation error for both methods is $O(h^3)$, whereas the global truncation error is $O(h^2)$. Given that the exact solution $y(t) = 1 + \frac{1}{4}t^2$ satisfies $y^{(3)} = 0$, explain why the distance between the numerical and exact solutions remains nonzero in these methods.

Stability of second-order methods

Stability of the explicit Heun and modified Euler methods can be considered similar to the stability of the Euler method (see Section 9.1). When $f(y) = -\lambda y$ is used for the linear ODE (9.10), the Heun method (9.20)–(9.21) transforms to the first-order difference equation:

$$y_{k+1} = \left(1 - h\lambda + \frac{1}{2}h^2\lambda^2\right) y_k, \tag{9.26}$$

because

$$p_{k+1} = (1 - h\lambda)y_k, \qquad y_{k+1} = y_k - \frac{1}{2}h\lambda\,(y_k + p_{k+1}).$$

By comparing the expression (9.26) with the Taylor series (9.11) for the exact solution $y_{k+1} = e^{-h\lambda}y_k$, you can confirm that the local truncation error of the explicit Heun method is $O(h^3)$ in agreement with the error formula (9.22). Similarly, the modified Euler method (9.23)–(9.24) transforms to the same first-order difference equation (9.26) because

$$p_{k+1/2} = \left(1 - \frac{1}{2}h\lambda\right)y_k, \qquad y_{k+1} = y_k - \frac{1}{2}h\lambda p_{k+1/2}.$$

Therefore, both Heun's and modified Euler's methods are stable when

$$\left|1 - h\lambda + \frac{1}{2}h^2\lambda^2\right| \leq 1. \tag{9.27}$$

The quadratic function $g(z) = 1 - z + z^2/2$ is convex and crosses the line $g(z) = 1$ at two points $z = 0$ and $z = 2$. Therefore, the explicit Heun and modified Euler methods have the same stability domain for $0 < h \leq 2/\lambda$ as the Euler method. Compared to the Euler method, these two methods present better convergence as $h \to 0$ but show no improvement in the stability of iterations for finite values of h.

Higher-order Runge–Kutta methods

When $n > 2$, the general solutions of Problem 9.2 have several free parameters. For instance, all third-order methods ($n = 3$) are obtained from a system of 6 equations for 8 unknown parameters, which exhibits a two-parameter solution [6], whereas all fourth-order methods ($n = 4$) are obtained from a system of 11 equations for 13 parameters, which also exhibits a two-parameter solution [17]. Among all available third-order and fourth-order Runge–Kutta methods, two particular methods are used widely in practical ODE solvers. These two numerical schemes generalize the Simpson integration rule (6.64), whose local truncation error is $O(h^5)$, according to the error formula (6.66).

The particular third-order Runge–Kutta method (with $n = 3$) takes the form:

$$\begin{aligned}
\mathbf{p}_{k+1/2} &= \mathbf{y}_k + \frac{1}{2}h\mathbf{f}(t_k, \mathbf{y}_k), \\
\mathbf{p}_{k+1} &= \mathbf{y}_k - h\mathbf{f}(t_k, \mathbf{y}_k) + 2h\mathbf{f}(t_{k+1/2}, \mathbf{p}_{k+1/2}), \\
\mathbf{y}_{k+1} &= \mathbf{y}_k + \frac{h}{6}\left[\mathbf{f}(t_k, \mathbf{y}_k) + 4\mathbf{f}(t_{k+1/2}, \mathbf{p}_{k+1/2}) + \mathbf{f}(t_{k+1}, \mathbf{p}_{k+1})\right],
\end{aligned}$$

where $\mathbf{p}_{k+1/2} \equiv \mathbf{p}_k^{(1)}$ and $\mathbf{p}_{k+1} \equiv \mathbf{p}_k^{(2)}$. The summation formula for $\mathbf{y}_{k+1}$ reduces to the Simpson integration rule (6.64) for the vector field $\mathbf{f}(t, \mathbf{y}) = \mathbf{f}(t)$, but the order $O(h^5)$ of the local truncation error of the Simpson rule is not preserved for a general vector field $\mathbf{f}(t,\mathbf{y})$ because the third-order Runge–Kutta solver has a local truncation error that is $O(h^4)$.

The particular fourth-order Runge–Kutta method (with $n = 4$) takes the form:

$$\mathbf{p}_{k+1/2} = \mathbf{y}_k + \frac{1}{2}h\mathbf{f}(t_k,\mathbf{y}_k),$$
$$\mathbf{q}_{k+1/2} = \mathbf{y}_k + \frac{1}{2}h\mathbf{f}(t_{k+1/2},\mathbf{p}_{k+1/2}),$$
$$\mathbf{p}_{k+1} = \mathbf{y}_k + h\mathbf{f}(t_{k+1/2},\mathbf{q}_{k+1/2}),$$
$$\mathbf{y}_{k+1} = \mathbf{y}_k + \frac{h}{6}\left[\mathbf{f}(t_k,\mathbf{y}_k) + 2\mathbf{f}(t_{k+1/2},\mathbf{p}_{k+1/2}) + 2\mathbf{f}(t_{k+1/2},\mathbf{q}_{k+1/2}) + \mathbf{f}(t_{k+1},\mathbf{p}_{k+1})\right],$$

where $\mathbf{p}_{k+1/2} \equiv \mathbf{p}_k^{(1)}$, $\mathbf{q}_{k+1/2} \equiv \mathbf{p}_k^{(2)}$, and $\mathbf{p}_{k+1} \equiv \mathbf{p}_k^{(3)}$. The summation formula for $\mathbf{y}_{k+1}$ reduces to the Simpson integration rule (6.64) for the vector field $\mathbf{f}(t,\mathbf{y}) = \mathbf{f}(t)$, and the order $O(h^5)$ of the local truncation error of the Simpson rule is preserved for a general vector field $\mathbf{f}(t,\mathbf{y})$ in the fourth-order Runge–Kutta ODE solver.

The significant achievement in accuracy of the higher-order Runge–Kutta methods compared to the lower-order Runge–Kutta methods is illustrated by the MATLAB script `Runge_Kutta_methods`. The script computes iterations and errors of the previous third-order and fourth-order Runge–Kutta methods in application to the ODE (9.2).

```
% Runge_Kutta_methods
h = 0.05; y0 = 1; T = 3; n = T/h;
t(1) = 0; % third-order and fourth-order Runge--Kutta methods
yRK3(1) = y0; yRK4(1) = y0;
for k = 1 : n
    t(k+1) = t(k) + h;
    fRK3 = -2*t(k)*yRK3(k); fRK4 = -2*t(k)*yRK4(k);
    pRK3 = yRK3(k) + h*fRK3/2; pRK4 = yRK4(k) + h*fRK4/2;
    gRK3 = -2*(t(k)+h/2)*pRK3; gRK4 = -2*(t(k)+h/2)*pRK4;
    qRK3 = yRK3(k) - h*fRK3 + 2*h*gRK3; qRK4 = yRK4(k) + h*gRK4/2;
    hRK4 = -2*(t(k)+h/2)*qRK4; rRK4 = yRK4(k) + h*hRK4;
    ffRK3 = -2*t(k+1)*qRK3; ffRK4 = -2*t(k+1)*rRK4;
    yRK3(k+1) = yRK3(k) + h*(fRK3+4*gRK3+ffRK3)/6;
    yRK4(k+1) = yRK4(k) + h*(fRK4+2*gRK4+2*hRK4+ffRK4)/6;
end
yExact = exp(-t.^2); % errors of the Runge--Kutta methods
eRK3 = abs(yRK3-yExact); normRK3 = max(eRK3)
eRK4 = abs(yRK4-yExact); normRK4 = max(eRK4)
plot(t,eRK3,'.b'); plot(t,eRK4,'.g');
```

Given the time step $h = 0.05$, the maximal error of the third-order Runge–Kutta method is of order 10^{-5}, while the maximal error of the fourth-order

Runge–Kutta method is of order 10^{-7}. The previous example suggests that the maximal error of the Euler method is of order 10^{-2} for the same value of h, while those of the Heun and modified Euler methods is of order 10^{-4}. Figure 9.4 shows the behavior of the third-order and fourth-order methods versus the evolution time. Comparison with Figure 9.3 shows that the third-order Runge–Kutta method resembles the behavior of the modified Euler method, while the fourth-order Runge–Kutta method resembles the behavior of the Heun method. It is remarkable how small the error of the fourth-order Runge–Kutta method is during the initial stage of the time evolution.

```
>> Runge_Kutta_methods

normRK3 =
    1.2391e-005

normRK4 =
    4.1760e-007
```

Exercise 9.5 Consider applications of the third-order and fourth-order Runge–Kutta methods to the ODE (9.5), where the singular initial data is replaced by the exact value of $y(t)$ at $t = t_0$ with small t_0. Use a power fit to show that, as the time step h gets smaller, the local error is $O(h^4)$ for the third-order method and $O(h^5)$ for the fourth-order method, while the global error is $O(h^3)$ for the third-order method and $O(h^4)$ for the fourth-order method.

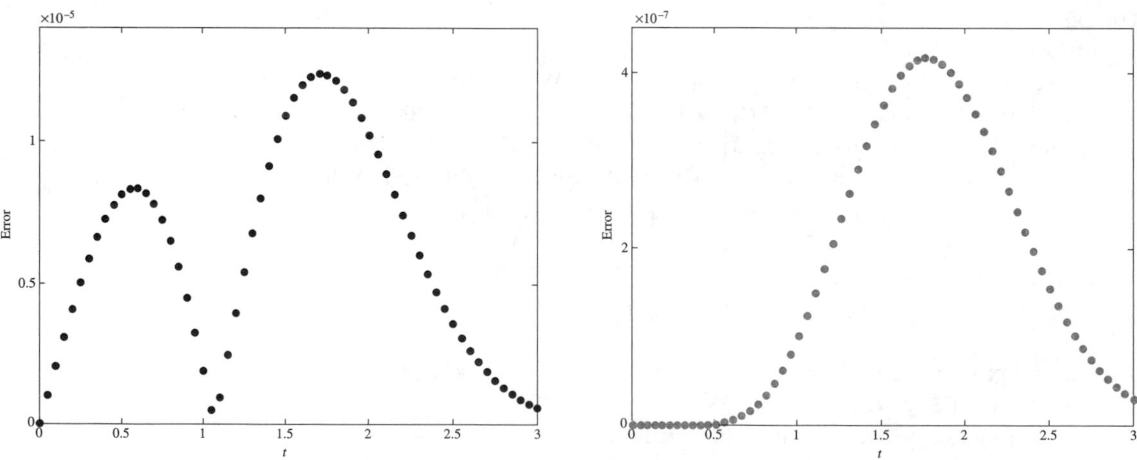

Figure 9.4 Global truncation errors for the explicit third-order (*left*) and fourth-order (*right*) Runge–Kutta methods for the ODE (9.2).

Convergence and stability

To illustrate convergence and stability of the third-order and fourth-order Runge–Kutta methods, we compute explicit iterations of both methods for the linear ODE (9.10). After simple computations with the linear vector field $f(y) = -\lambda y$, we obtain the first-order difference equation

$$y_{k+1} = \left(1 - h\lambda + \frac{1}{2}h^2\lambda^2 - \frac{1}{3!}h^3\lambda^3\right) y_k \tag{9.28}$$

for the third-order Runge–Kutta method and a similar equation

$$y_{k+1} = \left(1 - h\lambda + \frac{1}{2}h^2\lambda^2 - \frac{1}{3!}h^3\lambda^3 + \frac{1}{4!}h^4\lambda^4\right) y_k \tag{9.29}$$

for the fourth-order Runge–Kutta method. The difference equations (9.28) and (9.29) illustrate the correct convergence rate of the local truncation error in the third-order and fourth-order methods. In addition, you can find from the same equations the stability threshold when iterations become divergent. To do so, plot the cubic and quartic curves $g(z)$ associated with the first-order difference equations (9.28) and (9.29) versus $z = \lambda h$:

```
>> z = 0 : 0.01 : 3;
>> g1 = 1 - z + 0.5*z.^2 - z.^3/6;
>> g2 = 1 - z + 0.5*z.^2 - z.^3/6 + z.^4/24;
>> plot(z,g1,'b',z,g2,'g',z,ones(size(z)),':r',z,-ones(size(z)),':r');
```

Figure 9.5 shows the graphs of the two curves in the stability corridor $-1 \leq g(z) \leq 1$. You can see that the third-order Runge–Kutta method is stable for $0 < z \leq 2.5$, while the fourth-order Runge–Kutta method is stable for $0 < z \leq 2.785$. Although the stability interval increases with the higher-order Runge–Kutta methods, the improvement is insignificant compared to the price you pay on high complexity of the algorithm and a large number of additional computations of the vector field $\mathbf{f}(t, \mathbf{y})$. Therefore, we need a different strategy to extend the stability threshold in the family of the Runge–Kutta methods. For instance, implicit single-step ODE solvers have no constraints on the time step of numerical computations (see Section 9.5).

Exercise 9.6 Find numerical approximations of the solution of the ODE problem (9.2) using Heun's third-order, and fourth-order Runge–Kutta methods with $h = 0.6$ and compare the time when numerical instabilities of iterations are developed in each method.

Runge–Kutta methods for ODE systems

Although all previous ODE examples involved scalar first-order differential equations, the same explicit methods can be developed for systems of first-order differential equations. Moreover, if a scalar higher-order ODE is considered, it can be cast to a system of first-order ODEs by means of a stan-

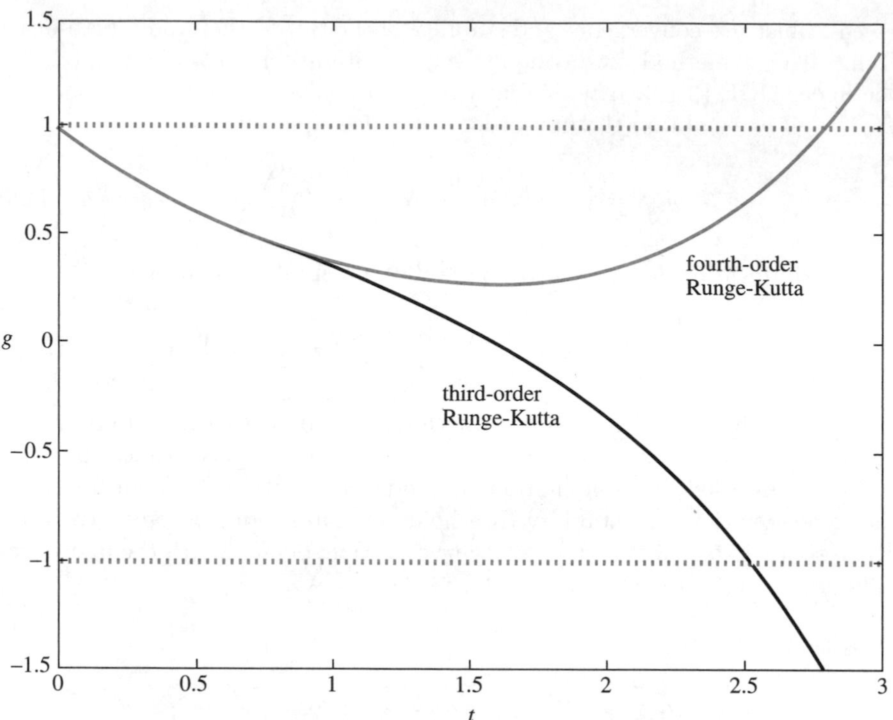

Figure 9.5 Stability corridor of the third-order and fourth-order Runge–Kutta methods.

dard vectorization: $x = x_1$, $\dot{x} = x_2$, and so forth. We illustrate the single-step Runge–Kutta methods on the example of the pendulum equation,

$$\ddot{x} + \sin x = 0, \qquad x(0) = x_0, \;\; \dot{x}(0) = y_0, \tag{9.30}$$

where (x_0, y_0) are initial values. Let $y = \dot{x}$ and transform the scalar second-order ODE (9.30) to the system of two first-order ODEs:

$$\frac{d}{dt}\begin{pmatrix} x \\ y \end{pmatrix} = \begin{pmatrix} y \\ -\sin x \end{pmatrix}, \tag{9.31}$$

where $x(0) = x_0$ and $y(0) = y_0$. The MATLAB script `pendulum` applies Euler's and Heun's methods to the pendulum equation (9.31) for two initial data points $x_0 = \frac{\pi}{4}$, $y_0 = 0$, and $x_0 = -\pi$, $y_0 = 0.05$.

```
% pendulum
h = 0.1; t = 0 : h : 30;
for j = 1 : 2
    xE(j,:) = zeros(size(t)); yE(j,:) = zeros(size(t));
    xH(j,:) = zeros(size(t)); yH(j,:) = zeros(size(t));
    if (j == 1)        % two initial conditions
        xE(j,1) = pi/4; yE(j,1) = 0;
        xH(j,1) = pi/4; yH(j,1) = 0;
    else
        xE(j,1) = -pi; yE(j,1) = 0.05;
        xH(j,1) = -pi; yH(j,1) = 0.05;
    end
    for k = 1 : (length(t) - 1)  % iterations of the two methods
        xE(j,k+1) = xE(j,k) + h*yE(j,k);
        yE(j,k+1) = yE(j,k) - h*sin(xE(j,k));
        pk = xH(j,k) + h*yH(j,k);
        qk = yH(j,k) - h*sin(xH(j,k));
        xH(j,k+1) = xH(j,k) + 0.5*h*(yH(j,k) + qk);
        yH(j,k+1) = yH(j,k) - 0.5*h*(sin(xH(j,k)) + sin(pk));
    end
    figure(j); plot(t,xE(j,:),'b',t,xH(j,:),'r');
end
```

Figure 9.6 shows the output of the MATLAB script **pendulum**. The initial condition $x_0 = \pi/4$, $y_0 = 0$ must correspond to the oscillatory motion of the

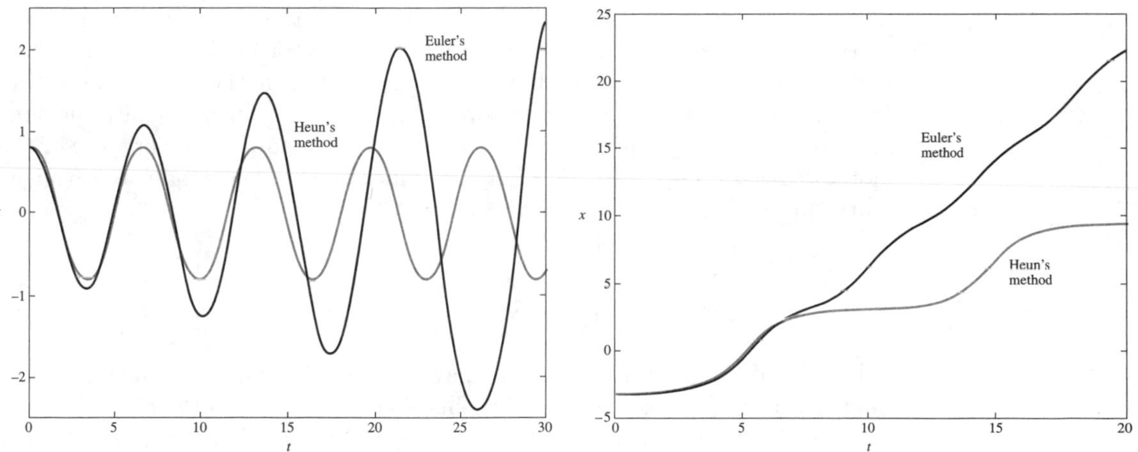

Figure 9.6 Numerical approximation of the solutions of the pendulum equation (9.30) with the explicit Euler and Heun methods for initial conditions $x_0 = \pi/4$, $y_0 = 0$ (*left*) and $x_0 = -\pi$, $y_0 = 0.05$ (*right*).

pendulum, when $x(t)$ is a periodic function of time. However, you can see from Figure 9.6 (*left*) that the Euler method introduces an artificial magnification of the oscillations so that the function $x(t)$ not only oscillates but also grows in time. On the other hand, the Heun method preserves the amplitude of oscillations of the solution $x(t)$, and it is thus more robust for the numerical approximations. The initial condition $x_0 = -\pi$, $y_0 = 0.05$ must correspond to the rotational motion of the pendulum, when the function $x(t)$ is growing in time. Figure 9.6 (*right*) shows that the Euler method displays a monotonic growth of $x(t)$, while the Heun method displays a growing and oscillating evolution of $x(t)$. The latter evolution is closer to the actual evolution of the pendulum according to the exact solution of the ODE (9.30).

More general families of the third-order and fourth-order Runge–Kutta methods, as well as their higher-order analogues, can be obtained from symbolic computational packages applied to solve Problem 9.2 with $n \geq 3$. Although the convergence rate increases for the higher-order Runge–Kutta methods, the higher-order methods are not popular because of two reasons. First, higher-order methods involve many computations of the vector field $\mathbf{f}(t, \mathbf{y})$ and this makes computations lengthy. Second, the coefficients in the summation rules (9.18) alternate in sign for large n and this leads to large round-off errors. These are the same reasons why the higher-order Newton–Cotes integration rules are not popular over the lower-order trapezoidal and Simpson's integration rules. For instance, you can see from Figure 9.4 (left) that the third-order Runge–Kutta method with sign alternation of the coefficients features a rapid growth of the computational error. The fourth-order Runge–Kutta method with the sign-definite coefficients is in fact the most popular method among the family of explicit Runge–Kutta ODE solvers. A useful modification of this method is referred to as the *Runge–Kutta–Fehlberg* method, which requires six computations of the vector field $\mathbf{f}(t, \mathbf{y})$ in two parallel applications of the fourth-order Runge–Kutta method. This algorithm enables one to control the truncation error, to adjust automatically the time step h during computations, and to achieve the fifth-order convergence of the global truncation error [6, 25]. These modifications of the single-step Runge–Kutta methods are considered in Section 9.3.

9.3 Adaptive Single-Step Solvers

Solutions of differential equations may behave differently at different time intervals. For instance, the Gaussian function $y(t) = e^{-t^2}$ in the exact solution of the ODE (9.2) changes rapidly from $y(0) = 1$ to small values of $y(t)$ on the initial time interval $t \in [0, 2]$, but it evolves slowly near the value $y = 0$ for the later times $t \geq 2$. It makes sense to use different time steps h for numerical computations at different time intervals so that the time step h is sufficiently small when the solution $y(t)$ changes rapidly and it is sufficiently large when the solution $y(t)$ changes slowly. In the former case, sufficient accuracy of numerical computations is achieved as a result of a small truncation error. In the latter case, numerical computations are not slowed down by unnecessary steps and the round-off error is small. Numerical methods for ODEs where the time step h and the error of computations are controlled are called *adaptive ODE solvers.*

Consider a single-step Runge–Kutta solver (see Section 9.2) and apply it twice with two different time steps. The order of the truncation error can be estimated by using Richardson extrapolation (see Section 6.4), and the time step h can be adjusted during the process of computations.

Error estimation

Let $O(h^{n+1})$ be the order of the local truncation error of the nth-order Runge–Kutta ODE solver in Problem 9.2. If $y_k = y(t_k)$ is an exact solution at t_k (which is only true for $k = 0$), then a single application of the Runge–Kutta ODE solver results in the numerical approximation y_{k+1} of the unknown solution $y(t_{k+1})$, such that

$$y(t_{k+1}) = y_{k+1} + \alpha_{n+1} h^{n+1}, \tag{9.32}$$

where α_{n+1} is a coefficient, which depends weakly on h in the limit of small h. Let us now apply the same Runge–Kutta ODE solver with the half time step $h/2$. Assuming that the coefficient α_{n+1} is constant in both computations, we can obtain another numerical approximation $\tilde{y}_{k+1}$ of the unknown solution $y(t_{k+1})$, such that

$$y(t_{k+1}) = y_{k+1} + \frac{1}{2^{n+1}} \alpha_{n+1} h^{n+1} + \frac{1}{2^{n+1}} \alpha_{n+1} h^{n+1} = \tilde{y}_{k+1} + \frac{1}{2^n} \alpha_{n+1} h^{n+1}. \tag{9.33}$$

By equating the two numerical approximations, we can estimate the order of the local truncation error at the time step k:

$$E_k = \alpha_{n+1} h^{n+1} = \frac{2^n}{2^n - 1} \left(\tilde{y}_{k+1} - y_{k+1} \right). \tag{9.34}$$

The MATLAB script `error_estimation` applies two computations of the fourth-order Runge–Kutta ODE solver to the ODE (9.3). Using the error estimation formula (9.34) with $n = 4$, you can estimate the local error at each step of the explicit single-step ODE solver. Alternatively, you can compute the distance between the numerical solution and the exact solution $y(t) = 1/(1-t)$.

```
% error_estimation
h = 0.01; k = 1; t(1) = 0; y(1) = 1; yy = 1; E(1) = 0;
 while (t(length(t)) < 0.9)
    t(k+1) = t(k) + h;
    k1 = y(k)^2;  % Runge-Kutta solver with h
```

```
    k2 = (y(k) + 0.5*h*k1)^2;
    k3 = (y(k) + 0.5*h*k2)^2;
    k4 = (y(k) + h*k3)^2;
    y(k+1) = y(k) + h*(k1+2*k2+2*k3+k4)/6;
    k1 = yy^2;    % Runge-Kutta solvers with h/2
    k2 = (yy + 0.25*h*k1)^2;
    k3 = (yy + 0.25*h*k2)^2;
    k4 = (yy + 0.5*h*k3)^2;
    yH = yy + h*(k1+2*k2+2*k3+k4)/12;
    k1 = yH^2;
    k2 = (yH + 0.25*h*k1)^2;
    k3 = (yH + 0.25*h*k2)^2;
    k4 = (yH + 0.5*h*k3)^2;
    yy = yH + h*(k1+2*k2+2*k3+k4)/12;
    E(k+1) = 16*(yy-y(k+1))/15;
    k = k+1;
end
yExact = 1./(1-t); Eexact = abs(yExact-y);
semilogy(t,E,'.b',t,Eexact,':r');
MaxError = max(abs(E-Eexact))
```

The output of the MATLAB script `error_estimation` is shown in Figure 9.7. The solid curve displays the distance between the exact and numerical solutions whereas the dots show the numerical estimation of the error from the error formula (9.34) with $n = 4$. Although the error formula (9.34) is based on the assumption that the previous value y_k is known exactly, the figure shows a good agreement between the approximate error formula (9.34) and the exact error, with the maximal absolute deviation being of order 10^{-8}. You can check that if the error bound is used with $n = 3$ or $n = 5$, then the maximal absolute deviation is two orders higher than it is predicted in Figure 9.7.

```
>> error_estimation

MaxError =   9.6813e-009
```

In practice, both computations with time steps h and $h/2$ should start at the same value y_k of the numerical approximation of the solution $y(t_k) \neq y_k$. The error formula (9.34) may not correspond to the actual distance between exact and numerical solutions because of the accumulating truncation error after several consequent steps. You can check that such comparison fails in the MATLAB script `error_estimation` if both computations start at the same value y_k. The following exercise corrects the comparison of the error estimation formula and the actual error.

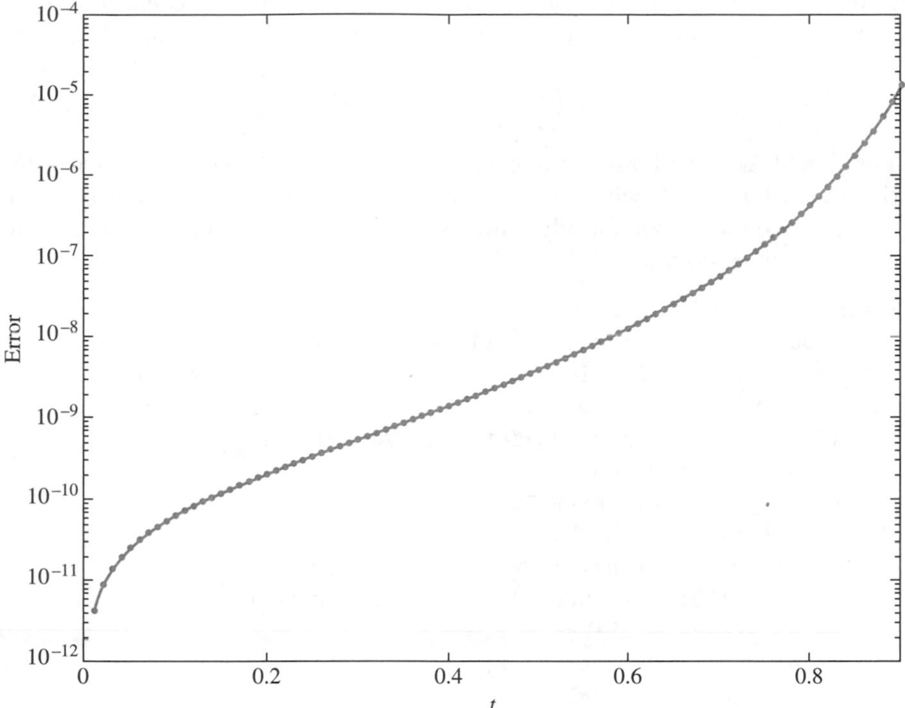

Figure 9.7 Estimation of the truncation error for the fourth-order Runge–Kutta solver in application to the ODE (9.3).

Exercise 9.7 For each step of the fourth-order Runge–Kutta ODE solver with $n = 4$ on $[t_k, t_{k+1}]$ starting with the same value of y_k, compute the error estimation (9.34) from two numerical approximations y_{k+1} and $\tilde{y}_{k+1}$. Compute the exact value of $y(t_{k+1})$ from the ODE (9.3) where the initial value is $y(t_k) = y_k$ and the time interval is $t_{k+1} - t_k = h$. Plot the error estimation and the exact distance between y_{k+1} and $y(t_{k+1})$ and comment on the accuracy of the error estimation formula (9.34).

Time step adjustment

If the truncation error of the Runge–Kutta ODE solver is controlled, we can assign a maximal acceptable error E_{tol} called the *tolerance* so that the time step h is adjusted after each iteration to maintain the same tolerance. By setting the local truncation error equal to the tolerance bound, we can define an optimal time step h_{opt} to perform the kth step:

$$\alpha_{n+1} h_{\text{opt}}^{n+1} = E_{\text{tol}}. \tag{9.35}$$

Combined with the error formula (9.34), we can eliminate the coefficient α_{n+1} and obtain the numerical estimation for the optimal time step h_{opt}:

$$h_{\text{opt}} = h \left(\frac{(2^n - 1)E_{\text{tol}}}{2^n \left(\tilde{y}_{n+1} - y_{n+1}\right)} \right)^{1/(n+1)}. \tag{9.36}$$

The MATLAB script `time_step_adjustment` computes the numerical solution of the ODE (9.3) with the fourth-order Runge–Kutta ODE solver, where the time step h is dynamically adjusted so that the local error matches the given tolerance bound (9.35).

```
% time_step_adjustment
h = 0.001; H(1) = h; k = 1; t(1) = 0; y(1) = 1;
E(1) = 0; Etol = 10^(-10); % initial error and tolerance
while (t(length(t)) < 0.99)
    k1 = y(k)^2;    % Runge-Kutta solver with h
    k2 = (y(k) + 0.5*h*k1)^2;
    k3 = (y(k) + 0.5*h*k2)^2;
    k4 = (y(k) + h*k3)^2;
    y(k+1) = y(k) + h*(k1+2*k2+2*k3+k4)/6;
    k1 = y(k)^2;    % Runge-Kutta solvers with h/2
    k2 = (y(k) + 0.25*h*k1)^2;
    k3 = (y(k) + 0.25*h*k2)^2;
    k4 = (y(k) + 0.5*h*k3)^2;
    yH = y(k) + h*(k1+2*k2+2*k3+k4)/12;
    k1 = yH^2;
    k2 = (yH + 0.25*h*k1)^2;
    k3 = (yH + 0.25*h*k2)^2;
    k4 = (yH + 0.5*h*k3)^2;
    yy = yH + h*(k1+2*k2+2*k3+k4)/12;
    E(k+1) = abs(16*(yy-y(k+1))/15); % error formula
    h = h*(Etol/E(k+1))^(1/5); % time step adjustment
    t(k+1) = t(k) + h;
    k1 = y(k)^2;    % new Runge-Kutta solver with h_opt
    k2 = (y(k) + 0.5*h*k1)^2;
    k3 = (y(k) + 0.5*h*k2)^2;
    k4 = (y(k) + h*k3)^2;
    y(k+1) = y(k) + h*(k1+2*k2+2*k3+k4)/6;
    H(k+1) = h;
    k = k+1;
end
yExact = 1./(1-t); Eexact = abs(yExact-y);
semilogy(t,E,'.b',t,Eexact,':r'); semilogy(t,H,'.b');
```

The output of the MATLAB script `time_step_adjustment` is shown in Figure 9.8. The tolerance bound is 10^{-10} and the initial time step is as small

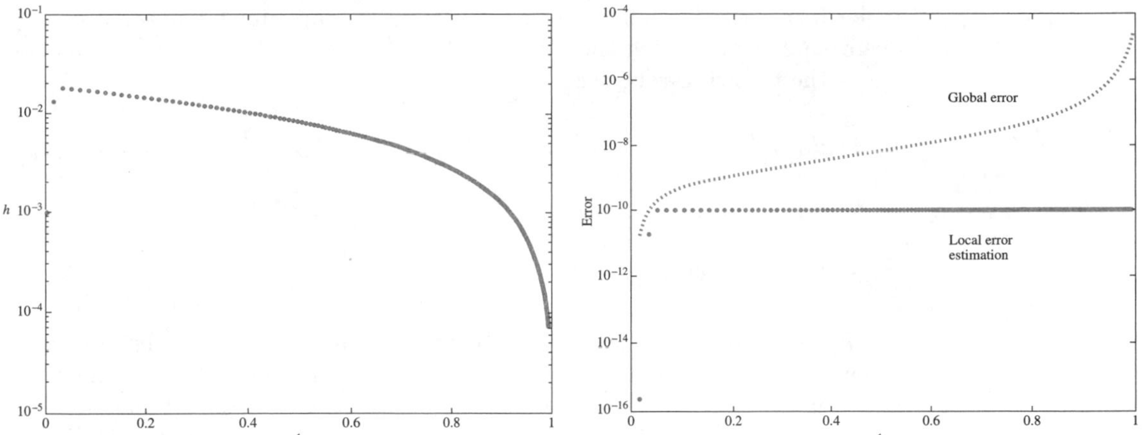

Figure 9.8 Dynamically adjusted time step (*left*) and the local and global errors (*right*) versus time t for the numerical solution of the ODE (9.3).

as $h = 0.001$. The left figure shows how the time step h is dynamically adjusted during the time evolution. The first two steps are performed with a smaller local truncation error than the given tolerance so that the time step is increased. The time step is monotonically decreased later since the solution $y(t)$ approaches the blow-up point at $t = 1$. Near the point $t = 1$, the solution increases rapidly and the accuracy becomes worse so that the time step is decreased drastically to maintain the same tolerance. More and more iterations of the Runge–Kutta ODE solvers are needed for further increments of time. The right figure shows the estimation of the local error versus time t. You can see that the tolerance of 10^{-10} for the local error is preserved throughout numerical computations. The dotted curve in the same figure shows that the global error (the distance from the exact solution) grows as t approaches the blow-up time $t = 1$. Therefore, the global error of the ODE solver is not controlled by the time step adjustment (9.36).

Exercise 9.8 Apply the algorithm with the time step adjustment and the estimation of the local truncation error to the Heun method in the context of the ODE problem (9.3). Plot the dynamically adjusted time step and the local and global errors versus time t.

From a practical viewpoint, some unnecessary computations can be avoided if the suggested optimal value h_{opt} is close to the original time step h. For instance, you can implement a simple rule that the time step is changed when the factor on the right-hand side of (9.36) (let us denote it by s) is either smaller than $\frac{3}{4}$ or larger than $\frac{3}{2}$ and it is left unchanged otherwise. If $s < \frac{3}{4}$, the time step h needs to be decreased, and if $s > \frac{3}{2}$, the time step h needs to be increased. In many cases, it is easier to implement computations by decreasing or increasing the time step by a factor of two. Therefore, you can adopt the following routine:

- If $s < \frac{3}{4}$, then $h_{\rm opt} = \frac{1}{2}h$
- If $\frac{3}{4} \leq s \leq \frac{3}{2}$, then $h_{\rm opt} = h$
- If $s > \frac{3}{2}$, then $h_{\rm opt} = 2h$

Additionally, you may worry about the round-off error if the time step becomes too small or too large, as a result of the possibility of underflow and overflow (see Section 1.7). You therefore need to control the range of values of h so that it is always kept in the interval $h_{\min} < h < h_{\max}$, where $h_{\min}$ and $h_{\max}$ depend on the floating point number system. If the time step adjustments push h beyond this interval, numerical computations must be stopped and alternative numerical methods need to be proposed.

Exercise 9.9 Modify the MATLAB script `time_step_adjustment` so that the practical routine is implemented to eliminate the unnecessary changes of the time step h and to control the range of its values.

The adaptive methods for the error estimation and the time step adjustment based on halving the time step require too many computations of the vector field $\mathbf{f}(t, \mathbf{y})$ compared to modern software algorithms. For instance, the Runge–Kutta–Fehlberg algorithm requires only 6 computations of the vector field [6], whereas the fourth-order Runge–Kutta method with time steps h and $h/2$ result in 11 computations of the vector field.

MATLAB ODE solvers

MATLAB implements two ODE solvers that are based on the explicit single-step Runge–Kutta methods with variable time step. The MATLAB function `ode23` implements the Runge–Kutta ODE solvers of the second and third orders that generalize the second-order Heun method. The MATLAB function `ode45` implements the Runge–Kutta ODE solvers of the fourth and fifth orders similar to the Runge–Kutta–Fehlberg method. Both the MATLAB ODE solvers require a minimum of three input arguments that are the name of the M-file with the ODE description, the span of the time interval, and the initial value for the dependent variable. The two output variables are the column vectors for times and the corresponding approximations of the solution $\mathbf{y}(t)$. If the span for the time interval contains only the starting and ending times, the output column vectors contain all numerical approximations after each time step of the iterative algorithm. If the span for the time interval contains a detailed partition between the starting and ending times, the output column vectors contain numerical approximations at the given partition.

The M-file function with the ODE description simply lists the right-hand side of the differential equation. For example, the following function ivpODE corresponds to the ODE (9.2), where we used Dy to denote the derivative of y (to avoid confusion with the transpose y'):

```
function Dy = ivpODE(t,y)
    Dy = -2*t.*y;
```

The M-file ivpODE for the ODE description is called in the main MATLAB script MATLAB_ODE.

```
% MATLAB_ODE
y0 = 1; tspan = 0 : 0.1 : 4;
[t,y] = ode45(@ivpODE,tspan,y0);
yExact = exp(-t.^2);
n = length(t)
Error = max(abs(y-yExact))
plot(t,yExact,'r',t,y,'.b');
```

The numerical approximations are computed on the equally spaced partition of the interval $[0, 4]$ with the time step $h = 0.1$. The output of the MATLAB script MATLAB_ODE shows the number of time steps and the maximal global error. You can see that the number of time steps $n = 41$ corresponds to the number of grid points in the given partition. The global error is of order 10^{-4}. The numerical approximation (dots) and the exact solution (solid curve) are shown in Figure 9.9.

```
>> MATLAB_ODE

n =
    41

Error =
    8.3739e-005
```

To improve the accuracy of the MATLAB ODE solvers, you need to specify the level of tolerance for the global error. This is possible with the fourth optional input argument in the MATLAB ODE functions. Not only the relative and absolute error tolerances can be specified in the optional fourth argument, but also the maximal time step, the suggested initial step, and other options. The MATLAB ODE functions also have a fifth optional input argument in which you can introduce parameters of the differential equations. The same list of parameters needs to be specified in the third optional input argument of the M-file for the ODE description, which is passed to the ODE solver. These additional properties are illustrated in the MATLAB

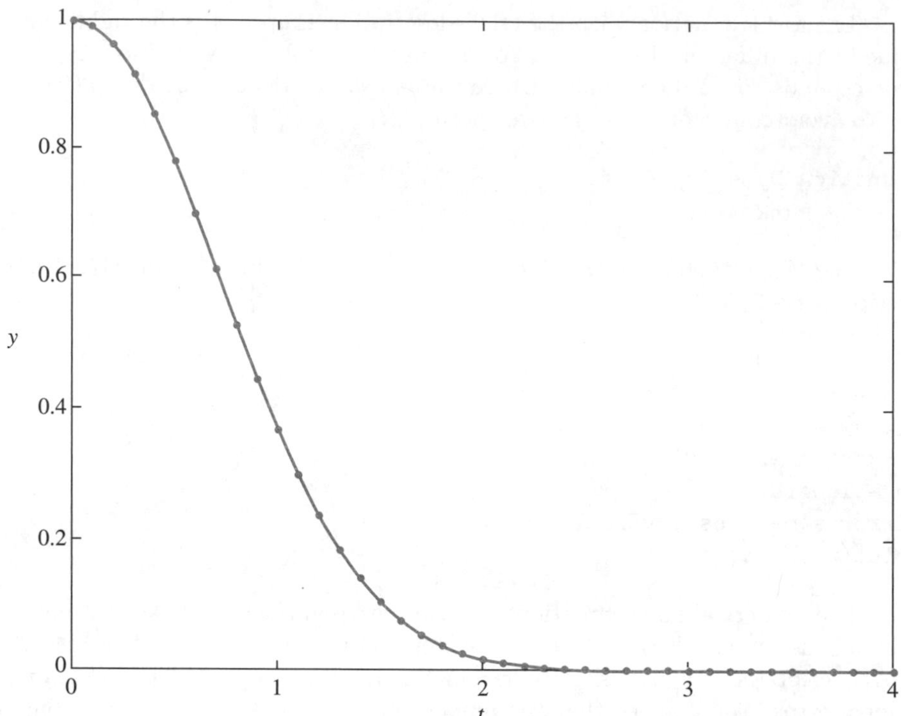

Figure 9.9 Numerical solution of the ODE (9.2) with the MATLAB ODE solver.

script MATLAB_ODE_options, which computes the numerical solution of the ODE (9.3).

```
% MATLAB_ODE_options
y0 = 1; tspan = [0, 0.98];
options=odeset('RelTol',10^(-6),'AbsTol',10^(-6));
p = 1; q = 0;         % options and parameters of the ODE solver
[t,y] = ode45(@ivp2ODE,tspan,y0,options,p,q);
n = length(t); h = t(2:n)-t(1:n-1);  % the dynamical time step
yExact = 1./(1-t); Error = max(abs(y-yExact)) % the global error
plot(t,yExact,'r',t,y,'.b'); plot(t(1:n-1),h,'.b');
```

The MATLAB script MATLAB_ODE_options uses the MATLAB function ivp2ODE, where the ODE (9.3) is coded with additional parameters p and q in the vector field $f(t,y) = py^2 + qt^2$ (the ODE (9.3) has $p = 1$ and $q = 0$).

```
function Dy = ivp2ODE(t,y,p,q)
    Dy = p*y.^2+q*t.^2;
```

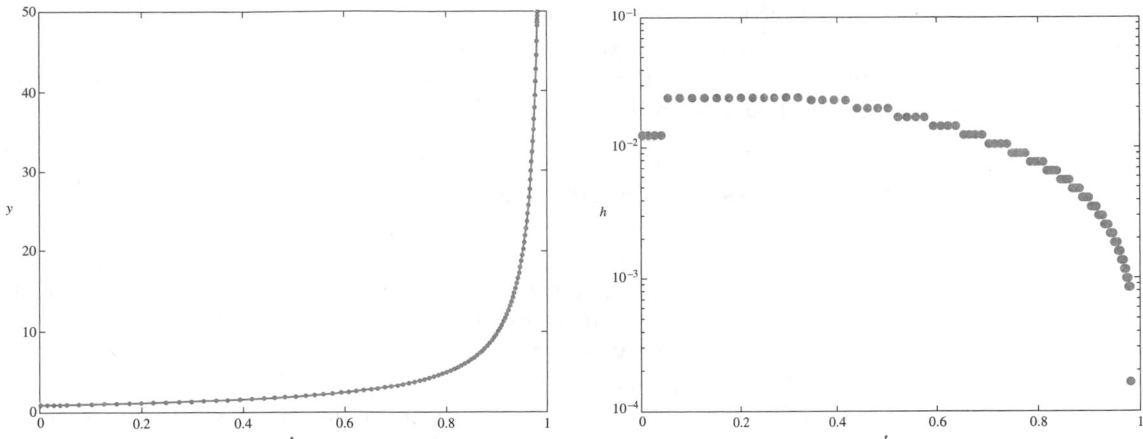

Figure 9.10 Numerical solution of the ODE (9.3) with the MATLAB ODE solver: exact (solid) and numerical (dots) solutions (*left*) and the time step (*right*).

The output of the MATLAB script `MATLAB_ODE_options` shows that the maximal global error is of order 10^{-3}. Although the tolerance level was specified at 10^{-6}, it only controls the level of the local error, while the global error can be much larger.

```
>> MATLAB_ODE_options

Error =   6.3657e-004
```

Figure 9.10 shows the numerical and exact solutions (*left*) and the dynamically adjusted time step h during iterations (*right*). You can see that the time step is adjusted to maintain the required level of tolerance, so that it is increased at the initial stage and decreased at the later stage when t approaches the blow-up time $t = 1$. This pattern of the MATLAB ODE solver `ode45` corresponds to the numerical solution in Figure 9.8 obtained with two computations of the fourth-order Runge–Kutta method for time steps h and $h/2$. You can see that the adjustment of the time step h is performed according to the previous practical routine: the time step is not adjusted if the estimation of the truncation error is close to the required level of tolerance.

Exercise 9.10 Compute the error of the numerical approximation of $y(t)$ at $t = 0.98$ of the ODE problem (9.3) with the MATLAB ODE solver `ode23` for different values of the tolerance. Plot the global error versus the tolerance in the logarithmic scale and characterize how the actual error depends on the prescribed tolerance.

Runge–Kutta methods are not the only ODE solvers that are widely used for numerical approximations of solutions of differential equations. Other methods such as the multistep Adams methods and implicit solvers are popular as well and are considered in the next two sections.

9.4 Multistep Adams Solvers

Single-step Runge–Kutta methods are designed to compute the numerical approximation of $\mathbf{y}(t)$ at t_{k+1} using the numerical approximation of $\mathbf{y}(t)$ at t_k and additional approximations of $\mathbf{y}(t)$ at interior points of the interval $[t_k, t_{k+1}]$. Multistep Adams methods incorporate the numerical approximations of $\mathbf{y}(t)$ at the previous times $t_k, t_{k-1}, \ldots, t_{k-m+1}$, where m is the order of the multistep method. This idea can be implemented efficiently in computer algorithms if the time step h is constant and the approximations of $\mathbf{y}_k$ are stored in the computer memory together with the values $\mathbf{f}_k = \mathbf{f}(t_k, \mathbf{y}_k)$, where $\mathbf{f}(t, \mathbf{y})$ is the vector field of the ODE (9.1). The advantage of this numerical algorithm is that only one computation of the vector function $\mathbf{f}(t, \mathbf{y})$ is required at each step of the iteration method.

Adams–Bashforth methods

Multistep Adams methods are classified as *explicit* and *implicit*. Explicit methods use the values $\mathbf{f}_k$, $\mathbf{f}_{k-1}$, ..., $\mathbf{f}_{k-m+1}$ in computations of $\mathbf{y}_{k+1}$ and are often referred to as the *Adams–Bashforth* methods. Implicit methods use the values $\mathbf{f}_{k+1}$, $\mathbf{f}_k$, ..., $\mathbf{f}_{k-m+2}$ in computations of $\mathbf{y}_{k+1}$ and are often referred to as the *Adams–Moulton* methods. Although both types of methods can be derived from the integral form (9.14) by integrating the interpolating polynomials for the function $\mathbf{f}(t, \mathbf{y}(t))$ (see Exercises 9.11 and 9.13 later), we prefer to use the Taylor series in Theorem 5.2 and the finite difference approximations of the numerical derivatives in Theorems 6.1 and 6.4 instead.

Problem 9.3 (Explicit m-Order Adams ODE Solver) Let $\mathbf{y}(t)$ be a smooth solution of the ODE (9.1) on $[t_k, t_{k+1}]$ starting with $\mathbf{y}(t_k) = \mathbf{y}_k$. Consider the Taylor series at $t = t_k$

$$\mathbf{y}(t_{k+1}) = \mathbf{y}(t_k) + h\mathbf{y}'(t_k) + \ldots + \frac{1}{m!}h^m \mathbf{y}^{(m)}(t_k) + \mathrm{O}(h^{m+1}) \tag{9.37}$$

and approximate the derivatives $\mathbf{y}^{(n)}(t_k)$ for $1 \leq n \leq m$ by the backward numerical derivatives through the points t_k, t_{k-1},...,t_{k-m} so that the local truncation error for the numerical approximation $\mathbf{y}_{k+1}$ of $\mathbf{y}(t_{k+1})$ is preserved as $\mathrm{O}(h^{m+1})$.

Because for each $\mathbf{y}^{(n)}(t_k)$, $1 \leq n \leq m$, there exists a backward numerical derivative through the points t_k,...,t_{k-m} with an $\mathrm{O}(h^{m-n+1})$ error, Problem 9.3 always has a solution. By construction, the global truncation error of the explicit m-order Adams method is $\mathrm{O}(h^m)$. As before, $\mathbf{y}(t_k)$ is not the same as $\mathbf{y}_k$ for $k \geq 1$ as a result of the accumulated truncation error, and this leads to a global error of one order higher than the local error.

When $m = 1$, the Taylor series (9.37) requires no additional computations and reduces to the single-step Euler method:

$$\mathbf{y}_{k+1} = \mathbf{y}_k + h\mathbf{f}_k, \tag{9.38}$$

which is described in Section 9.1. When $m = 2$, we can use the first-order backward difference approximation for the first derivative

$$\left.\frac{d}{dt}\mathbf{f}(t, \mathbf{y}(t))\right|_{t=t_k} = \frac{\mathbf{f}_k - \mathbf{f}_{k-1}}{h} + \mathrm{O}(h),$$

so that the explicit second-order Adams method becomes

$$\mathbf{y}_{k+1} = \mathbf{y}_k + \frac{1}{2}h\left(3\mathbf{f}_k - \mathbf{f}_{k-1}\right). \tag{9.39}$$

Continuing this process, we can use the second-order backward difference approximation for the first derivative and the first-order backward difference approximation for the second derivative

$$\left.\frac{d}{dt}\mathbf{f}(t, \mathbf{y}(t))\right|_{t=t_k} = \frac{3\mathbf{f}_k - 4\mathbf{f}_{k-1} + \mathbf{f}_{k-2}}{2h} + \mathrm{O}(h^2),$$
$$\left.\frac{d^2}{dt^2}\mathbf{f}(t, \mathbf{y}(t))\right|_{t=t_k} = \frac{\mathbf{f}_k - 2\mathbf{f}_{k-1} + \mathbf{f}_{k-2}}{h^2} + \mathrm{O}(h),$$

so that the explicit third-order Adams method becomes

$$\mathbf{y}_{k+1} = \mathbf{y}_k + \frac{1}{12}h\left(23\mathbf{f}_k - 16\mathbf{f}_{k-1} + 5\mathbf{f}_{k-2}\right). \tag{9.40}$$

Similarly, the explicit fourth-order Adams method takes the form:

$$\mathbf{y}_{k+1} = \mathbf{y}_k + \frac{1}{24}h\left(55\mathbf{f}_k - 59\mathbf{f}_{k-1} + 37\mathbf{f}_{k-2} - 9\mathbf{f}_{k-3}\right). \tag{9.41}$$

Exercise 9.11 Derive the explicit Adams methods (9.39)–(9.41) from integration of the integral equation (9.14), where $\mathbf{f}(t, \mathbf{y}(t))$ is interpolated by a polynomial through the points t_k, t_{k-1}, t_{k-2}, and t_{k-3}.

The MATLAB script `explicit_Adams` illustrates the iterations of the explicit second-order Adams method (9.39) for the ODE problem (9.2). Because the second-order difference equation (9.39) requires computations of the values of y_0 and y_1 before the start of iterations, the value of y_1 at t_1 has to be specified in addition to the initial value $y_0 = y(t_0)$. Because the exact solution

$y(t) = e^{-t^2}$ is available for the ODE problem (9.2), you can specify $y_0 = 1$ and $y_1 = e^{-t_1^2}$ from the exact solution. If the exact solution is unavailable (which is the motivation for solving the problem numerically to begin with), you can compute y_1 from a single-step second-order method (such as the Heun or modified Euler methods), which has the same order of the truncation error as the Adams method (9.39).

```
% explicit_Adams
t(1) = 0; y(1) = 1; T = 4;
nArray = 100 : 10 : 10000;
for j = 1 : length(nArray)  % approximations for different time steps
    n = nArray(j);
    h(j) = T/n;
    t(2) = t(1) + h(j);
    y(2) = exp(-t(2)^2);
    for k = 2 : n    % iterations of the second-order Adams method
            t(k+1) = t(k) + h(j);
            y(k+1) = y(k) - h(j)*(3*t(k)*y(k)-t(k-1)*y(k-1));
    end
    yExact = exp(-t.^2);
    Eloc(j) = abs(y(3)-yExact(3));
    Eglob(j) = abs(y(n+1)-yExact(n+1));
end
aLoc = polyfit(log(h),log(Eloc),1);  % power fit of local error
powerLoc = aLoc(1), EaprLoc=exp(aLoc(2))*h.^powerLoc;
aGlob=polyfit(log(h),log(Eglob),1);  % power fit of global error
powerGlob = aGlob(1), EaprGlob = exp(aGlob(2))*h.^powerGlob;
loglog(h,Eloc,'.b',h,EaprLoc,':r',h,Eglob,'.g',h,EaprGlob,':r')
```

The MATLAB script **explicit_Adams** plots the graphs of the local and global errors versus the time step h shown in Figure 9.11. The local error is defined after the first step at time $t_2 = 2h$. The global error is defined at the final time T. The power fits are shown by dotted curves, and the powers of the least-square fits are also displayed. The global error is $O(h^2)$ in agreement with the second order of the Adams method (9.39). The local error is $O(h^4)$, which is smaller than the predicted $O(h^3)$ error. The cancellation of the cubic power of h is a specific property of the particular ODE example (9.2).

```
>> explicit_Adams

powerLoc =
    3.9999

powerGlob =
    2.0141
```

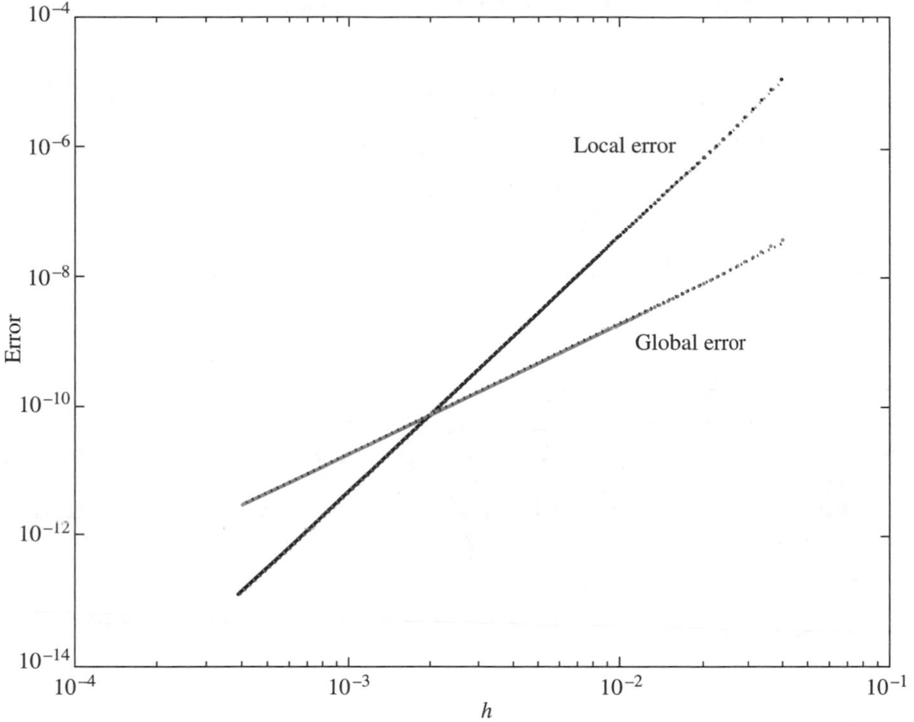

Figure 9.11 Numerical solution of the ODE (9.2) with the second-order explicit Adams ODE solver.

Exercise 9.12 Apply the second-order, third-order, and fourth-order explicit Adams methods to the ODE problem (9.3) and find the power fits of the local and global errors versus the time step h. Show that no cancellation occurs for the ODE problem (9.3) so that the local errors converge in accordance with the construction of Adams methods in Problem 9.3.

Convergence and stability

To consider stability of the explicit second-order Adams method (9.39), consider the linear ODE $y' = -\lambda y$. Substitution of $f(y) = -\lambda y$ into (9.39) results in the second-order difference equation

$$y_{k+1} = \left(1 - \frac{3}{2}\lambda h\right) y_k + \frac{1}{2}\lambda h y_{k-1}. \tag{9.42}$$

Since the second-order difference equation (9.42) has constant coefficients in k, we can reduce it to the quadratic equation with the substitution $y_k = q^k$, where q is the parameter to be determined. The quadratic equation for q takes the form

$$q^2 = (1 - 3z)q + z, \tag{9.43}$$

where $z = \lambda h/2$. The quadratic equation admits two roots

$$q_{1,2} = \frac{1 - 3z \pm \sqrt{(1 - 3z)^2 + 4z}}{2}.$$

The two roots $q_1(z)$ and $q_2(z)$ are plotted versus z by the MATLAB commands:

```
>> z = 0 : 0.01 : 1;
>> q1 = (1 - 3*z + sqrt((1 - 3*z).^2 + 4*z))/2;
>> q2 = (1 - 3*z - sqrt((1 - 3*z).^2 + 4*z))/2;
>> plot(z,q1,'r',z,q2,'b')
```

Figure 9.12 shows the behavior of the roots, from which it follows that one root is always stable between $0 < q < 1$. The other root is stable between

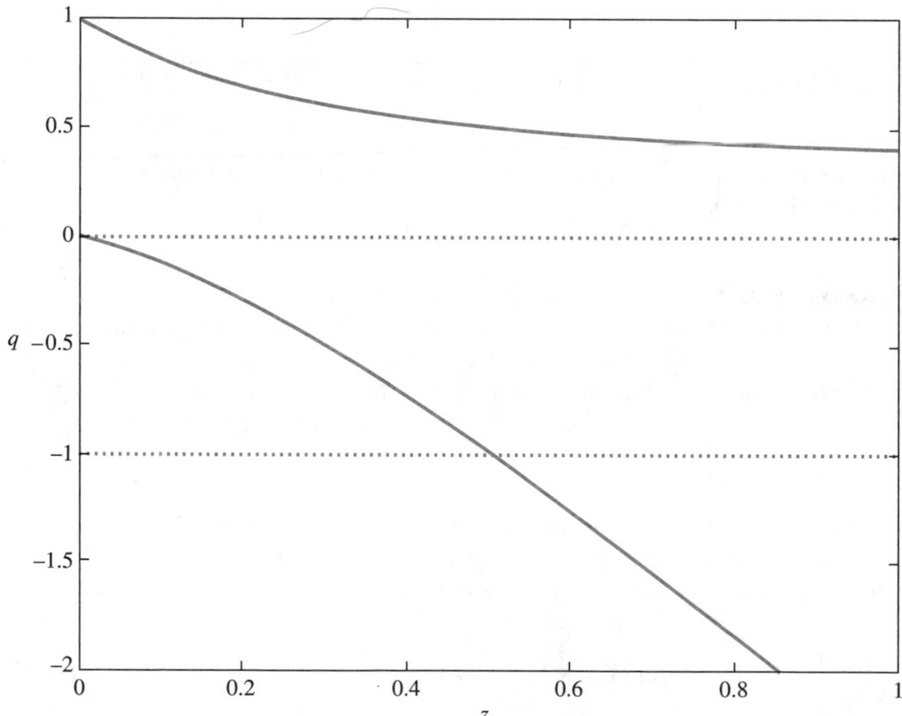

Figure 9.12 Roots of the quadratic equation (9.43) versus $z = \lambda h/2$.

$-1 < q < 0$ for $0 < \lambda h < 1$ and it is unstable with $q < -1$ for $\lambda h > 1$. When $h > 1/\lambda$, the second-order explicit Adams method generates a diverging sign-alternative sequence $\{y_k\}_{k=0}^{\infty}$ instead of the monotonically decaying sequence.

In comparison with the second-order Heun or the modified Euler method, the Adams method is less stable (the stability domain $0 < \lambda h < 1$ for the Adams method is narrower than the stability domain $0 < \lambda h < 2$ of the Euler method). An additional complication of the second-order difference equation (9.42) is that convergence of the truncation error as $h \to 0$ is not obvious. To consider convergence, we obtain a general solution of the second-order difference equation (9.42) as a linear combination of the two particular solutions

$$y_k = c_1 q_1^k + c_2 q_2^k, \qquad k \in \mathbb{N}, \tag{9.44}$$

where q_1, q_2 are two distinct roots of the quadratic equation (9.43) and c_1, c_2 are arbitrary coefficients. As $z \to 0$, the two roots are expanded as follows:

$$q_1 = 1 - 2z + 2z^2 + \mathrm{O}(z^3), \qquad q_2 = -z + \mathrm{O}(z^2), \tag{9.45}$$

where $z = \lambda h/2$. For $k > 3$, the term $q_2^k = \mathrm{O}(h^k)$ is beyond the truncation error of the second-order Adams method, and the explicit solution (9.44) with the expansion (9.45) produces the approximation

$$y_{k+1} = \left(1 - \lambda h + \frac{1}{2}\lambda^2 h^2 + \mathrm{O}(h^3)\right) y_k,$$

which is in agreement with Problem 9.3 that tells us that the local error of the second-order explicit Adams method is $\mathrm{O}(h^3)$.

The MATLAB script `instability_explicit_Adams` illustrates the long-term behavior of the Heun method (9.20)–(9.21) and the explicit second-order Adams method (9.39) for the ODE problem (9.2). Figure 9.13 shows numerical approximations (dark and bright dots) and the exact solution (dashed curve) for a given time step $h = 0.1$. Both explicit methods are unstable for sufficiently large time instance t, but the instability of the second-order Adams method occurs earlier than does the instability of the second-order Heun method. You also can see that the divergent sequence is sign-alternating for the second-order Adams method and it is sign-definite for the Heun method.

```
% instability_explicit_Adams
t(1) = 0; y(1) = 1; yH(1) = 1; T = 18; h = 0.1; n = T/h;
t(2) = t(1) + h; % first step (necessary for Adams method)
y(2) = exp(-t(2)^2); yH(2) = y(2);
for k = 2 : n
    t(k+1) = t(k) + h;
```

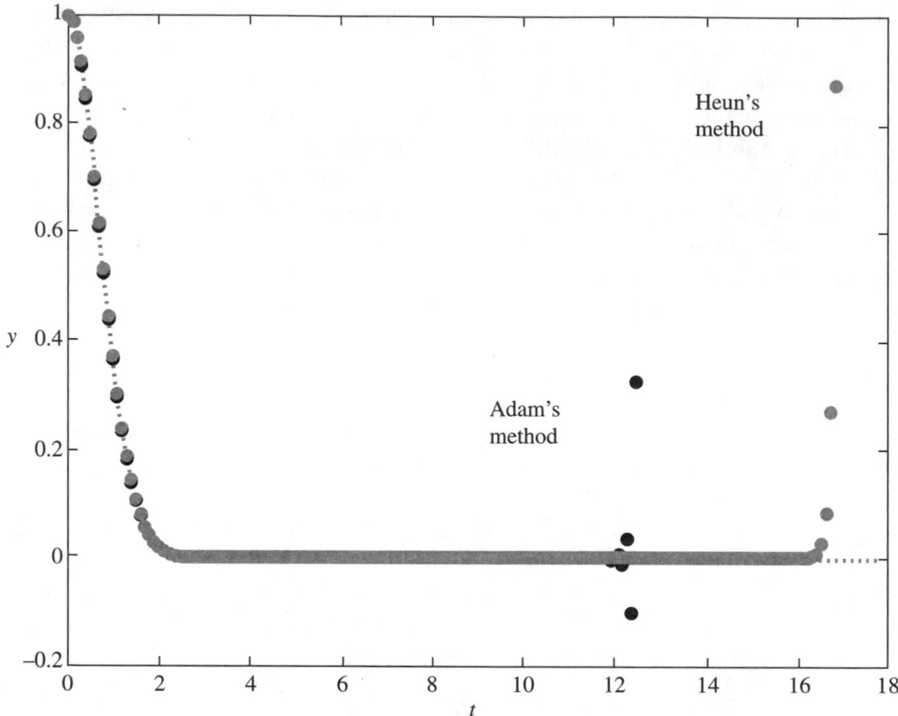

Figure 9.13 Instabilities of the Heun method (9.20)–(9.21) and the explicit Adams method (9.39) for the ODE problem (9.2) with $h = 0.1$.

```
    y(k+1) = y(k) - h*(3*t(k)*y(k)-t(k-1)*y(k-1));
    p = yH(k) - 2*h*t(k)*yH(k);
    yH(k+1) = yH(k) - h*(t(k)*yH(k)+t(k+1)*p);
end
yExact = exp(-t.^2);
plot(t,yExact,':r',t,y,'.b',t,yH,'.g');
```

Adams–Moulton methods

When the numerical approximation of $\mathbf{y}_{k+1}$ involves computation of the value $\mathbf{f}_{k+1}$, which is defined by the value of $\mathbf{y}_{k+1}$, we say that the ODE solver is implicit. Implicit methods have already appeared in Section 9.2, but their consideration was replaced with explicit Runge–Kutta methods in Problem 9.2. Here we construct implicit Adams methods, which can be used in pairs with the explicit Adams methods as predictor–corrector methods. More on implicit ODE solvers is given in Section 9.5.

Problem 9.4 (Implicit m-Order Adams ODE Solver) Let $\mathbf{y}(t)$ be a smooth solution of the ODE (9.1) on $[t_k, t_{k+1}]$ starting with $\mathbf{y}(t_k) = \mathbf{y}_k$. Consider Taylor series at $t = t_{k+1}$

$$\mathbf{y}(t_k) = \mathbf{y}(t_{k+1}) - h\mathbf{y}'(t_{k+1}) + \ldots + \frac{(-1)^m}{m!} h^m \mathbf{y}^{(m)}(t_{k+1}) + \mathrm{O}(h^{m+1}) \quad (9.46)$$

and approximate the derivatives $\mathbf{y}^{(n)}(t_{k+1})$ for $1 \le n \le m$ by the backward numerical derivatives through the points $t_{k+1}, t_k, \ldots, t_{k+1-m}$, so that the local truncation error for the numerical approximation $\mathbf{y}_{k+1}$ of $\mathbf{y}(t_{k+1})$ is preserved as $\mathrm{O}(h^{m+1})$.

It is obvious that the construction of implicit Adams methods is completely analogous to the construction of explicit Adams methods in Problem 9.3. When $m = 1$, the Taylor series (9.46) requires no additional computations and reduces to the implicit Euler method:

$$\mathbf{y}_{k+1} = \mathbf{y}_k + h\mathbf{f}_{k+1}, \quad (9.47)$$

which is explained in Section 9.5. When $m = 2$, we can use the first-order backward difference approximation for the first derivative

$$\left.\frac{d}{dt}\mathbf{f}(t, \mathbf{y}(t))\right|_{t=t_{k+1}} = \frac{\mathbf{f}_{k+1} - \mathbf{f}_k}{h} + \mathrm{O}(h),$$

so that the implicit second-order Adams method becomes

$$\mathbf{y}_{k+1} = \mathbf{y}_k + \frac{1}{2} h \left(\mathbf{f}_{k+1} + \mathbf{f}_k\right). \quad (9.48)$$

This method is nothing but the implicit Heun method (9.15) described in Section 9.2 in connection to the trapezoidal rule of numerical integration. Continuing this process, we can use the second-order backward difference approximation for the first derivative and the first-order backward difference approximation for the second derivative

$$\left.\frac{d}{dt}\mathbf{f}(t, \mathbf{y}(t))\right|_{t=t_{k+1}} = \frac{3\mathbf{f}_{k+1} - 4\mathbf{f}_k + \mathbf{f}_{k-1}}{2h} + \mathrm{O}(h^2),$$

$$\left.\frac{d^2}{dt^2}\mathbf{f}(t, \mathbf{y}(t))\right|_{t=t_{k+1}} = \frac{\mathbf{f}_{k+1} - 2\mathbf{f}_k + \mathbf{f}_{k-1}}{h^2} + \mathrm{O}(h),$$

so that the implicit third-order Adams method becomes

$$\mathbf{y}_{k+1} = \mathbf{y}_k + \frac{1}{12} h \left(5\mathbf{f}_{k+1} + 8\mathbf{f}_k - \mathbf{f}_{k-1}\right) \quad (9.49)$$

Similarly, the implicit fourth-order Adams method takes the form:

$$\mathbf{y}_{k+1} = \mathbf{y}_k + \frac{1}{24}h\left(9\mathbf{f}_{k+1} + 19\mathbf{f}_k - 5\mathbf{f}_{k-1} + \mathbf{f}_{k-2}\right). \tag{9.50}$$

Exercise 9.13 Derive the implicit Adams methods (9.48)–(9.50) from integration of the integral equation (9.14) where $\mathbf{f}(t, \mathbf{y}(t))$ is interpolated by the polynomial through the points t_{k+1}, t_k, t_{k-1}, and t_{k-2}.

Predictors and correctors

To avoid the root-finding algorithms for finding the value $\mathbf{y}_{k+1}$ in implicit Adams methods, we can use the explicit Adams method for a prediction of the value $\mathbf{p}_{k+1}$ at the time t_{k+1}. When the prediction is obtained, the value $\mathbf{f}_{k+1}$ can be computed as $\mathbf{f}(t_{k+1}, \mathbf{p}_{k+1})$, and the implicit Adams method of the corresponding order can be applied to find the correction of the value $\mathbf{y}_{k+1}$. These pairs of explicit and implicit Adams methods are known as the *predictor–corrector* methods.

One of the main advantages of the predictor–corrector methods is that only two computations of the vector field $\mathbf{f}(t, \mathbf{y})$ are required at each time step. Although the same amount of computations is required in the first-order and second-order single-step methods, more computations of the vector field are needed in higher-order Runge–Kutta methods. On the other hand, the Adams methods are not self-starting because $(m-1)$ values $\mathbf{y}_1, \ldots, \mathbf{y}_{m-1}$ must be approximated in addition to the initial value $\mathbf{y}_0$ to start iterations of the explicit m-order Adams method. Single-step Runge–Kutta methods of the corresponding order can be used for numerical approximations of these values.

The fourth-order explicit and implicit Adams methods applied to the ODE (9.2) are illustrated in the MATLAB script `predictor_corrector`. To simplify computations, we have used here the exact values for y_0, y_1, y_2, and y_3 from the exact solution $y(t) = e^{-t^2}$.

```
% predictor_corrector
t(1) = 0; y(1) = 1; f(1) = 0; T = 4;
nArray = 100 : 5 : 2000;
for j = 1 : length(nArray) % approximations for different time steps
    n = nArray(j); h(j) = T/n;
    t(2) = t(1) + h(j); y(2) = exp(-t(2)^2); f(2) = -2*t(2)*y(2);
    t(3) = t(2) + h(j); y(3) = exp(-t(3)^2); f(3) = -2*t(3)*y(3);
    t(4) = t(3) + h(j); y(4) = exp(-t(4)^2); f(4) = -2*t(4)*y(4);
    for k = 4 : n  % iterations of the predictor-corrector pair
        t(k+1) = t(k) + h(j);
        p = y(k) + h(j)*(55*f(k)-59*f(k-1)+37*f(k-2)-9*f(k-3))/24;
```

```
            ff = -2*t(k+1)*p;
            y(k+1) = y(k) + h(j)*(9*ff+19*f(k)-5*f(k-1)+f(k-2))/24;
            f(k+1) = -2*t(k+1)*y(k+1);
    end
    yExact = exp(-t.^2);
    Eloc(j) = abs(y(5)-yExact(5));
    Eglob(j) = abs(y(n+1)-yExact(n+1));
end
aLoc = polyfit(log(h),log(Eloc),1);  % power fit of local error
powerLoc = aLoc(1), EaprLoc = exp(aLoc(2))*h.^powerLoc;
aGlob=polyfit(log(h),log(Eglob),1); % power fit of global error
powerGlob = aGlob(1), EaprGlob = exp(aGlob(2))*h.^powerGlob;
loglog(h,Eloc,'.b',h,EaprLoc,':r',h,Eglob,'.g',h,EaprGlob,':r')
```

The MATLAB script `predictor_corrector` plots the graphs of the local and global errors versus the time step h, as shown in Figure 9.14. The local

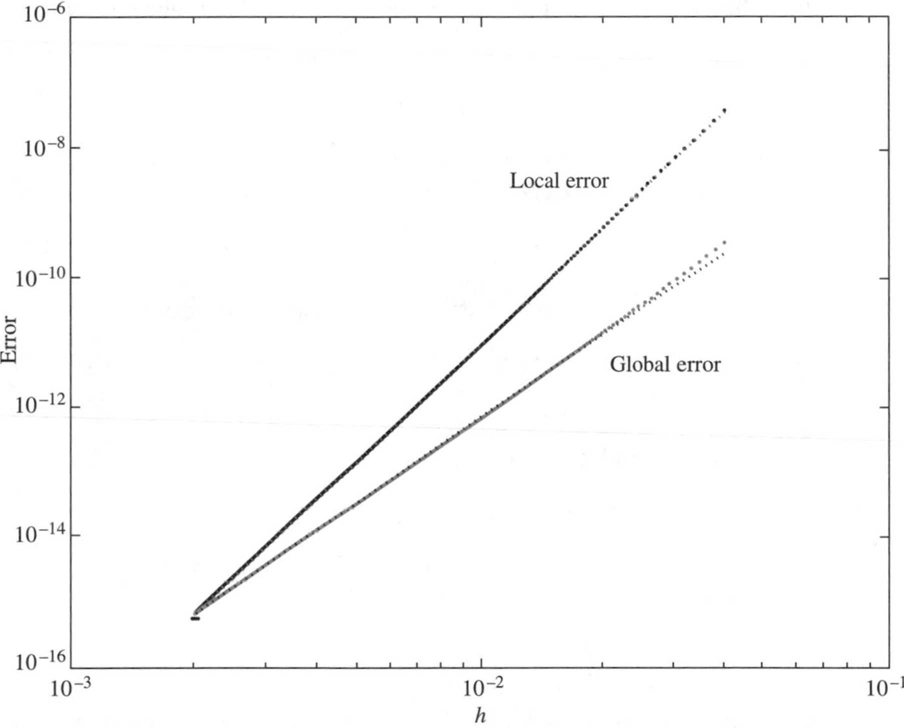

Figure 9.14 Numerical solution of the ODE (9.2) with the fourth-order predictor–corrector pair.

error is defined after the first step at time $t_5 = 5h$, while the global error is defined at the final time T. The power fits are shown by dotted curves, and the powers of the least-square fits are also displayed. The global error is $O(h^4)$ according to the fourth order of the predictor–corrector pair (9.41) and (9.50). The local truncation error is $O(h^6)$, which is smaller than the predicted $O(h^5)$ error. The cancellation of the fifth power of h is a specific property of the particular ODE example (9.2) (see Exercise 9.12).

```
>> predictor_corrector

powerLoc =
    5.9985

powerGlob =
    4.2844
```

Convergence and stability

Next, consider stability of the second-order predictor–corrector pair (9.39) and (9.48) in application to the linear ODE $y' = -\lambda y$. Substitution of $f(y) = -\lambda y$ into (9.39) and (9.48) results in the second-order difference equation

$$y_{k+1} = \left(1 - \lambda h + \frac{3}{4}\lambda^2 h^2\right) y_k - \frac{1}{4}\lambda^2 h^2 y_{k-1}. \tag{9.51}$$

Because the second-order difference equation (9.51) has constant coefficients in k, we can reduce it to the quadratic equation with the substitution $y_k = q^k$:

$$q^2 = (1 - 2z + 3z^2)q - z^2, \tag{9.52}$$

where $z = \lambda h/2$. The quadratic equation admits two roots

$$q_{1,2} = \frac{1 - 2z + 3z^2 \pm \sqrt{(1 - 2z + 3z^2)^2 - 4z^2}}{2}.$$

The two roots $q_1(z)$ and $q_2(z)$ are plotted versus z by the MATLAB commands:

```
>> z = 0 : 0.01 : 1.5;
>> q1 = (1 - 2*z + 3*z.^2 + sqrt((1 - 2*z + 3*z.^2).^2 - 4*z.^2))/2;
>> q2 = (1 - 2*z + 3*z.^2 - sqrt((1 - 2*z + 3*z.^2).^2 - 4*z.^2))/2;
>> plot(z,q1,'r',z,q2,'b')
```

Figure 9.15 shows the behavior of the roots. Two real roots for small z become complex-valued roots for $z_0 < z < 1$, and then reappear as real-valued roots for $z > 1$. Because the complex roots are complex conjugated and $q_1 q_2 = z^2$, we have $|q_1|^2 = z^2 < 1$ for complex roots, and they are always

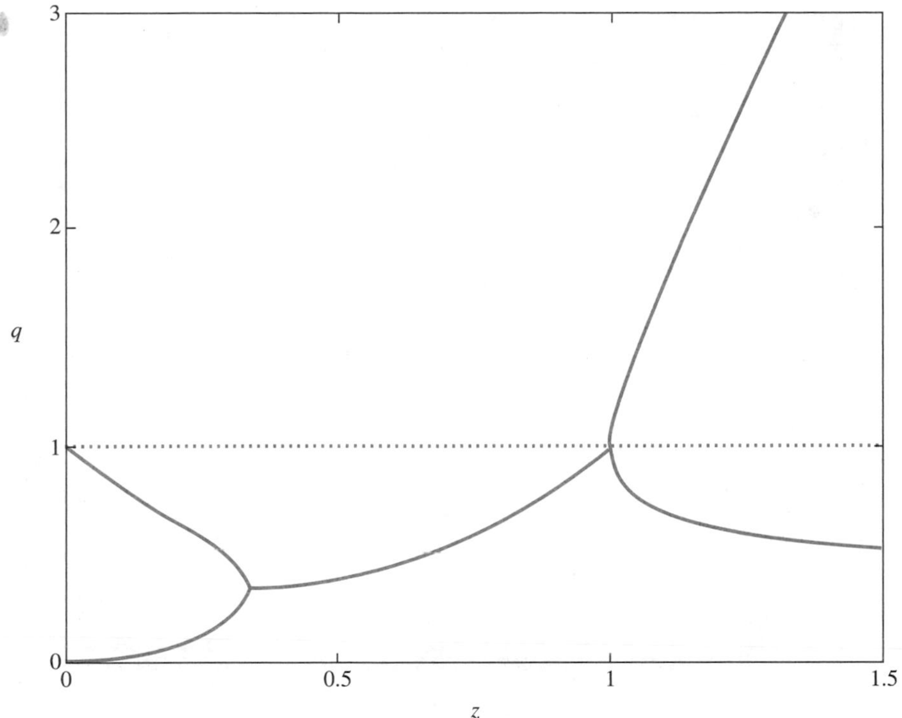

Figure 9.15 Roots of the quadratic equation (9.52) versus $z = \lambda h/2$.

stable. On the other hand, one real root for $z > 1$ is always greater than one, such that the method is unstable for $\lambda h > 2$. The stability domain $h < 2/\lambda$ of the second-order predictor–corrector pair is exactly the same as that of the Heun method (9.20)–(9.21). Because $q_1(z) > 1$ for $z > 1$, the unstable root leads to a monotonic divergence of the sequence $\{y_k\}_{k=0}^{\infty}$, similar to the behavior of the Heun method.

The MATLAB script `instability_implicit_Adams` illustrates the long-term behavior of the second-order predictor–corrector pair (9.39) and (9.48) and the modified Euler method (9.23)–(9.24) for the ODE problem (9.2). Figure 9.16 shows numerical approximations (bright and dark dots) and the exact solution (dashed curve) for a given time step $h = 0.1$. Although the instability of the second-order predictor–corrector pair occurs later than the instability of the explicit Adams method, it occurs earlier than the instabilities of the second-order Heun and modified Euler methods (compare Figures 9.13 and 9.16). You can also see that both divergent sequences are sign-definite in accordance with the preceding analysis.

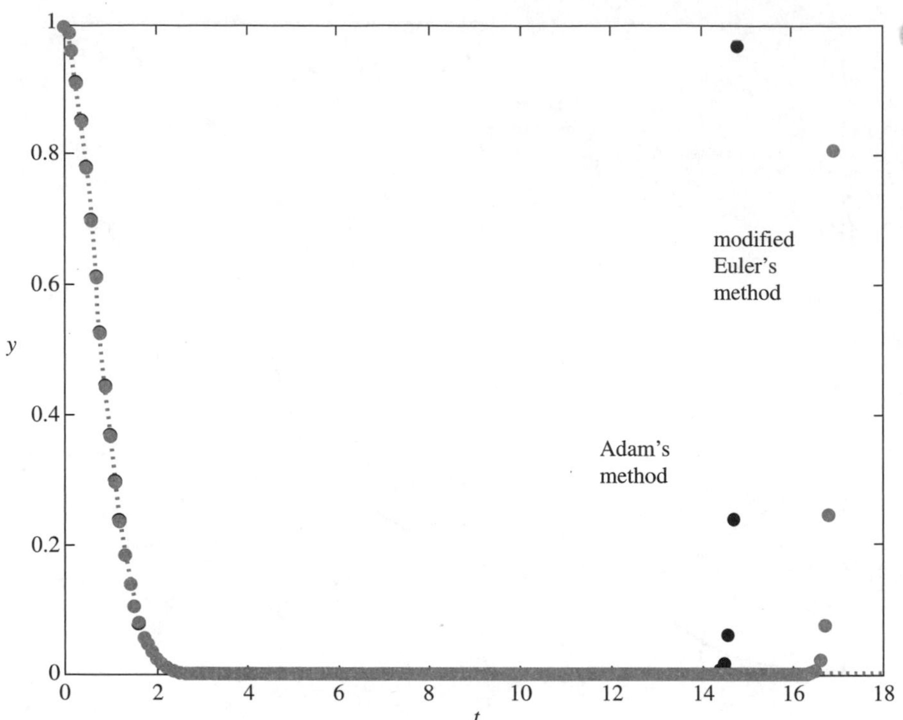

Figure 9.16 Instabilities of the predictor–corrector pair (9.39) and (9.48) and the modified Euler method (9.23)–(9.24) for the ODE problem (9.2) with $h = 0.1$.

```
% instability_implicit_Adams
t(1) = 0; y(1) = 1; yM(1) = 1; T = 18; h = 0.1; n = T/h;
t(2) = t(1) + h;     % first step
y(2) = exp(-t(2)^2); yM(2) = y(2);
for k = 2 : n
    t(k+1) = t(k) + h;
    p = y(k) - h*(3*t(k)*y(k)-t(k-1)*y(k-1));
    y(k+1) = y(k) - h*(t(k)*y(k) + t(k+1)*p);
    p = yM(k) - h*t(k)*yM(k);
    yM(k+1) = yM(k) - 2*h*(t(k)+h/2)*p;
end
yExact = exp(-t.^2);
plot(t,yExact,':r',t,y,'.b',t,yM,'.g');
```

Exercise 9.14 Consider the fourth-order predictor–corrector pair (9.41) and (9.50) and the fourth-order Runge–Kutta method. By applying both methods to the ODE problem (9.3), show numerically that the Adams method gives a smaller truncation error than the Runge–Kutta method does, although it is still of the same order. By applying both methods to the ODE problem $y' = -y$ with $y(0) = 1$, show numerically that the Adams method develops instabilities earlier than the Runge–Kutta method.

Error estimation

Because the time step h is constant in the Adams methods, any adjustment of the time step h in the error control procedure (see Section 9.3) requires computations of numerical approximations of $\mathbf{y}$ and $\mathbf{f}(t, \mathbf{y})$ at intermediate times. It is, however, easy to estimate the local truncation error at each time step of the predictor–corrector pair because the orders of both explicit and implicit Adams methods are taken to be the same.

For instance, the second-order Adams methods for a scalar ODE $y' = f(t, y)$ have the local truncation errors in the form [7]:

$$y(t_{k+1}) = p_{k+1} + \frac{5}{12}\alpha_3 h^3,$$
$$y(t_{k+1}) = y_{k+1} - \frac{1}{12}\tilde{\alpha}_3 h^3,$$

where α_3 and $\tilde{\alpha}_3$ are close to each other for small values of h. Neglecting the difference between the values of α_3 and $\tilde{\alpha}_3$, we obtain the error estimation formula

$$E_k = |y(t_{k+1}) - y_{k+1}| = \frac{1}{6}|y_{k+1} - p_{k+1}|. \tag{9.53}$$

The third-order Adams methods have the local truncation errors in the form [7]:

$$y(t_{k+1}) = p_{k+1} + \frac{3}{8}\alpha_4 h^4,$$
$$y(t_{k+1}) = y_{k+1} - \frac{1}{24}\tilde{\alpha}_4 h^4,$$

where α_4 and $\tilde{\alpha}_4$ are close to each other, such that

$$E_k = |y(t_{k+1}) - y_{k+1}| = \frac{1}{10}|y_{k+1} - p_{k+1}|. \tag{9.54}$$

Similarly, the fourth-order Adams methods have the local truncation errors [7]:

$$y(t_{k+1}) = p_{k+1} + \frac{251}{720}\alpha_5 h^5,$$
$$y(t_{k+1}) = y_{k+1} - \frac{19}{720}\tilde{\alpha}_5 h^5,$$

where α_5 and $\tilde{\alpha}_5$ are close to each other, such that

$$E_k = |y(t_{k+1}) - y_{k+1}| = \frac{19}{270}|y_{k+1} - p_{k+1}|. \tag{9.55}$$

The MATLAB script `error_estimation_Adams` illustrates the error estimation formula (9.53) for the second-order predictor–corrector pair in application to the ODE (9.3). For each time step from t_k to $t_{k+1} = t_k + h$, the exact solution of the ODE (9.3) with $y(t_k) = y_k$ is $y(t_{k+1}) = y_k/(1 - hy_k)$, which gives the exact value of the local error to be compared with the estimation formula (9.53).

```
% error_estimation_Adams
h = 0.01; t(1) = 0; y(1) = 1; yExact(1) = 1;
t(2) = h; y(2) = 1/(1-t(2)); yExact(2) = y(2); % no error for y_0 and y_1
E(1) = 0; E(2) = 0; k = 2;
while (t(length(t)) < 0.9)
    t(k+1) = t(k) + h;
    pp = y(k) + 0.5*h*(3*y(k)^2-y(k-1)^2);
    y(k+1) = y(k) + 0.5*h*(pp^2+y(k)^2);
    E(k+1) = abs(y(k+1)-pp)/6;      % error estimation formula
    yExact(k+1) = y(k)/(1-h*y(k));
    k = k+1;
end
Eexact = abs(yExact-y);                  % exact local error of a single step
semilogy(t,E,'.b',t,Eexact,':r');
```

The MATLAB script `error_estimation_Adams` plots the approximation of the local error by dots and the actual local error by a dashed curve, as shown in Figure 9.17. The graph shows that the error formula (9.53) deviates insignificantly from the actual error and this deviation becomes visible for larger time intervals of computations.

Time step adjustment

When the estimation of E_k exceeds a tolerance level E_{tol}, the time step h needs to be reduced. For instance, it can be halved if $E_k > \frac{3}{2}E_{\text{tol}}$. In this case, approximations of the solution $y(t)$ and the vector field $f(t, y)$ must be found at midpoints $t_{k-1/2}, t_{k-3/2}, \ldots, t_{k-m+1/2}$. To do so, polynomial interpolation of the same order as in the predictor–corrector pair can be used (see Section 5.2). When the estimation of E_k is smaller than a tolerance level E_{tol}, the time step h needs to be increased. If it is increased by a factor of two (for $E_k < \frac{3}{4}E_{\text{tol}}$), each second value of the numerical approximation of $y(t)$ and $f(t, y)$ can be dropped from the table of data values stored in the computer memory.

Exercise 9.15 Apply the fourth-order predictor–corrector pair to the ODE problem (9.3). Compute the estimation of the local truncation error (9.55)

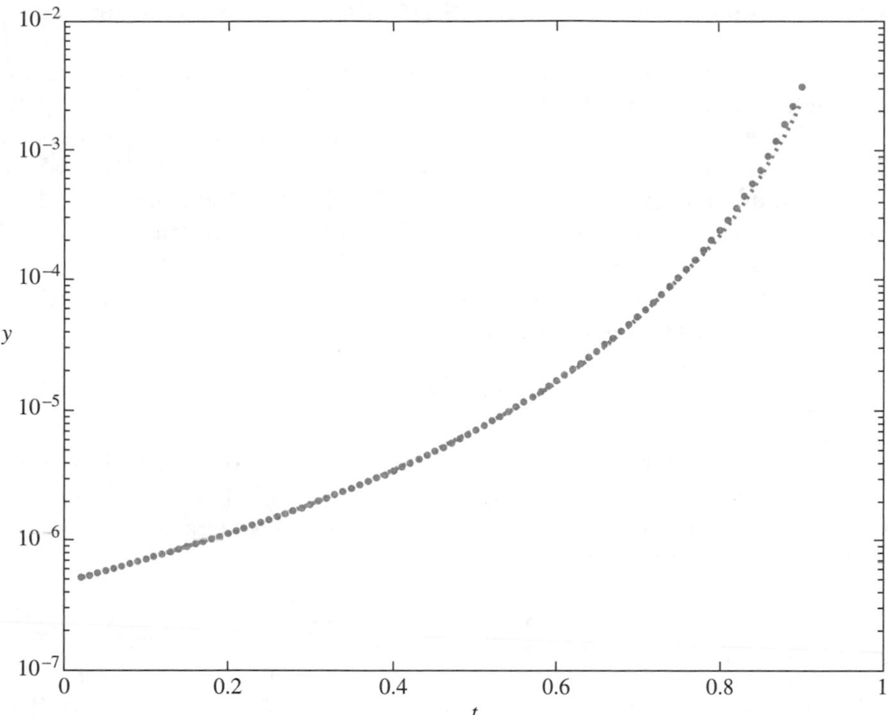

Figure 9.17 Estimation of the local truncation error from the second-order predictor–corrector pair.

and plot it versus time. Modify the script by adding an option to increase or decrease the time step h by the factor of 2 if the estimation of the local error is sufficiently smaller or larger than a given tolerance bound E_{tol}. Plot the dynamically adjusted time step h versus time t.

The MATLAB function `ode113` implements the Adams ODE solvers of a variable order from the 1st order to the 13th order. Details of the use of the MATLAB function `ode113` are the same as for the MATLAB functions `ode23` and `ode45` described in Section 9.3.

Exercise 9.16 Apply the MATLAB function `ode113` to the ODE problem (9.3) and plot the numerical approximation, the global truncation error, and the time step versus time. Compare the difference between the outputs of the ODE solvers `ode113` and `ode45` (see MATLAB script `MATLAB_ODE`).

9.5 Implicit Methods for Stiff Differential Equations

In Sections 9.2 and 9.4, we were concerned with the stability constraint for the single-step and multistep ODE solvers applied to the scalar linear ODE $y' = -\lambda y$, where $\lambda > 0$. You have seen that the time step of the explicit ODE solvers has to be sufficiently small in the interval $h \leq C/\lambda$, where C is constant. For instance, $C = 2$ for the Euler and Heun methods (9.8) and (9.20)–(9.21), $C = 2.785$ for the fourth-order Runge–Kutta method, and $C = 1$ for the explicit second-order Adams method (9.39).

Stiff ODE systems

The scalar ODE $y' = -\lambda y$ represents a normal mode of a general system of linear ODEs with constant coefficients

$$\frac{d\mathbf{y}}{dt} = \mathbf{A}\mathbf{y}, \qquad \mathbf{y} \in \mathbb{R}^n, \tag{9.56}$$

where $\mathbf{A}$ is a constant $n \times n$ matrix. If $\mathbf{A}$ is a diagonalizable matrix, then its eigenvectors $\{\mathbf{x}_1, \dots, \mathbf{x}_n\}$ for eigenvalues $\{\lambda_1, \dots, \lambda_n\}$ form a basis in $\mathbb{R}^n$ according to the representation (4.28). In this case, a general solution of the ODE system (9.56) takes the form

$$\mathbf{y}(t) = \zeta_1 \mathbf{x}_1 e^{\lambda_1 t} + \dots + \zeta_n \mathbf{x}_n e^{\lambda_n t}, \tag{9.57}$$

where $\zeta_1, \dots, \zeta_n$ are parameters of the linear combination. Each normal coordinates $z_j(t) = \zeta_j e^{\lambda_j t}$ solves the scalar ODE $\dot{z}_j = \lambda_j z_j$, $j = 1, \dots, n$. The linear system is *asymptotically* stable if all eigenvalues λ_j are negative (or have negative real part) so that $\lim_{t\to\infty} \mathbf{y}(t) = \mathbf{0}$ for any initial condition $\mathbf{y}(0) = \mathbf{y}_0 \in \mathbb{R}^n$. The stable linear ODE system (9.56) with large differences in eigenvalues $|\lambda_1|,\dots,|\lambda_n|$ is called a *stiff* ODE system.

For instance, consider a stiff two-component system in normal coordinates:

$$\dot{z}_1 = -\lambda_1 z_1, \quad \dot{z}_2 = -\lambda_2 z_2,$$

where $0 < \lambda_1 \ll \lambda_2$. Both components $z_1(t)$ and $z_2(t)$ decay to zero as $t \to \infty$, but the component $z_1(t)$ decays to zero much slower than the component $z_2(t)$ does. If the time step h satisfies the constraint $h < C/\lambda_2 \ll C/\lambda_1$, where C is defined earlier, then the stability constraints are satisfied for both components. However, it will take too many iterations to observe for any changes in the component $z_1(t)$. The accumulated errors grow with the number of iterations and may destroy validity of the numerical approximation. On the other hand, if the time step is sufficiently large to resolve the decay of the component $z_1(t)$, it may violate stability of iterations of the component $z_2(t)$, such that $C/\lambda_2 < h < C/\lambda_1$. In this case, the component $z_2(t)$ will grow because of numerical instability and will quickly destroy the validity of the numerical approximation. Therefore, stiff ODE systems cannot be solved with iterations of the explicit single-step or multistep ODE solvers, such as the Runge–Kutta and Adams methods.

The classical example of a stiff nonlinear ODE is the *Van-der-Pol equation*:

$$\ddot{y} - \mu(1 - y^2)\dot{y} + y = 0, \tag{9.58}$$

where $\mu > 0$ is a parameter. The scalar Van-der-Pol equation (9.58) can be rewritten in the standard vector form as follows:

$$\frac{d}{dt}\begin{pmatrix} y_1 \\ y_2 \end{pmatrix} = \begin{pmatrix} y_2 \\ \mu(1 - y_1^2)y_2 - y_1 \end{pmatrix}, \tag{9.59}$$

where $y_1 = y$ and $y_2 = \dot{y}$. The origin $(y_1, y_2) = (0, 0)$ is a unique critical point of the ODE system (9.59), where the vector field is zero. Linearization of the ODE system at the critical point $(0, 0)$ yields the linear ODE system (9.56) with $n = 2$ and

$$\mathbf{A} = \begin{pmatrix} 0 & 1 \\ -1 & \mu \end{pmatrix}.$$

The matrix $\mathbf{A}$ has two eigenvalues

$$\lambda_{\pm} = \frac{\mu \pm \sqrt{\mu^2 - 4}}{2},$$

which are also roots of the characteristic equation $\lambda^2 - \mu\lambda + 1 = 0$. When $|\mu| > 2$ and $|\mu|$ is large, both eigenvalues are real (either positive or negative) and the linear ODE system is stiff. Indeed, when μ is large, we can derive that $\lambda_+ = O(\mu)$ and $\lambda_- = O(\frac{1}{\mu})$ and the difference between λ_+ and λ_- grows for large values of μ.

The MATLAB script `explicit_vanderpol` computes two numerical solutions of the ODE system (9.59) with the explicit Euler method (9.8) for $h = 0.02$ and $h = 0.04$.

```
% explicit_vanderpol
t(1) = 0; y(1) = 0.1; z(1) = 0; mu = 10; T = 50;
hSpan = 0.02 : 0.02 : 0.04; % computations with h = 0.02 and h = 0.04
for j = 1 : 2
    h = hSpan(j); n = T/h;
    for k = 1 : n
        t(k+1) = t(k) + h;
        y(k+1) = y(k) + h*z(k);
        z(k+1) = z(k) + h*(mu*(1-y(k)^2)*z(k)-y(k));
    end
    if j == 1
        figure(1); plot(t,y,'.b'); hold on;
        figure(2); plot(y,z,'.r');
```

```
        else
            figure(1); plot(t,y,'.g'); hold off;
        end
    end
```

When $\mu = 10$, the Van-der-Pol system (9.59) is stiff with two eigenvalues $\lambda_+ \approx 10$ and $\lambda_- \approx 0.1$. Although both time steps fit into the stability interval $h \leq 2/\lambda_+$, Figure 9.18 (*left*) shows that the numerical solution with $h = 0.04$ is unstable on the time interval $[0, 50]$ and blows beyond the scale of the figure. The solution with $h = 0.02$ is stable at least on the computational interval [0,50]. Figure 9.18 (*right*) shows the phase plane (y_1, y_2) for the numerical approximation with $h = 0.02$. The behavior of the solution displays an alternating sequence of fast relaxations and slow evolutions, which is a typical feature of stiff systems.

Implicit methods

To deal with stiff ODE systems, we consider the implicit versions of the same single-step and multistep ODE solvers discussed in Sections 9.2 and 9.4. The ODE solver is said to be *implicit* if the vector field $\mathbf{f}(t, \mathbf{y})$ that defines the value of $\mathbf{y}_{k+1}$ is computed at t_{k+1} with $\mathbf{y}_{k+1}$.

The family of implicit multistep Adams methods is defined in Problem 9.4. Although the family is used in Section 9.4 in the context of the explicit predictor–corrector pair, it can be considered on its own, without additional simplifications. For instance, the implicit first-order Adams method (9.47) can be rewritten as

$$\mathbf{y}_{k+1} = \mathbf{y}_k + h\mathbf{f}(t_{k+1}, \mathbf{y}_{k+1}), \tag{9.60}$$

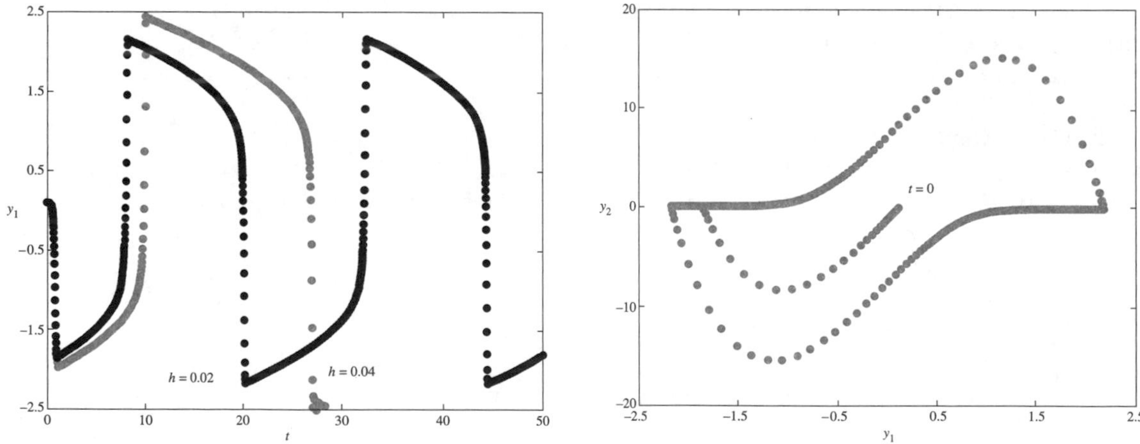

Figure 9.18 Numerical solution of the ODE system (9.59) with the explicit Euler method (9.8): $y_1 = y$ versus t (*left*) and phase plane $(y_1, y_2) = (y, \dot{y})$ (*right*).

which is equivalent to the implicit Euler method. Similar to the Euler method (9.8), it computes a numerical solution of the ODE $\mathbf{y}'(t) = \mathbf{f}(t, \mathbf{y}(t))$ by replacing the first derivative $\mathbf{y}'(t_k)$ with the backward difference approximation

$$D_1 \mathbf{y}(t_{k+1}) = \frac{\mathbf{y}_{k+1} - \mathbf{y}_k}{t_{k+1} - t_k}.$$

Because the truncation error of the implicit Euler method (9.60) is minus times the error of the explicit Euler method (9.8), convergence of both methods is the same in the limit of small h. Both methods are first order with respect to the global truncation error (see Section 9.1). On the other hand, the stability properties of these methods are different for finite values of h. Although the explicit Euler method can only produce a decaying solution to the ODE $y' = -\lambda y$ with $\lambda > 0$ in the interval $h < 2/\lambda$, the implicit Euler method produces such solutions for any $h > 0$. In this case, we say that the implicit Euler method is *unconditionally stable.* Indeed, substituting the vector field $f(y) = -\lambda y$ to the implicit Euler method (9.60), we obtain the first-order difference equation

$$y_{k+1} = \frac{y_k}{1 + \lambda h} = \left(1 - \lambda h + \mathrm{O}(h^2)\right) y_k,$$

from which it follows that the sequence $\{y_k\}_{k \in \mathbb{N}}$ is monotonically decreasing to zero as $k \to \infty$ for any $\lambda h > 0$. Similarly, the implicit second-order Adams method (9.48) can be rewritten as

$$\mathbf{y}_{k+1} = \mathbf{y}_k + \frac{1}{2} h \left(\mathbf{f}(t_k, \mathbf{y}_k) + \mathbf{f}(t_{k+1}, \mathbf{y}_{k+1})\right), \qquad (9.61)$$

which coincides with the implicit Heun method (9.15). It follows from the derivation of the implicit Heun method from the trapezoidal integration rule that the global truncation error has the second order (see Section 9.2). The stability properties of the implicit Heun method are again different from stability properties of the explicit Heun method. Although the explicit method is stable in the interval $h \leq 2/\lambda$, the implicit Heun method is unconditionally stable for any $h > 0$. Indeed, substituting the vector field $f(y) = -\lambda y$ in the implicit Heun method (9.61), we obtain the first-order difference equation

$$y_{k+1} = \frac{2 - \lambda h}{2 + \lambda h} y_k = \left(1 - \lambda h + \frac{1}{2}\lambda^2 h^2 + \mathrm{O}(h^3)\right) y_k,$$

from which it follows that the sequence $\{y_k\}_{k \in \mathbb{N}}$ is monotonically decreasing to zero as $k \to \infty$ for any $\lambda h > 0$.

We illustrate iterations of the implicit Euler and Heun methods in application to the ODE (9.2). Because the vector field $f(t, y) = -2ty$ has the *separable* form, the implicit methods can be rewritten as the explicit first-order difference equations. The MATLAB script `stability_implicit` computes numerical approximations of the solution.

```
% stability_implicit
t(1) = 0; yE(1) = 1; yH(1) = 1;
T = 18; h = 0.1; n = T/h;
for k = 1 : n   % iterations of two implicit methods
    t(k+1) = t(k) + h;
    yE(k+1) = yE(k)/(1+2*h*t(k+1));
    yH(k+1) = yH(k)*(1-h*t(k))/(1+h*t(k+1));
end
yExact = exp(-t.^2);
plot(t,yExact,':r',t,yE,'.b',t,yH,'.g');
errorE = max(abs(yE-yExact)), errorH = max(abs(yH-yExact))
```

The MATLAB script `stability_implicit` outputs the absolute values of the errors, which shows that the error of the implicit Heun method is one order smaller than that of the implicit Euler method, in agreement with the previous analysis. Iterations of both implicit methods are shown by black and colored dots in Figure 9.19. Both sequences converge to zero, similar to the exact solution $y(t) = e^{-t^2}$ shown by the dotted curve.

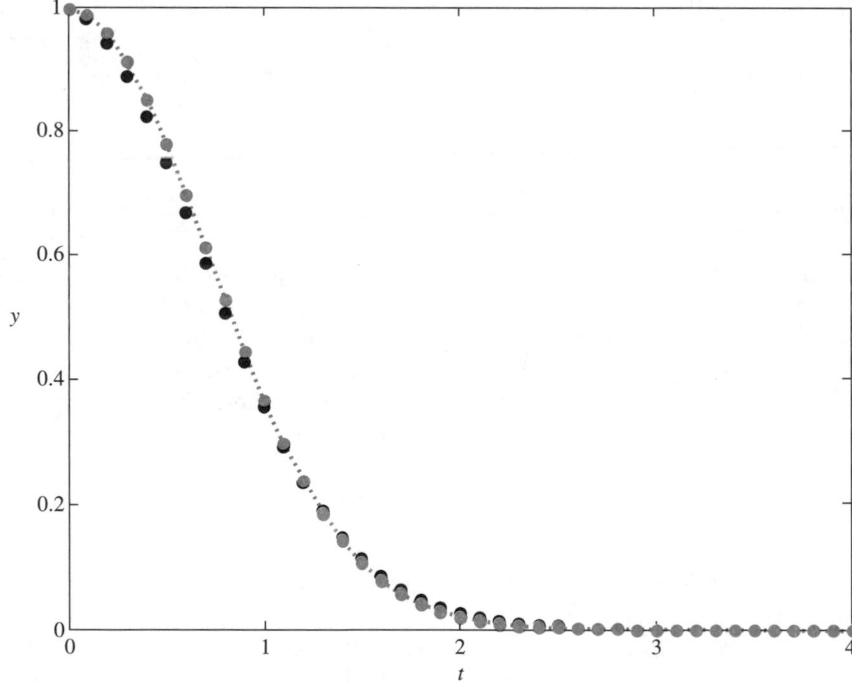

Figure 9.19 Numerical solution of the ODE system (9.2) with the implicit Euler and Heun methods (9.60) and (9.61).

```
>> stability_implicit

errorE =
    0.0302

errorH =
    0.0013
```

We can develop the idea of implicit ODE solvers beyond the simplest implicit Euler and Heun methods. For the family of multistep Adams methods in Problem 9.4, each nth-order method is implicit by the construction and no additional modifications are necessary. For the family of single-step Runge–Kutta methods in Problem 9.2, we can extend summations (9.18) to $i = j$, such that computations of $\mathbf{p}_k^{(j)}$ would require computations of $\mathbf{f}(t_k^{(j)}, \mathbf{p}_k^{(j)})$ for $j = 1, 2, \ldots, n$.

Exercise 9.17 Apply the implicit versions of the third-order Runge–Kutta and Adams methods to the ODE (9.2) and show that no constraints on stability of iterations occur in either method.

Semi-implicit methods

Although it is easy to rewrite the implicit ODE solver for a linear ODE as an explicit difference equation, many ODEs include nonlinear terms, for example, quadratic and cubic polynomials of the dependent variable $\mathbf{y}$. If the implicit methods cannot be simplified, every time step would require iterations of the root-finding algorithm. Bifurcations are then possible when iterations of the root-finding algorithm at the next time step converge to a different root compared to that at the previous time step. In many cases, *semi-implicit* methods allow us to avoid the necessity of using the root-finding algorithms. Because the order of the semi-implicit method must correspond to the order of the original implicit method, we shall discuss this modification only for the simplest case of the implicit Euler method (9.60). Let $\mathbf{v}_k = \mathbf{y}_{k+1} - \mathbf{y}_k$ and assume that the vector field $\mathbf{f}(t, \mathbf{y})$ is twice continuously differentiable in $\mathbf{y}$. Then,

$$\mathbf{f}(t_{k+1}, \mathbf{y}_{k+1}) = \mathbf{f}(t_{k+1}, \mathbf{y}_k) + D_{\mathbf{y}}\mathbf{f}(t_{k+1}, \mathbf{y}_k)\mathbf{v}_k + \mathrm{O}(\|\mathbf{v}_k\|^2), \tag{9.62}$$

where $D_{\mathbf{y}}\mathbf{f}(t_{k+1}, \mathbf{y}_k)$ is the Jacobian matrix for the vector field $\mathbf{f}(t, \mathbf{y})$. If the norm $\|\mathbf{v}_k\|$ is $\mathrm{O}(h)$, then the truncation error of the expansion (9.62) is comparable with the $\mathrm{O}(h^2)$ local truncation error of the implicit Euler method. Combining (9.60) and (9.62), we can rewrite the implicit Euler method as the semi-implicit Euler method:

$$(I - hD_{\mathbf{y}}\mathbf{f}(t_{k+1}, \mathbf{y}_k))\,(\mathbf{y}_{k+1} - \mathbf{y}_k) = h\mathbf{f}(t_{k+1}, \mathbf{y}_k), \tag{9.63}$$

where I is the identity matrix. The convergence properties of the implicit Euler method (9.60) are preserved in the semi-implicit Euler method (9.63). The unconditional stability of the implicit method is also preserved for the semi-implicit Euler method.

When the Van-der-Pol system (9.59) is considered, you can compute

$$\mathbf{f}(t,\mathbf{y}) = \begin{bmatrix} y_2 \\ \mu(1-y_1^2)y_2 - y_1 \end{bmatrix}, \qquad D_{\mathbf{y}}\mathbf{f}(t,\mathbf{y}) = \begin{bmatrix} 0 & 1 \\ -1-2\mu y_1 y_2 & \mu(1-y_1^2) \end{bmatrix},$$

such that the semi-implicit Euler method (9.63) leads to the linear system:

$$\begin{aligned} y_{1,k+1} - hy_{2,k+1} &= y_{1,k} \\ h(1+2\mu y_{1,k}y_{2,k})y_{1,k+1} + (1-h\mu+h\mu y_{1,k}^2)y_{2,k+1} &= y_{2,k} + 2h\mu y_{1,k}^2 y_{2,k}. \end{aligned}$$

The MATLAB script `implicit_vanderpol` computes the numerical solution of the ODE system (9.59) by solving the preceding linear system with $h = 0.04$ and $\mu = 10$. Figure 9.20 shows that the numerical solution is stable on the time interval $[0, 50]$ (compare with the solution of the explicit Euler method in Figure 9.18).

```
% implicit_vanderpol
t(1) = 0; y(1) = 0.1; z(1) = 0; mu = 10;
T = 50; h = 0.04; n =T/h;
for k = 1 : n  % iterations of the semi-implicit Euler method
```

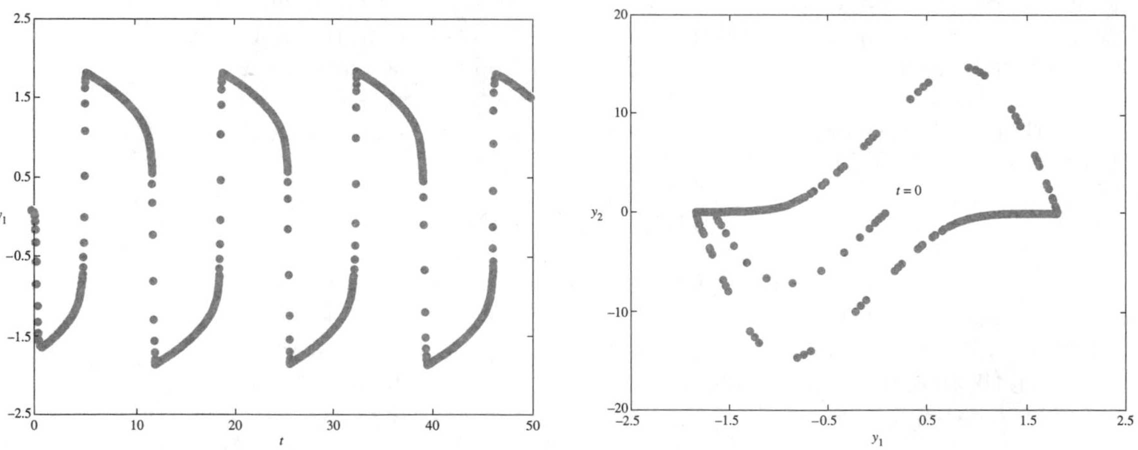

Figure 9.20 Numerical solution of the ODE system (9.59) with the semi-implicit Euler method (9.63).

```
    t(k+1) = t(k) + h;
    A = [1, -h ; h*(1+2*mu*y(k)*z(k)), (1-h*mu+h*mu*y(k)^2)];
    b = [y(k); z(k) + 2*h*mu*y(k)^2*z(k)];
    x = A\b; y(k+1) = x(1); z(k+1) = x(2);
end
figure(1); plot(t,y,'.b');
figure(2); plot(y,z,'.r');
```

The MATLAB functions `ode15s` and `ode23s` implement the implicit ODE solvers by using two different ideas. The function `ode15s` is based on backward differentiation of a variable order ranging from the first to the fifth orders. The function `ode23s` is based on the trapezoidal rule of the second and third orders. In some sense, the former solver generalizes the implicit Adams methods, whereas the latter solver generalizes the implicit Runge–Kutta methods. Implementations of these two MATLAB ODE functions are the same as those for the other MATLAB ODE functions described in Section 9.3.

Exercise 9.18 Apply the MATLAB functions `ode15s` and `ode23s` to the ODE system (9.59) and plot the numerical approximations on the phase plane. Compare the outputs of the ODE solvers `ode15s` and `ode23s` and the output of the MATLAB script `implicit_vanderPol`.

More general families of the ODE solvers can be designed beyond the single-step Runge–Kutta and multistep Adams methods. For instance, any polynomial interpolation of the function $\mathbf{f}(t, \mathbf{y}(t))$ in the integral form (9.14) through the points t_{k+1}, t_k, ..., t_{k-m} as well as through the intermediate points on the interval $[t_{k-m}, t_{k+1}]$ can lead to a new ODE solver. Similarly, any backward difference approximation of the higher-order derivatives of the Taylor series for the values of $\mathbf{f}(t_{k+1}, \mathbf{y}_{k+1})$ in (9.46) involving intermediate points will return a new ODE solver. These two approaches and their various mixtures were used for derivation of other ODE solvers, which are not mentioned in this book (see [6, 7] for other examples of the ODE solvers).

9.6 Summary and Notes

In this chapter, we study numerical approximations of solutions of the initial-value problems for ordinary differential equations:

- **Section 9.1:** We formulate the sufficient condition (Theorem 9.1) that guarantees existence of local solutions of the initial-value problem for ordinary differential equations (Problem 9.1). We explain how the solution to this problem can be approximated on a discrete set of time instances using the most elementary ODE solver, the Euler method. Convergence of the Euler method relies on the Taylor series (Theorem 9.2) and allows us to classify the Euler method as the first-order method in terms of its global truncation error. We identify a stability constraint on the time step of the Euler method for a linear ODE with constant coefficients.

- **Section 9.2:** Improvements of the Euler method are classified by transforming the differential equation to the equivalent integral form. Two improved methods follow from the trapezoidal and midpoint integration rules, namely, the Heun and modified Euler methods. A general family of the single-step Runge–Kutta ODE solvers is defined in Problem 9.2. We identify the errors and the stability constraints of the second-order, third-order, and fourth-order Runge–Kutta methods. We show how to apply the Runge–Kutta methods to the ODE systems with several components.

- **Section 9.3:** We show that all single-step methods can be made adaptive by estimating the numerical error and adjusting the time step accordingly. A particularly popular version of the adaptive Runge–Kutta methods is based on the combination of the fourth-order and fifth-order methods. Adaptive MATLAB ODE solvers are also discussed.

- **Section 9.4:** Two families of the multistep ODE solvers are constructed from the Taylor series and backward approximations of the numerical derivatives. The family of explicit Adams–Bashforth methods solves Problem 9.3, whereas the family of implicit Adams–Moulton methods solves Problem 9.4. These two families are used in the predictor–corrector pairs of the same order. We establish stability and convergence of the second-order pair from the roots of quadratic equations. Adaptive multistep methods are developed similarly to the adaptive single-step methods with the error estimation and the time step adjustment techniques.

- **Section 9.5:** We show how the stability constraint of explicit single-step and multistep methods can lead to the ill-posed numerical problem related to stiff ODE systems. Implicit methods with their unconditional stability are applied to the stiff linear systems to overcome this numerical difficulty. The stiff nonlinear systems can be approximated by an explicit modification of an implicit method called the semi-implicit method.

Existence, uniqueness, and stability of solutions of initial-value problems for ordinary differential equations are studied in courses in *Differential Equations* (the suggested text is [22]). Algorithms and truncation errors of the numerical approximations of solutions of ODEs are covered in *Numerical Analysis* courses (the suggested text is [7]). Advanced integrations methods that involve spectral and finite-element methods are covered in [9] (see also Chapters 11 and 12 of this text).

9.7 Exercises

1. Solve analytically the initial-value ODE problem

$$y' = \frac{1}{1+t^2} - 2y^2, \qquad y(0) = 0.$$

Approximate the solution with the Euler method on $[0, 10]$ with time steps $h = 0.1$ and $h = 0.01$. Plot the exact and numerical solutions on the same graph and the computational errors on the other graph, as a function of time.

2. Compute the value of $y(1)$ from the initial-value problem

$$y' = ty^2, \qquad y(0) = 1$$

using the Euler method with the time step h. Apply the Richardson extrapolation method and improve the numerical approximations by reductions of the time step by a factor of two. Plot the numerical error of the approximation of $y(1)$ versus the number of iterations of the Richardson algorithm.

3. Approximate the solution of the initial-value ODE problem

$$y' = (3t^2 - 2t + 1)y, \qquad y(0) = 1$$

with the Euler and Heun methods on $[0, 2]$ with the time step $h = 0.1$. Plot the distance between exact and numerical solutions versus t. Using two computations of $y(2)$ with $h = 0.1$ and $h = 0.05$, confirm that the global error of the Euler method is $O(h)$, whereas that of the Heun method is $O(h^2)$.

4. Approximate the solution of the initial-value ODE problem

$$y' = t^2 - y, \qquad y(0) = 2$$

with the modified Euler method on $[0, 1]$ with the time steps $h = 0.1$ and $h = 0.05$. Plot the distance between exact and numerical solutions versus t. Confirm that the ratio of the global errors at $t = 1$ is close to $2^2 = 4$.

5. Consider the second-order finite-difference approximations:

$$y'(t_k) \approx \frac{y(t_{k+1}) - y(t_{k-1})}{2h},$$

$$y'(t_{k-1}) \approx \frac{-y(t_{k+1}) + 4y(t_k) - 3y(t_{k-1})}{2h},$$

$$y'(t_{k+1}) \approx \frac{3y(t_{k+1}) - 4y(t_k) + y(t_{k-1})}{2h}.$$

For each approximation, develop a numerical method to solve the initial-value problem

$$y'(t) = y(t - y), \qquad y(0) = 1$$

and specify if (i) the method is single-step or multistep, (ii) the method is explicit or implicit, (iii) the method is stable or not. Plot the three approximations of the solution on the interval $[0, 2]$ and the exact solution on the same graph.

6. Consider the nonlinear oscillator equation

$$y'' + 2y - 3y^3 + y^5 = 0,$$

for three initial values $(y, y') = (0.5, 0)$, $(y, y') = (1.5, 0)$, and $(y, y') = (2.5, 0)$. Approximate the three numerical solutions with the fourth-order Runge–Kutta method and plot the solutions on the phase plane (y, y').

7. Repeat the previous exercise by using the central difference approximation

$$y''(t_k) \approx \frac{y(t_{k+1}) - 2y(t_k) + y(t_{k-1})}{h^2}.$$

Characterize convergence and stability of the corresponding ODE solver.

8. Approximate the solution of the initial-value ODE problem

$$y' = \frac{\sin t}{t} y, \qquad y(0) = 1$$

with the second-order, third-order, and fourth-order Runge–Kutta methods on $[0, 10]$ with time step $h = 0.1$. Plot the computation error of all three methods versus time on the same graph. Repeat the exercise with time step $h = 0.01$ and confirm that the global error at $t = 1$ reduces as 10^p, where p is the order of the method.

9. Use the explicit fourth-order Adams–Bashforth method for the numerical solution of the initial-value ODE problem

$$y' = 40 - 20y, \qquad y(0) = 0,$$

on the time interval $[0, 10]$. Use the sufficiently small time step h inside the stability constraint $0 < h < h_0$.

10. Repeat the previous exercise by using the implicit fourth-order Adams–Milton method and confirm the unconditional stability of the implicit method.

11. Derive the two-step explicit and implicit Adams methods with different step sizes $h_1 = t_1 - t_0$ and $h_2 = t_2 - t_1$. Study convergence of the new two-step Adams methods as $h = \sqrt{h_1^2 + h_2^2} \to 0$. Study stability of these methods for the linear ODE $y' = -\lambda y$.

12. Use the second-order predictor–corrector pair based on Adams methods and approximate the solution of the Lotka–Volterra system

$$\dot{y}_1 = 4y_1 - y_1 y_2,$$
$$\dot{y}_2 = -2y_2 + y_1 y_2$$

on $[0, 10]$ starting with $y_1(0) = 1$ and $y_2(0) = 4$. Plot the trajectory on the phase plane (y_1, y_2) with the time step $h = 0.1$. Repeat computations with $h = 0.05$ and $h = 0.025$ and plot the distance between two successive approximations of $y_1(t)$ versus t.

13. Use the error estimation and time step adjustment algorithms for the numerical solution of the initial-value problem

$$y' = 3t^2 y^2, \qquad y(0) = 1$$

with the modified Euler method on $[0, 1)$ and initial step $h = 0.1$. Plot the time step h versus time t. Terminate the algorithm when the time step h becomes smaller than 10^{-10}.

14. Repeat the previous exercise by using the second-order predictor–corrector pair based on Adams methods.

15. Find the exact solution of the third-order ODE

$$y''' + 4y'' + 5y' + 2y = 2t^2 + 10t + 8$$

starting with the initial values $y(0) = 1$, $y'(0) = -1$, and $y''(0) = 3$. Approximate the solution with the MATLAB ODE solvers `ode23`, `ode45`, `ode113`, and `ode15s`. Plot the numerical and exact solutions on the same graph. Plot the time step h versus time t for all approximations on the same graph.

16. Analyze stability of the explicit method

$$y_{k+1} = 4y_k - 3y_{k-1} - 2hf(t_{k-1}, y_{k-1})$$

for $f(y) = -\lambda y$. Apply the method to the initial-value ODE problem

$$y' = -y \log y, \qquad y(0) = 2$$

on the interval $[0, 3]$ with different time steps h and investigate the stability of the explicit method.

17. Analyze the stability of the implicit method known as Milne's method

$$y_{k+1} = y_{k-1} + \frac{h}{3}\left[f(t_{k-1}, y_{k-1}) + 4f(t_k, y_k) + f(t_{k+1}, y_{k+1})\right],$$

for $f(y) = -\lambda y$. Apply the method to the initial-value ODE problem

$$y' = y(3 - 4y + y^2), \qquad y(0) = 2$$

on the interval $[0, 5]$ with different time steps h and investigate the stability of Milne's method.

18. Consider the initial-value ODE problem

$$y' = -\lambda y + \frac{1}{1+t^2}, \qquad y(0) = 0$$

with $\lambda = -1$, $\lambda = -10$, and $\lambda = -100$. Apply the explicit and implicit Euler methods and the explicit and implicit Heun methods for approximation of the solution on $[0, 3]$ with time steps $h = 0.1$ and $h = 0.01$. Plot the exact and numerical solutions on the same graph and separate unstable and stable approximations. Plot the computational errors for the stable approximations on the same graph.

19. Use the semi-implicit Euler method for a numerical solution of the stiff system of differential equations

$$\begin{aligned} x' &= -x + xy, \qquad x(0) = 3 \\ y' &= x + 2x^2 - 10y, \qquad y(0) = 5 \end{aligned}$$

on the time interval $[0, 2]$ with time steps $h = 0.01$ and $h = 0.1$. Plot the components $x(t)$ and $y(t)$ of the numerical solutions on separate graphs.

20. Apply the implicit Heun method to the nonlinear ODE

$$y' = \cos^2 y - t^2, \qquad y(0) = 0,$$

using the root-finding algorithm at each time step. Approximate the solution on $[0, 3]$ with time steps $h = 0.1$ and $h = 0.2$. Plot the numerical solutions and confirm the second-order convergence and unconditional stability of the implicit Heun method.

CHAPTER 10

Boundary-Value Problems for ODEs and PDEs

WHEREAS SOLUTIONS OF initial-value problems for ordinary differential equations (ODEs) exist whenever the vector field for the differential equation is sufficiently smooth, boundary-value problems may have no solutions if the boundary conditions are inconsistent. Nevertheless, many boundary-value problems arise in the context of physical and engineering problems where the existence of solutions is suggested by the governing laws of physics. Although the proof of existence of solutions can be complicated, it is legitimate to develop numerical approximations and graphical visualizations of the solutions before analyzing the properties of the given boundary-value problem, consistent with general strategy adopted in our numerical laboratory.

Many specific results on solutions of boundary-value problems for ordinary and partial differential equations (PDEs) cannot be formulated as general theorems. Moreover, no general numerical recipe is available for numerical solutions of nonlinear differential equations. Experience and knowledge of various numerical routines are valuable assets in the design of new numerical algorithms for solutions of boundary-value problems.

This chapter covers the simplest numerical approximations of solutions of boundary-value problems associated with ODEs and PDEs, as well as convergence, stability, and robustness of the numerical algorithms.

10.1 Finite-Difference Methods for ODEs

A boundary-value problem for an ODE consists of a system of differential equations of the form

$$\frac{d\mathbf{y}}{dx} = \mathbf{f}(x, \mathbf{y}), \qquad x \in [a, b], \tag{10.1}$$

where $\mathbf{y} : \mathbb{R} \to \mathbb{R}^d$ is a vector-valued function satisfying certain conditions at the endpoints of the interval $[a, b]$, and $\mathbf{f} : \mathbb{R} \times \mathbb{R}^d \to \mathbb{R}^d$ is a vector field. When the condition is of the form $\mathbf{y}(a) = \mathbf{y}_a$ without any other specification at the point $x = b$, this reduces to the initial-value problem. We obtain a boundary-value problem when some components of $\mathbf{y}(a)$ are not specified, while the corresponding components are specified at $\mathbf{y}(b)$. This apparently innocuous modification, however, alters the character of the problem in much deeper ways. For instance, no result as general as Theorem 9.1 is available for this boundary-value problem.

For this reason, we focus on the following simplified problem.

Problem 10.1 (Boundary-Value Second-Order ODE Problem) Find a twice continuously differentiable function $y(x)$ on the interval $[a, b]$ that solves the second-order ODE

$$y'' = F(x, y, y'), \qquad a < x < b, \tag{10.2}$$

subject to the boundary conditions

$$y(a) = \alpha, \qquad y(b) = \beta, \tag{10.3}$$

where $F(x, y, y')$ is continuous in x and continuously differentiable in (y, y') and (α, β) are constants.

Conditions of the form (10.3), that is, given in terms of the function $y(x)$ itself, are called *Dirichlet boundary conditions.* By contrast, a modified version of the preceding problem consists of imposing conditions on the derivative function $y'(x)$, that is,

$$y'(a) = \gamma, \qquad y'(b) = \delta, \tag{10.4}$$

which are called *Neumann boundary conditions.* In some cases, a boundary-value problem can be specified by a combination of Dirichlet and Neumann boundary conditions.

It is easy to see how the scalar second-order differential equation (10.2) can be viewed as a special case of the vector-valued first-order differential equation (10.1). All we need to do is define the vector

$$\mathbf{y}(x) = \begin{bmatrix} y_1(x) \\ y_2(x) \end{bmatrix} := \begin{bmatrix} y(x) \\ y'(x) \end{bmatrix} \tag{10.5}$$

and consider the problem

$$\begin{aligned} y_1'(x) &= y_2(x) \\ y_2'(x) &= F(x, y_1, y_2), \end{aligned} \tag{10.6}$$

where $F : \mathbb{R}^3 \to \mathbb{R}$ is a scalar function of three variables. In this setting, Dirichlet boundary conditions of the form (10.3) correspond to conditions on the first component of the vector $\mathbf{y}$ in (10.5), while Neumann boundary conditions of the form (10.4) correspond to conditions on its second component.

Some general existence results are available for Problem 10.1. For instance, when $F(x, y, y')$ is linear in y and y', Equation (10.2) becomes a linear second-order ODE, which takes the most general form

$$y'' = p(x)y' + q(x)y + r(x), \qquad a < x < b, \tag{10.7}$$

where $p(x)$, $q(x)$, and $r(x)$ are given continuous functions on $[a, b]$. A sufficient condition for existence of a solution of the linear boundary-value problem is given in the next theorem.

Theorem 10.1 *If $p(x)$, $q(x)$, and $r(x)$ are continuous functions on $[a, b]$ and $q(x) > 0$ on $[a, b]$, then the second-order ODE (10.7) with boundary conditions (10.3) always has a solution.*

We note that the linear boundary-value problem may admit a solution even when the sufficient condition of Theorem 10.1 is not satisfied.

The boundary-value problem (10.2)–(10.3) can be solved with the MATLAB function `bvp4c`. The function has three input variables: the name of the M-file specifying a 2-by-2 ODE system of the form (10.6), the name of the M-file specifying the boundary values, and an initial approximation of the solution prepared with the MATLAB function `bvpinit`. The output variable of `bvp4c` returns the solution in the form $[y(x), y'(x)]$, which can be evaluated at any given grid with the MATLAB function `deval`. These MATLAB functions are illustrated for the linear ODE:

MATLAB BVP solver

$$-y'' + x^2 y = \frac{1 - 3x^2}{(1 + x^2)^2}, \qquad -10 < x < 10, \tag{10.8}$$

subject to the Dirichlet boundary conditions $y(-10) = y(10) = 0$. By Theorem 10.1, the linear boundary-value problem (10.8) has a solution, which you can approximate numerically with the MATLAB function `bvp4c`. First, you need to put (10.8) into the form (10.6), that is,

$$y_1'(x) = y_2(x) \tag{10.9}$$

$$y_2'(x) = x^2 y(x) - \frac{1 - 3x^2}{(1 + x^2)^2}. \tag{10.10}$$

You then code this system in the following MATLAB function:

```
function dydx = myODE(x,y)
% function lbvODE defines the linear ODE
    dydx = [ y(2)
           x.^2.*y(1)-(1-3*x.^2)./(1+x.^2).^2];
```

Notice that dydx is a vector with two components corresponding to (10.9) and (10.10), respectively. Next, we prepare the MATLAB function specifying the boundary conditions:

```
function res = myboundary(ya,yb)
% function lbvbc defines the separated boundary conditions
    res = [ ya(1); yb(1) ];
```

At first sight, this seems like an awkward way to code the boundary conditions. The arguments ya and yb correspond to the vectors $\mathbf{y}_a$ and $\mathbf{y}_b$, which give the boundary conditions at the endpoints a and b of an interval $[a, b]$ for problem (10.1). For Dirichlet boundary conditions, you need only specify the first component of these vectors, namely, ya(1) and yb(1) in the preceding function. If you happened to be interested in Neumann boundary conditions, then you would have to specify the terms ya(2) and yb(2) instead. Next, the use of the variable res in the preceding function refers to the fact that MATLAB interprets each boundary condition in the form of a "residual" at the boundary. This means that boundary conditions are specified by setting the components of the vector res equal to zero. In this example, you have ya(1)=0 and yb(1)=0, corresponding to the desired conditions $y(-10) = 0$ and $y(10) = 0$. Alternatively, an expression of the form res=[ya(1)-3; yb(1)+1] would correspond to the boundary conditions $y(-10) = 3$ and $y(10) = -1$.

Finally, you can create a MATLAB function specifying the initial approximations $y(x) = 1/(1 + x^2)$ and $y'(x) = -2x/(1 + x^2)^2$:

```
function yinit = myinitial(x)
% function lbvinit defines the starting approximation
    yinit = [1./(1+x.^2); -2*x./(1+x.^2).^2];;
```

These functions are used in the MATLAB script BVP_solver. A unique solution of the inhomogeneous problem (10.8) is approximated numerically on a sufficiently dense grid. The numerical approximation is shown in Figure 10.1.

```
% BVP_solver
L = 10; xinit = linspace(-L,L,21);
initial_sol = bvpinit(xinit,@myinitial); % call of "bvpinit"
sol = bvp4c(@myODE,@myboundary,initial_sol); % call of "bvp4c"
xint = linspace(-L,L,1001);
yint = deval(sol,xint); % evaluation of the numerical solution
plot(xint,yint(1,:));
```

Finite-difference method for ODEs

All linear and many nonlinear boundary-value problems can be solved with the *finite difference method*, which is based on the *closed* numerical differentiation described in Section 6.2. The finite-difference method is very different from the *shooting method* explained in Section 10.2, which uses the ODE

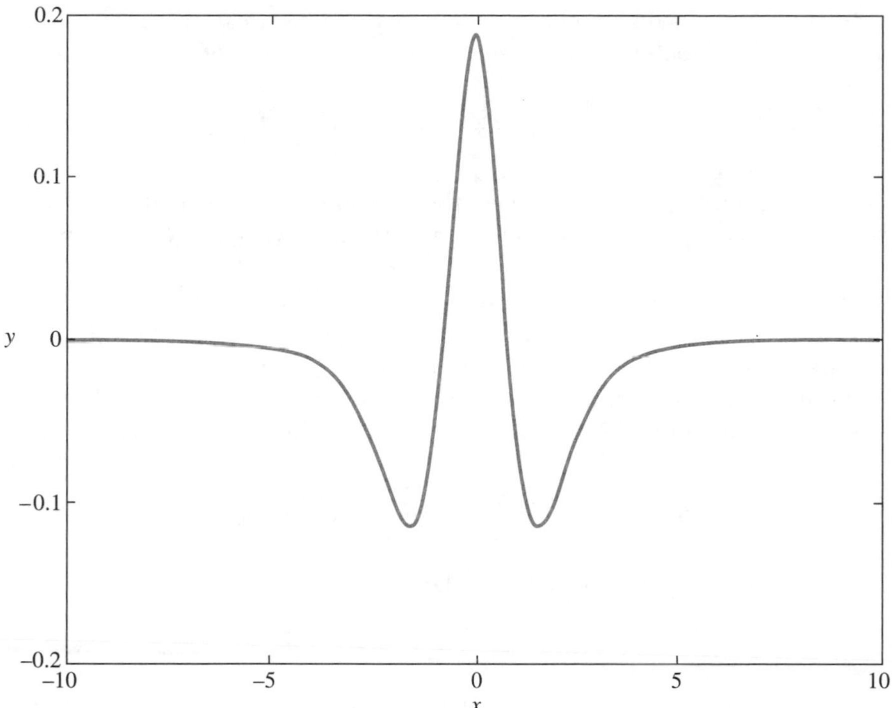

Figure 10.1 Solution of the boundary-value problem (10.8) by the MATLAB solver.

solver for the initial-value problem and the root-finding algorithms. Compared to the shooting method, the finite-difference method is more robust in the sense that it can be easily applied to higher-order differential equations and for partial differential equations.

We start by replacing the continuous interval $[a, b]$ with the equally spaced (*uniform*) discrete grid $x_1, x_2, \ldots, x_{n+1}$, where $x_1 = a$ and $x_{n+1} = b$. Since the interval is divided into n equal subintervals, the step size of the grid is $h = (b - a)/n$. The function $y(x)$ in the ODE (10.2) is represented by the vector $[y_1, y_2, \ldots, y_{n+1}]$, where the end values $y_1 = \alpha$ and $y_{n+1} = \beta$ are known from the boundary conditions (10.3). Given the values of $y(x)$ at the discrete grid, a function $y(x)$ on $[a, b]$ can be interpolated with polynomial or other smooth functions.

At each interior grid point x_k for $k = 2, 3, \ldots, n$, the first and second derivatives of $y(x)$ can be expressed by the second-order central numerical derivatives, described in Section 6.1. Using this numerical procedure, we replace the ODE (10.2) with the system of nonlinear equations:

$$\frac{y_{k+1} - 2y_k + y_{k-1}}{h^2} = F\left(x_k, y_k, \frac{y_{k+1} - y_{k-1}}{2h}\right), \qquad k = 2, 3, \ldots, n. \quad (10.11)$$

Since $y_1 = \alpha$ and $y_{n+1} = \beta$ are given, the system (10.11) is closed in the sense that $(n-1)$ equations correspond to $(n-1)$ unknown values at the interior grid points.

For the *linear* boundary-value problem, we can develop an explicit finite-difference method. On the uniform discrete grid, the functions $p(x)$, $q(x)$, and $r(x)$ in the ODE (10.7) are evaluated as finite-dimensional vectors, similar to the unknown solution $y(x)$. Using the same discretization procedure, the differential equation (10.7) becomes the system of linear equations:

$$\left(1 + \frac{hp_k}{2}\right) y_{k-1} - \left(2 + h^2 q_k\right) y_k + \left(1 - \frac{hp_k}{2}\right) y_{k+1} = h^2 r_k. \qquad (10.12)$$

The linear system (10.12) is put into the vector-matrix form:

$$\mathbf{A}\mathbf{y}_{\text{in}} = \mathbf{b} + \alpha \mathbf{b}_1 + \beta \mathbf{b}_{n+1}, \qquad (10.13)$$

where $\mathbf{y}_{\text{in}} = [y_2, y_3, \ldots, y_n]$ is an unknown vector, $\mathbf{A}$ is the coefficient matrix

$$\mathbf{A} = \begin{bmatrix} -(2+h^2q_2) & (1-\frac{hp_2}{2}) & 0 & \ldots & 0 & 0 \\ (1+\frac{hp_3}{2}) & -(2+h^2q_3) & (1-\frac{hp_3}{2}) & \ldots & 0 & 0 \\ \vdots & \vdots & \vdots & \ddots & \vdots & \vdots \\ 0 & 0 & 0 & \ldots & (1+\frac{hp_n}{2}) & -(2+h^2q_n) \end{bmatrix},$$

and $\mathbf{b}, \mathbf{b}_1, \mathbf{b}_{n+1}$ are the right-hand side vectors

$$\mathbf{b} = h^2 \begin{bmatrix} r_2 \\ r_3 \\ \vdots \\ r_n \end{bmatrix}, \quad \mathbf{b}_1 = -\left(1 + \frac{hp_2}{2}\right) \begin{bmatrix} 1 \\ 0 \\ \vdots \\ 0 \end{bmatrix}, \quad \mathbf{b}_{n+1} = -\left(1 - \frac{hp_n}{2}\right) \begin{bmatrix} 0 \\ 0 \\ \vdots \\ 1 \end{bmatrix}.$$

The truncation error of the central numerical derivatives in the system (10.12) for sufficiently smooth functions $y(x)$ is $O(h^2)$. If the matrix $\mathbf{A}$ is nonsingular and well conditioned, the inverse matrix $\mathbf{A}^{-1}$ is computed up to round-off errors. Therefore, the numerical approximation of $y(x)$ at the interior grid points has the same order $O(h^2)$ and the finite-difference solution $\mathbf{y}_{\text{in}}$ converges quadratically in h to the exact solution $y(x)$ at the grid points $\{x_k\}_{k=1}^{n+1}$, when the step size h is reduced to zero.

The finite-difference numerical solution is illustrated for the linear boundary-value problem:

$$-y'' + \left(1 - \gamma\, \mathrm{sech}^2 x\right) y = \mathrm{sech} x, \qquad x \in \mathbb{R}, \tag{10.14}$$

where γ is a parameter and $y(x)$ decays to zero as $|x| \to \infty$. Because the problem is defined on an infinite interval, we use the transformation $z = \tanh x$ to map $\mathbb{R}$ into $[-1, 1]$. After the transformation, the differential equation (10.14) is rewritten in z:

$$y'' - \frac{2z}{(1-z^2)} y' + \frac{\gamma(1-z^2) - 1}{(1-z^2)^2} y = \frac{-1}{(1-z^2)^{3/2}}, \qquad -1 < z < 1, \tag{10.15}$$

subject to

$$\lim_{z \to \pm 1} y(z) = 0. \tag{10.16}$$

The problem is singular at the endpoints $z = \pm 1$, but the finite-difference numerical solution is computed at the interior grid points $-1 < z < 1$. The numerical solution $y(z)$ is then transformed back to the solution $y(x)$ by the inverse transformation $x = \tanh^{-1} z$. The solution of the linear system (10.13) for the problem (10.15) is computed in the MATLAB script `linear_bvp`. Figure 10.2 (*left*) shows the solution $y(x)$ on $x \in \mathbb{R}$ for $\gamma = 1$. The solution is a smooth positive single-humped function, which resembles the right-hand side of the ODE (10.14).

```
% linear_bvp
h = 0.01; z = -1 : h : 1;
n = length(z)-1; zz = z(2:n); gamma = 1;
% building and solving the linear system
 A1 = -2*diag(ones(n-1,1))+diag(ones(n-2,1),1)+diag(ones(n-2,1),-1);
 A2 = -h^2*diag((1 - gamma*(1-zz.^2))./(1-zz.^2).^2);
 A3 = -h*diag(zz(1:n-2)./(1-zz(1:n-2).^2),1);
 A4 = +h*diag(zz(2:n-1)./(1-zz(2:n-1).^2),-1);
 A = A1 + A2 + A3 + A4;
b = -h^2*((1-zz.^2).^(3/2))';
yIn = A\b; xIn = atanh(zz)';
zInt = linspace(-1,1,11001);
xInt = atanh(zInt(2:length(zInt)-1));
yInt = spline(zz,yIn,zInt);
yInt = yInt(2:length(yInt)-1);
plot(xInt,yInt,'r',xIn,yIn,'.g');
```

Continuing the same MATLAB script, you can compute the solution $y(x)$ for different values of γ in the interval $1 \leq \gamma \leq 3$. Figure 10.2 (*right*) shows

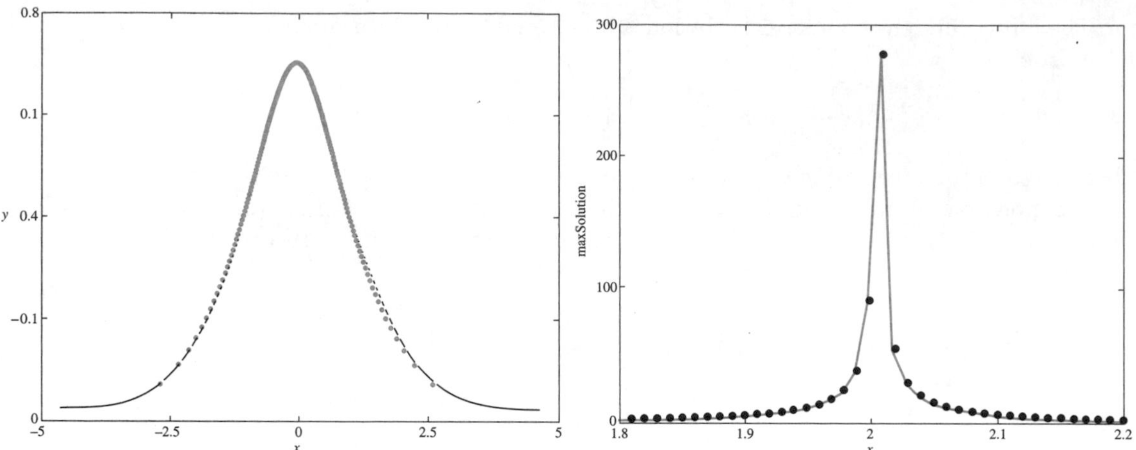

Figure 10.2 The numerical approximation of the solution of the ODE (10.14) with $\gamma = 1$ (*left*) and its maximum as a function of γ (*right*).

the maximum of the solution $y(x)$ versus γ. A problem occurs near $\gamma = 2$, where the maximum becomes too large. The problem is explained by the fact that the homogeneous equation

$$-y'' + \left(1 - 2\,\mathrm{sech}^2 x\right) y = 0$$

has a nontrivial solution $y(x) = \mathrm{sech} x$. By the Fredholm Alternative for inhomogeneous systems, the solution $y(x)$ of the ODE (10.14) with $\gamma = 2$ does not decay as $|x| \to \infty$. When boundary conditions (10.16) are enforced, the solution of the ODE (10.15) blows up. The numerical approximation of $y(x)$ for finite values of h still exists, but its magnitude becomes large.

```
% linear_bvp (continued)
gamma = 1 : 0.01 : 3;
for k = 1 : length(gamma)
    gammak = gamma(k);
    yIn = A\b;
    maxSolution(k) = max(abs(yIn));
end
plot(gamma,maxSolution,'.',gamma,maxSolution);
```

Another example enables us to study the error of the numerical approximation in the finite-difference solution of the linear boundary-value problem. The linear equation

$$y'' = \frac{2x}{1+x^2}y' - \frac{2}{1+x^2}y + 1 + x^2, \qquad 0 < x < 1, \tag{10.17}$$

subject to

$$y(0) = 2, \qquad y(1) = \frac{5}{3}, \tag{10.18}$$

has the exact solution:

$$y(x) = \frac{1}{6}x^4 - \frac{3}{2}x^2 + x + 2. \tag{10.19}$$

The absolute error of the numerical approximation can be computed from the maximal distance between the exact and numerical solutions at the interior grid points. The MATLAB script `error_linear_bvp` computes the best power fit of the maximal error versus the step size h.

```
% error_linear_bvp
n = 10:10:400;
for k = 1 : length(n)
    nn = n(k); hh = 1/nn; h(k) = hh;
    x = linspace(0,1,nn+1); xx = x(2:nn);
    A1 = -2*diag(ones(nn-1,1))+diag(ones(nn-2,1),1)+diag(ones(nn-2,1),-1);
    A2 = 2*hh^2*diag(1./(1+xx.^2))-hh*diag(xx(1:nn-2)./(1+xx(1:nn-2).^2),1);
    A3 = hh*diag(xx(2:nn-1)./(1+xx(2:nn-1).^2),-1);
    A = A1 + A2 + A3;
    b = hh^2*(1+xx.^2)';
    b(1) = b(1)-(1 + hh*xx(1)/(1+xx(1)^2))*2;
    b(nn-1) = b(nn-1)-(1 - hh*xx(nn-1)/(1+xx(nn-1)^2))*5/3;
    u = A\b; y = [ 2; u; 5/3]';
    yExact = x.^4/6-3*x.^2/2+x+2;
    Error(k) = max(abs(y - yExact));
end
a = polyfit(log(h),log(Error),1); power = a(1)  % best power fit
Int = linspace(h(1),h(length(h)),101);
ErrorInt=exp(a(2)+a(1)*log(hInt));
plot(h,Error,'.b',hInt,ErrorInt,':r');
```

The output of the MATLAB script `error_linear_bvp` shows that the maximal error is $O(h^2)$. Figure 10.3 shows the graph of the maximal error and its best power fit versus h.

```
>> error_linear_bvp

power =
    1.9996
```

Few comments are in place here. The matrix $\mathbf{A}$ has a tridiagonal structure because of the second-order central numerical derivatives used. Although it

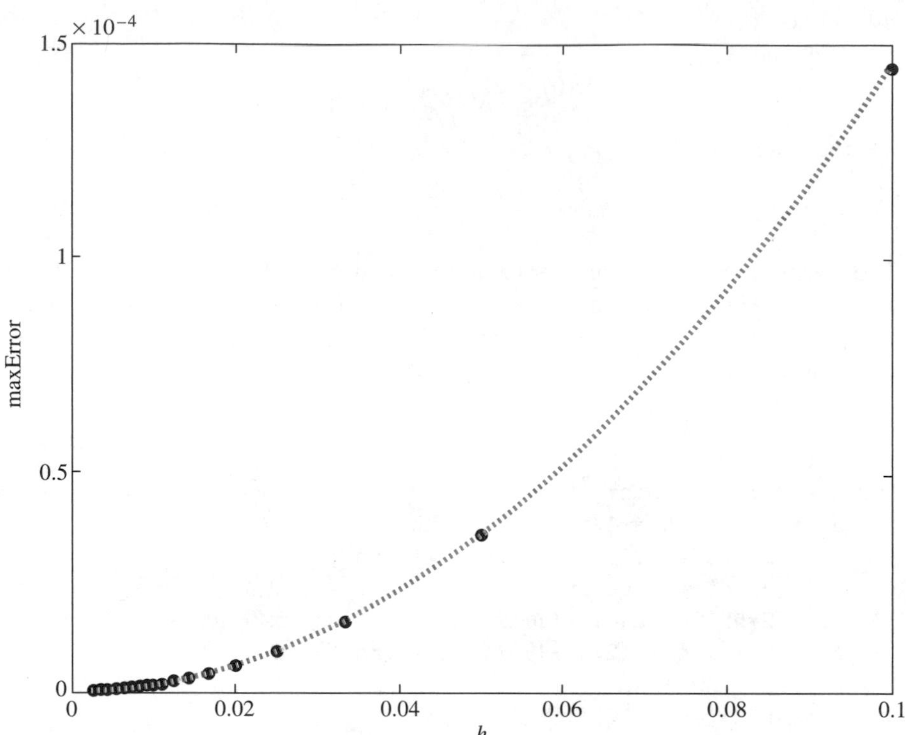

Figure 10.3 Numerical error of the finite-difference solution of the linear equation (10.17) versus step size h.

is nonsymmetric for nonconstant values of $p(x)$, it approaches a symmetric matrix with diagonally dominant coefficients in the limit of small h. When the matrix $\mathbf{A}$ is invertible, fast algorithms of numerical linear algebra can be employed to compute solutions for the linear system (10.13). However, the matrix $\mathbf{A}$ could be singular for finite values of h, in which case the homogeneous linear system $\mathbf{A}\mathbf{y}_{\text{in}} = \mathbf{0}$ has a non-trivial solution. When it happens, the solution of the inhomogeneous problem (10.13) subject to the prescribed boundary conditions may not exist. Recall that the linear system (10.13) represents the ODE (10.7) after the numerical discretization (10.12).

The boundary values (10.3) are effectively incorporated into the tridiagonal linear system (10.13) as the vectors $\mathbf{b}_1$ and $\mathbf{b}_{n+1}$. If the fourth-order central numerical derivatives are invoked on the uniform discrete grid, the linear system (10.13) is not closed unless the fourth-order central numerical derivatives at the first and last interior grid points $x = x_2$ and $x = x_n$ are replaced by equivalent forward and backward differences. These differences involve coupling y_2 with y_3, y_4, y_5, and coupling y_n with y_{n-1}, y_{n-2}, y_{n-3}, resulting in a coefficient matrix that has no five-diagonal structure. Therefore, second-order central differences are best suited for second-order boundary-value ODE problem.

Because the numerical error in computations of y_k for $k = 2, 3, \ldots, n$ is $O(h^2)$, the interpolation of a smooth function $y(x)$ through the discrete values is supposed to have the same accuracy. Interpolations of higher accuracy do not generally reduce the error of the finite-difference solution. It is clear that the discretization of the linear boundary-value problem (10.7) into the linear system (10.13) is not unique. For instance, the values p_k can be replaced by the mean value of p_{k-1} and p_{k+1} within the same second-order accuracy.

Neumann conditions

It is easy to incorporate other boundary conditions in the finite-difference solution of the linear boundary-value problem, such as Neumann boundary conditions:

$$y'(a) = \gamma, \qquad y'(b) = \delta. \tag{10.20}$$

Boundary values for derivatives (10.20) can be incorporated in two equivalent ways: (1) the second-order forward and backward differences can be used at the endpoints, and (2) virtual grid points can be added beyond the endpoints. In the first method, the derivatives at the endpoints give additional relations for the values y_1 and y_{n+1} at the endpoints in terms of the values y_k, $k = 2, 3, \ldots, n$ at the interior grid points:

$$\begin{aligned} -y_3 + 4y_2 - 3y_1 &= 2h\gamma, \\ y_{n-1} - 4y_n + 3y_{n+1} &= 2h\delta. \end{aligned}$$

When y_1 and y_{n+1} are eliminated from the remaining system of linear equations (10.12), the tridiagonal structure (10.13) is preserved, with α and β replaced by γ and δ, and the elements of the first and last rows of the coefficient matrix $\mathbf{A}$ and the vectors $\mathbf{b}_1$ and $\mathbf{b}_{n+1}$ are modified as follows:

$$\mathbf{A} = \begin{bmatrix} -\frac{2}{3}(1 - hp_2) - h^2 q_2 & \frac{2}{3}(1 - hp_2) & 0 & \ldots & 0 & 0 \\ (1 + \frac{hp_3}{2}) & -(2 + h^2 q_3) & (1 - \frac{hp_3}{2}) & \ldots & 0 & 0 \\ \vdots & \vdots & \vdots & \ddots & \vdots & \vdots \\ 0 & 0 & 0 & \ldots & \frac{2}{3}(1 + hp_n) & -\frac{2}{3}(1 + hp_n) - h^2 q_n \end{bmatrix}$$

and

$$\mathbf{b}_1 = \frac{2h}{3}\left(1 + \frac{hp_2}{2}\right)\begin{bmatrix} 1 \\ 0 \\ \vdots \\ 0 \end{bmatrix}, \qquad \mathbf{b}_{n+1} = -\frac{2h}{3}\left(1 - \frac{hp_n}{2}\right)\begin{bmatrix} 0 \\ 0 \\ \vdots \\ 1 \end{bmatrix}.$$

In the second method, the virtual grid points $x_0 = a - h$ and $x_{n+2} = b + h$ are added beyond the closed grid $a \leq x \leq b$ and the central difference approximation are extended for the boundary conditions (10.20):

$$y_2 - y_0 = 2h\gamma, \qquad y_{n+2} - y_n = 2h\delta.$$

The linear system is then extended to include two additional equations at the endpoints while the artificial values y_0 and y_{n+2} are eliminated from the system. As a result, the linear system (10.13) is set for the vector $\mathbf{y} = [y_1, y_2, \ldots, y_{n+1}]$, whereas the coefficient matrix and the right-hand side vectors are

$$\mathbf{A} = \begin{bmatrix} -(2+h^2q_1) & 2 & 0 & \ldots & 0 & 0 \\ (1+\frac{hp_2}{2}) & -(2+h^2q_2) & (1-\frac{hp_2}{2}) & \ldots & 0 & 0 \\ \vdots & \vdots & \vdots & \ddots & \vdots & \vdots \\ 0 & 0 & 0 & \ldots & 2 & -(2+h^2q_{n+1}) \end{bmatrix}$$

and

$$\mathbf{b} = h^2 \begin{bmatrix} r_1 \\ r_2 \\ \vdots \\ r_{n+1} \end{bmatrix}, \quad \mathbf{b}_1 = 2h\left(1 + \frac{hp_1}{2}\right)\begin{bmatrix} 1 \\ 0 \\ \vdots \\ 0 \end{bmatrix}, \quad \mathbf{b}_{n+1} = -2h\left(1 - \frac{hp_{n+1}}{2}\right)\begin{bmatrix} 0 \\ 0 \\ \vdots \\ 1 \end{bmatrix}.$$

The two methods are illustrated on the example of a homogeneous linear equation

$$-y'' + \cos x\, y = \lambda y, \qquad 0 < x < 2\pi, \tag{10.21}$$

subject to the Neumann boundary conditions $y'(0) = y'(2\pi) = 0$. This boundary-value problem defines the spectrum of *Mathieu's* equation associated with a periodic potential. The parameter λ is interpreted as an eigenvalue and the corresponding non-zero solution of the homogeneous boundary-value problem is interpreted as an eigenvector. MATLAB script `eigenvalue_lbv` computes approximations for the three smallest eigenvalues with the corresponding eigenvectors by two different methods. The maximal value of the components of eigenvectors is normalized by one.

```
% eigenvalue_lbv
x = linspace(0,2*pi,101); h = x(2) - x(1); n = length(x)-1;
                % computations of the first method
A1 = 2*diag(ones(n-1,1))-diag(ones(n-2,1),1)-diag(ones(n-2,1),-1);
A2 = h^2*diag(cos(x(2:n)));
A1(1,1) = 2/3; A1(n-1,n-1) = 2/3;
A1(1,2) = -2/3; A1(n-1,n-2) = -2/3;
```

```
A = A1 + A2; [V,D] = eig(A);
[lambda, ind] = sort(diag(D)'/h^2);
 disp('The smallest three eigenvalues in the first method are:');
lambda1 = lambda(1:3)
                   % computations of the second method
A1 = 2*diag(ones(n+1,1))-diag(ones(n,1),1)-diag(ones(n,1),-1);
A1(1,2) = -2; A1(n+1,n) = -2;
A2 = h^2*diag(cos(x)); A = A1 + A2;
[V,D] = eig(A); [lambda, ind] = sort(diag(D)'/h^2);
 disp('The smallest three eigenvalues in the second method are:');
lambda1 = lambda(1:3) figure(1);
plot(lambda,zeros(size(lambda)),'.');
                   % normalization of eigenvectors
vector1 = V(:,ind(1))'; vector1 = vector1/max(abs(vector1));
vector2 = V(:,ind(2))'; vector2 = vector2/max(abs(vector2));
vector3 = V(:,ind(3))'; vector3 = vector3/max(abs(vector3));
figure(2); plot(x,vector1,'b',x,vector2,'g',x,vector3,'r');
```

The output of the script **eigenvalue_lbv** shows that both methods give similar results for the three smallest eigenvalues with the difference of 10^{-5} in their values. Figure 10.4 shows all values of λ and the first three normalized eigenvectors computed from the second method. Note that the eigenvector for the smallest eigenvalue has no zeros on the interval $[0, 2\pi]$, whereas the eigenvectors for the second and third eigenvalues have one and two zeros, respectively. This fact is well-known for eigenvectors of the Sturm–Liouville eigenvalue problem in *Partial Differential Equations.*

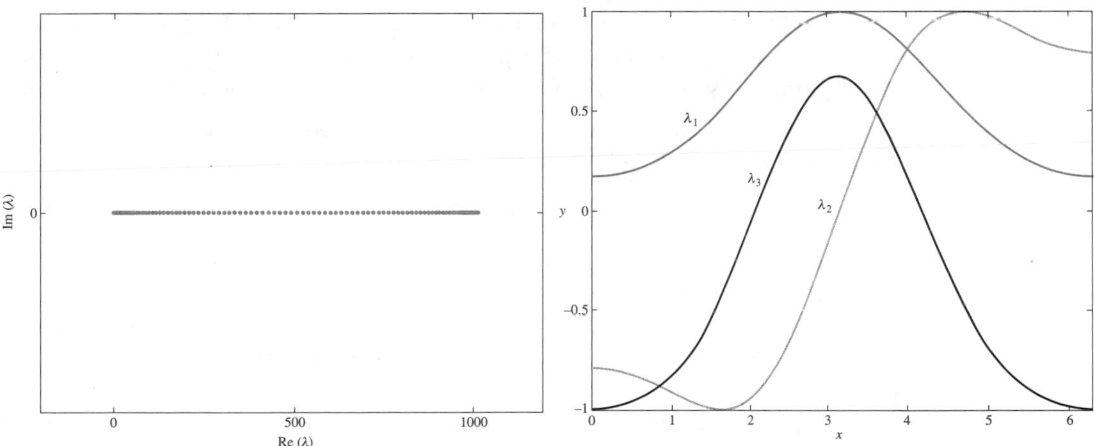

Figure 10.4 Eigenvalues λ for the ODE (10.21) (*left*). The first three normalized eigenvectors (*right*).

```
>> eigenvalue_lbv

The smallest three eigenvalues in the first method are:

lambda1 =
  -0.37857648343995   0.59461047979905   1.29273905853575

The smallest three eigenvalues in the second method are:

lambda1 =
  -0.37857794209418   0.59462590450130   1.29277868344366
```

Nonlinear ODEs

For *nonlinear* boundary-value problems, the system of nonlinear equations (10.11) cannot be solved in general. However, an iterative solution can be constructed with the root-finding algorithms for vector-valued nonlinear functions described in Section 8.2. In this formulation, the nonlinear system (10.11) is represented in the vector form:

$$\mathbf{f}(\mathbf{y}) = \mathbf{0}, \tag{10.22}$$

where $\mathbf{y} = [y_2, y_3, \ldots, y_n]$ and $\mathbf{f} = [f_2, f_3, \ldots, f_n]$, so that

$$f_k = y_{k+1} - 2y_k + y_{k-1} - h^2 F\left(x_k, y_k, \frac{y_{k+1} - y_{k-1}}{2h}\right), \quad k = 2, 3, \ldots, n.$$

The boundary values $y_1 = \alpha$ and $y_{n+1} = \beta$ must be used again in the first and last equations of the system. The matrix-vector Newton–Raphson method takes the form:

$$\mathbf{y}^{(m+1)} = \mathbf{y}^{(m)} - \mathbf{J}^{-1}(\mathbf{y}^{(m)})\mathbf{f}(\mathbf{y}^{(m)}), \tag{10.23}$$

where $\mathbf{y}^{(m)}$ is the mth approximation of the solution and $\mathbf{J}(\mathbf{y})$ is the Jacobian matrix of the vector function $\mathbf{f}(\mathbf{y})$. If the sequence $\{\mathbf{y}^{(m)}\}_{m=0}^{\infty}$ converges to a fixed point $\mathbf{y}_*$, so that $\lim_{m\to\infty} \mathbf{y}^{(m)} = \mathbf{y}_*$ and $\mathbf{f}(\mathbf{y}_*) = \mathbf{0}$, the iterative loop can be stopped when the distance between two successive iterations fits into the tolerance bound:

$$\|\mathbf{y}^{(m+1)} - \mathbf{y}^{(m)}\| \leq \text{tolerance}. \tag{10.24}$$

Although the matrix-vector Newton–Raphson method converges absolutely near any fixed point of the nonlinear system (10.22), the convergence domain for a particular fixed point $\mathbf{y}_*$ could be quite narrow. As a result, it can be difficult to guess a suitable starting approximation $\mathbf{y}_0$ for a convergent sequence $\{\mathbf{y}^{(m)}\}_{m=0}^{\infty}$.

An illustrative example of a nonlinear boundary-value problem is given by *Duffing's* equation,

$$y'' + 2y(1 - y^2) = 0, \qquad x \in \mathbb{R}, \tag{10.25}$$

subject to the boundary conditions:

$$\lim_{x \to -\infty} y(x) = -1, \qquad \lim_{x \to \infty} y(x) = 1. \qquad (10.26)$$

The exact solution of the nonlinear problem exists in closed form as $y(x) = \tanh x$. Using the finite-difference approximation, you can truncate the computational domain on $[-L, L]$ for large $L \gg 1$ and represent the system of nonlinear equations in the form (10.22), where

$$f_k = y_{k+1} - 2y_k + y_{k-1} + 2h^2 y_k(1 - y_k^2).$$

The Jacobian matrix is written in the form,

$$J_{k,l}(\mathbf{y}) = \delta_{k,k+1} - 2\delta_{k,k} + \delta_{k,k-1} + 2h^2(1 - 3y_k^2)\delta_{k,k}, \quad 2 \le k, l \le n,$$

where $\delta_{k,l}$ is the Kronecker symbol defined as $\delta_{k,l} = 0$ for $k \neq l$ and $\delta_{k,k} = 1$. You can start iterations with the initial approximation $y_k^{(0)} = \tanh x_k$. The MATLAB script `nonlinear_bvp` computes the successive iterations and checks the convergence criterion (10.24).

```
% nonlinear_bvp
L = 8; x = linspace(-L,L,201);
h = x(2) - x(1); n = length(x)-1;
y = tanh(x)'; y(1) = -1; y(n+1) = 1; % starting iteration
tolerance = 10^(-10); solutionNorm = 1;
k = 0; maxNumber = 100;
while ((solutionNorm > tolerance) & (k < maxNumber)) % iterations
    J1 = -2*diag(ones(n-1,1))+diag(ones(n-2,1),1)+diag(ones(n-2,1),-1);
    J2 = 2*h^2*diag(1-3*y(2:n).^2); J = J1 + J2;
    F = y(3:n+1)-2*y(2:n)+y(1:n-1)+2*h^2*y(2:n).*(1-y(2:n).^2);
    yy = y(2:n) - inv(J)*F; yy = [-1 ; yy ; 1];
    solutionNorm = max(abs(y-yy));
    y = yy; k = k + 1;
    distance(k) = solutionNorm;
end
figure(1); plot(distance); figure(2);
plot(x,y','.b',x,tanh(x),':r');
```

Figure 10.5 (*left*) shows the distance between two successive iterations (10.24) versus the number of iterations. You can see that the iterations do not converge but oscillate around the fixed point. The graph of the solution $y(x)$ after 100 iterations is shown in Figure 10.5 (*right*). Although the graph resembles the exact solution $y(x) = \tanh x$, it is clear that the iterative loop can be terminated only if the tolerance is large enough. There are two main reasons for limitations of the finite-difference method applied to the nonlinear problem (10.25)–(10.26): (1) the computational domain $\mathbb{R}$ is truncated as

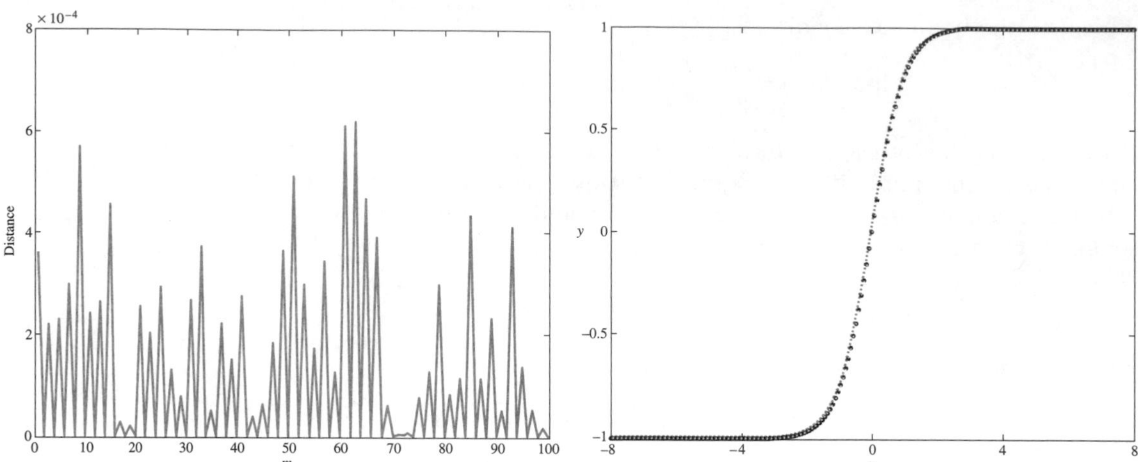

Figure 10.5 Distance between two successive iterations of the Newton–Raphson method versus the number of iterations for the ODE (10.25) (*left*). Finite-difference solution of the nonlinear boundary-value problem (10.25)–(10.26) after truncation and discretization (dots) and the exact solution (dashed curve) (*right*).

$[-L, L]$, and (2) the boundary conditions are enforced at the endpoints $x = \pm L$ instead of the limits $x \to \pm\infty$. Additional complications are induced by the discretization of the interval $[-L, L]$ and by iterations of the Newton–Raphson root-finding algorithm for nonlinear matrix-vector systems.

Exercise 10.1 Apply the transformation $z = \tanh x$ to the nonlinear boundary-value problem (10.25)–(10.26) and develop the Newton–Raphson iterative method for the resulting system of nonlinear equations. Observe a better convergence of iterations on the finite computational interval $[-1, 1]$.

Boundary-value problems for higher-order ordinary differential equations can also be solved with the finite-difference method. For instance, the higher-order derivatives can be replaced by the central-difference approximations of the same second order. More boundary conditions are set for a higher-order differential equation, and the second-order central difference for higher derivatives involves more points. As a result, all boundary values can be incorporated into a closed system of algebraic equations at the interior points of the discrete grid.

10.2 Shooting Methods for ODEs

Although the finite-difference methods are very robust for boundary-value problems associated with differential equations, they are not accurate even if the multipoint numerical derivatives are used to replace the continuous second-order and higher-order derivatives. In this section, we present an alternative method that uses the ODE initial-value solvers in applications to the boundary-value problems. This method is called the *shooting method* and provides better accuracy and control of the numerical solution, with the obvious limitation of being applicable only to ordinary differential equations. Although this is certainly a shortcoming, the shooting method is very popular because of its simplicity both in theory and implementation. Other accurate methods that are more difficult to learn and to code are covered in Sections 11.3 and 12.3 in the context of spectral and finite-element techniques.

Let us consider again the second-order boundary-value problem (10.2)–(10.3). It is clear from Theorem 9.1 that the initial-value problem for the second-order ODE (10.2) with a smooth function $F(x, y, y')$ admits a unique solution $y(x)$ if the two values for $y(x)$ and $y'(x)$ are specified in any point, say, at $x = a$. Since $y(a) = \alpha$ is given, we only lack the data for $y'(a)$, which can be viewed as a parameter of the initial-value problem:

$$y(a) = \alpha, \qquad y'(a) = s. \tag{10.27}$$

Let us denote the corresponding solution of the ODE (10.2) with the initial value (10.27) by $y(x; s)$. The function $y(x; s)$ is well defined on the entire interval $[a, b]$ if the solution $y(x; s)$ at a given s lies in the domain of definition of the ODE (10.2). When the solution $y(x; s)$ is obtained, the value of $y(b; s)$ can be compared against the boundary value β using the *displacement function*:

$$f(s) = y(b; s) - \beta. \tag{10.28}$$

When $f(s_*) = 0$ for some $s = s_*$, the solution of the ODE (10.2) with the initial value (10.27) is simultaneously a solution of the same ODE with the boundary value (10.3). Therefore, to find s_* we only need to run a root-finding algorithm and iterate along a sequence of approximations $\{s_n\}_{n \geq 0}$ using numerical computations of the solution $y(x; s)$ until either the sequence $\{s_n\}_{n \geq 0}$ converges to a root s_* or the solution $y(x; s_n)$ jumps outside the domain of definition of the ODE (10.2) for some n. This procedure explains why it is called the shooting method: we shoot the solution $y(x; s)$ from the left end $x = a$ with an approximate slope $s = s_n$ and, depending on how the target is hit at the right end $x = b$, correct the slope $y'(a) = s$ in the next approximation $s = s_{n+1}$.

Consider the linear second-order ODE

$$xy'' + y' = 0, \qquad 1 < x < 2, \tag{10.29}$$

subject to the boundary values $y(1) = 1$ and $y(2) = 3$, for which the shooting method can be implemented analytically. Indeed, a general solution of the ODE (10.29) is

$$y(x) = A \log x + B,$$

where A and B are arbitrary constants. From the boundary conditions, you find that the values $B = 1$ and $A = 2/\log 2$ give the exact solution of the boundary-value problem. From the initial conditions of the shooting method $y(1) = 1$ and $y'(1) = s$, you can find that $B = 1$ and $A = s$ in the solution $y(x; s)$, after which the function $f(s)$ is computed explicitly as $f(s) = s \log 2 - 2$. Because the function $f(s)$ is linear in s, only one solution $y(x; s)$ of the boundary-value problem exists with $s = 2/\log 2$.

Linear shooting method

The root-finding problem for a scalar function $f(s)$ is considered in Section 8.1. The Newton–Raphson and secant methods represent unconditionally stable iterations that converge if the initial approximation s_0 is sufficiently close to the root s_*. In particular, the *Newton–Raphson* method runs iterations from s_n to s_{n+1} based on the linear approximation of the function $f(s)$. If the second-order ODE (10.2) is linear in the form (10.7), then the function $f(s)$ is linear in s and the Newton method converges in a single iteration if the round-off error is controllably small. Moreover, we can use the principle of the linear superposition for the linear ODE (10.7) to construct the function $f(s)$ and look for its zeros. Indeed, let us represent a solution of the linear ODE (10.7) in the form:

$$y(x; s) = u(x) + sv(x), \tag{10.30}$$

where $u(x)$ is a unique solution of the linear inhomogeneous ODE (10.7) with the initial values $u(a) = \alpha$, $u'(a) = 0$ and $v(x)$ is a unique solution of the linear homogeneous ODE (10.7) with $r(x) \equiv 0$ and initial conditions $v(a) = 0$, $v'(a) = 1$. In terms of these two linearly independent solutions, we can compute $f(s) = u(b) + sv(b) - \beta$ and observe that a single iteration of the linear shooting method immediately produces a unique solution of the boundary-value problem:

$$s = s_* = \frac{\beta - u(b)}{v(b)}, \tag{10.31}$$

which is valid if $v(b) \neq 0$. If $v(b) = 0$, the linear shooting method fails. In this case, the homogeneous equation has a nontrivial solution satisfying $v(a) = v(b) = 0$.

We illustrate the linear shooting method for the linear ODE (10.8), which is generalized here on a general computation interval $[-L, L]$:

$$-y'' + x^2 y = \frac{1 - 3x^2}{(1 + x^2)^2}, \qquad -L < x < L, \tag{10.32}$$

subject to the Dirichlet boundary conditions $y(-L) = y(L) = 0$. The linear shooting method is coded in the MATLAB script `shooting_linear_ODE`, where the Heun method is used as the ODE initial-value solver.

```
% shooting_linear_ODE
h = 0.1; L = input('Enter the half-length of the interval L = ');
x = -L : h : L;
y(1) = 0; u(1) = 0; % (y,u) - components of u(x)
z(1) = 0; v(1) = 1; % (z,v) - components of v(x)
for k = 1 : length(x)-1   % iterations of the ODE solver
        yp = y(k) + h*u(k);
        up = u(k) + h*(x(k)^2*y(k)+(3*x(k)^2-1)/((1+x(k)^2)^2));
        y(k+1) = y(k) + 0.5*h*(u(k)+up);
        uu1 = x(k)^2*y(k)+(3*x(k)^2-1)/((1+x(k)^2)^2);
        uu2 = x(k+1)^2*yp+(3*x(k+1)^2-1)/((1+x(k+1)^2)^2);
        u(k+1) = u(k) + 0.5*h*(uu1 + uu2);
        zp = z(k) + h*v(k);
        vp = v(k) + h*x(k)^2*z(k);
        z(k+1) = z(k) + 0.5*h*(v(k)+vp);
        v(k+1) = v(k) + 0.5*h*(x(k)^2*z(k)+x(k+1)^2*zp);
end
s = -y(length(x))/z(length(x)); % linear shooting method
y = y + s*z; % resulting solution
if (abs(max(y)) < 1000)
    plot(x,y,'b');
else
    disp('ODE solvers fails.');
end
```

When the MATLAB script shooting_linear_ODE is executed with $L = 5$, the linear shooting method produces a smooth solution $y(x)$, shown in Figure 10.6. This solution is similar to the one shown in Figure 10.1 for the same problem with $L = 10$. However, when the shooting method is applied with $L = 10$, it produces an invalid solution with large values of $y(x)$. This invalid solution is generated by the instabilities of the Heun method used to solve the linear ODE (10.32). Therefore, the shooting method can suffer from instabilities of the initial-value ODE solvers used in computations of the particular solutions $u(x)$ and $v(x)$ even if the iteration of the linear shooting method (10.30) is constructed explicitly. To correct the situation, you need to choose a different ODE solver for the linear ODE (10.32), for example, the implicit ODE solvers explained in Section 9.5.

```
>> shooting_linear_ODE
Enter the half-length of the interval L = 5

>> shooting_linear_ODE
Enter the half-length of the interval L = 10
ODE solvers fails.
```

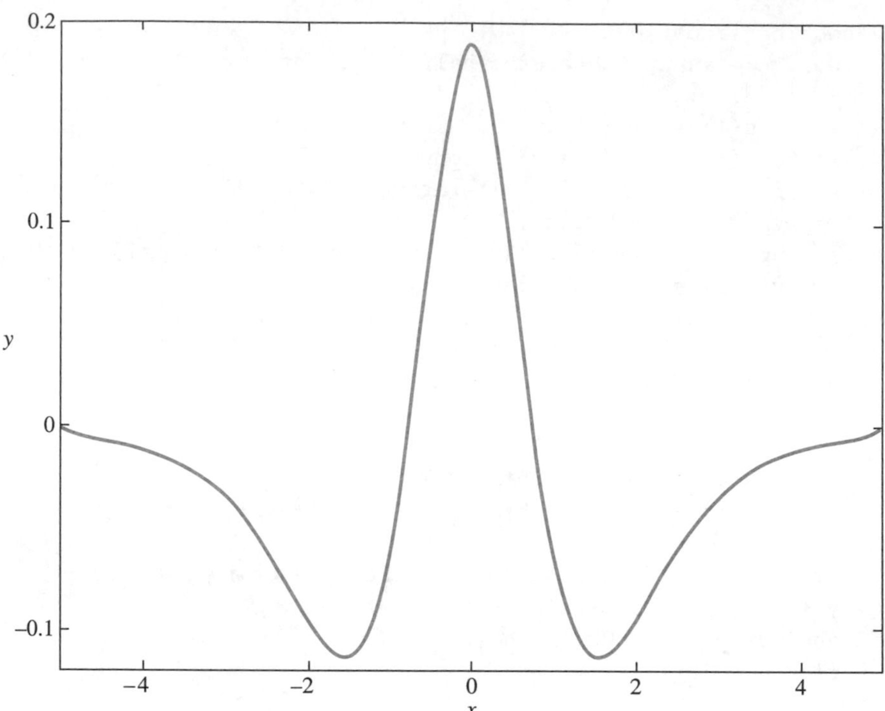

Figure 10.6 The solution of the linear ODE (10.32) with the linear shooting method for $L = 5$.

Exercise 10.2 Use the implicit second-order Adams method for the linear ODE (10.32) and approximate the solution on the interval $-L < x < L$ with $L = 10$.

Newton–Raphson method

The shooting method for the nonlinear ODE (10.2) does not generally converge in a single iteration. Moreover, the linear superposition principle is no longer available and we need to implement the Newton–Raphson algorithm for finding zeros of $f(s) = 0$. However, we can formulate this algorithm in terms of two solutions $u(x)$ and $v(x)$ constructed for a particular value $s = s_n$ at some $n \geq 0$. The function $u(x)$ solves the nonlinear inhomogeneous ODE (10.2) with the initial value (10.27), for example:

$$u'' = F(x, u, u'), \qquad a < x < b, \tag{10.33}$$

subject to the initial values $u(a) = \alpha$ and $u'(a) = s$. The function $v(x)$ solves the linearized homogeneous ODE

$$v'' = F_2(x)v + F_3(x)v', \qquad a < x < b, \tag{10.34}$$

where

$$F_2(x) = \frac{\partial F}{\partial u}(x, u, u'), \quad F_3(x) = \frac{\partial F}{\partial u'}(x, u, u'),$$

subject to the initial values $v(a) = 0$ and $v'(a) = 1$. It is clear from the construction that if $f(s) = u(b) - \beta$, then $f'(s) = v(b)$, where both functions $u(x)$ and $v(x)$ depend on s and have to be computed separately for different values of $s = s_n$. If both $f(s_n)$ and $f'(s_n)$ are known, the Newton–Raphson method produces a sequence of approximations $\{s_n\}_{n\geq 0}$ by using the first-order difference equation

$$s_{n+1} = s_n - \frac{f(s_n)}{f'(s_n)}, \qquad n \geq 0. \tag{10.35}$$

Because the Newton–Raphson method (10.35) converges unconditionally, the shooting method approximates a solution of Problem 10.1 under the conditions that (1) the solution $y(x)$ exists, (2) the initial condition $s = s_0$ is sufficiently close to the root $s = s_*$, and (3) the ODE solver is numerically accurate, for instance, being unconditionally stable.

We illustrate the shooting method for the nonlinear ODE

$$y'' + \frac{1}{x}y' - y + y^3 = 0, \qquad 0 < x < L, \tag{10.36}$$

which describes radially symmetric bound states in two spatial dimensions. We are interested in a large computational domain (e.g., $L = 10$) subject to the mixed boundary conditions $y'(0) = 0$ and $y(L) = 0$. The shooting parameter is then $y(0) = s$ and the Taylor series approximation near $x = 0$ leads to the solution:

$$y(x) = s + \frac{s(1 - s^2)}{4}x^2 + \mathrm{O}(x^4).$$

The value of s must be found from the boundary condition for $y(1)$. Because the nonlinear ODE (10.36) may admit several solutions, the starting approximation for s may be crucial for convergence of iterations. The MATLAB script `shooting_nonlinear_ODE` gives details of implementations of the shooting method with the Heun method as the ODE initial-value solver.

```
% shooting_nonlinear_ODE
h = 0.01; x = 0 : h : 10; m = 0; ss = 0;
s = input('Enter starting value for s = ');  % user input
while (abs(s-ss) > 10^(-10)) & (abs(s-ss) < 100) & (m < 1000)
    ss = s; y(1) = s; u(1) = 0;        % (y,u) - components of u(x)
    y(2) = s + s*(1-s^2)*x(2)^2/4;     % Taylor series approximation
    u(2) = s*(1-s^2)*x(2)/2;           % for the first step
    z(1) = 1; v(1) = 0;                % (z,v) - components of v(x)
```

```
        z(2) = 1 + (1-3*s^2)*x(2)^2/4;
        v(2) = (1-3*s^2)*x(2)/2;
        for k = 2 : length(x)-1          % iterations of the ODE solver
            yp = y(k) + h*u(k);
            up = u(k) - h*(u(k)/x(k)-y(k)+y(k)^3);
            y(k+1) = y(k) + 0.5*h*(u(k)+up);
            u(k+1) = u(k) - 0.5*h*(u(k)/x(k)-y(k)+y(k)^3+up/x(k+1)-yp+yp^3);
            zp = z(k) + h*v(k);
            vp = v(k) - h*(v(k)/x(k)-z(k)+3*y(k)^2*z(k));
            z(k+1) = z(k) + 0.5*h*(v(k)+vp);
            vv = v(k)/x(k)-z(k)+3*y(k)^2*z(k)+vp/x(k+1)-vp+3*yp^2*zp;
            v(k+1) = v(k) - 0.5*h*vv;
        end
        s = ss - y(length(x))/z(length(x)); m = m+1;
        if (m == 1)
            plot(x,y,':r'); hold on;
        end
    end
    if (abs(s-ss) < 100) & (m < 1000)
        fprintf('The shooting method converges in %d iterations\n',m);
        plot(x,y,'b');
    else
        disp('The shooting method fails.');
    end
```

When the MATLAB script `shooting_nonlinear_ODE` is executed with several starting values of s, you find that convergence may not be observed for all values of s. The shooting method converges for $s = 1.5$ and the starting (dashed curve) and final (solid curve) approximations are shown in Figure 10.7. The resulting solution has no zeros on $0 < x < L$ and it is referred to as the ground state of the nonlinear ODE (10.36). However, the shooting method fails for $s = 2$, because of a failure of the ODE initial-value solver to approximate the solution near the unstable critical point $(0, 0)$.

```
>> shooting_nonlinear_ODE
Enter starting value for s = 1.5
    The shooting method converges in 33 iterations

>> shooting_nonlinear_ODE
Enter starting value for s = 2
    The shooting method fails.
```

Secant method

A possible alternative of the Newton–Raphson method is the secant method, which avoids the construction of the solution $v(x)$ in the linearized equation (10.34). Indeed, if two approximations of $y(x; s)$ and $f(s)$ are found

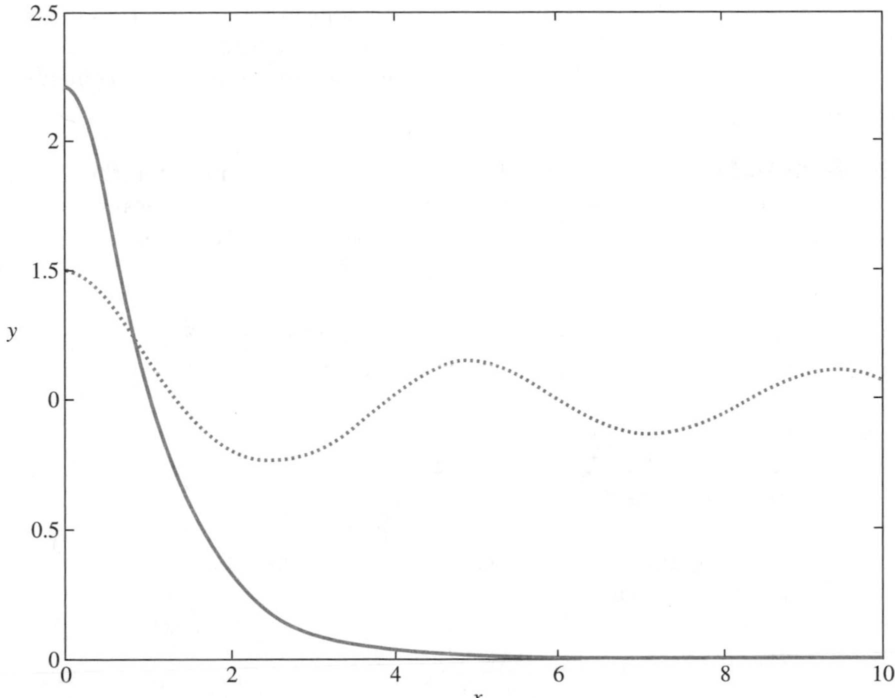

Figure 10.7 The ground state solution (solid curve) of the nonlinear ODE (10.36) and the starting approximation (dashed curve) of the nonlinear shooting method.

for $s = s_0$ and $s = s_0 + \Delta s$ with small Δs, then the derivative $f'(s_0)$ can be approximated by the forward-difference approximation:

$$Df(s_0) = \frac{f(s_0 + \Delta s) - f(s_0)}{\Delta s},$$

so that the first iteration can be performed by

$$s_1 = s_0 - \frac{\Delta s f(s_0)}{f(s_0 + \Delta s) - f(s_0)}.$$

In the subsequent iterations, the backward-difference approximation can be used to approximate the derivative $f'(s_n)$, so that further iterations of the secant method can be defined by the two-step difference equation:

$$s_{n+1} = s_n - \frac{(s_n - s_{n-1})f(s_n)}{f(s_n) - f(s_{n-1})}, \qquad n \geq 0. \tag{10.37}$$

Similar to the Newton–Raphson method, the secant method converges unconditionally, but the rate of convergence is slower for a simple root s_* of $f(s)$ (see Section 8.1). On the other hand, the method can be easily implemented with fewer *flops*.

Exercise 10.3 Show that the function $f(s)$ is linear in s for the linear second-order ODE (10.7) so that the Newton method (10.35) converges in a single iteration at $n = 0$, while the secant method (10.37) converges in two iterations.

The secant method applied to the linear ODE (10.32) with $L = 5$ is coded in the MATLAB script `shooting_secant_method`.

```
% shooting_secant_method
h = 0.1; x = -5 : h : 5; m = 0; ss = s;
s = input('Enter starting value for s = ');
y(1) = 0; u(1) = s;          % (y,u) - components of y(x)
for k = 1 : length(x)-1      % iterations of the ODE solver
        yp = y(k) + h*u(k);
        up = u(k) + h*(x(k)^2*y(k)+(3*x(k)^2-1)/((1+x(k)^2)^2));
        y(k+1) = y(k) + 0.5*h*(u(k)+up);
        uu1 = x(k)^2*y(k)+(3*x(k)^2-1)/((1+x(k)^2)^2);
        uu2 = x(k+1)^2*yp+(3*x(k+1)^2-1)/((1+x(k+1)^2)^2);
        u(k+1) = u(k) + 0.5*h*(uu1 + uu2);
end
yy = y(length(x)); s = s + h;
while (abs(s-ss) > 10^(-10)) % iterations of the secant method
    m = m + 1; y(1) = 0; u(1) = s;
    for k = 1 : length(x)-1   % iterations of the ODE solver
        yp = y(k) + h*u(k);
        up = u(k) + h*(x(k)^2*y(k)+(3*x(k)^2-1)/((1+x(k)^2)^2));
        y(k+1) = y(k) + 0.5*h*(u(k)+up);
        uu1 = x(k)^2*y(k)+(3*x(k)^2-1)/((1+x(k)^2)^2);
        uu2 = x(k+1)^2*yp+(3*x(k+1)^2-1)/((1+x(k+1)^2)^2);
        u(k+1) = u(k) + 0.5*h*(uu1 + uu2);
    end
    if (m == 1)
        s = ss - h*yy/(y(length(x))-yy);
    else
        sN = s - (s - ss)*y(length(x))/(y(length(x))-yy);
        yy = y(length(x)); ss = s; s = sN;
    end
end
fprintf('The secant method converges in %d iterations.',m);
```

The output of the secant method is independent of the starting value for s, and it converges in two iterations. The resulting solution $y(x)$ is not different from the one shown in Figure 10.6.

```
>> shooting_secant_method
Enter starting value for s = 0
    The secant method converges after 2 iterations.
```

Eigenvalue problems

The shooting method can be used for problems where the shooting parameter occurs explicitly in the ODE rather than in the boundary conditions. We illustrate this modification of the shooting method for the *Bessel* equation:

$$x^2 y'' + xy' + \lambda x^2 y = 0, \qquad 0 < x < 1, \tag{10.38}$$

subject to the boundary conditions $y'(0) = 0$ and $y(1) = 0$. Since the ODE (10.38) is linear, any solution can be multiplied by a nonzero constant and remains a solution, so that we can always obtain a solution normalized by $y(0) = 1$. Therefore, for each λ there exists a unique solution of the initial-value problem with $y(0) = 1$ and $y'(0) = 0$, while the additional boundary condition $y(1) = 0$ over-determines the problem. It is only for isolated values of λ called *eigenvalues* that the over-determined initial-boundary-value problem has a unique solution $y(x)$. The exact solution of the ODE (10.38) with $y(0) = 1$ and $y'(0) = 0$ is the Bessel function of zero order $y(x) = J_0(\sqrt{\lambda}x)$ so that all eigenvalues λ are given by roots of $J_0(\sqrt{\lambda}) = 0$. Plotting the Bessel function $J_0(\sqrt{\lambda})$ for $0 < \lambda < 140$, one can approximate the first four eigenvalues from the roots of $J_0(\sqrt{\lambda}) = 0$, as shown in Figure 10.8.

```
>> x = 0 : 1 : 140;
>> plot(x,BesselJ(0,sqrt(x)),'b',x,zeros(size(x)),':r')
```

You can compute eigenvalues of the boundary-value problem (10.38) numerically by using the shooting method, with the value $y(1)$ depending on the parameter λ. The MATLAB script `shooting_eigenvalue` gives details of the implementation of the corresponding shooting method. The ODE solver is the explicit Heun method and since the first step involves a singular value at $x = 0$, you need to use a power series approximation $y(x) = 1 - \lambda x^2/4 + O(x^4)$ of the solution $y(x)$ near $x = 0$.

```
% shooting_eigenvalue
h = 0.01; x = 0 : h : 1; m = 0; aa = 0;
a = input('Enter an initial approximation for eigenvalue = ');
while (abs(a-aa) > 10^(-10)) & (m < 1000)
    aa = a;
    y(1) = 1; u(1) = 0;                         % (y,u) - components of u(x)
```

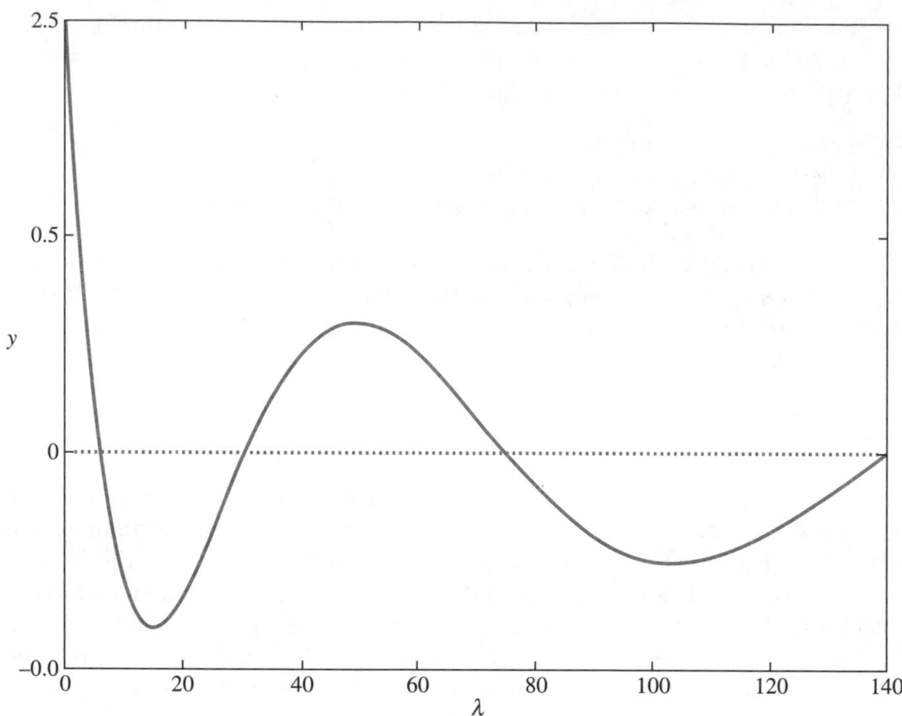

Figure 10.8 The first four roots of the Bessel function $J_0(\sqrt{\lambda}) = 0$.

```
    y(2) = 1 - a*x(2)^2/4; u(2) = -a*x(2)/2; % power series approximation
    z(1) = 0; v(1) = 0;                      % (z,v) - components of v(x)
    z(2) = -x(2)^2/4; v(2) = -x(2)/2;
    for k = 2 : length(x)-1                  % Heun's method
        yp = y(k) + h*u(k); up = u(k) - h*(u(k)/x(k)+a*y(k));
        y(k+1) = y(k) + 0.5*h*(u(k)+up);
        u(k+1) = u(k) - 0.5*h*(u(k)/x(k)+a*y(k)+up/x(k+1)+a*yp);
        zp = z(k) + h*v(k); vp = v(k) - h*(v(k)/x(k)+a*z(k)+y(k));
        z(k+1) = z(k) + 0.5*h*(v(k)+vp);
        v(k+1) = v(k) - 0.5*h*(v(k)/x(k)+a*z(k)+vp/x(k+1)+a*zp+y(k)+yp);
    end
    a = aa - y(length(x))/z(length(x));       % shooting method
    m = m+1;
end
fprintf('Final approximation of eigenvalue = %f\n',a)
fprintf('The shooting method converges in %d iterations\n',m);
plot(x,y);                                    % plotting eigenfunction
```

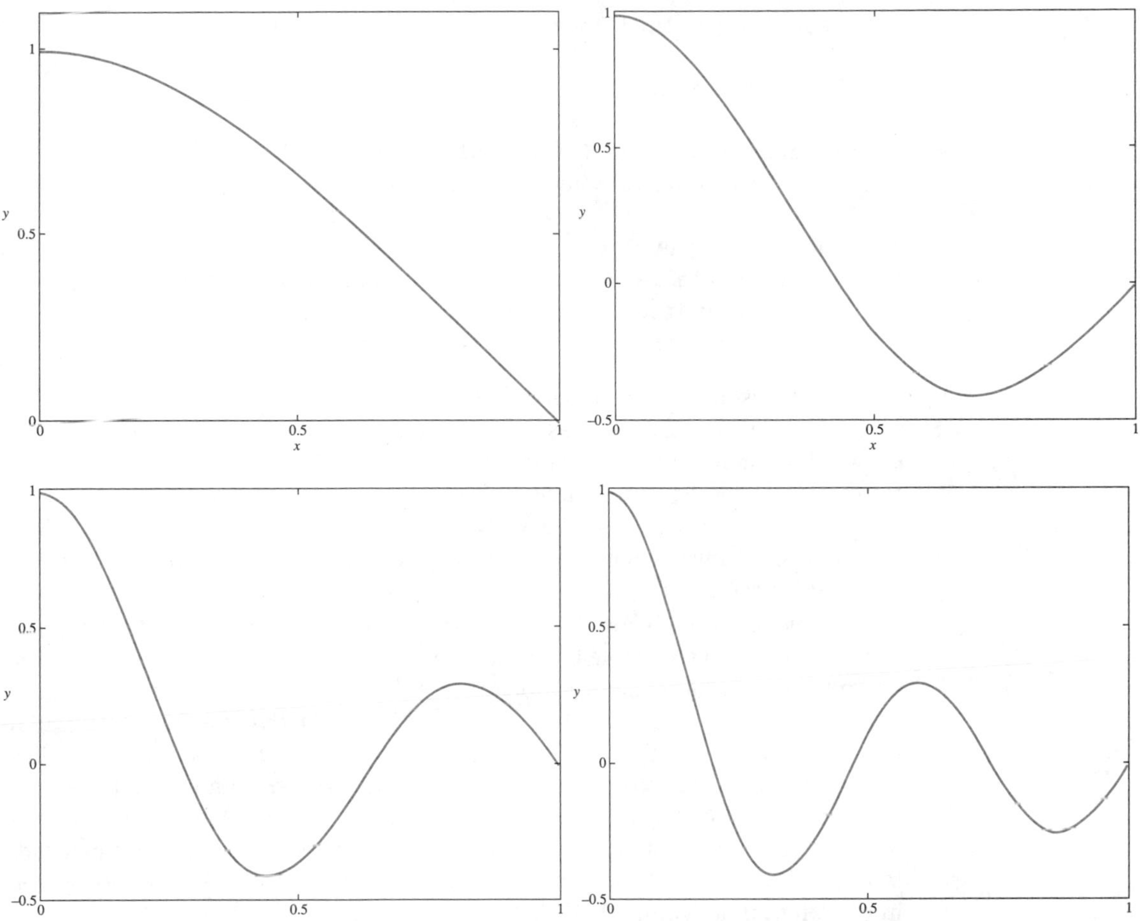

Figure 10.9 Four eigenfunctions of the Bessel equation (10.38).

Executing the script several times for different inputs, you can find good starting values for the first four smallest eigenvalues by trial and error. The corresponding eigenfunctions are plotted in Figure 10.9.

```
>> shooting_eigenvalue
 Enter an initial approximation for eigenvalue = 5
 Final approximation of eigenvalue = 5.782936
 The shooting method converges in 5 iterations

>> shooting_eigenvalue
 Enter an initial approximation for eigenvalue = 20
 Final approximation of eigenvalue = 30.451708
```

```
 The shooting method converges in 6 iterations

>> shooting_eigenvalue
 Enter an initial approximation for eigenvalue = 60
 Final approximation of eigenvalue = 74.743196
 The shooting method converges in 5 iterations

>> shooting_eigenvalue
 Enter an initial approximation for eigenvalue = 120
 Final approximation of eigenvalue = 138.502165
 The shooting method converges in 5 iterations
```

Comparison of the numerical values from the output of the shooting method with numerical values from the roots of $J_0(\sqrt{\lambda}) = 0$ in Figure 10.8 shows that these numerical values are identical. Also note that the eigenfunctions satisfy the Sturm–Liouville theory: the ground state for the smallest eigenvalue has no zeros on the space interval $(0, 1)$, while the eigenfunction for the nth eigenvalue sorted in increasing order has exactly $(n - 1)$ zeros on the space interval $(0, 1)$.

Shooting methods for eigenvalue problems can be generalized for higher-order ODEs with separable boundary conditions, when some boundary values are given on the left end of the domain of definition and some boundary values are given on the right end. In the case of linear ODEs, the resulting scalar function $f(\lambda)$ is a determinant of the matrix of fundamental solutions called the *Evans function*. Extensions to the nonlinear problems and boundary-value problems without the eigenvalue parameter are also possible but would take us too far adrift. We should only remark that no extensions of shooting methods to multidimensional PDE problems have been developed so far or implemented in existing numerical methods.

10.3 Finite-Difference Methods for Parabolic PDEs

Partial differential equations, or simply PDEs, arise in fundamental laws of physics for modeling of various processes, such as mechanical vibrations of strings and heat conductivity of metal rods. PDEs involve functions of more than one independent variable, for example, functions of time variable t and space variables x, y, and z. Only a few of the simplest boundary-value problems for linear PDEs can be solved analytically in a closed form. These solutions are derived with the techniques of separation of variables, the linear superposition principle, and the generalized Fourier series, which are typically covered in *Partial Differential Equations* courses. Because these analytical methods have limited applicability, we need to consider a numerical approximation of the solutions of PDEs. The simplest yet robust method for such numerical approximations is based on the extension of the finite-difference method for boundary-value ODE problems (see Section 10.1). We apply this method to a general mixed (initial-value in time and boundary-value in space) PDE problem on a finite interval, which involves first-order derivatives in time variable and second-order derivatives in space variable.

Problem 10.2 (Parabolic PDE Problem) Find a function $u(x,t)$ satisfying the second-order PDE

$$u_t = u_{xx} + F(x,t,u,u_x), \qquad a < x < b, \ \ t > 0, \tag{10.39}$$

subject to the Dirichlet boundary values

$$u(a,t) = \alpha(t), \qquad u(b,t) = \beta(t), \qquad t > 0, \tag{10.40}$$

and the Cauchy initial value

$$u(x,0) = u_0(x), \qquad a < x < b. \tag{10.41}$$

Here u_t, u_{xx} denote partial derivatives, all functions are supposed to be as smooth as needed, and the consistent conditions

$$u_0(a) = \alpha(0), \qquad u_0(b) = \beta(0),$$

are assumed to be satisfied.

A fundamental example of the partial differential equation (10.39) is the *one-dimensional heat equation*:

$$u_t = u_{xx} + F(x,t), \tag{10.42}$$

where $F(x,t)$ and $u(x,t)$ are referred to as the heat source and temperature distribution, respectively. The heat equation (10.42) has many important applications in heat transfer, diffusivity theory, combustion waves, and financial mathematics.

Finite-difference method for PDEs

The closed domain of definition of Problem 10.2 is the semi-infinite vertical strip:

$$D = \left\{(x,t) \in \mathbb{R}^2 : a \le x \le b, \ t \ge 0\right\}.$$

Following the ideas of the finite-difference method, we introduce a mesh of equally spaced vertical and horizontal grid levels:

$$D_{\text{mesh}} = \{(x_k,t_l), \ \ k = 0,1,\ldots,n, \ \ l = 0,1,\ldots\}, \tag{10.43}$$

where the discrete grid of x-values is defined by the spatial step size $h = (b-a)/n$ so that $x_k = a+kh$, $k = 0,1,\ldots,n$, while the discrete grid of t-values is defined by the time step τ so that $t_l = l\tau$, $l = 0,1,\ldots$. A smooth function of two variables $u(x,t)$ is approximated numerically at the discrete points of the mesh D_{mesh} as $u_{k,l}$. A two-dimensional interpolation through the values $u_{k,l}$ at the mesh points (x_k, t_l) can be used to recover the function $u(x,t)$ in the entire domain D. It is important to understand that $u_{k,l}$ is only an approximation for the value $u(x_k, t_l)$ subject to truncation and round-off errors of numerical computations. The central problem of *Numerical Analysis* is to measure the error between the exact and numerical values of the solution surface $u(x,t)$ and to establish the convergence criterion under which the truncation error reduces to zero as the step size h and the time step τ become negligibly small.

The finite-difference method for PDEs replaces continuous first and second derivatives with finite differences and reduces the partial differential equation (10.39) in the interior of the domain D to a set of algebraic equations. The algebraic system is then solved by algorithms of numerical linear algebra. Three methods of finite-difference approximations are commonly used for the heat equation (10.42): (1) explicit, (2) implicit, and (3) Crank–Nicolson.

Explicit method

The *explicit* method is equivalent to the Euler method at the time level t_l, $l = 0, 1, \dots$ for each interior grid point x_k, $k = 1, 2, \dots, n-1$:

$$
\begin{aligned}
u_{k,l+1} &= u_{k,l} + \tau u_t(x_k, t_l) + \mathrm{O}(\tau^2) \\
&= u_{k,l} + \tau \left(u_{xx}(x_k, t_l) + F(x_k, t_l)\right) + \mathrm{O}(\tau^2) \\
&= u_{k,l} + \tau \left(\frac{u_{k+1,l} - 2u_{k,l} + u_{k-1,l}}{h^2} + F_{k,l} \right) + \mathrm{O}(\tau^2, h^2),
\end{aligned}
$$

where the central-difference approximation is used for u_{xx} and the relation $F_{k,l} = F(x_k, t_l)$ is exact. Starting with the exact initial values $u_{k,0} = u_0(x_k)$ and incorporating the exact boundary values $u_{0,l} = \alpha(t_l)$ and $u_{n,l} = \beta(t_l)$, we can perform an explicit step from t_l to t_{l+1} and recover the values of $u_{k,l}$ at the interior mesh points in D_{mesh}. The global truncation error of the numerical approximation at any fixed final time $t = T > 0$ is one order larger in τ compared to the local error, that is, it is $\mathrm{O}(\tau, h^2)$.

Implicit method

The *implicit* method is equivalent to the backward Euler ODE solver:

$$
\begin{aligned}
u_{k,l+1} &= u_{k,l} + \tau u_t(x_k, t_{l+1}) + \mathrm{O}(\tau^2) \\
&= u_{k,l} + \tau \left(u_{xx}(x_k, t_{l+1}) + F(x_k, t_{l+1})\right) + \mathrm{O}(\tau^2) \\
&= u_{k,l} + \tau \left(\frac{u_{k+1,l+1} - 2u_{k,l+1} + u_{k-1,l+1}}{h^2} + F_{k,l+1} \right) + \mathrm{O}(\tau^2, h^2).
\end{aligned}
$$

The time step from t_l to t_{l+1} is performed by solving a linear system for the unknown values of $u_{k,l+1}$ at the interior mesh points in D_{mesh}. The global truncation error of the implicit method remains the same as in the explicit method, that is, it is $\mathrm{O}(\tau, h^2)$.

The *Crank–Nicolson* method is equivalent to the implicit Heun ODE solver:

Crank–Nicolson method

$$\begin{aligned}
u_{k,l+1} &= u_{k,l} + \frac{\tau}{2}\left(u_t(x_k,t_l) + u_t(x_k,t_{l+1})\right) + O(\tau^3) \\
&= u_{k,l} + \frac{\tau}{2}\left(u_{xx}(x_k,t_l) + F(x_k,t_l) + u_{xx}(x_k,t_{l+1}) + F(x_k,t_{l+1})\right) + O(\tau^3) \\
&= u_{k,l} + \frac{\tau}{2}\left(\frac{u_{k+1,l} - 2u_{k,l} + u_{k-1,l}}{h^2} + \frac{u_{k+1,l+1} - 2u_{k,l+1} + u_{k-1,l+1}}{h^2}\right) \\
&\quad + \frac{\tau}{2}\left(F_{k,l} + F_{k,l+1}\right) + O(\tau^3, h^2).
\end{aligned}$$

The time step from t_l to t_{l+1} is performed by solving a linear system for the unknown values of $u_{k,l+1}$ at the interior mesh points in D_{mesh}. The global truncation error of the Crank–Nicolson is $O(\tau^2, h^2)$, symmetrically with respect to time step τ and step size h.

All three finite-difference methods combine the single-step ODE solvers for initial-value problems and finite-difference ODE solvers for boundary-value problems. The resulting finite-difference equations are obtained after truncation of the remainder terms in the preceding linear equations.

At each time level t_l, all values of $u_{k,l+1}$ except for the boundary values $u_{0,l+1}$ and $u_{n,l+1}$ are unknown. Therefore, the finite-difference equations are best represented in the form where the unknown values are written on the left-hand side of the linear system and all known values are written on the right-hand side of the system. This representation is formulated in matrix-vector notations, which are suitable for numerical computations.

Let $r = \tau/h^2$. The explicit method is rewritten in the form:

$$u_{k,l+1} = (1-2r)u_{k,l} + r\left(u_{k+1,l} + u_{k-1,l}\right) + \tau F_{k,l}, \tag{10.44}$$

or in the matrix-vector form:

$$\mathbf{u}_{l+1} = \mathbf{A}_1 \mathbf{u}_l + \tau \mathbf{F}_l + r\mathbf{B}_l, \tag{10.45}$$

where

$$\mathbf{A}_1 = \begin{bmatrix} 1-2r & r & 0 \ldots 0 & 0 \\ r & 1-2r & r \ldots 0 & 0 \\ \vdots & \vdots & \vdots \ddots \vdots & \vdots \\ 0 & 0 & 0 \ldots r & 1-2r \end{bmatrix},$$

and

$$\mathbf{u}_l = \begin{bmatrix} u_{1,l} \\ u_{2,l} \\ \vdots \\ u_{n-1,l} \end{bmatrix}, \quad \mathbf{F}_l = \begin{bmatrix} F_{1,l} \\ F_{2,l} \\ \vdots \\ F_{n-1,l} \end{bmatrix}, \quad \mathbf{B}_l = \begin{bmatrix} u_{0,l} \\ 0 \\ \vdots \\ u_{n,l} \end{bmatrix}.$$

The implicit method is rewritten in the form:

$$(1+2r)u_{k,l+1} - r\left(u_{k+1,l+1} + u_{k-1,l+1}\right) = u_{k,l} + \tau F_{k,l+1}, \tag{10.46}$$

or in the matrix-vector form:

$$\mathbf{A}_2 \mathbf{u}_{l+1} = \mathbf{u}_l + \tau \mathbf{F}_{l+1} + r\mathbf{B}_{l+1}, \tag{10.47}$$

where

$$\mathbf{A}_2 = \begin{bmatrix} 1+2r & -r & 0 & \dots & 0 & 0 \\ -r & 1+2r & -r & \dots & 0 & 0 \\ \vdots & \vdots & \vdots & \ddots & \vdots & \vdots \\ 0 & 0 & 0 & \dots & -r & 1+2r \end{bmatrix},$$

and $\mathbf{u}_l$, $\mathbf{F}_l$ and $\mathbf{B}_l$ are the same as in the explicit method. Finally, the Crank–Nicolson method is rewritten in the form:

$$\begin{aligned} 2(1+r)u_{k,l+1} - r\left(u_{k+1,l+1} + u_{k-1,l+1}\right) &= 2(1-r)u_{k,l} \\ &+ r\left(u_{k+1,l} + u_{k-1,l}\right) + \tau\left(F_{k,l} + F_{k,l+1}\right), \end{aligned} \tag{10.48}$$

or in the matrix-vector form:

$$\mathbf{A}_4 \mathbf{u}_{l+1} = \mathbf{A}_3 \mathbf{u}_l + \tau\left(\mathbf{F}_l + \mathbf{F}_{l+1}\right) + r\left(\mathbf{B}_l + \mathbf{B}_{l+1}\right), \tag{10.49}$$

where

$$\mathbf{A}_3 = \begin{bmatrix} 2(1-r) & r & 0 & \dots & 0 & 0 \\ r & 2(1-r) & r & \dots & 0 & 0 \\ \vdots & \vdots & \vdots & \ddots & \vdots & \vdots \\ 0 & 0 & 0 & \dots & r & 2(1-r) \end{bmatrix},$$

$$\mathbf{A}_4 = \begin{bmatrix} 2(1+r) & -r & 0 & \dots & 0 & 0 \\ -r & 2(1+r) & -r & \dots & 0 & 0 \\ \vdots & \vdots & \vdots & \ddots & \vdots & \vdots \\ 0 & 0 & 0 & \dots & -r & 2(1+r) \end{bmatrix},$$

and $\mathbf{u}_l$, $\mathbf{F}_l$, and $\mathbf{B}_l$ are the same as in the explicit method. In all cases, the matrices $\mathbf{A}_1$, $\mathbf{A}_2$, $\mathbf{A}_3$, and $\mathbf{A}_4$ are tridiagonal. They are all diagonally dominant and invertible for small values of r so that direct and iterative algorithms of numerical linear algebra (see Sections 2.5 and 2.6) provide fast computational bases for iterations in time.

The next example gives comparative analysis of the three finite-difference methods in application to the heat equation (10.42) on the space interval $[0,1]$ with $F(x) = \sin(3\pi x)$, $u_0(x) = \sin(\pi x)$, and $\alpha(t) = \beta(t) \equiv 0$. The heat equation (10.42) has the exact solution:

$$u(x,t) = e^{-\pi^2 t} \sin(\pi x) + \frac{1}{9\pi^2}\left(1 - e^{-9\pi^2 t}\right) \sin(3\pi x). \qquad (10.50)$$

The solution $u(x,t)$ describes a quick relaxation of the initial condition $u(x,0) = \sin(\pi x)$ to the time-independent solution $\lim_{t\to\infty} u(x,t) = (9\pi^2)^{-1} \sin(3\pi x)$, which is supported by the time-independent source $F(x) = \sin(3\pi x)$.

```
% heat_equation
h = 0.1; x = 0 : h : 1; % grid in space
n = length(x) - 1;
tau = 0.005; t = 0 : tau : 0.6;  % grid in time
m = length(t)-1; m2 = 1+m/2; r = tau/h^2;
U1 = zeros(length(x),length(t)); % solution matrices for three methods
U1(:,1)=sin(pi*x'); U2 = U1; U3 = U1;
f = sin(3*pi*x)';  % iterations of the three linear systems
A1=(1-2*r)*diag(ones(1,n-1))+r*(diag(ones(1,n-2),1)+diag(ones(1,n-2),-1));
A2=(1+2*r)*diag(ones(1,n-1))-r*(diag(ones(1,n-2),1)+diag(ones(1,n-2),-1));
A3=(1-r)*diag(ones(1,n-1))+0.5*r*(diag(ones(1,n-2),1)+diag(ones(1,n-2),-1));
A4=(1+r)*diag(ones(1,n-1))-0.5*r*(diag(ones(1,n-2),1)+diag(ones(1,n-2),-1));
for k = 1 : m
    U1(2:n,k+1) = A1*U1(2:n,k) + tau*f(2:n);
    U2(2:n,k+1) = inv(A2)*(U2(2:n,k) + tau*f(2:n));
    U3(2:n,k+1) = inv(A4)*(A3*U3(2:n,k) + tau*f(2:n));
end
T = 0.3; xInt = 0 : 0.01 : 1; % plotting solutions at t = 0.3
Uex=exp(-pi^2*T)*sin(pi*xInt)+(1-exp(-9*pi^2*T))*sin(3*pi*xInt)/(9*pi^2);
figure(1); hold on;
plot(x',U1(:,m2),'r',x',U2(:,m2),'b',x',U3(:,m2),'g',xInt,Uex,':k');
plot(x',U1(:,m2),'*r',x',U2(:,m2),'*b',x',U3(:,m2),'*g');
[T,X] = meshgrid(t,x);  % plotting truncation errors versus time
Uex=exp(-pi^2*T).*sin(pi*X)+(1-exp(-9*pi^2*T)).*sin(3*pi*X)/(9*pi^2);
E1=max(abs(U1-Uex)); E2=max(abs(U2-Uex)); E3=max(abs(U3-Uex));
figure(2); plot(t,E1,'r',t,E2,'b',t,E3,'g');
```

When the MATLAB script `heat_equation` is executed, it produces two figures. Figure 10.10 (*left*) shows the value of $u(x,t)$ at $t = 0.3$, computed from the exact solution (10.50) and the three finite-difference methods. The truncation error is minimal in the Crank–Nicolson method and it is comparable between the explicit and implicit methods. Figure 10.10 (*right*) shows the graph of the maximal distance between exact and numerical solutions

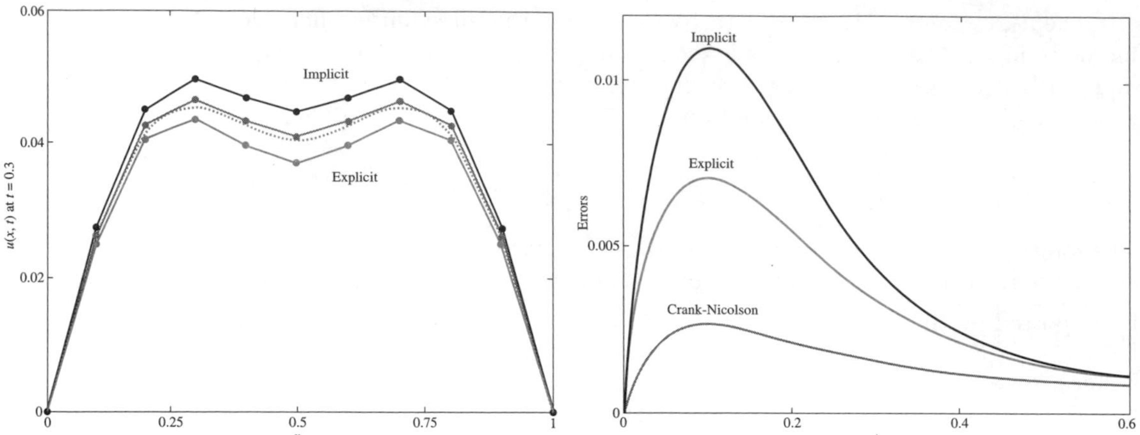

Figure 10.10 The exact solution (dotted curve) and the numerical solutions of the explicit, implicit, and Crank–Nicolson methods at $t = 0.3$ (*left*). The maximal distance between the exact and numerical solutions versus time t (*right*).

versus time. The maximal distance grows linearly in t at the beginning stage but relaxes to zero for larger values of t. Relaxation of the maximal distance is correlated with relaxation of the entire solution to the time-independent limiting solution.

Exercise 10.4 Consider the heat equation (10.42) with the exact solution (10.50). Plot the ∞-norm of the difference between exact and numerical solutions for a fixed value of time $t = T$ versus τ and h. Verify numerically that convergence of the explicit and implicit methods is linear in τ and quadratic in h, while convergence of the Crank–Nicolson method is quadratic both in τ and h.

Stability of iterations

Finite-difference methods for numerical solutions of partial differential equations can be surprisingly inappropriate for numerical approximations. The main problem with finite-difference methods (especially with explicit iteration schemes) is that they may magnify the numerical round-off noise.

We can investigate instabilities of numerical finite-difference methods by employing the separation of variables, the linear superposition principle, and the discrete Fourier transform. For simplicity, we consider the explicit method (10.44) for the homogeneous heat equation (10.42) with $F(x,t) \equiv 0$ and $\alpha(t) = \beta(t) \equiv 0$. The explicit scheme takes the form

$$u_{k,l+1} = (1-2r)u_{k,l} + r(u_{k+1,l} + u_{k-1,l}) \tag{10.51}$$

for $k = 1, 2, \ldots, n-1$ and $l = 0, 1, \ldots$, subject to the boundary conditions $u_{0,l} = u_{n,l} = 0$ and initial condition $u_{k,0} = u_0(x_k)$. Freeze the time level $t = t_l$ and expand the values of $u_{k,l}$ in the discrete Fourier sine-transform:

$$u_{k,l} = \sum_{j=1}^{n} a_{j,l} \sin\left(\frac{\pi k j}{n}\right), \qquad k = 0, 1, \ldots, n. \tag{10.52}$$

The boundary conditions $u_{0,l} = u_{n,l} = 0$ are satisfied for any $l \geq 0$. Because of the linear superposition principle, we can consider the time iterations of each term in the sum (10.52) separately. Hence, we can substitute $u_{k,l} = a_{j,l}\sin(\kappa_j k)$ with $\kappa_j = \frac{\pi j}{n}$ into the explicit method (10.51) and obtain

$$a_{j,l+1}\sin(\kappa_j k) = (1-2r)a_{j,l}\sin(\kappa_j k) + r a_{j,k}\left(\sin(\kappa_j(k+1)) + \sin(\kappa_j(k-1))\right).$$

Because of the trigonometric formula,

$$\sin(\kappa(k+1)) + \sin(\kappa(k-1)) = 2\cos(\kappa)\sin(\kappa k),$$

the factor $\sin(\kappa_j k)$ cancels out, and we obtain a first-order difference equation in l for fixed j:

$$a_{j,l+1} = Q_j a_{j,l}, \tag{10.53}$$

where

$$Q_j = 1 - 2r + 2r\cos(\kappa_j). \tag{10.54}$$

Because the factor Q_j is l-independent, the amplitude $a_{j,l}$ of the Fourier mode $\sin(\kappa_j k)$ changes during iterations in l, according to the powers of the factor Q_j:

$$a_{j,l} = Q_j^l a_{j,0}, \qquad l = 0, 1, \ldots$$

The amplitude $a_{j,l}$ grows in l if $|Q_j| > 1$ and it is bounded or decaying if $|Q_j| \leq 1$. Therefore, the explicit method is stable when

$$|Q_j| \leq 1, \qquad j = 1, 2, \ldots, n. \tag{10.55}$$

Because $Q_j < 1$ for $r > 0$ in (10.54), the stability constraint (10.55) can be rewritten as follows:

$$1 - 4r\sin^2\left(\frac{\pi j}{2n}\right) \geq -1, \qquad j = 1, 2, \ldots, n, \tag{10.56}$$

which results in the *conditional stability* of the explicit method for $0 < r \leq \frac{1}{2}$. When $r > \frac{1}{2}$, the first unstable Fourier mode corresponds to $j = n$, which is responsible for a pattern of time-growing space-alternative sequence of $u_{k,l}$.

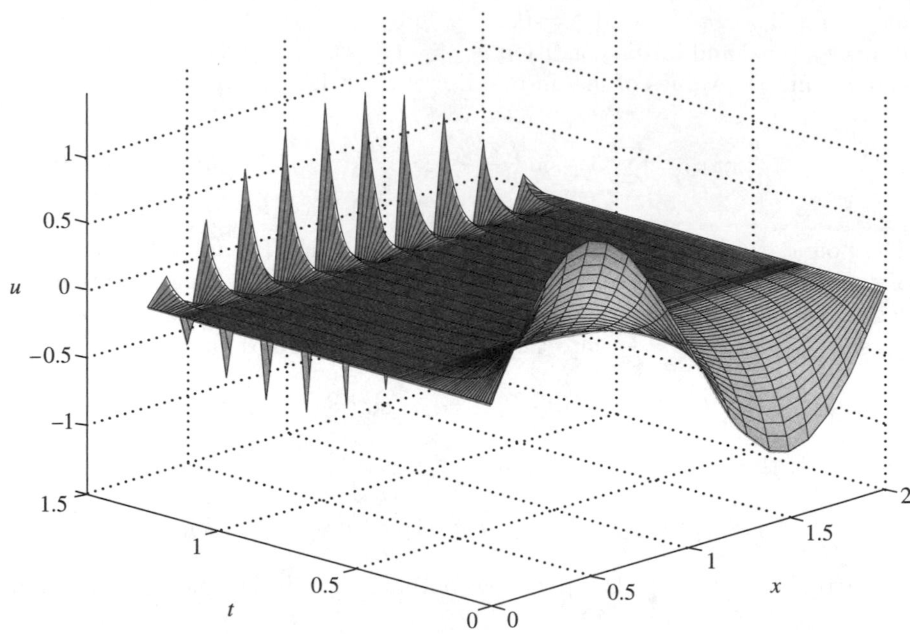

Figure 10.11 Instabilities of the explicit finite-difference method.

We illustrate the instability of the explicit finite-difference method (10.51) for an example with $r = 0.55$ for the space interval $[0, 1]$ and the initial condition $u_0 = \sin(\pi x)$. The exact solution $u(x, t) = e^{-\pi^2 t} \sin(\pi x)$ describes smooth decay of an initial condition $u_0(x)$ to the zero solution $\lim_{t\to\infty} u(x, t) = 0$. The output of the MATLAB script `instability_explicit-method` is shown in Figure 10.11. The expected smooth decay is destroyed by the noise that grows rapidly as a result of dynamical instabilities of the explicit method.

```
% instability_explicit_method
h = 0.1; x = 0 : h : 2; n = length(x) - 1;
 tau = 0.55*0.01; t = 0 : tau : 1.28; r = tau/h^2;
U1 = zeros(length(x),length(t)); U1(:,1) = sin(pi*x');
A1=(1-2*r)*diag(ones(1,n-1))+r*(diag(ones(1,n-2),1)+diag(ones(1,n-2),-1));
for k = 1 : (length(t)-1)
    U1(2:n,k+1) = A1*U1(2:n,k); % explicit method
end
[T,X] = meshgrid(t,x); surf(X,T,U);
```

Using the same analysis, we can easily prove *unconditional* stability of the implicit and Crank–Nicolson finite-difference methods. Consider the implicit method (10.46) for the homogeneous problem:

$$(1+2r)u_{k,l+1} - r(u_{k-1,l+1} + u_{k+1,l+1}) = u_{k,l}. \tag{10.57}$$

When $u_{k,l} = a_{j,l}\sin(\kappa_j k)$ is substituted for an individual Fourier mode with $\kappa_j = \frac{\pi j}{n}$ and $j = 1, 2, \ldots, n$, the implicit method (10.57) reduces to the first-order difference equation (10.53) with

$$Q_j = \frac{1}{1 + 2r - 2r\cos(\kappa_j)}. \tag{10.58}$$

For any $r > 0$ and any $j = 1, 2, \ldots, n$, we have $0 < Q_j < 1$, which implies that the implicit method is unconditionally stable. Similarly, the Crank–Nicolson method (10.48) for the homogeneous problem:

$$2(1+r)u_{k,l+1} - r\left(u_{k+1,l+1} + u_{k-1,l+1}\right) = 2(1-r)u_{k,l} + r\left(u_{k+1,l} + u_{k-1,l}\right) \tag{10.59}$$

reduces to the first-order difference equation (10.53) with

$$Q_j = \frac{1 - r + r\cos(\kappa_j)}{1 + r - r\cos(\kappa_j)}. \tag{10.60}$$

For any $r > 0$ and any $j = 1, 2, \ldots, n$, we have again $0 < Q_j < 1$, which implies that the Crank–Nicolson method is unconditionally stable.

Convergence of iterations

Analysis of convergence and stability of the finite-difference method provides a useful tool to control accuracy of the numerical solutions. If the step size h of the discrete spatial grid is decreased, the truncation error of the central-difference numerical derivatives reduces to zero as $O(h^2)$. For instance, if h is halved, then the truncation error is quartered. However, the stability of the explicit method depends on $r = \tau/h^2$, that is, the time step τ must be adjusted accordingly, to restore the stability of the explicit iterations. This implies that if h is halved, then τ needs to be quartered, and the computational complexity increases eight times. No adjustment of the time step τ is required in the implicit and Crank–Nicolson methods because of their unconditional stability. The difference between these two methods is in the convergence rate for the truncation error: $O(\tau)$ in the implicit method versus $O(\tau^2)$ in the Crank–Nicolson method. To reduce the error by a factor of four when the step size h is halved, the time step τ must be quartered in the implicit method and halved in the Crank–Nicolson method.

We illustrate the convergence of the three finite-difference methods for the homogeneous heat equation on the space interval $[0, 1]$ with the initial condition $u_0(x) = \sin(2\pi x)$. The exact solution is $u(x,t) = e^{-4\pi^2 t}\sin(2\pi x)$. The MATLAB script `error_heat_equation` compares the exact solution with the three numerical solutions for two step sizes $h = 0.1$ and $h = 0.05$. The time step is adjusted as $\tau = 0.1h^2$ in the explicit method, and so $r = 0.1$ is preserved. On the other hand, the time step is adjusted as $\tau = 0.1h$ in the implicit and Crank–Nicolson methods so that $r = 1$ for $h = 0.1$ and $r = 2$ for $h = 0.05$.

```
% error_heat_equation
T = 0.1;
for h = 0.05 : 0.05 : 0.1
    x = 0 : h : 1; n = length(x) - 1;
    tau1 = 0.1*h^2; t1 = 0 : tau1 : T;
    m1 = length(t1)-1; r1 = tau1/h^2;
    U1 = zeros(length(x),length(t1));
    tau2 = 0.1*h; t2 = 0 : tau2 : T;
    m2 = length(t2)-1; r2 = tau2/h^2;
    U2 = zeros(length(x),length(t2));
    U3 = zeros(length(x),length(t2));
    U1(:,1) = sin(2*pi*x'); U2(:,1) = sin(2*pi*x');
    U3(:,1) = sin(2*pi*x'); d1 = diag(ones(1,n-1));
    d2 = diag(ones(1,n-2),1) + diag(ones(1,n-2),-1);
    A1 = (1-2*r1)*d1 + r1*d2; A2 = (1+2*r2)*d1 - r2*d2;
    A3 = (1-r2)*d1 + 0.5*r2*d2; A4 = (1+r2)*d1 - 0.5*r2*d2;
    for k = 1 : m1
        U1(2:n,k+1) = A1*U1(2:n,k);
    end
    for k = 1 : m2
        U2(2:n,k+1) = inv(A2)*U2(2:n,k);
        U3(2:n,k+1) = inv(A4)*A3*U3(2:n,k);
    end
    Uex = exp(-4*pi^2*T)*sin(2*pi*x);
    if (h == 0.05)
        E1a = max(abs(U1(:,m1+1)-Uex'));
        E2a = max(abs(U2(:,m2+1)-Uex'));
        E3a = max(abs(U3(:,m2+1)-Uex'));
        fprintf('E1a = %f, E2a = %f, E3a = %f\n',E1a,E2a,E3a);
    elseif (h == 0.1)
        E1b = max(abs(U1(:,m1+1)-Uex'));
        E2b = max(abs(U2(:,m2+1)-Uex'));
        E3b = max(abs(U3(:,m2+1)-Uex'));
        fprintf('E1b = %f, E2b = %f, E3b = %f\n',E1b,E2b,E3b);
        r1 = E1b/E1a; r2 = E2b/E2a; r3 = E3b/E3a;
        fprintf('r1 = %f, r2 = %f, r3 = %f\n',r1,r2,r3);
    end
end
```

The output of the MATLAB script `error_heat_equation` shows the quadratic convergence for the explicit method, the linear convergence of the implicit method, and the quadratic convergence of the Crank–Nicolson method. The rate of convergence is estimated from the ratio of the global truncation errors in two computations. The ratio is close to 2 when the convergence is linear, and it is close to $2^2 = 4$ when the convergence is quadratic. The quadratic convergence of the explicit method is supported by the increased computational complexity because $\tau \sim h^2$. The rates of convergence of the implicit and Crank–Nicolson methods agree with the theoretical predictions of $O(\tau, h^2)$ and $O(\tau^2, h^2)$, respectively, where $\tau \sim h$.

```
>> error_heat_equation

E1a = 0.000254, E2a = 0.008695, E3a = 0.000386

E1b = 0.001005, E2b = 0.019079, E3b = 0.001542

r1 = 3.958203, r2 = 2.194354, r3 = 4.000137
```

Nonlinear heat equation

Although the analysis of convergence and stability is much more complicated when the function $F(x, t, u, u_x)$ in the PDE (10.39) is nonlinear in u, the finite-difference methods can be extended to nonlinear problems. For illustration, we consider the nonlinear heat equation

$$u_t = u_{xx} + u(1-u), \qquad -L < x < L, \ \ t > 0, \tag{10.61}$$

subject to the boundary conditions

$$u(-L, t) = 1, \qquad u(L, t) = 0, \tag{10.62}$$

and the initial condition

$$u(x, 0) = \frac{1}{2}\left(1 - \tanh x\right). \tag{10.63}$$

The initial condition corresponds to the transition front (called a *kink*) between the equilibrium state $u = 1$ for large negative x and the equilibrium state $u = 0$ for large positive x. When L is large (say, $L = 20$), the truncation of the solution on the bounded interval $[-L, L]$ introduces a small error, which is invisible in numerical simulations. The MATLAB script `nonlinear_heat_equation` shows an application of the explicit method to the nonlinear heat equation (10.61). The stability constraint on iterations of the explicit method is chosen from the linear part of the nonlinear equation (10.61).

```
% nonlinear_heat_equation
h = 0.1; L = 20; T = 5;
x = -L : h : L; n = length(x) - 1;
e1 = zeros(n-1,1); e1(1) = 1; % boundary conditions
tau = 0.5*h^2; t = 0 : tau : T;
m = length(t)-1; r = tau/h^2;
U = zeros(length(x),length(t)); % initial conditions
U(1,:) = ones(1,m+1); U(:,1)=0.5*(1 - tanh(x'));
A1=(1-2*r)*diag(ones(1,n-1))+r*(diag(ones(1,n-2),1)+diag(ones(1,n-2),-1));
for k = 1 : (length(t)-1)  % explicit method
    U(2:n,k+1) = A1*U(2:n,k) + tau*U(2:n,k).*(1-U(2:n,k)) + r*e1;
end
[T,X] = meshgrid(t,x); surf(X,T,U);
```

When the MATLAB script `nonlinear_heat_equation` is executed, it produces the solution surface $u(x,t)$ for $x \in [-20, 20]$ and $t \in [0, 5]$. The solution surface is shown in Figure 10.12 from the side (*left*) and the top (*right*). Despite limitations of the explicit method, the numerical iterations recover an accurate solution of the nonlinear heat equation (10.61), which describes a steady propagation of the kink to the right from the equilibrium state $u = 1$ for large negative x to the equilibrium state $u = 0$ for large positive x.

Note that applications of implicit methods to the nonlinear heat equation (10.61) would involve the necessity to solve a system of nonlinear equations at each time step of the iterative methods. Similar to the implicit methods for ODEs (see Section 9.5), semi-implicit methods could simplify iterations of the finite-difference methods for nonlinear PDEs.

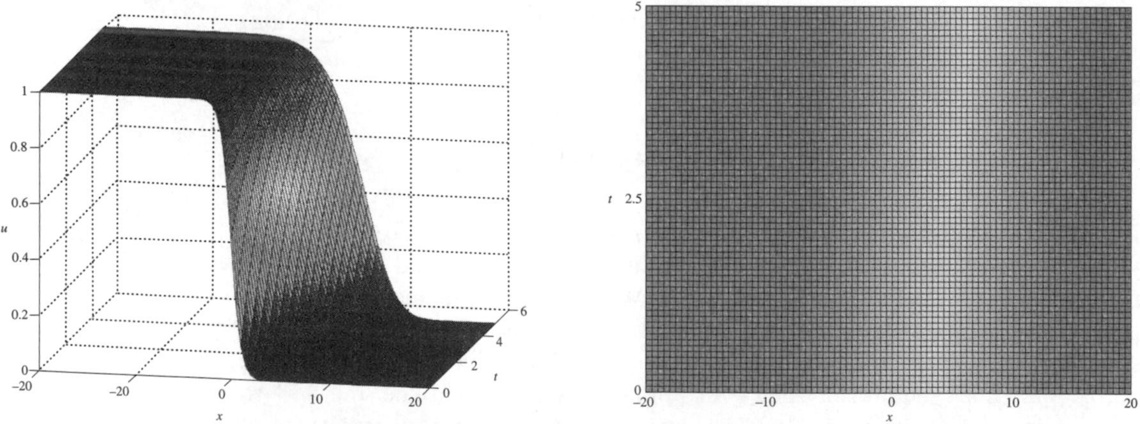

Figure 10.12 Numerical solution of the nonlinear heat equation (10.61) by using the explicit method. Solution surface (*left*). Contour plot (*right*).

Although we have constructed finite-difference methods for the heat equation (10.42) as combinations of single-step ODE solvers in time and central-difference ODE solvers in space, the same methods can be recovered from the finite-difference approximations of numerical derivatives. The first-order time derivative is replaced by the forward-difference numerical derivative in the explicit method and by the backward-difference numerical derivative in the implicit method, while the second-order space derivative is computed by the central-difference approximation at t_l and t_{l+1}, respectively. In the Crank–Nicolson method, the first-order time derivative is evaluated at the midpoint $t = t_l + \frac{1}{2}\tau$ by the central-difference formula, while the second-order space derivative at the midpoint is approximated by the mean value of these derivatives at the two adjacent time levels $t = t_l$ and $t = t_{l+1}$.

10.4 Finite-Difference Methods for Hyperbolic PDEs

Many problems arising in mathematical physics are described by partial differential equations with second-order derivatives both in time and space variables. For instance, one of the central law of physics, Newton's second law, states that the acceleration of a point mass, which is the second-order time derivative of its coordinate, is proportional to the resulting force. In the case of the tension force acting on a string, for instance, the coordinate is proportional to the vertical displacement of the string while the force is proportional to the second-order space derivative of the string displacement. We shall develop a simple modification of the finite-difference method to the partial differential equation with the second-order derivatives both in time and space variables.

Problem 10.3 (Hyperbolic PDE Problem) Find a function $u(x,t)$ satisfying the second-order PDE

$$u_{tt} = u_{xx} + F(x,t,u,u_t,u_x), \qquad a < x < b, \quad t > 0, \tag{10.64}$$

subject to the Dirichlet boundary values (10.40) and the Cauchy initial values

$$u(x,0) = u_0(x), \quad u_t(x,0) = v_0(x) \qquad a < x < b. \tag{10.65}$$

A fundamental example of the second-order partial differential equation (10.64) is the *one-dimensional wave equation*:

$$u_{tt} = u_{xx} + F(x,t), \tag{10.66}$$

where $F(x,t)$ and $u(x,t)$ are referred to as the external wave source and wave displacement, respectively. The wave equation (10.66) has many important applications in water waves, electromagnetic waves, acoustic waves, and elasticity theory.

Explicit and implicit methods

Because partial derivatives in time and space have equivalent roles in the wave equation (10.66), the finite-difference method can be developed directly by replacing the continuous derivatives by the finite-difference approximations. In this approach, it is not necessary to write the scalar second-order PDE (10.66) as a first-order system and to use the ODE solvers for first-order systems. Instead, the second-order partial derivatives can be replaced by the second-order central-difference numerical derivatives so that the wave equation (10.66) reduces to the system of linear algebraic equations. Two modifications of the finite-difference method are used: (1) explicit and (2) implicit. In the *explicit* method, the second derivatives are centered at the particular interior point of the numerical mesh (10.43):

$$\frac{u_{k,l+1} - 2u_{k,l} + u_{k,l-1}}{\tau^2} = \frac{u_{k+1,l} - 2u_{k,l} + u_{k-1,l}}{h^2} + F(x_k, t_l). \qquad (10.67)$$

In the *implicit* method, the spatial second derivatives are replaced by the mean value of the spatial second derivatives centered at the previous time layer t_{l-1} and the future time level t_{l+1}. The truncation order for this averaging procedure is $O(\tau^2)$ as the central-difference approximation for the second-order time derivative. On the other hand, time iterations now become implicit because the unknown values $u_{k,l+1}$ occur both on the left-hand side and right-hand side of the system:

$$\frac{u_{k,l+1} - 2u_{k,l} + u_{k,l-1}}{\tau^2} = F(x_k, t_l)$$
$$+\frac{u_{k+1,l+1} - 2u_{k,l+1} + u_{k-1,l+1} + u_{k+1,l-1} - 2u_{k,l-1} + u_{k-1,l-1}}{2h^2}. \qquad (10.68)$$

Defining the *Courant number* $\gamma = \tau/h$, the explicit method is rewritten in the form:

$$u_{k,l+1} = 2(1-\gamma^2)u_{k,l} + \gamma^2\left(u_{k+1,l} + u_{k-1,l}\right) - u_{k,l-1} + \tau^2 F_{k,l}, \qquad (10.69)$$

or in the matrix-vector form:

$$\mathbf{u}_{l+1} = \mathbf{A}_1\mathbf{u}_l - \mathbf{u}_{l-1} + \tau^2\mathbf{F}_l + \gamma^2\mathbf{B}_l, \qquad (10.70)$$

where

$$\mathbf{A}_1 = \begin{bmatrix} 2(1-\gamma^2) & \gamma^2 & 0 & \dots & 0 & 0 \\ \gamma^2 & 2(1-\gamma^2) & \gamma^2 & \dots & 0 & 0 \\ \vdots & \vdots & \vdots & \ddots & \vdots & \vdots \\ 0 & 0 & 0 & \dots & \gamma^2 & 2(1-\gamma^2) \end{bmatrix},$$

and

$$\mathbf{u}_l = \begin{bmatrix} u_{1,l} \\ u_{2,l} \\ \vdots \\ u_{n-1,l} \end{bmatrix}, \qquad \mathbf{F}_l = \begin{bmatrix} F_{1,l} \\ F_{2,l} \\ \vdots \\ F_{n-1,l} \end{bmatrix}, \qquad \mathbf{B}_l = \begin{bmatrix} u_{0,l} \\ 0 \\ \vdots \\ u_{n,l} \end{bmatrix}.$$

The implicit method is rewritten in the form:

$$\begin{aligned} 2(1+\gamma^2)u_{k,l+1} - \gamma^2\left(u_{k+1,l+1} + u_{k-1,l+1}\right) = 4u_{k,l} + \tau^2 F_{k,l} \\ -2(1+\gamma^2)u_{k,l-1} + \gamma^2\left(u_{k+1,l-1} + u_{k-1,l-1}\right), \end{aligned} \quad (10.71)$$

or in the matrix-vector form:

$$\mathbf{A}_2\mathbf{u}_{l+1} = 4\mathbf{u}_l - \mathbf{A}_2\mathbf{u}_{l-1} + \tau^2\mathbf{F}_l + \gamma^2\left(\mathbf{B}_{l-1} + \mathbf{B}_{l+1}\right), \quad (10.72)$$

where

$$\mathbf{A}_2 = \begin{bmatrix} 2(1+\gamma^2) & -\gamma^2 & 0 & \dots & 0 & 0 \\ -\gamma^2 & 2(1+\gamma^2) & -\gamma^2 & \dots & 0 & 0 \\ \vdots & \vdots & \vdots & \ddots & \vdots & \vdots \\ 0 & 0 & 0 & \dots & -\gamma^2 & 2(1+\gamma^2) \end{bmatrix}.$$

Initial time step

Both iterative methods are classified as *two-step* discrete maps because the values of $\mathbf{u}_0$ and $\mathbf{u}_1$ are needed to obtain the values of $\mathbf{u}_2$, $\mathbf{u}_3$ at the future time levels. Since the initial condition $u(x,0) = u_0(x)$ in (10.65) gives only $\mathbf{u}_0$, we have to express $\mathbf{u}_1$ from the initial condition $u_t(x,0) = v_0(x)$. This relationship can be obtained by using the finite-difference methods for Neumann boundary conditions (see Section 10.1). Let $t = t_{-1}$ be a virtual time level and $\mathbf{u}_{-1}$ be the corresponding vector. Using the central-difference approximation for the first derivative $u_t(x,0)$, we find that

$$\mathbf{u}_{-1} = \mathbf{u}_1 - 2\tau\mathbf{v}_0, \quad (10.73)$$

where $\mathbf{v}_0$ is computed from $v_0(x)$ at the interior grid points x_k for $k = 1, 2, \dots, n-1$. Extending the finite-difference equations to the level $l = 0$, we eliminate the value of $\mathbf{u}_{-1}$ from the system and obtain the first time step separately. The first step for the explicit method is given by

$$\mathbf{u}_1 = \frac{1}{2}\mathbf{A}_1\mathbf{u}_0 + \tau\mathbf{v}_0 + \frac{1}{2}\tau^2\mathbf{F}_0 + \frac{1}{2}\gamma^2\mathbf{B}_0, \quad (10.74)$$

while that for the implicit method is given by

$$\mathbf{A}_2\mathbf{u}_1 = 2\mathbf{u}_0 + \tau\mathbf{A}_2\mathbf{v}_0 + \frac{1}{2}\tau^2\mathbf{F}_0 + \frac{1}{2}\gamma^2\left(\mathbf{B}_{-1} + \mathbf{B}_1\right). \quad (10.75)$$

After the first step is performed and both $\mathbf{u}_0$ and $\mathbf{u}_1$ are found, the two-step discrete maps (10.69) and (10.71) can be used to obtain values for $\mathbf{u}_2$, $\mathbf{u}_3$, and so on.

Convergence and stability

Both the explicit and implicit methods have truncation errors that are $O(\tau^2, h^2)$. Therefore, both methods converge as $\tau \to 0$ and $h \to 0$ as second-order numerical methods. The two methods are, however, different with respect to the stability properties. After an example, we show that the explicit method is *conditionally* stable for $\gamma^2 \leq 1$, while the implicit method is *unconditionally* stable for any γ.

Consider the homogeneous wave equation (10.66) on the space interval $[0, 1]$ with $F(x,t) \equiv 0$, $\alpha(t) = \beta(t) \equiv 0$, $u_0(x) = \sin(\pi x)$, and $v_0(x) = \sin(2\pi x)$. The exact solution to the wave equation (10.66) has the form:

$$u(x,t) = \cos(\pi t)\sin(\pi x) + \frac{1}{2\pi}\sin(2\pi t)\sin(2\pi x). \tag{10.76}$$

This solution describes a standing wave that oscillates on the finite interval space $[0, 1]$. The MATLAB script `wave_equation` produces computations of the explicit and implicit methods with step size $h = 0.01$ and time step $\tau = 0.0125$.

```
% wave_equation
h = 0.01; x = 0 : h : 1; % grid in space
nX = length(x)-2; x1 = x(2:nX+1);
tau = 0.0125; r = tau/h; t = 0 : tau : 1; % grid in time
nT = length(t); U1 = zeros(nX,nT);
U1(:,1) = sin(pi*x1'); U2 = U1; % initial conditions
 d1 = diag(ones(1,nX));
 d2 = diag(ones(1,nX-1),1) + diag(ones(1,nX-1),-1);
A1 = 2*(1-r^2)**+r^2*d2; A2 = 2*(1+r^2)*d1 - r^2*d2;
U1(:,2)=0.5*A1*U1(:,1)+tau*sin(2*pi*x1');  % first steps
U2(:,2)=2*inv(A2)*U2(:,1)+tau*sin(2*pi*x1');
for k = 3 : nT  % further steps
        U1(:,k) = A1*U1(:,k-1)-U1(:,k-2);
        U2(:,k) = 4*inv(A2)*U2(:,k-1)-U2(:,k-2);
end
U1=[zeros(1,length(t));U1;zeros(1,length(t))];
U2=[zeros(1,length(t));U2;zeros(1,length(t))];
 [X,T1] = meshgrid(x,t(1:(nT-1-12*2)/2));
figure(1); surf(X,T1,U1(:,1:(nT-1-12*2)/2)');
 [X,T] = meshgrid(x,t);
figure(2); surf(X,T,U2');
```

The graphical output of the MATLAB script `wave_equation` (Figure 10.13) shows that the explicit method generates instabilities if $\gamma = 1.25$.

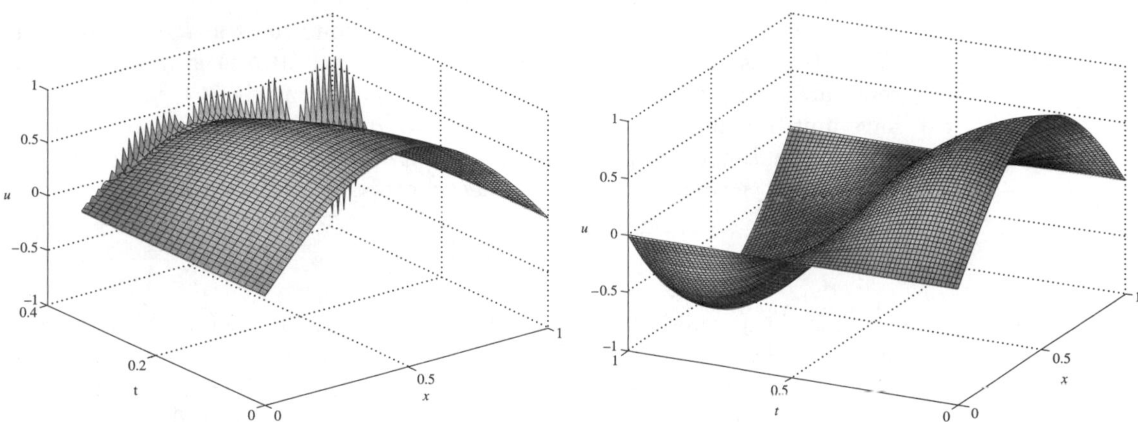

Figure 10.13 Numerical instabilities of the explicit method for the wave equation (10.66) (*left*). Stable numerical solution with the implicit method (*right*).

The implicit method produces an accurate solution of the homogeneous wave equation (10.66) for the same γ.

We study the stability of the explicit method for the homogeneous problem with $F(x,t) \equiv 0$ and $\alpha(t) = \beta(t) \equiv 0$. Let $u_{k,l}$ be represented in the form of a discrete Fourier transform with an individual Fourier mode $u_{k,l} = a_{j,l}\sin(\kappa_j k)$, where $\kappa_j = \frac{\pi j}{n}$ and $j = 1, 2, \ldots, n$. The explicit method (10.69) reduces to the second-order difference equation:

$$a_{j,l+1} = 2(1 - \gamma^2 + \gamma^2 \cos\kappa_j)a_{j,l} - a_{j,l-1}. \tag{10.77}$$

Similar to the analysis of the second-order difference equations in Section 9.4, we reduce the second-order difference equation (10.77) to the first-order difference equation (10.53) with the l-independent factor Q_j. The two possible values for Q_j follow from the quadratic equation:

$$Q_j^2 - 2(1 - \gamma^2 + \gamma^2 \cos\kappa_j)Q_j + 1 = 0. \tag{10.78}$$

Two roots $Q_j^{\pm}$ of the quadratic equation (10.78) satisfy the relations:

$$Q_j^+ Q_j^- = 1, \qquad Q_j^+ + Q_j^- = 2(1 - \gamma^2 + \gamma^2 \cos\kappa_j), \tag{10.79}$$

and they can be found explicitly from the quadratic formula:

$$Q_j^{\pm} = 1 - 2\gamma^2 \sin^2\frac{\kappa_j}{2} \pm 2\gamma \sin\frac{\kappa_j}{2}\sqrt{\gamma^2 \sin^2\frac{\kappa_j}{2} - 1}. \tag{10.80}$$

If the two roots are distinct, the general solution of the two-step discrete map (10.77) is represented by the linear superposition principle,

$$a_{j,l} = Q_j^{+l}\alpha_j + Q_j^{-l}\beta_j, \tag{10.81}$$

where the constants α_j, β_j are found from the conditions at $l = 0$ and $l = 1$. When the discriminant of the quadratic equation (10.78) is negative, both roots are complex conjugate. By the first equality in (10.79), the two roots have unit modulus and the Fourier mode $a_{j,l}$ does not grow but oscillates in l. This situation corresponds to the stability of the Fourier mode in the time evolution of the explicit finite-difference method. The discriminant is negative when

$$\gamma^2 \sin^2 \frac{\kappa_j}{2} < 1,$$

which is always valid when $\gamma^2 < 1$. On the other hand, when the discriminant of the quadratic equation is strictly positive, the two roots are real and distinct. As a result of the first equality in (10.79), one of the roots has modulus greater than one. Therefore, the Fourier mode $a_{j,l}$ grows in l (when coefficients α_j and β_j in the representation (10.81) are nonzero). It is clear that the discriminant is strictly positive for $j = n$ when $\gamma^2 > 1$, which indicates the most unstable Fourier mode at $\kappa_n = \pi$. In the marginal case $\gamma = 1$, the two roots $Q_j^{\pm} = e^{\pm i\kappa_j}$ are distinct and stable for all κ_j except for $\kappa_n = \pi$, when the two roots coincide at $Q_n^{\pm} = -1$ and the general solution of the second-order difference equation (10.77) is no longer (10.81). Although the general solution in this case grows linearly in l, we can neglect this weak instability since the Fourier mode $\sin(\kappa_n k) = \sin(\pi k)$ is identically zero for all $k \in \mathbb{Z}$. Thus, the marginal case $\gamma = 1$ is considered to be stable if the boundary conditions in Problem 10.3 are Dirichlet.

Exercise 10.5 Consider the implicit method (10.71) for the homogeneous wave equation and find the second-order difference equation for the Fourier mode $u_{k,l} = a_{j,l} \sin(\kappa_j k)$, where $\kappa_j = \frac{\pi j}{n}$. Find and solve the quadratic equation for Q_j in the first-order difference equation (10.53). Prove that the implicit method (10.71) is unconditionally stable for any γ.

When the step size h is reduced by the factor of two, the time step of both methods needs to be adjusted for stability and convergence. If the time step τ is also reduced by the factor of two, then the truncation error is reduced by the factor of four (because the method is second order in h and τ), while the value of γ is then preserved. The same final time $t = T$ is then reached with four times more computations. This adjustment of the time step is necessary in the explicit method to preserve stability of iterations. On the other hand, the implicit method can operate with the original time step because of its unconditional stability for any γ, but the truncation error may reduce slower than by the factor of four.

Consider the same example of the homogeneous wave equation (10.66), which admits the exact solution (10.76). The MATLAB script `error_wave_equation` estimates the rate of convergence of the global truncation error for the explicit and implicit methods by performing two computations with step sizes h and $2h$ and fixed values of $\gamma = 0.5$.

```
% error_wave_equation
T = 1; r = 0.5;
for h = 0.01 : 0.01 : 0.02
    x = 0 : h : 1; nX = length(x)-2;
    x1 = x(2:nX+1); tau = r*h;
    t = 0 : tau : T; nT = length(t);
    U1 = zeros(nX,nT); U1(:,1) = sin(pi*x1');
    U2 = U1; d1 = diag(ones(1,nX));
    d2 = diag(ones(1,nX-1),1)+diag(ones(1,nX-1),-1);
    A1 = 2*(1-r^2)*d1 + r^2*d2; A2 = 2*(1+r^2)*d1 - r^2*d2;
    U1(:,2) = 0.5*A1*U1(:,1)+tau*sin(2*pi*x1');
    U2(:,2) = 2*inv(A2)*U2(:,1)+tau*sin(2*pi*x1');
    for k = 3 : nT
        U1(:,k) = A1*U1(:,k-1)-U1(:,k-2);
        U2(:,k) = 4*inv(A2)*U2(:,k-1)-U2(:,k-2);
    end
    Uexact = cos(pi*T)*sin(pi*x1')+sin(2*pi*T)*sin(2*pi*x1')/(2*pi);
    if (h == 0.01)
        E1a = max(abs(Uexact-U1(:,nT)));
        E2a = max(abs(Uexact-U2(:,nT)));
        fprintf('E1a = %f, E2a = %f\n',E1a,E2a);
    elseif (h == 0.02)
        E1b = max(abs(Uexact-U1(:,nT)));
        E2b = max(abs(Uexact-U2(:,nT)));
        fprintf('E1b = %f, E2b = %f\n',E1b,E2b);
        R1 = E1b/E1a; R2 = E2b/E2a;
        fprintf('R1 = %f, R2 = %f\n',R1,R2);
    end
end
```

When the MATLAB script `error_wave_equation` is executed, it shows that the ratio of truncation errors is close to $2^2 = 4$. These numerical results confirm the second-order convergence of the explicit and implicit methods for the wave equation with an $O(\tau^2, h^2)$ truncation error.

```
>> error_wave_equation

E1a = 0.000123, E2a = 0.000370

E1b = 0.000493, E2b = 0.001478

R1 = 3.996624, R2 = 3.993796
```

Exact finite-difference solution

A remarkable cancelation exists for the explicit method, when it is applied to the homogeneous wave equation (10.66) with $\gamma = 1$ ($\tau = h$). The exact solution of Problem 10.3 for the homogeneous wave equation (10.66) is written in the *D'Alambert form*:

$$u(x,t) = \frac{1}{2}\left[u_0(x-t) + u_0(x+t)\right] + \frac{1}{2}\int_{x-t}^{x+t} v_0(s)ds. \qquad (10.82)$$

Let $\tau = h$ and evaluate $u(x,t)$ at the mesh points (10.43):

$$\begin{aligned} u_{k,l+1} + u_{k,l-1} &= \frac{1}{2}\left[u_0(x_k - t_{l+1}) + u_0(x_k + t_{l+1}) + u_0(x_k - t_{l-1}) + u_0(x_k + t_{l-1})\right] \\ &\quad + \frac{1}{2}\left(\int_{x_k - t_{l+1}}^{x_k + t_{l+1}} + \int_{x_k - t_{l-1}}^{x_k + t_{l-1}}\right) v_0(s)ds \\ &= \frac{1}{2}\left[u_0(x_{k-1} - t_l) + u_0(x_{k+1} + t_l) + u_0(x_{k+1} - t_l) + u_0(x_{k-1} + t_l)\right] \\ &\quad + \frac{1}{2}\left(\int_{x_{k-1} - t_l}^{x_{k+1} + t_l} + \int_{x_{k+1} - t_l}^{x_{k-1} + t_l}\right) v_0(s)ds = u_{k+1,l} + u_{k-1,l}. \end{aligned}$$

Therefore, the explicit method with $\gamma = 1$ reproduces an *exact* solution of the homogeneous wave equation, and the truncation error is identically zero. This cancelation of the truncation error is illustrated in the previous example with a relatively large step size $h = 0.1$.

```
% exact_solution
T = 1; h = 0.1; x = 0 : h : 1;
nX = length(x)-2; x1 = x(2:nX+1);
tau = h; t = 0 : tau : T; nT = length(t);
U1 = zeros(nX,nT); U1(:,1) = sin(pi*x1');
A1 = diag(ones(1,nX-1),1) + diag(ones(1,nX-1),-1);
U1(:,2) = 0.5*A1*U1(:,1) + tau*sin(2*pi*x1');
for k = 3 : nT
        U1(:,k) = A1*U1(:,k-1)-U1(:,k-2);
end
Uexact = cos(pi*T)*sin(pi*x1')+sin(2*pi*T)*sin(2*pi*x1')/(2*pi);
```

When the MATLAB script `exact_solution` is executed, you can see that the error is of the order of 10^{-16}, which corresponds to the level of machine precision:

```
>> exact_solution

Error =
    4.4409e-016
```

Nonlinear wave equation

The finite-difference methods can be applied to a general nonlinear second-order PDE (10.64). For instance, consider the nonlinear wave equation

$$u_{tt} - u_{xx} + u(1-u^2) = 0, \qquad -L < x < L, \;\; t > 0, \tag{10.83}$$

subject to the periodic boundary conditions

$$u(-L,t) = u(L,t), \qquad u_x(-L,t) = u_x(L,t)$$

and to the initial conditions

$$u(x,0) = e^{-x^2}, \qquad u_t(x,0) = 0.$$

Let $L = 10$ and the discrete grid be defined by $x_k = -L + (k-1)h$, where $h = 2L/n$. The periodic boundary conditions can be implemented by closing the discrete grid in a loop so that $u_{0,l} = u_{n,l}$ and $u_{1,l} = u_{n+1,l}$. As a result, at each time level $t = t_l$ there exists n unknown values of $u_{k,l+1}$ for $k = 1, 2, \ldots, n$, which are found from the explicit method:

$$u_{k,l+1} = 2(1-\gamma^2)u_{k,l} + \gamma^2\left(u_{k+1,l} + u_{k-1,l}\right) - u_{k,l-1} - \tau^2 u_{k,l}(1 - u_{k,l}^2), \tag{10.84}$$

for $k = 1, 2, \ldots, n$, subject to $u_{0,l} = u_{n,l}$ and $u_{n+1,l} = u_{1,l}$. The linear part of the explicit method is stable for $\gamma \le 1$.

```
% nonlinear_wave_equation
h = 0.1; x = -10 : h : 10;
nX = length(x)-1; x1 = x(1:nX);
tau = 0.01; r = tau/h;
t = 0 : tau : 10; nT = length(t);
U = zeros(nX,nT); U(:,1) = exp(-x1'.*x1'); % initial condition
d1 = diag(ones(1,nX));
d2 = diag(ones(1,nX-1),1) + diag(ones(1,nX-1),-1);
A = 2*(1-r^2)*d1 +r^2*d2; A(1,nX) = r^2; A(nX,1) = r^2;
U(:,2) = 0.5*A*U(:,1); % first step
for k = 3 : nT     % explicit method
        U(:,k) = A*U(:,k-1)-U(:,k-2)+tau^2*U(:,k-1).*(U(:,k-1).^2-1);
end
U = [ U; U(1,:) ];
[X,T] = meshgrid(x,t);
surf(X,T,U');
```

When the MATLAB script `nonlinear_wave_equation` is executed, it produces the solution surface shown in Figure 10.14 from the side (*left*) and top (*right*). The numerical solution describes a generation of radiative waves escaping the localized initial hump centered at $x = 0$ to the boundaries $x =$

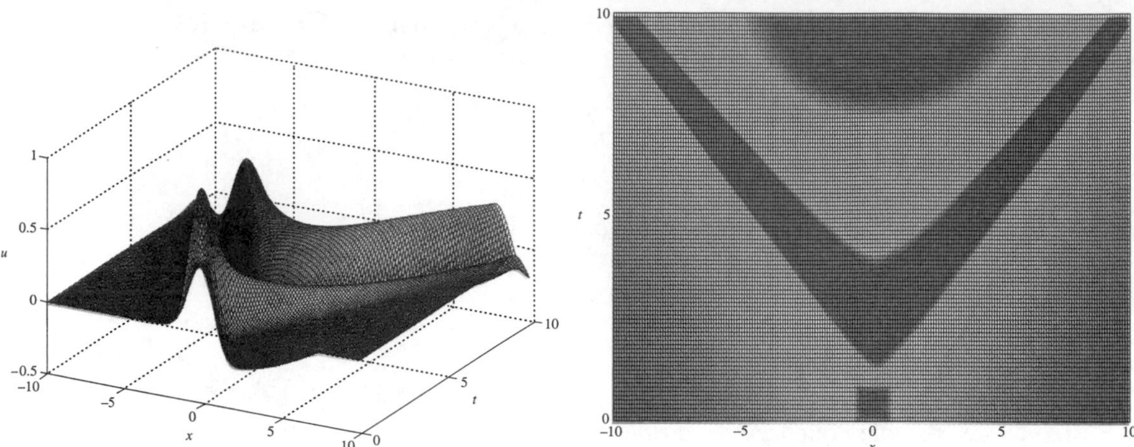

Figure 10.14 Numerical solution of the nonlinear wave equation (10.83) by using the explicit method with $\gamma = 0.1$. Solution surface (*left*). Contour plot (*right*).

$\pm L$. The radiative waves later reappear back to the computational domain because of periodic reflections from the boundaries.

Applications of the implicit finite-difference methods to the nonlinear wave equations are limited by the root-finding algorithms for large systems of nonlinear equations. Semi-implicit methods avoid these difficulties, if the nonlinear terms are defined at the present time level t_l and the linear terms are discretized at the future and past time levels t_{l+1} and t_{l-1}. By solving the linear part with numerical linear algebra, we can define an explicit iterative scheme, which has better stability properties compared to the explicit method. Nevertheless, the semi-implicit methods may suffer from instabilities caused by the nonlinear terms of the nonlinear PDEs.

10.5 Finite-Difference Methods for Elliptic PDEs

In many physical problems such as heat conductivity, time evolution of a solution is observed only at the transient stage, during which the solution relaxes asymptotically to a time-independent state. If the object is localized in a bounded domain, the time-independent state represents a balance between external forces or sources and the boundary conditions from exterior objects. The finite-difference method based on discretization of spatial derivatives can also be applied to boundary-value PDE problems even if no explicit time evolution occurs. We consider, in particular, a two-dimensional second-order PDE problem with separated boundary conditions.

Problem 10.4 (Elliptic PDE Problem) Find a twice continuously differentiable function $u(x, y)$ in a bounded domain $D \subset \mathbb{R}^2$, such that $u(x, y)$ solves the second-order PDE

$$u_{xx} + u_{yy} + F(x, y, u, u_x, u_y) = 0, \qquad (x, y) \in D, \tag{10.85}$$

subject to the Dirichlet boundary values:

$$u = \phi(x, y), \qquad (x, y) \in \partial D, \tag{10.86}$$

where $F(x, y, u, u_x, u_y)$ and $\phi(x, y)$ are continuous functions.

The fundamental example of the partial differential equation (10.85) is the *Poisson equation*:

$$u_{xx} + u_{yy} + F(x, y) = 0. \tag{10.87}$$

When $F(x, y) = 0$, the Poisson equation is referred to as the *Laplace equation*. The Poisson and Laplace equations have many important applications in electrostatics, potential motion of fluids, and elasticity theory.

Finite-difference solution

We consider the boundary-value PDE problem in a simple case, when the domain D is a rectangle,

$$D = \{(x, y) \in \mathbb{R}^2 : \ a_x < x < b_x, \ a_y < y < b_y\}. \tag{10.88}$$

For open rectangular domains, a mesh of equally spaced vertical and horizontal grid levels can be introduced similar to the mesh (10.43):

$$D_{\text{mesh}} = \{(x_k, y_m), \ \ k = 1, \ldots, n_x - 1, \ \ m = 1, \ldots, n_y - 1\}, \tag{10.89}$$

where the discrete grids are defined by the step sizes $h_x = (b_x - a_x)/n_x$ and $h_y = (b_y - a_y)/n_y$ so that $x_k = a_x + kh_x$, $k = 1, \ldots, n_x - 1$ and $y_m = a_y + mh_y$, $m = 1, \ldots, n_y - 1$. A smooth function of two variables $u(x, y)$ is approximated numerically at the discrete points of the mesh D_{mesh} by $u_{k,m}$, with the second derivatives at the point (x_k, y_m) being replaced with the second-order central numerical derivatives:

$$\frac{u_{k+1,m} - 2u_{k,m} + u_{k-1,m}}{h_x^2} + \frac{u_{k,m+1} - 2u_{k,m} + u_{k,m-1}}{h_y^2}$$
$$+F\left(x_k, y_m, u_{k,l}, \frac{u_{k+1,m} - u_{k-1,m}}{2h_x}, \frac{u_{k,m+1} - u_{k,m-1}}{2h_y}\right) = 0. \tag{10.90}$$

It is clear that the set of nonlinear equations at each node of the two-dimensional mesh (10.89) contains as many equations as there exist unknown variables $u_{k,m}$. The boundary values of $u_{k,m}$ at $k = 0$, $k = n_x$, $m = 0$, and $m = n_y$ known from the boundary conditions (10.86) are incorporated in equations at the nodes of the first interior layer of the discrete mesh (10.89).

It is, however, unclear how to solve the nonlinear system (10.90), which has no banded structure and has a huge size. Even if the function $F(x, y, u, u_x, u_y)$ is linear in u, u_x, and u_y, the numerical solutions of large linear systems are not feasible because of the storage problems, long computation times, and accumulating round-off errors. For instance, we can form a structure of the system (10.90) by concatenating rows of the matrix $u_{k,m}$ into a one-dimensional vector and representing all equations on the unknown $u_{k,m}$ as a matrix-vector product. If the discrete mesh has 100 interior x-values and 100 interior y-values, the vector for unknown values $u_{k,m}$ has size $100 \times 100 = 10,000$ and the coefficient matrix has size $10,000 \times 10,000$, which is far beyond the current capacity of MATLAB memory allocation.

Direct method

If the discrete mesh is sparse, however, a feasible numerical solution of the system (10.90) can be obtained by direct methods of numerical linear algebra (see Section 2.5). The next example illustrates the solution of the Poisson equation:

$$u_{xx} + u_{yy} + \sin(\pi x)\sin(\pi y) = 0, \qquad 0 < x < 1,\ 0 < y < 1, \tag{10.91}$$

subject to the Dirichlet boundary data:

$$u(x,0) = \sin(\pi x), \quad u(x,1) = 0, \quad u(0,y) = \sin(2\pi y), \quad u(1,y) = 0. \tag{10.92}$$

There exists an exact solution of the boundary-value problem:

$$u(x,y) = \sin(\pi x)\frac{\sinh(\pi(1-y))}{\sinh(\pi)} + \frac{\sinh(2\pi(1-x))}{\sinh(2\pi)}\sin(2\pi y) - \frac{1}{2\pi^2}\sin(\pi x)\sin(\pi y).$$

You can construct a numerical approximation of the solution for a sparse grid with $h_x = h_y = 0.25$. A closed system of 9 linear equations for 9 unknown values of $u_{k,m}$ at the interior mesh points is obtained from the system (10.90). The MATLAB script `Poisson_equation` solves a linear system associated with the 9×9 coefficient matrix and the 9×1 constant term from the given values at the boundary and the source.

```
% Poisson_equation
% u = [u11, u21, u31, u12, u22, u32, u13, u23, u33]
A = [-4,1,0,1,0,0,0,0,0; 1,-4,1,0,1,0,0,0,0; 0,1,-4,0,0,1,0,0,0; ...
    1,0,0,-4,1,0,1,0,0; 0,1,0,1,-4,1,0,1,0; 0,0,1,0,1,-4,0,0,1; ...
    0,0,0,1,0,0,-4,1,0; 0,0,0,0,1,0,1,-4,1; 0,0,0,0,0,1,0,1,-4 ];
h = 0.25; x = [0.25,  0.5, 0.75];
b1 = sin(pi*x); b2 = sin(2*pi*x); % boundary conditions
b=-[b1(1)+b2(1);b1(2);b1(3);b2(2);0;0;b2(3);0;0];
c1=-h^2*sin(pi*x')*sin(pi*x(1)); c2=-h^2*sin(pi*x')*sin(pi*x(2));
c3=-h^2*sin(pi*x')*sin(pi*x(3)); % sources
c = [ c1; c2; c3 ]; r = b + c;
u = A\r; % numerical solver
```

```
u1 = u(1:3)'; u2 = u(4:6)'; u3 = u(7:9)';
x = 0 : 0.25 : 1; y = 0 : 0.25 : 1;
[X,Y] = meshgrid(x,y); % solution surface
u0 = sin(pi*x); u1 = [b2(1),u1,0]; u2 = [b2(2),u2,0];
u3 = [b2(3),u3,0]; u4 = zeros(1,5);
Z = [u0;u1;u2;u3;u4]; mesh(X,Y,Z);
```

When the MATLAB script `Poisson_equation` is executed, it outputs Figure 10.15, which displays a numerical approximation of the preceding exact solution. Since the mesh (10.89) is sparse, the solution surface is pretty rough.

Iterative methods

Because the solution surface has a low resolution if the discrete mesh is sparse, direct methods of solutions of the system (10.90) are rarely used. Instead, iterative methods can be developed similarly to the iterative algorithms of numerical linear algebra (see Section 2.6). To systematically cover these algorithms in the context of Problem 10.4, consider embedding the boundary-value problem (10.85) into the initial-value problem in time t:

$$u_t = u_{xx} + u_{yy} + F(x, y, u, u_x, u_y), \qquad (x, y) \in D, \quad t > 0, \qquad (10.93)$$

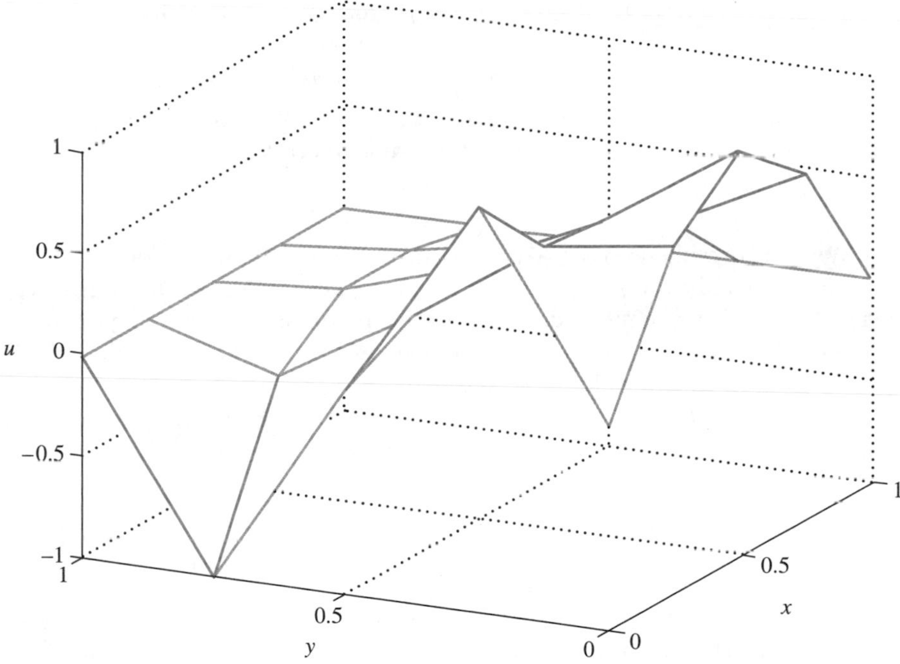

Figure 10.15 Solution of the Poisson equation with a direct finite-difference method.

subject to the Dirichlet boundary values

$$u(x,y,t) = \phi(x,y), \qquad (x,y) \in \partial D, \ \ t > 0, \tag{10.94}$$

and the Cauchy initial value

$$u(x,y,0) = u_0(x,y), \qquad (x,y) \in D. \tag{10.95}$$

It is clear that if the solution $u(x,y,t)$ of the PDE (10.93) with the initial value (10.95) converges to the time-independent limit as t gets large, the asymptotic solution $u_\infty(x,y) = \lim_{t\to\infty} u(x,y,t)$ satisfies the boundary-value problem (10.85)–(10.86). This happens for the two-dimensional heat equation

$$u_t = u_{xx} + u_{yy} + F(x,y), \tag{10.96}$$

which reduces to the Poisson equation (10.87) as $t \to \infty$. The following convergence result is proved in courses in *Partial Differential Equations.*

Theorem 10.2 *Let D be a bounded simply connected domain with a smooth boundary ∂D. Let $F(x,y)$, $u_0(x,y)$, and $\phi(x,y)$ be continuous functions on their domains of definition. There exists a unique limit $u_\infty(x,y) = \lim_{t\to\infty} u(x,y,t)$ of the initial-value problem (10.96) with (10.94)–(10.95), which is independent of $u_0(x,y)$. The function $u_\infty(x,y)$ is a twice continuously differentiable solution of the boundary-value problem (10.87) with boundary data (10.86).*

With this result, we can take $u_0(x,y) = 0$ in the time integration of the two-dimensional heat equation (10.96) and approximate the solution $u(x,y,t)$ with a finite-difference method. Two iterative methods are commonly used for a numerical solution of (10.96): (1) the relaxation (explicit) method, and (2) the alternative direction (implicit) method.

Relaxation methods

In the *relaxation* method, the time integration of the two-dimensional heat equation (10.96) is based on the Euler method:

$$\frac{u_{k,m}^{l+1} - u_{k,m}^{l}}{\tau} = \frac{u_{k+1,m}^{l} - 2u_{k,m}^{l} + u_{k-1,m}^{l}}{h_x^2} + \frac{u_{k,m+1}^{l} - 2u_{k,m}^{l} + u_{k,m-1}^{l}}{h_y^2} + F_{k,m},$$

where the superscript refers to the approximation at a discrete time level $t = t_l$ and the starting approximation is $u_{k,m}^0 = 0$ for all $(k,m) \in D_{\text{mesh}}$. The boundary values of $u_{k,m}^l$ are used in numerical computations at the first layer of interior nodes in the mesh (10.89). The relaxation method is similar to the explicit method for the one-dimensional heat equation (see Section 10.3).

If a numerical solution of the time-independent equation (10.87) is the ultimate goal, the iterative algorithm can be run up to sufficiently many iterations. The iterative loop stops when the distance between two successive iterations becomes smaller than a tolerance bound. The truncation error of the finite-difference solution of the boundary-value problem (10.87) is then $O(h_x^2, h_y^2)$.

Explicit iterations of the relaxation method can be rewritten in the form:

$$u_{k,m}^{l+1} = (1-\omega)u_{k,m}^{l} + \frac{\tau}{h_x^2}\left(u_{k+1,m}^{l} + u_{k-1,m}^{l}\right) + \frac{\tau}{h_y^2}\left(u_{k,m+1}^{l} + u_{k,m-1}^{l}\right) + \tau F_{k,m},$$

where

$$\omega = 2\tau\left(\frac{1}{h_x^2} + \frac{1}{h_y^2}\right).$$

When $\omega = 1$, the explicit method recovers the *Jacobi* iterative method for solutions of the linear systems (see Section 2.6). The explicit method is referred to as the *underrelaxation* method for $0 < \omega < 1$ and as the *overrelaxation* method for $\omega > 1$.

Similar to algorithms of numerical linear algebra, there exists a useful modification of the relaxation method. When one-step iterations are performed from the level t_l to the level t_{l+1}, the updated values of $u_{k,m}^{l+1}$ can be used instead of the previous values of $u_{k,m}^{l}$. For instance, if the new values of $u_{k,m}^{l+1}$ are computed with a double loop that runs by incrementing values of (k, m), then the updated values of $u_{k-1,m}^{l+1}$ and $u_{k,m-1}^{l+1}$ are evaluated at the time when the values of $u_{k,m}^{l+1}$ are being computed. The explicit iteration scheme of the modified relaxation method can then be rewritten in the form:

$$u_{k,m}^{l+1} = (1-\omega)u_{k,m}^{l} + \frac{\tau}{h_x^2}\left(u_{k+1,m}^{l} + u_{k-1,m}^{l+1}\right) + \frac{\tau}{h_y^2}\left(u_{k,m+1}^{l} + u_{k,m-1}^{l+1}\right) + \tau F_{k,m}.$$

When $\omega = 1$, the explicit method recovers the *Gauss–Seidel* iterative method (see Section 2.6). Similar to the extensions of the Jacobi method, the explicit method is referred to as the underrelaxation method for $0 < \omega < 1$ and the overrelaxation method for $\omega > 1$.

Stability of relaxation methods

Stability analysis shows that the explicit method based on the Jacobi iterations is unstable for $\omega > 1$, implying that the overrelaxation method cannot be useful. On the other hand, convergence analysis shows that the global truncation error of the time iterations of the two-dimensional heat equation (10.96) is $O(\omega)$. If you do not care about accuracy of the intermediate computations, the fastest convergence of the explicit method based on Jacobi iterations occurs for the largest time step inside the stability domain, that is, for $\omega = 1$.

Exercise 10.6 Consider the explicit method based on Jacobi iterations for the homogeneous equations

$$u_{k,m}^{l+1} = (1-\omega)u_{k,m}^l + \frac{\tau}{h_x^2}\left(u_{k+1,m}^l + u_{k-1,m}^l\right) + \frac{\tau}{h_y^2}\left(u_{k,m+1}^l + u_{k,m-1}^l\right)$$

with the homogeneous boundary conditions $u_{0,m}^l = u_{n_x,m}^l = u_{k,0}^l = u_{k,n_y}^l = 0$. By using the discrete Fourier transform, prove that the Jacobi iterations are stable for $\omega \leq 1$ and unstable for $\omega > 1$. Illustrate the analytical result with numerical computations of any solution that starts with nonzero values of $u_{k,m}^0$.

A modification of the same stability analysis shows that the explicit method based on the Gauss–Seidel iterations is unstable for $\omega > 2$ so that the relaxation method makes sense for $\omega \leq 2$. Moreover, it is shown in [29] that the fastest convergence occurs for $\omega = \omega_*$, where

$$\omega_* = \frac{2}{1+\sqrt{1-\rho^2}}, \qquad \rho = \frac{h_y^2 \cos\frac{\pi}{n_x} + h_x^2 \cos\frac{\pi}{n_y}}{h_x^2 + h_y^2}.$$

Because $0 < \rho < 1$ for $n_x \geq 3$ and $n_y \geq 3$, the fastest convergence occurs for overrelaxation iteration methods with $\omega = \omega_* > 1$.

The relaxation method is illustrated on the example of the Poisson equation (10.91) with boundary values (10.92). The MATLAB script **relaxation_method** performs iterations of the explicit method with $\omega = 0.5$ (underrelaxation), $\omega = 1$ (Gauss–Seidel), and $\omega = \omega_*$ (overrelaxation) on the mesh of 50×50 nodes within the tolerance bound of 10^{-6}.

```
% relaxation_method
n = 50; h = 1/n; x = 0 : h : 1; y = 0 : h : 1;
for j = 1 : 3  % three explicit iterations
    if (j == 1)
        omega = 0.5;
    elseif (j == 2)
        omega = 1;
    else
        omega = 2/(1+sin(pi/n));
    end
    u = zeros(n+1,n+1);  % boundary values
    u(1,:) = sin(pi*x); u(:,1) = sin(2*pi*y');
    l = 0; nor = 1;
    while( (nor > 10^(-6)) & (l < 10000))
        l = l + 1; uu = u;
        for k = 2 : n-1
            for m = 2 : n-1
```

```
                uu = u(k-1,m) + u(k+1,m) + u(k,m-1) + u(k,m+1);
                uu = uu + h^2*sin(pi*x(k))*sin(pi*y(k));
                u(k,m) = (1-omega)*u(k,m)+0.25*omega*uu;
            end
        end
        nor = max(max(abs(u-uu)));
    end
    fprintf('Number of iterations for omega = %f is %d\n',omega,l);
end
[X,Y] = meshgrid(x,y); mesh(X,Y,u);
```

When the MATLAB script **relaxation_method** is executed, it outputs the number of iterations required for convergence of each iterative procedure. It is clear that the overrelaxation method with $\omega = \omega_*$ converges faster than the Gauss–Seidel method and the underrelaxation method with $\omega = 0.5$. Figure 10.16 shows a smooth solution surface $u(x, y)$ in the resulting solution obtained with the overrelaxation method for $\omega = \omega_*$, the smoothness of which is controlled by the size of the mesh (10.89).

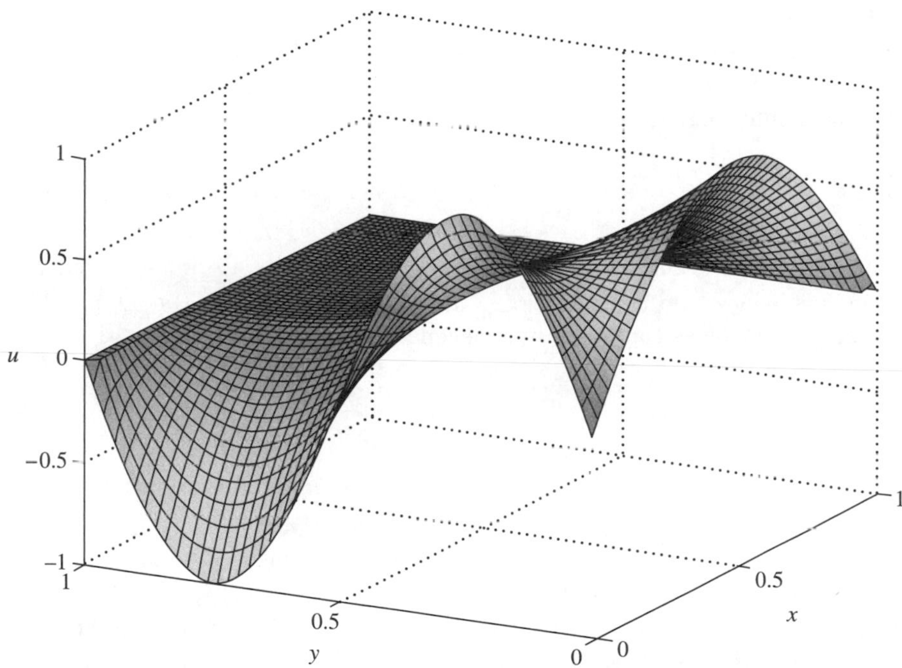

Figure 10.16 Solution of the Poisson equation with the overrelaxation method.

```
>> relaxation_method
   Number of iterations for omega = 0.500000 is 4560
   Number of iterations for omega = 1.000000 is 1778
   Number of iterations for omega = 1.881838 is 106
```

Exercise 10.7 Show numerically that the truncation error of the relaxation method is $O(h_x^2, h_y^2)$ in the example of the Poisson equation (10.91).

Alternative direction method

In the *alternative direction* method, the time integration of the two-dimensional heat equation (10.96) is based on the implicit Euler method with two modifications: (1) while one spatial second-order derivative is defined at the future time level t_{l+1}, the other one is computed at the present time level t_l; and (2) the choice of x and y spatial second-order derivatives is alternated in time iterations. With these two modifications, we can solve the linear system of equations by inverting matrices with tridiagonal (banded) structure and to preserve the second-order accuracy both in space grid size (h_x, h_y) and in time step τ. These modifications imply that the alternative direction method is similar to the Crank–Nicolson method for the one-dimensional heat equation (see Section 10.3). The method can be useful if the intermediate dynamics of the two-dimensional heat equation (10.96) is a subject of interest, in addition to the time-independent solution of the Poisson equation (10.87).

Two steps of the alternative direction method change the values $u_{k,m}^l$ at the time level t_l to the values $u_{k,m}^{l+2}$ at the time level t_{l+2}. The first step involves the linear inhomogeneous system:

$$u_{k,m}^{l+1} - \frac{\tau}{h_x^2}\left(u_{k+1,m}^{l+1} - 2u_{k,m}^{l+1} + u_{k-1,m}^{l+1}\right) = u_{k,m}^l + \frac{\tau}{h_y^2}\left(u_{k,m+1}^l - 2u_{k,m}^l + u_{k,m-1}^l\right) + \tau F_{k,m},$$

where $m = 1, 2, \ldots, n_y - 1$, the values $u_{0,m}^{l+1}$ and $u_{n_x,m}^{l+1}$ are known from the boundary values, and the values $u_{k,m}^l$ are known for any $(k, m) \in D_{\text{mesh}}$. The second step involves the linear inhomogeneous system:

$$u_{k,m}^{l+2} - \frac{\tau}{h_y^2}\left(u_{k,m+1}^{l+2} - 2u_{k,m}^{l+2} + u_{k,m-1}^{l+2}\right) = u_{k,m}^{l+1} + \frac{\tau}{h_x^2}\left(u_{k+1,m}^{l+1} - 2u_{k,m}^{l+1} + u_{k-1,m}^{l+1}\right) + \tau F_{k,m},$$

where $k = 1, 2, \ldots, n_x - 1$, the values $u_{k,0}^{l+2}$ and u_{k,n_y}^{l+2} are known from the boundary values, and the values of $u_{k,m}^{l+1}$ are now available for any $(n, m) \in D_{\text{mesh}}$. At each iteration of the two-step method, you need to run only one loop through the column vectors $\mathbf{u}_m^l = [u_{1,m}^l, u_{2,m}^l, \ldots, u_{n_x-1,m}^l]$ and another loop through the row vectors $\mathbf{u}_k^{l+1} = [u_{k,1}^{l+1}, u_{k,2}^{l+1}, \ldots, u_{k,n_y-1}^{l+1}]$. The method has an $O(\tau^2, h_x^2, h_y^2)$ global truncation error. As many other implicit methods for linear systems, the alternative direction method is unconditionally stable for any τ and (h_x, h_y).

Exercise 10.8 Consider the homogeneous version of the alternative direction method and prove that the method is unconditionally stable for any τ and (h_x, h_y). Confirm the analytical result by increasing the values of τ in numerical computations of a solution that starts with nonzero values of $u^0_{k,m}$.

We illustrate the alternative direction method for the example of the Poisson equation (10.91). The MATLAB script `alternative_direction_method` performs iterations of the implicit method with $\omega = 0.5$ on the mesh of 50×50 nodes within the tolerance bound of 10^{-6}.

```
% alternative_direction_method
n = 50; h = 1/n; omega = 0.5; x = 0 : h : 1; y = 0 : h : 1;
d1 = diag(ones(1,n-1)); d2 = diag(ones(1,n-2),1)+diag(ones(1,n-2),-1);
A = (1+2*omega)*d1 - omega*d2; u = zeros(n+1,n+1);
u(1,:) = sin(pi*x); u(:,1) = sin(2*pi*y'); % boundary values
uN = u; l = 0; nor = 1;
while( (nor > 10^(-6)) & (l < 10000))
    l = l + 1;
    uu = u;
     for m = 2 : n
        u1 = (1-2*omega)*uu(2:n,m)+omega*(uu(2:n,m-1)+uu(2:n,m+1));
        u1 = u1 + omega*uu(1,m)*eye(n-1,1);
        u1 = u1 + omega*h^2*sin(pi*x(m))*sin(pi*y(2:n)');
        u(2:n,m) = inv(A)*u1;
    end
    for k = 2 : n
        u2 = (1-2*omega)*u(k,2:n)'+omega*(u(k-1,2:n)'+uu(k+1,2:n)');
        u2 = u2 + omega*u(k,1)*eye(n-1,1);
        u2 = u2 + omega*h^2*sin(pi*y(k))*sin(pi*x(2:n)');
        uN(k,2:n) = (inv(A)*u2)';
    end
    nor = max(max(abs(uN-uu)));
    u = uN;
end
fprintf('Number of iterations for omega = %f is %d\n',omega,l);
[X,Y] = meshgrid(x,y); mesh(X,Y,u);
```

When the MATLAB script `alternative_direction_method` is executed, it produces the solution surface in Figure 10.17. Since each iteration of the two-step method involves computations of two loops through solutions of the linear inhomogeneous systems, the convergence speed to the time-independent solution is slow. However, the transient behavior of the two-dimensional heat equation is approximated more accurately compared to the relaxation method.

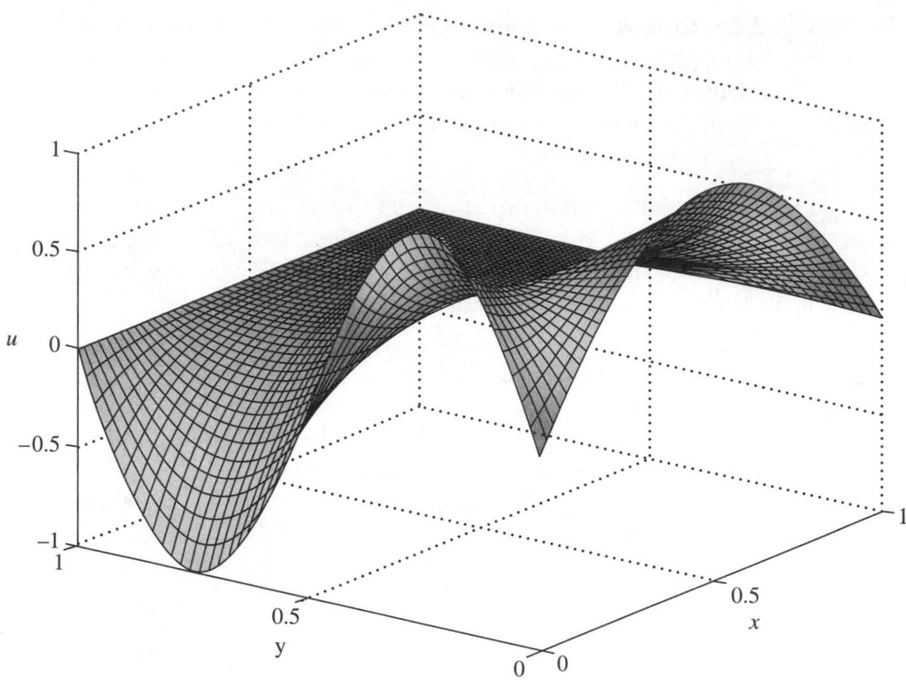

Figure 10.17 Solution of the Poisson equation with the alternative direction method.

```
>> alternative_direction_method

Number of iterations for omega = 0.500000 is 1293
```

Nonlinear elliptic problem

The relaxation method can also be used to reduce a nonlinear elliptic boundary-value problem. For instance, consider the nonlinear elliptic eigenvalue problem on a large square:

$$u_{xx} + u_{yy} + u^3 = \lambda u, \qquad -L < x < L, \;\; -L < y < L, \tag{10.97}$$

subject to the Dirichlet boundary values

$$u(x, -L) = u(x, L) = u(-L, y) = u(L, y) = 0,$$

where λ is a parameter. Besides the trivial solution $u(x, y) = 0$, the nonlinear problem (10.97) has a set of nonzero solutions, which includes the *ground state*

$u(x,y) > 0$. To exclude the trivial solution from the iterative method and to define the values of λ, we impose a constraint on the solution

$$\int_{-L}^{L}\int_{-L}^{L} u^2(x,y)dxdy = P_0 > 0. \qquad (10.98)$$

The starting iteration in the form of the Gaussian pulse $u(x,y) = e^{-(x^2+y^2)/2}$ is used to approximate the positive (ground state) solution $u(x,y)$. The MATLAB script **ground_state** performs iterations of the relaxation method with $\omega = 0.5$ under a renormalization of the squared norm (10.98) with $P_0 = 1$. This renormalization is similar to the power method for eigenvalues (see Section 4.6). The discrete mesh with 50-by-50 nodal points is used on the squared domain with $L = 20$.

```
% ground_state
n = 50; L = 20; omega = 0.5; h = (2*L)/n;
x = -L : h : L; y = -L : h : L;
[X,Y] = meshgrid(x,y); % starting approximation
u = exp(-X.^2-Y.^2);
u(1,:) = 0; u(n+1,:) = 0; u(:,1) = 0; u(:,n+1) = 0;
l = 0; nor = 1;
while( (nor > 10^(-8)) & (l < 100000))
    l = l + 1; uu = u;
    for k = 2 : n
        for m = 2 : n
            w = u(k-1,m) + u(k+1,m) + u(k,m-1) + u(k,m+1) + h^2*u(k,m)^3;
            u(k,m) = (1-omega)*u(k,m)+0.25*omega*w;
        end
    end
    power = sum(sum(u.^2));
    u = u/sqrt(power);
    nor = max(max(abs(u-uu)));
end
fprintf('Number of iterations = %f is %d\n',omega,l);
Z = u; mesh(X,Y,Z);
ux = (u(3:n+1,:)-u(1:n-1,:))/(2*h); % partial derivatives
uy = (u(:,3:n+1)-u(:,1:n-1))/(2*h); % Rayleigh quotient
lambda=(sum(sum(u.^4))-sum(sum(ux.^2))-sum(sum(ux.^2)))/sum(sum(u.^2))
```

When the MATLAB script **ground_state** is executed, it converges to the ground state, which is shown in Figure 10.18. The solution surface with the normalized power (10.98) corresponds to a particular value of λ, which can be computed from the Rayleigh quotient,

$$\lambda = \frac{\int_{-L}^{L}\int_{-L}^{L}\left(u^4 - u_x^2 - u_y^2\right)dxdy}{\int_{-L}^{L}\int_{-L}^{L} u^2 dxdy},$$

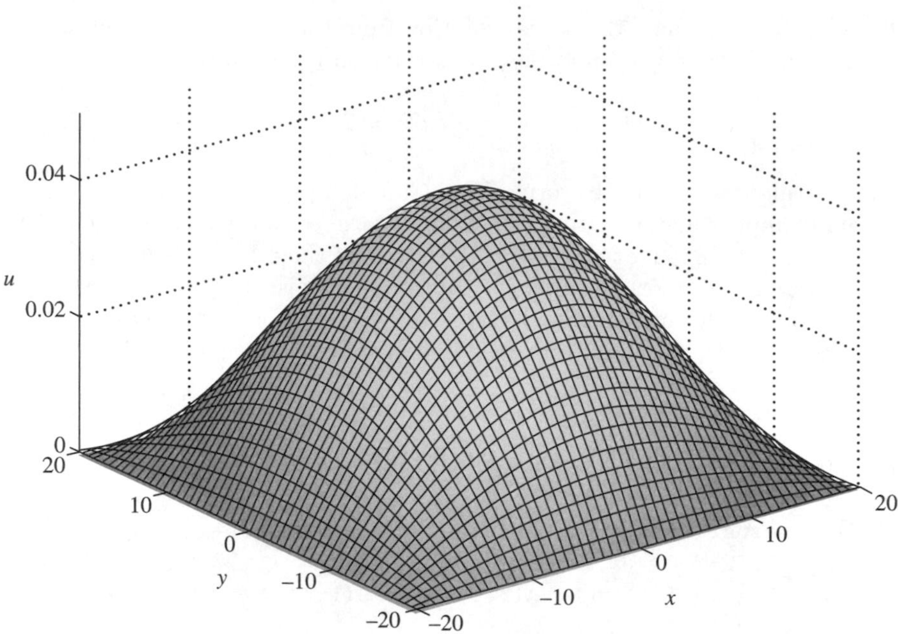

Figure 10.18 Solution of the nonlinear elliptic equation (10.97) with the relaxation method.

after the partial derivatives are discretized by using central differences.

```
>> ground_state

Number of iterations = 0.500000 is 2891

lambda =
    -0.0109
```

Although finite-difference methods are the simplest methods to solve differential equations of mathematical physics, they have many limitations caused by low accuracy, slow rate of convergence, and ineffective memory management. More complicated algorithms can be more efficient, while satisfying the same objective of constructing a numerical approximation of solutions of differential equations. While finite-difference methods originate from the polynomial interpolation and finite-difference approximations of numerical derivatives, other methods such as spectral and finite-element methods originate from the trigonometric and piecewise polynomial approximations, respectively. These two methods are covered in the last two chapters of the book.

10.6 Summary and Notes

In this chapter, we obtain approximate solutions of the boundary-value problems for ordinary and partial differential equations.

- **Section 10.1:** We formulate a sufficient condition (Theorem 10.1) for existence of solutions of the second-order boundary-value ODE problem (Problem 10.1). We introduce the finite-difference method for numerical approximations of the linear boundary-value ODE problem by replacing the continuous derivatives with their central-difference approximations. We show that finite-difference methods are robust with respect to different boundary conditions, such as Dirichlet, Neumann, and mixed boundary conditions. These methods are also applied to the nonlinear boundary-value ODE problem after the Newton–Raphson iteration method is used for root-finding.

- **Section 10.2:** We show that the shooting method is useful to solve the boundary-value ODE problem from the solutions of the initial-value ODE solvers. The linear boundary-value ODE problem is solved with a single iteration. The shooting method relies on the Newton–Raphson and secant algorithms for root finding of a scalar function. The same method is applied for approximation of eigenvalues in the linear eigenvalue problem for second-order ordinary differential equations.

- **Section 10.3:** Finite-difference methods are applied to parabolic partial differential equations (Problem 10.2). Three methods, namely, explicit, implicit, and Crank–Nicolson methods, are developed for the linear heat equation, and their convergence and stability are analyzed. The explicit method is also developed for the nonlinear parabolic PDE.

- **Section 10.4:** Finite-difference methods are applied to hyperbolic PDEs (Problem 10.3). We develop explicit and implicit methods for the linear wave equation and analyze their convergence and stability. We show that the explicit method recovers the exact solution of the homogeneous wave equation. The solutions to nonlinear hyperbolic PDEs are also approximated by the explicit method.

- **Section 10.5:** Finite-difference methods are applied to elliptic PDEs (Problem 10.4). We show that direct methods based on numerical solutions of linear systems produce a sparse solution. On the other hand, iterative methods based on iterations of the linear systems are related to the explicit finite-difference method for a two-dimensional parabolic PDE. The correspondence between time-independent elliptic and time-dependent parabolic PDEs is established in Theorem 10.2. Two methods, namely, relaxation and alternative direction methods, are developed for the linear two-dimensional heat equation. The relaxation method is also extended for the nonlinear elliptic PDE.

Finite-difference and shooting methods for ordinary differential equations are covered in [25]. Existence, uniqueness, and stability of solutions of initial-value and boundary-value problems for partial differential equations are studied in [27]. The suggested reference for truncation errors and stability of iterations of finite-difference methods for partial differential equations is [8].

10.7 Exercises

1. Consider the linear boundary-value ODE problem

$$-y'' + y' + y = x(1-x), \qquad 0 < x < 1,$$

subject to the Neumann boundary conditions $y'(0) = y'(1) = 0$. Approximate the solution with the central-difference method for step sizes $h = 0.1$ and $h = 0.05$. Confirm that the truncation error at each point of the discrete grid is $O(h^2)$. Use the Richardson extrapolation and compute a more accurate approximation of the solution $y(x)$. Plot the exact solution and the three numerical approximations on the same graph.

2. Consider the linear boundary-value ODE problem

$$y'' + x^2 y = 1, \qquad 0 < x < 1,$$

subject to the Dirichlet boundary conditions $y(0) = y(1) = 0$. Use the finite-difference (with the second-order central differences) and linear shooting (with the fourth-order Runge–Kutta ODE solver) methods and find two numerical approximations of the solution of the boundary-value problem on a grid of 100 data points. Plot the two solutions on the same graph on the interval $[0, 1]$.

3. Consider the linear eigenvalue problem

$$y'' + x^2 y = \lambda y, \qquad 0 < x < 1,$$

subject to the Dirichlet boundary conditions $y(0) = y(1) = 0$. Use the finite-difference method and find numerical approximations of 100 eigenvalues and eigenfunctions of the problem. Show that the value $\lambda = 0$ is not an eigenvalue in the finite-difference method.

4. Consider the Bessel equation

$$y'' + \frac{1}{x} y' + \lambda y = 0, \qquad 0 < x < 1,$$

subject to $y'(0) = 0$ and $y(1) = 0$, where λ is a parameter. Use the finite-difference method and approximate the three smallest eigenvalues λ. Construct and plot the eigenfunctions corresponding to these eigenvalues. Verify that the numerical approximations of eigenvalues λ converge quadratically as step size h goes to zero.

5. Use the fourth-order central-difference method and find a numerical approximation of the solution of the boundary-value problem

$$y'' + \cos xy = \cos(2x), \qquad 0 < x < \pi,$$

subject to the periodic boundary conditions $y(x+\pi) = y(x)$. Plot these solutions on the same graph for $[0, \pi]$.

6. Consider the *Liouville equation*

$$-y'' + e^y = 0, \qquad 0 < x < 1,$$

subject to the boundary conditions $y(0) = y(1) = 0$. Use the finite-difference method and set up the system of nonlinear equations at the grid points. Approximate solutions of the nonlinear system with the vector Newton–Raphson method for $h = 0.1$.

7. Consider the ground state solution $y(x)$ of the cubic elliptic ODE problem:

$$y'' + \frac{1}{x}y' + y - y^3 = 0, \qquad 0 < x < L,$$

subject to the boundary conditions $y(0) = 0$ and $y(L) = 1$ for large $L = 10$. Use the shooting method combined with two ODE solvers: the explicit Heun method and the fourth-order Runge–Kutta method. Plot the two solutions for $h = 0.1$ and observe that the ∞-norm of the distance between two solutions is smaller for $h = 0.02$.

8. Use the MATLAB function `odebvp` and approximate a numerical solution of the *Euler equation*

$$x^2y'' - 2xy' + 2y = 0, \qquad 1 < x < 2,$$

with $y(1) = 0$ and $y(2) = 1$. Use the least-square power fit for the graph of the computational error at the interior grid points versus step size h.

9. Use the Crank–Nicolson method and approximate the solution of the inhomogeneous heat equation

$$u_t = u_{xx} + \sin(2x), \qquad 0 < x < \pi, \;\; t > 0,$$

subject to $u(0,t) = u(\pi, 0) = 0$ and $u(x, 0) = \sin(10x)$. Use the equally spaced grid with step size $h = \pi/10$ and time step $\tau = 0.1$ on the space interval $[0, 1]$ and time interval $[0, 10]$ and plot the solution surface.

10. Consider the two-step explicit method for the heat equation:

$$u_{k,l+1} = 2r\left(u_{k-1,l} - 2u_{k,l} + u_{k+1,l}\right) + u_{k,l-1}.$$

Reduce the finite-difference equation to the second-order difference equation for the Fourier mode $u_{k,l} = a_{j,l}\sin(\kappa_j k)$, where $\kappa_j = \frac{\pi j}{n}$. Find a quadratic equation for Q_j and solve it with the quadratic formula. Prove that the preceding explicit method is *unconditionally unstable* for any $r > 0$.

11. Solve the heat equation

$$u_t = u_{xx}, \qquad 0 < x < 1, \;\; 0 < t < 1,$$

subject to the initial condition $u(x,0) = 0$ and boundary conditions $u(0,t) = t^2$ and $u(1,t) = -t^2$. Use the explicit, implicit, and Crank–Nicolson methods with step size $h = 0.1$ and time step $\tau = 0.5h^2$. Plot the computational error surface for each of the numerical approximations.

12. Consider the nonlinear heat equation

$$u_t = u_{xx} - u(1-u^2), \qquad -1 < x < 1, \;\; t > 0,$$

subject to the boundary conditions $u(-1,t) = u(1,t) = 0$ and the initial condition $u(x,0) = 1 - x^2$. Approximate the solution with the explicit method on the discrete grid with step size $h = 0.1$ and time step $\tau = 0.5h^2$. Plot the solution $u(x,t)$ versus x for times $t = k$, $k \in \mathbb{N}$.

13. Approximate the steady-state solution $y(x) = \lim_{t\to\infty} u(x,t)$ to the previous exercise with the shooting method and plot the ∞-norm for the distance between $u(x,t)$ and $y(x)$ versus t.

14. Use the explicit method to approximate the solution of the wave equation

$$u_{tt} = u_{xx}, \qquad 0 < x < 1, \;\; t > 0,$$

subject to the Neumann boundary conditions $u_x(0,t) = u_x(1,t) = 0$ and initial conditions $u(x,0) = 0$ and $u_t(x,0) = \cos(2\pi x)$. Use the equally spaced grid with step size $h = 0.1$ and time step $\tau = 0.1$ on the space interval $[0,1]$ and time interval $[0,1]$. Confirm with repeated computations that the finite-difference method is unstable for $\tau > h = 0.1$, so that the Neumann boundary conditions did not alter the stability domain of the explicit method.

15. Consider the nonlinear *Boussinesq equation*

$$u_{tt} = u_{xx} - u_{xxxx} + (u^2)_{xx}, \qquad 0 < x < 1, \;\; t > 0,$$

subject to the periodic boundary conditions $u(x,t) = u(x+1,t)$ and the initial condition $u(x,0) = \sin(2\pi x)$ and $u_t(x,0) = 0$. Apply the explicit finite-difference method and a compute numerical approximation of the solution with $h = 0.1$ and $\tau = 0.01$. Find the convergence rate and the stability interval for the explicit finite-difference method.

16. Consider the *advection equation*

$$u_t + u_x = 0, \qquad 0 < x < L, \;\; t > 0$$

and derive the forward-time backward-space finite-difference method

$$u_{k,l+1} = (1-\gamma)u_{k,l} + \gamma u_{k-1,l},$$

where $\gamma = \tau/h$. Using the complex discrete Fourier mode $u_{k,l} = a_{j,l}e^{i\kappa_j k}$, where $\kappa_j = \frac{\pi j}{n}$, find the complex-valued factor Q_j in the first-order difference equation $a_{j,l+1} = Q_j a_{j,l}$. Prove that the finite-difference method is stable for $0 < \gamma \leq 1$ and unstable for $\gamma > 1$. Illustrate these results for the boundary condition $u(0,t) = \sin(3\pi x)$ and initial condition $u(x,0) = \sin(\pi x)$ on $L = 1$, $h = 0.1$, and two values of $\tau = 0.1$ and $\tau = 0.2$.

17. Consider the same advection equation from the previous exercise and derive the backward-time central-space finite-difference method:

$$\gamma u_{k+1,l+1} + 2u_{k,l+1} - \gamma u_{k-1,l+1} = 2u_{k,l},$$

where $\gamma = \tau/h$. Using the complex discrete Fourier mode $u_{k,l} = a_{j,l}e^{i\kappa_j k}$, where $\kappa_j = \frac{\pi j}{n}$, find the complex valued factor Q_j in the first-order difference equation $a_{j,l+1} = Q_j a_{j,l}$. Prove that the finite-difference method is *unconditionally stable* for any $\gamma > 0$. Illustrate these results for the boundary condition $u(0,t) = \sin(3\pi x)$ and initial condition $u(x,0) = \sin(\pi x)$ on $L = 1$, $h = 0.1$, and $\tau = 0.2$.

18. Consider the boundary-value problem for the Poisson equation:

$$u_{xx} + u_{yy} = x(1-x)y(1-y), \qquad 0 < x < 1, \;\; 0 < y < 1,$$

subject to the homogeneous Neumann boundary conditions on the boundary of the unit square. Use the equally spaced grid with the step size $h_x = h_y = 0.25$ and compute the direct solution of the linear inhomogeneous system in the finite-difference method.

19. Implement the overrelaxation method based on Gauss–Seidel iterations with an optimal parameter ω to the Poisson equation

$$u_{xx} + u_{yy} = \sin^2 x + \sin^2 y, \qquad 0 < x < \pi, \;\; 0 < y < \pi,$$

subject to the homogeneous Dirichlet boundary conditions on the boundary of the square. Plot the numerical solution surface with $h_x = h_y = 0.1$.

20. Consider the Helmholtz equation

$$-u_{xx} - u_{yy} + u = \pi^2 \sin(\pi x)\sin(\pi y), \qquad 0 < x < 1, \;\; 0 < y < 1,$$

subject to the homogeneous Dirichlet boundary conditions on the boundary of the unit square. Use the alternative direction method and obtain a numerical approximation of the stationary solution with $h_x = h_y = 0.1$. Confirm that the computational error is $\mathrm{O}(h^2)$, where $h_x = h_y$.

CHAPTER **11**

Spectral Methods

TRIGONOMETRIC APPROXIMATION AND INTERPOLATION are the simplest but most fundamental *spectral methods.* These methods offer an effective numerical solution to the problems of function approximation, calculus of derivatives, and differential equations. Trigonometric functions such as $\cos(kx)$, $\sin(kx)$, and e^{ikx}, with integer k, represent an arbitrary function $f(x)$ on a finite interval $[0, 2\pi]$ better than the power functions x^k. This surprising feature is a result of the oscillatory behavior of the periodic trigonometric functions. If the function $f(x)$ is sufficiently smooth, the coefficients of the trigonometric functions with larger values of k in its representation are much smaller than coefficients with smaller values of k. As a result, a sum of a few trigonometric functions gives an adequate numerical representation of the given function $f(x)$ that is free of numerical artifacts such as polynomial wiggle for power functions with large exponents n.

This chapter covers the construction, properties, and errors of trigonometric approximation and interpolation, as well as their applications to numerical approximations of solutions of ordinary and partial differential equations.

11.1 Trigonometric Approximation and Interpolation

Whereas polynomial approximation and interpolation (see Sections 5.2 and 5.6) fit a given set of data points to a linear combination of power functions, trigonometric approximation and interpolation represent the data points by a linear combination of orthogonal trigonometric functions.

Problem 11.1 (Trigonometric Approximation) Let $f(x)$ be a function with a finite number of jump discontinuities on the interval $0 \le x \le L$ and let $f_m(x)$ be the partial trigonometric sum in the form

$$f_m(x) = \frac{1}{2}a_0 + \sum_{j=1}^{m} a_j \cos\left(\frac{2\pi jx}{L}\right) + b_j \sin\left(\frac{2\pi jx}{L}\right), \qquad 0 \le x \le L. \quad (11.1)$$

Find the coefficients $\{a_j\}_{j=0}^m$ and $\{b_j\}_{j=1}^m$ that minimize the square error

$$\int_0^L [f(x) - f_m(x)]^2 \, dx. \tag{11.2}$$

Strictly speaking, this problem can be formulated for the more general class of *square integrable* functions, but since this involves subtle definitions from the theory of Lesbegue integration, we focus on functions with jump discontinuities at a finite number of points, which covers most common practical applications.

Trigonometric approximation

The following solution of Problem 11.1 is typically presented in courses in *Partial Differential Equations.*

Theorem 11.1 *There exists a unique solution of Problem 11.1 with*

$$a_j = \frac{2}{L} \int_0^L f(x) \cos\left(\frac{2\pi j x}{L}\right) dx, \qquad j = 0, 1, ..., m, \tag{11.3}$$

$$b_j = \frac{2}{L} \int_0^L f(x) \sin\left(\frac{2\pi j x}{L}\right) dx, \qquad j = 1, 2, ..., m. \tag{11.4}$$

For any small $\epsilon > 0$, there exists $m \geq 1$ such that

$$\int_0^L [f(x) - f_m(x)]^2 \, dx \leq \epsilon^2. \tag{11.5}$$

A typical example of the trigonometric approximation is the representation of the sign function on the interval $[-1, 1]$

$$f(x) = \text{sign}(x) = \begin{cases} +1, & 0 < x < 1 \\ -1, & -1 < x < 0 \end{cases} \tag{11.6}$$

by the trigonometric sum of sinusoidal functions

$$f_m(x) = \sum_{j=1}^m \frac{4}{\pi(2j-1)} \sin \pi(2j-1)x, \qquad -1 \leq x \leq 1. \tag{11.7}$$

The partial sum $f_m(x)$ is an odd periodic function with the period $L = 2$, which satisfies the Dirichlet boundary conditions $f_m(0) = f_m(1) = 0$. Notice that the original function $f(x)$ has two jump discontinuities at the points

$x = 0$ and $x = 1$. Theorem 11.1 guarantees that the distance (11.5) can be reduced with larger values of m for any given small $\epsilon^2 > 0$. Indeed, the error

$$\int_{-1}^{1} [f(x) - f_m(x)]^2 \, dx = 2 - \frac{16}{\pi^2} \sum_{j=1}^{m} \frac{1}{(2j-1)^2}$$

can be reduced below any small number ϵ^2, since the sum $\sum_{j=1}^{m} \frac{1}{(2j-1)^2}$ converges absolutely to $\frac{\pi^2}{8}$ as m goes to infinity.

Convergence of trigonometric approximation

If the function $f(x)$ is sufficiently smooth, the partial sum $f_m(x)$ converges *pointwise* to $f(x)$ at each point $x \in [0, L]$ as $m \to \infty$. Moreover, for sufficiently smooth $f(x)$ there exists a uniform bound on the absolute distance $|f(x) - f_m(x)|$ for all $x \in [0, L]$, which converges to zero as $m \to \infty$. In this case, we say that $f_m(x)$ converges to $f(x)$ *uniformly* on $[0, L]$. In general, the absolute distance between $f(x)$ and $f_m(x)$ can be estimated in terms of powers of m depending on the smoothness of $f(x)$.

Theorem 11.2 *Let $f(x)$ be k-times continuously differentiable on $[0, L]$ for some $k \geq 1$ and let the periodic boundary conditions $f(0) = f(L)$, $f'(0) = f'(L)$, ..., $f^{(k)}(0) = f^{(k)}(L)$ be satisfied. Then, there exists constant $C_k > 0$ such that*

$$\sup_{0 \leq x \leq L} |f(x) - f_m(x)| \leq \frac{C_k}{m^{k+1}}, \qquad m \geq 1. \tag{11.8}$$

If the function $f(x)$ has jump discontinuities either in the interior points $0 < x < L$ or at the endpoints $x = 0$ and $x = L$ (when $f(0) \neq f(L)$), the partial sum $f_m(x)$ converges pointwise to the mean value of $f(x)$ at the jump discontinuity. In this case, the convergence is nonuniform and *Gibbs oscillations* near the discontinuity points arise resulting in the nonvanishing local error as $m \to \infty$. For the sign function (11.6) with the jump discontinuity at $x = 0$, the partial sum $f_m(x)$ overshoots the value $f(x) = 1$ for small positive x for any $m \geq 1$. As $m \to \infty$, the local error shifts toward the point $x = 0$, but it does not vanish [9].

Trigonometric interpolation

Following the trigonometric approximation, the trigonometric interpolation can be introduced in two different ways. In the first method, the partial sum (11.1) is discretized on the uniform grid with constant step size to match the given set of y-values and the solution of the resulting linear system represents discretizations of integrals in the coefficients (11.3)–(11.4). In the second method, the same discretizations of the integrals (11.3)–(11.4) are recovered from analysis of eigenvalues and eigenvectors of the difference eigenvalue problems. Whereas the first method extends the linear systems of polynomial interpolation in Section 5.2, the second method is closely related to calculus of differences for numerical derivatives in Section 6.2.

Problem 11.2 (Trigonometric Interpolation) Let a discrete set of data points (x_1, y_1), (x_2, y_2), ..., (x_{n+1}, y_{n+1}) be defined on the uniform grid of equally spaced x-values

$$x_k = \frac{(k-1)L}{n}, \qquad k = 1, 2, \ldots, n+1, \tag{11.9}$$

subject to the periodic boundary condition $y_{n+1} = y_1$. When $n = 2m$, find the interpolant $F_m(x)$ in the form,

$$F_m(x) = \frac{1}{2}a_0 + \sum_{j=1}^{m-1} a_j \cos\left(\frac{2\pi j x}{L}\right) + b_j \sin\left(\frac{2\pi j x}{L}\right) + \frac{1}{2}a_m \cos\left(\frac{2\pi m x}{L}\right), \tag{11.10}$$

such that

$$F_m(x_k) = y_k, \qquad k = 1, 2, \ldots, n. \tag{11.11}$$

We have adopted several conventions in Problem 11.2. We enumerate the coefficients a_j from $j = 0$ to $j = m$ to be consistent with the trigonometric sum (11.1) in Problem 11.1. On the other hand, we enumerate the discrete grid from $k = 1$ to $k = n+1$ to keep consistent with Problem 5.1. The reason why we have defined the coefficient a_m with the factor $\frac{1}{2}$ will be clear from the solution of the system (11.11). Furthermore, we have dropped the coefficient b_m since it is not defined on the discrete grid (11.9) where $\sin(\pi(k-1)) = 0$ for any $k \in \mathbb{N}$. The set of equations (11.11) is truncated at $k = n$ since the equation for $k = n+1$ is redundant as a result of periodic boundary conditions $y_{n+1} = y_1$. Finally, we shall consider only the case of even values of n. Problem 11.2 can be extended to the case of odd values of n, but all formulas need to be rewritten with simple modifications.

The system of equations (11.10)–(11.11) is equivalent to the linear algebraic system for $(m+1)$ coefficients $\{a_j\}_{j=0}^{m}$ and $(m-1)$ coefficients $\{b_j\}_{j=1}^{m-1}$:

$$y_k = \frac{1}{2}a_0 + \sum_{j=1}^{m-1} a_j \cos\left(\frac{\pi j(k-1)}{m}\right) + b_j \sin\left(\frac{\pi j(k-1)}{m}\right) + \frac{1}{2}a_m \cos\left(\pi(k-1)\right), \tag{11.12}$$

where $k = 1, 2, \ldots, 2m$.

The MATLAB solver \ (backslash) gives a unique solution to the system (11.12), assuming that the linear system is neither singular nor ill-conditioned. The next example presents a numerical solution for the sign function (11.6), extended from the symmetric interval $[-1, 1]$ to the fundamental interval $[0, 2]$

by periodic continuation. The MATLAB script trigonometric_sums performs computations of the coefficients $\{a_j\}_{j=0}^{m}$ and $\{b_j\}_{j=1}^{m-1}$ of Problem 11.2. These coefficients are compared with the exact values (11.7) for the coefficients $\{b_j\}_{j=1}^{m}$ of Problem 11.1.

```
% trigonometric_sums
x = -1 : 0.25 : 1; y = sign(x);
n = length(x) - 1; m = n/2; L = 2;
y(1) = 0; y(n+1) = 0;  % continuation of data to x in [0,L]
xx = [x(m+1:n),x(1:m)]'; yy=[y(m+1:n),y(1:m)]';
A = 0.5*ones(n,1);  % building the coefficient matrix
for j = 1 : (m-1)
        A = [ A, cos(2*pi*j*xx/L), sin(2*pi*j*xx/L) ];
end
A = [ A, 0.5*cos(2*pi*m*xx/L) ];
c = A\yy;  % Fourier coefficients of interpolation
a = [c(1);c(2:2:n)]', b = [0;c(3:2:n);0]'
bb = 0; jj = 1;  % Fourier coefficients of approximation
for j = 1 : m-1
    if (jj == 1)
        bb = [bb,4/(pi*j)]; jj = 0;
    else
        bb = [bb,0]; jj = 1;
    end
end
xInt = -1: 0.002 : 1;
yInt = 0.5*a(1)*ones(1,length(xInt));
for j = 1 : (m-1)    % trigonometric interpolation
    yInt = yInt + a(1+j)*cos(2*pi*j*xInt/L) + b(1+j)*sin(2*pi*j*xInt/L);
end
yInt = yInt + 0.5*a(m+1)*cos(2*pi*m*xInt/L);
yyInt = zeros(1,length(xInt)); % trigonometric approximation
for j = 1 : (m-1)
    yyInt = yyInt + bb(1+j)*sin(2*pi*j*xInt/L);
end
plot(x,y,'.g',xInt,yInt,'b',xInt,yyInt,':r');
```

When the MATLAB script trigonometric_sums is executed, its output shows that all coefficients $\{a_j\}_{j=0}^{m}$ are of the order of machine precision. This is explained by the fact that the function $f(x)$ is odd in x. On the other hand, the coefficients $\{b_j\}_{j=1}^{m-1}$ of the trigonometric interpolation are different from the coefficients $\{b_j\}_{j=1}^{m}$ of the trigonometric approximation. Therefore, the functions $F_m(x)$ and $f_m(x)$ give generally different representations of the original function $f(x)$. Figure 11.1 plots functions $F_m(x)$ (solid curve) and

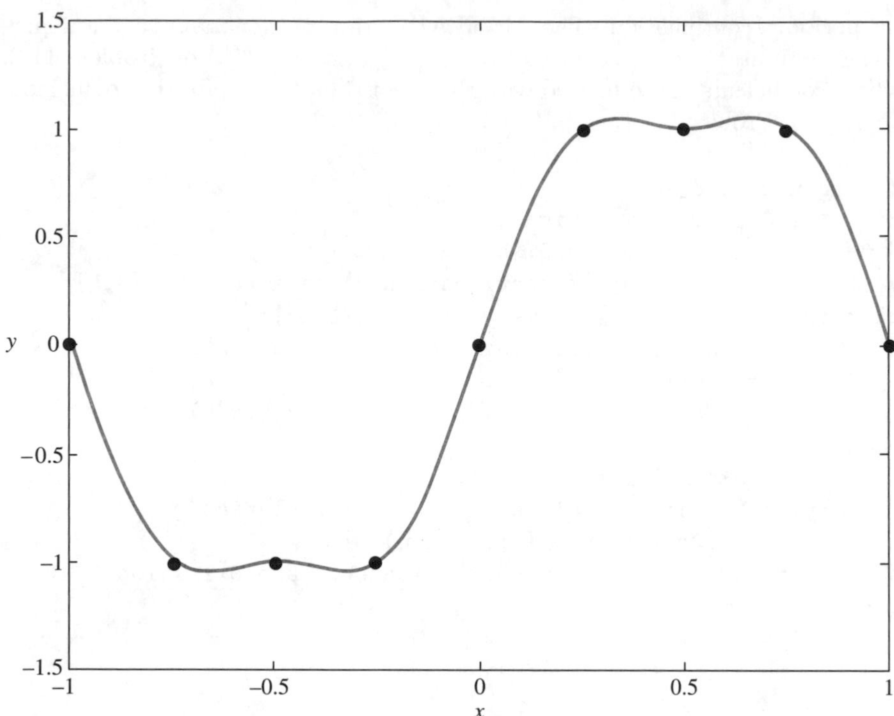

Figure 11.1 Trigonometric interpolation $F_m(x)$ (solid curve), trigonometric approximation $f_m(x)$ (dashed curve), and the original function $f(x)$ (dots).

$f_m(x)$ (dashed curve) with respect to the original function $f(x)$ (dots). The trigonometric approximation displays Gibbs oscillations at the jump discontinuities $x = 0$ and $x = \pm 1$, while trigonometric interpolation fits much closer to the given data points.

```
>> trigonometric_sums

a =
  1.0e-015 *
    0.1248   -0.0236    0.0565   -0.0883   -0.0139

b =
         0    1.2071    0.0000    0.2071         0

bb =
         0    1.2732         0    0.4244
```

Exercise 11.1 Compute the trigonometric approximation $f_m(x)$ for the sign function (11.6) with different values of m and prove numerically that the local truncation error for $f_m(x)$ does not reduce with larger values of m but shifts toward the jump point $x = 0$. Plot the maximum error for the trigonometric approximation versus m.

Exercise 11.2 Repeat Exercise 11.1 for the trigonometric interpolation $F_m(x)$.

To ensure that the linear system (11.12) is neither singular nor ill-conditioned, compute the product of the coefficient matrix **A** and its transposed matrix **A′** in the previous example. The result is surprising: the product matrix is diagonal, implying that **A** can be made into an orthogonal matrix afterward by appropriate scaling of its columns.

```
>> P = A'*A

P =
    Columns 1 to 6
    2.0000    0.0000         0   -0.0000    0.0000         0
    0.0000    4.0000    0.0000    0.0000   -0.0000   -0.0000
         0    0.0000    4.0000   -0.0000         0    0.0000
   -0.0000    0.0000   -0.0000    4.0000    0.0000    0.0000
    0.0000   -0.0000         0    0.0000    4.0000   -0.0000
         0   -0.0000    0.0000    0.0000   -0.0000    4.0000
   -0.0000    0.0000   -0.0000   -0.0000    0.0000    0.0000
         0         0   -0.0000    0.0000    0.0000   -0.0000

   Columns 7 to 8
   -0.0000         0
    0.0000         0
   -0.0000   -0.0000
   -0.0000    0.0000
    0.0000    0.0000
    0.0000   -0.0000
    4.0000   -0.0000
   -0.0000    2.0000
```

The diagonal entries of `A'*A` are found to match $m = 4$, except for the first and last entries that match $\frac{m}{2} = 2$. By using these normalization factors and the orthogonality of columns of **A**, we obtain the exact solution of Problem 11.2.

Theorem 11.3 *There exists a unique solution $F_m(x)$ of Problem 11.2 for any $m \geq 1$:*

$$a_j = \frac{1}{m}\sum_{k=1}^{n} y_k \cos\left(\frac{\pi j(k-1)}{m}\right), \qquad j = 0, 1, \ldots, m, \tag{11.13}$$

$$b_j = \frac{1}{m}\sum_{k=1}^{n} y_k \sin\left(\frac{\pi j(k-1)}{m}\right), \qquad j = 1, \ldots, m-1. \tag{11.14}$$

Because the first and last terms in the interpolating function $F_m(x)$ in (11.10) are defined with the factor $\frac{1}{2}$, the summation formulas (11.13)–(11.14) are defined uniformly for any j. By using the vector dot products for fast vector-matrix computations of summation formulas, you can replace the MATLAB solver for the linear system (11.12) in the previous example by the summation formulas (11.13)–(11.14). The result is, of course, the same.

```
>> for j = 0 : m
       a(j+1) = yy'*cos(2*pi*j*xx/L)/m;
       b(j+1) = yy'*sin(2*pi*j*xx/L)/m;
   end
>> a

ans =
  1.0e-016 *
         0         0    0.4020         0         0

>> b

ans =
         0    1.2071    0.0000    0.2071    0.0000
```

Discrete Fourier transform

The fast Fourier transform is a popular algorithm for fast computations of the discrete Fourier transforms (11.13)–(11.14). It is implemented in most software packages and computational libraries. MATLAB operates with two basic functions: `fft` and `ifft` for the discrete Fourier transforms and their various modifications. The fast Fourier transform operates with the complex form of the trigonometric interpolation:

$$y_k = \frac{1}{n}\sum_{j=0}^{n-1} c_j e^{i\pi j(k-1)/m}, \qquad k = 1, 2, \ldots, n, \tag{11.15}$$

$$c_j = \sum_{k=1}^{n} y_k e^{-i\pi j(k-1)/m}, \qquad j = 0, 1, \ldots, n-1, \tag{11.16}$$

where for $j = 1, 2, \ldots, m-1$

$$c_j = m(a_j - ib_j), \qquad c_{n-j} = m(a_j + ib_j) = c_{-j},$$

and for $j = 0$ and $j = m$

$$c_0 = ma_0, \qquad c_m = ma_m.$$

The coefficients $\{a_j\}_{j=0}^{m}$ and $\{b_j\}_{j=1}^{m-1}$ can be found from the coefficients $\{c_j\}_{j=0}^{m}$ by

$$a_j = \frac{\text{Re}(c_j)}{m}, \qquad b_j = -\frac{\text{Im}(c_j)}{m},$$

where c_0 and c_m are real. We note that the relation $c_{n-j} = c_{-j}$ can be derived from the simple identity

$$e^{i\pi j(k-1)/m} = e^{i\pi(j-n)(k-1)/m}. \tag{11.17}$$

The latter modification allows us to move coefficients with negative indices $\{c_j\}_{j=-m}^{-1}$ to coefficients with positive indices $\{c_j\}_{j=m}^{n-1}$. The fast Fourier transform is based on the complex discrete Fourier transform with $n = 2^N$, where $N \in \mathbb{N}$. See [6] for details of the computational algorithm. The fast Fourier transform requires $n \log n$ computational operations versus n^2 operations of the discrete Fourier transform.

Exercise 11.3 Apply the MATLAB function `fft` to the function $f(x) = e^{-x^2}$ on $[-10, 10]$ with $n = 2^N$ and $n = 2^N - 2$ data points and compute the CPU time of each application. Show that there is no difference between the CPU time of the two operations even when the values of N becomes larger.

Finishing the previous example, we give yet another equivalent method for computing coefficients of the trigonometric interpolation:

```
>> c = fft(yy);
>> a = (real(c(1:m+1))/m)'

aF =
     0     0     0     0     0

>> b = (-imag(c(1:m+1))/m)'

bF =
          0    1.2071          0    0.2071          0
```

The formula (11.15) is referred to as the *inverse discrete Fourier transform*, whereas the formula (11.16) is referred to as the *discrete Fourier transform*. You can confirm that the two operations are inverse to each other in the sense that the successive use of (11.16) and (11.15) restores the same set of data values y_k, $k = 1, 2, \ldots, n$.

```
>> yy = ifft(c)'

yy =
        Columns 1 to 6
        0     1.0000     1.0000     1.0000          0    -1.0000
        Column 7 to 8
        -1.0000    -1.0000
```

The pair of direct and inverse transforms is used in the *pseudo-spectral method*, where the solution of a time-dependent problem is computed by means of iterations and each iteration involves the discrete Fourier transform of the unknown function, its derivatives, and its multiplications.

11.2 Errors of Trigonometric Interpolation

The summation formulas (11.13)–(11.14) can be viewed as an application of the trapezoidal integration rule to computations of the continuous Fourier integrals (11.3)–(11.4) over the equally spaced discrete grid of x-values. (The trapezoidal rule is discussed in Section 6.5.) It is surprising that the approximate trapezoidal rule recovers the exact summation formula found from the orthogonality of the coefficient matrix of the linear system (11.12). This miraculous property can be explained by the method of spectral decompositions from *Linear Algebra* (eigenvalues are covered in Section 4.1). Moreover, the same method can be extended to construction of other spectral interpolations, such as the polynomial interpolation with orthogonal (for example, Chebyshev and Legendre) polynomials [24].

Let us discuss the principal difference between the trigonometric approximation in Problem 11.1 and the trigonometric interpolation in Problem 11.2. Problem 11.1 operates on the infinite-dimensional space of continuous functions $f(x)$ defined on a finite interval $[0, L]$ with the periodic boundary conditions. On the other hand, Problem 11.2 operates on the finite-dimensional space of data points $y_1, y_2, \ldots, y_{n+1}$ defined on the grid of equally spaced values $x_1, x_2, \ldots, x_{n+1}$ with the boundary conditions $y_{n+1} = y_1$. As a result, we may view the linear system (11.12) as a unique decomposition of the vector $\mathbf{y} = [y_1, y_2, \ldots, y_n] \in \mathbb{R}^n$ over the set of basis vectors in $\mathbb{R}^n$, where the unknown values of $a_0, a_1, \ldots, a_m$ and $b_1, b_2, \ldots, b_{m-1}$ for $n = 2m$ are coordinates of the decomposition. The discrete trigonometric functions in the system (11.12) represent an orthogonal set of eigenvectors of a linear eigenvalue problem that builds a particular basis in $\mathbb{R}^n$.

Eigenvalue problems

To understand the linear eigenvalue problem for discrete trigonometric functions, recall the continuous theory of the Sturm–Liouville eigenvalue problem.

Theorem 11.4 *Consider the boundary-value problem on the finite interval,*

$$u''(x) + \lambda u(x) = 0, \qquad 0 < x < L, \tag{11.18}$$

subject to the periodic boundary conditions $u(L) = u(0)$ and $u'(L) = u'(0)$. The problem (11.18) has a complete orthogonal set of eigenfunctions $u_j(x) = e^{i2\pi jx/L}$, $j \in \mathbb{Z}$, which corresponds to the set of eigenvalues $\lambda_j = \left(\frac{2\pi j}{L}\right)^2$ and satisfies the orthogonality relations

$$\int_0^L u_j(x)\bar{u}_l(x)dx = L\delta_{j,l}, \tag{11.19}$$

where $\delta_{j,l}$ is the Kroneker symbol. Any square integrable function $f(x)$ can be represented as the complex Fourier series

$$f(x) = \sum_{j\in\mathbb{Z}} c_j u_j(x), \qquad c_j = \frac{1}{L}\int_0^L f(x)\bar{u}_j(x)dx. \tag{11.20}$$

Replacing the continuous second derivative in the differential equation (11.18) with the second-order central difference (see Section 6.2), we obtain the difference eigenvalue problem:

$$u_{k+1} - 2u_k + u_{k-1} + h^2\lambda u_k = 0, \qquad 1 \le k \le n, \tag{11.21}$$

subject to the periodic boundary conditions $u_{n+1} = u_1$ and $u_0 = u_N$. Here, h is the step size of the spatial discretization, such that $h = L/n$. The difference eigenvalue problem (11.21) has a complete set of exact solutions for eigenvectors $\mathbf{u} = [u_1, u_2, \ldots, u_n]$ and eigenvalues λ:

$$u_k = e^{i2\pi j(k-1)/n}, \qquad \lambda = \frac{4}{h^2}\sin^2\left(\frac{\pi j}{n}\right), \qquad 0 \le j \le n-1. \tag{11.22}$$

The eigenvalue for $j = 0$ is simple, while all other eigenvalues are double. The validity of the solution (11.22) can be verified from (11.21) by the direct substitution and the use of the elementary trigonometric identity:

$$e^{i\theta} - 2 + e^{-i\theta} = 2\cos\theta - 2 = -4\sin^2\frac{\theta}{2}.$$

If the difference equation (11.21) is viewed as the matrix eigenvalue problem for a three-banded coefficient matrix, the set of eigenvectors and eigenvalues can be found with the MATLAB eigenvalue solver `eig`. These computations are illustrated in the MATLAB script `eigenvalues_eigenvectors`.

```
% eigenvalues_eigenvectors
n = 101;
A=2*diag(ones(1,n))-diag(ones(1,n-1),1)-diag(ones(1,n-1),-1);
A(1,n) = -1; A(n,1) = -1;
[V,D] = eig(A); lambda = diag(D);
lambda=[lambda(1:2:n);lambda(n-1:-2:2)]; % reorganize eigenvalues
kappa = pi*(0:1:n)/n; lambdaTheory = 4*sin(kappa).^2;
plot(kappa,lambdaTheory,'r:',kappa(1:n),lambda','b.')
x = (0:1:n-1)'/(n-1); v1 = V(:,1)/max(V(:,1)); % first five eigenvectors
v2=V(:,2)/max(V(:,2)); v3 = V(:,3)/max(V(:,3));
v4=V(:,4)/max(V(:,4)); v5 = V(:,5)/max(V(:,5));
plot(x,v1,'b',x,v2,'g',x,v3,'g:',x,v4,'m',x,v5,'m:');
```

When the MATLAB script `eigenvalues_eigenvectors` is executed, it computes eigenvalues of the difference eigenvalue problem (11.21) and plots them together with the exact solution (11.22) in Figure 11.2 (*left*). Figure 11.2 (*right*) shows graphs of eigenvectors for the smallest five eigenvalues starting with the ground state $u_k = 1$ and alternating through even $u_k = \cos(2\pi jk/n)$ and odd $u_k = \sin(2\pi jk/n)$ eigenfunctions (11.22).

The three-banded coefficient matrix in the difference eigenvalue problem (11.21) is symmetric. The eigenvalues λ are all real, while the eigenvectors are

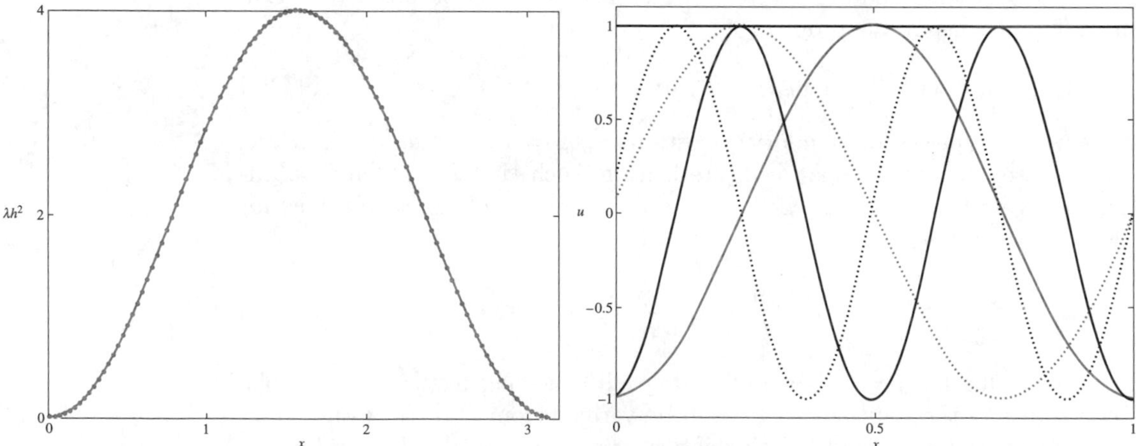

Figure 11.2 Eigenvalues (*left*) and the first five eigenvectors (*right*) for the difference eigenvalue problem (11.21).

the same as the complex exponentials of the inverse discrete Fourier transform (11.15). By adding complex numbers on the unit circle,

$$\sum_{k=1}^{n} e^{2\pi i(j-l)(k-1)/n} = n\delta_{j,l},$$

we recover the discretization of the continuous inner products (11.19). All these facts follow from general results in *Linear Algebra*. In particular, Theorem 4.6 implies that any vector $\mathbf{x} \in \mathbb{R}^n$ is uniquely represented over the orthogonal basis of eigenvectors $\mathbf{u}_1, \mathbf{u}_2, \ldots, \mathbf{u}_n$ of an n-by-n symmetric (self-adjoint) matrix, such that

$$\mathbf{x} = x_1\mathbf{u}_1 + x_2\mathbf{u}_2 + \ldots + x_n\mathbf{u}_n, \tag{11.23}$$

where the coordinates $x_1, x_2, \ldots, x_n$ are found from the projection formulas

$$x_j = \frac{\langle \mathbf{x}, \mathbf{u}_j \rangle}{\langle \mathbf{u}_j, \mathbf{u}_j \rangle}, \qquad 1 \le j \le n. \tag{11.24}$$

The pair of decomposition formulas (11.23)–(11.24) becomes the pair of discrete Fourier transforms (11.15)–(11.16). Whereas the discrete Fourier transform (11.16) corresponds to the second-order trapezoidal rule to the continuous Fourier integrals (11.3)–(11.4), the difference eigenvalue problem (11.21) corresponds to the second-order central-difference approximation of the differential eigenvalue problem (11.18). These facts explain the "miracle" of orthogonality of the coefficient matrix of the linear system (11.12).

Convergence of trigonometric interpolation

Suppose now that the data points (x_1, y_1), (x_2, y_2), ..., (x_{n+1}, y_{n+1}) correspond to the continuous function $f(x)$ evaluated on the discrete grid (11.9). The function $F_m(x)$ is a trigonometric interpolant for the function $f(x)$ on the interval $[0, L]$. The distance between these two functions defines the *truncation error* of the trigonometric interpolation. Given that we investigate in detail the truncation error of the polynomial interpolation in Section 5.5, we can now ask if the trigonometric interpolation performs a better job for a representation of the function $f(x)$. It is proved in *Numerical Analysis* that the trigonometric interpolant has an exponentially small truncation error in terms of the number of data points provided that the function $f(x)$ is extended into an analytic function off the real interval $[0, L]$.

Theorem 11.5 *Let $f(x)$ be a periodic function on $[0, L]$, which can be extended to an analytic function in the complex strip $|\mathrm{Im}(z)| < a$ with $|u(x + iy)| \le c$ uniformly in the rectangle $[0, L] \times [-a, a]$, where $a > 0$ and $c > 0$. Then, for any sufficiently large m there exists $C > 0$ such that*

$$\max_{0 \le x \le L} |f(x) - F_m(x)| \le Ce^{-am}. \tag{11.25}$$

This result is referred to as *spectral accuracy* of spectral methods, which we now illustrate with several examples. In the first example, we consider the Runge function $f(x) = 1/(1+25x^2)$ on the interval $[-1, 1]$ when the polynomial interpolation features the polynomial wiggle (see Section 5.5). The MATLAB script `Runge_trig_interpolation` computes the trigonometric interpolation of the Runge function for $n = 4, 8, 12$ subintervals.

```
% Runge_trig_interpolation
for n = 4 : 4 : 12
    x = linspace(-1,1,n+1);
    y = 1./(1+25*x.^2);
    m = n/2; L = 2;
    xx = [x(m+1:n),x(1:m)]';
    yy = [y(m+1:n),y(1:m)]';
    for j = 0 : m   % coefficients of interpolation
        a(j+1) = 2*yy'*cos(2*pi*j*xx/L)/n;
        b(j+1) = 2*yy'*sin(2*pi*j*xx/L)/n;
    end
    xInt = -1 : 0.001 : 1;
    yInt = 0.5*a(1)*ones(1,length(xInt));
    for j = 1 : (m-1) % trigonometric interpolant
        yInt = yInt + a(1+j)*cos(2*pi*j*xInt/L) + b(1+j)*sin(2*pi*j*xInt/L);
    end
    yInt = yInt + 0.5*a(m+1)*cos(2*pi*m*xInt/L);
    plot(xInt,yInt,x,y,'.');
end
yExact = 1./(1+25*xInt.^2);
plot(xInt,yExact,':r');
```

The output of the MATLAB script `Runge_trig_interpolation` is shown in Figure 11.3. No polynomial wiggle for the Runge function occurs in the trigonometric interpolation for large values of m.

The next example computes the truncation error of the trigonometric interpolation for two functions $f(x) = 1/(1+25x^2)$ and $f(x) = 1/(1+x^2)$ on the interval $[-5, 5]$ versus the number m, where $n = 2m$. The MATLAB script `error_trig_interpolation` contains code for computations of the errors for the two functions.

```
% error_trig_interpolation
for m = 1 : 300
    x = linspace(-5,5,2*m+1);
    y1 = 1./(1 + 25*x.^2);
    y2 = 1./(1 + x.^2);
    n= 2*m; L = 10;
    xx = [x(m+1:n),x(1:m)]';
```

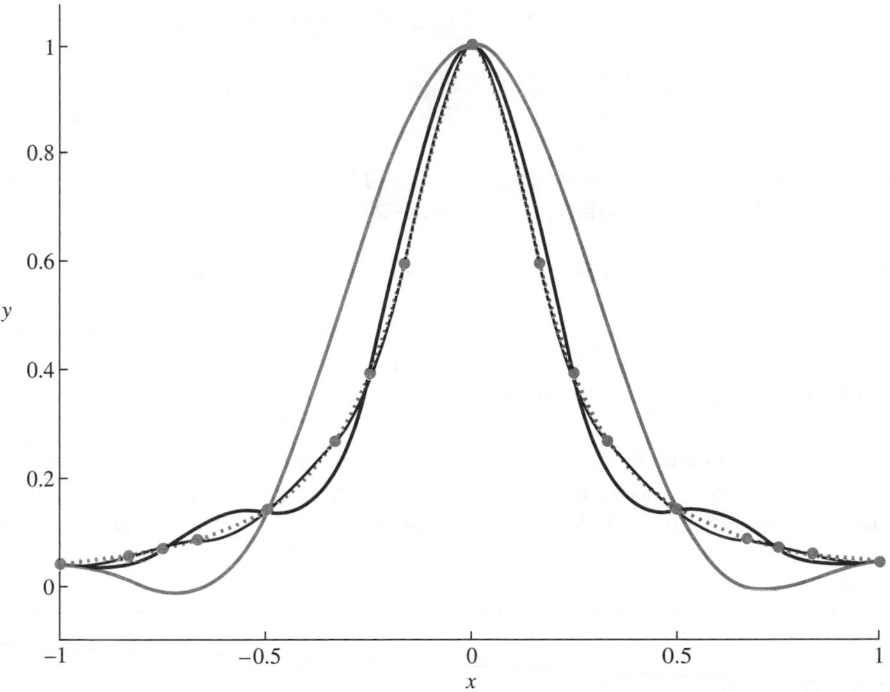

Figure 11.3 Trigonometric interpolation of the Runge function.

```
yy1 = [y1(m+1:n),y1(1:m)]';
yy2 = [y2(m+1:n),y2(1:m)]';
for j = 0 : m
    a1(j+1) = 2*yy1'*cos(2*pi*j*xx/L)/n;
    b1(j+1) = 2*yy1'*sin(2*pi*j*xx/L)/n;
    a2(j+1) = 2*yy2'*cos(2*pi*j*xx/L)/n;
    b2(j+1) = 2*yy2'*sin(2*pi*j*xx/L)/n;
end
xInt = linspace(-5,5,1001);
yInt1 = 0.5*a1(1)*ones(1,length(xInt));
for j = 1 : (m-1)
    yInt1 = yInt1 + a1(1+j)*cos(2*pi*j*xInt/L);
    yInt1 = yInt1 + b1(1+j)*sin(2*pi*j*xInt/L);
end
yInt1 = yInt1 + 0.5*a1(m+1)*cos(2*pi*m*xInt/L);
yInt2 = 0.5*a2(1)*ones(1,length(xInt));
for j = 1 : (m-1)
    yInt2 = yInt2 + a2(1+j)*cos(2*pi*j*xInt/L);
```

```
        yInt2 = yInt2 + b2(1+j)*sin(2*pi*j*xInt/L);
    end
    yInt2 = yInt2 + 0.5*a2(m+1)*cos(2*pi*m*xInt/L);
    yExact1 = 1./(1 + 25*xInt.^2);
    yExact2 = 1./(1 + xInt.^2);
    Error1(m) = max(abs(yExact1-yInt1));
    Error2(m) = max(abs(yExact2-yInt2));
end
hold on; semilogy(Error1,'b.'); semilogy(Error2,'m.');
```

The output of the MATLAB script `error_trig_interpolation` is shown in Figure 11.4. For smaller values of m, the logarithm of the total error reduces linearly in m for both functions. Because the function $f(x) = 1/(1 + x^2)$ has a pole at $a = 1$ and the function $f(x) = 1/(1 + 25x^2)$ has a pole at $a = \frac{1}{5}$, the logarithm of the total error drops faster for the first function in agreement with Theorem 11.5. For larger values of m, the logarithmic errors for both functions decay much slower than the linear decay predicted

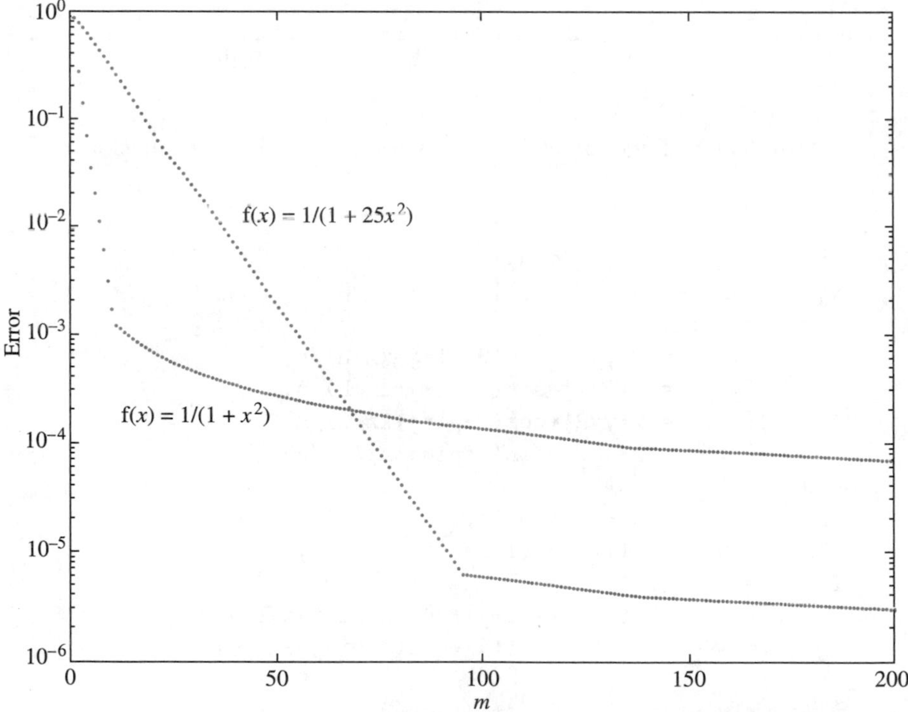

Figure 11.4 Maximum error versus the number m of the trigonometric interpolation.

by the theorem. The slow decay of the total error is caused by the influence of the round-off error, which increases with larger values of m. Compared to the slow decay of the truncation error and the growth of round-off error in the polynomial interpolation for large m (see Section 5.5), the trigonometric interpolation gives a much more robust algorithm for numerical representation of the function $f(x)$.

Similar to Theorem 11.2, the error of the trigonometric interpolation is $O(m^{-k-1})$ if the function $f(x)$ has k continuous derivatives (see convergence theorems in [28]). In this case, the efficiency of the trigonometric interpolation becomes comparable to the efficiency of the piecewise polynomial interpolation (see Section 12.1).

Exercise 11.4 Compute the maximum truncation error for the trigonometric interpolation $F_m(x)$ of the function $f(x) = \sqrt{1-x^2}$ on the interval $[-1, 1]$ versus the number m. Show that the maximum truncation error does not satisfy the bound (11.25) because $f(x)$ is not a real analytic function in the end points $x = \pm 1$. Because $f(x)$ is not differentiable at $x = \pm 1$, show that the maximum error is $O(m^{-1})$ as m increases.

Since we have assumed periodic boundary conditions at the endpoints of the interpolation interval $[0, L]$, improvements in the trigonometric interpolation are possible by reducing the step size h and simultaneously increasing the number of data points n. Unlike the case of polynomial interpolation, it is impossible to reduce the step size h and the interpolating interval $[0, L]$ simultaneously because it results in the violation of the boundary conditions on the function $f(x)$. In addition, you must be careful when no specific boundary conditions for the function $f(x)$ are implied by the interpolation problem. Interpolations based on other orthogonal functions (such as Chebyshev and Legendre interpolations) could be more appropriate for representation of functions without specific boundary conditions [24].

11.3 Trigonometric Methods for Differential Equations

In the problems where spectral accuracy of Theorem 11.5 can be reached, spectral methods give a rapidly convergent approximation. In these problems, the Galerkin method becomes accurate with very few terms in the trigonometric sum (11.1), while the collocation method can be applied with very few grid points in the uniform grid (11.9). These methods would work for various problems, including numerical solutions of boundary-value problems for differential equations. We shall develop trigonometric approximation and interpolation for the simplest boundary-value problems, leaving the general theory of spectral methods and their applications for more specialized texts [9, 24, 28].

Consider the heat equation (10.42) as the basic example of Problem 10.2. In particular, consider the boundary-value problem

$$u_t = u_{xx} + f(x), \qquad 0 < x < 1, \quad t > 0, \tag{11.26}$$

subject to the Dirichlet boundary conditions $u(0,t) = u(1,t) = 0$ and the initial condition $u(x,0) = g(x)$, where $f(x)$ and $g(x)$ are given functions on $[0,1]$. For instance, you can take a particular example of these functions as

$$f(x) = 10x(1-x), \qquad g(x) = 0.2\sin(3\pi x).$$

Using trigonometric approximation for odd functions on the symmetric interval $[-1,1]$ (see Problem 11.1 and Theorem 11.1 with $L = 2$), the function $f(x)$ is expanded into the trigonometric sum,

$$f_m(x) = \sum_{j=1}^{m} b_j \sin(\pi j x), \qquad 0 \le x \le 1, \tag{11.27}$$

where

$$b_j = 2\int_0^1 f(x)\sin(\pi j x)dx, \qquad 1 \le j \le m. \tag{11.28}$$

If $f(x) = 10x(1-x)$, the explicit expression for b_j is

$$b_j = \frac{40(1-(-1)^j)}{\pi^3 j^3}.$$

Because the function $f(x)$ is continuously differentiable on $[0,1]$, Theorem 11.2 states that the partial sum $f_m(x)$ converges to $f(x)$ pointwise and uniformly on $[0,1]$ as $m \to \infty$, while the difference between $f_m(x)$ and $f(x)$ is $O(m^{-2})$. The same order follows also from the explicit expression for b_j since the sum $\sum_{j=m+1}^{\infty} \frac{1}{j^3}$ is $O(m^{-2})$. Because the truncation error does not decay exponentially, the trigonometric approximation is not spectrally accurate. This inaccuracy is explained by the fact that if the function $f(x)$ is extended as an odd function from $[0,1]$ to $[-1,1]$, and then it is continued periodically from $[-1,1]$ to $\mathbb{R}$ with period $L = 2$, then the second derivative $f''(x)$ has jump discontinuities at the points $x = 0$ and $x = 1$. As a result, the function $f(x)$ cannot be analytically continued in the complex plane.

Galerkin method The Galerkin method is based on the approximation of the solution $u(x,t)$ of the time evolution problem (11.26) by the trigonometric sum:

$$u(x,t) = \sum_{j=1}^{m} u_j(t)\sin(\pi j x), \qquad 0 \le x \le 1, \tag{11.29}$$

which satisfies the boundary conditions $u(0,t) = u(1,t) = 0$. The initial values for $u_j(0)$ follow from the initial condition $u(x,0) = g(x)$ expanded into the same trigonometric sum with

$$u_j(0) = 2\int_0^1 g(x)\sin(\pi j x)dx, \qquad 1 \le j \le m. \tag{11.30}$$

When $g(x) = 0.2\sin(3\pi x)$, the explicit expression for $u_j(0)$ is $u_j(0) = 0.2\delta_{3,j}$, where $\delta_{i,j}$ is the Kronecker symbol. By substituting (11.27) and (11.29) into the heat equation (11.26), we find the uncoupled systems of ordinary differential equations,

$$\frac{du_j}{dt} = -\pi^2 j^2 u_j(t) + b_j, \qquad 1 \le j \le m. \tag{11.31}$$

The exact solution of the ODE system (11.31) is available in analytic form:

$$u_j(t) = u_j(0)e^{-\pi^2 j^2 t} + \frac{b_j}{\pi^2 j^2}\left(1 - e^{-\pi^2 j^2 t}\right), \qquad 1 \le j \le m, \tag{11.32}$$

where $u_j(0)$ are given. Alternatively, initial-value ODE solvers can be applied to the numerical solution of the ODE system (11.31). In either case, the trigonometric sum (11.29) becomes the numerical approximation of the solution surface $u(x,t)$. The MATLAB script Galerkin_method shows details of the Galerkin method supplemented by the fourth-order Runge–Kutta method for solutions of the ODE system (11.31) (see Section 9.2).

```
% Galerkin_method
m = 5; j = 1 : m;
b = 40*(1 - (-1).^j)./(pi^3*j.^3);  % inhomogeneous term
u = zeros(1,m); u(3) = 0.2;          % initial condition
T = 0.5; t = linspace(0,T,101); dt = t(2)-t(1); % time grid
uSpectrum(1,:) = u;
for k = 1 : length(t)-1            % fourth-order Runge-Kutta method
    k1 = -pi^2*j.^2.*u+b; u1 = u + 0.5*dt*k1;
    k2 = -pi^2*j.^2.*u1+b; u2 = u + 0.5*dt*k2;
    k3 = -pi^2*j.^2.*u2+b; u3 = u + dt*k3;
    k4 = -pi^2*j.^2.*u3+b;
    u = u + dt*(k1+2*k2+2*k3+k4)/6;
    uSpectrum(k+1,:) = u;
end
x = linspace(0,1,101);
U = zeros(length(t),length(x));
```

```
for k = 1 : length(t)              % solution surface
   for jj = 1 : m
        U(k,:) = U(k,:) + uSpectrum(k,jj)*sin(pi*jj*x);
   end
end
[X,T] = meshgrid(x,t); mesh(X,T,U);
```

When the MATLAB script Galerkin_method is executed, it computes the numerical approximation of the solution surface $u(x,t)$ by the trigonometric sum (11.29) with $m = 5$ for $0 \leq t \leq 0.5$. Figure 11.5 shows the solution surface $u(x,t)$, which displays a transformation of the initial condition $u(x,0) = g(x)$ to the time-independent solution of the boundary-value problem (11.26):

$$u_\infty(x) = \lim_{t\to\infty} u(x,t) = \frac{5}{6}x(x^3 - 2x^2 + 1). \tag{11.33}$$

Except for the initial time $t = 0$, there exists an error between the exact solution $u(x,t)$ of the heat equation (11.26) and the trigonometric sum (11.29). The numerical error is caused by two main sources. The first source of the numerical error is a discretization of the numerical solution of the ODE system

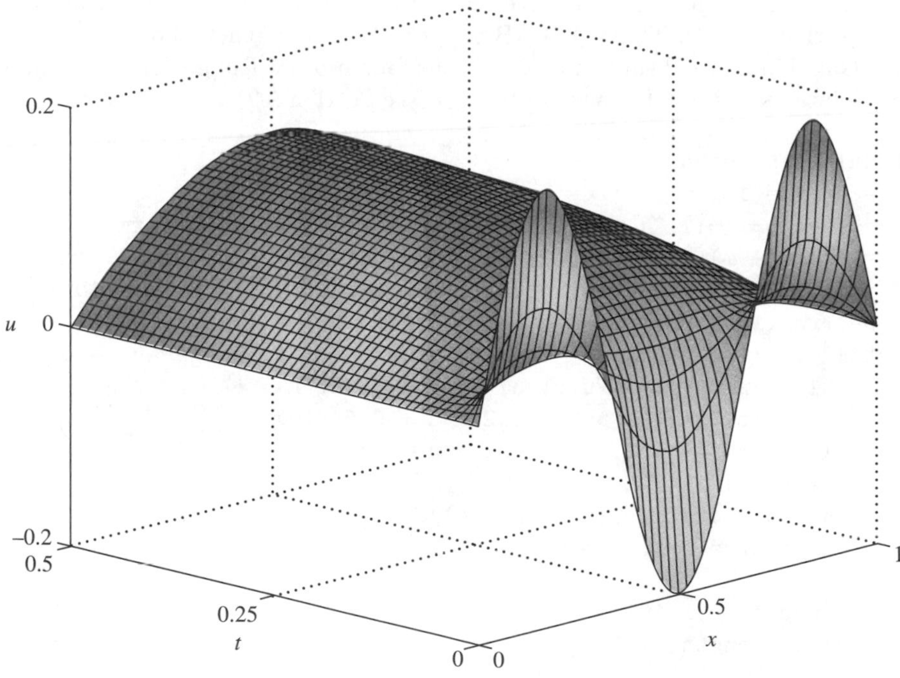

Figure 11.5 Solution of the heat equation (11.26) by the Galerkin method.

(11.31). Although the exact solution (11.32) is available in this particular case, the numerical ODE solver is often the only available tool for computations of solutions of the ODE systems. Unfortunately, the ODE system (11.31) becomes stiff for large values of m because the decay rate $\lambda_j = \pi^2 j^2$ becomes larger with larger values of j. As a result, all explicit ODE solvers develop instabilities for larger values of j, while implicit ODE solvers may produce a stable numerical solution of the ODE systems (see Section 9.5). Even if the ODE solver is implicit and stable, it leads to a numerical error that depends on the time step τ and converges to zero as a power function of τ. For instance, the global truncation error of the fourth-order Runge–Kutta method is $O(\tau^4)$ (see Section 9.2).

The other source of the numerical error is a truncation of the trigonometric sum (11.29). This source is controlled by Theorem 11.2 on the trigonometric approximation. In the particular problem (11.26) with $f(x) = 10x(1-x)$, the limiting stationary solution $u_\infty(x)$ is the polynomial of degree four (11.33). If $u_\infty(x)$ is extended to the entire axis as an odd periodic function with period $L = 2$, it has jump discontinuities in the fourth derivative, implying that the corresponding trigonometric sum (11.29) has an $O(m^{-4})$ error as $m \to \infty$. Since $u(x,0) = g(x) = \sin(3\pi x)$ is represented by the trigonometric sum exactly, the truncation error of the representation for $u(x,t)$ coincides with that for $u_\infty(x)$ for any fixed value of $t > 0$. If $f(x)$ were represented by a trigonometric sum (11.27) with spectral accuracy, then the solution $u(x,t)$ would be represented by the sum (11.29) with spectral accuracy, too.

Exercise 11.5 Use the exact solution (11.32) for the ODE system (11.31) and show that the error between the exact solution $u(x,t)$ in the Fourier series form and the trigonometric sum (11.29) for a fixed value of $t > 0$ is $O(m^{-4})$ for $f(x) = 10x(1-x)$ and is exponentially small in m for $f(x) = \sin(\pi x)$.

Collocation method

By using trigonometric interpolation, we develop the collocation method, which is based on the numerical approximation of the solution $u(x,t)$ of the PDE (11.26) at the uniform grid

$$x_k = \frac{(k-1)}{(m+1)}, \qquad 1 \le k \le m+2. \tag{11.34}$$

Compared to the formalism in Problem 11.2, the trigonometric sum (11.29) extends the sum (11.10) by increasing the index m by one. In addition, the cosine terms in the sum (11.10) are identically zero as the function $u(x,t)$ is extended into a periodic odd function on the x-axis with the period $L = 2$. The initial values for $u_j(0)$ in the trigonometric sum (11.29) follow from the trigonometric interpolation of the initial condition $u(x,0) = g(x)$ by

$$g(x_k) = \sum_{j=1}^{m} u_j(0) \sin(\pi j x_k), \qquad 2 \le k \le m+1,$$

with the inversion formula

$$u_j(0) = \frac{2}{m+1} \sum_{k=2}^{m+1} g(x_k) \sin(\pi j x_k), \qquad 1 \leq j \leq m. \tag{11.35}$$

Compared to the sum (11.10), the summation over $m+3 \leq k \leq n-1$ is not performed, since the function $g(x)$ is extended into a periodic odd function on values of x_k beyond the range $1 \leq k \leq m+2$. Similarly, the source term in the heat equation (11.26) is approximated at the grid points (11.34) by

$$f(x_k) = \sum_{j=1}^{m} b_j \sin(\pi j x_k), \qquad 2 \leq k \leq m+1,$$

with the inversion formula

$$b_j = \frac{2}{m+1} \sum_{k=2}^{m+1} f(x_k) \sin(\pi j x_k), \qquad 1 \leq j \leq m. \tag{11.36}$$

When the trigonometric sums for $u(x,t)$ and $f(x)$ are substituted into the heat equation (11.26), the same system of uncoupled ordinary differential equations (11.31) arises. The ODE system can be solved with an initial-value ODE solver such as the fourth-order Runge–Kutta method. The MATLAB script `collocation_method` shows details of the collocation method for the functions $f(x)$ and $g(x)$, while the ODE solver is coded similarly to the MATLAB script `Galerkin_method`.

```
% collocation_method
m = 5; x = linspace(0,1,m+2);
f = 10*x.*(1-x); % source term
for j = 1 : m
    b(j) = 2*f*sin(pi*j*x')/(m+1);
end
g = 0.2*sin(3*pi*x); % initial condition
for j = 1 : m
    u(j) = 2*g*sin(pi*j*x')/(m+1);
end
j = 1:m; T = 0.5;
t = linspace(0,T,101); dt = t(2)-t(1);
uSpectrum(1,:) = u;
for k = 1 : length(t)-1  % time iterations
     k1 = -pi^2*j.^2.*u+b; u1 = u + 0.5*dt*k1;
     k2 = -pi^2*j.^2.*u1+b; u2 = u + 0.5*dt*k2;
     k3 = -pi^2*j.^2.*u2+b; u3 = u + dt*k3;
     k4 = -pi^2*j.^2.*u3+b;
```

```
    u = u + dt*(k1+2*k2+2*k3+k4)/6;
    uSpectrum(k+1,:) = u;
end
x = linspace(0,1,101);
U = zeros(length(t),length(x));
for k = 1 : length(t)   % solution surface
   for jj = 1 : m
        U(k,:) = U(k,:) + uSpectrum(k,jj)*sin(pi*jj*x);
   end
end
[X,T] = meshgrid(x,t); mesh(X,T,U);
```

When the MATLAB script `collocation_method` is executed, it computes the numerical approximation of the solution surface $u(x,t)$, which is shown in Figure 11.6. Both the transient process and the limiting solution look similar to ones modelled by the Galerkin method (see Figure 11.5).

Errors of spectral methods

Let us now compare the truncation errors of the Galerkin and collocation methods for numerical approximation of the limiting solutions $u_\infty(x)$ of the

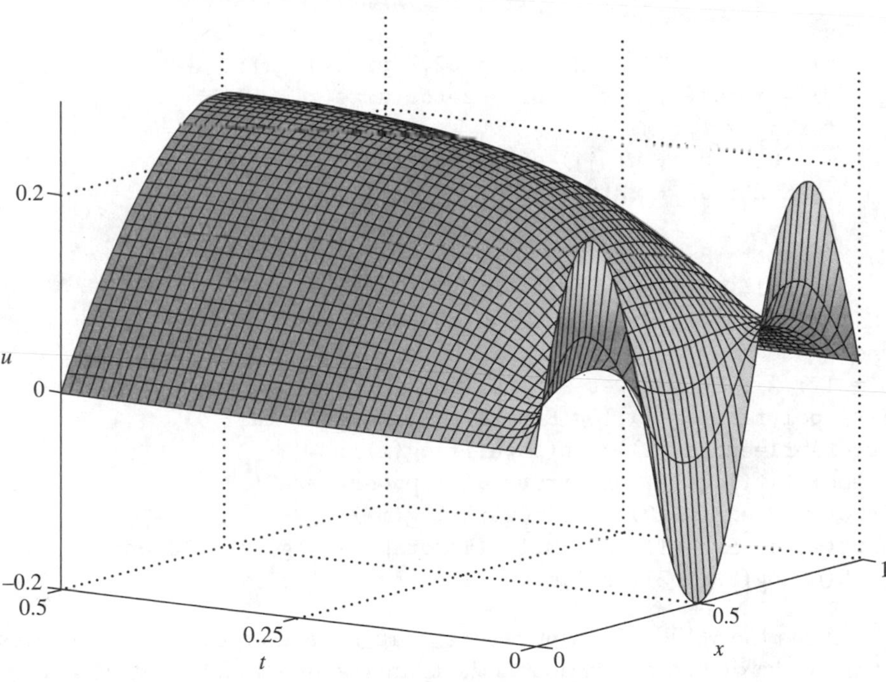

Figure 11.6 Solution of the heat equation (11.26) by the collocation method.

heat equation (11.26). The limiting solution $u_\infty(x)$ solves the boundary-value ODE problem

$$u_{xx} + f(x) = 0, \qquad 0 < x < 1, \tag{11.37}$$

subject to the Dirichlet boundary conditions $u(0) = u(1) = 0$. In both methods, the solution $u(x)$ is approximated by the trigonometric sum (11.29) with the time-independent coefficients $u_j = b_j/(\pi^2 j^2)$, $j = 1, \ldots, m$. The only difference occurs in the computations of the values of b_j. These values are computed from continuous integrals (11.28) in the Galerkin method, while they are computed from the discrete sum (11.36) in the collocation method. Numerical solutions of the ODE problem (11.37) with $f(x) = 10x(1-x)$ are computed in the MATLAB script errors_trig_methods by both methods and the truncation errors are found from the exact solution (11.33).

```
% errors_trig_methods
M = 100; x = linspace(0,1,1001);
for m = 1 : 2 : M
    j = 1 : m; b1 = 40*(1 - (-1).^j)./(pi^3*j.^3);
    xx = linspace(0,1,m+2); f = 10*xx.*(1-xx);
    for jj = 1 : m
        b2(jj) = 2*f*sin(pi*jj*xx')/(m+1);
    end
    u1 = b1./(pi^2*j.^2); u2 = b2./(pi^2*j.^2);
    U1 = zeros(size(x)); U2 = zeros(size(x));
    for jj = 1 : m
        U1 = U1 + u1(jj)*sin(pi*jj*x);
        U2 = U2 + u2(jj)*sin(pi*jj*x);
    end
    Uexact = 5*x.*(x.^3-2*x.^2+1)/6;
    Error1((m+1)/2) = sqrt(sum((U1-Uexact).^2));
    Error2((m+1)/2) = sqrt(sum((U2-Uexact).^2));
end
m = 1 : 2 : M; % power fit for the error dependence
a1 = polyfit(log(m),log(Error1),1); power1 = a1(1)
ErrorApr1=exp(a1(2))*exp(power1*log(m));
a2=polyfit(log(m),log(Error2),1); power2 = a2(1)
ErrorApr2=exp(a2(2))*exp(power2*log(m));
plot(m,log(Error1),'b.',m,log(ErrorApr1),':g'); hold on;
plot(m,log(Error2),'r.',m,log(ErrorApr2),':y')
```

When the MATLAB script errors_trig_methods is executed, the errors of the Galerkin and collocation methods are computed and shown as dots in Figure 11.7. The best power fits for the two data sets are also computed and plotted by the dotted curves. You can see from the power fits that the error

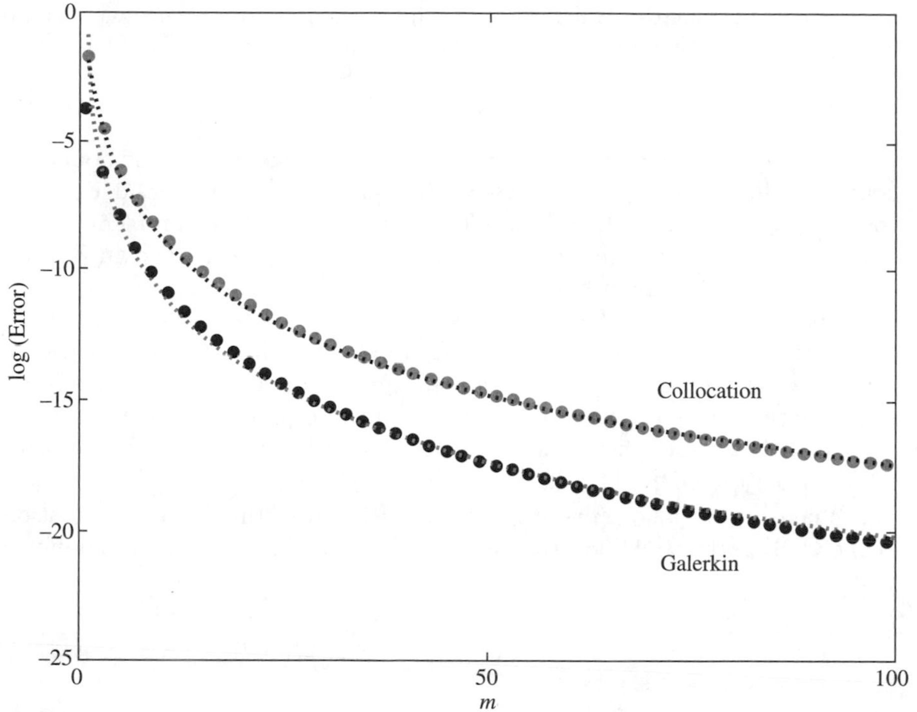

Figure 11.7 Truncation errors for the numerical solutions of the ODE problem (11.37) with the Galerkin and collocation methods

for either numerical solution is $O(m^{-4})$ because the solution $u_\infty(x)$ in (11.32) has the jump discontinuities in the fourth derivative. You can also see from Figure 11.7 that the Galerkin method has a smaller truncation error compared to the collocation method with the same rate of convergence.

```
>> errors_trigonometric_methods

power1 =
    -3.9924

power2 =
    -3.6631
```

The heat equation (11.26) is one of the simplest problems for applications of spectral methods. Other linear and nonlinear differential equations offer more challenges for the Galerkin and collocation methods. We review these challenges when attempting to solve the boundary-value ODE problem

$$-u'' + f(x)u = g(x), \qquad 0 < x < 2\pi, \tag{11.38}$$

where $f(x+2\pi) = f(x)$ and $g(x+2\pi) = g(x)$ subject to the periodic boundary conditions $u(2\pi) = u(0)$ and $u'(2\pi) = u'(0)$. In particular, we consider $f(x) = \cos x$ and $g(x) = \sin x$. (Eigenvalues of the linear eigenvalue problem $-u'' + \cos xu = \lambda u$ are approximated with the finite-difference method in Section 10.1, where the Neumann boundary conditions $u'(0) = u'(2\pi) = 0$ are used.) If no periodic solution of the homogeneous ODE $-u'' + f(x)u = 0$ exists, the inhomogeneous boundary-value problem (11.38) admits a unique periodic solution $u(x)$, which can be approximated numerically by using trigonometric sums. Both trigonometric approximation and interpolation lead to a linear system with a full coefficient matrix, compared to the diagonal system that follows from the ODE (11.37).

Advanced collocation method

When the trigonometric interpolation is used in the numerical solution of the ODE problem (11.38), the interval $[0, 2\pi]$ is represented by the uniform grid

$$x_k = \frac{2\pi(k-1)}{n}, \qquad k = 1, 2, \ldots, n+1, \tag{11.39}$$

where n is even. The periodic function $u(x)$ is then represented by the complex trigonometric sum

$$u(x) = \frac{1}{n} \sum_{j=-m+1}^{m-1} c_j e^{ijx} + \frac{1}{2n}\left(c_{-m}e^{-imx} + c_m e^{imx}\right), \tag{11.40}$$

where the set of coefficients $\{c_j\}_{j=-m}^{m}$ is computed from the set of function values $\{u_k\}_{k=1}^{n}$ with $u_k = u(x_k)$ by the inversion formula

$$c_j = \sum_{k=1}^{n} u_k e^{-ijx_k}, \qquad -m \le j \le m. \tag{11.41}$$

The complex trigonometric sum (11.40) is obtained from the trigonometric sum (11.10) in Problem 11.2. The coefficients $\{c_j\}_{j=-m}^{m}$ are defined by the same formulas as those given later in (11.15)–(11.16) but no reflection to positive indices $c_{n-j} = c_{-j}$ is used. Proceeding with the collocation method for a numerical solution of the ODE problem (11.38), we define the functions $f(x)$ and $g(x)$ at the collocation points (11.39) and meet the obstacle that the linear problem for the set of coefficients $\{u_k\}_{k=1}^{n}$ is not closed because the derivative terms $u''(x_k)$ are not defined in terms of the values $\{u_k\}_{k=1}^{n}$. A solution to this obstacle is constructed in [28]. According to this solution, we define the interpolation function $S(x)$ that passes through points of the discrete delta function $S(x_k) = \delta_{k,1}$. Using (11.41), we obtain that $c_j = 1$ for all $-m \le j \le m$, leading to

$$\begin{aligned} S(x) &= \frac{1}{n}\Big(1 + e^{ix} + e^{-ix} + \ldots + e^{i(m-1)x} + e^{-i(m-1)x} + \cos(mx)\Big) \\ &= \frac{1}{n}\left(\frac{1-e^{imx}}{1-e^{ix}} + \frac{1-e^{-imx}}{1-e^{-ix}} + \cos(mx) - 1\right) \\ &= \frac{\sin(mx)\sin x}{n(1-\cos x)} = \frac{\sin(mx)\cos(x/2)}{2m\sin(x/2)}. \end{aligned}$$

Using the function $S(x)$, the trigonometric interpolation is written in the form

$$u(x) = \sum_{k=1}^{n} u_k S(x - x_k), \qquad (11.42)$$

so that the first and second derivatives of $u(x)$ at the grid point $x = x_k$ are expressed in the matrix form

$$u'(x_k) = \sum_{l=1}^{n} (\mathbf{D}_1)_{k,l} u_l, \qquad u''(x_k) = \sum_{l=1}^{n} (\mathbf{D}_2)_{k,l} u_l,$$

where $\mathbf{D}_1$ and $\mathbf{D}_2$ are symmetric matrices obtained from the first and second derivatives of $S(x)$ in the form [28]:

$$(\mathbf{D}_1)_{i,j} = \begin{cases} 0, & i = j, \\ \frac{1}{2}(-1)^{i-j}\cot(\pi(i-j)/n), & i \neq j \end{cases}$$

and

$$(\mathbf{D}_2)_{i,j} = \begin{cases} -\frac{n^2}{12} - \frac{1}{6}, & i = j, \\ -\frac{(-1)^{i-j}}{2\sin^2(\pi(i-j)/n)}, & i \neq j \end{cases}$$

By eliminating $u''(x_k)$ from the ODE (11.38) at the grid point $x = x_k$, we can close the system of linear equations for the set $\{u_k\}_{k=1}^{n}$, solve it with MATLAB linear algebra, and display the solution $u(x)$.

Advanced Galerkin method

When the trigonometric approximation is used in the numerical solution of the ODE problem (11.38), the periodic function $u(x)$ is represented by the complex trigonometric sum

$$u(x) = \sum_{j=-m}^{m} c_j e^{ijx}, \qquad (11.43)$$

where the set of coefficients $\{c_j\}_{j=-m}^{m}$ is computed from the continuous function $u(x)$ by the inversion formula

$$c_j = \frac{1}{2\pi}\int_0^{2\pi} u(x) e^{-ijx}, \qquad -m \le j \le m. \qquad (11.44)$$

The functions $f(x)$ and $g(x)$ can be represented exactly by

$$\cos x = \frac{e^{ix} + e^{-ix}}{2}, \qquad \sin x = \frac{e^{ix} - e^{-ix}}{2i}.$$

Proceeding with the Galerkin method for a numerical solution of the ODE problem (11.38), we substitute the trigonometric sum (11.43) into the ODE (11.38) and meet the obstacle that the set of coefficients $\{c_j\}_{j=-m}^{m}$ is not closed because the product term $f(x)u(x)$ generates Fourier terms $e^{i(m+1)x}$ and $e^{-i(m+1)x}$ beyond the truncation order of the sum (11.43). If the explicit form for $f(x)$ and $g(x)$ is used, the linear system for the set $\{c_j\}_{j=-m}^{m}$ takes the explicit form of the difference equation

$$j^2 c_j + \frac{1}{2}\left(c_{j+1} + c_{j-1}\right) = \frac{1}{2i}\left(\delta_{j,1} - \delta_{j,-1}\right), \qquad -m \le j \le m. \tag{11.45}$$

Assuming that the coefficients c_j become smaller with larger values of j, we can truncate the linear system (11.45) beyond the terms of $-m \le j \le m$. Because $c_j \neq 0$ for $|j| \ge m+1$, the truncation of the linear system (11.45) introduces an additional truncation error to the numerical approximation of $u(x)$ by the trigonometric sum (11.43). Because of the truncation procedure, the linear system (11.45) becomes closed and can be solved with MATLAB linear algebra.

The corresponding computations of the collocation and Galerkin methods are coded in the MATLAB script `bvp_trig_methods`. Two numerical approximations of the solution $u(x)$ obtained by the collocation and Galerkin methods are shown graphically in Figure 11.8.

```
% bvp_trig_methods
n = 100; x = linspace(0,2*pi,n+1);
f = diag(cos(x(1:n))); % collocation method
g = sin(x(1:n))';
D = -diag(ones(1,n))*(n^2/12+1/6); % matrix for second derivatives
for j = 1 : n-1
    for k = j+1 : n
        D(j,k) = 0.5*(-1)^(j-k-1)/(sin(pi*(j-k)/n))^2;
        D(k,j) = D(j,k);
    end
end
A = -D + f; u = A\g; u(n+1) = u(1);
m = n/2; j = -m : m;     % Galerkin method
A = diag(j.^2) + 0.5*(diag(ones(n,1),1) + diag(ones(n,1),-1));
gg = zeros(n+1,1); % inhomogeneous term
gg(m) = -1/(2*i); gg(m+2) = 1/(2*i);
c = A\gg; % solution in Fourier space
xx = linspace(0,2*pi,1001); uu = zeros(1,1001);
```

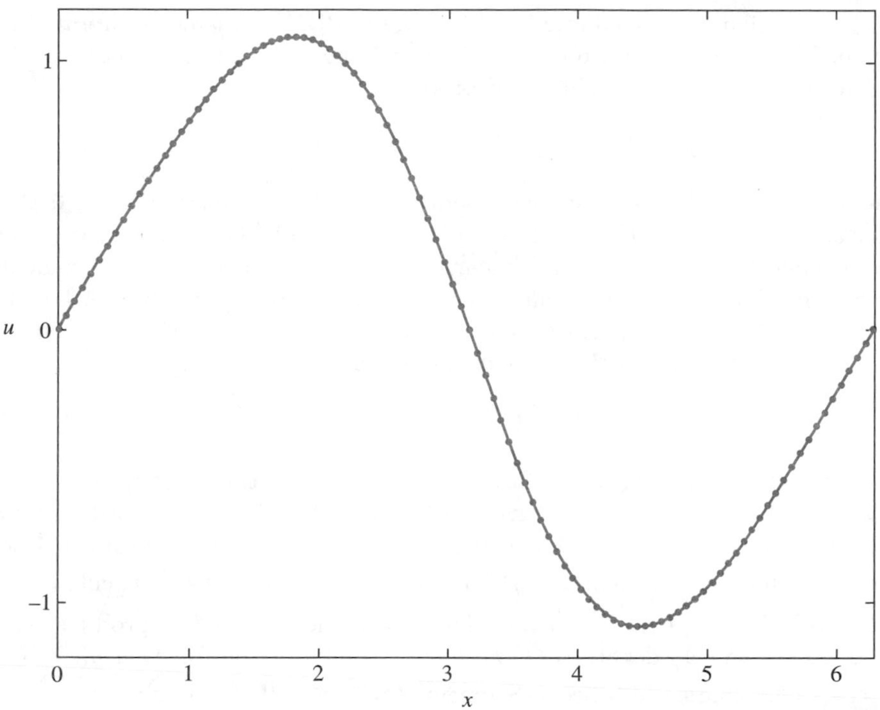

Figure 11.8 Numerical approximations of the solution $u(x)$ of the boundary-value problem (11.38) by the collocation method (dots) and by the Galerkin method (solid).

```
for k = 1 : n+1    % solution in physical space
    uu = uu + c(k)*exp(i*j(k)*xx);
end
plot(x,u,'.b',xx,real(uu),'r');
```

Exercise 11.6 Show numerically that the ∞-norm of the difference between the numerical approximations of the collocation and Galerkin methods reduces with larger values of m. Plot the ∞-norm of the computational error versus $n = 2m$ and fit the dependence with a power law.

Pseudospectral method

The boundary-value ODE problem (11.38) leads to the full coefficient matrix in either the collocation or Galerkin method. Another popular spectral method called *the pseudospectral method* overcomes this obstacle and produces a diagonal linear system for the price of an iterative method. The pseudospectral method iterates numerical approximations of the boundary-value problem

(11.38) similar to how elliptic boundary-value PDE problems are iterated by embedding the elliptic problem in the parabolic problem (see Section 10.5). Consider the boundary-value PDE problem

$$u_t = u_{xx} - f(x)u + g(x), \qquad 0 < x < 2\pi, \quad t > 0, \tag{11.46}$$

subject to the periodic boundary conditions $u(0,t) = u(2\pi,t)$ and $u_x(0,t) = u_x(2\pi,t)$ and the initial condition $u(x,0) = u_0(x)$. The stationary (time-independent) solutions of the time-evolution PDE problem (11.46) coincide with the solutions of the boundary-value ODE problem (11.38). After the time discretization is performed with the explicit Euler method, we obtain the iterative rule that defines a sequence of functions $\{u_k(x)\}_{k=0}^{\infty}$:

$$u_{k+1}(x) = u_k(x) + \tau \left(u_k''(x) - f(x)u_k(x) + g(x)\right). \tag{11.47}$$

When time iterations are performed, the given functions $u_k(x)$, $f(x)$, and $g(x)$ can be computed on the grid points (11.39) and all terms of the iterative scheme (11.47) can be diagonalized using the trigonometric sums (11.40) with the inversion formula (11.41). Let $c_j^{(k)}$ denote Fourier coefficients of the approximation $u_k(x)$, $b_j^{(k)}$ denote Fourier coefficients of the product term $f(x)u_k(x)$, and a_j denote the Fourier coefficients of the function $g(x)$. The iterative scheme (11.47) becomes diagonal in terms of the coefficients $c_j^{(k)}$:

$$c_j^{(k+1)} = c_j^{(k)}(1 - \tau j^2) + \tau(a_j - b_j^{(k)}), \qquad -m \le j \le m. \tag{11.48}$$

Since the problem (11.46) is linear, the iterative procedure (11.48) converges to a solution from any starting approximation $u_0(x)$, if it converges at all. Therefore, we can start the iterative procedure with $u_0(x) = 0$. As in all other iterative methods, iterations can be stopped when the distance between two successive iterations becomes smaller than a given tolerance. The MATLAB script `pseudospectral_method` performs computations of the pseudospectral method for the time-evolution PDE problem (11.46).

```
% pseudospectral_method
n = 50; m = n/2; x = linspace(0,2*pi,n+1);
f = cos(x(1:n)); g = sin(x(1:n)); % inhomogeneous terms
j = -m : m;  % discrete Fourier transform for g(x)
for jj = 1 : length(j)
    a(jj) = g*exp(-i*j(jj)*x(1:n)');
end
c = zeros(size(a)); % initial approximation
for k = 1 : n
    u(k) = real(c(2:n)*exp(i*(-m+1:m-1)'*x(k)));
    u(k) = (u(k)+(c(1)*exp(-i*m*x(k))+c(n+1)*exp(i*m*x(k)))/2)/n;
end
```

```
tau = 0.001; du = 1; toler = 10^(-9);
count = 0; term = 100000; % pseudospectral method
while ((du > toler) & (count < term))
      ff = f.*u;  % discrete transform for f(x) u_k(x)
      for jj = 1 : length(j)
           b(jj) = ff*exp(-i*j(jj)*x(1:n)');
      end
      cc = c + tau*(-j.^2.*c-b+a);
      for k = 1 : n % inverse transform for u_{k+1}(x)
            uu(k) = real(cc(2:n)*exp(i*(-m+1:m-1)'*x(k)));
            uu(k) = (uu(k)+(cc(1)*exp(-i*m*x(k))+cc(n+1)*exp(i*m*x(k)))/2)/n;
      end
      du = max(abs(uu-u)); c = cc;
      u = uu; count = count + 1;
end
fprintf('The algorithm converges after %d iterations\n',count);
u(n+1) = u(1); plot(x,u,'.');
```

When the MATLAB script pseudospectral_method is executed, it computes the sequence of numerical approximations $\{u_k(x)\}_{k=0}^{k_{\text{term}}}$ that starts with $u_0(x) = 0$ and terminates at $k = k_{\text{term}}$ when the distance between two successive iterations becomes smaller than the tolerance 10^{-9}. The number of iterations is displayed. The solution $u(x)$ at the grid points (11.39) is shown graphically in Figure 11.9. It looks similar to the solution obtained by the collocation and Galerkin methods in Figure 11.8.

```
>> pseudospectral_method

The algorithm converges after 15058 iterations
```

The use of explicit single-step methods such as the Euler method in pseudospectral methods can limit their applicability as a result of instabilities of explicit ODE solvers. When the iterations (11.48) are considered with $f(x) = g(x) = 0$, it is clear that the Euler method is stable only if

$$1 - j^2\tau > -1, \qquad -m \le j \le m,$$

such that $\tau < 2/m^2$. When the number of terms m in the trigonometric sum (11.40) grows, the time step τ becomes smaller and the number of iterations it takes to reach the same level of tolerance grows. This property results in slow convergence and large round-off error of the pseudospectral method.

There are several ways to improve the convergence and stability of the pseudospectral methods. For stability, implicit methods such as the implicit Euler method can be used for a numerical approximation of the solution $u(x)$.

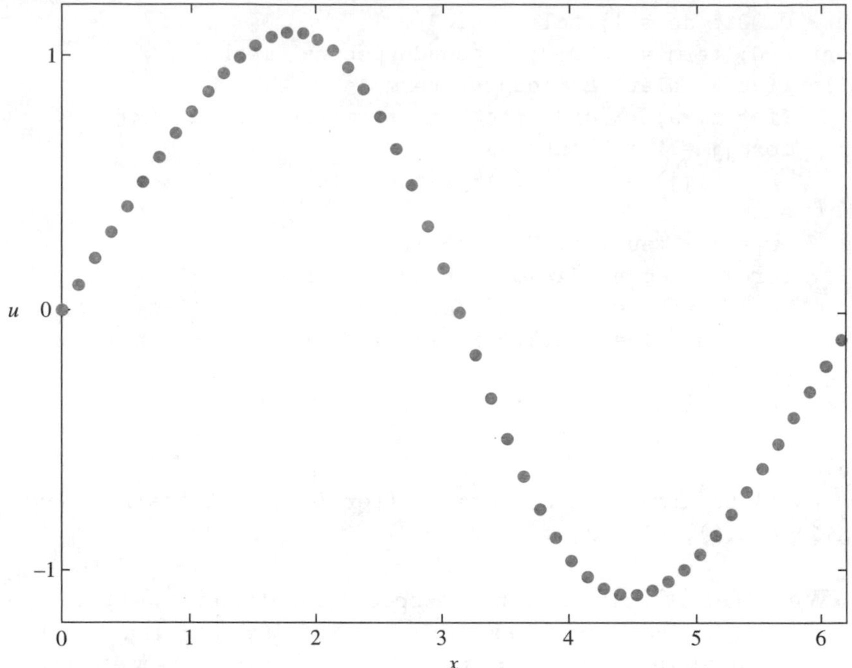

Figure 11.9 Numerical approximation of the solution $u(x)$ of the boundary-value problem (11.38) by the pseudospectral method.

Then, the number m can be arbitrarily large with no effects on stability of iterations. For convergence, we can avoid diagonalizing each term in the iterative scheme (11.47), which requires computations of the direct and inverse discrete Fourier transforms at each iteration of the time-evolution scheme. When the second derivatives of the solution $u_k(x)$ are approximated at the grid points (11.39) with the matrix $\mathbf{D}_2$ obtained from the representation (11.42) (as in the collocation method), the iterative scheme can be closed for the vector of the approximation $u_k(x)$ evaluated at the grid points (11.39). When the convolution sum is truncated beyond the terms $|j| \geq m + 1$ (as in the Galerkin method), the iterative scheme can be closed for the vector of the Fourier coefficients $c_j^{(k)}$ computed for the approximation $u_k(x)$.

When the nonlinear differential equations are considered, pseudospectral methods are more useful compared to the direct collocation and Galerkin methods. For instance, the power terms u^2 and u^3 as well as their derivatives can be computed from the kth iteration $u_k(x)$ directly, by using the pair of direct and inverse discrete Fourier transforms. Iterative methods become explicit methods suitable for numerical approximations of solutions of the time-independent nonlinear boundary-value PDE problem.

11.4 Summary and Notes

In this chapter, we studied properties of trigonometric interpolation and approximation and their applications to numerical solutions of ordinary and partial differential equations.

- **Section 11.1:** Trigonometric approximation is formulated in Problem 11.1. The solution to this problem is given in Theorem 11.1, while convergence of the approximation for smooth functions is described in Theorem 11.2. Trigonometric interpolation is formulated in Problem 11.2. The solution to this problem is given in Theorem 11.3 and is implemented in the pair of discrete Fourier transforms.

- **Section 11.2:** Discrete eigenvalue problems for difference equations are analyzed in connection to continuous eigenvalue problems for differential operators (Theorem 11.4). The error of trigonometric interpolation for real analytic functions is described in Theorem 11.5. Analysis of convergence of trigonometric interpolation shows that no Runge phenomenon can occur for trigonometric sums.

- **Section 11.3:** Two spectral methods originate from applications of the trigonometric approximation and interpolation, namely, the Galerkin and collocation methods. The simplest (diagonal) application of these methods is described in the example of the inhomogeneous heat equation. Errors of spectral methods and details of their numerical implementations are discussed for two examples of the boundary-value problems for second-order ordinary differential equations. An additional pseudo-spectral method is described in the context of the time-evolution PDE problem that embeds the boundary-value ODE problem.

Trigonometric approximations and Sturm–Liouville eigenvalue problems are presented in [27]. Trigonometric interpolations and spectral accuracy are covered in [28]. Applications of spectral methods to numerical approximations of solutions of ordinary and partial differential equations are treated in [9].

11.5 Exercises

1. Compute the Fourier series, the trigonometric approximant $f_m(x)$, and the trigonometric interpolant $F_m(x)$ for the function

$$f(x) = 1 - |x|, \qquad -1 < x < 1,$$

which is extended periodically with the period $L = 2$. Plot the functions $f_m(x)$, $F_m(x)$, and $f(x)$ on $[-1, 1]$ for $m = 10$. Plot the L^2-norm of the error E of the trigonometric approximation and interpolation versus the number of terms m in log-log scale and find the power fits in the dependencies of E versus m. Compare the power fits with Theorem 11.2.

2. Repeat the previous exercise for the function $f(x)$ on $(0,1)$ reflected antisymmetrically on $(-1,0)$ and extended periodically with the period $L = 2$. Observe the Gibbs phenomenon by plotting the local error of the trigonometric interpolation and approximation on $[-1,1]$ for different values of m.

3. Compute the trigonometric interpolation $F_m(x)$ of the function

$$f(x) = \frac{1 + 2\sin x}{3 - 2\cos x}, \qquad 0 \le x \le 2\pi$$

and plot it on $[0, 2\pi]$ for $m = 5$. Find the power fit for the 2-norm of the error E versus the number of terms m. Compare the power fit with Theorem 11.5.

4. Write the MATLAB function `[a] = LSsine(x,y,n)`, where **x** and **y** are the column vectors of m elements, $0 < n \le m$, and **a** is the column vector of n elements in the sine interpolation

$$F_{\text{sine}}(x) = a_1 \sin(\pi x) + a_2 \sin(2\pi x) + \ldots + a_n \sin(n\pi x).$$

Apply this function for the data points related to the function $f(x) = x(1-x)e^{-x}$ on $[0,1]$ with $n = m$ and plot the 2-norm of the error versus n.

5. Use the MATLAB function `fft` and compute the discrete Fourier transform of the function $f(x) = x(1-x^2)$ on $[0,1]$. Plot the trigonometric interpolant $F_m(x)$ on a dense grid of data points on $[-1,1]$. Explain why the resulting function is even in x. Modify the function $f(x)$ so that the same computation produces an odd function $F_m(x)$ on $[-1,1]$.

6. Find eigenvalues and eigenvectors for the discrete eigenvalue problem related to the set of orthogonal Hermite polynomials:

$$u_{k+1} - 2u_k + u_{k-1} - hx_k(u_{k+1} - u_{k-1}) + h^2\lambda u_k = 0, \qquad 1 \le k \le n,$$

subject to the Dirichlet boundary conditions $u_0 = u_{n+1} = 0$, where the points $\{x_k\}_{k=1}^{n+1}$ represent a uniform grid on $[-L, L]$ and L is sufficiently large, say, $L = 10$. Plot the distribution of eigenvalues and the first five eigenvectors of the Hermite difference eigenvalue problem.

7. Consider the integral representation of the Bessel function

$$J_0(x) = \frac{1}{\pi}\int_0^{\pi} e^{ix\cos t}\,dt.$$

For a set of equally spaced grid points on the x-interval $[0,5]$, replace $e^{ix\cos t}$ for $t \in [0,\pi]$ with the trigonometric interpolant $F_m(t)$, integrate the function $F_m(t)$ analytically on $[0,\pi]$, and compute the Bessel function $J_0(x)$.

8. Consider the boundary-value ODE problem

$$y'' + x^2 y = 1, \qquad 0 < x < 1,$$

subject to the Dirichlet boundary conditions $y(0) = y(1) = 0$. Construct the numerical approximations of the solution based on the Galerkin and collocation methods. Find the power fits of the total square error of the numerical approximations versus the step size h on $[0,1]$.

9. Consider the linear eigenvalue problem

$$y'' + x^2 y = \lambda y, \qquad 0 < x < 1,$$

subject to the Dirichlet boundary conditions $y(0) = y(1) = 0$. Approximate the spectrum of eigenvalues by truncating the trigonometric approximation for $m = 100$ terms. Show that the value $\lambda = 0$ is not the eigenvalue in the Galerkin method. Repeat the exercise by replacing derivatives with matrices in the collocation method on the $n = 100$ equally spaced grid points.

10. Consider the Hill equation

$$-y'' + \cos xy = \lambda y, \qquad 0 < x < 2\pi$$

subject to the periodic boundary conditions $y(x + 2\pi) = y(x)$. Expand the solution in the trigonometric series and truncate the system at $m = 100$ terms in the Galerkin method. Plot the spectrum of eigenvalues and the first five eigenfunctions for the smallest eigenvalues. Illustrate that the number of zeros of $y(x)$ on $(0, 2\pi)$ increases in the same ascending order as the eigenvalues are sorted.

11. Repeat the previous exercise with the antiperiodic boundary conditions $y(x + 2\pi) = -y(x)$. Show that each pair of eigenvalues with the antiperiodic eigenfunctions is located between each pair of eigenvalues with the periodic eigenfunctions and vice versa.

12. Consider the linear Schrödinger equation

$$iu_t = u_{xx}, \qquad 0 < x < 1, \ t > 0$$

for complex-valued function $u(x,t)$, subject to the Dirichlet boundary conditions $u(0,t) = u(1,t) = 0$ and the initial condition

$u(x,0) = e^{-x^2+ix^2}$. Construct the numerical approximation of the solution surface $u(x,t)$ for $x \in [0,1]$ and $t \in [0,10]$ based on the Galerkin and collocation methods supplemented with the exact ODE integration. Study how the 2-norm of the error converges as $h \to 0$ for different fixed values of t.

13. Consider the wave equation

$$u_{tt} = u_{xx} + \cos(2x), \qquad 0 < x < \pi, \;\; t > 0,$$

subject to the Neumann boundary conditions $u_x(0,t) = u_x(\pi,t) = 0$ and the initial conditions $u(x,0) = u_t(x,0) = 0$. Construct the numerical approximations of the solution surface $u(x,t)$ for $x \in [0,\pi]$ and $t \in [0,3]$ based on the Galerkin and collocation methods and the Heun method. Show the two solution surfaces and the exact solution of the wave equation.

14. Consider the *Burgers* equation

$$u_t = uu_x + u_{xx}, \qquad 0 < x < 2\pi, \quad t > 0,$$

subject to the periodic boundary conditions $u(0,t) = u(2\pi,t)$ and $u_x(0,t) = u_x(2\pi,t)$ and the initial condition $u(x,0) = e^{-x^2}$. Compute the numerical approximation for $t \in [0,3]$ by using the pseudospectral method based on the direct and inverse discrete Fourier transforms. Study how convergence and stability of the pseudospectral method depend on step size h and time step τ.

15. Repeat the previous exercise with the pseudospectral method based on the collocation method supplemented by the matrix representation of the first and second derivatives of $u(x,t)$ in x. Compare convergence and stability between the two pseudospectral methods.

CHAPTER 12

Splines and Finite Elements

ALTHOUGH TRIGONOMETRIC INTERPOLATION offers an accurate numerical solution to the interpolation problem, it also may have some shortcomings. The discrete grid of x-values has to be equally spaced to enforce all nice properties of trigonometric interpolation, such as the orthogonality and convergence of discrete trigonometric functions. On the other hand, although polynomial interpolation is valid on nonequally spaced discrete grids, it may develop a polynomial wiggle. There exists an alternative method to overcome the limitations of both trigonometric and polynomial interpolations. If the entire interpolation interval is decomposed into smaller intervals connected at the given data points, the degree of interpolating polynomials can be reduced to avoid the polynomial wiggle. This idea leads to the *spline interpolation*, which is a basis for the *finite-element method*, a useful tool in numerical approximations of solutions of boundary-value problems for differential equations.

This chapter covers the construction, properties, and errors of spline and Hermite interpolations and applications of finite elements to numerical approximations of solutions of ordinary differential equations.

12.1 Spline Interpolation

We refer to the polynomial of a low degree between two adjacent data points as *a spline* and to the grid point that connects two adjacent splines as *a breaking point.* The simplest *linear* splines between each two subsequent breaking points coincide with the piecewise linear interpolation for the given set of data points. Starting with *quadratic* and *cubic* polynomials, you will see a difference between spline interpolation and piecewise polynomial interpolation. The former match the first-order, second-order, and higher-order derivatives of the resulting curve at the breaking points, whereas the latter prepares independent polynomials between each two subsequent breaking points. Similar to the uniform polynomial interpolation, spline and piecewise polynomial interpolations with higher-order (fourth, fifth, etc.) polynomials may develop polynomial wiggles. As a result, such higher-order interpolation is used less often.

Let the set of $(n+1)$ data points (x_1, y_1), (x_2, y_2), ..., (x_{n+1}, y_{n+1}) be ordered in the ascending order of distinct breaking points

$$x_1 < x_2 < \ldots < x_{n+1} \tag{12.1}$$

(if some grid points x_k and x_{k+1} coincide, the vertical spline can be used between different data values y_k and y_{k+1}). We now define a sequence of spline interpolation problems and describe algorithms of the exact solution to each problem.

Linear spines

Problem 12.1 (Linear Splines) Connect the given set of data points using linear splines $S_k(x)$ between two data points (x_k, y_k) and (x_{k+1}, y_{k+1}), that is,

$$S_k(x) = y_k + m_k(x - x_k), \qquad x_k \leq x \leq x_{k+1}, \qquad k = 1, 2, \ldots, n, \tag{12.2}$$

where m_k is the slope of $S_k(x)$ at $x = x_k$.

According to the linear interpolation (5.14) between two data points, a unique solution of Problem 12.1 is given in the explicit form:

$$m_k = D_1[y_k, y_{k+1}] = \frac{y_{k+1} - y_k}{x_{k+1} - x_k}, \tag{12.3}$$

where $D_1[y_k, y_{k+1}]$ is the first-order divided difference. If the data points (x_k, y_k) and (x_{k+1}, y_{k+1}) are connected by a smooth function $y = f(x)$, the slope (12.3) represents the *forward-difference approximation* for $f'(x_k)$, the *backward-difference approximation* for $f'(x_{k+1})$, or the *central-difference approximation* for $f'\left(\frac{x_k + x_{k+1}}{2}\right)$ (see Section 6.1).

The MATLAB script `linear_splines` illustrates computations of linear splines for the MATLAB demo function `humps`. The function `humps` returns a smooth function with two local maxima near the points $x = 0.3$ and $x = 0.9$. In computations of the first-order divided differences (12.3), you can use the MATLAB function `diff`, which takes the vector of data values and returns the vector of differences between two subsequent data values. The output of MATLAB script `linear_splines` is shown in Figure 12.1.

```
% linear_splines
x = -1 : 0.2 : 3; n = length(x)-1; y = humps(x);
y0 = y(1:n); m = diff(y)./diff(x); % slopes of linear splines
xInt = -1 : 0.001 : 3;
for j = 1 : length(xInt) % intervals between breaking points
    if xInt(j) ~= x(n+1)
        iInt(j) = sum(x <= xInt(j));
    else
        iInt(j) = n;
    end
end
yInt = y0(iInt) + m(iInt).*(xInt-x(iInt));
yEx = humps(xInt);
plot(x,y,'b.',xInt,yInt,'g',xInt,yEx,'r:');
```

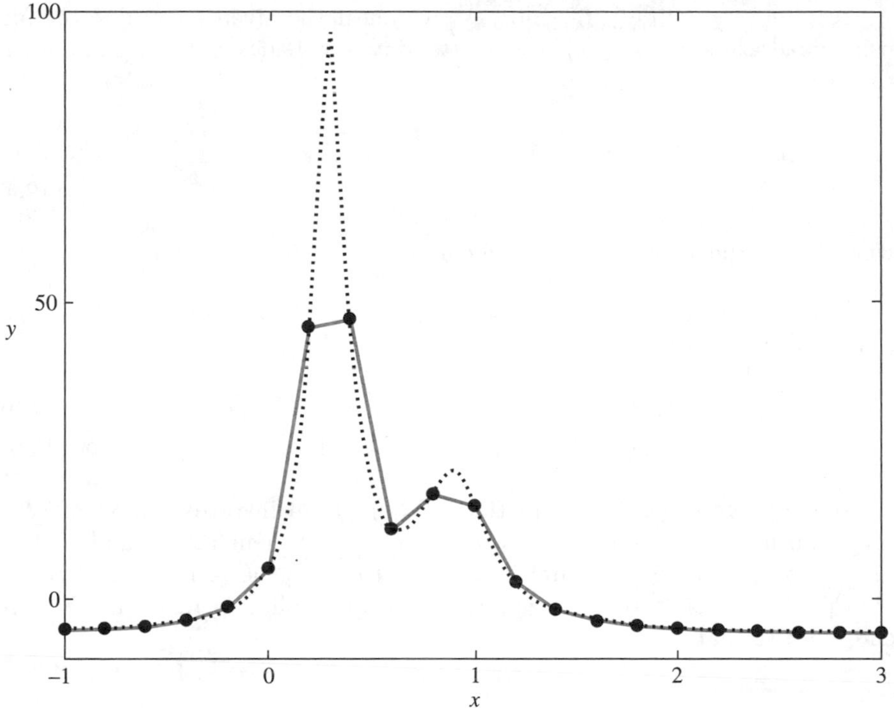

Figure 12.1 Example of linear spline interpolation.

By construction, linear splines display a continuous but nondifferentiable curve with sharp corners at the breaking points, where the slopes $\{m_k\}_{k=1}^{n}$ jump. The MATLAB function `plot` connects all available data points by straight lines, that is, it produces the linear spline interpolation by default. The graphical output of Figure 12.1 is recovered with a single call of the plot function on the same set of data points.

```
>> plot(x,y,'b.',x,y,'g',xInt,yEx,'r:');
```

In many practical problems, linear spline interpolation can be used for rough and quick approximation of the unknown function at the intermediate data points. However, because the linear spline interpolation produces a non-smooth curve, it is rarely useful for a graphical visualization. Quadratic and cubic splines are more useful for smooth graphical visualizations and more accurate numerical interpolation of the unknown functional dependencies.

Quadratic splines

Problem 12.2 (Quadratic Splines) Connect the given set of data points using quadratic splines $S_k(x)$ between two data points (x_k, y_k) and (x_{k+1}, y_{k+1}), that is,

$$S_k(x) = y_k + m_k(x - x_k) + \frac{1}{2}\kappa_k(x - x_k)^2, \quad x_k \le x \le x_{k+1}, \quad k = 1, 2, \ldots, n, \tag{12.4}$$

where m_k is the slope and κ_k is the curvature of $S_k(x)$ at $x = x_k$.

The quadratic splines define a continuously differentiable curve on the interpolation interval, that is,

$$S_k(x_{k+1}) = y_{k+1}, \qquad k = 1, 2, \ldots, n \tag{12.5}$$

$$S'_k(x_{k+1}) = S'_{k+1}(x_{k+1}), \qquad k = 1, 2, \ldots, n-1. \tag{12.6}$$

The system of $(2n - 1)$ equations (12.5)–(12.6) involves the set of $(2n)$ unknown parameters $\{(m_k, \kappa_k)\}_{k=1}^{n}$. Therefore, Problem 12.2 is underdetermined and one more constraint is required to make the problem *well posed.* One possibility is to require that the quadratic spline is linear at the left endpoint, so that

$$\kappa_1 = 0. \tag{12.7}$$

Quadratic splines with the condition (12.7) are referred to as *the natural quadratic splines*. A unique solution of Problem 12.2 with the natural quadratic spline condition (12.7) is found by using an effective computational algorithm, which results in the first-order difference equation for slopes $\{m_k\}_{k=1}^{n+1}$, where $m_{n+1} = S'_n(x_{n+1})$ is defined similarly to the other slopes. Since the derivative $S'_k(x)$ of quadratic splines is a linear function, the linear spline given by $S'_k(x)$ passes through two data points (x_k, m_k) and (x_{k+1}, m_{k+1}) with unknown values of m_k and m_{k+1}. As a result, the constraints (12.6) are satisfied by the relation

$$\kappa_k = \frac{m_{k+1} - m_k}{x_{k+1} - x_k}, \qquad k = 1, 2, \ldots, n. \tag{12.8}$$

Although the value m_{n+1} is not defined in Problem 12.2, it is needed in (12.8) for $k = n$. With the account of relations (12.8), the continuity constraints (12.5) result in the first-order difference equation for slopes $\{m_k\}_{k=1}^{n+1}$:

$$m_{k+1} = -m_k + 2D_1[y_k, y_{k+1}], \qquad k = 1, 2, \ldots, n, \tag{12.9}$$

where the first-order divided difference $D_1[y_k, y_{k+1}]$ is defined in (12.3). The first-order difference equation (12.9) is well posed if the starting value m_1 is defined. When the natural spline condition (12.7) is used, the relation

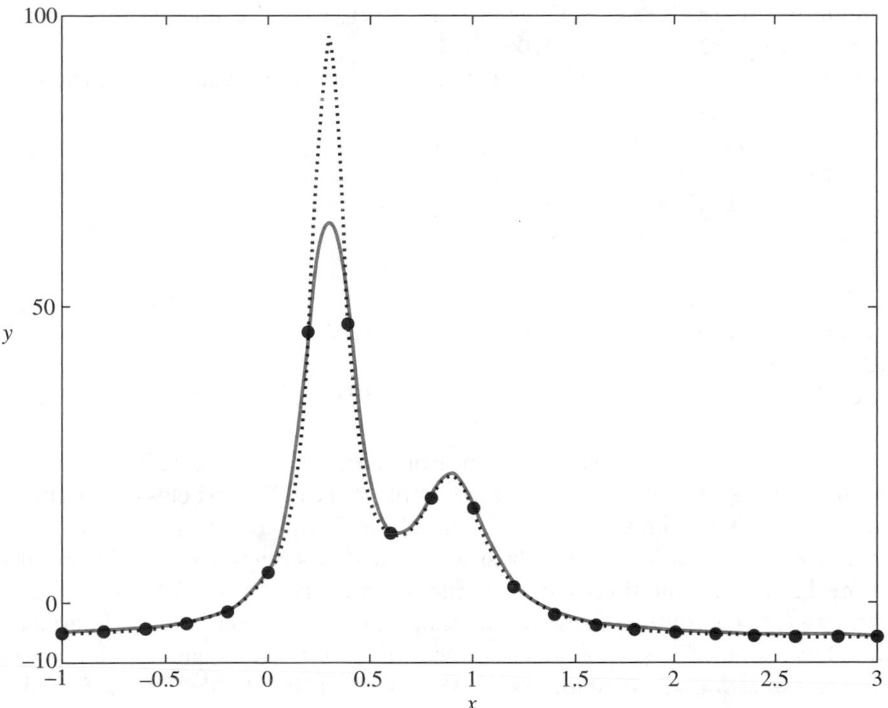

Figure 12.2 Example of quadratic spline interpolation.

$m_2 = m_1$ must be added, and the first iteration of the difference equation (12.9) with $k = 1$ leads to the obvious solution:

$$m_1 = m_2 = D_1[y_1, y_2] = \frac{y_2 - y_1}{x_2 - x_1}.$$

With these explicit values, the set of other slopes $\{m_k\}_{k=3}^{n+1}$ is found uniquely from the first-order difference equation (12.9) for $k = 2, 3, \ldots, n$. The previous example is modified to illustrate the computational algorithm of quadratic spline interpolation. The output of the MATLAB script `quadratic_splines` is shown in Figure 12.2.

```
% quadratic_splines
x = -1 : 0.2 : 3; y = humps(x); n = length(y)-1; y0 = y(1:n);
firstDifference = (y(2:n+1)-y(1:n))./(x(2:n+1)-x(1:n));
m(1) = firstDifference(1); m(2) = m(1); % slopes
for k = 2 : n
    m(k+1) = -m(k)+2*firstDifference(k);
end
```

```
kappa = (m(2:n+1)-m(1:n))./(x(2:n+1)-x(1:n)); % curvatures
m = m(1:n); xInt = -1 : 0.001 : 3;
for j = 1 : length(xInt)  % intervals between breaking points
    if xInt(j) ~= x(n+1)
        iInt(j) = sum(x <= xInt(j));
    else
        iInt(j) = n;
    end
end
xx = xInt-x(iInt);
yInt=y0(iInt)+m(iInt).*xx+0.5*kappa(iInt).*xx.^2;
yEx=humps(xInt);
plot(x,y,'b.',xInt,yInt,'g',xInt,yEx,'r:');
```

Compared to the linear spline interpolation (see Figure 12.1), quadratic splines display a continuously differentiable curve. Nevertheless, the curvatures jump at the breaking points, and these jumps could be visible to the human eye. In addition, many physical applications are modeled with second-order differential equations such that the second derivatives of physical quantities (such as coordinates or wave functions) need to be continuous. Therefore, in many applications, quadratic splines are not smooth enough and cubic splines are required to enhance smoothness. Additional cubic terms in cubic splines are chosen to ensure that the curvatures are continuous at the breaking points.

Cubic splines

Problem 12.3 (Cubic Splines) Connect the given set of data points using cubic splines $S_k(x)$ between two data points (x_k, y_k) and (x_{k+1}, y_{k+1}), that is,

$$S_k(x) = y_k + m_k(x - x_k) + \frac{1}{2}\kappa_k(x - x_k)^2 + \frac{1}{6}\gamma_k(x - x_k)^3, \quad x_k \le x \le x_{k+1}, \tag{12.10}$$

where m_k is the slope, κ_k is the curvature, and γ_k is the third derivative of $S_k(x)$ at $x = x_k$.

The cubic splines define a twice continuously differentiable curve on the interpolation interval, that is, three sets of constraints are satisfied:

$$S_k(x_{k+1}) = y_{k+1}, \qquad k = 1, 2, \ldots, n \tag{12.11}$$

$$S'_k(x_{k+1}) = S'_{k+1}(x_{k+1}), \qquad k = 1, 2, \ldots, n-1 \tag{12.12}$$

$$S''_k(x_{k+1}) = S''_{k+1}(x_{k+1}), \qquad k = 1, 2, \ldots, n-1. \tag{12.13}$$

The cubic spline interpolation is an underdetermined problem because the system of $(3n - 2)$ equations (12.11)–(12.13) involves the set of $(3n)$

unknown parameters $\{(m_k, \kappa_k, \gamma_k)\}_{k=1}^{n}$. Two more conditions are needed and the *natural cubic spline* is defined by the zero curvatures at both endpoints:

$$\kappa_1 = 0, \qquad \kappa_{n+1} = 0, \tag{12.14}$$

where $\kappa_{n+1} = S_n''(x_{n+1})$ is defined similarly to the other curvatures. The natural cubic spline has inflection points at $x = x_1$ and $x = x_{n+1}$, but it is generally nonlinear because of cubic terms at the endpoints. Other endpoint conditions are also used for cubic spline interpolation. See [4] for a full description of cubic splines.

The solution of Problem 12.3 repeats the steps of the solution of Problem 12.2. Since the second derivative $S_k''(x)$ of cubic splines is a linear function, the linear spline given by $S_k''(x)$ passes through two data points (x_k, κ_k) and (x_{k+1}, κ_{k+1}) with unknown values of κ_k and κ_{k+1}. As a result, the constraints (12.13) are satisfied by the relation for the third derivative terms γ_k:

$$\gamma_k = \frac{\kappa_{k+1} - \kappa_k}{x_{k+1} - x_k}, \qquad k = 1, 2, \ldots, n. \tag{12.15}$$

Although the value κ_{n+1} is not defined in Problem 12.3, it is needed in (12.15) for $k = n$. With the account of relations (12.15), the constraints (12.12) are satisfied by the relation for the slope terms m_k:

$$m_k = D_1[y_k, y_{k+1}] - \frac{1}{6}(2\kappa_k + \kappa_{k+1})(x_{k+1} - x_k), \tag{12.16}$$

where the first-order divided difference $D_1[y_k, y_{k+1}]$ is defined in (12.3). With the account of relations (12.15) and (12.16), the final set of constraints (12.11) results in the second-order difference equation for curvatures $\{\kappa_k\}_{k=1}^{n+1}$:

$$\left(\frac{x_k - x_{k-1}}{x_{k+1} - x_{k-1}}\right)\kappa_{k-1} + 2\kappa_k + \left(\frac{x_{k+1} - x_k}{x_{k+1} - x_{k-1}}\right)\kappa_{k+1} = 6D_2[y_{k-1}, y_k, y_{k+1}], \tag{12.17}$$

where $k = 2, 3, \ldots, n$, and $D_2[y_{k-1}, y_k, y_{k+1}]$ is the second-order divided difference defined by

$$D_2[y_{k-1}, y_k, y_{k+1}] = \frac{1}{x_{k+1} - x_{k-1}}\left(\frac{y_{k+1} - y_k}{x_{k+1} - x_k} - \frac{y_k - y_{k-1}}{x_k - x_{k-1}}\right).$$

The second-order divided difference $D_2[y_{k-1}, y_k, y_{k+1}]$ equals one-half of the central-difference approximation for the second derivative, as we explain in Section 6.2. According to the endpoint conditions (12.14), the system of second-order difference equations (12.17) is completed by the boundary conditions $\kappa_1 = \kappa_{n+1} = 0$. In this case, the linear system can be written in the matrix-vector form

$$\mathbf{A}\boldsymbol{\kappa} = 6\mathbf{d}_2, \tag{12.18}$$

where

$$\mathbf{A} = \begin{bmatrix} 2 & \frac{x_3-x_2}{x_3-x_1} & 0 & \dots 0 \\ \frac{x_3-x_2}{x_4-x_2} & 2 & \frac{x_4-x_3}{x_4-x_2} & \dots 0 \\ 0 & \frac{x_4-x_3}{x_5-x_3} & 2 & \dots 0 \\ \vdots & \vdots & \vdots & \ddots \vdots \\ 0 & 0 & 0 & \dots 2 \end{bmatrix}$$

and

$$\boldsymbol{\kappa} = \begin{bmatrix} \kappa_2 \\ \kappa_3 \\ \vdots \\ \kappa_n \end{bmatrix}, \qquad \mathbf{d}_2 = \begin{bmatrix} D_2[y_1, y_2, y_3] \\ D_2[y_2, y_3, y_4] \\ \vdots \\ D_2[y_{n-1}, y_n, y_{n+1}] \end{bmatrix}.$$

Under the ordering (12.1), the coefficient matrix **A** is diagonally dominant and a unique solution to the linear system (12.18) always exists. The MATLAB script `cubic_splines` illustrates computations of the cubic splines for the MATLAB demo function `humps`. The output of the MATLAB script `cubic_splines` is shown in Figure 12.3.

```
% cubic_splines
x = -1 : 0.2 : 3; y = humps(x); n = length(y)-1;
h = x(2:n+1)-x(1:n); % curvatures
A=2*diag(h(1:n-1))+2*diag(h(2:n))+diag(h(2:n-1),1)+diag(h(2:n-1),-1);
b = 6*((y(3:n+1)-y(2:n))./h(2:n)-(y(2:n)-y(1:n-1))./h(1:n-1));
kappa = A\b'; kappa = [0;kappa;0]';
y0 = y(1:n);  % slopes and third derivatives
m=(y(2:n+1)-y(1:n))./h(1:n)-h(1:n).*(kappa(2:n+1)+2*kappa(1:n))/6;
gamma = (kappa(2:n+1)-kappa(1:n))./h(1:n);
xInt = -1 : 0.001 : 3;
for j = 1 : length(xInt) % intervals between breaking points
    if xInt(j) ~= x(n+1)
        iInt(j) = sum(x <= xInt(j));
    else
        iInt(j) = n;
    end
end
xx = xInt-x(iInt);
yInt=y0(iInt)+m(iInt).*xx+0.5*kappa(iInt).*xx.^2+gamma(iInt).*xx.^3/6;
yEx = humps(xInt);
plot(x,y,'b.',xInt,yInt,'g',xInt,yEx,'r:');
```

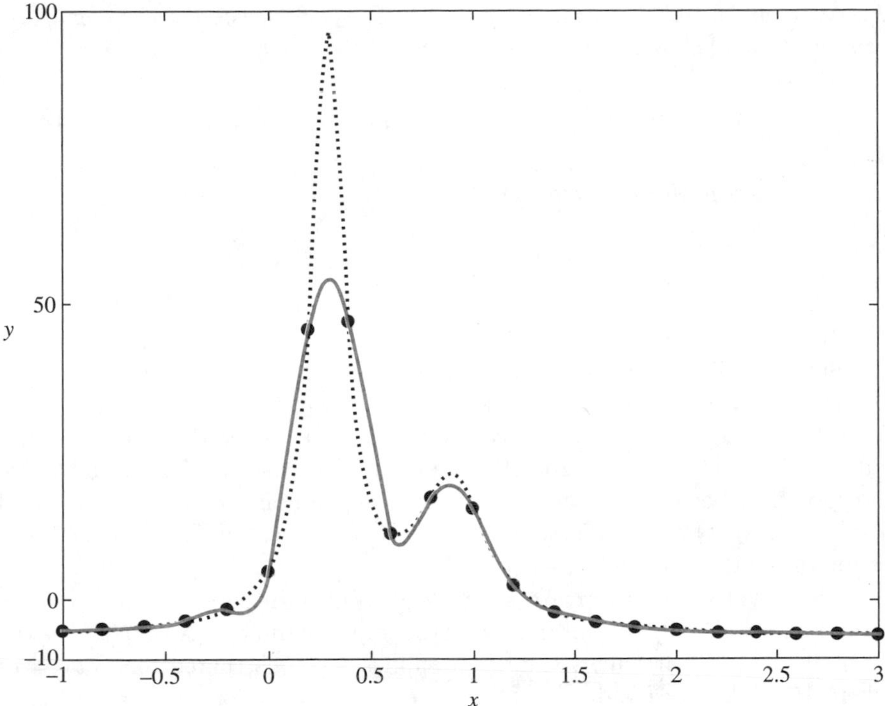

Figure 12.3 Example of cubic spline interpolation.

Exercise 12.1 Replace the natural cubic spline conditions (12.14) with the Neumann boundary conditions $m_1 = m_{n+1} = 0$ and set up the linear system for the set of curvatures $\{\kappa_k\}_{k=1}^{n+1}$.

Errors of splines

Although the cubic spline interpolation provides a smooth curve with continuous curvatures, the distance between the original MATLAB function **humps** and the cubic splines (which is the truncation error of the cubic spline interpolation) becomes larger than the truncation error of the quadratic spline interpolation (compare Figures 12.2 and 12.3). A polynomial wiggle of the cubic spline interpolation becomes visible, and the graphical disagreement is the price to pay for continuity of curvatures in the cubic splines. The following result from *Numerical Analysis* characterizes the truncation error of spline interpolation.

Theorem 12.1 *Let $f(x)$ be $(m+1)$-times continuously differentiable on $[a,b]$ and $S^{(m)}(x)$ be the linear ($m=1$), quadratic ($m=2$), or cubic ($m=3$) splines through the uniform grid of $(n+1)$ data points (12.1) with $x_1 = a$,*

$x_{n+1} = b$, and equal step size $h = (b-a)/n$. Let $f(x)$ satisfy the same endpoint conditions as the corresponding splines $S^{(m)}(x)$. Then,

$$\sup_{a \le x \le b} |f(x) - S^{(m)}(x)| \le \frac{1}{(m+1)!} C_m M_{m+1} h^{m+1}, \tag{12.19}$$

where C_m is a positive constant and

$$M_{m+1} = \max_{a \le x \le b} |f^{(m+1)}(x)|.$$

Since the linear spline is just a piecewise linear interpolation between adjacent data points, C_1 has the same value as in the upper bound (5.27) for $n = 1$, that is, $C_1 = \frac{1}{4}$. The constants C_2 and C_3 for the quadratic and cubic splines are different from those in the piecewise quadratic and cubic interpolations because parameters of splines depend on the global behavior of the function $f(x)$ on the discrete grid. See [1] for proofs of Theorem 12.1 and estimates on the constants C_m.

The MATLAB script `error_splines` computes the constants $C_{1,2,3}$ in the error bounds (12.19) by working with the simplest functions of $f(x)$ on the interval $[0, 1]$. In particular, the linear splines are constructed for $f(x) = x^2$ when $M_2 = 2$, the quadratic splines are constructed for $f(x) = x^3$ when $M_3 = 6$, and the cubic splines are constructed for $f(x) = x^3(1 - x/2)$ when $M_4 = 12$. The function $f(x) = x^3$ satisfies the natural spline condition (12.7) at $x = 0$, whereas the function $f(x) = x^3(1 - x/2)$ satisfies the natural spline conditions (12.14) at $x = 0$ and $x = 1$.

```
% error_splines
x = 0 : 0.1 : 1; h = x(2)-x(1); n = length(x)-1;
xInt = 0 : 0.001 : 1; %  % intervals between breaking points
for j = 1 : length(xInt)
    if xInt(j) ~= x(n+1)
        iInt(j) = sum(x <= xInt(j));
    else
        iInt(j) = n;
    end
end
xx = xInt-x(iInt); y1 = x.^2; y0 = y1(1:n);
m = diff(y1)./diff(x); % linear splines
yInt1 = y0(iInt) + m(iInt).*xx;
yEx1 = xInt.^2; % estimation of constant C1
C1 = max(yInt1-yEx1)/h^2
y2 = x.^3; y0 = y2(1:n);
firstDifference=(y2(2:n+1)-y2(1:n))./(x(2:n+1)-x(1:n));
```

```
m(1)=firstDifference(1); m(2) = m(1); for k = 2 : n
    m(k+1) = -m(k)+2*firstDifference(k);
end
kappa = (m(2:n+1)-m(1:n))./(x(2:n+1)-x(1:n));
m = m(1:n); % estimation of constant C2
yInt2=y0(iInt)+m(iInt).*xx+0.5*kappa(iInt).*xx.^2;
yEx2 = xInt.^3;
C2 = max(yInt2-yEx2)/h^3
y3 = x.^3.*(1-x/2);
h = x(2:n+1)-x(1:n);
A=2*diag(h(1:n-1))+2*diag(h(2:n))+diag(h(2:n-1),1)+diag(h(2:n-1),-1);
b = 6*((y3(3:n+1)-y3(2:n))./h(2:n)-(y3(2:n)-y3(1:n-1))./h(1:n-1));
kappa = A\b'; kappa = [0;kappa;0]';
y0 = y3(1:n); % estimation of constant C3
m=(y3(2:n+1)-y3(1:n))./h(1:n)-h(1:n).*(kappa(2:n+1)+2*kappa(1:n))/6;
gamma = (kappa(2:n+1)-kappa(1:n))./h(1:n);
yInt3=y0(iInt)+m(iInt).*xx+0.5*kappa(iInt).*xx.^2+gamma(iInt).*xx.^3/6;
yEx3 = xInt.^3.*(1-xInt/2);
C3 = 2*max(yInt3-yEx3)/h(1)^4
```

The output of the MATLAB script `error_splines` shows that $C_1 = \frac{1}{4}$ in accordance with the exact value obtained from the Taylor series, whereas C_2 and C_3 are of the same magnitude.

```
>> error_splines

C1 =
    0.2500

C2 =
    0.3849

C3 =
    0.1571
```

Exercise 12.2 Using linear splines ($m = 1$) for $f(x) = x^2$, natural quadratic splines ($m = 2$) for $f(x) = x^3$, and natural cubic splines ($m = 3$) for $f(x) = x^3(1 - x/2)$ on the interval $[0, 1]$, confirm numerically that the truncation error of the spline interpolation $S^{(m)}$ is $O(h^{m+1})$.

By slightly modifying the MATLAB script `error_splines`, you can check that the error of the natural quadratic spline is nonzero even for the quadratic function $f(x) = x^2$ because this function violates the condition for the natural quadratic splines (12.7) with $f''(0) = 2 \neq 0$. On the other hand, if

$m_1 = 0$ is used in the first-order difference equation (12.9), which agrees with the given value $f'(0) = 0$, the quadratic splines recover the quadratic function $f(x) = x^2$ exactly, up to machine precision. The MATLAB script `error_quadratic_splines` computes quadratic splines from the first-order difference equation (12.9) supplemented by the condition $m_1 = 0$. (The first part of the MATLAB code is omitted because it repeats codes of the MATLAB script `error_splines`.)

```
% error_quadratic_splines
y2 = x.^2; y0 = y2(1:n);
firstDifference=(y2(2:n+1)-y2(1:n))./(x(2:n+1)-x(1:n)); m(1) = 0;
for k = 1 : n
    m(k+1) = -m(k)+2*firstDifference(k);
end
kappa = (m(2:n+1)-m(1:n))./(x(2:n+1)-x(1:n));
m = m(1:n); % quadratic splines
xx = xInt-x(iInt);
yInt2=y0(iInt)+m(iInt).*xx+0.5*kappa(iInt).*xx.^2; yEx2 = xInt.^2;
C2 = 3*max(yInt2-yEx2)/h^3
```

The output of the MATLAB script `error_quadratic_splines` shows the round-off error of the order of 10^{-13}.

```
>> error_quadratic_splines

C2 =
    6.6613e-013
```

Similarly, you can check that the error of the natural cubic spline (and the constant C_3) is larger for the quartic function $f(x) = x^4$ because the second condition for the natural cubic splines (12.14) is violated with $f''(1) = 12 \neq 0$. If the second-order difference equation (12.17) is closed by the end-point conditions $\kappa_1 = 0$ and $\kappa_{n+1} = 12$, the error of spline interpolation becomes as small as that for the natural cubic spline for the function $f(x) = x^3(1 - x/2)$. The MATLAB script `error_cubic_splines` computes cubic splines from the second-order difference equation (12.17) supplemented by the conditions $\kappa_1 = 0$ and $\kappa_{n+1} = 12$. (The first part of the MATLAB code is omitted because it repeats code of the MATLAB script `error_splines`.)

```
% error_cubic_splines
y3 = x.^4; h = x(2:n+1)-x(1:n);
A=2*diag(h(1:n-1))+2*diag(h(2:n))+diag(h(2:n-1),1)+diag(h(2:n-1),-1);
b = 6*((y3(3:n+1)-y3(2:n))./h(2:n)-(y3(2:n)-y3(1:n-1))./h(1:n-1));
b(n-1) = b(n-1) - 12*h(n);
kappa = A\b'; kappa = [0;kappa;12]';
```

```
y0 = y3(1:n);      % cubic splines
m=(y3(2:n+1)-y3(1:n))./h(1:n)-h(1:n).*(kappa(2:n+1)+2*kappa(1:n))/6;
gamma = (kappa(2:n+1)-kappa(1:n))./h(1:n);
xx = xInt-x(iInt);
yInt3=y0(iInt)+m(iInt).*xx+0.5*kappa(iInt).*xx.^2+gamma(iInt).*xx.^3/6;
yEx3 = xInt.^4;
C3=max(yInt3-yEx3)/h(1)^4

>> error_cubic_splines

C3 =
     0.0053
```

It follows from the output of the MATLAB script `error_cubic_splines` that the constant C_3 in Theorem 12.1 is not universal because it depends on the function $f(x)$ and the endpoint conditions even if $f^{(4)}(x)$ is constant on the interval $[a, b]$.

Exercise 12.3 Consider the natural quadratic spline ($m = 2$) for $f(x) = x^2$ and natural cubic splines ($m = 3$) for $f(x) = x^4$ on the interval $[0, 1]$ and confirm numerically that the truncation error of the spline interpolation $S^{(m)}$ is $O(h^2)$ in both cases.

The main improvement of the spline interpolation compared to the polynomial interpolation is that the convergence of the interpolant $S^{(m)}(x)$ to the given function $f(x)$ is guaranteed by the boundness of the constant M_{m+1} if the function $f(x)$ is $(m + 1)$ times continuously differentiable on $[a, b]$ (see Theorem 12.1). As a result, the error bound (12.19) converges to zero if $h \to 0$. The causes for polynomial wiggle (such as the unbounded growth of higher derivatives of $f(x)$ as $m \to \infty$) are absent from the spline interpolation. The MATLAB script `Runge_splines` illustrates the smooth behavior of the cubic spline interpolation for the Runge function $f(x) = 1/(1+25x^2)$ on the interval $[-1, 1]$.

```
% Runge_splines
for n = 4 : 4 : 12
    x = linspace(-1,1,n+1); y = 1./(1+25*x.^2);
    h = x(2:n+1)-x(1:n);
    A = 2*diag(h(1:n-1))+2*diag(h(2:n))+diag(h(2:n-1),1)+diag(h(2:n-1),-1);
    b = 6*((y(3:n+1)-y(2:n))./h(2:n)-(y(2:n)-y(1:n-1))./h(1:n-1));
    kappa = A\b'; kappa = [0;kappa;0]';
    y0 = y(1:n);
    m = (y(2:n+1)-y(1:n))./h(1:n)-h(1:n).*(kappa(2:n+1)+2*kappa(1:n))/6;
    gamma = (kappa(2:n+1)-kappa(1:n))./h(1:n);
    xInt = linspace(-1,1,1001);
```

```
    for j = 1 : length(xInt)
        if xInt(j) ~= x(n+1)
            iInt(j) = sum(x <= xInt(j));
        else
            iInt(j) = n;
        end
    end
    xx = xInt-x(iInt);
    yInt = y0(iInt)+m(iInt).*xx+0.5*kappa(iInt).*xx.^2+gamma(iInt).*xx.^3/6;
    plot(xInt,yInt,x,y,'*');
end
yExact = 1./(1+25*xInt.^2);
plot(xInt,yExact,':r');
```

The output of the MATLAB script Runge_splines is shown in Figure 12.4. The solid color, dotted color, and solid black lines correspond to cubic splines with $n = 4, 8$, and 12, respectively. When n increases, the cubic spline interpolation gets closer to the Runge function $f(x)$ and displays no polynomial wiggle.

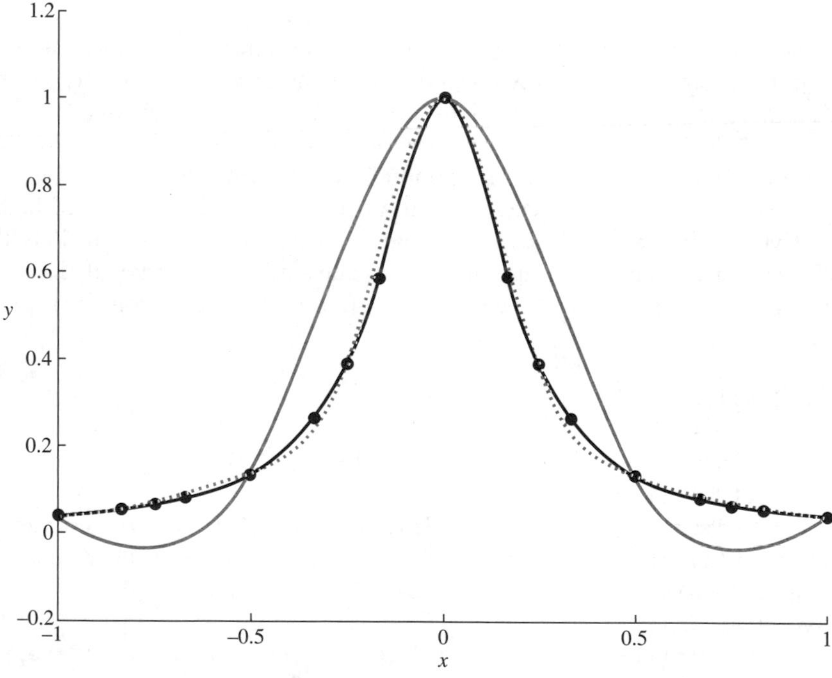

Figure 12.4 Cubic spline interpolation of the Runge function.

The spline interpolation features no polynomial wiggle, which is similar to the trigonometric interpolation (see Section 11.2). On the other hand, the spline interpolation may not be as good as the trigonometric interpolation. If the function $f(x)$ admits an analytic extension off the real axis, the convergence of spline interpolation is only polynomial in h, while the trigonometric interpolation provides the exponential convergence (see Theorem 11.5).

If the upper bound M_m does not grow with larger values of m, then the convergence is faster for cubic splines and the truncation error for cubic splines is smaller than that for linear and quadratic splines. On the other hand, if the upper bound M_m grows, the truncation error for quadratic splines at a given value of h may be smaller than that for cubic splines. The example of such a situation is seen by comparing Figures 12.2 and 12.3.

12.2 Hermite Interpolation

There are many other kinds of piecewise polynomial interpolation. We can try constructing a piecewise polynomial interpolation by breaking the set of $(n+1)$ data points into subsets of $(m+1)$ adjacent data points and interpolating each subset with a polynomial $p_m(x)$ of degree m. This procedure is used in composite integration rules, as we explain in Section 6.6. See [15] for analysis of the piecewise polynomial interpolation and its truncation error. Alternatively, we can try modifying a set of cubic splines with a set of other polynomials such as the cubic Hermite polynomials. These cubic Hermite polynomials allow us to provide not only continuity of the first-order derivatives of the curve but also to match the set of prescribed slopes. This interpolation is appropriate when both the function values and the derivative values are available, for example, from the experimental data.

Cubic Hermite interpolation

Problem 12.4 (Cubic Hermite Interpolation) Connect the given set of data points and their slopes (x_1, y_1, m_1), (x_2, y_2, m_2), ..., $(x_{n+1}, y_{n+1}, m_{n+1})$ using cubic Hermite polynomials between two data points (x_k, y_k, m_k) and $(x_{k+1}, y_{k+1}, m_{k+1})$, that is,

$$S_k(x) = y_k + m_k(x - x_k) + \alpha_k(x - x_k)^2 + \beta_k(x - x_k)^2(x - x_{k+1}), \quad (12.20)$$

where α_k and β_k are parameters, $x_k \le x \le x_{k+1}$, and $k = 1, 2, \ldots, n$.

The cubic Hermite interpolation defines a continuously differentiable curve on the interpolation interval, such that two sets of constraints are met:

$$S_k(x_{k+1}) = y_{k+1}, \qquad S'_k(x_{k+1}) = m_{k+1}, \qquad k = 1, 2, \ldots, n. \quad (12.21)$$

The set of $(2n)$ coefficients $\{\alpha_k, \beta_k\}_{k=1}^{n}$ is found from the set of $(2n)$ conditions (12.21), which require that the function $S_k(x)$ remains continuous and smooth across the breaking point $x = x_{k+1}$. When the cubic Hermite polynomial (12.20) is substituted into the set of equations (12.21), the coefficients $\{(\alpha_k, \beta_k)\}_{k=1}^{n}$ are found explicitly in the form

$$\alpha_k = \frac{1}{x_{k+1} - x_k} \left(D_1[y_k, y_{k+1}] - m_k \right), \tag{12.22}$$

$$\beta_k = \frac{1}{(x_{k+1} - x_k)^2} \left(m_k + m_{k+1} - 2D_1[y_k, y_{k+1}] \right), \tag{12.23}$$

where the first-order divided difference $D_1[y_k, y_{k+1}]$ is defined by (12.3). Although cubic Hermite interpolation produces a continuously differentiable function, it may have jump discontinuities in the second derivatives at the breaking points.

We illustrate the cubic Hermite interpolation on the same example of the MATLAB demo function **humps** that is used in Section 12.1. The values of the slopes of the demo function at the breaking points can be approximated by the central differences computed with a small step size h_0 (central differences are explained in Section 6.1). The MATLAB script `cubic_Hermite_interpolation` shows details of the cubic Hermite interpolation.

```
% cubic_Hermite_interpolation
x = -1 : 0.2 : 3; n = length(x)-1;
y = humps(x); h0 = 0.01;
m = (humps(x+h0)-humps(x-h0))/(2*h0); % slopes of the function "humps"
h = diff(x); y0 = y(1:n); m0 = m(1:n);
alpha = (diff(y)./h - m0)./h; % coefficients of Hermite polynomials
beta=(m(2:n+1)+m(1:n)-2*diff(y)./h)./(h.^2);
xInt = -1 : 0.001 : 3;      % evaluations of Hermite polynomials
for j = 1 : length(xInt)
    if xInt(j) ~= x(n+1)
        iInt(j) = sum(x <= xInt(j));
    else
        iInt(j) = n;
    end
end
xx = xInt-x(iInt); xx1 = xInt-x(iInt+1);
yInt=y0(iInt)+m0(iInt).*xx+alpha(iInt).*xx.^2+beta(iInt).*xx.^2.*xx1;
yEx = humps(xInt); plot(x,y,'b.',xInt,yInt,'g',xInt,yEx,'r:');
```

The output of the MATLAB script `cubic_Hermite_interpolation` is shown on Figure 12.5. The cubic Hermite interpolation represents the two humps of the demo function **humps** much better than the cubic spline interpolation (compare Figures 12.3 and 12.5).

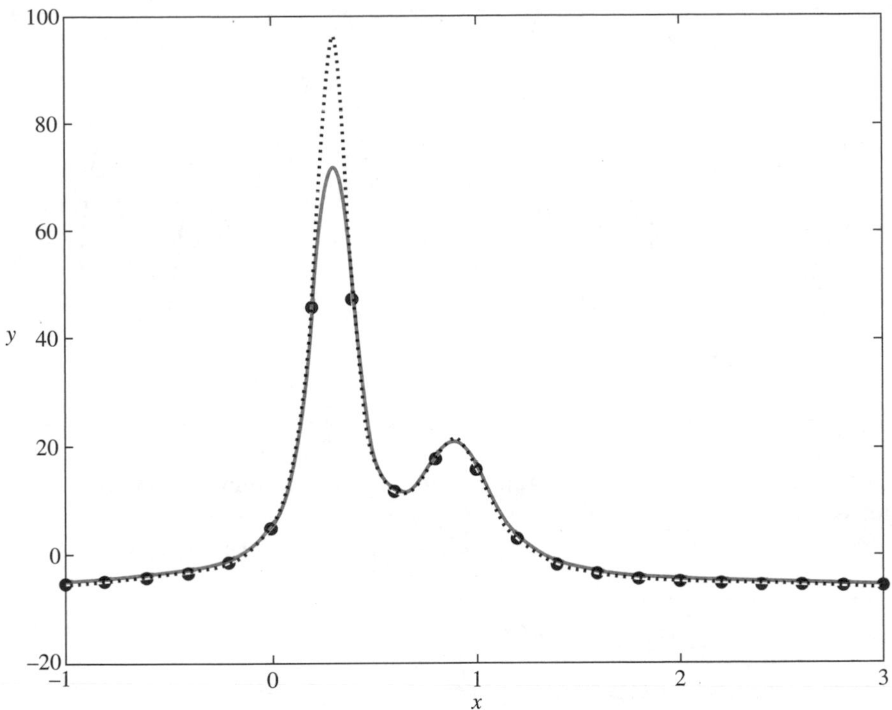

Figure 12.5 Example of cubic Hermite interpolation.

Hermite polynomials versus cubic splines

The cubic Hermite interpolation works better than the cubic spline interpolation in many other cases because it uses more information on the functional dependence $y = f(x)$ (slopes at the discrete grid points are known in addition to the data values). Another example of the advantages of the cubic Hermite interpolation is represented by a rapidly oscillating function when few data points are given. In particular, consider the function

$$f(x) = \cos(10\pi x) + 2\cos(12\pi x)$$

on the interval $[0, 1]$ with the step size $h = 0.125$. The MATLAB script **rapidly_oscillating_functions** evaluates both cubic spline and Hermite interpolations and outputs them in Figure 12.2.

```
% rapidly_oscillating_functions
x = 0:0.125:1; y = cos(10*pi*x)+2*cos(12*pi*x);
m=-10*pi*sin(10*pi*x)-24*pi*sin(12*pi*x);
xInt = 0 : 0.001 : 1;
for j = 1 : length(xInt)
    if xInt(j) ~= x(n+1)
```

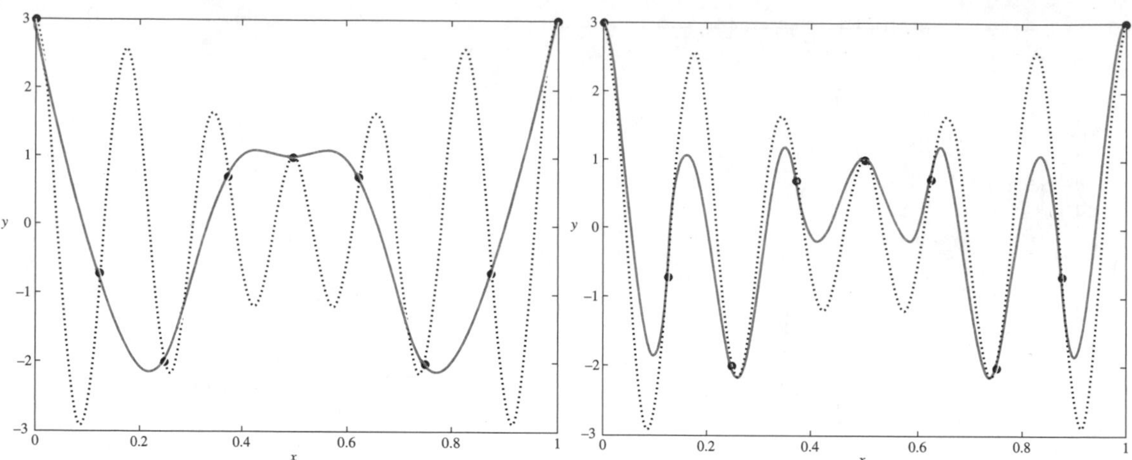

Figure 12.6 Cubic spline interpolation (*left*) versus cubic Hermite interpolation (*right*) for a rapidly oscillating function (dotted curve).

```
        iInt(j) = sum(x <= xInt(j));
    else
        iInt(j) = n;
    end
end
xx = xInt-x(iInt);
n=length(x)-1; h=diff(x);
            % cubic Hermite polynomials
y0 = y(1:n); m0 = m(1:n);
alpha=(diff(y)./h - m0)./h;
beta=(m(2:n+1)+m(1:n)-2*diff(y)./h)./(h.^2);
yInt=y0(iInt)+m0(iInt).*xx+alpha(iInt).*xx.^2+beta(iInt).*xx.^2.*(xInt-x(iInt+1));
yEx=cos(10*pi*xInt)+2*cos(12*pi*xInt);
plot(x,y,'b.',xInt,yInt,'g',xInt,yEx,'r:');
            % cubic splines
A=2*diag(h(1:n-1))+2*diag(h(2:n))+diag(h(2:n-1),1)+diag(h(2:n-1),-1);
b = 6*((y(3:n+1)-y(2:n))./h(2:n)-(y(2:n)-y(1:n-1))./h(1:n-1));
kappa=A\b'; kappa = [0;kappa;0]';
m=diff(y)./h-h.*(kappa(2:n+1)+2*kappa(1:n))/6;
gamma=diff(kappa)./h(1:n);
yInt=y0(iInt)+m(iInt).*xx+0.5*kappa(iInt).*xx.^2+gamma(iInt).*xx.^3/6;
plot(x,y,'b.',xInt,yInt,'g',xInt,yEx,'r:');
```

You can see in Figure 12.2 that the cubic spline interpolation produces a very inadequate curve, whereas the cubic Hermite interpolation predicts a correct behavior of the curve of the given function.

Similar to Theorem 12.1, the truncation error of the cubic Hermite interpolation can be bounded from above.

Errors of Hermite interpolation

Theorem 12.2 *Let $f(x)$ be 4 times continuously differentiable on $[a,b]$ and $S(x)$ be the cubic Hermite interpolant through the uniform grid of $(n+1)$ data points (12.1)* with $x_1 = a$, $x_{n+1} = b$, and equal step size $h = (b-a)/n$. Then,

$$\sup_{x\in[a,b]} |f(x) - S(x)| \leq \frac{1}{4!} C M_4 h^4, \tag{12.24}$$

where C is a positive constant and $M_4 = \max_{a\leq x\leq b} |f^{(4)}(x)|$.

The error formula (12.24) is similar to the error formula (12.19) for $m = 3$ since the cubic Hermite polynomials interpolate cubic polynomials exactly but introduce the truncation error because of the fourth derivative of the function $f(x)$. The MATLAB script `error_Hermite` computes the truncation error of the cubic Hermite interpolation for the functions $f(x) = x^k$, $k = 2, 3, 4$ on the interval $[0, 1]$.

```
% error_Hermite
xInt = linspace(0,1,1001);
for k = 2 : 4
    yEx = xInt.^k;
    for n = 1 : 50
        x = linspace(0,1,n+1); Step(n) = x(2)-x(1);
        y = x.^k; m = k*x.^(k-1); h = diff(x);
        y0 = y(1:n); m0 = m(1:n);
        alpha = (diff(y)./h - m0)./h;
        beta = (m(2:n+1)+m(1:n)-2*diff(y)./h)./(h.^2);
        for j = 1 : length(xInt)
            if xInt(j) ~= x(n+1)
                iInt(j) = sum(x <= xInt(j));
            else
                iInt(j) = n;
            end
        end
        xx = xInt-x(iInt);
        yInt = y0(iInt) + m0(iInt).*xx + alpha(iInt).*xx.^2;
        yInt = yInt + beta(iInt).*xx.^2.*(xInt-x(iInt+1));
        Error(n) = max(abs(yInt - yEx));
    end
    ErrorMax = max(Error)
```

```
    if (k == 4)
        a = polyfit(log(Step),log(Error),1);
        power = a(1)
    end
end
```

When the MATLAB script `error_Hermite` is executed, it computes the maximal distance between the power functions $f(x) = x^k$, $k = 2, 3, 4$ and the cubic Hermite polynomial $S(x)$ on the interval $[0, 1]$. The output shows that the maximal error is of the order of machine precision for $k = 2$ and $k = 3$. It is nonzero for the quartic function $f(x) = x^4$, when the maximal error (which occurs for $n = 1$ and $h = 1$) gives the value $C = 0.0625$ of the constant in the error bound (12.24). The power fit of the dependence of error versus the step size h shows that the error is $O(h^4)$, in full agreement with the error-bound (12.24). We note that the standard theorem on Hermite interpolation claims that the error is proportional to the $(2n+2)$-th derivative of the function $f(x)$ [7], which is identically zero for $f(x) = x^4$ and $n \geq 2$.

```
>> error_Hermite

ErrorMax =
    1.1102e-016

ErrorMax =
    2.2204e-016

ErrorMax =
    0.0625

power =
    4.0001
```

The error formula (12.24) shows that the Hermite interpolation does not lead to polynomial wiggle if the fourth derivative of the function $f(x)$ is bounded on the interval $[a, b]$.

Exercise 12.4 Compute the cubic Hermite interpolation to the Runge function $f(x) = 1/(1 + 25x^2)$ on the interval $[-1, 1]$. Show that the polynomial wiggle does not occur when the number of data points n is increased.

MATLAB interpolation

The linear and cubic splines and the cubic Hermite interpolation are produced by the MATLAB function `interp1`, where the fourth input argument allows a user to select the linear spline (default) by `linear`, the cubic spline by `spline`, and the cubic Hermite interpolation by `cubic`. The script

`MATLAB_interpolation` computes and plots the cubic spline and Hermite interpolations of the bell-shaped hyperbolic function $f(x) = \text{sech}^2 x$.

```
% MATLAB_interpolation
x = -3:1:3; y = (cosh(x)).^(-2);
yD = -2*sinh(x).*(cosh(x)).^(-3);
xInt = -3 : 0.001 : 3; yInt = (cosh(xInt)).^(-2);
yInt1 = interp1(x,y,xInt,'spline'); % cubic spline interpolation
yInt2 = interp1(x,y,xInt,'cubic'); % cubic Hermite interpolation
plot(x,y,'r.',xInt,yInt,'r:',xInt,yInt1,'b',xInt,yInt2,'g');
```

The output of the MATLAB script `MATLAB_interpolation` is shown in Figure 12.7. Comparison with the exact function (dotted curve) shows that the cubic spline interpolation (colored curve) is the best fit near the central peak of the bell-shaped function, while the cubic Hermite interpolation (black curve) is the best fit near the tails of the bell-shaped function.

The MATLAB function `interp1` with the third argument `'spline'` and `'cubic'` are rewritten as the MATLAB functions `spline` and `pchip`,

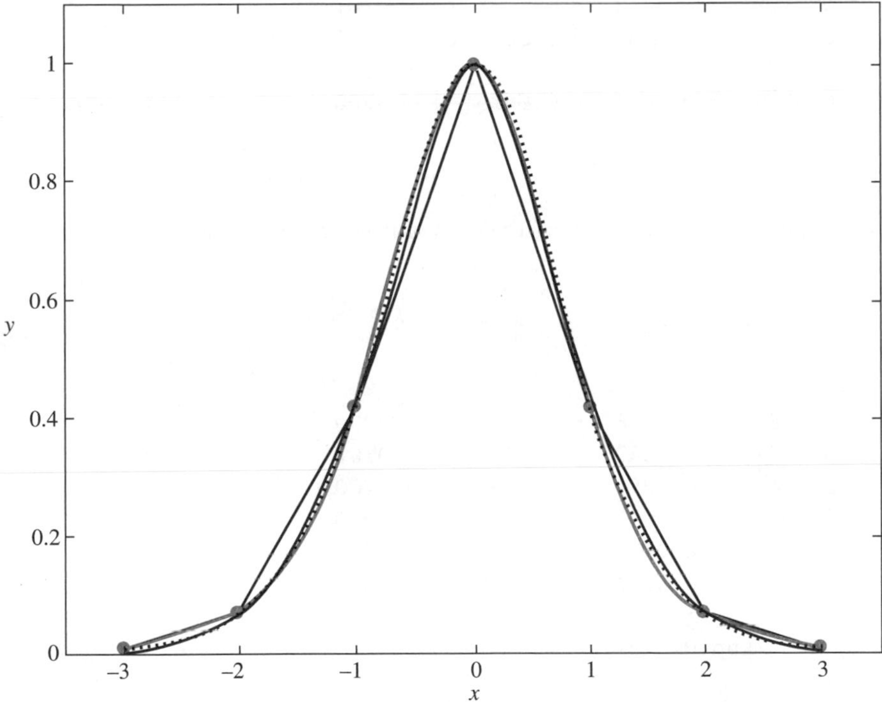

Figure 12.7 MATLAB cubic spline and Hermite interpolations of the bell-shaped function $f(x) = \text{sech}^2 x$.

respectively. The last two functions are useful for a semianalytical representation of piecewise cubic polynomials. If n is the number of piecewise polynomials and m is the degree of each polynomial, the function **unmpkk** applied to the functions **spline** and **pchip** returns the vector of breaking points and the matrix of coefficients of piecewise polynomials with n rows and $m+1$ columns (sorted in the MATLAB order of coefficients as in Section 5.1). This representation can be used in the two MATLAB functions **mkpp** and **ppval** for the evaluation of cubic Hermite polynomials at any set of x-values. Continuing the previous example, we illustrate the use of these MATLAB functions for representation, evaluation, and plotting of the polynomial interpolations.

```
>> [P1,R1] = unmkpp(spline(x,y)) % cubic spline interpolation

P1 =
    -3     -2     -1      0      1      2      3

R1 =
    0.0869   -0.1166    0.0904    0.0099
    0.0869    0.1443    0.1181    0.0707
   -0.4926    0.4051    0.6675    0.4200
    0.4926   -1.0726         0    1.0000
   -0.0869    0.4051   -0.6675    0.4200
   -0.0869    0.1443   -0.1181    0.0707

>> SS1 = mkpp(P1,R1);
>> [P2,R2] = unmkpp(pchip(x,y)) % cubic Hermite interpolation

P2 =
    -3     -2     -1      0      1      2      3

R2 =
   -0.0180    0.0788         0    0.0099
   -0.1591    0.4048    0.1036    0.0707
   -0.7240    0.8680    0.4360    0.4200
    0.7240   -1.3040         0    1.0000
    0.1591   -0.0723   -0.4360    0.4200
    0.0180    0.0247   -0.1036    0.0707

>> SS2 = mkpp(P2,R2);
```

After the analytical data of the cubic interpolating polynomials are stored on a computer, you can try evaluating the values of the interpolating polynomials at any point, for example, at $x_0 = 0.5$.

```
>> x0 = 0.5; yExact = (cosh(x0))^(-2)

yExact =
   0.78644773296593

>> y1 = ppval(SS1,x0)

y1 =
   0.79342277554157

>> y2 = ppval(SS2,x0)

y2 =
   0.76449214273183
```

We illustrate the computations of the first derivative of the cubic spline and Hermite interpolations in the MATLAB script derivatives_splines.

```
% derivatives_splines
x = -3:1:3; y = (cosh(x)).^(-2);
yDer=-2*sinh(x).*(cosh(x)).^(-3);
xInt = -3 : 0.001 : 3; yEx = (cosh(xInt)).^(-2);
yDerEx=-2*sinh(xInt).*(cosh(xInt)).^(-3);
            % semi-analytic computations of polynomials
[P1,R1] = unmkpp(spline(x,y));
[P2,R2] = unmkpp(pchip(x,y));
            % semi-analytic computations of their derivatives
[n,m] = size(R1);
for j = 1 : n
    Rder1(j,:) = polyder(R1(j,:));
end
yDerInt1 = ppval(mkpp(P1,Rder1),xInt);
[n,m] = size(R2);
 for j = 1 : n
    Rder2(j,:) = polyder(R2(j,:));
end
yDerInt2 = ppval(mkpp(P2,Rder2),xInt);
plot(x,yDer,'r.',xInt,yDerEx,'r:',xInt,yDerInt1,'g',xInt,yDerInt2,'k');
```

The output of the MATLAB script derivatives_splines is shown in Figure 12.8. You can see from the figure that the cubic spline interpolation provides continuous first and second derivatives, whereas the cubic Hermite interpolation provides continuous first derivatives and jump discontinuities in the second derivatives. It is also clear that the MATLAB function spline does not use the conditions for *natural cubic splines* because the slopes of

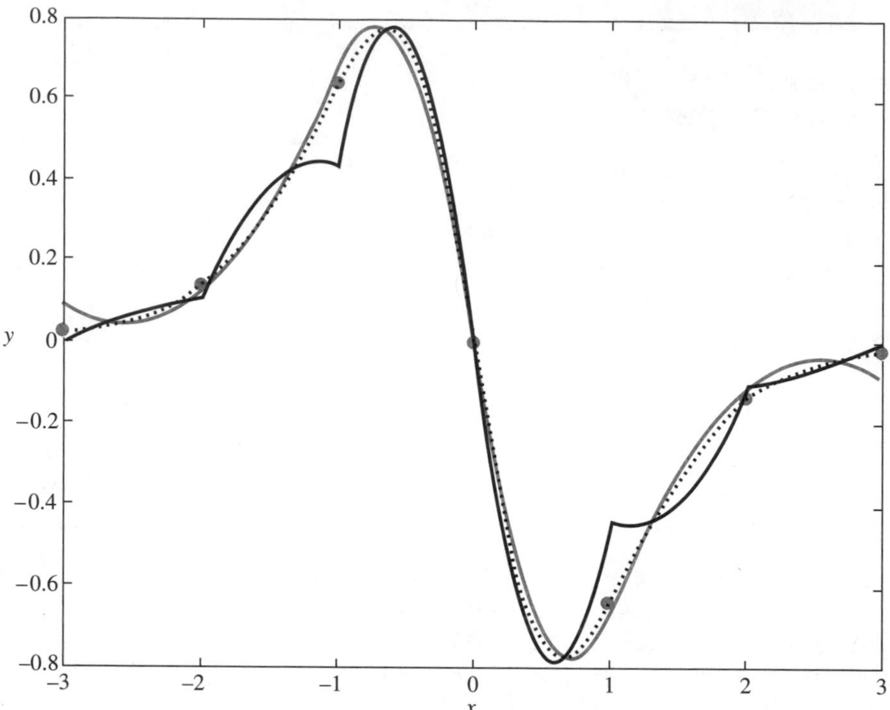

Figure 12.8 The first derivative of the cubic spline and Hermite interpolations for the bell-shaped function $f(x) = \mathrm{sech}^2 x$.

the first derivative of cubic splines at the endpoints are nonzero. In addition, the MATLAB function `pchip` does not require the values of slopes to be specified, but it approximates the slopes from the given set of data points by a modified finite-difference method. See the book [19], where the creator of MATLAB explains modifications of standard cubic spline and cubic Hermite interpolations in the MATLAB functions `spline` and `pchip`.

Exercise 12.5 Apply the MATLAB functions `spline` and `pchip` to the function $f(x) = e^{-x^2}$ on the interval $[-2, 2]$ and show that the computation error is $\mathrm{O}(h^4)$ for both modifications of the cubic spline and cubic Hermite interpolations.

12.3 Finite Elements for Differential Equations

Spectral methods for solutions of ordinary and partial differential equations are based on specific partitions of the computational domain. For instance, the collocation method based on the trigonometric interpolation relies on a uniform grid of equally spaced grid points (see Section 11.3). Although finite-difference methods are not limited by the constant step size, the uniform grids are commonly used for control of convergence of the finite-difference approximations of numerical derivatives (see Section 10.1).

The finite-element method remedies the problem of specific or uniform partitions of the computational domain with any user-defined partition. This method is particularly important for two-dimensional and three-dimensional domains with curved boundaries. A number of software algorithms offer implementations of the finite-element method, in particular the FEMLAB (Finite Element Method Laboratory), which originated from the MATLAB PDE toolbox and developed into a rapidly growing computational software industry.

Rayleigh–Ritz problem

Let us explain the main idea of the finite-element method and the relations between the finite elements and the spline interpolation. Although the multidimensional boundary-value PDE problems are the primary target for applications of the finite-element method, we work in the context of a one-dimensional problem (e.g., for an ODE) where many technical details of the presentation can be avoided. In particular, we shall consider the linear second-order ODE problem

$$y''(x) + q(x)y = r(x), \qquad a < x < b, \tag{12.25}$$

subject to the Dirichlet boundary conditions

$$y(a) = y(b) = 0, \tag{12.26}$$

where $q(x)$ and $r(x)$ are given continuous functions on $[a, b]$. The boundary-value ODE problem (12.25)–(12.26) can be reformulated as the variational *Rayleigh–Ritz* problem, introduced in the following result.

Theorem 12.3 *Let $y(x)$ be a twice continuously differentiable solution of the boundary-value ODE problem (12.25)–(12.26). Then, $y(x)$ is an extremal value of the energy functional*

$$E[y] = \int_a^b \left[\left(\frac{dy}{dx} \right)^2 - q(x)y^2 + 2r(x)y \right] dx, \tag{12.27}$$

in the sense that

$$\frac{d}{d\epsilon} E\left[y(x) + \epsilon z(x)\right] \bigg|_{\epsilon=0} = 0,$$

for any smooth function $z(x)$ on $[a, b]$ with $z(a) = z(b) = 0$.

Indeed, a straightforward computation shows that

$$E[y+\epsilon z]-E[y] = -2\epsilon \int_a^b \left[y'' + q(x)y - r(x)\right] z(x)dx + 2\epsilon y'(x)z(x)\Big|_{x=a}^{x=b} + \mathrm{O}(\epsilon^2),$$

where the linear terms in ϵ vanish if $y(x)$ is a solution of the ODE (12.25) and $z(a) = z(b) = 0$. The converse may not be true since the extremal value of $E[y]$ may be represented by the function $y(x)$ with jump discontinuities in itself or its derivatives, which is not a classical (twice continuously differentiable) solution of the ODE (12.25). However, if the extremal value of $E[y]$ is represented by a twice continuously differentiable function $y(x)$ on $[a,b]$ with fixed values at the end points (12.26), then the function $y(x)$ is necessarily a classical solution of the ODE (12.25).

A similar correspondence between the variational Rayleigh–Ritz problem and solutions of ODEs may involve higher-order derivatives of the function $y(x)$. Although this correspondence can be exploited in many different directions, we shall mainly use Theorem 12.3 for a numerical approximation of the solution $y(x)$ of the boundary-value problem (12.25)–(12.26) by using the extremal value $y(x)$ of the energy functional $E[y]$.

Exercise 12.6 Show that the cubic spline interpolant $y = S_k(x)$ on the elementary interval $[x_k, x_{k+1}] \subset [a,b]$ gives the minimal value of the energy functional

$$E_k = \int_{x_k}^{x_{k+1}} \left[y''(x)\right]^2 dx.$$

Show that the cubic spline interpolant $y = S(x)$ on the entire interval $[a,b]$ gives the minimal value of the total energy $E = \sum_{k=1}^n E_k$ but $y(x)$ is not a classical solution of the relevant ODE.

Basis functions

Let the set of grid points $\{x_k\}_{k=0}^{n+1}$ represent the partition of the continuous interval $[a,b]$ with $x_0 = a$, $x_{n+1} = b$, and $x_0 < x_1 < \ldots < x_{n+1}$. The partition may not be uniform and, in particular, the grid points can accumulate around some points of the interval, for example, at the singular points of the second-order ODE. Consider a set of basis functions associated with the given partition $\{\phi_k(x)\}_{k=1}^n$ such that

- the function $\phi_k(x)$ is piecewise continuous on $[a,b]$ for any $k = 1, 2, \ldots, n$
- $\phi_k(x_k) = 1$ and $\phi_k(x_l) = 0$ for any $l \neq k$

A function $y(x)$ in the energy functional $E[y]$ can be approximated by a linear combination of basis functions,

$$y(x; y_1, \ldots, y_n) = \sum_{k=1}^n y_k \phi_k(x), \qquad a \le x \le b, \tag{12.28}$$

where the values $y_k \equiv y(x_k; y_1, \ldots, y_n)$ for $k = 1, 2, \ldots, n$ are parameters of the representation (12.28). If $y(x; y_1, \ldots, y_n)$ gives an extremal value of $E[y]$, then the function $E(y_1, \ldots, y_n) \equiv E[y(x; y_1, \ldots, y_n)]$ is extremal in n variables $(y_1, \ldots, y_n)$, which implies that

$$\frac{\partial}{\partial y_k} E(y_1, \ldots, y_n) = 0, \qquad k = 1, 2, \ldots, n. \tag{12.29}$$

The extremal property (12.29) for the energy functional (12.27) leads to a system of linear equations on variables $y_1, \ldots, y_n$. The coefficients of the linear system are expressed by integrals that can be either computed analytically in some simple cases or approximated numerically, as in Section 6.5. Note that the linear system that follows from the extremal property (12.29) is similar to the linear system derived for the least-square polynomial approximation in Section 5.6 and for the trigonometric approximation in Section 11.1. This analogy suggests that the finite-element method can be regarded as the spline *approximation* of the function $y(x)$.

The set of basic functions $\{\phi_k(x)\}_{k=1}^n$ in the representation (12.28) can be fairly arbitrary. Lagrange interpolation is based on the choice of Lagrange polynomials as the basis functions (see Section 5.3). Trigonometric interpolation is based on the choice of periodic sine functions as the basis functions (see Section 11.1). Both the Lagrange polynomials and the periodic sine functions are defined on the entire interval $[a, b]$.

Finite elements

The *finite-elements* method originates from the choice of piecewise continuous polynomials as the basis functions. The piecewise continuous polynomials are localized near the corresponding grid points. The simplest piecewise polynomials are linear splines in the form

$$\begin{aligned} \phi_k(x) &= \frac{x - x_{k-1}}{x_k - x_{k-1}}, \qquad x_{k-1} \le x \le x_k, \\ \phi_k(x) &= \frac{x_{k+1} - x}{x_{k+1} - x_k}, \qquad x_k \le x \le x_{k+1}, \end{aligned} \tag{12.30}$$

and $\phi_k(x) = 0$ beyond the interval $x_{k-1} \le x \le x_{k+1}$. When the representation (12.28) is expanded, it becomes nothing but the linear spline interpolation constructed in Section 12.1, such that $y = S_k(x)$ on the interval $x_k \le x \le x_{k+1}$, where the linear spline $S_k(x)$ between data points (x_k, y_k) and (x_{k+1}, y_{k+1}) is given by the function (12.2) with the slope (12.3).

When the finite-element representation (12.28) is substituted into the energy functional (12.27) and the extremal property (12.29) for a solution $y(x)$ is invoked, the unknown set of parameters $\{y_k\}_{k=1}^n$ is found from the linear system of equations

$$\sum_{j=1}^{n} y_j \int_a^b \left[\phi_k'(x)\phi_j'(x) - q(x)\phi_k(x)\phi_j(x)\right] dx + \int_a^b r(x)\phi_k(x) dx = 0, \tag{12.31}$$

where $k = 1, 2, \ldots, n$. The linear system (12.31) can be cast to the matrix-vector form

$$\mathbf{A}\mathbf{y} = \mathbf{b}, \tag{12.32}$$

where $\mathbf{y} = [y_1, \ldots, y_n]$ and the elements of $\mathbf{A}$ and $\mathbf{b}$ are given by

$$\begin{aligned} a_{kj} &= \int_a^b \left[\phi_k'(x)\phi_j'(x) - q(x)\phi_k(x)\phi_j(x)\right] dx, \\ b_k &= -\int_a^b r(x)\phi_k(x)dx. \end{aligned}$$

Because of the localization properties (12.30) of the linear finite elements $\phi_k(x)$, the matrix $\mathbf{A}$ is tridiagonal with the only nonzero elements at $j = k$ and $j = k \pm 1$:

$$\begin{aligned} a_{kk} &= \frac{1}{h_{k-1}} + \frac{1}{h_k} - \frac{1}{h_{k-1}^2}\int_{x_{k-1}}^{x_k} q(x)(x - x_{k-1})^2 dx - \frac{1}{h_k^2}\int_{x_k}^{x_{k+1}} q(x)(x - x_{k+1})^2 dx, \\ a_{k,k-1} &= -\frac{1}{h_{k-1}} + \frac{1}{h_{k-1}^2}\int_{x_{k-1}}^{x_k} q(x)(x - x_{k-1})(x - x_k)dx = a_{k-1,k}, \end{aligned}$$

where $h_k = x_{k+1} - x_k$ is defined for $k = 0, 1, \ldots, n$. The constant term is also computed in the explicit form:

$$b_k = -\frac{1}{h_{k-1}}\int_{x_{k-1}}^{x_k} r(x)(x - x_{k-1})dx + \frac{1}{h_k}\int_{x_k}^{x_{k+1}} r(x)(x - x_{k+1})dx.$$

When the integrals are approximated with the trapezoidal rule (see Section 6.5), the coefficients a_{kj} and b_k are simplified to the form:

$$\begin{aligned} a_{kk} &= \frac{1}{h_{k-1}} + \frac{1}{h_k} - \frac{1}{2}q_k(h_k + h_{k-1}), \\ a_{k,k-1} &= -\frac{1}{h_{k-1}} = a_{k-1,k}, \\ b_k &= -\frac{1}{2}r_k(h_k + h_{k-1}), \end{aligned}$$

where $r_k = r(x_k)$ and $q_k = q(x_k)$. In this approximation, the linear system (12.31) coincides with the linear system obtained in the finite-difference method, when the second derivative of the second-order ODE (12.25) is approximated with the central-difference numerical derivatives (see Section 6.1). In the case of the uniform grid with $h = h_k$, the linear system is rewritten in the explicit form

$$\frac{y_{k-1} - 2y_k + y_{k+1}}{h^2} + q_k y_k = r_k, \qquad k = 1, 2, \ldots, n. \tag{12.33}$$

This system coincides with the linear system (10.12) in Section 10.1.

Exercise 12.7 Use Simpson's rule for numerical integration and find explicit expressions for the coefficients $a_{kk}, a_{k,k-1}$, and b_k. Simplify the linear system for the uniform grid with equal step size h.

If the functions $q(x)$ and $r(x)$ are given analytically, the integrals in the expressions for a_{kk}, $a_{k,k-1}$, and b_k can be computed analytically so that the linear system (12.31) can be closed without additional truncations. We illustrate these computations with the example of the functions $q(x) = \cos x$ and $r(x) = \sin x$ on the interval $[0, 2\pi]$ subject to the Dirichlet boundary conditions (12.26). (The example of the boundary-value problem (12.25) with $q(x) = \cos x$ and $r(x) = \sin x$ subject to the periodic boundary conditions on $y(x)$ is considered in Section 11.3, using the Galerkin and collocation methods.) By analytic computations, we obtain the coefficients of the matrix $\mathbf{A}$ and the constant term $\mathbf{b}$ in the explicit form:

$$a_{kk} = \left(\frac{1}{h_{k-1}} + \frac{1}{h_k}\right)(1 - 2\cos x_k) + \frac{2(\sin x_k - \sin x_{k-1})}{h_{k-1}^2} - \frac{2(\sin x_k - \sin x_{k+1})}{h_k^2},$$

$$a_{k,k-1} = -\frac{1}{h_{k-1}}(1 - \cos x_{k-1} - \cos x_k) - \frac{2(\sin x_k - \sin x_{k-1})}{h_{k-1}^2} = a_{k-1,k},$$

$$b_k = \frac{\sin x_{k-1} - \sin x_k}{h_{k-1}} + \frac{\sin x_{k+1} - \sin x_k}{h_k}.$$

We illustrate the finite element method in the MATLAB script `finite_elements`. Since the linear finite elements (12.30) are equivalent to the linear spline interpolation, the function $y(x)$ is reconstructed from the set of data values $\{y_k\}_{k=0}^{n+1}$ with $y_0 = y_{n+1} = 0$ using the MATLAB graphical function `plot`. The output for the numerical solution $y(x)$ of the boundary-value problem is shown in Figure 12.9.

```
% finite_elements
n = 40; x = linspace(0,2*pi,n+1);
q = cos(x); r = sin(x);
h = x(2:n+1)-x(1:n); % linear system for linear finite elements
d1 = (1./h(1:n-1)+1./h(2:n)).*(1-2*q(2:n));
d1=d1+2*(r(2:n)-r(1:n-1))./(h(1:n-1).^2)-2*(r(2:n)-r(3:n+1))./(h(2:n).^2);
d2=-(1-q(2:n-1)-q(3:n))./h(2:n-1)-2*(r(3:n)-r(2:n-1))./(h(2:n-1).^2);
d3=-(1-q(2:n-1)-q(3:n))./h(2:n-1)+2*(r(2:n-1)-r(3:n))./(h(2:n-1).^2);
A = diag(d1) + diag(d2,-1) + diag(d3,+1);
b=(r(1:n-1)-r(2:n))./h(1:n-1)+(r(3:n+1)-r(2:n))./h(2:n);
y = A\b'; y = [0,y',0]; % plotting finite-element approximation
plot(x,y,'.r',x,y,'b');
```

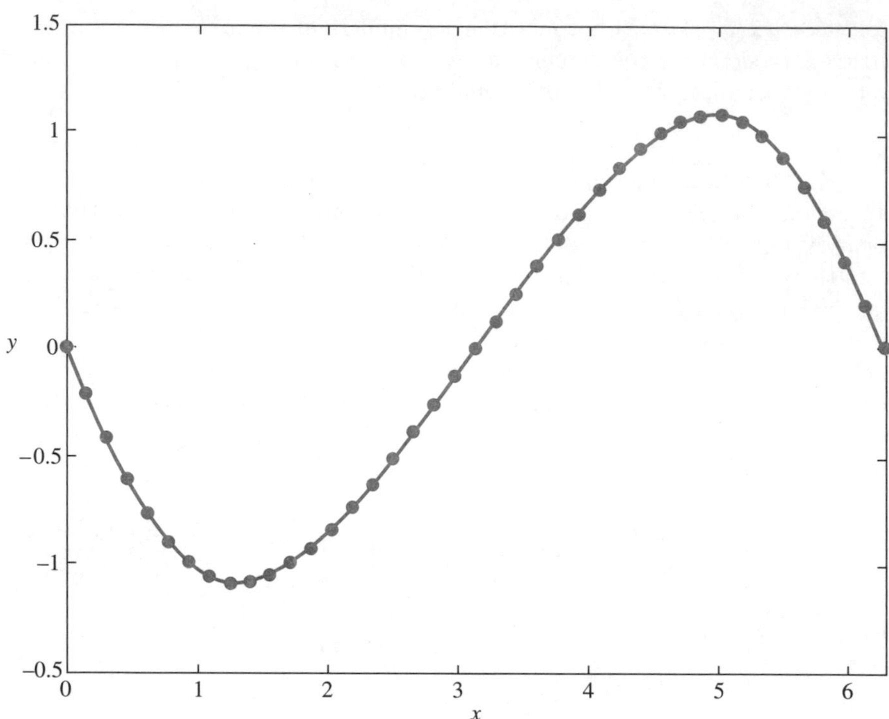

Figure 12.9 Finite-element approximation of the solution of the boundary-value problem (12.25) with $q(x) = \cos x$ and $r(x) = \sin x$.

Errors of finite elements

If the discrete grid is uniform with the constant step size h and the trapezoidal rule is used for numerical approximation of the integrals, the linear systems (10.12) and (12.33) are identical, and the truncation errors of the solution $y(x)$ obtained with the linear finite elements (12.30) and with the second-order central differences are the same. By error analysis of central differences (see Section 6.2), the maximal difference between the exact solution at the grid points $y(x_k)$ and the numerical approximations y_k is $\mathrm{O}(h^2)$. It is surprising, however, that if the exact values of integrals are used in the finite-element method instead of the numerical trapezoidal rule, the truncation error of the finite-element method becomes much smaller than that of the finite-difference method. We illustrate this property in the example of the boundary-value problem

$$y'' = x^2, \qquad 0 < x < 1,$$

with $y(0) = y(1) = 0$, which admits an exact solution

$$y(x) = \frac{1}{12} x(x^3 - 1).$$

Since $q(x) = 0$ and $r(x) = x^2$, elements of the matrix $\mathbf{A}$ and the constant term $\mathbf{b}$ are computed exactly:

$$a_{kk} = \frac{1}{h_{k-1}} + \frac{1}{h_k} - \frac{1}{2}q_k(h_k + h_{k-1}),$$
$$a_{k,k-1} = -\frac{1}{h_{k-1}} = a_{k-1,k},$$
$$b_k = -\frac{1}{12}\left(x_{k+1}^3 - x_{k-1}^3 + x_k(x_{k+1}^2 - x_{k-1}^2) + x_k^2(x_{k+1} - x_{k-1})\right).$$

The MATLAB script `error_finite_elements` computes numerical approximations of both the finite-element and finite-difference methods on the uniform grid and compares the maximal distance between the exact and numerical solutions.

```
% error_finite_elements
for n = 2 : 50;
    x = linspace(0,1,n+1); yExact = x.*(x.^3-1)./12;
    h = x(2:n+1)-x(1:n); h(n-1) = h(1);
                % finite-element method
    d1 = (1./h(1:n-1)+1./h(2:n)); d2 = -1./h(2:n-1);
    A = diag(d1) + diag(d2,-1) + diag(d2,+1);
    b = -(x(3:n+1).^3-x(1:n-1).^3+x(2:n).*(x(3:n+1).^2-x(1:n-1).^2))/12;
    b = b-x(2:n).^2.*(x(3:n+1)-x(1:n-1))/12;
    y = A\b'; y = [0,y',0];
    errFEM(n-1) = max(abs(y-yExact));
                % finite-difference method
    A = 2*diag(ones(1,n-1)) - diag(ones(1,n-2),1) - diag(ones(1,n-2),-1);
    b = -hh(n-1)^2*x(2:n).^2;
    y = A\b'; y = [0,y',0];
    errFDM(n-1) = max(abs(y-yExact));
end
a = polyfit(log(hh),log(errFDM),1); % power fit
power = a(1), MaxErrFDM = max(errFDM), MaxErrFEM = max(errFEM)
```

The output of the MATLAB script `error_finite_elements` shows that the finite-difference method constructed with the second-order central-difference numerical derivatives has an $O(h^2)$ truncation error, whereas the finite-element method constructed with the linear finite elements recovers the exact solution up to machine precision.

```
>> error_finite_elements

power =
    1.9891
```

```
MaxErrFDM =
    0.0052

MaxErrFEM =
    1.0755e-015
```

Exercise 12.8 Consider the boundary-value problem (12.25) with power functions $q(x) = x^k$ and $r(x) = x^m$ for $k, m \in \mathbb{N}$ and find the lowest values of k and m, when the numerical approximation obtained with the linear finite elements has a nonzero truncation error.

Eigenvalue problems

When multiple sources of the truncation errors occur, rough approximations resulting from a particular truncation procedure can prevent getting good accuracy in other truncation procedures. You have seen already that the use of the trapezoidal integration rule reduces the finite-element method to the finite-difference method. We shall consider another example, namely, the linear eigenvalue problem

$$y'' + \lambda y = 0, \qquad 0 < x < \pi, \tag{12.34}$$

subject to the Dirichlet boundary values $y(0) = y(\pi) = 0$. The exact eigenvalues λ and the corresponding eigenfunctions $y(x)$ are

$$\lambda = k^2, \qquad y(x) = \sin(kx), \qquad k \in \mathbb{N}.$$

To construct a numerical approximation of the eigenfunctions and eigenvalues of the problem (12.34) by using the finite-element method, we consider the Rayleigh–Ritz functional

$$E[y] = \int_0^{\pi} \left[\left(\frac{dy}{dx} \right)^2 - \lambda y^2 \right] dx, \tag{12.35}$$

where λ is a Lagrange multiplier. The extremal values of the Rayleigh–Ritz functional (12.35) are approximated by the finite-element representation (12.28) if the variables $\{y_k\}_{k=1}^n$ satisfy the matrix eigenvalue problem:

$$\mathbf{A}\mathbf{y} + \lambda \mathbf{y} = \mathbf{0},$$

where coefficients of the matrix $\mathbf{A}$ are the same as earlier for $q(x) = 0$. If the grid is uniform with constant step size h, the matrix eigenvalue problem coincides with that of the finite difference method (compare with the system (11.21) in Section 11.2). Each eigenvalue converges to the squared integer values with an $O(h^2)$ truncation error. The MATLAB script `eigenvalues_finite_elements` illustrates computations of the first three eigenvalues and the power laws of the truncation errors.

```
% eigenvalues_finite_elements
k = 0;
for n = 100 : 10 : 300;
    k = k+1; h(k) = pi/(n-1);
    A = 2*diag(ones(1,n))-diag(ones(1,n-1),1)-diag(ones(1,n-1),-1);
    [V,D] = eig(A/h(k)^2); lambda = diag(D);
    lam1(k) = abs(lambda(1)-1);
    lam2(k) = abs(lambda(2)-4);
    lam3(k) = abs(lambda(3)-9);
end
a1=polyfit(log(h),log(lam1),1); power1 = a1(1)
a2=polyfit(log(h),log(lam2),1); power2 = a2(1)
a3=polyfit(log(h),log(lam3),1); power3 = a3(1)
```

The output of the MATLAB script `eigenvalues_finite_elements` shows a surprising result. The truncation error for each of the three smallest eigenvalues is $O(h)$. This apparent contradiction with the theory of the finite-element (finite-difference) method is explained by the use of the MATLAB function `eig`, which computes eigenvalues of the matrix with a numerical routine. The routine used in the MATLAB eigenvalue solver introduces an additional error, which changes the truncation error of the finite-element method from $O(h^2)$ to $O(h)$.

```
>> eigenvalues_finite_elements

power1 =
    0.9839

power2 =
    0.9872

power3 =
    0.9928
```

Nonlinear finite elements can also be constructed as extensions of the linear finite elements (12.30). For instance, quadratic and cubic spline interpolations can be replaced by the quadratic and cubic spline approximations, when the finite elements are given by the quadratic and cubic polynomials interpolating through few data points and preserving smoothness of the finite elements and the continuity of its curvatures (in the case of cubic splines). See [5] for the comprehensive studies of finite elements.

Finite elements are applied to two-dimensional and three-dimensional boundary-value PDE problems as well as to time-dependent PDE problems. In the first case, the simplest finite elements are triangles and rectangles in two dimensions and pyramids and parallelepipeds in three dimensions. In the second case, the variables y_k in the finite-element representation (12.28) are functions of time and the extremal property for the energy functional results in the derivation of the system of ODEs for time evolution of the variables $y_k(t)$, which is then solved with the initial-value ODE solvers. The text [8] gives a good introduction to multidimensional and time-dependent finite elements with examples in the MATLAB PDE Toolbox.

12.4 Summary and Notes

In this chapter, we studied properties of the spline and Hermite interpolations and their applications to numerical computations of solutions of differential equations:

- **Section 12.1:** Three basic spline interpolations are formulated in Problem 12.1 for linear splines, Problem 12.2 for quadratic splines, and Problem 12.3 for cubic splines. Solutions of these problems are developed on the basis of difference equations for parameters of the spline interpolations supplemented by additional endpoint conditions. Errors of linear, quadratic, and cubic splines are described in Theorem 12.1 and illustrated in the context of smooth functions satisfying or violating the required endpoint conditions. We show that no Runge phenomenon can occur in the spline interpolation.

- **Section 12.2:** The cubic Hermite interpolation is formulated in Problem 12.4, which admits an immediate explicit solution. We compare cubic Hermite and spline interpolations by means of numerical examples. The error of the cubic Hermite interpolation is described in Theorem 12.2. We perform semianalytic operations with spline and Hermite interpolations using relevant MATLAB functions.

- **Section 12.3:** We discuss applications of finite elements to numerical approximations of solutions of differential equations in the context of the Rayleigh–Ritz variational method (Theorem 12.3). The method is known as the finite-element method, which generalizes the finite-difference method. Implementations and errors of this method are discussed in the context of the boundary-value and linear eigenvalue problems for ordinary differential equations.

Spline and Hermite interpolations are covered in [7], while applications of finite-element methods to numerical approximations of solutions of differential equations are reviewed in [9].

12.5 Exercises

1. Compute linear, quadratic, and cubic splines for the functions

$$f = \frac{1}{1+x^4}$$

on the equally spaced grid on $[0, 1]$ with the step size $h = 0.1$. Plot the three spline interpolants on the same graph and compare the 2-norm of the error of the spline interpolations.

2. Construct the piecewise quadratic and quadratic spline interpolations of the function $f = e^{-|x|}$ on $[-2, 2]$ with the step size $h = 0.25$. Plot the two interpolants on the same graph and compare the 2-norm of the error of the polynomial interpolations.

3. Consider the function $f(x) = \cos x$ on $[0, 2\pi]$ with the step size $h = \pi/2$. Compute the natural cubic spline and the cubic Hermite interpolations and plot them on the same graph. Compare the 2-norm of the error of the interpolations.

4. Derive the formula for the integral of the linear, quadratic, and cubic spline interpolants $S_k(x)$ on the elementary interval $[x_k, x_{k+1}] \subset [a, b]$. Compute the total integral of the spline interpolant for the function

$$f(x) = \frac{1}{\sqrt{1-x^2}}$$

on $[0, 1]$ and compare with the exact computation.

5. Compute the first-order and second-order derivatives of the natural quadratic and cubic splines for the function $f(x) = xe^{-x}$ with $n = 10$ data points on the interval $[0, 2]$. Confirm that the quadratic splines have continuous first derivatives at the breaking points, while the cubic splines have continuous second derivatives at the breaking points.

6. Use the MATLAB function `spline` and compare `S=spline(x,y)` for the given set of data points x and y. Use the MATLAB function `max` and write a MATLAB function that computes the maximum jump in the third derivative of the spline S. Repeat the same exercise for the cubic Hermite interpolant in `S = cubic(x,y)` and compute the maximum jump in the second derivative of the interpolant.

7. Consider a polynomial in the form

$$p(x) = a_1 + a_2 x + a_3 x^2 + a_4 x(x-1) + a_5 (x-1)^2.$$

Find coefficients of the interpolant from the conditions that $p(x)$ equals a function $f(x)$ at $x = 0$, $x = 0.5$, and $x = 1$ and the derivative $p'(x)$

equals the derivative $f'(x)$ at $x = 0$ and $x = 1$. Plot the interpolant for the function $f(x) = e^{-2x} \sin(5x)$.

8. Consider a not-a-knot cubic spline, where the natural interpolation conditions (12.14) are replaced by the conditions $S_1(x_{1.5}) = y_{1.5}$ and $S_n(x_{n+0.5}) = y_{n+0.5}$, where the two additional data points $(x_{1.5}, y_{1.5})$ and $(x_{n+0.5}, y_{n+0.5})$ are midpoints on the intervals $[x_1, x_2]$ and $[x_n, x_{n+1}]$. Set up the linear system and construct the not-a-knot cubic spline for the function $y = \log x$ on $[1, 2]$ with $h = 0.2$.

9. Compute the first-order and second-order derivatives of the cubic Hermite interpolation of the function $f(x) = xe^{-x}$ on a uniform grid of the interval $[0, 2]$. Confirm that the cubic Hermite interpolation in the form (12.20) has continuous first derivatives at the breaking points but jump discontinuities of the second derivatives.

10. Construct the cubic Hermite interpolation for the function $f(x) = 1/x$ on $[a, 1]$ and plot the error of the interpolation versus x for $h = 0.1$ and $a = 0.5$. Find the power fit of the 2-norm of the error E versus the step size h for $a = 0.5$, $a = 0.1$, and $a = 0.01$. Find the convergence rate as $a \to 0$.

11. Construct the piecewise cubic, natural cubic spline, and cubic Hermite interpolations of the function $f = \log x$ on the interval $[1, 3]$ with the step size $h = 0.5$. Plot the three interpolants on the same graph and compare the total errors of the interpolations.

12. Consider the cubic Hermite interpolation of the data points (x_k, y_k) from the function $y = x^{1/3}$ on $[0.1, 2]$ with the step size $h = 0.1$. Use the second-order difference approximation for the slopes m_k at the grid points and compute the cubic Hermite interpolation. Plot the computational error versus x.

13. Consider the boundary-value ODE problem

$$-y'' + y = 2 \sin x, \qquad 0 < x < \pi,$$

subject to the Dirichlet boundary conditions $y(0) = y(\pi) = 0$. Approximate the numerical solution with the linear finite-element method on the uniform grid with $h = \pi/10$. Plot the total error versus h and find the convergence rate from the power least-square fit.

14. Consider the boundary-value problem for the Legendre equation

$$(1 - x^2)y'' - 2xy' + y = x^4, \qquad -1 < x < 1,$$

such that the solution $y(x)$ is bounded at the regular singular points $x = 1$ and $x = -1$. The boundary-value problem has the exact solution

$$y = -\frac{1}{95}(5x^4 + 12x^2 - 24).$$

Construct numerical approximations of the solution $y(x)$ by using the finite-element and finite-difference methods and compare their total errors.

15. Consider the linear eigenvalue problem

$$y'' + \frac{2}{x}y' + \lambda y = 0, \qquad 1 < x < 2,$$

subject to the Dirichlet boundary conditions $y(1) = y(2) = 0$. Exact values for λ and $y(x)$ can be found from the substitution $y(x) = u(x)/x$, where $u(x)$ solves $u'' + \lambda u = 0$ subject to the same Dirichlet boundary conditions. Construct numerical approximations of the first three eigenvalues and eigenfunctions by using the finite-element method supplemented with the trapezoidal rule. Plot the computational error in finding the first three eigenvalues versus the step size h and confirm the quadratic convergence of the finite-element method.

Bibliography

[1] J. H. E. Ahlberg, E. N. Nilson, and J. L. Walsh. *The Theory of Splines and their Applications.* New York: Academic Press, 1967.

[2] S. Axler, *Linear Algebra Done Right.* New York: Springer, 1997.

[3] R. Barrett, M. Berry, T. F. Chan, J. Demmel, J. Donato, J. Dongarra, V. Eijkhout, R. Pozo, C. Romine, and H. Van der Vorst. *Templates for the Solution of Linear Systems: Building Blocks for Iterative Methods,* 2nd ed. Philadelphia: SIAM, 1994.

[4] C. de Boor. *A Practical Guide to Splines.* New York: Springer–Verlag,1978.

[5] D. S. Burnett. *Finite Element Analysis: From Concepts to Applications.* Reading, MA: Addison–Wesley, 1987.

[6] S. C. Chapra and R. P. Canale. *Numerical Methods for Engineers.* New York: McGraw-Hill, 2006

[7] J. F. Epperson. *An Introduction to Numerical Methods and Analysis.* New York: John Wiley & Sons, 2002.

[8] C. F. Gerald and P. O. Wheatley. *Applied Numerical Analysis.* Boston: Pearson Education, 2004.

[9] D. Gottlieb and S. A. Orszag. *Numerical Analysis of Spectral Methods: Theory and Applications.* Philadelphia: SIAM, 1977.

[10] P. R. Halmos. *Finite–dimensional Vector Spaces.* 2nd ed. New York: Springer, 1974.

[11] D. Hanselman and B. Littlefield. *Mastering MATLAB 6: A Comprehensive Tutorial and Reference.* Upper Saddle River, NJ: Prentice Hall, 2001.

[12] M. T. Heath. *Scientific Computing: An Introductory Survey.* 2nd ed. New York: McGraw-Hill, 2002.

[13] K. Janich. *Linear Algebra.* New York: Springer-Verlag, 1994.

[14] P. D. Lax. *Linear Algebra.* New York: John Wiley & Sons, 1997.

[15] P. Linz and R. L. C. Wang. *Exploring Numerical Methods.* Boston: Jones and Bartlett Publishers, 2003.

[16] C. F. Van Loan. *Introduction to Scientific Computing.* Upper Saddle River, NJ: Prentice Hall, 2000.

[17] J. H. Mathews and K. D. Fink. *Numerical Methods Using Matlab.* Upper Saddle River, NJ: Prentice Hall, 1999.

[18] J. E. Marsden and A. J. Tromba. *Vector Calculus.* New York: W. H. Freeman and Company, 2003.

[19] C. B. Moler. *Numerical Computing with MATLAB.* Philadelphia: SIAM, 2004.

[20] W. K. Nicholson. *Linear Algebra with Applications.* New York: McGraw-Hill Ryerson, 2006.

[21] S. R. Otto and J. P. Denier. *An Introduction to Programming and Numerical Methods in MATLAB.* London: Springer-Verlag, 2005.

[22] M. W. Hirsch, S. Smale, and R. L. Devaney. *Differential Equations, Dynamical Systems, and an Introduction to Chaos.* San Diego: Academic Press, 2004.

[23] A. Ralson and P. Rabinowitz. *A First Course in Numerical Analysis.* New York: McGraw-Hill, 1978.

[24] W. Gautschi. *Orthogonal Polynomials. Computation and Approximation.* Oxford: Oxford University Press, 2004.

[25] R. J. Schilling and S. L. Harris. *Applied Numerical Methods for Engineers Using Matlab and C.* Pacific Grove, CA: Brooks/Cole, 2000.

[26] J. Stewart. *Calculus: Early Transcendentals.* 6th ed. Belmont, CA: Thomson-Brooks/Cole, 2008.

[27] W. A. Strauss. *Partial Differential Equations: An Introduction.* New York: John Wiley & Sons, 1992.

[28] L. N. Trefethen. *Spectral Methods in Matlab.* Philadelphia: SIAM, 2000.

[29] D. Watkins. *Fundamentals of Matrix Computations.* New York: John Wiley & Sons, 1991.

[30] J. H. Wilkinson. *The Algebraic Eigenvalue Problem.* New York: Oxford University Press, 1965.

Subject Index

MATLAB Index

New MATLAB Scripts and Functions

p56

p57 Functional spaces. ∞ dimension – Functional analysis – truncation.

p60 – Derivative Maps